D0820066

COLORADO
MOUNTAIN
COLLEGE

Mechanics of Materials

Give me matter, and I will construct a world out of it.

IMMANUEL KANT

Mechanics of Materials

Anthony Bedford and Kenneth M. Liechti
University of Texas at Austin

Prentice Hall
Upper Saddle River, NJ 07458

Library of Congress Cataloging-in-Publication Data

Bedford, A.
Mechanics of materials / Anthony Bedford and Kenneth M. Liechti.
p. cm.
Includes bibliographical references and index.
ISBN 0-201-89552-8
1. Strength of materials. I. Liechti, K. M. II. Title.

TA405.B414 2000
620.1′12–dc21
00-029129

Vice-President and Editorial Director, ECS: Marcia Horton
Acquisitions Editor: Eric Svendsen
Editorial Assistant: Kristen Blanco
Associate Editor: Joe Russo
Marketing Manager: Danny Hoyt
Production Editor: Kerry Reardon
Executive Managing Editor: Vince O'Brien
Managing Editor: David A. George
Art Manager: Gus Vibal
Art Editor: Xiaohong Zhu
Art Management Support: Julie Nazario
Cover Art Director: Jayne Conte
Line Art: Titan Digital
Photo Research: Abby Reip
Copy Editors: Barbara Zeiders and Martha Ocker
Page Layout and Composition: ICC
Cover Art: Fallingwater, Courtesy of Super Stock, Inc.
Cover Design: Bruce Kenselaar
Manufacturing Buyer: Pat Brown
Assistant Vice President of Production and Manufacturing: David W. Riccardi

Printed in the United States of America

10 9 8 7 6 5 4 3 2 1

ISBN 0-201-89552-8

Prentice-Hall International (UK) Limited, *London*
Prentice-Hall of Australia Pty. Limited, *Sydney*
Prentice-Hall Canada Inc., *Toronto*
Prentice-Hall Hispanoamericana, *S.A., Mexico*
Prentice-Hall of India Private Limited, *New Delhi*
Prentice-Hall of Japan, Inc., *Tokyo*
Pearson Education Asia Pte. Ltd., *Singapore*
Editora Prentice-Hall do Brasil, Ltda., *Rio de Janeiro*

PHOTO CREDITS

Page 2, Bill Hedrich, Hedrich-Blessing; page 57, Beech Aircraft Corporation; page 60, Robert Laberge, Allsport Photography (USA, Inc.); page 94, Professor Roy E. Olsen; pages 136 and 137, Ride & Drive Magazine; page 176, NASA Headquarters; page 186, Richard Pasley, Stock Boston; page 220, Don Morley, Stone; page 237, Stone; page 239, Textron Inc.; page 240, Dale Boyer, Stone; pages 242, 243, and 288, NASA Headquarters; pages 248, 249, and 285, French Government Tourist Office; page 256, Corbis; page 291, NASA/Glenn Research Center; pages 292 and 293, Julius Shulman; pages 322 and 323, Brian Parker, Tom Stack & Associates; pages 438 and 439, Bob Rowan, Corbis; pages 476 and 477, Lindsay Hebberd, Corbis; pages 502 and 503, Corbis; page 535, Robert Nichols, AP/Wide World Photos; pages 28, 106, 121, 145, 169, 324, 404, 405, 440, and 447, authors.

CONTENTS

ABOUT THE AUTHORS

Anthony Bedford is Professor of Aerospace Engineering and Engineering Mechanics at the University of Texas at Austin. He received his B.S. degree at the University of Texas at Austin, his M.S. degree at the California Institute of Technology, and his Ph.D. degree at Rice University in 1967. He has industrial experience at Douglas Aircraft Company and at TRW, where he did structural dynamics and trajectory studies for the Apollo program. He has been on the faculty of the University of Texas at Austin since 1968.

Dr. Bedford's main professional activity has been education and research in engineering mechanics. He is author or coauthor of papers on the mechanics of composite materials and mixtures and four books, including *Engineering Mechanics: Statics* and *Engineering Mechanics: Dynamics* published by Addison Wesley Longman. From 1973 until 1983 he was a consultant to Sandia National Laboratories, Albuquerque, New Mexico.

He is a licensed professional engineer and a member of the American Society for Engineering Education, the Society for Engineering Science, the American Academy of Mechanics, and the Society for Natural Philosophy.

Kenneth Liechti is Professor of Aerospace Engineering and Engineering Mechanics at the University of Texas at Austin and holds the E. P. Schoch Professorship in Engineering. He received his B.Sc. in Aeronautical Engineering at Glasgow University and M.S. and Ph.D. degrees in Aeronautics at the California Institute of Technology. He gained industrial experience at General Dynamics Fort Worth Division prior to joining the faculty of the University of Texas at Austin in 1982.

Dr. Liechti's main areas of teaching and research are in the mechanics of materials and fracture mechanics. He is the author or coauthor of papers on interfacial fracture, fracture in adhesively bonded joints, and the nonlinear behavior of polymers. He has consulted on fracture problems with several companies.

He is a fellow of the American Society of Mechanical Engineers and a member of the Society for Experimental Mechanics, the American Academy of Mechanics, and the Adhesion Society. He is an associate editor of the journal Experimental Mechanics.

PREFACE

Mechanics of materials is concerned with the internal forces and deformations of objects that result from the external forces acting on them. This book appears in a time of transition for education in the mechanics of materials. The traditional course in "strength of materials" that long formed an important part of the engineering curriculum had as one of its primary goals acquainting students with the details of many analytical and empirical solutions that could be applied to structural design. This reliance on a catalog of results has lessened as the finite element method has become commonly available for stress analysis. Another important development is that current research in mechanics of materials is beginning to bring to reality the long-held dream of the merger of continuum solid mechanics and materials science into a unified field. For these reasons, the principle emphasis in the first course in mechanics of materials is becoming oriented toward helping students understand the theoretical foundations, especially the concepts of stress and strain, the stress-strain relations including the meaning of isotropy, and criteria for failure and fracture.

In Chapter 1 we provide an extensive review of statics, including problems, that the instructor may choose to cover or simply have students read. In reviewing distributed loads, we lay the groundwork for our definitions in Chapter 2 of the traction vector and the normal and shear stresses. In Chapter 2 we also introduce the longitudinal and shear strains in terms of changes in infinitesimal material elements. In Chapters 3 and 4 we cover bars subjected to axial and torsional loads and introduce the definitions of the elastic and shear moduli. With these examples as motivation, in Chapters 5 and 6 we introduce the general states of stress and strain and their transformations and the stress-strain relations for linearly elastic materials. In Chapters 7–10 we discuss stresses in beams, deformations of beams, and the buckling of columns. We introduce energy methods in Chapter 11. In Chapter 12 we discuss failure criteria for general states of stress and introduce modern fracture mechanics, which we believe should become an integral part of the first course in mechanics of materials.

Most of the topics in Chapters 1 through 10 and selected topics from Chapters 11 and 12 can be covered in a typical one-semester course. An important decision in teaching the mechanics of materials is whether to introduce the general states of stress and strain before or after discussing applications. We have chosen a compromise based on our teaching experience. Bars subjected to axial load and torsion are first discussed to provide students with simple examples of normal and shear stresses and extensional and shear strains. The states of stress and strain are then discussed, followed by stresses in beams and beam deflections. Instructors preferring to introduce the general states of stress and strain in the beginning should cover Chapters 1, 2, 5, and 6 before covering Chapters 3 and 4.

The first course in mechanics of materials prepares students for subsequent courses in structural analysis, structural dynamics, and advanced mechanics of deformable media. In comparison to courses in statics and dynamics, it is an interesting challenge for instructors and students. Engineering students begin the study of statics and dynamics having had some prior experience with the basic concepts in high school and college physics courses. In contrast, the core concepts in mechanics of materials are new to most students. It is therefore essential that the textbook used introduce and explain these concepts with great care and reinforce them with many examples. This has been our objective in writing this book. Our approach is to present the material as we do in the classroom: emphasizing understanding of the basic principles of mechanics of materials and demonstrating them with examples drawn from contemporary engineering applications and design.

Features

- **Examples That Teach** We continue reinforcing the problem-solving skills students have learned in their introductory mechanics courses. Separate *Strategy* sections preceding most examples and selected problems teach students how to approach and solve problems in engineering. What principles apply? What must be determined, and in what order? Many examples conclude with *Discussion* sections that indicate ways of checking and interpreting answers, point out interesting features of the solution, or suggest alternative methods of solution.
- **Free-Body Diagrams** Correct and consistent use of free-body diagrams is the most essential skill that students of mechanics must acquire. We review the steps involved in drawing free-body diagrams in Chapter 1 and emphasize their use throughout the book. Many of our figures are designed to teach how free-body diagrams are chosen and drawn:

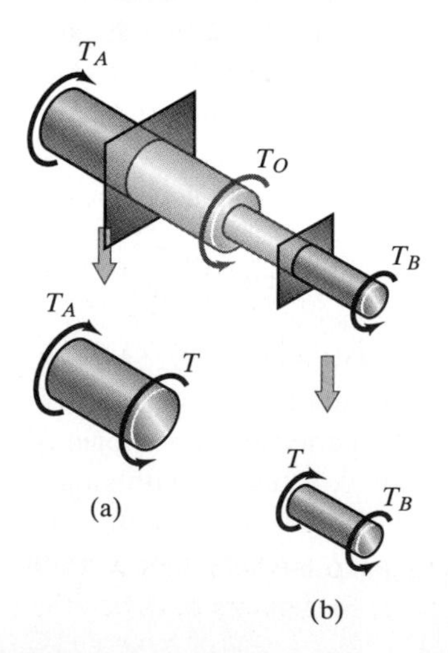

(a)

(b)

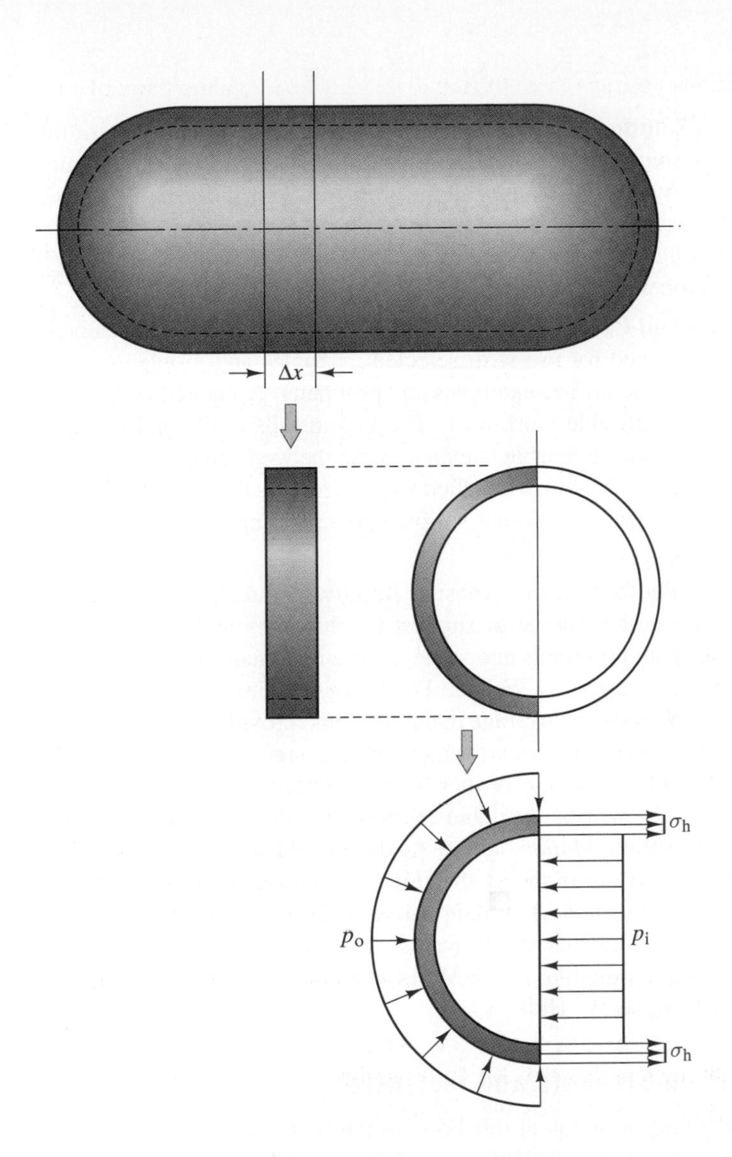

- **Design** The Accrediting Board for Engineering and Technology (ABET), as well as many practicing engineers, strongly encourage the introduction of design throughout the engineering curriculum. By expressing many of our examples and problems in terms of engineering design, we demonstrate the use of mechanics of materials within the larger context of engineering practice. Design problems are designated by blue problem numbers. Chapters 3, 4, 5, and 8 contain sections explicitly addressing the design of particular structural elements, and in Chapter 12 we discuss failure and fracture criteria used in structural design.

Supplements

We have developed and are developing supplements to the book that will contribute to students' understanding as well as enable instructors to enrich their presentations:

- **Instructor's Solution Manual with CD** Prepared by Michael May of the University of South Carolina–Aiken, this manual for the instructor contains complete step-by-step solutions to all the problems. It includes the associated art as well as essential free body diagrams. The accompanying CD contains a selection of figures from the text in the form of PowerPoint slides as well as Adobe Acrobat pdf files of all the figures.
- **Mathcad Student CD** The CD included with this text includes 15 Mathcad worksheets designed for use with selected examples and problems at the option of the instructor. The examples and problems are marked with an M indicating the applicable worksheet. The CD installs easily and contains the Mathcad 8 engine to enable students to use the worksheets on their own computers. Users with Matlab installed who prefer it as their equation solver will be able to download Matlab scripts corresponding to the worksheets from the website.
- **Website** The Bedford/Liechti website, http://www.prenhall.com/bedford, will serve as a further source of support for this book and is intended as a resource for both professors and students. It will contain: *Design Projects* These problems, created by Wallace Fowler of the University of Texas at Austin, extend the design coverage in the book and provide additional applications of mechanics of materials to engineering design. *Concept Animations* Animations will be created to further help illustrate concepts in mechanics of materials. They will be available to view on-line as well as download for class presentation. *Matlab Scripts* As described above, the website will contain Matlab versions of the Mathcad worksheets on the student CD. *Syllabus Builder* The site will contain a free tool to help instructors create and maintain personal course web pages. As described above, the website will contain Matlab versions of the Mathcad worksheets on the student CD, prepared by Douglas W. Hull.

Commitment to Students and Instructors

We have ensured the accuracy of this book to the best of our ability. We have solved every example and problem in an effort to confirm that their answers are correct and that they are of an appropriate level of difficulty. Suzanne Mescan of Progressive Publishing Alternatives carefully checked the entire manuscript. Any errors that remain are our responsibility. We welcome communication from students and instructors concerning needed corrections or improvements. Our address is Department of Aerospace Engineering and Engineering Mechanics, University of Texas at Austin, Austin, Texas 78712 or abedford@mail.utexas.edu.

Acknowledgements

A book of this kind results from a collaboration by many people. We have learned mechanics from our own teachers, colleagues, and students. The following faculty at other institutions critically reviewed the manuscript and gave us many insightful suggestions based on their knowledge of mechanics of materials and how it can best be taught:

From *Prentice Hall:*

William McCarthy, *New Mexico State University;* Harish P. Cherukuri, *University of North Caroliana;* Wendy Taniwangsa, *Santa Clara University;* Darren L. Hitt, *University of Vermont;* Wen S. Chan, *University of Texas at Arlington;* Manoj Chopra, *University of Central Florida;* Sidney Thompson, *University of Georgia;* Alfred Striz, *University of Oaklahoma*

From *Addison Wesley:*

James Casey, *University of California, Berkeley;* Hidenori Murakami, *University of California at San Diego;* Ahmet S. Cakmak, *Princeton University;* Ellen M. Arruda, *University of Michigan;* James H. Williams, Jr., *Massachusetts Institute of Technology*

The initial inspiration for this book was a suggestion by Editor Stuart Johnston, then at Addison Wesley. After a long planning phase supported by Addison Wesley Longman editors Rob Merino, Michael Slaughter, and Chuck Iossi, the project began in earnest when we started working with Editor Eric Svendsen of Prentice Hall. The book could not have been published in its present form without his astute editorial help and organizational skill. We are also grateful for his efforts in arranging for the development of the supplements. Many other talented and agreeable people at Prentice Hall and elsewhere helped us and made important contributions to the book, especially Kristen Blanco, John Hayball, Martha Ocker, Kerry Reardon, Joe Russo, Barbara Zeiders, and Xiaohong Zhu.

Mechanics of Materials

Frank Lloyd Wright's Fallingwater (1938), one of the most influential architectural and structural designs of the twentieth century, has cantilevered decks of reinforced concrete extending over a waterfall.

CHAPTER 1

Introduction

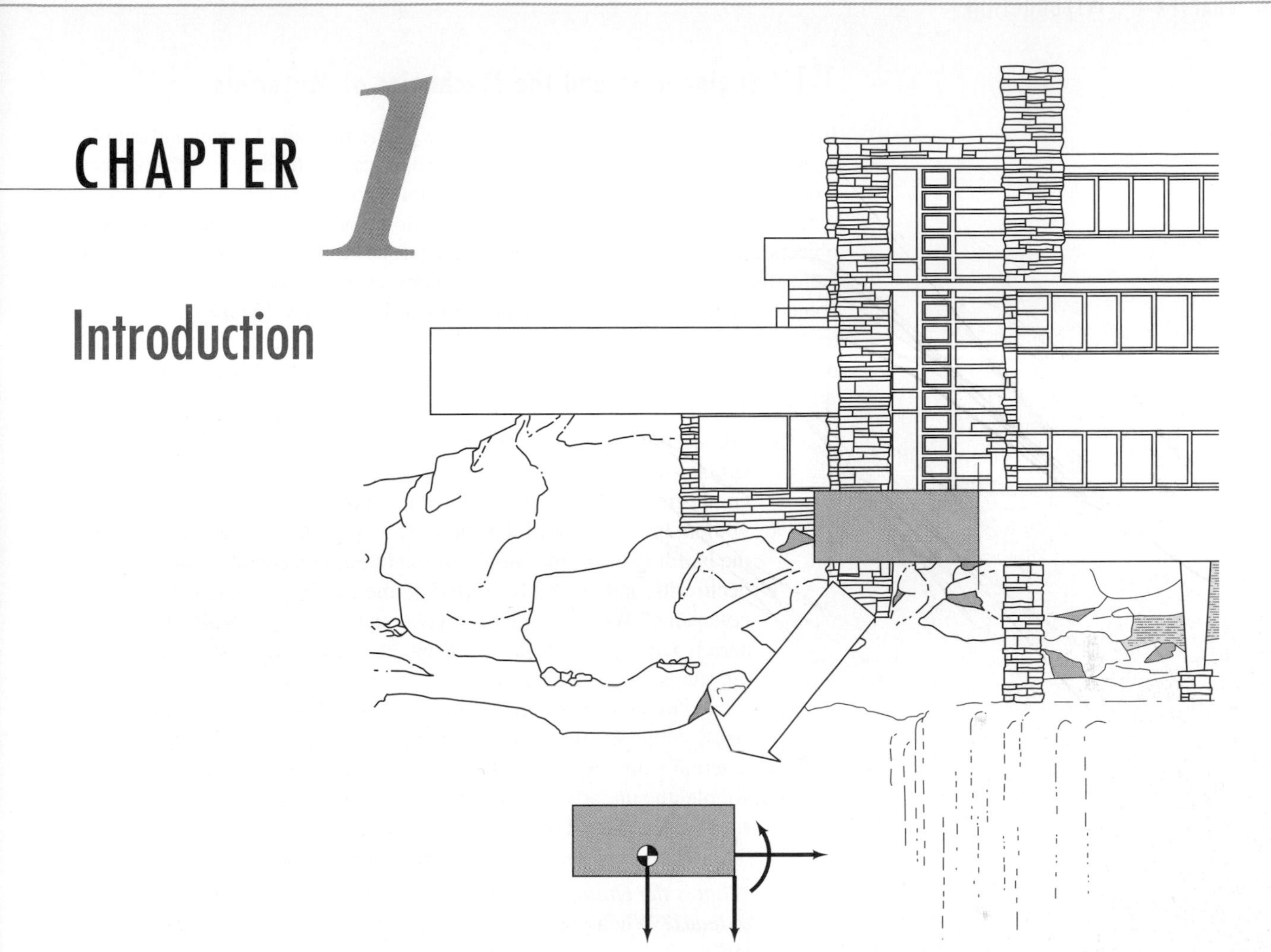

The diagram shows the internal forces and moment exerted on a portion of a cantilevered deck. The mechanics of materials is concerned with the study and analysis of the internal forces within materials and the deformations resulting from those forces. It is one of the sciences underlying the design of any device that must support loads, from the simplest machines and tools to complex vehicles and structures such as the masterpiece Fallingwater. In this chapter we discuss the central questions addressed in the mechanics of materials and review topics with which you must be familiar to begin studying this subject.

1-1 | Engineering and the Mechanics of Materials

In the study of statics you were concerned with the analysis of forces and couples acting on objects in equilibrium. In doing so, it was tacitly assumed that objects would indeed support the forces and couples to which they were subjected without collapsing. Furthermore, deformations, or changes in dimensions of objects due to external loads, were disregarded. It was assumed that objects were effectively rigid, or at least that their deformations were so small that they could be neglected. In the mechanics of materials we develop the concepts and tools of analysis necessary to examine the ability of objects to support loads and determine the resulting deformations. We examine three fundamental questions:

1. ***Will an object or structure support the loads to which it is subjected?*** This is the principal question faced by structural design engineers. Will the parts of a machine perform their functions without breaking? Will the frame of a building support the weight of the building itself, the weight of the building's contents and occupants, and the loads exerted on the building by winds without collapsing? Will an airplane's wing support the gravitational and aerodynamic forces to which it will be subjected? To show how these questions are answered, we introduce the state of stress, which is related to the forces within a material. We show how the states of stress within an object determine whether it will support specified external loads. Although the examples we present are relatively simple, the underlying procedure applies to all structural design: The state of stress must be determined throughout a structure to ensure that it does not exceed the capacities of the materials used.
2. ***What is the change in shape, or deformation, of an object subjected to loads?*** When you pull on a rubber band, it stretches. Riding in an airliner, you can see the wings flex when they are subjected to loads by turbulent air. Any object deforms when it is subjected to loads. We introduce the state of strain, which is related to the state of deformation in the neighborhood of a given point of a material. For the model of materials called linearly elastic materials, we present the relationships between the state of stress and the state of strain and show how these relationships are used in determining the deformations of simple objects. Design engineers must often be concerned with how objects change shape due to the loads acting on them—for example, the members of a linkage must not be allowed to deform so much that the functioning of the linkage is affected—but engineers determine deformations for another very important reason. Doing so enables them to solve statically indeterminate problems.
3. ***What can be done if the external loads on an object cannot be determined by using the equilibrium equations?*** In statics you saw examples of statically indeterminate objects, for which the number of unknown reactions on the free-body diagram exceeded the number of

independent equilibrium equations. Such problems, which are very common in engineering, cannot be solved using the methods of statics alone. We will show that by supplementing the equilibium equations with the relationships between the loads acting on an object and its deformation, the unknown reactions can be determined.

Before we begin answering these questions in Chapter 2, we must discuss units and review concepts from statics.

1-2 | Units and Numbers

We use both the SI and U.S. Customary systems of units in examples and problems. We summarize these systems in this section.

International System of Units

Length is measured in meters (m), mass in kilograms (kg), and time in seconds (s). These are the base units of the SI system. Force is measured in newtons (N). One newton is the force required to give an object of 1 kilogram mass an acceleration of 1 meter per second squared. The relationship between an object's mass m in kilograms and its weight W at sea level in newtons is $W = mg$, where $g = 9.81\ \text{m/s}^2$ is the acceleration due to gravity at sea level.

Pressures and stresses are usually expressed in SI units in terms of newtons per square meter (N/m^2), which are called pascals (Pa). Occasionally they are expressed in terms of bars (1 bar $= 10^5$ Pa).

To express quantities in SI units by numbers of convenient size, multiples of units are indicated by prefixes. The most common prefixes, their abbreviations, and the multiples they represent are shown in Table 1-1. For example, 1 GPa (gigapascal) is 1×10^9 pascals.

U.S. Customary Units

Length is measured in feet (ft), force in pounds (lb), and time in seconds (s). These are the base units of the U.S. Customary system. The unit of mass is a

Table 1-1 Common prefixes used in SI units and the multiples they represent

Prefix	Abbreviation	Multiple
nano-	n	10^{-9}
micro-	μ	10^{-6}
milli-	m	10^{-3}
kilo-	k	10^{3}
mega-	M	10^{6}
giga-	G	10^{9}

derived unit, the slug, which is the mass of material accelerated at 1 foot per second squared by a force of 1 pound. The relationship between an object's mass m in slugs and its weight W at sea level in pounds is $W = mg$, where $g = 32.2$ ft/s^2. We frequently use the inch (1 ft = 12 in.) and kilopound (1 kip = 1000 lb).

Pressures and stresses are expressed in U.S. Customary Units in terms of pounds per square foot (lb/ft^2 or psf) or pounds per square inch (lb/in^2 or psi). We will also use thousands of pounds per square foot (kip/ft^2 or ksf) and thousands of pounds per square inch (kip/in^2 or ksi).

In some engineering applications an alternative unit of mass called the pound mass (lbm) is used, which is the mass of material having a weight of 1 pound at sea level. The weight at sea level of an object that has a mass of 1 slug is $W = mg = (1 \text{ slug})(32.2 \text{ ft/s}^2) = 32.2$ lb, so 1 lbm = 1/32.2 slug. When the pound mass is used, a pound of force is usually denoted by the abbreviation lbf. We do not use the pound mass.

Use of Numbers

We treat numbers given in problems as exact values: If a problem states that a quantity equals 32.2, you can assume that its value is 32.200.... We express intermediate results and answers in the examples and the answers to the problems to at least three significant digits. In your own calculations, try to avoid round-off errors that occur if you round off intermediate results when making a series of calculations. Maintain as much accuracy as you can by retaining values in your calculator.

1-3 Review of Statics

The fundamental concepts of mechanics of materials that we discuss in this book—the states of stress and strain in materials and their relationships—apply in both static and dynamic situations. However, for simplicity in this introductory treatment we consider only materials and objects that are in equilibrium. In this section we review some of the concepts from statics with which you need to be familiar to understand developments in subsequent chapters.

Free-Body Diagrams

The *free-body diagram*, an essential tool in mechanics, is simply a drawing of a particular object showing the external forces and couples acting on it. The object is said to be *isolated*, or *freed*, from its surroundings. *External* forces and couples are those exerted by objects not included in the free-body diagram. For example, suppose that a man is standing on a floor. In a free-body diagram of the man, his weight, exerted by the earth, and the forces exerted on his feet by the floor are external forces and would be shown in his free-body diagram. But if he exerts forces on his hands by pressing them together, those are internal

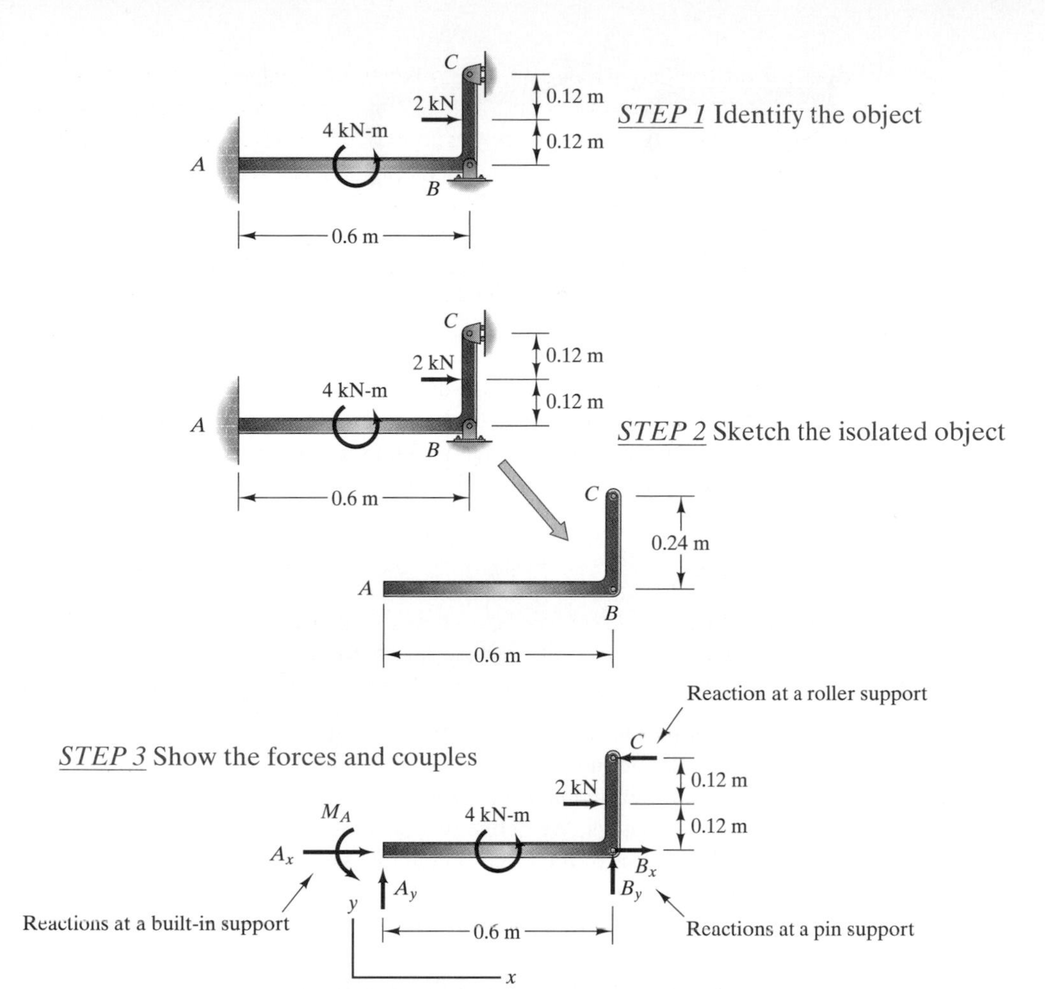

FIGURE 1-1 Steps in drawing a free-body diagram.

forces and would not be shown. Drawing a free-body diagram involves three steps, which we illustrate in Fig. 1-1:

1. ***Identify the object you want to isolate.*** Your choice will often be dictated by particular forces and couples you want to determine. In Fig. 1-1 our objective is to draw a free-body diagram of the L-shaped bar. The bar is subjected to two loads—a 2-kN force and a 4-kN-m couple—and has built-in, pin and roller supports.
2. ***Draw a sketch of the object isolated from its surroundings.*** In Fig. 1-1 we carry out this step by drawing a sketch of the bar isolated from its supports.
3. ***Show the external forces and couples acting on the isolated object and label them.*** In Fig. 1-1 we complete the free-body diagram of the bar by adding the loads and the reactions exerted by the supports to the sketch of the isolated bar.

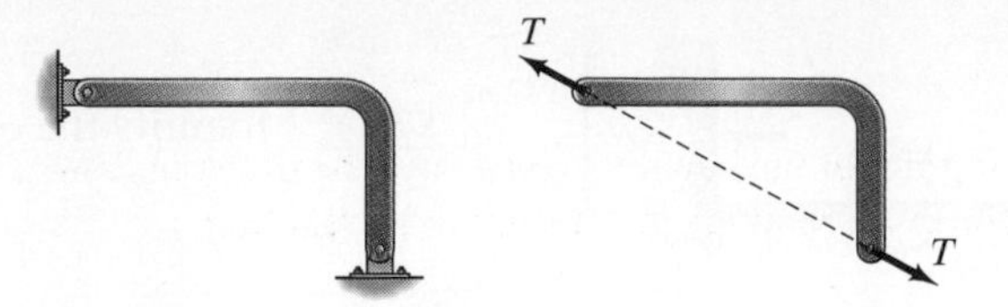

FIGURE 1-2 A two-force member.

Equilibrium

We say that an object is in *equilibrium* during an interval of time if it is stationary or is in steady translation relative to an inertial reference frame. If an object is in equilibrium, the sum of the external forces acting on it and the sum of the moments about any point due to the external forces and couples acting on it are zero:

$$\Sigma \mathbf{F} = \mathbf{0}, \tag{1-1}$$

$$\Sigma \mathbf{M}_{\text{any point}} = \mathbf{0}. \tag{1-2}$$

In some situations you can use the equilibrium equations to determine unknown forces and couples acting on objects (see Example 1-1).

A *system of forces and moments* is simply some particular set of forces and moments due to couples. We define two systems of forces and moments to be *equivalent* if the sums of the forces in the two systems are equal and the sums of the moments about any point due to the two systems are equal. Notice that if the equilibrium equations (1-1) and (1-2) hold for a given system of forces and moments, they also hold for any equivalent system.

A *two-force member* is an object that is subjected to two forces acting at different points and no couples. If a two-force member is in equilibrium, the two forces must be equal in magnitude, opposite in direction, and have the same line of action (Fig. 1-2).

EXAMPLE 1-1

The beam in Fig. 1-3 supports a 100-kN force and a 600-kN-m couple. Determine the reactions at the built-in support A.

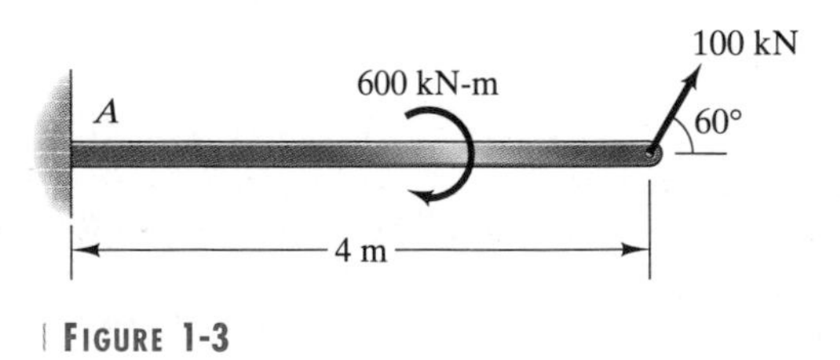

FIGURE 1-3

Strategy

We will draw a free-body diagram of the beam by isolating it from the built-in support and apply the equilibrium equations to determine the reactions.

Solution

We draw the free-body diagram of the beam in Fig. (a). The terms A_x and A_y are the components of the force and M_A is the couple exerted on the beam by the support.

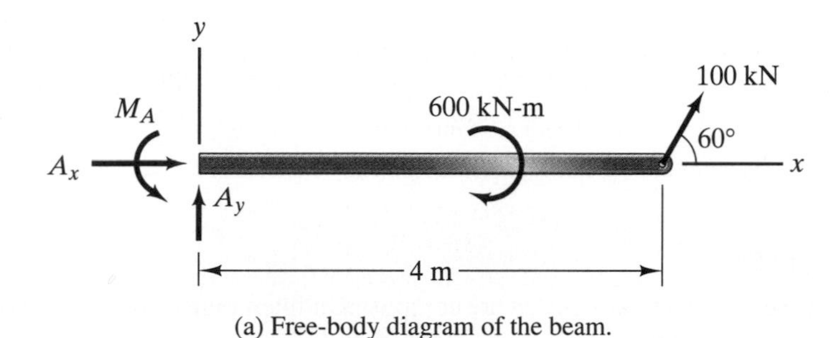

(a) Free-body diagram of the beam.

From the equilibrium equations

$$\Sigma F_x = A_x + 100\cos 60^\circ = 0,$$

$$\Sigma F_y = A_y + 100\sin 60^\circ = 0,$$

$$\Sigma M_{\text{point } A} = M_A - 600 + (4)(100\sin 60^\circ) = 0,$$

we determine the reactions $A_x = -50$ kN, $A_y = -86.6$ kN, and $M_A = 253.6$ kN-m.

Discussion

If we modify this beam by placing a roller support at the right end [Fig. (b)], it becomes statically indeterminate. There are now four unknown reactions [Fig. (c)], and we can write only three independent equilibrium equations:

$$\Sigma F_x = A_x = 0,$$

$$\Sigma F_y = A_y + B = 0,$$

$$\Sigma M_{\text{point } A} = M_A - 600 + 4B = 0.$$

We cannot determine the reactions A_y, M_A, and B from these equations. In Chapter 9 we solve problems such as this by using the equilibrium equations together with equations that relate the loads and reactions on the beam to its lateral deformation.

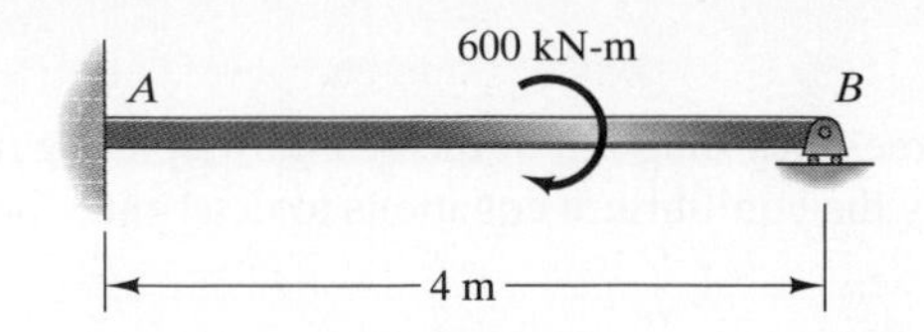

(b) Placing a roller support at the right end of the beam.

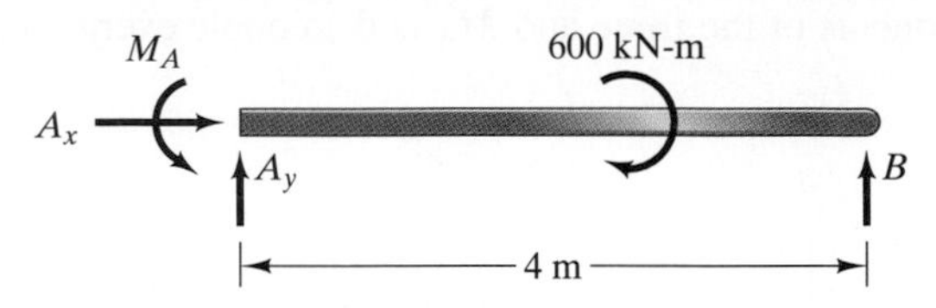

(c) There are four unknown reactions.

Structures

Here we consider structures that are composed of interconnected parts, or *members*. We call a structure a *truss* if it consists entirely of two-force members. A typical truss is made up of straight bars connected at their ends by pins, is loaded by forces at joints where the members are connected, and is supported at joints (Fig. 1-4). Because the members of a truss are two-force members, they are subjected only to axial forces. In Example 1-2 we review methods for determining the axial forces in the members of a statically determinate truss.

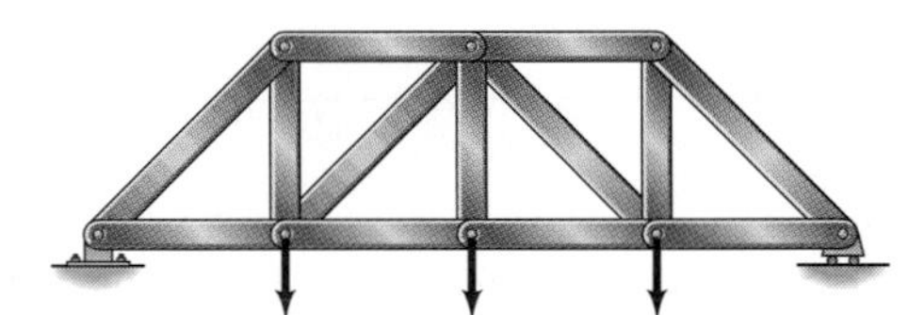

FIGURE 1-4 Howe bridge truss. Notice that it is loaded and supported at its joints.

A structure of interconnected members that does not satisfy the definition of a truss is called a *frame* if it is designed to remain stationary and support loads, and a *machine* if it is designed to move and exert loads. Such structures are analyzed to determine the reactions at the connections between members by drawing the free-body diagrams of the individual members. We review this process for a statically determinate frame in Example 1-3.

EXAMPLE 1-2

Determine the axial force in member *BH* of the Howe truss in Fig. 1-5 and indicate whether the member is in tension or compression.

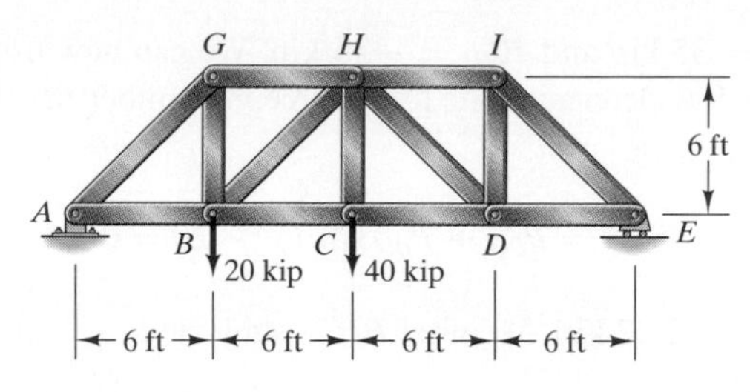

FIGURE 1-5

Strategy

We will apply the method of joints, using equilibrium to analyze carefully chosen joints until we have determined the axial force in member BH.

Solution

We first draw the free-body diagram of the entire truss [Fig. (a)]. From the equilibrium equations for this free-body diagram, we determine that the reactions at the supports are $A_x = 0$, $A_y = 35$ kip, and $E = 25$ kip.

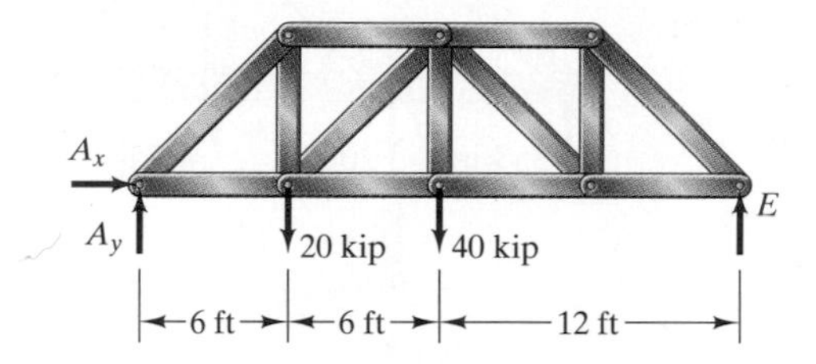

(a) Free-body diagram of the entire truss.

Applying the method of joints, we first draw the free-body diagram of joint A [Fig. (b)], because it is subjected to a known force, the 35-kip reaction at A, and only two unknown forces, the axial forces in members AB and AG. From the equilibrium equations

$$\Sigma F_x = P_{AB} + P_{AG} \cos 45^\circ = 0,$$

$$\Sigma F_y = 35 + P_{AG} \sin 45^\circ = 0,$$

we obtain $P_{AB} = 35$ kip and $P_{AG} = -49.5$ kip. (The signs of these answers indicate that member AB is in tension and member AG is in compression.) Now that we know P_{AG}, the free-body diagram of joint G has only two unknown forces [Fig. (c)]. From the equilibrium equations

$$\Sigma F_x = P_{GH} - P_{AG} \sin 45^\circ = 0,$$

$$\Sigma F_y = -P_{BG} - P_{AG} \cos 45^\circ = 0,$$

we obtain $P_{BG} = 35$ kip and $P_{GH} = -35$ kip. We can now use the free-body diagram of joint B to determine the axial force in member BH [Fig. (d)]. From the equation

$$\Sigma F_y = P_{BG} + P_{BH} \sin 45^\circ - 20 = 0,$$

we obtain $P_{BH} = -21.2$ kip. Member BH is subjected to a compressible axial force of 21.2 kip.

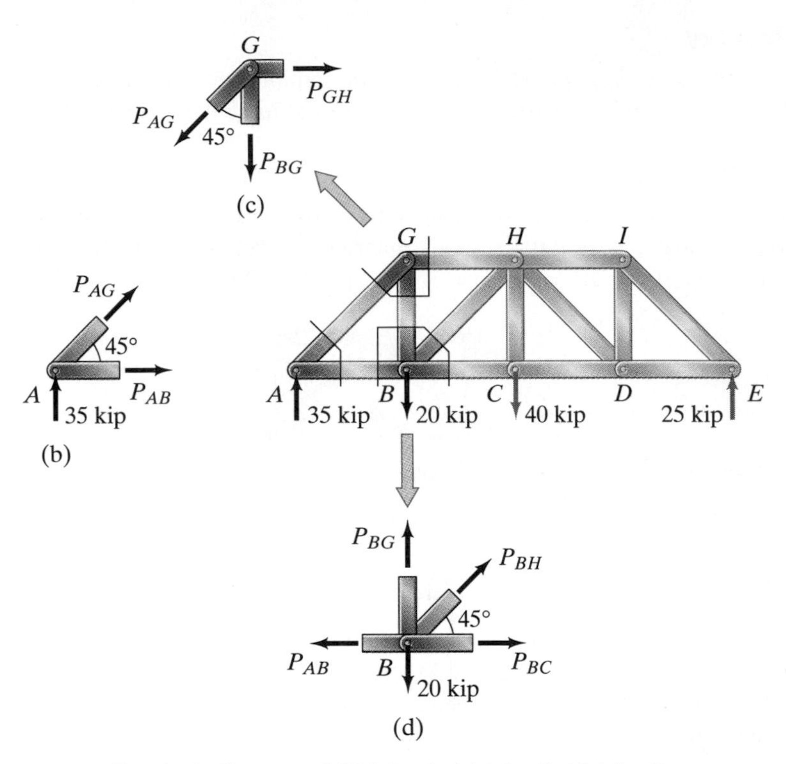

Free-body diagrams of (b) joint A; (c) joint G; (d) joint B.

Discussion

We can also determine the axial force in member BH by applying the method of sections. In Fig. (e) we isolate a section of the truss by "cutting" members BC, BH, and GH. From the equilibrium equation

$$\Sigma F_y = 35 - 20 + P_{BH} \sin 45^\circ = 0,$$

we obtain $P_{BH} = -21.2$ kip. The method of sections can sometimes provide the information you need much more easily than the method of joints, but you can't always find a section that works.

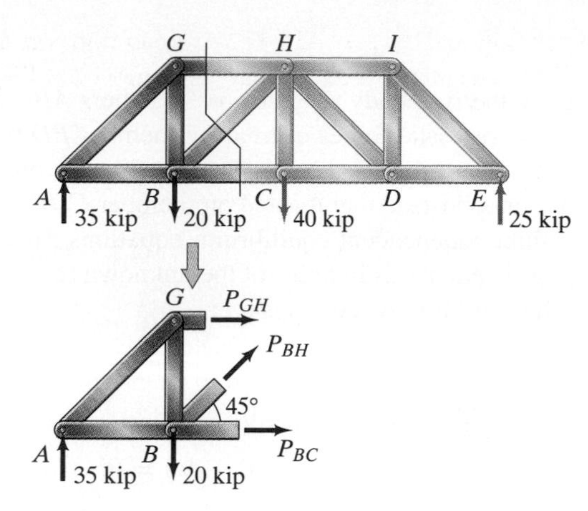

(e) Section suitable for determining the axial force in member *BH*.

EXAMPLE 1-3

The frame in Fig. 1-6 supports a suspended mass $m = 20$ kg. Determine the reactions on its members.

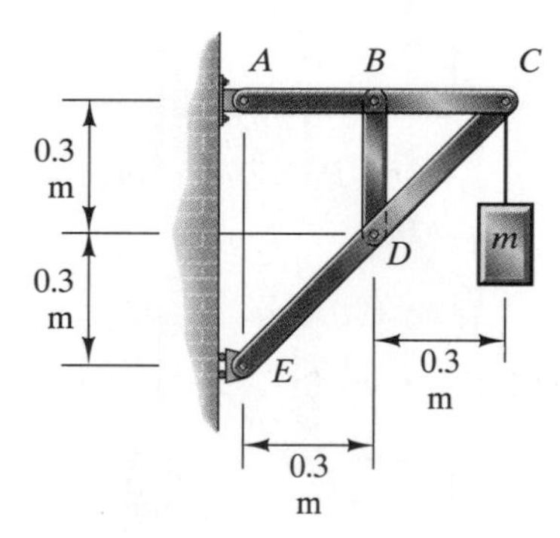

FIGURE 1-6

Strategy

We will draw the free-body diagrams of the individual members and apply the equilibrium equations to each member to determine the reactions. Notice that member *BD* is a two-force member. We can take advantage of this observation to simplify the free-body diagrams of members *ABC* and *CDE*—member *BD* exerts equal and opposite forces at *B* and *D* that are parallel to member *BD*—and we do not need to write equilibrium equations for member *BD*.

Solution

In Fig. (a) we draw the free-body diagrams of members *ABC* and *CDE*. We denote the equal and opposite forces exerted by member *BD* by P. Also, we have assumed that the force exerted by the weight mg acts on member *ABC*. We could have assumed instead that it acted on member *CDE*.

We can write three independent equilibrium equations for each free-body diagram, obtaining six equations in terms of the unknown reactions A_x, A_y, P, C_x, C_y, and E. The equilibrium equations are:

Member ABC:

$$\Sigma F_x = A_x + C_x = 0,$$

$$\Sigma F_y = A_y + C_y - P - mg = 0,$$

$$\Sigma M_{\text{point } A} = -0.3P + 0.6C_y - 0.6mg = 0.$$

Member CDE:

$$\Sigma F_x = E - C_x = 0,$$

$$\Sigma F_y = P - C_y = 0,$$

$$\Sigma M_{\text{point } E} = 0.3P + 0.6C_x - 0.6C_y = 0.$$

Setting $m = 20$ kg and $g = 9.81$ m/s^2 and solving, we obtain $A_x = -196.2$ N, $A_y = 196.2$ N, $P = 392.4$ N, $C_x = 196.2$ N, $C_y = 392.4$ N, and $E = 196.2$ N.

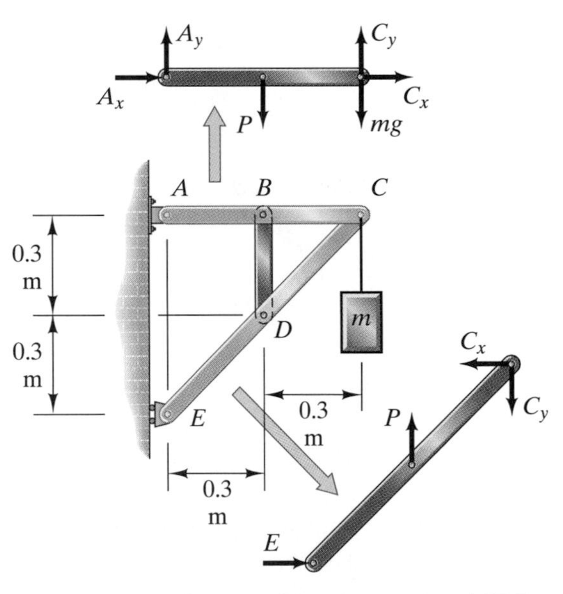

(a) Free-body diagrams of members *ABC* and *CDE*.

Centroids

The coordinates of the centroid of an area A in the x–y plane (Fig. 1-7) are

$$\bar{x} = \frac{\int_A x\, dA}{\int_A dA}, \qquad \bar{y} = \frac{\int_A y\, dA}{\int_A dA}.$$

The centroid of a composite area consisting of parts 1, 2, . . . whose centroid locations are known (Fig. 1-8) can be determined from the relations

$$\bar{x} = \frac{\bar{x}_1 A_1 + \bar{x}_2 A_2 + \cdots}{A_1 + A_2 + \cdots}, \qquad \bar{y} = \frac{\bar{y}_1 A_1 + \bar{y}_2 A_2 + \cdots}{A_1 + A_2 + \cdots}.$$

A "hole" or cutout can be treated as a negative area (see Example 1-4).

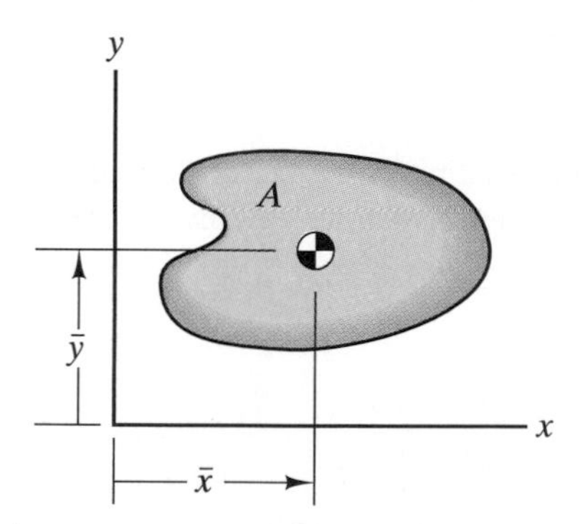

FIGURE 1-7 Coordinates of the centroid of A.

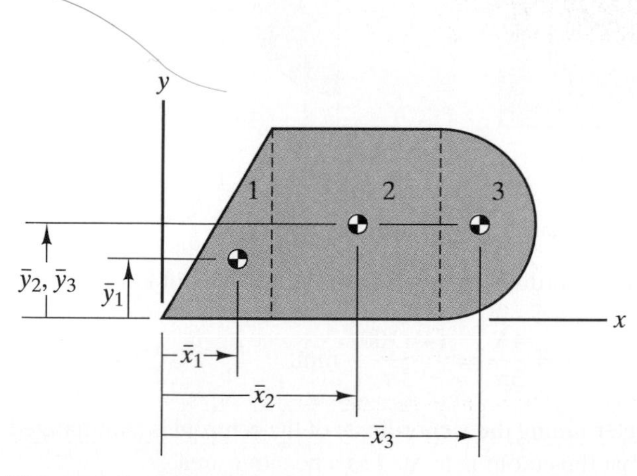

FIGURE 1-8 Composite area showing the coordinates of the centroids of the parts.

EXAMPLE 1-4

Determine the x coordinate of the centroid of the area in Fig. 1-9.

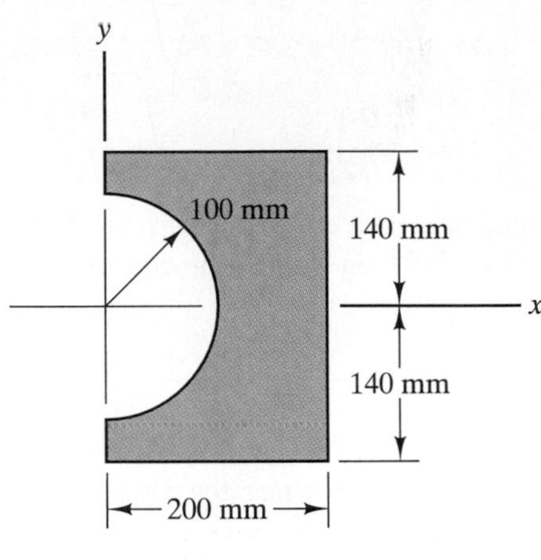

FIGURE 1-9

Solution

We will treat the area as a composite area consisting of the rectangle without the semicircular cutout and the area of the cutout, which we call parts 1 and 2 [Fig. (a)].

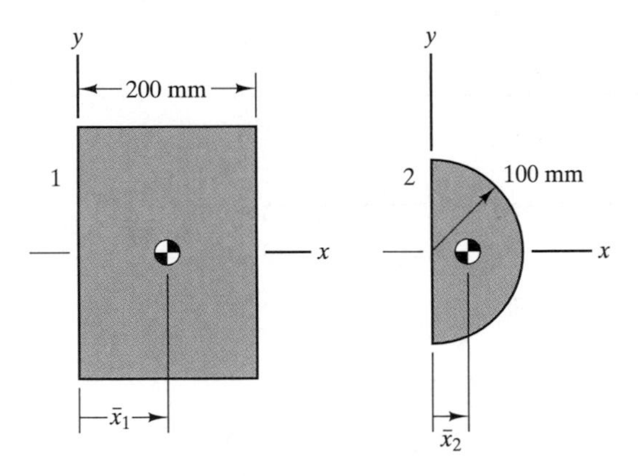

(a) Rectangle and semicircular cutout.

From Appendix D, the x coordinate of the centroid of the cutout is

$$\bar{x}_2 = \frac{4R}{3\pi} = \frac{(4)(100)}{3\pi} \text{ mm.}$$

The information for determining the x coordinate of the centroid is summarized in Table 1-2. Notice that the cutout is treated as a negative area.

Table 1-2 Information for determining $\bar{x}$

	$\bar{x}_i$ (mm)	A_i (mm^2)	$\bar{x}_i A_i$ (mm^3)
Part 1 (rectangle)	100	(200)(280)	(100)[(200)(280)]
Part 2 (cutout)	$\frac{(4)(100)}{3\pi}$	$-\frac{1}{2}\pi(100)^2$	$-\frac{(4)(100)}{3\pi}\left[\frac{1}{2}\pi(100)^2\right]$

The x coordinate of the centroid is

$$\bar{x} = \frac{\bar{x}_1 A_1 + \bar{x}_2 A_2}{A_1 + A_2} = \frac{(100)[(200)(280)] - [(4)(100)/3\pi]\left[\frac{1}{2}\pi(100)^2\right]}{(200)(280) - \frac{1}{2}\pi(100)^2}$$

$$= 122.4 \text{ mm}.$$

Distributed Forces

To express a force distributed along a line, we define a function w, called the *loading curve,* such that the *downward* force exerted on each infinitesimal element dx of the line is $w\,dx$ (Fig. 1-10). For example, the force exerted by a floor of a building on one of the horizontal beams supporting the floor is distributed along the length of the beam. This type of distributed load can also

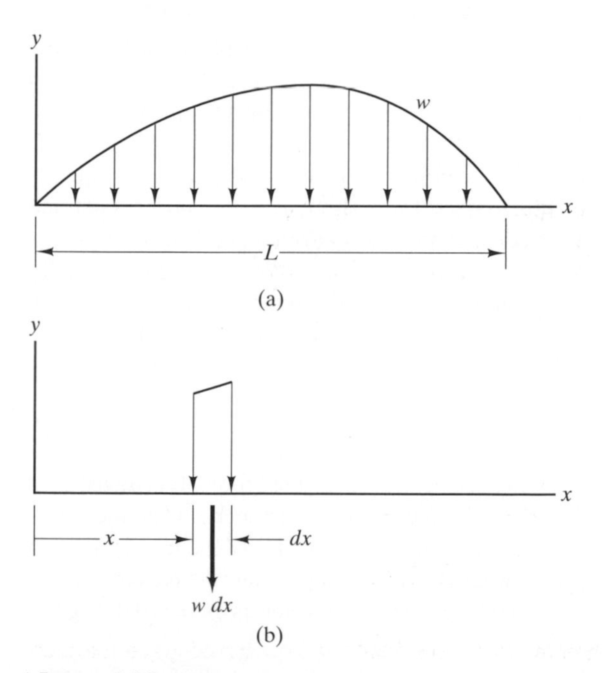

FIGURE 1-10 (a) Force distributed along the x axis. (b) The force on each element dx is $w\,dx$.

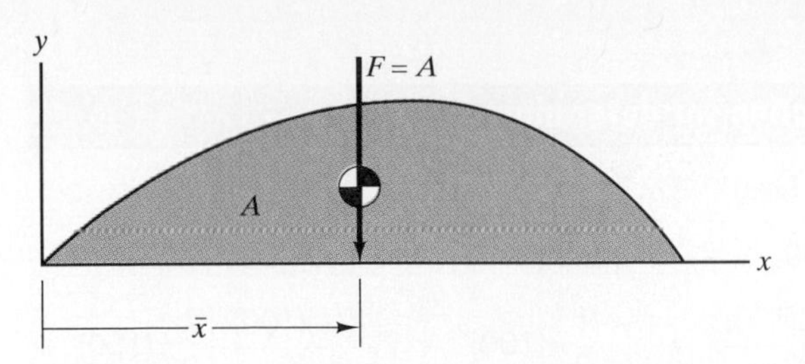

FIGURE 1-11 Representing a distributed load by an equivalent force.

be used to model a large number of discrete loads acting along a line, such as the forces exerted by the wheels of the traffic on a bridge.

The total downward force exerted by the distributed load on a portion L of the x axis is

$$F = \int_L w\, dx,$$

and the total clockwise moment about the origin is

$$M = \int_L xw\, dx.$$

The integral determining F indicates that the total force equals the "area" A between the loading curve and the x axis. Also, the moment due to F about the origin equals the total moment M due to the distributed load if F acts at the centroid of A (Fig. 1-11). Therefore, the force $F = A$ placed with its line of action through the centroid of A is equivalent to the distributed load. In Example 1-5 we review an equilibrium problem involving a beam subjected to a distributed load.

The force exerted by pressure on an object submerged in a stationary liquid is distributed over the surface of the object (Fig. 1-12a). The pressure p is defined such that the normal force exerted on an infinitesimal element dA of the object's surface is $p\, dA$ (Fig. 1-12b). The pressure increases with depth, so it is usually necessary to integrate to determine the total force exerted on a finite part of an object's surface. The total normal force on a plane surface of area A is

$$F = \int_A p\, dA.$$

An object's weight is another example of a distributed force. It is distributed over the volume of the object. The weight density γ of a material is defined such that the weight of an infinitesimal element of volume dV of the material at sea level is $\gamma\, dV$. The relationship between the weight density and the mass density ρ is $\gamma = \rho g$. For example, the mass density of tungsten is $\rho = 19.4\ \text{Mg/m}^3$ (megagrams per cubic meter), so its weight density is

$$\gamma = (19.4 \times 10^3\ \text{kg/m}^3)(9.81\ \text{m/s}^2) = 190 \times 10^3\ \text{N/m}^3.$$

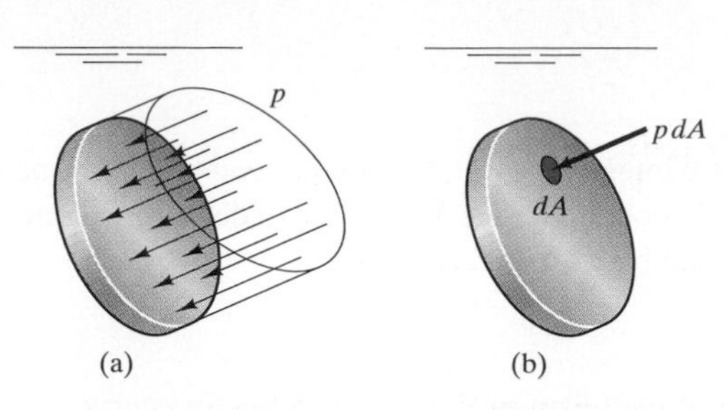

FIGURE 1-12 (a) Pressure is a distributed load. (b) The normal force on an element dA is $p\,dA$.

Table 1-3 Distributed loads

Type of distributed load	Domain	Force on an element
Distributed load on a beam	Line	$w\,dx$
Pressure	Area	$p\,dA$
Weight	Volume	$\gamma\,dV$

An object's total weight W is obtained by integrating the weight density over its volume,

$$W = \int_V \gamma\,dV,$$

and if the object is homogeneous ($\gamma =$ constant), its total weight is $W = \gamma V$.

We have discussed loads distributed over lines, areas, and volumes (Table 1-3). In each case the load is characterized by a function that determines the force acting on an infinitesimal element of its domain of definition. In Chapter 2 we show that the stresses in materials are described in terms of distributed loads defined exactly the same way as these familiar examples.

EXAMPLE 1-5

The beam in Fig. 1-13 is subjected to a triangular distributed load whose value at the right end of the beam is 120 lb/ft. What are the reactions at A and B?

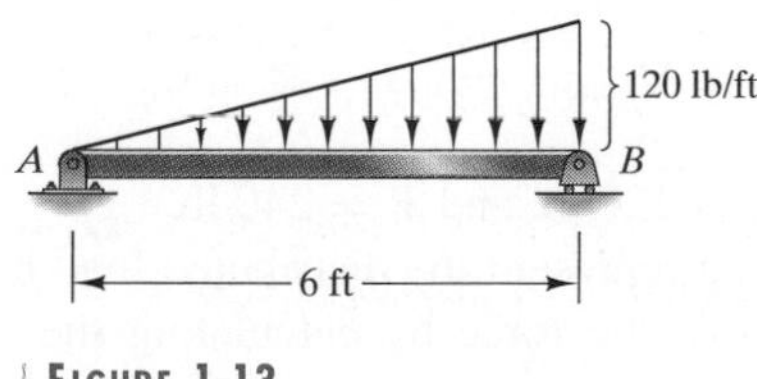

FIGURE 1-13

Strategy

We can determine the reactions by integrating the distributed load to determine the total force and moment it exerts on the beam. We will also determine them by representing the distributed load by an equivalent force.

Solution

We draw the free-body diagram of the beam in Fig. (a). The loading curve is a straight line whose value is 120 lb/ft at $x = 6$ ft. Therefore, the function w is

$$w = \left(\frac{120 \text{ lb/ft}}{6 \text{ ft}}\right) x = 20x \text{ lb/ft}.$$

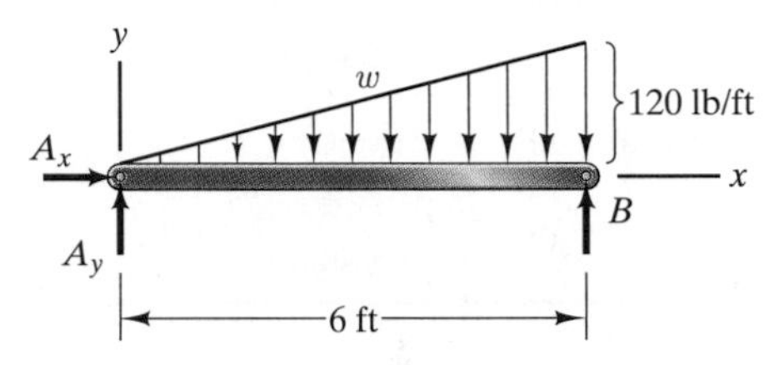

(a) Free-body diagram of the beam.

The total downward force exerted by the distributed load is

$$F = \int_0^6 w \, dx = \int_0^6 20x \, dx$$
$$= 360 \text{ lb},$$

and the total clockwise moment about the origin is

$$M = \int_0^6 xw \, dx = \int_0^6 20x^2 \, dx$$
$$= 1440 \text{ ft-lb}.$$

From the equilibrium equations

$$\Sigma F_x = A_x = 0,$$
$$\Sigma F_y = A_y + B - F = 0,$$
$$\Sigma M_{\text{point } A} = 6B - M = 0,$$

we obtain $A_x = 0$, $A_y = 120$ lb, and $B = 240$ lb.

Alternatively, we can represent the distributed load by an equivalent force [Fig. (b)]. We determine the force by calculating the triangular area under the loading curve,

$$F = \tfrac{1}{2}(120\ \text{lb/ft})(6\ \text{ft}) = 360\ \text{lb},$$

and we place the force at the centroid of the triangle, $\bar{x} = \tfrac{2}{3}(6\,\text{ft}) = 4$ ft.

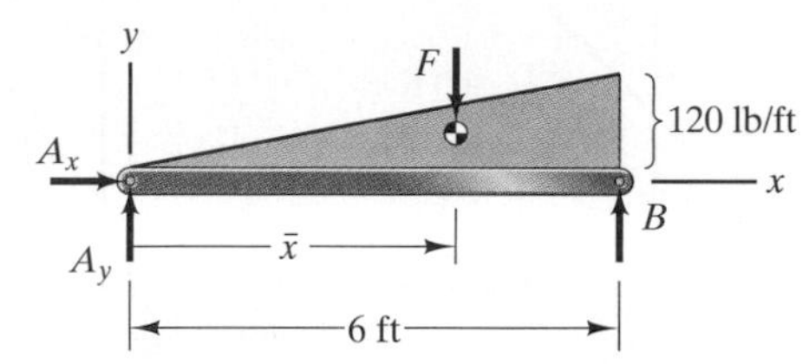

(b) Representing the distributed load by an equivalent force.

The equilibrium equations are

$$\Sigma F_x = A_x = 0,$$

$$\Sigma F_y = A_y + B - F = 0,$$

$$\Sigma M_{\text{point } A} = 6B - \bar{x}F = 0,$$

and we again obtain $A_x = 0$, $A_y = 120$ lb, and $B = 240$ lb.

PROBLEMS

1-3.1. The beam is subjected to a force and a couple. **(a)** Draw the free-body diagram of the beam. **(b)** Determine the reactions at A and B.

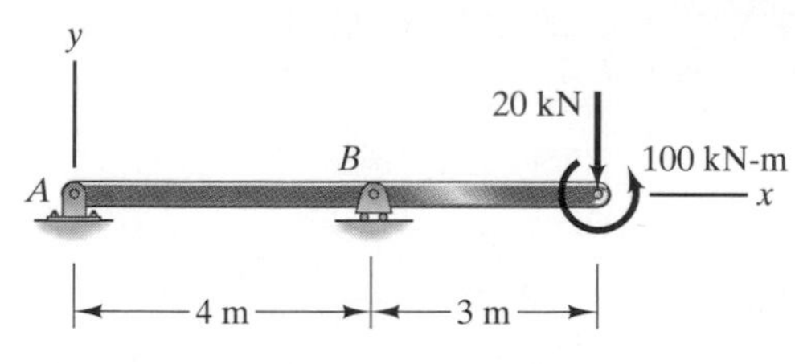

PROBLEM 1-3.1

1-3.2. In Problem 1-3.1, determine the reactions at A and B if the 100 kN-m couple acts on the beam at A. Compare your answers to the answers to Problem 1-3.1.

1-3.3. **(a)** Draw the free-body diagram of the bar. **(b)** Determine the reactions at the built-in support A.

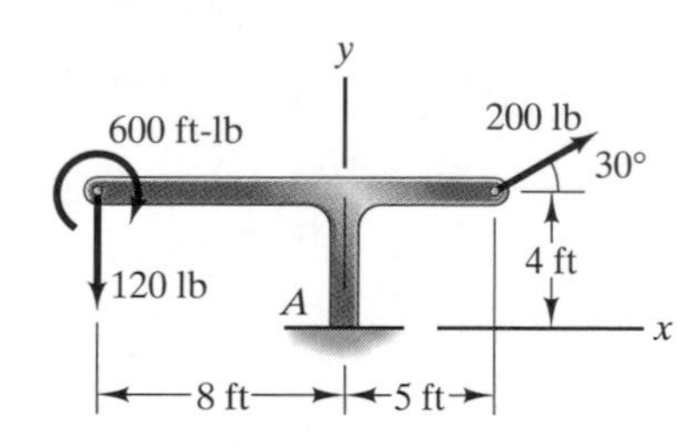

PROBLEM 1-3.3

1-3.4. In Problem 1-3.3, suppose that you want to change the 5-ft dimension so that the couple exerted on the beam by the built-in support is zero. What is the required dimension?

1-3.5. Determine the reactions exerted on the L-shaped bar by its supports at A and B.

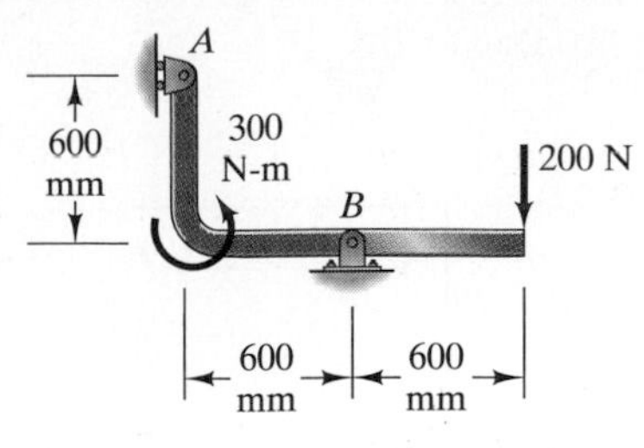

PROBLEM 1-3.5

1-3.6. The mass of the isolated portion of one of Fallingwater's decks is 14,700 kg. Determine the forces P and V and the couple M.

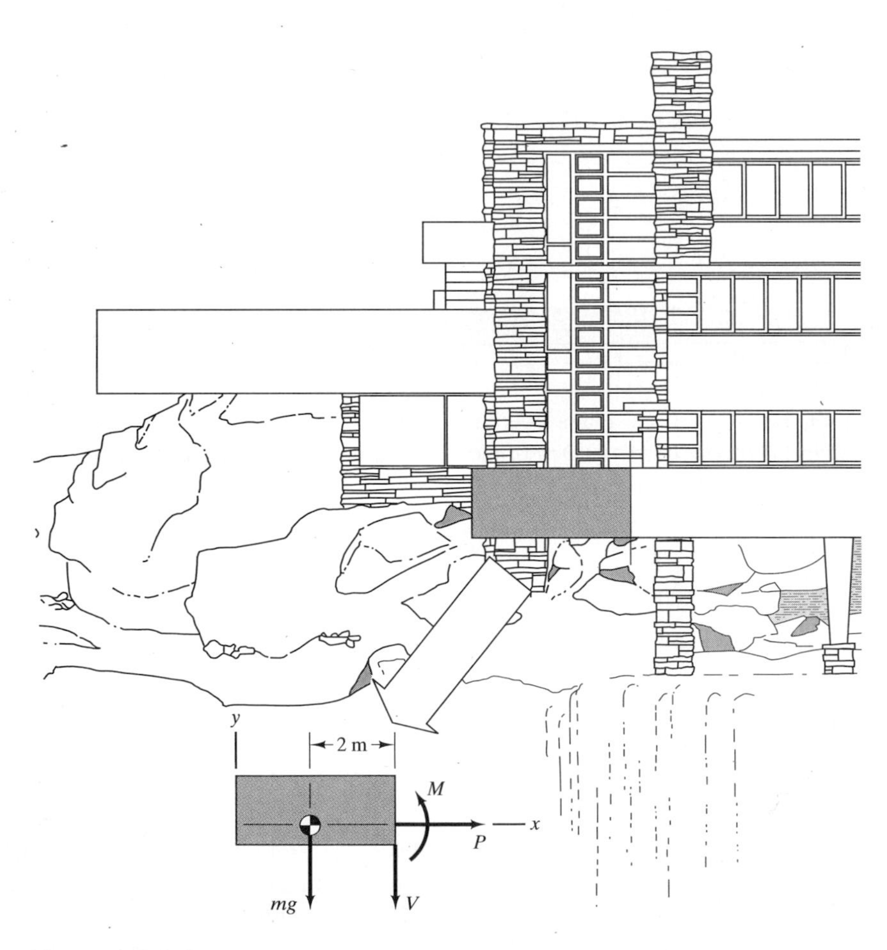

PROBLEM 1-3.6

1-3.7. The two systems of forces and moments (a) and (b) are equivalent. Determine the magnitude of the force F, the angle α, and the magnitude of the clockwise couple C.

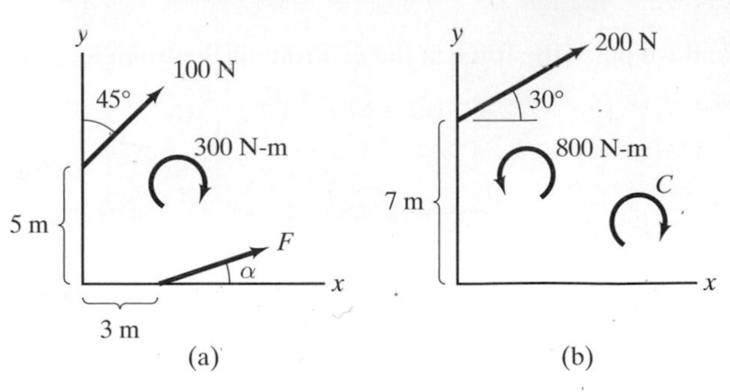

PROBLEM 1-3.7

1-3.8. If you represent the equivalent systems of forces and moments in Problem 1-3.7 by a new equivalent system consisting of a single force **R** acting at the origin of the coordinate system and a couple **M**, what are the force **R** and the couple **M**?

1-3.9. The suspended mass $m = 20$ kg. Determine the axial force in the bar AB and indicate whether it is in tension or compression. (*Strategy:* The bar is a two-force member. Draw the free-body diagram of joint B.)

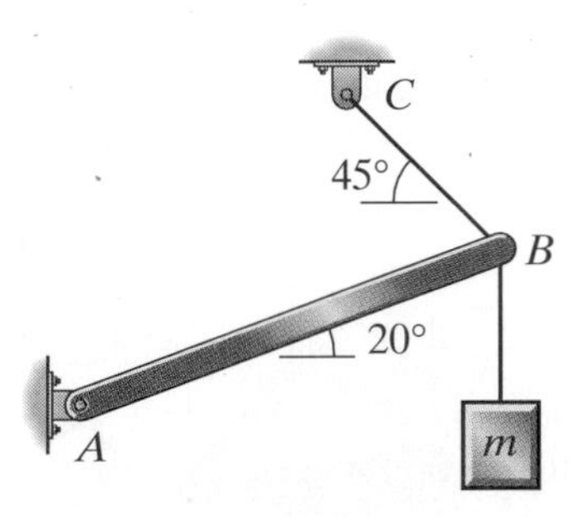

PROBLEM 1-3.9

1-3.10. If the bar AB in Problem 1-3.9 will safely support a compressive load of 400 N, what is the largest mass m that can be supended as shown?

1-3.11. Write the equilibrium equations for the entire truss shown in Fig. 1-5 and confirm that $A_x = 0$, $A_y = 35$ kip, and $E = 25$ kip.

1-3.12. Determine the magnitude of the axial force in member DH of the truss shown in Fig. 1-5 and indicate whether it is in tension or compression.

1-3.13. The force $F = 2$ kip. Determine the axial loads in members AB and AC of the truss and indicate whether they are in tension (T) or compression (C).

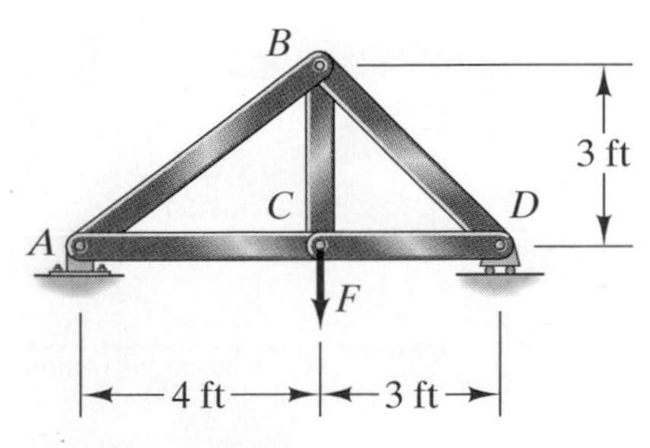

PROBLEM 1-3.13

1-3.14. The members of the truss shown in Problem 1-3.13 will safely support a tensile axial load of 4 kip and a compressive axial load of 2 kip. Based on these criteria, what is the largest safe value of F?

1-3.15. The suspended mass $m = 200$ kg. Determine the magnitudes of the axial forces in members BC, BD, and CD and indicate whether they are in tension (T) or compression (C).

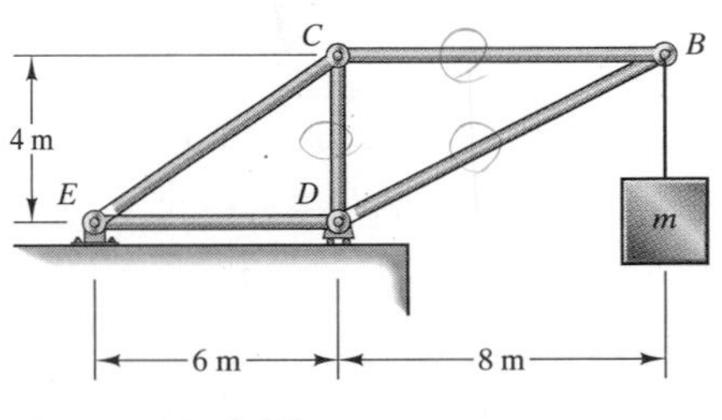

PROBLEM 1-3.15

1-3.16. For the truss in Problem 1-3.15, determine the magnitudes of the axial forces in members CE and DE and indicate whether they are in tension (T) or compression (C).

1-3.17. The loads $F_B = 40$ kN, $F_C = 60$ kN, and $F_D = 20$ kN. Determine the magnitudes of the axial forces in members BC, CG, and GH and indicate whether they are in tension (T) or compression (C).

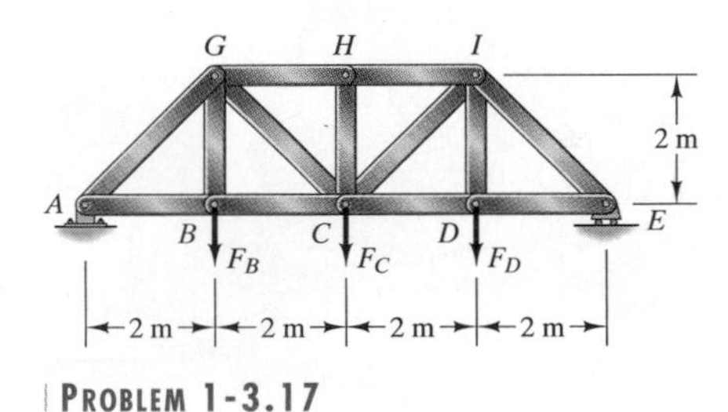

PROBLEM 1-3.17

1-3.18. Determine the reactions on member CDE of the frame.

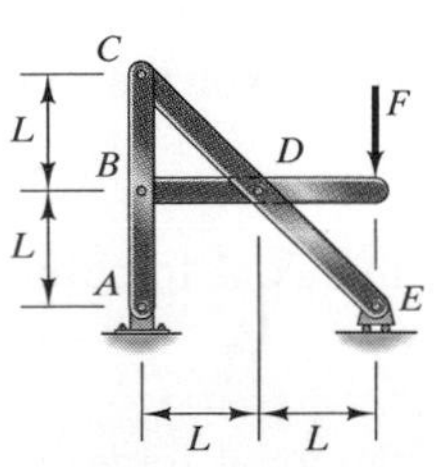

PROBLEM 1-3.18

1-3.19. The bucket of the front-end loader is supported by a pin support at C and the hydraulic actuator AB. If the mass of the bucket is 180 kg and the system is stationary, what is the axial load in the hydraulic actuator?

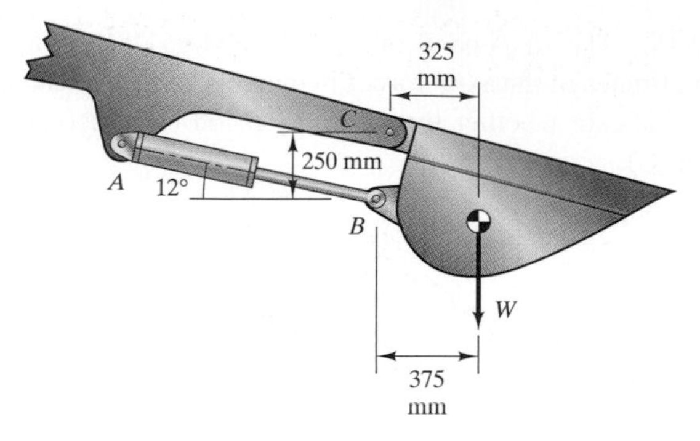

PROBLEM 1-3.19

1-3.20. The suspended crate weighs 2000 lb and the angle $\alpha = 30°$. If you neglect the weight of the crane's boom, what is the axial force in the hydraulic cylinder BC?

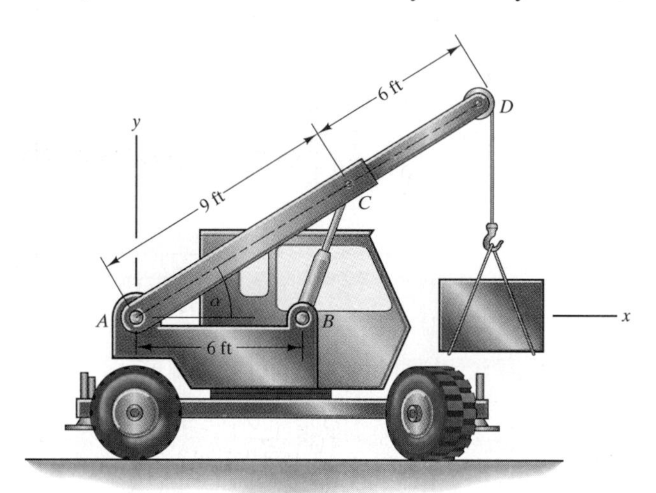

PROBLEM 1-3.20

1-3.21. In Problem 1-3.20, determine the x and y components of the reaction exerted on the crane's boom by the pin support A.

1-3.22. Determine the axial force in member BE of the frame.

1-3.23. Determine the axial force in member CF of the frame in Problem 1-3.22.

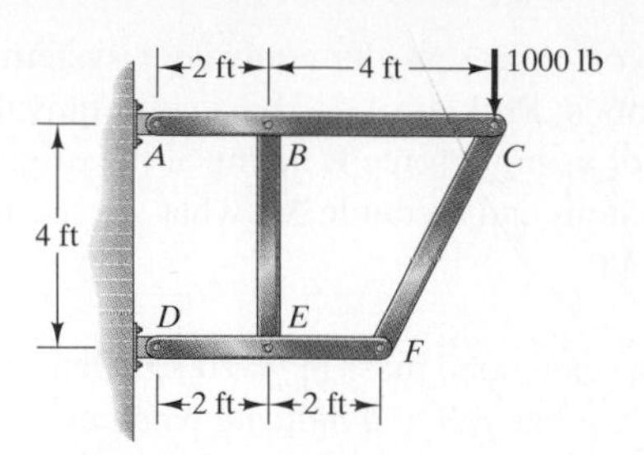

PROBLEM 1-3.22

1-3.24. The system shown supports half of the weight of the 680-kg excavator. If the system is stationary, what is the axial load in member AB?

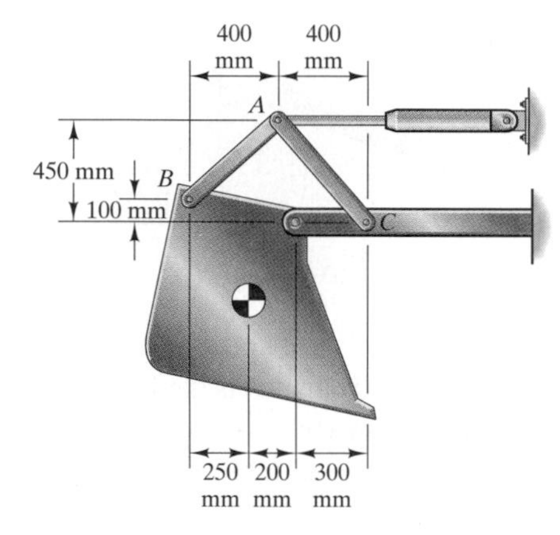

PROBLEM 1-3.24

1-3.25. Determine the axial load in member AC of the system in Problem 1-3.24.

1-3.26. A person applies the 150-N forces shown to the handles of the pliers. Determine the axial force in the link AB.

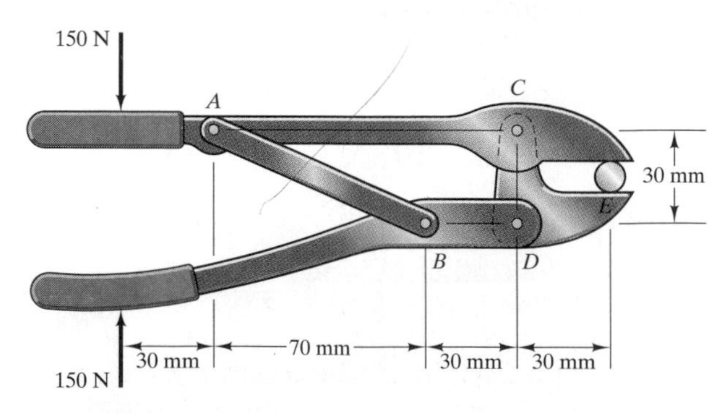

PROBLEM 1-3.26

1-3.27. In Problem 1-3.26, determine the x and y components of the reaction exerted on the upper handle AC of the pliers at C.

1-3.28. The radius of gear A is 0.13 m and the radius of gear B is 0.08 m. If the system is in equilibrium and the torque $T_A = 80$ N-m, what is the torque T_B?

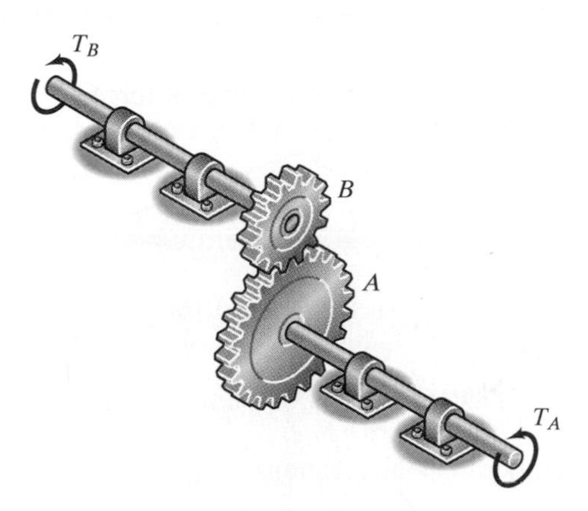

PROBLEM 1-3.28

1-3.29. The force **F** exerted on the bar is 20**i**−20**j**−10**k** (lb). Determine the reactions exerted on the bar at the built-in support O.

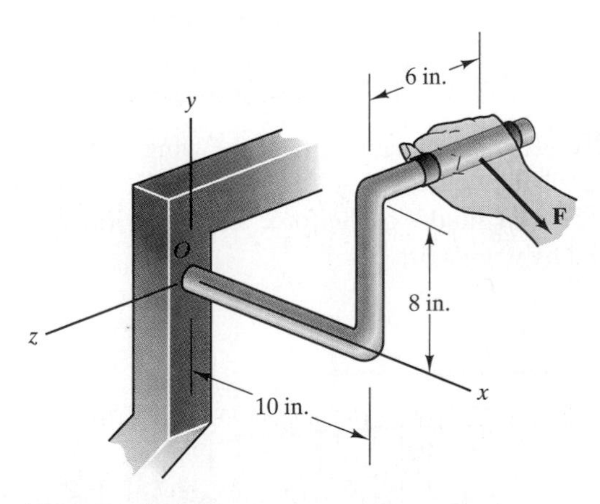

PROBLEM 1-3.29

1-3.30. The force **F** exerted on the bar in Problem 1-3.29 is 20**i** − 20**j** − 10**k** (lb). If you represent this force by an equivalent system consisting of a force **R** acting at the origin of the coordinate system and a couple **M**, what are the force **R** and the couple **M**? Compare your answers to the answers to Problem 1.3-29.

1-3.31. An axial force of magnitude P acts on the beam. If you represent the force by an equivalent system consisting of a force **F** acting at the origin O and a couple **M**, what are **F** and **M**?

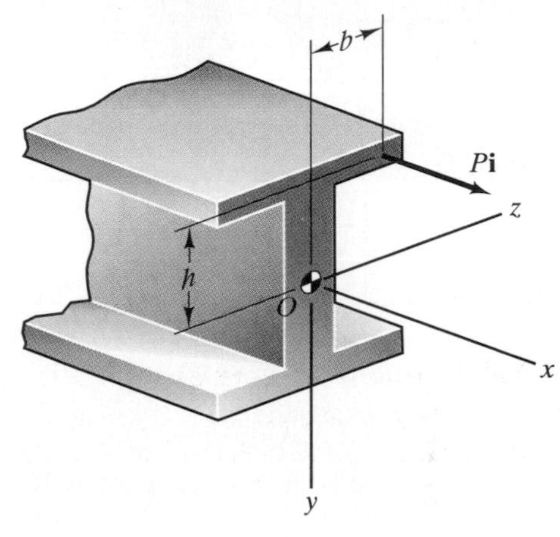

PROBLEM 1-3.31

1-3.32. Determine the x and y coordinates of the centroid of the area.

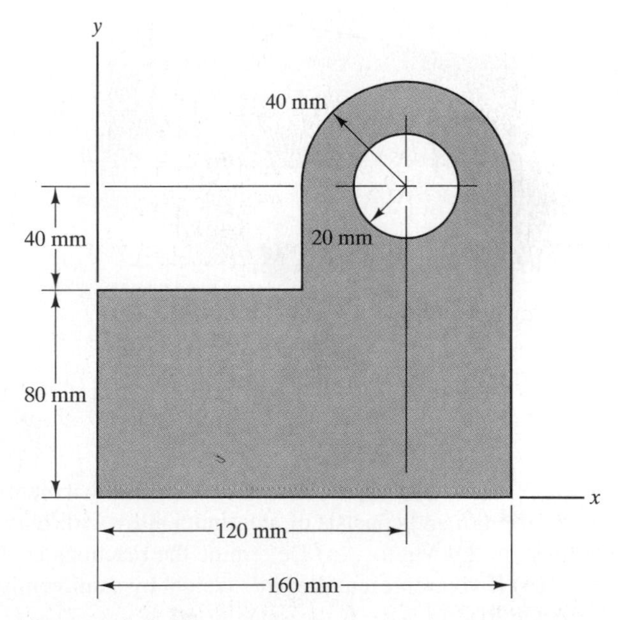

PROBLEM 1-3.32

1-3.33. The steel plate is homogeneous, of uniform thickness, and weighs 10 lb. **(a)** Determine the x coordinate of the plate's center of mass. **(b)** What are the reactions at A and B?

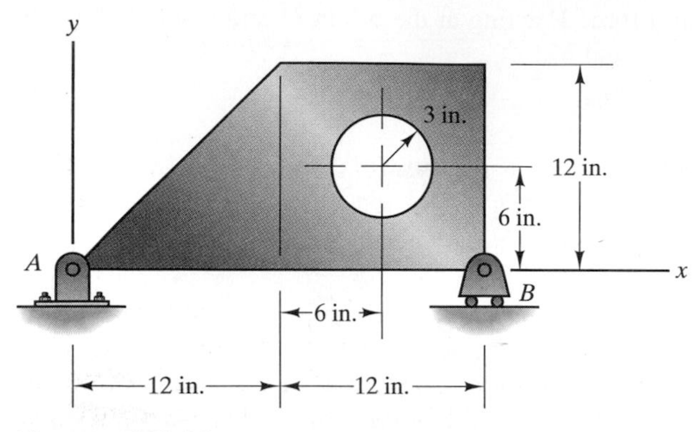

PROBLEM 1-3.33

1-3.34. The area of the homogeneous plate is 10 ft². The vertical reactions on the plate at A and B are 80 lb and 84 lb, respectively. Suppose that you want to equalize the reactions at A and B by drilling a 1-ft-diameter hole in the plate. At what horizontal distance from A should the center of the hole be placed?

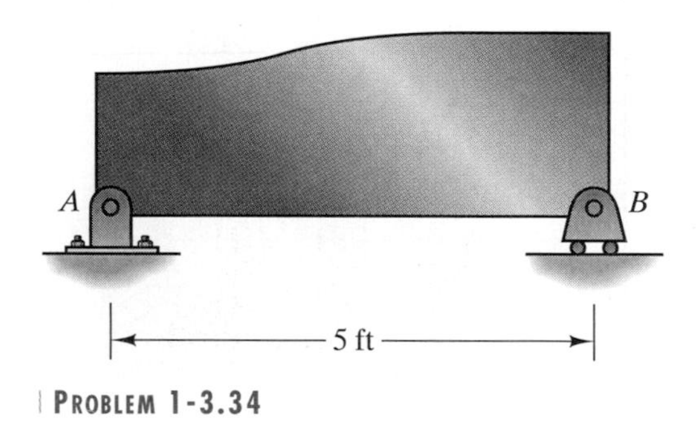

PROBLEM 1-3.34

1-3.35. In Problem 1-3.34, what are the reactions at A and B after the hole is drilled in the plate?

1-3.36. The beam has a circular cross section with a diameter of 100 mm and consists of aluminum alloy with mass density $\rho = 2.9$ Mg/m³. **(a)** Determine the reactions at A and B. **(b)** If you represent the bar's weight by a uniformly distributed load w, what is the value of w?

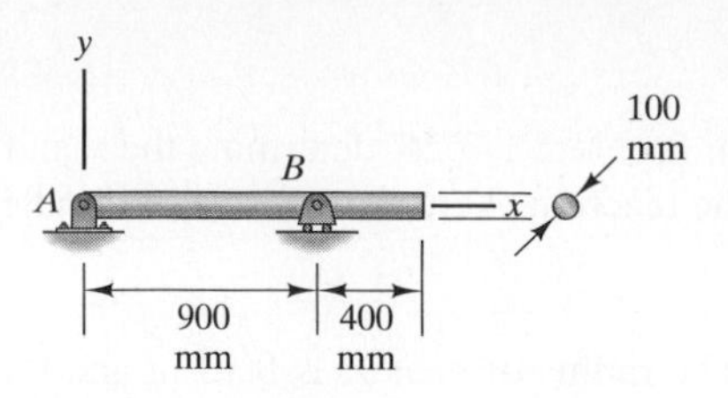

PROBLEM 1-3.36

1-3.37. Determine the reactions at A and B. (*Strategy:* Treat the distributed load as two triangular distributed loads and represent each one by an equivalent force.)

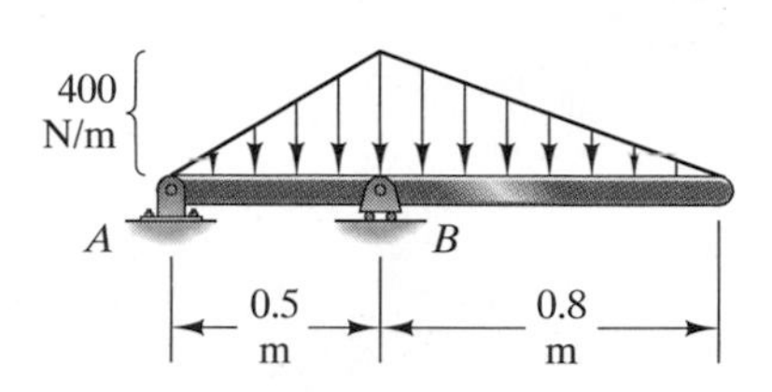

PROBLEM 1-3.37

1-3.38. Determine the reactions at A and B.

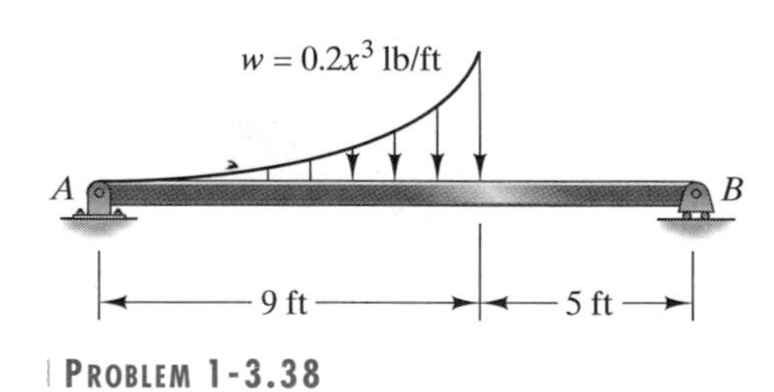

PROBLEM 1-3.38

1-3.39. The aerodynamic lift of the wing is described by the distributed load $w = -300(1 - 0.04x^2)^{1/2}$ N/m. Determine the magnitudes of the force and the moment about R exerted by wing's lift.

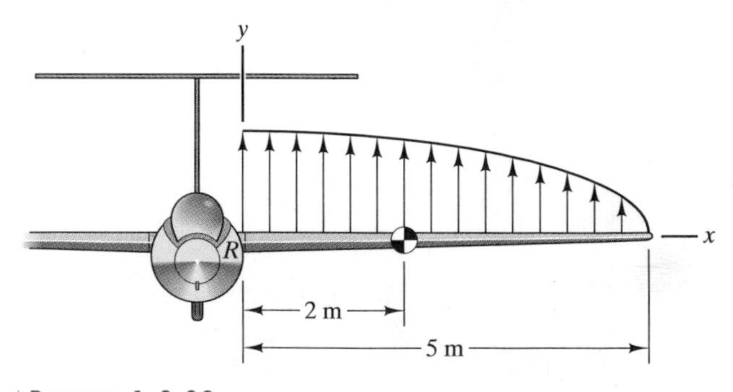

PROBLEM 1-3.39

1-3.40. The mass of the wing in Problem 1-3.39 is 27 kg and its center of mass is located 2 m to the right of the wing root R. Including the effects of the wing's lift, determine the reactions exerted on the wing at R where it is built in to the fuselage.

1-3.41. Determine the reactions at A and C.

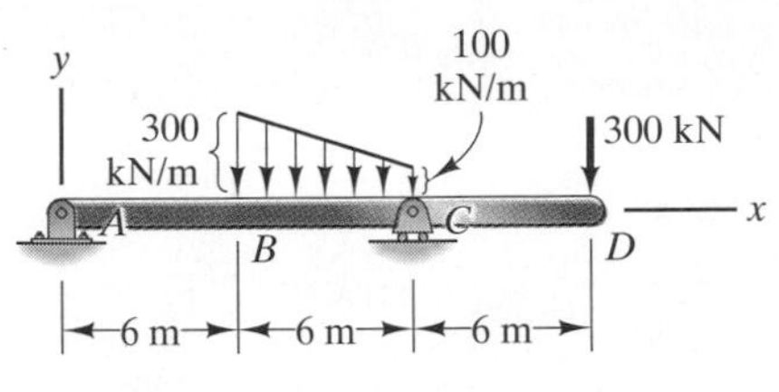

PROBLEM 1-3.41

1-3.42. A plane surface in the x–y plane of area A is subjected to a uniform pressure p_0. Show that this distributed load can be represented by an equivalent force of magnitude p_0A whose line of action passes through the centroid of A. (*Strategy:* Write integral expressions for the total force and the total moment about the origin due to the uniform pressure and use the definitions of the centroid of an area.)

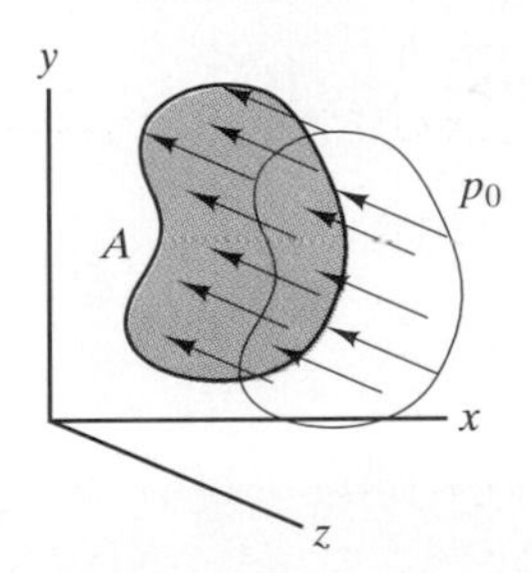

PROBLEM 1-3.42

1-3.43. The beam is subjected to a distributed couple c, defined such that each infinitesimal element dx of the beam is acted upon by a counterclockwise couple $c\,dx$. If $c = c_0 =$ constant from $x = 0$ to $x = L$, determine the reactions at A and B.

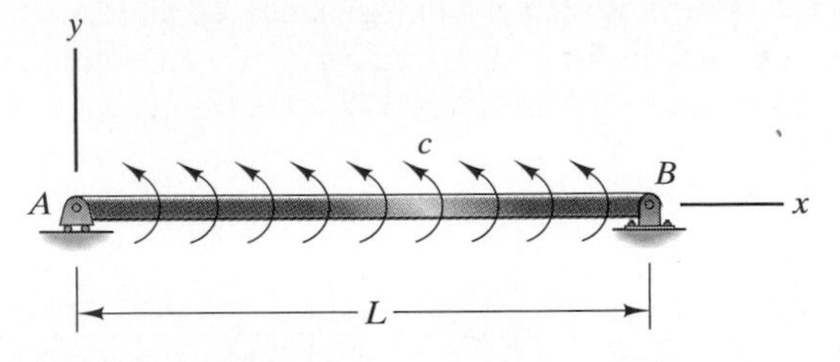

PROBLEM 1-3.43

1-3.44. If the distributed couple acting on the beam in Problem 1-3.43 is given by the equation $c = (x/L)c_0$, where c_0 is a constant, what are the reactions at A and B?

1-3.45. A pile is being slowly pushed into the ground by a vertical force F. The friction of the ground exerts a distributed axial force q, defined such that each infinitesimal element dx of the pile is acted upon by an axial force $q\,dx$. If $q = 400(1 - 0.4x^2/L^2)$ lb/ft and $L = 30$ ft, determine the force F at the instant shown. (The pile's weight and the force exerted by the ground at the bottom of the pile are negligible in comparison to the frictional force exerted along the pile's length.)

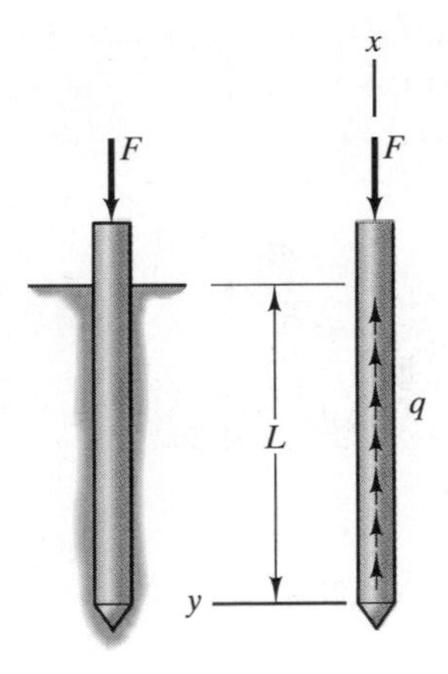

PROBLEM 1-3.45

This highway bridge over the Colorado River near Austin, Texas is supported by cables connected to steel arches that span the river.

CHAPTER 2

Measures of Stress and Strain

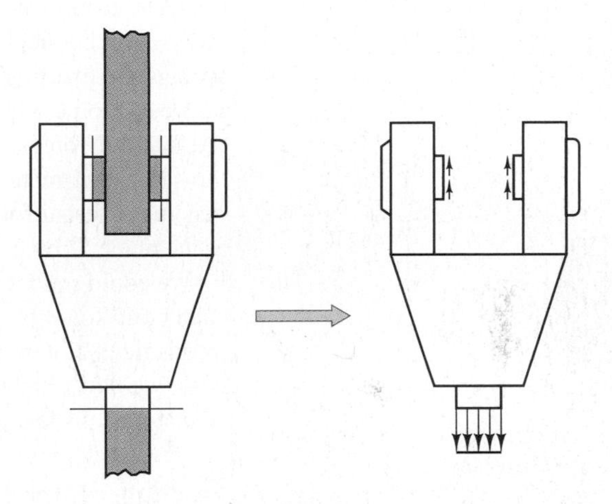

The diagram shows the normal stresses in one of the bridge's cables and the shear stresses in the pin connecting the cable to the arch. In this chapter we define stresses and strains, which are measures quantifying the internal forces and deformations in materials. We need these measures to assess the capability of structural elements such as the bridge's cables and arches to support loads and to determine the deformations that result from those loads.

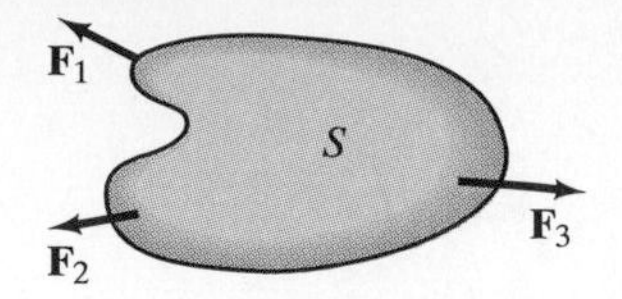

FIGURE 2-1 Sample of material subjected to forces.

2-1 | Stresses

Suppose that we subject a sample S of some solid material, such as iron, to the system of forces shown in Fig. 2-1. We assume that the sample is in equilibrium, and that its weight is negligible compared to the other external forces, so that $\mathbf{F}_1 + \mathbf{F}_2 + \mathbf{F}_3 = 0$. Let us pass an imaginary plane $\mathcal{P}$ through the sample, dividing it into parts S' and S'', and isolate part S' (Fig. 2-2). Let A be the area of intersection of the plane $\mathcal{P}$ with the sample. It is clear from Fig. 2-2b that part S' cannot be in equilibrium under the action of forces $\mathbf{F}_1$ and $\mathbf{F}_2$ alone. Part S'' must exert forces at the plane $\mathcal{P}$ that keep part S' in equilibrium.

What are those forces? They are literally the forces that hold the material together. Iron in its solid state consists of atoms "connected" to neighboring atoms by chemical bonds, electromagnetic forces that we may visualize as springs. By exerting external forces on the sample of iron, we alter the distances between atoms, changing the forces exerted by the bonds. The bond forces are internal forces within the material. But when we hypothetically separate the sample of material into two parts, the bond forces at the separating plane become external forces on the individual parts. It is these bond forces exerted at the plane $\mathcal{P}$ by part S'' which keep part S' in equilibrium (Fig. 2-3).

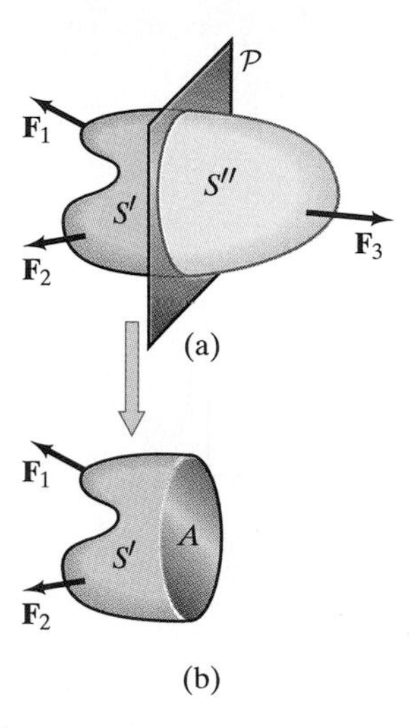

FIGURE 2-2 (a) The plane $\mathcal{P}$ divides the sample into parts S' and S''. (b) Isolating part S'.

We could model the forces acting on part S' at the plane $\mathcal{P}$ by representing each bond force by a vector as shown in Fig. 2-3, but this approach would be impractical. Not only would there be an unwieldy number of vectors to contend with, we don't know the actual arrangement of the atoms in any given sample of material. Instead, *we will represent the forces as a distributed load.*

Traction, Normal Stress, and Shear Stress

We define a vector-valued function $\mathbf{t}$, the *traction,* such that the force exerted on each infinitesimal element dA of the area A is $\mathbf{t}\,dA$ (Fig. 2-4). Notice the similarity between the definition of $\mathbf{t}$ and the definition of the pressure p in a stationary liquid (Fig. 1-12). The pressure is a scalar function, but here we must use a vector-valued function because the forces on A may act in any direction.

Since the value of $\mathbf{t}$ at a point on A is a vector, we can resolve it into components normal and tangential to A (Fig. 2-5). The scalar normal component σ is called the *normal stress* at the point, and the scalar tangential component τ is called the *shear stress* at the point. The normal stress σ is defined to be positive if it points outward, or away from the material, and is then said to be *tensile*. If σ is negative, the normal stress points toward the material and is said

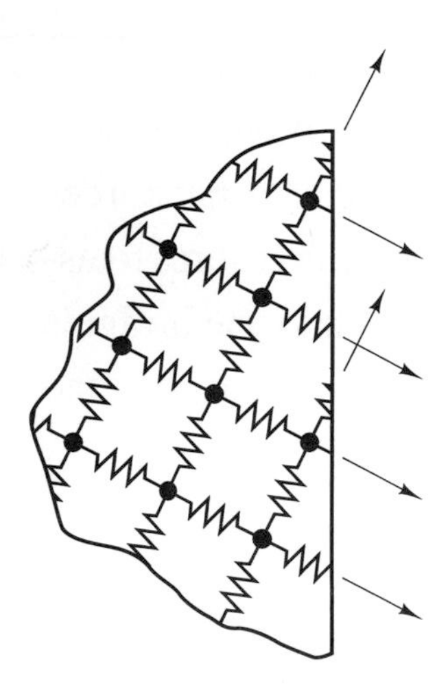

FIGURE 2-3 Bond forces at the plane $\mathcal{P}$.

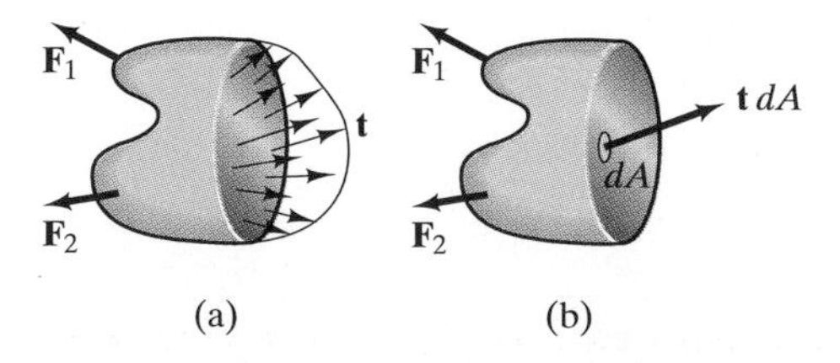

FIGURE 2-4 (a) Representing the forces on A by a distributed load. (b) The force on an element dA is $\mathbf{t}\,dA$.

to be *compressive*. The force exerted on an element dA is $\sigma\, dA$ in the normal direction and $\tau\, dA$ in the tangential direction.

The products $\mathbf{t}\, dA$, $\sigma\, dA$, and $\tau\, dA$ are forces, so the dimensions of $\mathbf{t}$, σ, and τ are force/area. In SI units, the traction and the normal and shear stresses are normally expressed in pascals (Pa), which are newtons per square meter, and in U.S. Customary units they are normally expressed in pounds per square inch (psi) or pounds per square foot (psf). We also use kips per square foot (ksf) and kips per square inch (ksi).

We introduced the definitions of the traction and the normal and shear stresses using a sample of iron in equilibrium as an example, but the same definitions are used to describe the internal forces in other media, and they need not be in equilibrium. The sample with which we began, Fig. 2-1, could model a different solid material, a liquid, or a gas, and it could be in an arbitrary state of motion. Depending on the medium, the internal forces represented may be quite different in nature from the bond forces between the atoms in iron. For example, the internal forces in a gas arise from impacts between the atoms or molecules of the gas resulting from their thermal motions. Nevertheless, we can represent the forces by a distributed load in exactly the same way (Fig. 2-4). The same definitions of the traction and its components are used in both solid and fluid mechanics to represent the internal forces in materials. In the special case of a liquid or gas at rest, the normal stress $\sigma = -p$, where p is the pressure, and the shear stress $\tau = 0$ (see Fig. 1-12). Recall that the normal stress σ is defined to be positive in tension, whereas the pressure p is defined to be positive in compression.

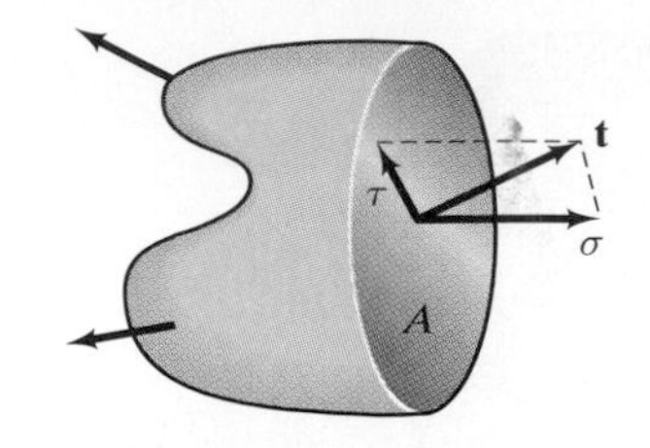

FIGURE 2-5 The components of **t** normal and tangential to A are the normal and shear stresses.

Average Stresses

The traction **t** or its components, the normal stress σ and shear stress τ, allow us in principle to represent the forces acting on area A in Fig. 2-5. But how can we determine their values as functions of position on A for a given sample of material and state of motion? This determination, evaluating the *stress distribution* for a given plane $\mathcal{P}$ within a material, is called *stress analysis*. In the following chapters we discuss the stress distributions in simple structural elements subjected to loads. In this section our goal is less ambitious: We want to determine the average stress at a plane when an object is in equilibrium.

The average value of the traction **t** on A is defined by

$$\mathbf{t}_{\text{av}} = \frac{1}{A}\int_A \mathbf{t}\, dA, \qquad \textbf{(2-1)}$$

where the subscript A on the integral sign signifies that integration is carried out over the entire area A. Therefore, we can express the total force exerted on the area A in terms of the average traction (Fig. 2-6):

$$\int_A \mathbf{t}\, dA = \mathbf{t}_{\text{av}} A.$$

The components of the average traction normal and tangential to A are the average normal and shear stresses on A (Fig. 2-7). The total normal and tangential forces exerted on A are $\sigma_{\text{av}} A$ and $\tau_{\text{av}} A$, respectively.

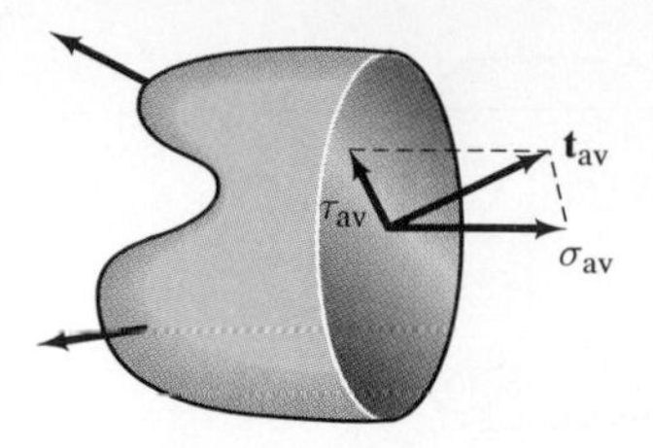

FIGURE 2-7 Decomposing the average traction into the average normal and shear stresses.

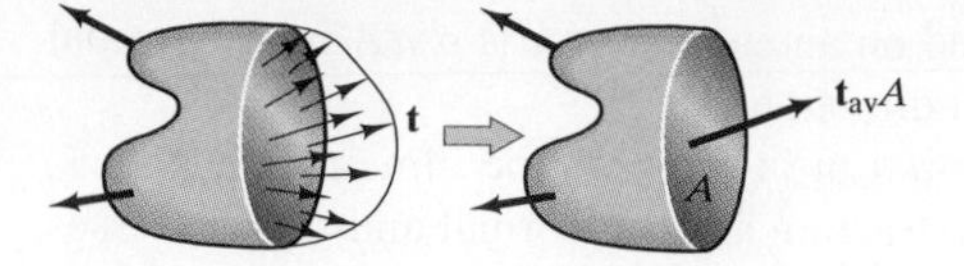

FIGURE 2-6 Expressing the total force exerted on A by the traction distribution in terms of the average traction.

If we know the external forces acting on an object in equilibrium, we can determine the average normal and shear stresses acting on an arbitrary plane $\mathcal{P}$. Two particular examples are important in applications.

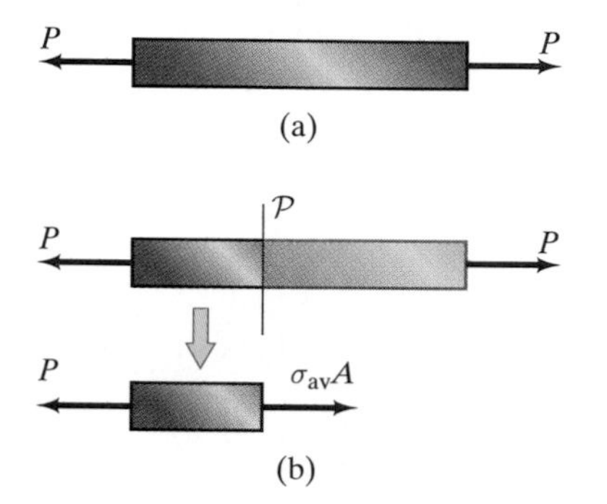

FIGURE 2-8 (a) Subjecting a bar to axial loads. (b) Obtaining a free-body diagram by passing a plane through the bar perpendicular to its axis.

AVERAGE NORMAL STRESS IN AN AXIALLY LOADED BAR

A straight bar whose cross section is uniform throughout its length is said to be *prismatic*. (The familiar triangular glass prism is a prismatic bar.) Figure 2-8a shows a prismatic bar subjected to axial forces parallel to its axis. In Fig. 2-8b we pass a plane $\mathcal{P}$ perpendicular to the bar's axis and isolate part of the bar. We show the force exerted by the average normal stress at $\mathcal{P}$, where A is the bar's cross-sectional area. We know that the average shear stress at $\mathcal{P}$ is zero because there is no external force tangential to $\mathcal{P}$. The sum of the forces equals zero,

$$\sigma_{av}A - P = 0,$$

so we can solve for the average normal stress:

$$\sigma_{av} = \frac{P}{A}.$$

We see that the average normal stress is proportional to the external axial load and inversely proportional to the bar's cross-sectional area. Notice that the value of the average normal stress does not depend on the location of the plane $\mathcal{P}$ along the bar's axis. We discuss the analysis of stresses in axially loaded bars fully in Chapter 3.

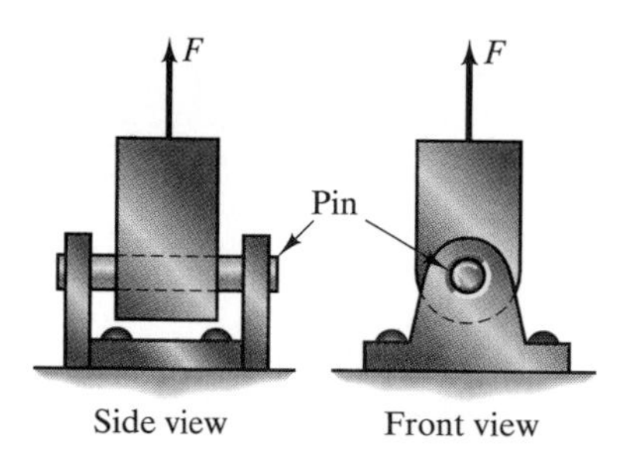

FIGURE 2-9 Two views of a pin support.

AVERAGE SHEAR STRESS IN A PIN

The pin support in Fig. 2-9 holds a bar subjected to an axial load F. The support consists of a bracket supporting the cylindrical pin, which passes through a hole in the bar. In Fig. 2-10 we obtain a free-body diagram by passing two planes through the pin perpendicular to its axis, one on each side of the supported bar. We show the forces resulting from the average shear stresses in the pin at the two cutting planes, where A is the pin's cross-sectional area. From the equilibrium equation

$$F - 2\tau_{av}A = 0,$$

we obtain the average shear stress:

$$\tau_{av} = \frac{F}{2A}.$$

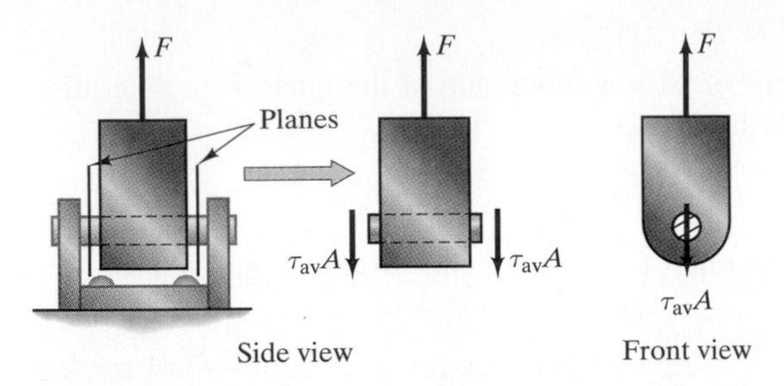

FIGURE 2-10 Passing two planes through the pin to determine the average shear stress.

These two examples demonstrate the ease with which we can determine the average stresses on a given plane within an object. But knowing the average stress on a plane is of limited usefulness in design, because it is the maximum, not the average, values of the normal and shear stresses that determine whether material failure will occur. Depending on the nature of the actual distributions of stress, the maximum stresses may be much greater than their average values. Moreover, it is not sufficient to determine the maximum stresses for a particular plane. The stresses acting on every plane must be considered. We examine the problem of determining maximum normal and shear stresses in Chapter 5.

There is one circumstance in which knowing the average stress on a given plane is useful in design. If experiments or analyses are first used to establish the relationship between the largest safe load on a given structural element and the value of the average normal or shear stress on a given plane, design can be carried out on that basis. For example, bolts made of a particular material that are to be used for a specific application can be tested to determine the largest safe value of average shear stress they will support.

EXAMPLE 2-1

The truss in Fig. 2-11 supports a 10-kN force. Its members are solid cylindrical bars of 40-mm radius. Determine the average normal and shear stresses on the plane $\mathcal{P}$.

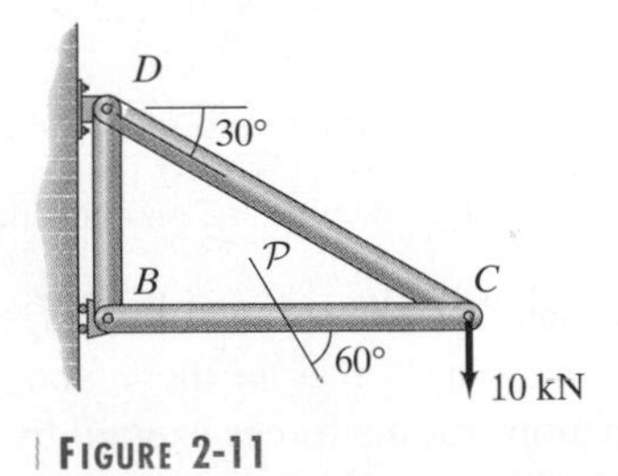

FIGURE 2-11

Strategy

We must first determine the axial force in member BC, which we can do by drawing the free-body diagram of joint C. We can then use the free-body diagram of the part of member BC on either side of the plane $\mathcal{P}$ to determine the average normal and shear stresses.

Solution

We draw the free-body diagram of joint C of the truss in Fig. (a). From the equilibrium equations

$$\Sigma F_x = -P_{BC} - P_{CD} \cos 30^\circ = 0,$$
$$\Sigma F_y = P_{CD} \sin 30^\circ - 10 = 0,$$

we obtain $P_{BC} = -17.3$ kN, $P_{CD} = 20$ kN. Member BC is subject to a compressive axial load of 17.3 kN.

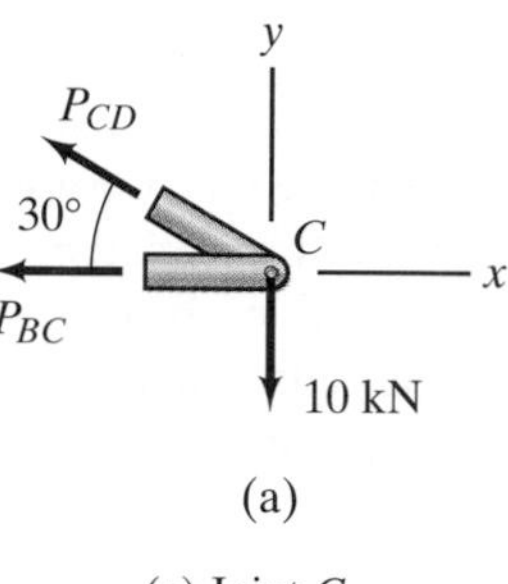

(a) Joint C.

In Fig. (b) we isolate the part of member BC to the left of the plane $\mathcal{P}$ and complete the free-body diagram by showing the forces exerted by the average normal and shear stresses.

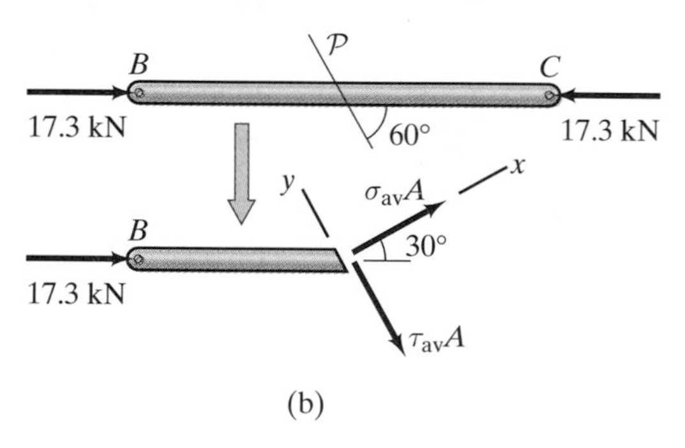

(b) Isolating the part of member BC on one side of plane $\mathcal{P}$.

By aligning a coordinate system normal and tangential to the plane $\mathcal{P}$ as shown in Fig. (b) and summing forces in the x and y directions, we obtain simple equilibrium equations for the forces exerted by the stresses:

$$\Sigma F_x = \sigma_{av} A + 17.3 \cos 30^\circ = 0,$$
$$\Sigma F_y = -\tau_{av} A - 17.3 \sin 30^\circ = 0.$$

We find that $\sigma_{av} A = -15$ kN and $\tau_{av} A = -8.66$ kN. The area A is the intersection of the plane $\mathcal{P}$ with the bar, the area upon which the average stresses σ_{av} and τ_{av} act. The relationship between A and the cross-sectional area of the cylindrical bar of 0.04-m radius is

$$A \cos 30^\circ = \pi(0.04)^2,$$

so $A = \pi(0.04)^2/\cos 30^\circ = 0.00580 \text{ m}^2$. Solving for the average normal and shear stresses, we obtain $\sigma_{av} = -2.58$ MPa and $\tau_{av} = -1.49$ MPa.

Discussion

The negative value of σ_{av} tells us that the average normal stress is compressive. Although we have established a convention for the positive direction of the normal stress, we have not yet done so for the shear stress. In drawing the free-body diagram in Fig. (b), we chose the direction of the stress τ_{av} arbitrarily, and the negative value we obtained indicates that it acts in the opposite direction.

EXAMPLE 2-2

In Fig. 2-12 detailed views of joint C of the truss are shown. The joint has a pin of 20-mm radius. What is the average shear stress in the pin?

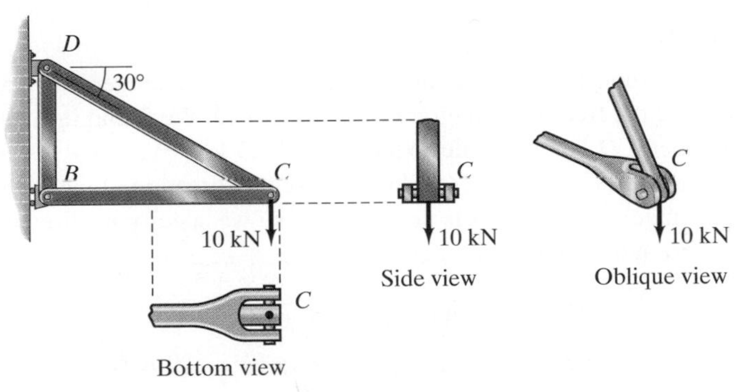

FIGURE 2-12

Strategy

By passing planes through the pin on both sides of member CD, we can isolate either member BC or CD and obtain a free-body diagram from which we can determine the average shear stress in the pin. Since the 10-kN external load acts on member CD, we will obtain a simpler free-body diagram by isolating member BC.

Solution

To draw the free-body diagram of member BC, we must know the axial force to which it is subjected. In the solution to Example 2-1, we found that member BC

has a compressive axial load of 17.3 kN. In Fig. (a) we show the bottom view of member BC with member CD still attached. By passing planes through the pin on both sides of member CD, we obtain the free-body diagram of member BC in Fig. (b). The 17.3-kN compressive axial load is supported by the average shear stresses in the pin.

From the equilibrium equation

$$17.3 - 2\tau_{av} A = 0,$$

we obtain $\tau_{av} A = 8.66$ kN. The pin's cross-sectional area is $A = \pi(0.02)^2 = 0.00126\ \mathrm{m}^2$, so the average shear stress in the pin is $\tau_{av} = 6.89$ MPa.

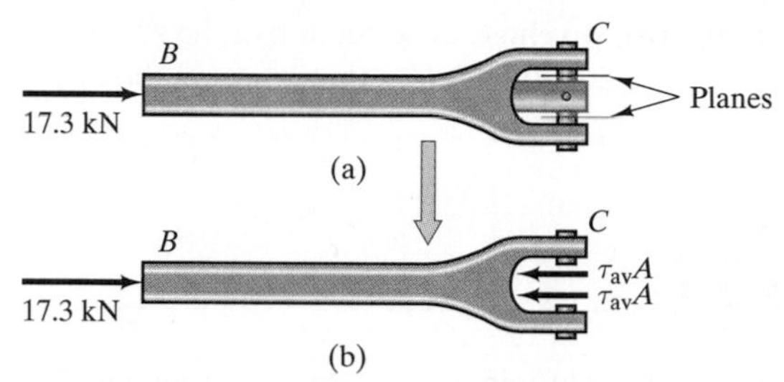

(a) Passing planes through the pin to isolate member BC. (b) Free-body diagram of member BC showing the forces exerted on the pin.

Discussion

We should demonstrate that we can also determine the average shear stress in the pin by using the free-body diagram of member CD. We found in Example 2-1 that member CD has a tensile axial load of 20 kN. We show its free-body diagram (in side view) in Fig. (c). In this case the direction of the force exerted by the average shear stresses is not obvious, so we specify its direction by the angle β. From the equilibrium equations

$$\Sigma F_x = 2\tau_{av} A \cos\beta - 20\cos 30^\circ = 0,$$
$$\Sigma F_y = 2\tau_{av} A \sin\beta + 20\sin 30^\circ - 10 = 0,$$

we again obtain $\tau_{av} A = 8.66$ kN.

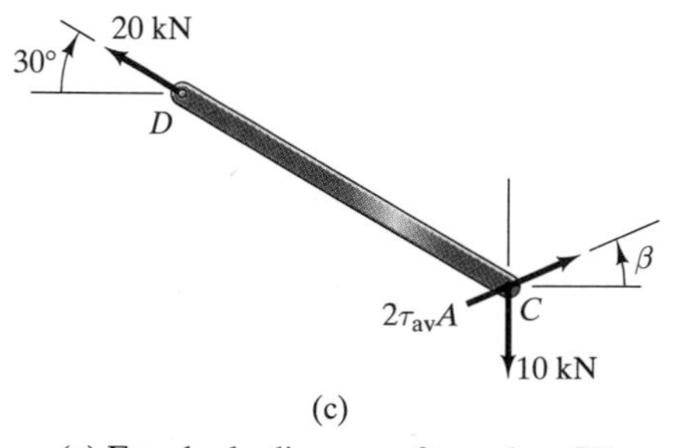

(c) Free-body diagram of member CD.

EXAMPLE 2-3

The arrangement in Fig. 2-13 is designed to cut a circular "blank" from the plate by exerting a sufficient force F on the punch. The plate is of thickness t and the punch and the matching hole beneath the plate are of diameter D. For a given force F on the punch, what average shear stress is induced in the plate?

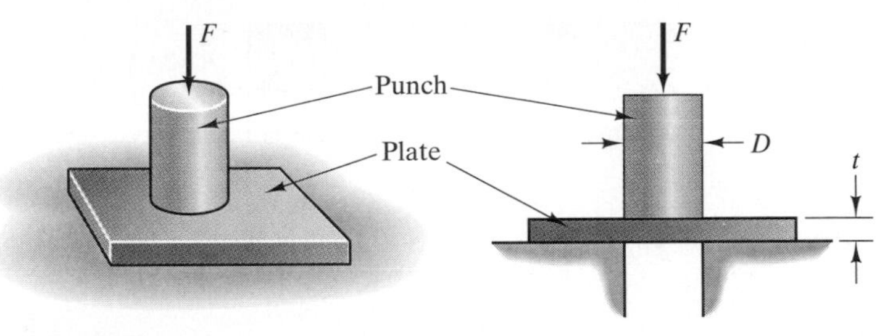

FIGURE 2-13

Strategy

To determine the average shear stress induced in the plate we obtain a free-body diagram by making a cylindrical "cut" of diameter D through the plate directly below the punch [Fig. (a)]. The force F on the free-body diagram is supported by vertical shear stress on the surface of the part of the plate we cut by the cylinder. We can use equilibrium to determine the average shear stress.

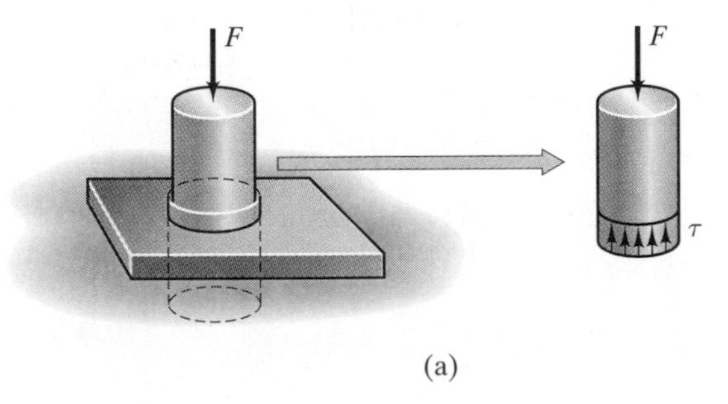

(a) Cutting the plate with a cylindrical surface.

Solution

The surface area of the part of the plate included in the free-body diagram on which the shear stress acts is equal to the product of the thickness t of the plate and the circumference πD. Summing the forces on the free-body diagram,

$$\tau_{\text{av}}(t)(\pi D) - F = 0,$$

we find that the average shear stress is $\tau_{\text{av}} = F/\pi t D$.

EXAMPLE 2-4

The normal stress acting on the rectangular cross section A in Fig. 2-14 is given by the equation $\sigma = 200y^2$ lb/in^2. Determine **(a)** the maximum normal stress; **(b)** thc total normal force; **(c)** the average normal stress.

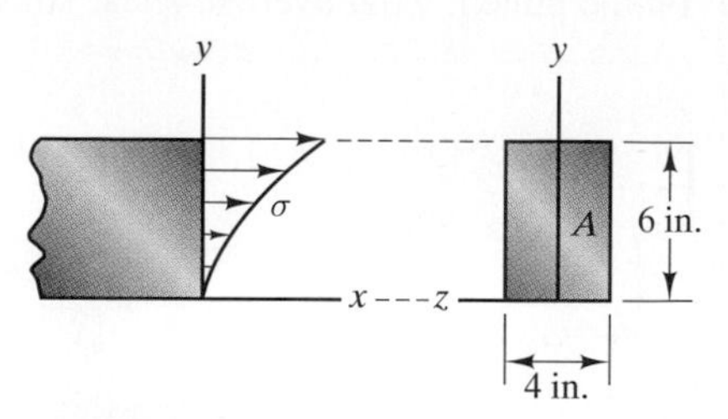

FIGURE 2-14

Strategy

(**a**) The value of σ increases monotonically with increasing y, so its maximum value occurs at $y = 6$ in. (**b**) The normal stress σ depends only upon y, so we can integrate to determine the total normal force by using an element of area dA in the form of a horizontal strip [Fig. (a)]. (**c**) The average stress equals the total normal force divided by A.

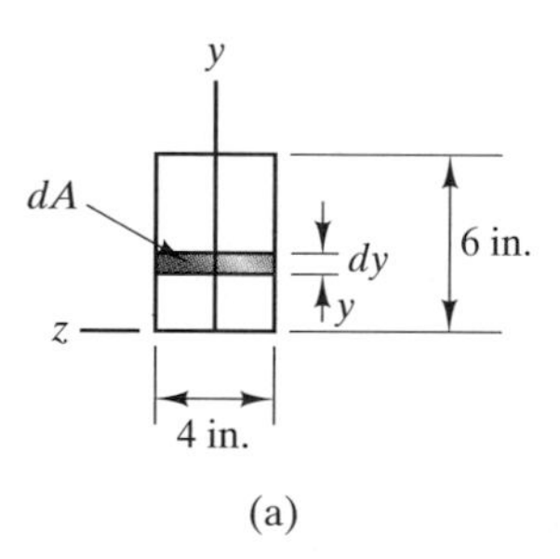

(a) Integration element dA.

Solution

(**a**) The maximum normal stress is

$$\sigma_{\max} = (200)(6)^2 = 7200 \text{ lb/in}^2.$$

(**b**) Using the strip element dA in Fig. (a), the total normal force is

$$\begin{aligned}\int_A \sigma\, dA &= \int_0^6 (200y^2)(4dy) \\ &= 800\left[\frac{y^3}{3}\right]_0^6 \\ &= 57{,}600 \text{ lb}.\end{aligned}$$

(**c**) The average stress is

$$\sigma_{av} = \frac{1}{A}\int_A \sigma\, dA$$

$$= \frac{1}{(4)(6)}(57{,}600)$$

$$= 2400 \text{ lb/in}^2.$$

Discussion

Notice that the maximum normal stress is three times the average normal stress. This illustrates why knowledge of the average stress is usually insufficient for design. The actual distribution of the stress must be known to determine its maximum value.

2-2 | Strains

In this section we consider what happens to a sample of material when it is subjected to external loads. From everyday experiments—stretching rubber bands, bending popsicle sticks—we are all familiar with the fact that objects can *deform,* or change shape, under the action of loads. Our ultimate objective is to be able to determine the deformations of simple structural elements resulting from given loads. For now, we merely want to introduce quantities that describe an object's change in shape.

If we put ourselves in the place of the pioneers of this subject, the task before us seems very difficult. How can we even describe the shape of a given object analytically, much less a change in its shape? To make the problem tractable, we don't approach it in this global way, asking "What is the object's new shape?" Instead, we begin by considering what happens to the material near a single point of the object.

Extensional Strain

Consider a sample of a solid material such as the piece of iron with which we began discussing stresses. We can imagine taking a pen and drawing an infinitesimal line dL somewhere within the material (Fig. 2-15a). Suppose that we then subject the sample to some set of loads, causing it to undergo a deformation. What happens to the imaginary line? We don't know, since we don't know what the loads are and cannot yet predict the effects those loads have on the material, but the line of original length dL may contract or stretch to some new length dL' and may also change direction (Fig. 2-15b). The *extensional strain* ϵ is a measure of the change in length of the line dL. It is defined to be the change in length divided by the original length:

$$\epsilon = \frac{dL' - dL}{dL}. \qquad \textbf{(2-2)}$$

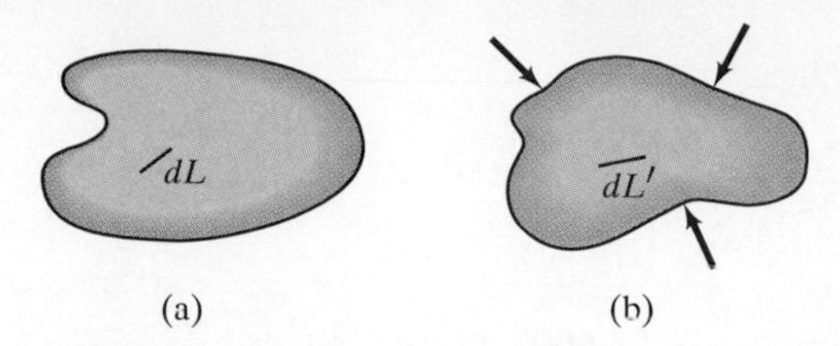

FIGURE 2-15 (a) Infinitesimal line within a sample of material. (b) Sample and line after a deformation.

For a given point of the material and a given direction (the direction of the line dL), ϵ measures how much the material contracts or stretches in the subsequent deformation. Notice that ϵ is negative if the material contracts and positive if it stretches. Since ϵ is a change in length divided by a length, it is dimensionless.

It is important to realize that the value of the extensional strain corresponds to a given point and a given direction in the material prior to the deformation. To define it, we had to choose a point in the material to place the line dL and choose a direction for dL. In general, the value of the extensional strain may vary from point to point in a material and may be different in different directions. For example, a material may contract in one direction but stretch in a different direction. Also, notice that the extensional strain measures the contraction or stretch of the material relative to some initial state, which we call the *reference state*.

If we know the value of ϵ corresponding to a given point and a given direction prior to the deformation, what does that tell us? We know that if we draw an infinitesimal line of length dL in the material located at that point and in that direction, its change in length resulting from the deformation is $dL' - dL = \epsilon\, dL$, so its new length is

$$dL' = (1 + \epsilon)\, dL. \tag{2-3}$$

We can determine the new length of a finite line if we know the value of the extensional strain ϵ in the direction tangent to the line at each point of the line. Let L be the length of a finite line in a sample of material in a reference state (Fig. 2-16a). We obtain the length L' of the line in a deformed state (Fig. 2-16b) by integrating Eq. (2-3),

$$L' = \int_L (1 + \epsilon)\, dL = L + \int_L \epsilon\, dL, \tag{2-4}$$

where the subscript L on the integral signs means that the integrations are carried out over the entire length of the line. We denote the change in length of a finite line by δ:

$$\delta = L' - L = \int_L \epsilon\, dL. \tag{2-5}$$

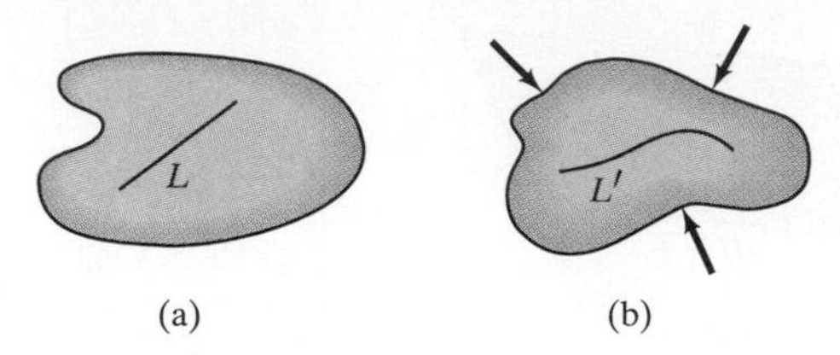

FIGURE 2-16 (a) Finite line within a sample of material. (b) Sample and line after a deformation.

If the value of ϵ is uniform (constant) along the line, the change in length is

$$\delta = L' - L = \epsilon L \qquad \text{(if } \epsilon \text{ is constant along } L\text{).} \qquad \textbf{(2-6)}$$

Notice that Eqs. (2-4)–(2-6) apply even if the line is not straight in the reference state as long as ϵ is the extensional strain in the direction tangent to the line at each point.

Shear Strain

The extensional strain tells us how much a material contracts or stretches in a given direction at a given point of a material. We will find in subsequent chapters that another type of strain is also useful for describing the deformation of a material. Let's return to our sample of material and imagine drawing two perpendicular infinitesimal lines dL_1 and dL_2 somewhere within a reference state (Fig. 2-17a). The angle between the two lines is 90°, or $\pi/2$ radians. After the material undergoes a deformation, these two lines may no longer be perpendicular. Let us denote the angle in radians between the two lines after the deformation as $\pi/2 - \gamma$ (Fig. 2-17b). The angle γ will be positive, zero, or negative if the angle between the two lines decreases, remains the same, or increases, respectively.

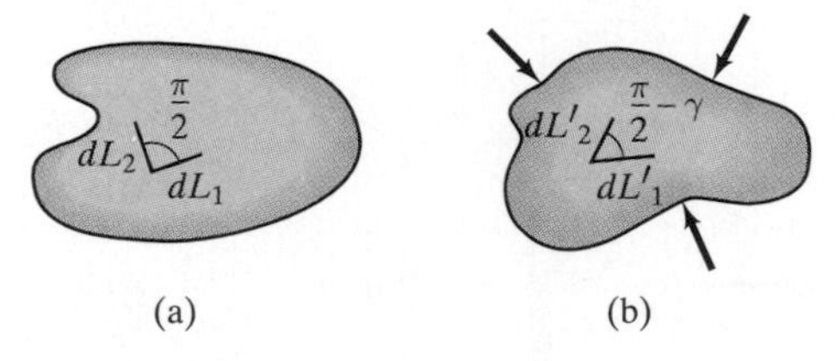

FIGURE 2-17 (a) Perpendicular infinitesimal lines at a point in the reference state. (b) The lines in a deformed state.

The angle γ is called the *shear strain* referred to the directions dL_1 and dL_2. Thus the shear strain is a measure of the change in the angle between two particular lines that were perpendicular in the reference state. In general, its value may vary from point to point in a material and may be different for different directions of the lines dL_1 and dL_2. Since γ is an angle in radians, the shear strain is dimensionless.

If we consider an infinitesimal rectangle in the reference state with sides dL_1 and dL_2 (Fig. 2-18a), the shear strain referred to the directions dL_1 and dL_2 tells

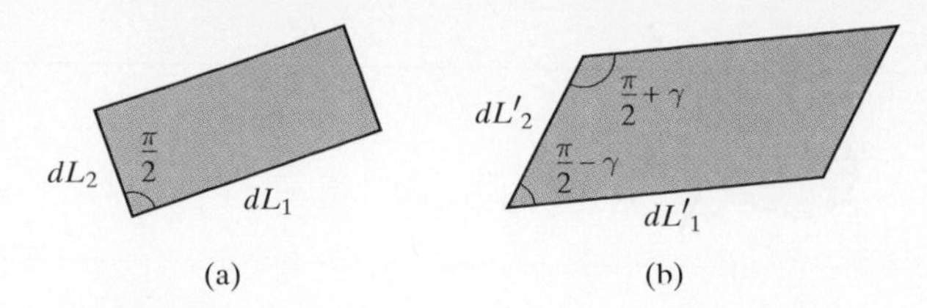

FIGURE 2-18 (a) Rectangle in the reference state. (b) Shear strain causes the rectangle to become a parallelogram.

us the angles between the sides of the resulting parallelogram in the deformed state (Fig. 2-18b).

Notice that in defining the extensional and shear strains we have made no assumption about the state of motion of the material. In the examples and problems presented in this introductory treatment of mechanics of materials, we assume that the material in its deformed state is in equilibrium. But our definitions of strain are not limited to that situation. The deformed state can instead represent the state at an instant in time of a material undergoing an arbitrary motion.

EXAMPLE 2-5

A bar has length $L = 2$ m in the unloaded state. **(a)** In Fig. 2-19a, the bar is subjected to axial loads which increase its length to 2.04 m. If the extensional strain ϵ in the direction of the bar's axis is assumed to be uniform throughout the bar's length, what is ϵ? **(b)** In Fig. 2-19b, the bar is suspended from one end, causing its length to increase to 2.01 m. The resulting extensional strain in the x direction is given by the equation $\epsilon = ax/L$, where a is a constant. What is the value of a?

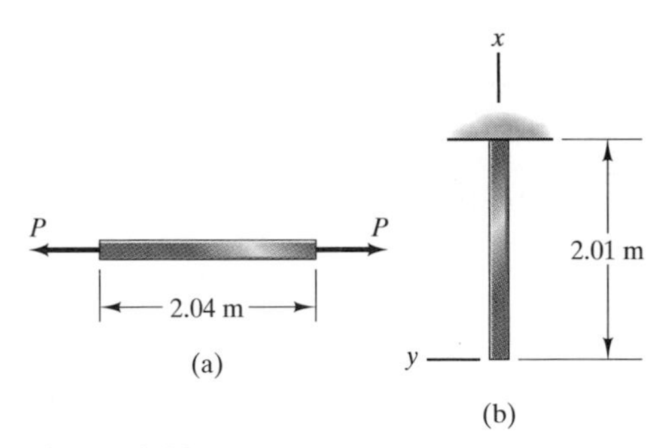

FIGURE 2-19

Strategy

In part (a) the extensional strain is uniform, so the bar's change in length δ in the deformed state is given in terms of its reference length L and the strain ϵ by Eq. (2-6). We know the change in length, so we can solve for ϵ. In part (b) the strain is a function of distance along the bar's axis, so we must use Eq. (2-5).

By substituting the change in length and the given equation for ϵ, we can solve for the constant a.

Solution

(**a**) From Eq. (2-6),

$$\begin{aligned} \delta = L' - L = \epsilon L : \\ 2.04 - 2 = \epsilon(2) \\ 0.04 = \epsilon(2). \end{aligned}$$

We obtain $\epsilon = 0.02$.
(**b**) From Eq. (2-5), the bar's change in length is

$$\delta = L' - L = \int_L \epsilon \, dL :$$

$$2.01 - 2 = \int_0^2 \frac{ax}{2} \, dx$$

$$0.01 = \frac{a}{2} \left[\frac{x^2}{2} \right]_0^2 = a.$$

The constant $a = 0.01$.

Discussion

In case (b) we gave the form of the equation for the distribution of extensional strain along the axis of a suspended bar. In Chapter 3 we relate the extensional strain to the normal stress in a linearly elastic bar subjected to axial load. You will then be able to show that the extensional strain is a linear function of x in a suspended prismatic bar.

EXAMPLE 2-6

Consider the infinitesimal rectangle in Fig. 2-20a at a point of a material in a reference state. The diagram in a deformed state is shown in Fig. 2-20b. Let the extensional strains in the dL_1 and dL_2 directions be ϵ_1 and ϵ_2, and let the shear strain referred to the dL_1 and dL_2 directions be γ. What is the extensional strain in the direction of the diagonal dL?

Strategy

From the extensional strains in the dL_1 and dL_2 directions we can determine the lengths dL_1' and dL_2'. From the shear strain we know the angles between the sides of the parallelogram in the deformed state (see Fig. 2-18). By applying the law of cosines to one of the oblique triangles comprising the parallelogram, we can determine the length dL'.

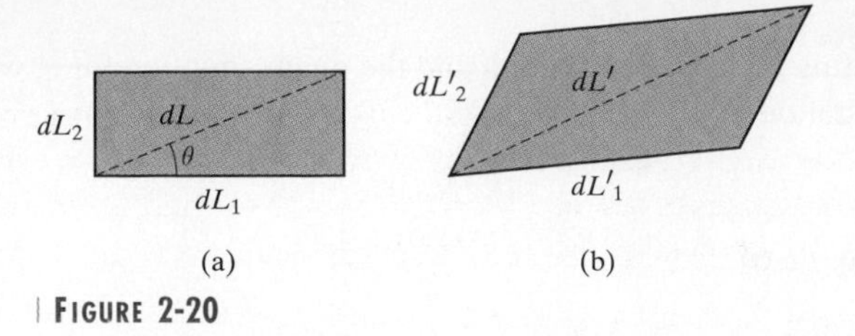

| FIGURE 2-20

Solution

The lengths dL_1' and dL_2' are

$$dL_1' = (1 + \epsilon_1)\, dL_1 = (1 + \epsilon_1)\, dL \cos\theta,$$
$$dL_2' = (1 + \epsilon_2)\, dL_2 = (1 + \epsilon_2)\, dL \sin\theta.$$

Consider the upper triangle of the parallelogram in the deformed state [Fig. (c)].

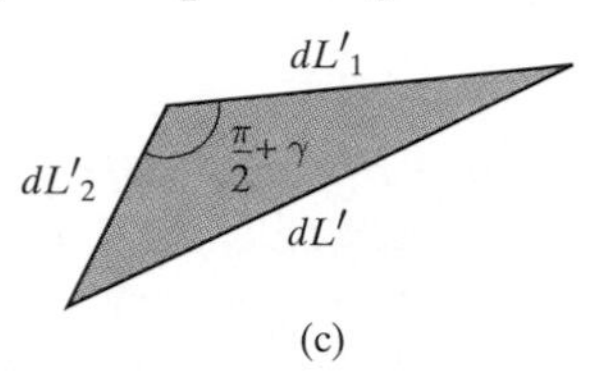

(c)

(c) Upper triangle of the element.

From the law of cosines,

$$(dL')^2 = (dL_1')^2 + (dL_2')^2 - 2\, dL_1'\, dL_2' \cos\left(\frac{\pi}{2} + \gamma\right).$$

Substituting our expressions for dL_1' and dL_2' into this equation, we obtain

$$(dL')^2 = \left[(1 + \epsilon_1)^2 \cos^2\theta + (1 + \epsilon_2)^2 \sin^2\theta - 2(1 + \epsilon_1)(1 + \epsilon_2) \cos\theta \sin\theta \cos\left(\frac{\pi}{2} + \gamma\right)\right] (dL)^2.$$

The extensional strain in the direction of the diagonal dL is therefore

$$\epsilon = \frac{dL' - dL}{dL}$$
$$= \sqrt{(1 + \epsilon_1)^2 \cos^2\theta + (1 + \epsilon_2)^2 \sin^2\theta - 2(1 + \epsilon_1)(1 + \epsilon_2) \cos\theta \sin\theta \cos\left(\frac{\pi}{2} + \gamma\right)} - 1.$$

Discussion

This example demonstrates why the shear strain is needed for determining the deformation near a point of a material. The extensional strains in the dL_1 and dL_2 directions are not sufficient to determine the extensional strain in the dL direction. We also need the value of the shear strain.

Chapter Summary

In this chapter we have introduced fundamental measures of the stresses and strains in a material. We defined average stresses on an area and showed how to evaluate them in particular cases. In Chapter 3 we apply these definitions to structural applications involving axially loaded bars.

Traction, Normal Stress, and Shear Stress

The *traction* **t** [Fig. (a)] is defined such that the force exerted on each infinitesimal element dA of the area A is $\mathbf{t}\,dA$. The normal component σ is the *normal stress* and the tangential component τ is the *shear stress*. The normal stress σ is defined to be positive if it points outward, or away from the material, and is then said to be *tensile*. If σ is negative, the normal stress points toward the material and is said to be *compressive*. The dimensions of **t**, σ, and τ are force/area.

The average value of the traction **t** on the area A in Fig. (a) is

$$\mathbf{t}_{\mathrm{av}} = \frac{1}{A}\int_A \mathbf{t}\,dA. \qquad \text{Eq. (2-1)}$$

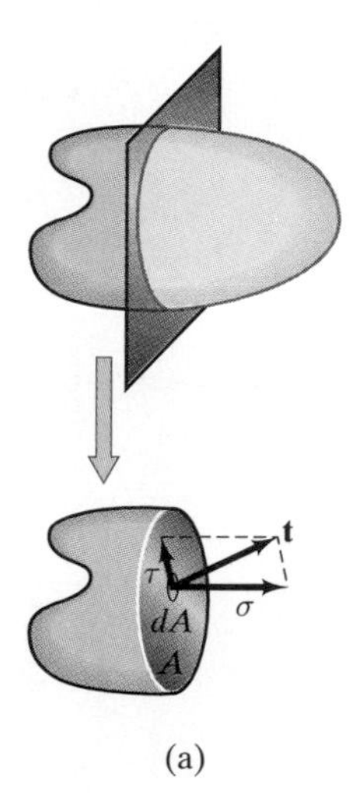

(a)

The components of the average traction normal and tangential to A are the average normal stress σ_{av} and average shear stress τ_{av}. The total normal and tangential forces exerted on A are $\sigma_{\mathrm{av}}A$ and $\tau_{\mathrm{av}}A$, respectively.

Extensional and Shear Strains

Consider an infinitesimal line dL in a sample of material in a reference state [Fig. (b)]. Let the length of the line in a deformed state be dL' [Fig. (c)]. The *extensional strain* ϵ is defined by

$$\epsilon = \frac{dL' - dL}{dL}. \qquad \text{Eq. (2-2)}$$

(b) (c)

Let L be the length of a finite line within the material in a reference state. The length L' in a deformed state is

$$L' = L + \int_L \epsilon\,dL, \qquad \text{Eq. (2-4)}$$

where ϵ is the extensional strain in the direction tangent to the line at each point. The change in length of the line is denoted by δ:

$$\delta = L' - L = \int_L \epsilon\,dL. \qquad \text{Eq. (2-5)}$$

If ϵ is uniform along the line, the change in length is

$$\delta = L' - L = \epsilon L. \qquad \text{Eq. (2-6)}$$

Consider two perpendicular infinitesimal lines dL_1 and dL_2 in a sample of material in a reference state [Fig. (d)]. Let the angle in radians between the two lines in a deformed state be denoted by $\pi/2 - \gamma$ [Fig. (e)]. The angle γ is the *shear strain* referred to the directions dL_1 and dL_2.

(d) (e)

If an infinitesimal rectangle in the reference state has sides dL_1 and dL_2 [Fig. (f)], the shear strain referred to the directions dL_1 and dL_2 determines the angles between the sides of the resulting parallelogram in the deformed state [Fig. (g)].

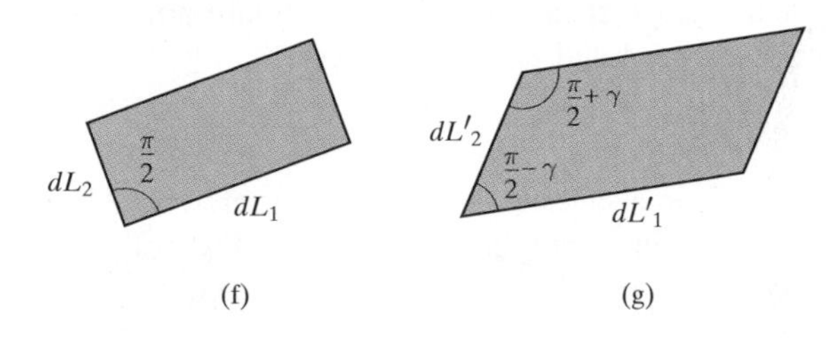

(f) (g)

PROBLEMS

2-1.1. The prismatic bar has a circular cross section with 50-mm radius and is subjected to 4-kN axial loads. Determine the average normal stress at the plane $\mathcal{P}$.

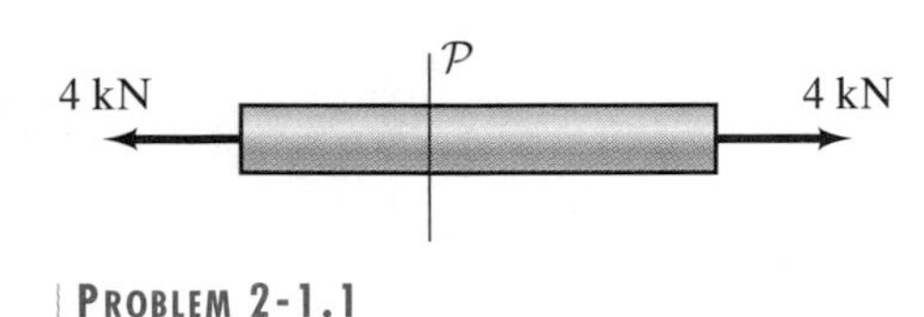

PROBLEM 2-1.1

2-1.2. In Problem 2-1.1, what is the average shear stress at the plane $\mathcal{P}$?

2-1.3. The prismatic bar has cross-sectional area $A = 30$ in^2 and is subjected to axial loads. Determine the average normal stress **(a)** at plane $\mathcal{P}_1$; **(b)** at plane $\mathcal{P}_2$.

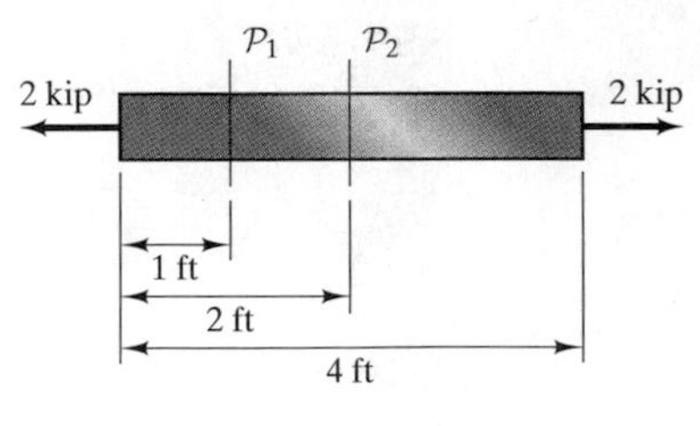

PROBLEM 2-1.3

2-1.4. The prismatic bar has a solid circular cross section with 2-in. radius. Determine the average normal stress **(a)** at plane $\mathcal{P}_1$; **(b)** at plane $\mathcal{P}_2$.

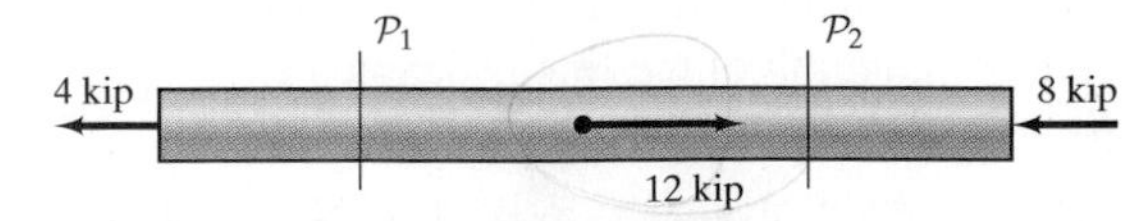

PROBLEM 2-1.4

2-1.5. A prismatic bar with cross-sectional area A is subjected to axial loads. Determine the average normal and shear stresses at the plane $\mathcal{P}$ if $A = 0.02\ \text{m}^2$, $P = 4$ kN, and $\theta = 25°$.

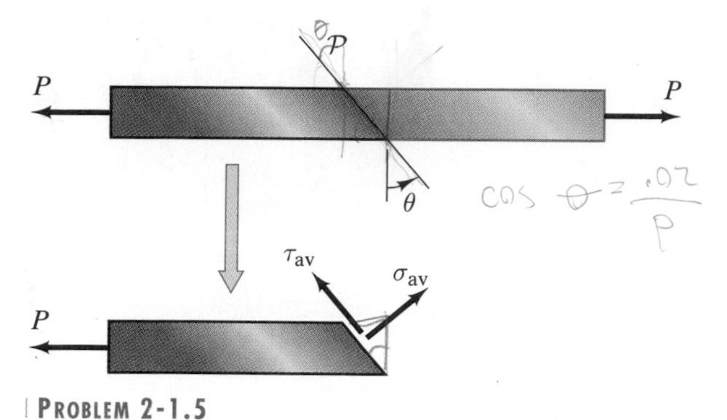

PROBLEM 2-1.5

2-1.6. Suppose that the prismatic bar shown in Problem 2-1.5 has cross-sectional area $A = 0.024\ \text{m}^2$. If the angle $\theta = 35°$ and the average normal stress on the plane $\mathcal{P}$ is $\sigma_{av} = 200$ kPa, what are τ_{av} and the axial force P?

2-1.7. For the prismatic bar in Problem 2-1.5, derive equations for the average normal and shear stresses at the plane $\mathcal{P}$ as functions of θ.

2-1.8. The prismatic bar has a solid circular cross section with 30-mm radius. It is suspended from one end and is loaded only by its own weight. The mass density of the homogeneous material is 2800 kg/m^3. Determine the average normal stress at the plane $\mathcal{P}$, where x is the distance from the bottom of the bar in meters. (*Strategy:* Draw a free-body diagram of the part of the bar below the plane $\mathcal{P}$.)

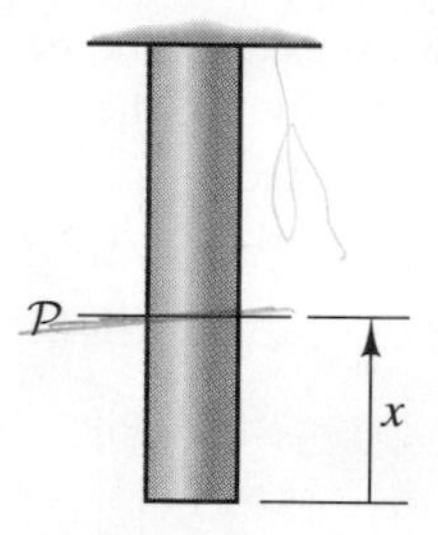

PROBLEM 2-1.8

2-1.9. The bar is made of material 1 in. thick. Its width varies linearly from 2 in. at its left end to 4 in. at its right end. If the axial load $P = 200$ lb, what is the average normal stress **(a)** at plane $\mathcal{P}_1$; **(b)** at plane $\mathcal{P}_2$?

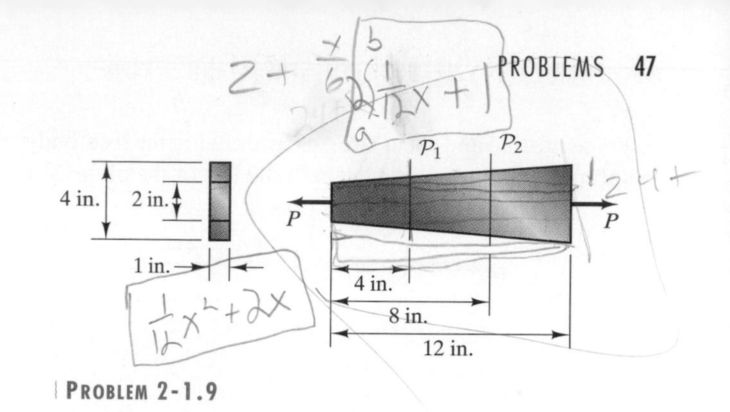

PROBLEM 2-1.9

2-1.10. If the average normal stress at plane $\mathcal{P}_1$ of the bar described in Problem 2-1.9 is $\sigma_{av} = 300$ psi, what are the axial load P and the average normal stress at plane $\mathcal{P}_2$?

2-1.11. The bar described in Problem 2-1.9 is suspended from its wide end and is loaded only by its own weight. The weight density of the homogeneous material is 0.32 lb/in^3. Determine the average normal stress at the plane $\mathcal{P}$, where x is the distance from the bottom of the bar in inches.

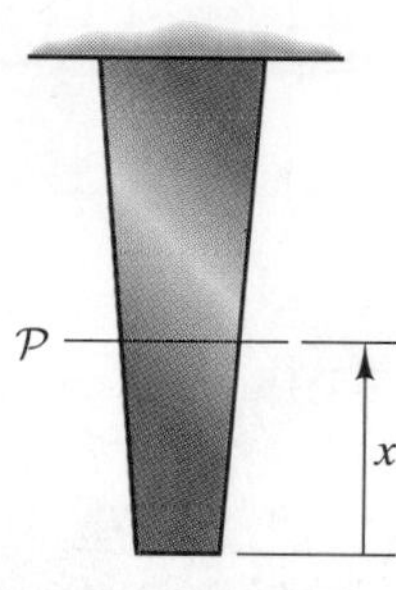

PROBLEM 2-1.11

2-1.12. The cantilever beam has cross-sectional area $A = 12\ \text{in}^2$. What are the average normal stress and the magnitude of the average shear stress at the plane $\mathcal{P}$?

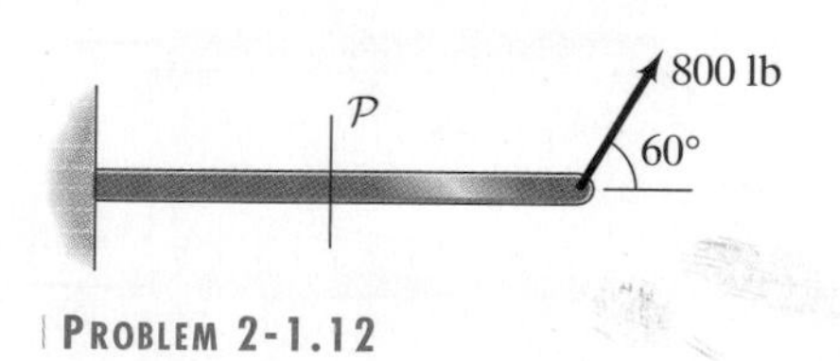

PROBLEM 2-1.12

2-1.13. The beam has cross-sectional area $A = 0.0625\ \text{m}^2$. What are the average normal stress and the magnitude of the average shear stress at the plane $\mathcal{P}$? (*Strategy:* Draw the free-body diagram of the entire beam and determine the reactions at the pin and roller supports. Then determine the average normal and shear stresses by drawing the free-body diagram of the part of the beam to the left of the plane $\mathcal{P}$.)

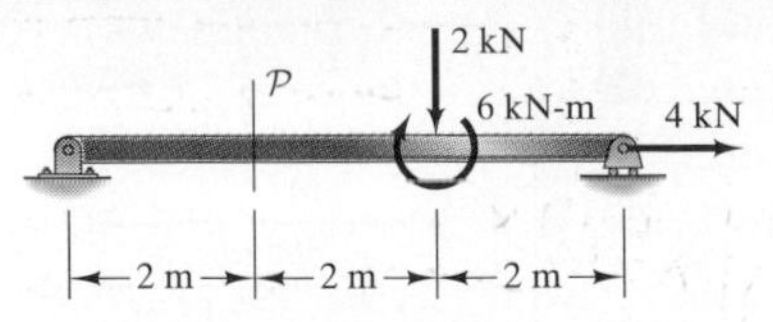

PROBLEM 2-1.13

2-1.14. Determine the average normal stress and the magnitude of the average shear stress at the plane $\mathcal{P}$ of the beam in Problem 2-1.13 by drawing the free-body diagram of the part of the beam to the right of the plane $\mathcal{P}$ and compare your answers to those of Problem 2-1.13.

2-1.15. The beam has cross-sectional area $A = 0.1\ \text{m}^2$. What are the average normal stress and the magnitude of the average shear stress at the plane $\mathcal{P}$?

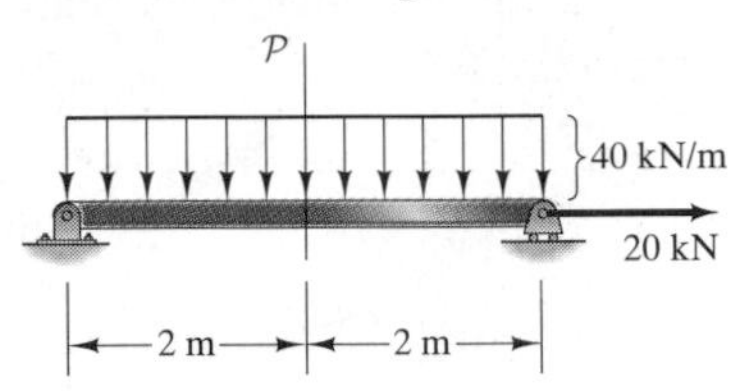

PROBLEM 2-1.15

2-1.16. In Problem 2-1.15, what are the average normal stress and the magnitude of the average shear stress at the plane $\mathcal{P}$ if the plane is 1 m from the left end of the beam?

2-1.17. The beams have cross-sectional area $A = 60\ \text{in}^2$. What are the average normal stress and the magnitude of the average shear stress at the plane $\mathcal{P}$ in cases (a) and (b)?

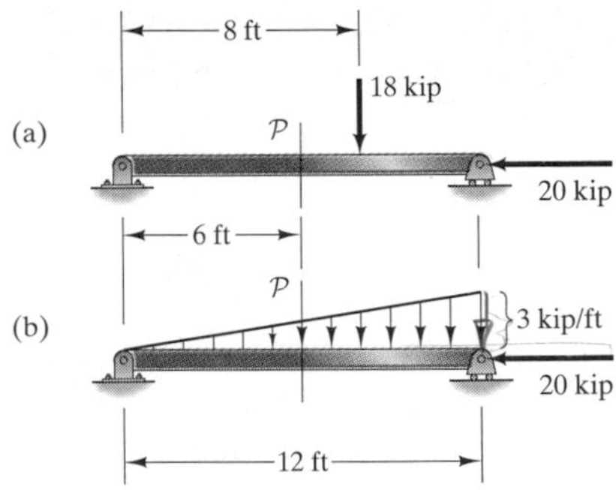

PROBLEM 2-1.17

2-1.18. Figure (a) is a diagram of the bones and biceps muscle of a person's arm supporting a mass. Figure (b) is a biomechanical model of the arm in which the biceps muscle AB is represented by a bar with pin supports. The suspended mass is $m = 2$ kg and the weight of the forearm is 9 N. If the cross-sectional area of the tendon connecting the biceps to the forearm at A is $28\ \text{mm}^2$, what is the average normal stress in the tendon?

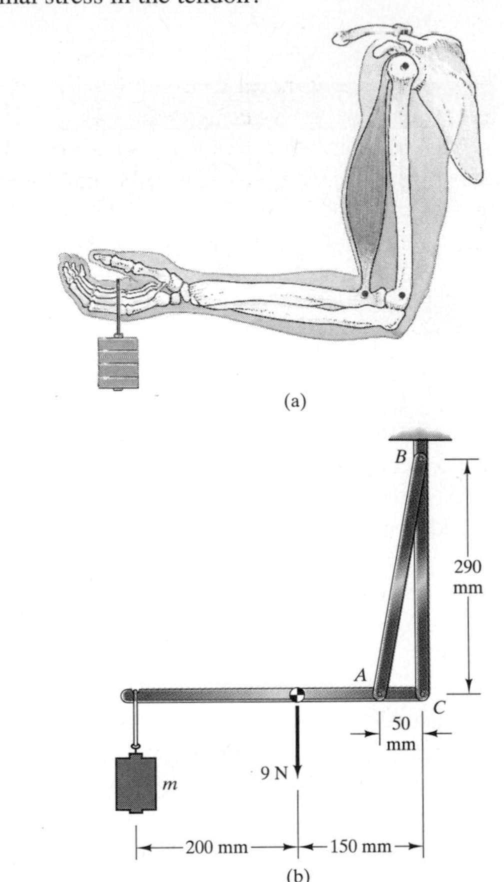

PROBLEM 2-1.18

2-1.19. The force **F** exerted on the bar is $20\mathbf{i} - 20\mathbf{j} - 10\mathbf{k}$ (lb). The plane $\mathcal{P}$ is parallel to the y–z plane and is 5 in.

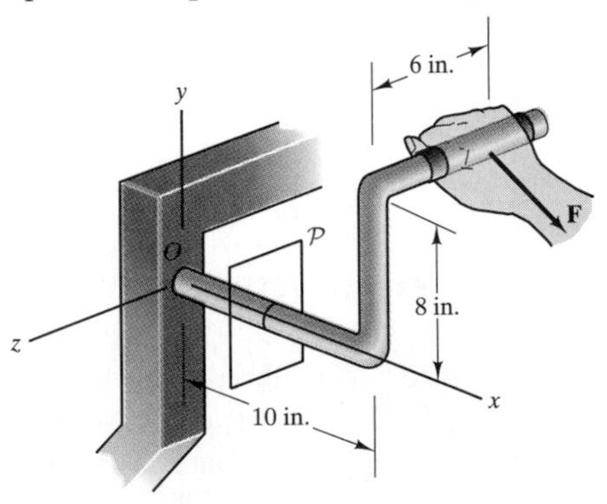

PROBLEM 2-1.19

from the origin O. The bar's cross-sectional area at $\mathcal{P}$ is $0.65\ \text{in}^2$. What is the average normal stress in the bar at $\mathcal{P}$?

2-1.20. In Problem 2-1.19, what is the magnitude of the average shear stress in the bar at $\mathcal{P}$?

2-1.21. The plane $\mathcal{P}$ is parallel to the y–z plane of the coordinate system. The cross-sectional area of the tennis racquet at $\mathcal{P}$ is $400\ \text{mm}^2$. Including the force exerted on the racquet by the ball and inertial effects, the total force on the racquet above the plane $\mathcal{P}$ is $35\mathbf{i} - 16\mathbf{j} - 85\mathbf{k}$ (N). What is the average normal stress on the racquet at $\mathcal{P}$?

PROBLEM 2-1.21

2-1.22. In Problem 2-1.21, what is the magnitude of the average shear stress on the racquet at $\mathcal{P}$?

2-1.23. The prismatic bar AB has cross-sectional area $A = 0.01\ \text{m}^2$. If the force $F = 6$ kN, what is the average normal stress at the plane $\mathcal{P}$?

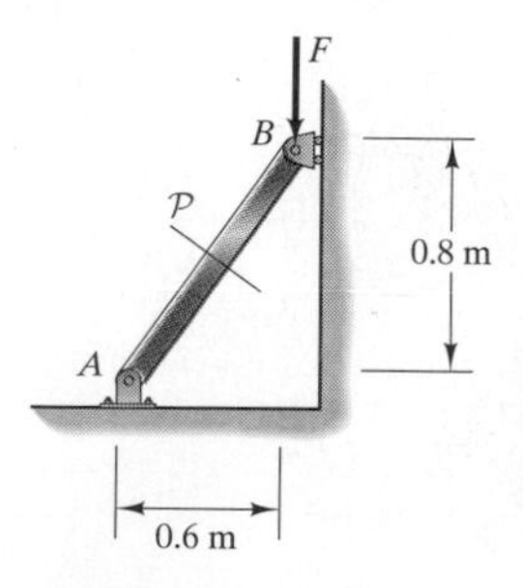

PROBLEM 2-1.23

2-1.24. The prismatic bar AB in Problem 2-1.23 will safely support an average compressive normal stress of 1.2 MPa on the plane $\mathcal{P}$. Based on this criterion, what is the largest downward force F that can safely be applied?

2-1.25. Two views of the support at point A of the prismatic bar AB in Problem 2-1.23 are shown. The support has a pin 40 mm in diameter. If the force $F = 8$ kN, what is the magnitude of the average shear stress in the pin?

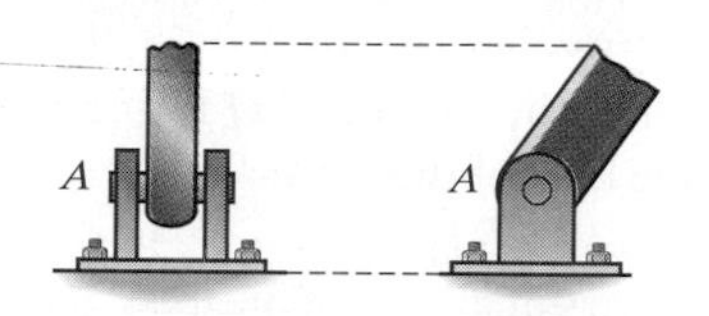

PROBLEM 2-1.25

2-1.26. The fixture shown connects a 50-mm-diameter bridge cable to a flange that is attached to the bridge. A 60-mm-diameter circular pin connects the fixture to the flange. If the average normal stress in the cable is $\sigma_{av} = 120$ MPa, what average shear stress τ_{av} must the pin support?

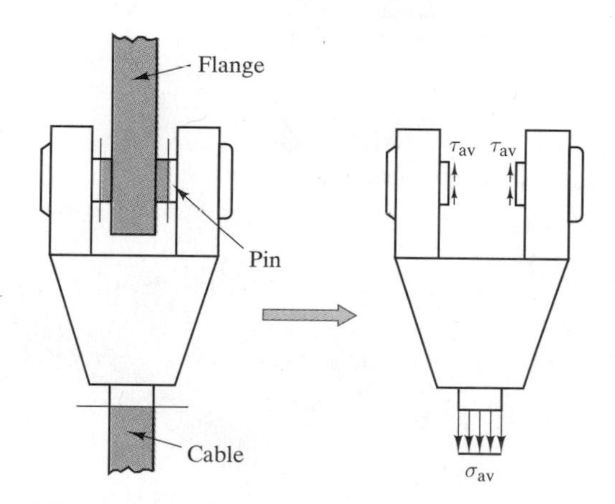

PROBLEM 2-1.26

2-1.27. Consider the fixture shown in Problem 2-1.26. The cable will safely support an average normal stress of 700 MPa and the circular pin will safely support an average shear stress of 220 MPa. Based on these criteria, what is the largest tensile load the cable will safely support?

2-1.28. The truss is made of prismatic bars with cross-sectional area $A = 0.25\ \text{ft}^2$. Determine the average normal stress in member BE acting on a plane perpendicular to the axis of the member.

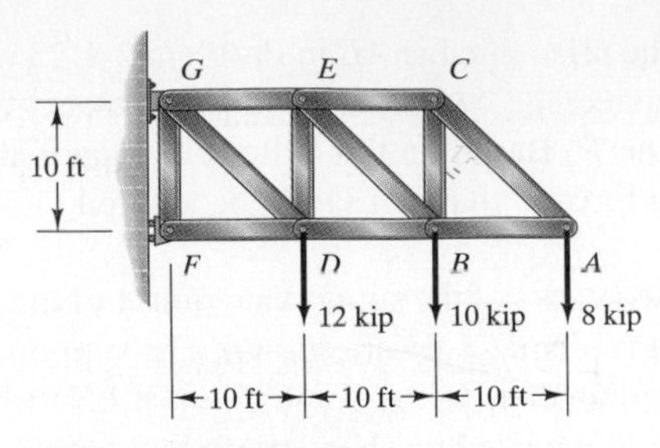

| PROBLEM 2-1.28

2-1.29. For the truss in Problem 2-1.28, determine the average normal stress in member *BD* acting on a plane perpendicular to the axis of the member.

2-1.30. Three views of joint *A* of the truss in Problem 2-1.28 are shown. The joint is supported by a cylindrical pin 2 in. in diameter. What is the magnitude of the average shear stress in the pin?

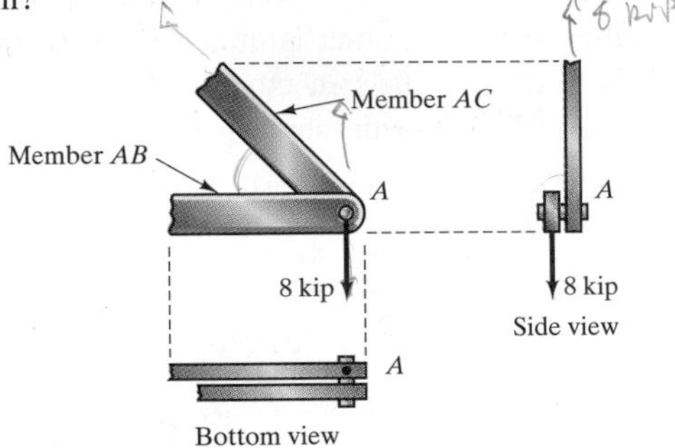

| PROBLEM 2-1.30

2-1.31. The top view of pin *A* of the pliers is shown. The cross-sectional area of the pin is 4.5 mm². What is the average shear stress in the pin when 150-N forces are applied to the pliers as shown?

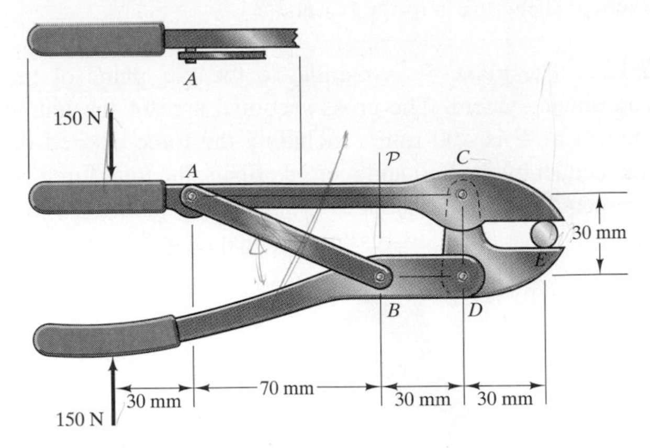

| PROBLEM 2-1.31

2-1.32. In Problem 2-1.31 the vertical plane $\mathcal{P}$ is 30 mm to the left of C. The cross-sectional area of member AC of the pliers at the plane $\mathcal{P}$ is 50 mm². Determine the average normal stress and the magnitude of the average shear stress at $\mathcal{P}$ when 150-N forces are applied to the pliers as shown.

2-1.33. The suspended crate weighs 2000 lb and the angle $\alpha = 30°$. The top view of the pin support *A* of the

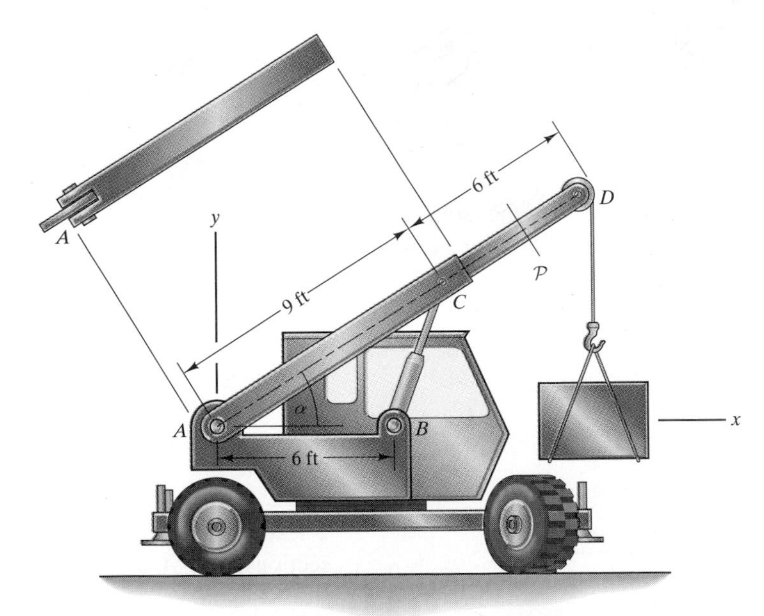

| PROBLEM 2-1.33

crane's boom is shown. The cross-sectional area of the pin is 23 in^2. What is the average shear stress in the pin?

2-1.34. In Problem 2-1.33, the plane $\mathcal{P}$ is 3 ft from end D of the crane's boom and is perpendicular to the boom. The cross-sectional area of the boom at $\mathcal{P}$ is 15 in^2. Determine the average normal stress and the magnitude of the average shear stress in the boom at $\mathcal{P}$.

2-1.35. The jaws of the bolt cutter are connected by two links AB. The cross-sectional area of each link is 750 mm^2. What average normal stress is induced in each link by the 90-N forces exerted on the handles?

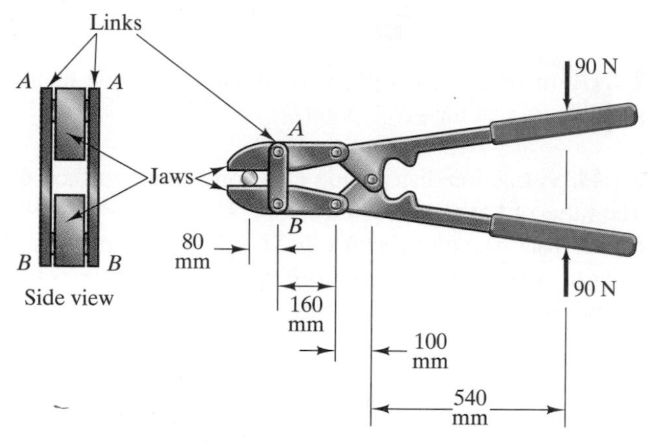

PROBLEM 2-1.35

2-1.36. The pins connecting the links AB to the jaws of the bolt cutter in Problem 2-1.35 are 20 mm in diameter. What average shear stress is induced in the pins by the 90-N forces exerted on the handles?

2-1.37. Three rectangular boards are glued together and subjected to axial loads as shown. What is the average shear stress on each glued surface?

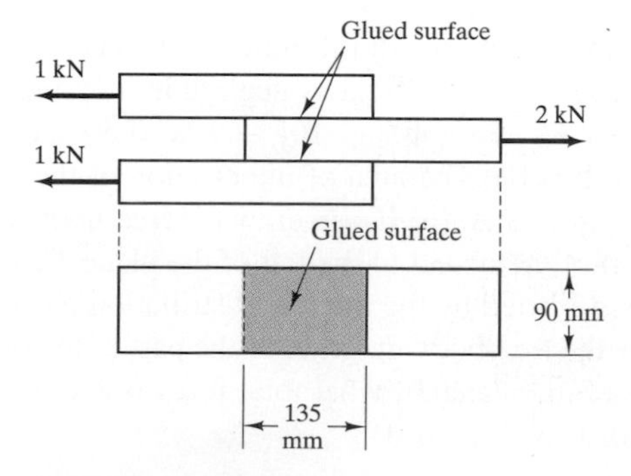

PROBLEM 2-1.37

2-1.38. Two boards with 4 in. × 4 in. square cross sections are mitered and glued together as shown. If the axial forces $P = 600$ lb, what average shear stress must the glue support?

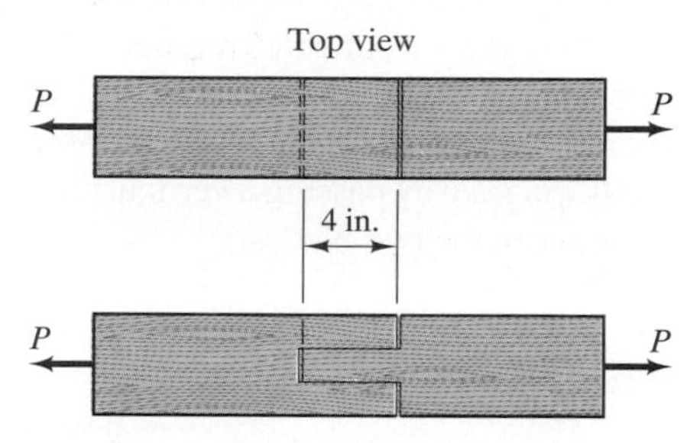

PROBLEM 2-1.38

2-1.39. A $\frac{1}{8}$-in.-diameter punch is used to cut blanks out of a $\frac{1}{16}$-in.-thick plate of aluminum. If an average shear stress of 20,000 psi must be induced in the plate to create a blank, what force F must be applied?

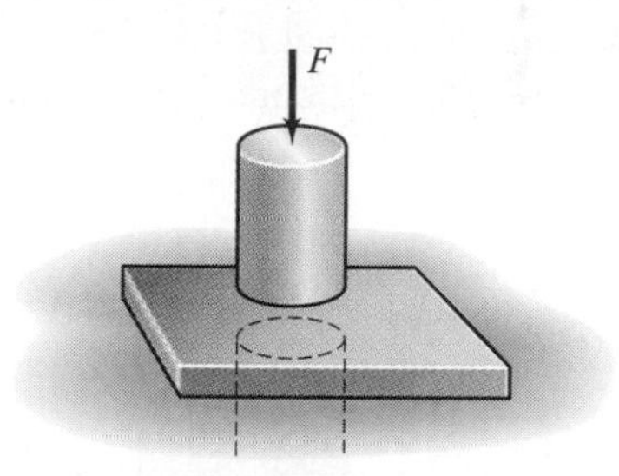

PROBLEM 2-1.39

2-1.40. Two pipes are connected by bolted flanges. The bolts are 20 mm in diameter. One pipe has a built-in support and the other is subjected to a torque $T = 6$ kN-m about its axis. Estimate the resulting average shear stress in each bolt. Why is this result an estimate?

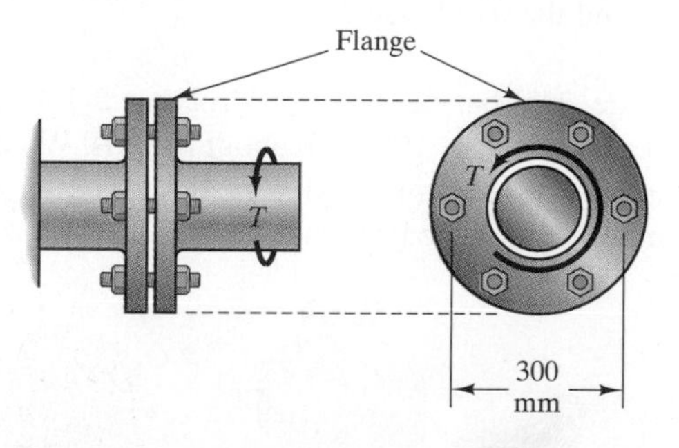

PROBLEM 2-1.40

2-1.41. The bolts in Problem 2-1.40 will each safely support an average shear stress of 130 MPa. Based on this criterion, estimate the largest safe torque T that can be applied.

2-1.42. "Shears" such as the familiar scissors have two blades that subject a material to shear stress. For the shearing process shown, draw a suitable free-body diagram and determine the average shear stress the blades exert on the sheet of material of thickness t and width b. (*Strategy:* Obtain a free-body diagram by passing a vertical plane through the material between the two blades.)

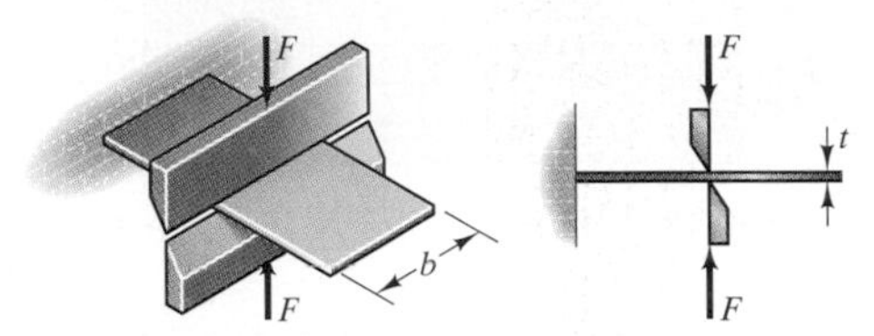

| PROBLEM 2-1.42

2-1.43. A 2-in.-diameter cylindrical steel bar is attached to a 3-in.-thick fixed plate by a cylindrical rubber grommet. If the axial load $P = 60$ lb, what is the average shear stress on the cylindrical surface of contact between the bar and the grommet?

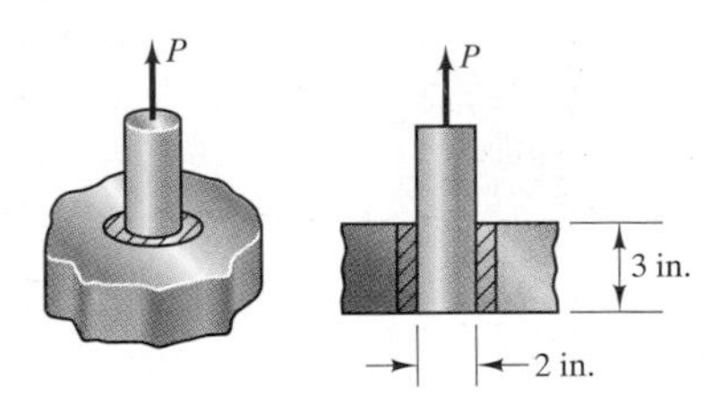

| PROBLEM 2-1.43

2-1.44. The outer diameter of the cylindrical rubber grommet in Problem 2-1.43 is 3.5 in. What is the average shear stress on the cylindrical surface of contact between the grommet and the fixed plate?

2-1.45. The steel bar described in Problem 2-1.43 is subjected to a torque $T = 100$ in-lb about its axis. What is the average shear stress on the cylindrical surface of contact between the bar and the grommet?

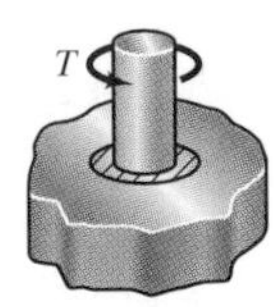

| PROBLEM 2-1.45

2-1.46. A traction distribution $\mathbf{t}$ acts on a plane surface A. The value of $\mathbf{t}$ at a given point on A is $\mathbf{t} = 45\mathbf{i} + 40\mathbf{j} - 30\mathbf{k}$ (kPa). The unit vector $\mathbf{i}$ is perpendicular to A and points away from the material. What is the normal stress σ at the given point?

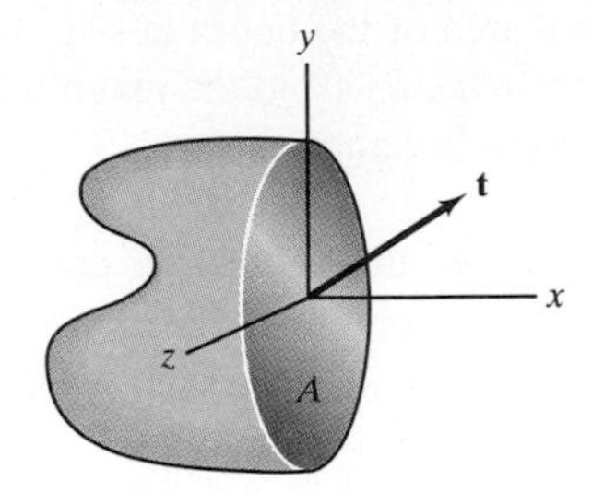

| PROBLEM 2-1.46

2-1.47. In Problem 2-1.46, what is the magnitude of the shear stress τ at the given point?

2-1.48. A traction distribution $\mathbf{t}$ acts on a plane surface A. The value of $\mathbf{t}$ at a given point on A is $\mathbf{t} = 3000\mathbf{i} - 2000\mathbf{j} + 6000\mathbf{k}$ (psi). The unit vector $\mathbf{e} = \frac{6}{7}\mathbf{i} + \frac{3}{7}\mathbf{j} + \frac{2}{7}\mathbf{k}$ is perpendicular to A and points away from the material. What is the normal stress σ at the given point?

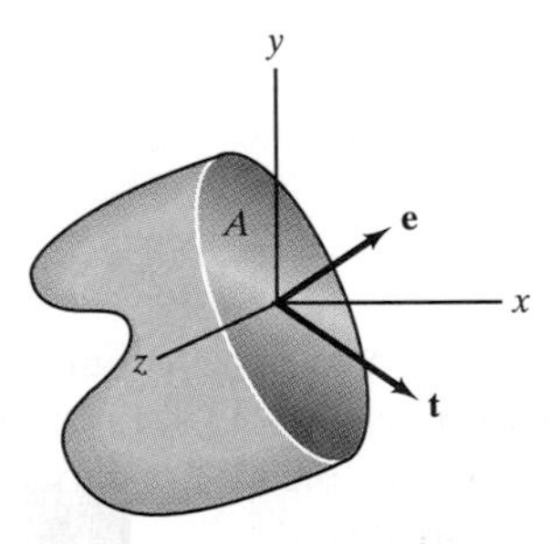

| PROBLEM 2-1.48

2-1.49. In Problem 2-1.48, what is the magnitude of the shear stress τ at the given point?

2-1.50. An object in equilibrium is subjected to three forces. (The object's weight is negligible in comparison.) The forces $\mathbf{F}_1 = -40\mathbf{i} + 10\mathbf{j} - 10\mathbf{k}$ (kN) and $\mathbf{F}_2 = -30\mathbf{i} + 15\mathbf{k}$ (kN). The area of intersection of the plane $\mathcal{P}$ with the object is A. **(a)** If you draw the free-body diagram of the part of the object to the left of the plane $\mathcal{P}$, what total force is exerted by the traction distribution on A? **(b)** If you draw the free-body diagram of the part of the object to the right of the plane $\mathcal{P}$, what total force is exerted by the traction distribution on A?

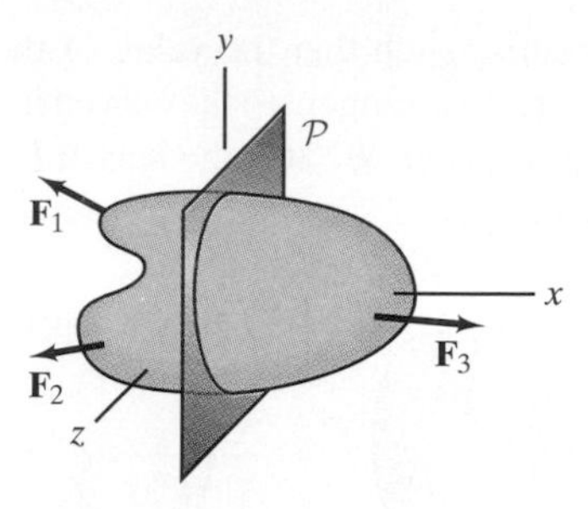

PROBLEM 2-1.50

2-1.51. The plane $\mathcal{P}$ in Problem 2-1.50 is perpendicular to the x axis and the area of intersection of $\mathcal{P}$ with the object is $A = 0.04\ \text{m}^2$. If you draw the free-body diagram of the part of the object to the left of the plane $\mathcal{P}$, determine **(a)** the average traction on A; **(b)** the average normal stress on A.

2-1.52. The rectangular cross section A is subjected to the traction distribution

$$\mathbf{t} = (2 - 1.5y^2)\mathbf{i} + 3y^2\mathbf{j} + (1 + 2y^2)\mathbf{k}\ (\text{MPa}).$$

What is the average normal stress on A?

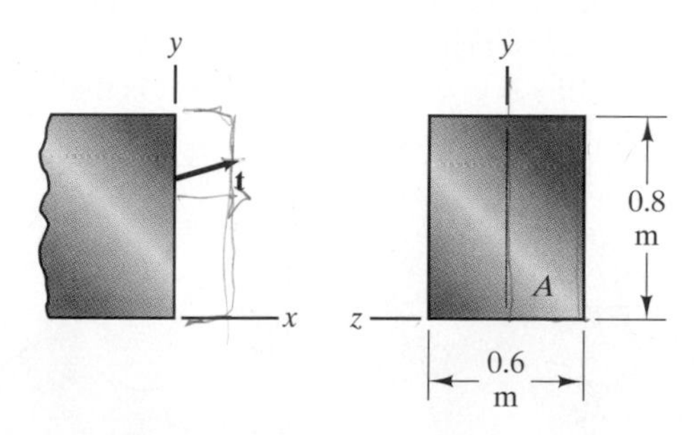

PROBLEM 2-1.52

2-1.53. In Problem 2-1.52, what is the magnitude of the average shear stress on A?

2-1.54. Suppose that the rectangular cross section A in Problem 2-1.52 is subjected to the traction distribution

$$\mathbf{t} = (2 - 1.5z^2)\mathbf{i} + 3z^2\mathbf{j} + (1 + 2z^2)\mathbf{k}\ (\text{MPa}).$$

What is the average traction $\mathbf{t}_{\text{av}}$ on A?

2-1.55. The rectangular cross section A is subjected to a normal stress given by the equation $\sigma = ay$, where a is a constant. At $y = 0.8$ m, $\sigma = 20$ MPa. Determine **(a)** the total normal force exerted on A; **(b)** the average normal stress on A; **(c)** the ratio of the maximum normal stress on A to the average normal stress.

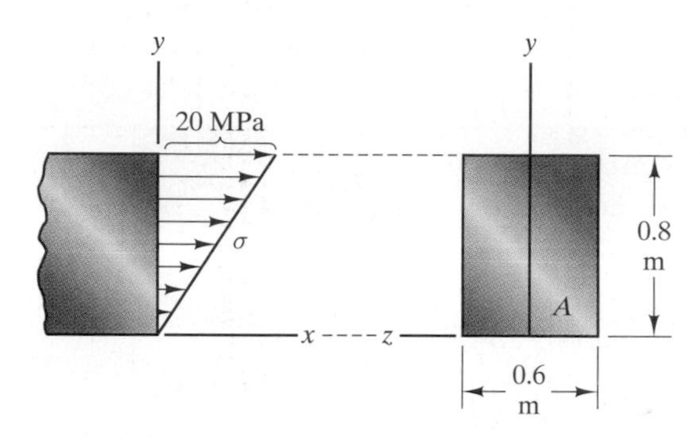

PROBLEM 2-1.55

2-1.56. A beam with rectangular cross section is subjected to a counterclockwise couple M at its left end and at its right end is subjected to a normal stress distribution $\sigma = ay$, where a is a constant. The origin of the coordinate system is at the center of the beam's cross section and the beam is in equilibrium. **(a)** What total force is exerted on the beam by the normal stress distribution? **(b)** Show that the constant $a = 12M/bh^3$. [*Strategy:* To do part (b), calculate the total moment about the z axis. The force exerted on each infinitesimal element dA of the cross section by the normal stress distribution is $\sigma\, dA$, so the moment about the z axis due to the stress distribution is $-(\int_A y\sigma\, dA)\mathbf{k}$.]

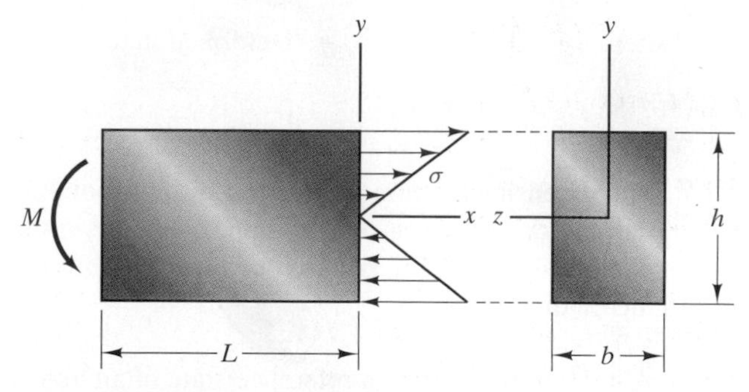

PROBLEM 2-1.56

2-1.57. The rectangular cross section A is subjected to a shear stress in the y-axis direction given by the equation

$$\tau = \frac{6V}{bh^3}\left(\frac{h^2}{4} - y^2\right),$$

where V is a constant. Determine **(a)** the total force exerted on A; **(b)** the average shear stress on A; **(c)** the ratio of the maximum shear stress on A to the average shear stress.

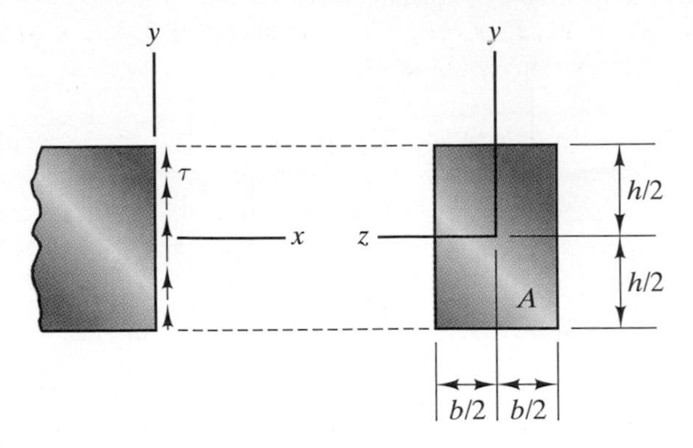

PROBLEM 2-1.57

2-2.1. A line of length dL at a particular point of a material in a reference state has length $dL' = 1.2dL$ in a deformed state. What is the extensional strain corresponding to that particular point and the direction of the line dL?

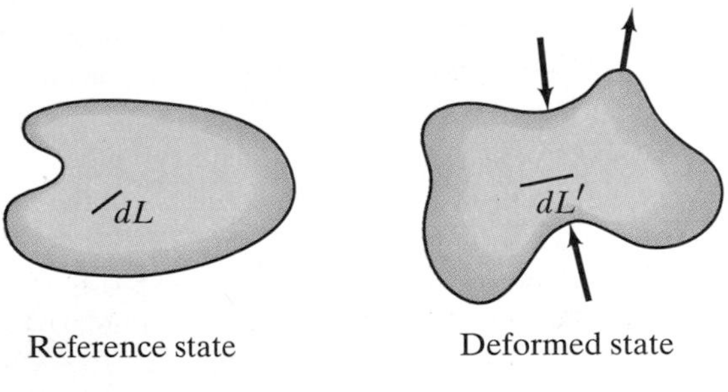

PROBLEM 2-2.1

2-2.2. The extensional strain corresponding to a point of a material and the direction of a line of length dL in the reference state is $\epsilon = 0.15$. What is the length dL' of the line in the deformed state?

2-2.3. A straight line within a reference state of an object is 50 mm long. In a deformed state, the line is 54 mm long. If the extensional strain ϵ in the direction tangent to the line is uniform thoughout the line's length, what is ϵ?

2-2.4. The length of the curved line within the material in the reference state is $L = 0.2$ m. The material then undergoes a deformation such that the value of the extensional strain ϵ in the direction tangent to the curved line is $\epsilon = 0.03$ at each point of the line. What is the length L' of the line in the deformed state?

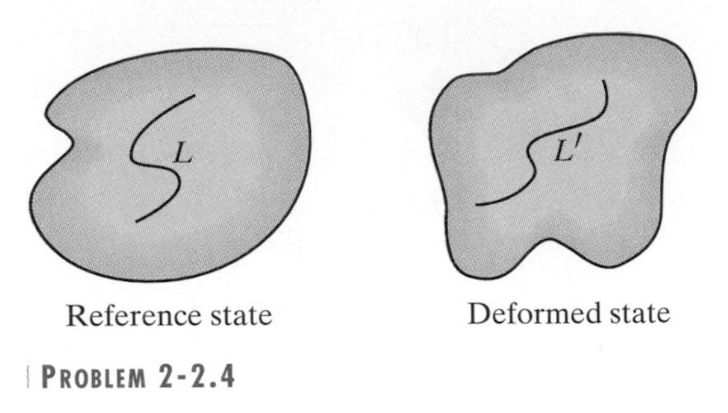

PROBLEM 2-2.4

2-2.5. The length of the curved line L in the reference state shown in Problem 2-2.4 is 0.5 m. Its length in the deformed state is $L' = 0.488$ m. If the value of the extensional strain ϵ in the direction tangent to the line is the same at each point of the line, what is ϵ?

2-2.6. The coordinate s measures distance along the curved line in the reference state. The length of the line is $L = 0.2$ m. The material then undergoes a deformation such that the value of the extensional strain ϵ in the direction tangent to the curved line is $\epsilon = 0.03 + 2s^2$. What is the length L' of the line in the deformed state?

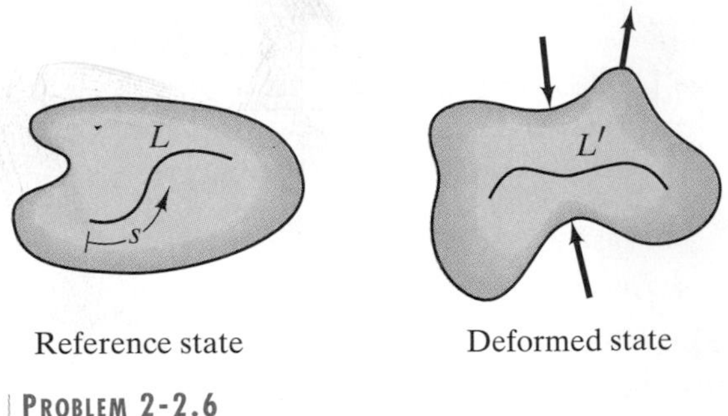

PROBLEM 2-2.6

2-2.7. In Problem 2-2.6, suppose that the material undergoes a deformation that induces an extensional strain tangent to the line given by the equation $\epsilon = 0.01[1 + (s/L)^3]$. What is the length L' of the line in the deformed state?

2-2.8. A 4-ft prismatic bar is subjected to axial forces that increase its length to 4.025 ft. If you assume that the extensional strain ϵ in the direction of the bar's axis is uniform (constant) along its axis, what is ϵ?

2-2.9. Suppose that you subject a 2-m prismatic bar to compressive axial forces that cause a uniform extensional strain $\epsilon = -0.003$ in the axial direction. What is the bar's length in the deformed state?

2-2.10. A prismatic bar is subjected to loads that cause uniform axial strains $\epsilon = 0.002$ in its left half and $\epsilon = -0.004$ in its right half. What is the resulting change in length of the 28-in. bar?

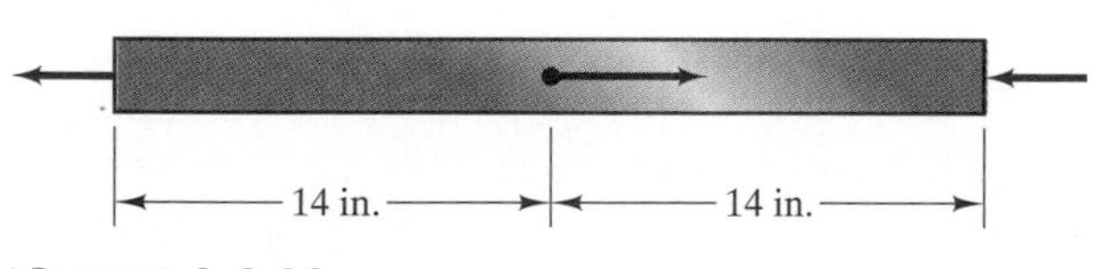

PROBLEM 2-2.10

2-2.11. The prismatic bar in Problem 2-2.10 is subjected to loads that cause a uniform axial strain $\epsilon_L = 0.006$ in its left half and a uniform axial strain ϵ_R in its right half. As a result, the length of the 28-in. bar increases by 0.032 in. What is ϵ_R?

2-2.12. In a shock-wave experiment, the left side of a 100-mm-thick plate of steel is subjected to a constant velocity of 1.5 km/s to the right at time $t = 0$. As a result, a shock wave travels across the plate with a constant velocity $U > 1.5$ km/s. To the right of the shock wave, the material

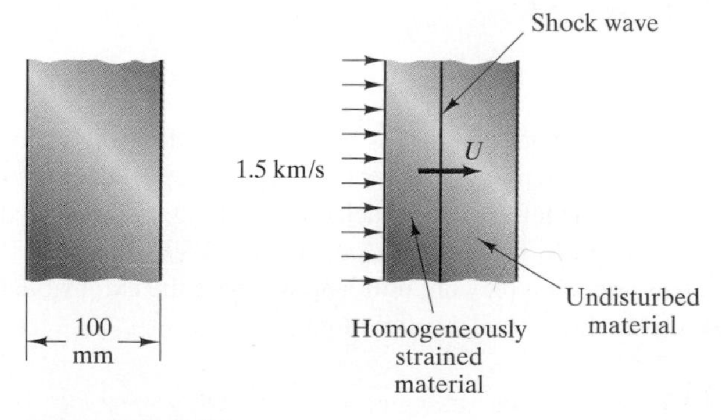

PROBLEM 2-2.12

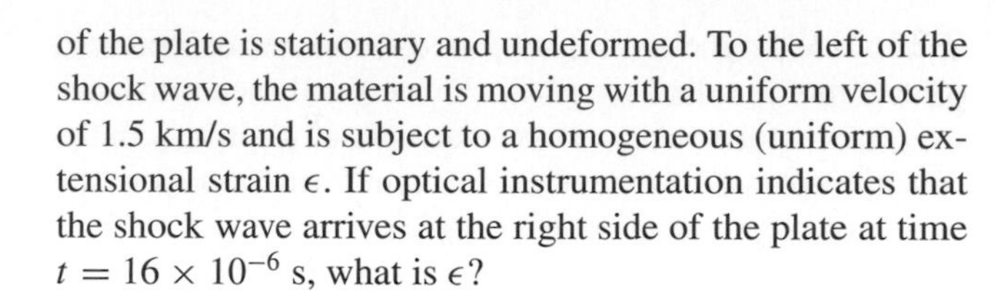

of the plate is stationary and undeformed. To the left of the shock wave, the material is moving with a uniform velocity of 1.5 km/s and is subject to a homogeneous (uniform) extensional strain ϵ. If optical instrumentation indicates that the shock wave arrives at the right side of the plate at time $t = 16 \times 10^{-6}$ s, what is ϵ?

2-2.13. In Problem 2-2.12, suppose that the time at which the shock wave arrives at the right side of the plate is unknown, but an embedded strain gauge indicates that the homogeneous extensional strain to the left of the shock wave is $\epsilon = -0.3$. What is the velocity U of the shock wave?

2-2.14. When it is unloaded, the nonprismatic bar is 12 in. long. The loads cause axial strain given by the equation $\epsilon = 0.04/(x + 12)$, where x is the distance from the left end of the bar in inches. What is the change in length of the bar?

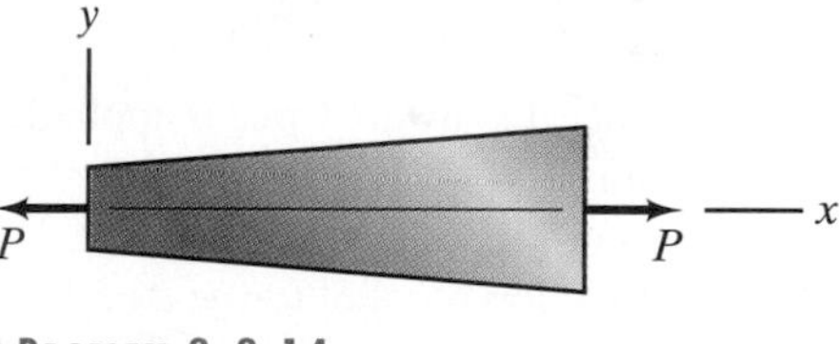

PROBLEM 2-2.14

2-2.15. The force F causes point B to move downward 0.002 m. If you assume the resulting extensional strain ϵ parallel to the axis of the bar AB is uniform along the bar's length, what is ϵ?

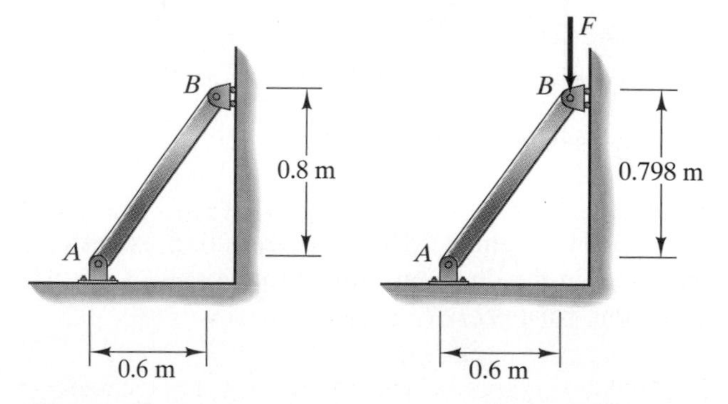

PROBLEM 2-2.15

2-2.16. When the truss is subjected to the vertical force F, joint A moves a distance $v = 0.3$ m vertically and a distance $u = 0.1$ m horizontally. If the extensional strain ϵ_{AB} in the direction parallel to member AB is uniform thoughout the length of the member, what is ϵ_{AB}?

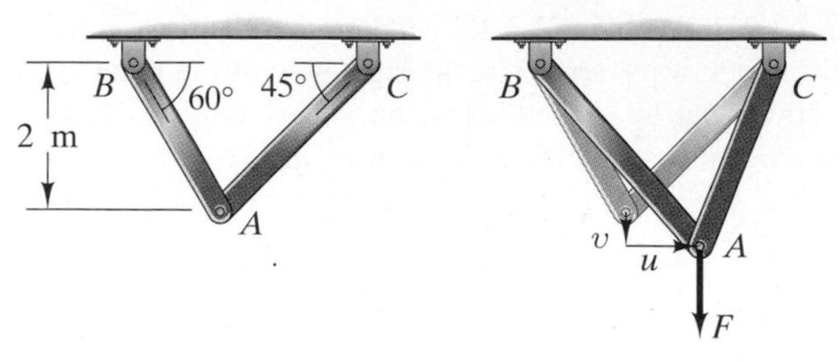

PROBLEM 2-2.16

2-2.17. In Problem 2-2.16, if the extensional strain ϵ_{AC} in the direction parallel to member AC is uniform thoughout the length of the member, what is ϵ_{AC}?

2-2.18. Suppose that a downward force is applied at point A of the truss, causing point A to move 0.360 in. downward and 0.220 in. to the left. If the resulting extensional strain ϵ_{AB} in the direction parallel to the axis of bar AB is uniform, what is ϵ_{AB}?

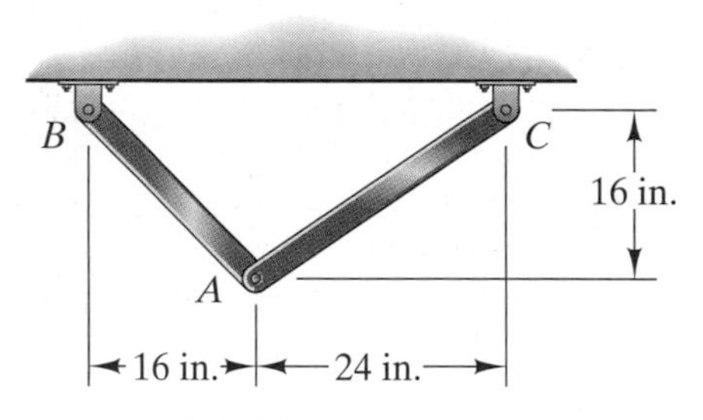

PROBLEM 2-2.18

2-2.19. In Problem 2-2.18, if the resulting extensional strain ϵ_{AC} in the direction parallel to the axis of bar AC is uniform, what is ϵ_{AC}?

2-2.20. A steel tube (a) has an outer radius $r = 20$ mm. The tube is then pressurized, increasing its outer radius to $r' = 20.04$ mm (b). What is the resulting extensional strain of the bar's outer circumference in the direction tangent to the circumference?

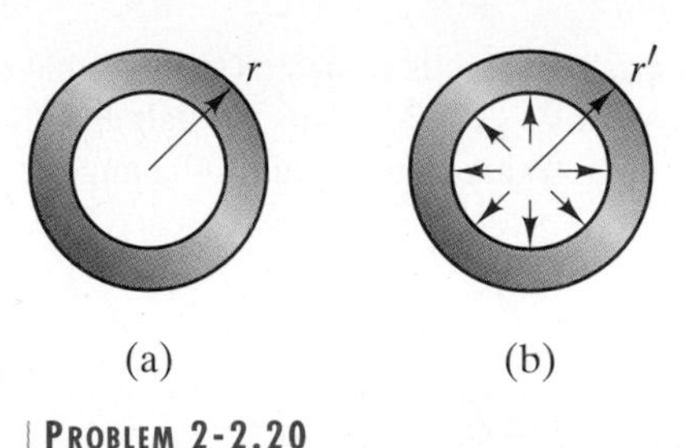

PROBLEM 2-2.20

2-2.21. The angle between two infinitesimal lines dL_1 and dL_2 that are perpendicular in a reference state is 120° in a deformed state. What is the shear strain at this point corresponding to the directions dL_1 and dL_2?

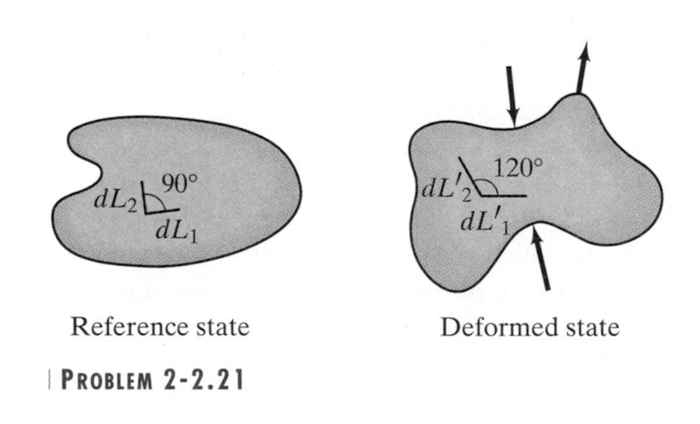

PROBLEM 2-2.21

2-2.22. When the airplane's wing is unloaded (the reference state), the perpendicular lines L_1 and L_2 on the upper surface of the left wing are each 600 mm long. In the loaded state shown, L_1 is 600.2 mm long and L_2 is 595 mm long. If you assume that they are uniform, what are the extensional strains in the L_1 and L_2 directions?

PROBLEM 2-2.22

2-2.23. In Problem 2-2.22, the angle between the lines L_1 and L_2 at the point where they intersect is 90.2° in the loaded state. What is the shear strain referred to the directions L_1 and L_2 at that point?

2-2.24. Two infinitesimal lines dL_1 and dL_2 are shown in a reference state and in a deformed state. (The lines dL_1, dL_2, dL'_1, and dL'_2 are contained in the x–y plane.) What is the shear strain at this point corresponding to the directions dL_1 and dL_2?

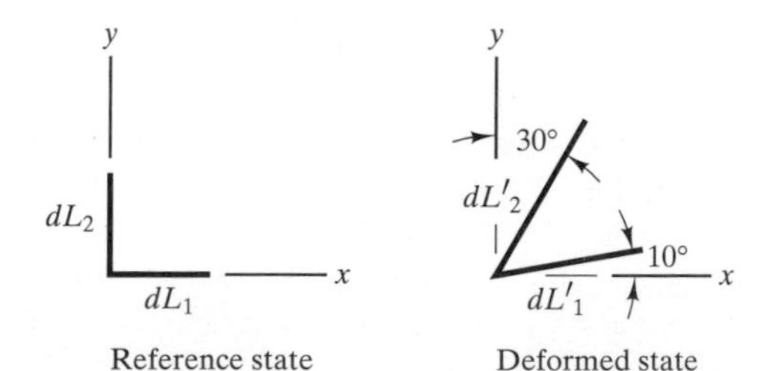

PROBLEM 2-2.24

2-2.25. In the upper part of Fig. (a) a cylindrical bar with 12-m length and 1-m radius has a built-in support at the left end. Axial and circumferential lines are drawn on the bar's surface. The lower part of Fig. (a) shows the bar's curved surface "unwrapped," illustrating the geometry of the axial and circumferential lines. The infinitesimal line dL_1 lies on an axial line and the infinitesimal line dL_2 lies on a circumferential line. A torque T is then applied about the axis of the bar at its right end, causing the end to rotate [Fig. (b)]. The bar's length and radius do not change when T is applied. The resulting geometry of the axial and circumferential lines is shown in the lower part of Fig. (b); in this unwrapped view they remain straight but are no longer orthogonal. The resulting shear strain referred to the directions dL_1 and dL_2 is 0.13. Use this value to determine the angle through which the end of the bar rotates.

2-2.26. In Problem 2-2.25, determine the angle through which the end of the bar rotates if the shear strain referred to the directions dL_1 and dL_2 is 0.2.

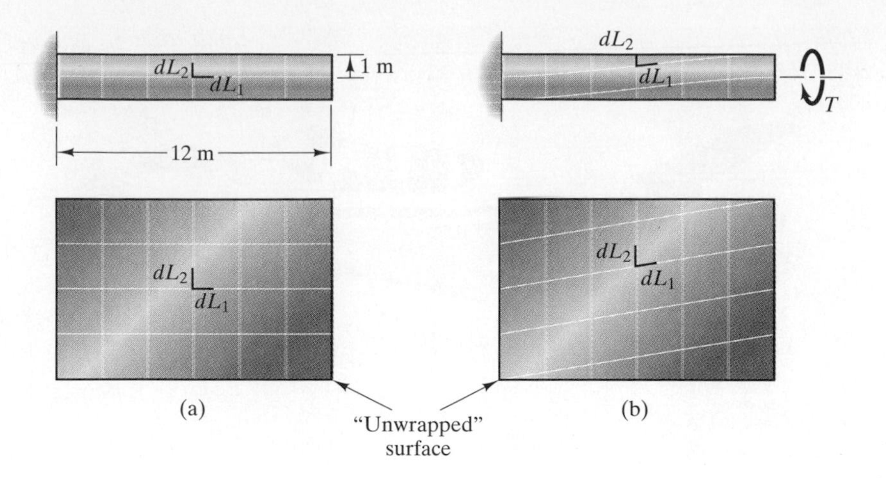

PROBLEM 2-2.25

2-2.27. Two infinitesimal lines dL_1 and dL_2 within a material are parallel to the x and y axes in a reference state [Fig. (a)]. After a motion and deformation of the material, dL_1 points in the direction of the unit vector $\mathbf{e}_1 = 0.667\mathbf{i} + 0.667\mathbf{j} + 0.333\mathbf{k}$, and dL_2 points in the direction of the unit vector $\mathbf{e}_2 = -0.408\mathbf{i} + 0.816\mathbf{j} - 0.408\mathbf{k}$ [Fig. (b)]. What is the shear strain referred to the directions dL_1 and dL_2?

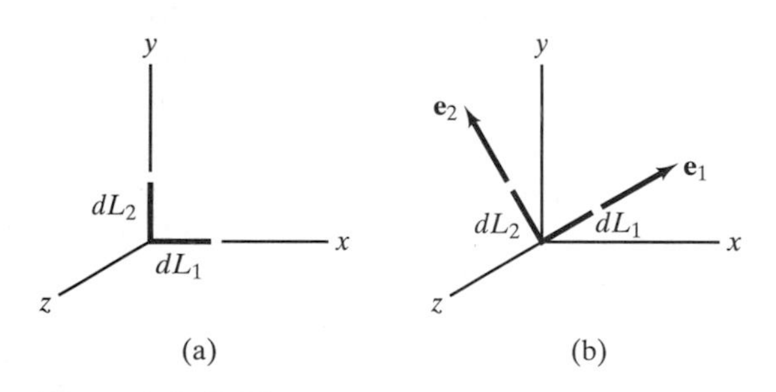

PROBLEM 2-2.27

2-2.28. In Problem 2-2.27, suppose that after the motion and deformation of the material, dL_1 points in the direction of the unit vector $\mathbf{e}_1 = 0.667\mathbf{i} + 0.667\mathbf{j} + 0.333\mathbf{k}$, and dL_2 points in the direction of the unit vector $\mathbf{e}_2 = -0.514\mathbf{i} + 0.686\mathbf{j} + 0.514\mathbf{k}$. What is the shear strain referred to the directions dL_1 and dL_2?

2-2.29. An infinitesimal rectangle at a point in a reference state of a material is shown. In a deformed state the extensional strains in the dL_1 and dL_2 directions are $\epsilon_1 = 0.04$ and $\epsilon_2 = -0.02$ and the shear strain referred to the dL_1 and dL_2 directions is $\gamma = 0.02$. What is the extensional strain in the dL direction? (See Example 2-6.)

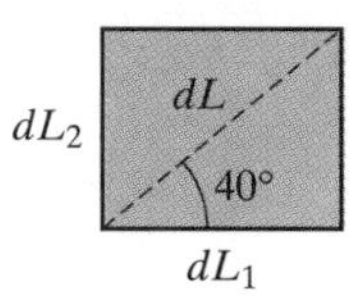

PROBLEM 2-2.29

2-2.30. For the infinitesimal rectangle at a point in a reference state of a material shown in Problem 2-2.29, suppose that in a deformed state the extensional strains in the dL_1, dL_2, and dL directions are $\epsilon_1 = 0.030$, $\epsilon_2 = 0.020$, and $\epsilon = 0.038$. What is the shear strain referred to the dL_1 and dL_2 directions?

2-2.31. An infinitesimal rectangle at a point in a reference state of a material becomes the parallelogram shown in a deformed state. Determine **(a)** the extensional strain in the dL_1 direction; **(b)** the extensional strain in the dL_2 direction; **(c)** the shear strain corresponding to the dL_1 and dL_2 directions.

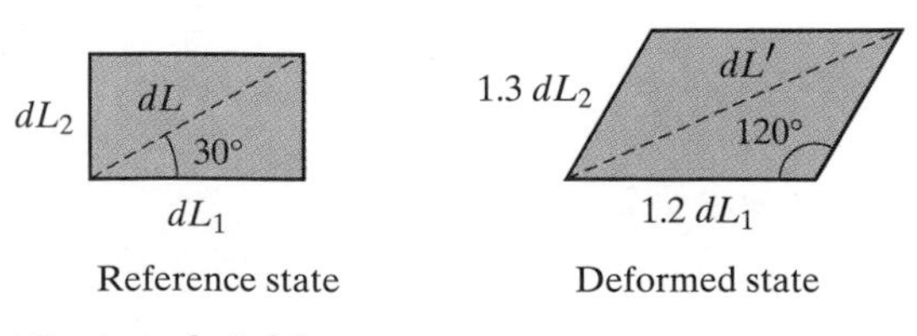

PROBLEM 2-2.31

2-2.32. For the infinitesimal rectangle in Problem 2-2.31, show that the length of the diagonal dL in the deformed state is $dL' = 1.476dL$.

2-2.33. In Example 2-6 we derived an equation for the extensional strain ϵ in the direction of an infinitesimal line dL. If the strains ϵ, ϵ_1, ϵ_2, and γ are "small," meaning that products of the strains are negligible in comparison to the strains themselves, show that the equation for ϵ simplifies to

$$\epsilon = \epsilon_1 \cos^2\theta + \epsilon_2 \sin^2\theta + \gamma \sin\theta \cos\theta.$$

Strategy: Begin with the equation

$$(dL')^2 = \left[(1+\epsilon_1)^2 \cos^2\theta + (1+\epsilon_2)^2 \sin^2\theta - 2(1+\epsilon_1) \times (1+\epsilon_2) \cos\theta \sin\theta \cos\left(\frac{\pi}{2}+\gamma\right)\right](dL)^2.$$

Write the left side in terms of ϵ using the relation $dL' = (1+\epsilon)dL$ and express the term $\cos\left(\frac{\pi}{2}+\gamma\right)$ as a Taylor series:

$$\cos\left(\frac{\pi}{2}+\gamma\right) = \cos\frac{\pi}{2} - \left(\sin\frac{\pi}{2}\right)\gamma - \frac{1}{2}\left(\cos\frac{\pi}{2}\right)\gamma^2 + \cdots.$$

Then neglect terms in the equation involving products of the strains.

The front wheels of the Formula 1 car are attached to the car's frame by a light and aerodynamically efficient structure of axially loaded steel rods.

CHAPTER 3

Axially Loaded Bars

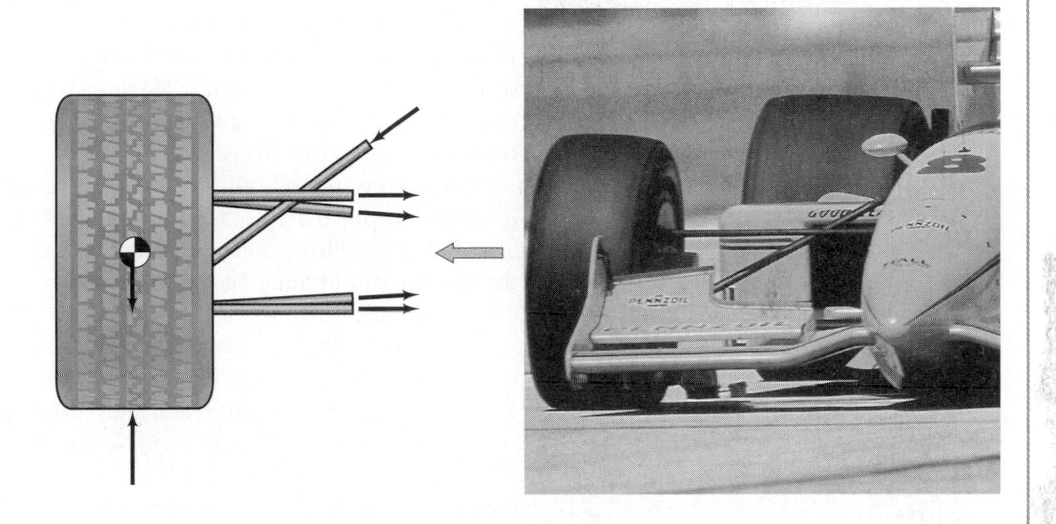

The figure shows the axial loads in the truss supporting the car's wheel. Truss structures—widely used to support bridges, buildings, vehicles, and other mechanical devices—are frameworks of bars that support loads by subjecting their members to axial loads. Exerting axial loads on a bar and measuring the resulting change in its length is a common method of testing the response of materials to stress. Not only is this test relatively straightforward to carry out, it subjects the material to a very simple and easily determined distribution of stress. Because of this simplicity, and the great importance of axially loaded bars from the standpoint of applications, we analyze them in detail in this chapter.

3-1 | Stresses in Prismatic Bars

In Chapter 2 we determined average normal and shear stresses in axially loaded bars. We emphasized, however, that it is the actual distributions of stress, not their average values, that usually must be known for design. We now describe these distributions.

Stresses on Perpendicular Planes

We first consider the stresses acting on planes perpendicular to the axis of an axially loaded bar. Suppose that we could somehow load a prismatic bar by applying uniform normal stresses σ at its ends (Fig. 3-1). Then it can be shown that the stress distribution on *every* plane perpendicular to the axis of the bar consists of the same uniform normal stress and no shear stress (Fig. 3-2). In Fig. 3-3 we draw the free-body diagram of an element of the bar having rectangular faces, two of which are perpendicular to the bar's axis. The faces perpendicular to the axis are subjected to the uniform normal stress σ and the other faces are free of stress.

The proof of these results, beyond our scope in this elementary treatment, involves showing that no other distribution of stress can satisfy both the boundary conditions (the tractions on the ends and the traction-free lateral surfaces of the bar) and equilibrium. They also depend upon the material having a property called isotropy, which we discuss in Chapter 6.

So we know the stress distribution on planes perpendicular to the axis of an axially loaded prismatic bar, but only if the bar is loaded in a very idealized way. What does that tell us about real axially loaded bars? To answer this question we first need to demonstrate that applying a uniform normal stress to the end of a bar is equivalent to applying a single force at the centroid of its cross section. (Recall that we define two systems of forces and moments to be equivalent if

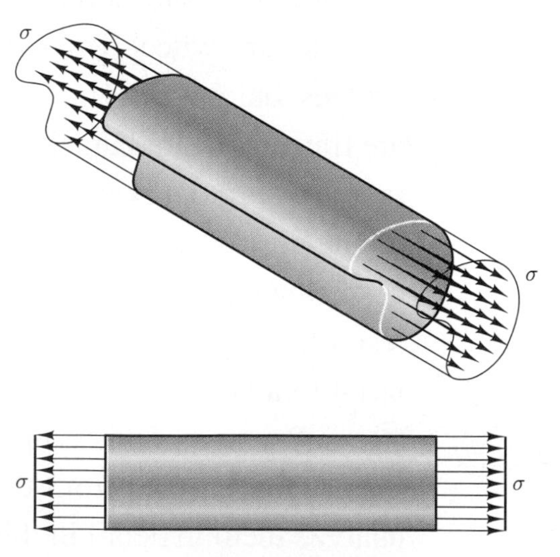

FIGURE 3-1 Loading a bar by normal tractions at its ends.

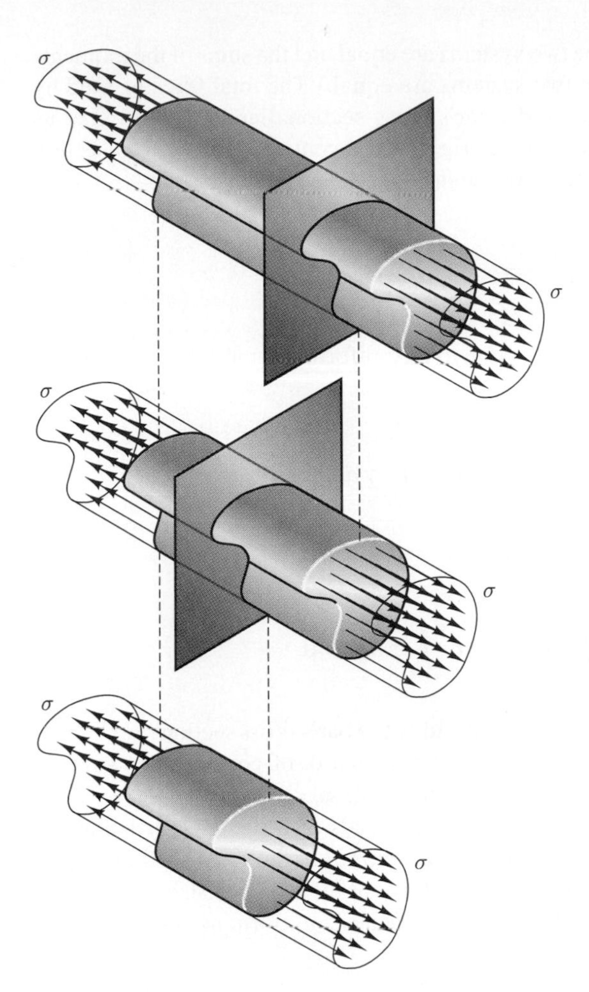

FIGURE 3-2 The same uniform normal stress acts on every plane perpendicular to the bar's axis.

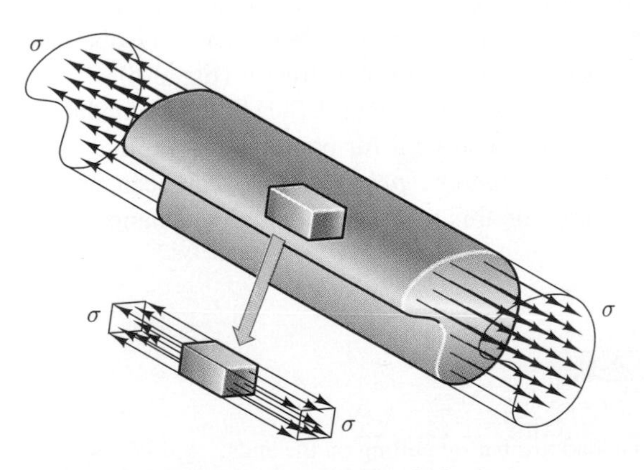

FIGURE 3-3 Isolating an element within the bar.

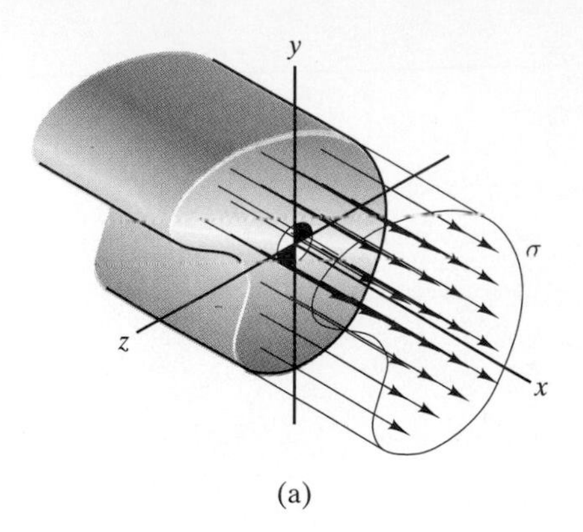

(a)

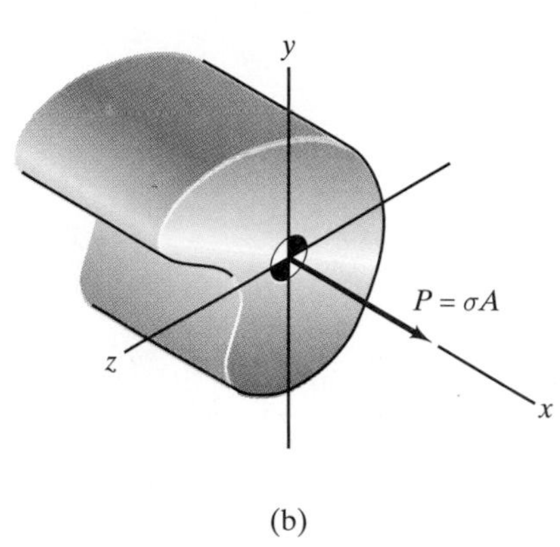

(b)

FIGURE 3-4 (a) Coordinate system with its origin at the centroid. (b) The force P acting at the centroid is equivalent to the uniform stress σ.

the sums of the forces in the two systems are equal and the sums of the moments about any point due to the two systems are equal.) The total force exerted by a uniform normal stress σ on the bar's cross-sectional area A is σA. Let us place a coordinate system with its origin at the centroid of the cross section (Fig. 3-4a), so that the y and z coordinates of the centroid are zero:

$$\bar{y} = \frac{\int_A y\,dA}{A} = 0, \qquad \bar{z} = \frac{\int_A z\,dA}{A} = 0.$$

Then the moment about the z axis due to the uniform normal stress distribution is zero,

$$\int_A -y\sigma\,dA = -\sigma \int_A y\,dA = 0,$$

and the moment about the y axis is zero,

$$\int_A z\sigma\,dA = \sigma \int_A z\,dA = 0.$$

If we place a force $P = \sigma A$ at the centroid of the bar's cross section (Fig. 3-4b), the moments due to P about the y and z axes are, of course, zero, so P is equivalent to the uniform normal stress distribution.

Now suppose that we apply loads to the ends of a prismatic bar that are equivalent to axial forces P applied at the centroid of the cross section, but do it in some realizable way (Fig. 3-5). On planes at increasing distances from the ends of the bar, the stress distribution approaches a uniform normal stress

$$\sigma = \frac{P}{A}. \qquad \textbf{(3-1)}$$

No matter how we apply the loads, the stress distribution at axial distances greater than a few times the bar's width is approximately the same one obtained by subjecting the ends of the bar to uniform normal stresses (Fig. 3-6).

This result was stated by Barré de Saint-Venant (1797–1886), who conducted experiments with rubber bars to demonstrate it for particular cases. In a more general form it is known as *Saint-Venant's principle,* and has been proven analytically in recent years. Based on this result, we assume in examples and

FIGURE 3-5 Applying an axial load to a bar by pulling on the ends.

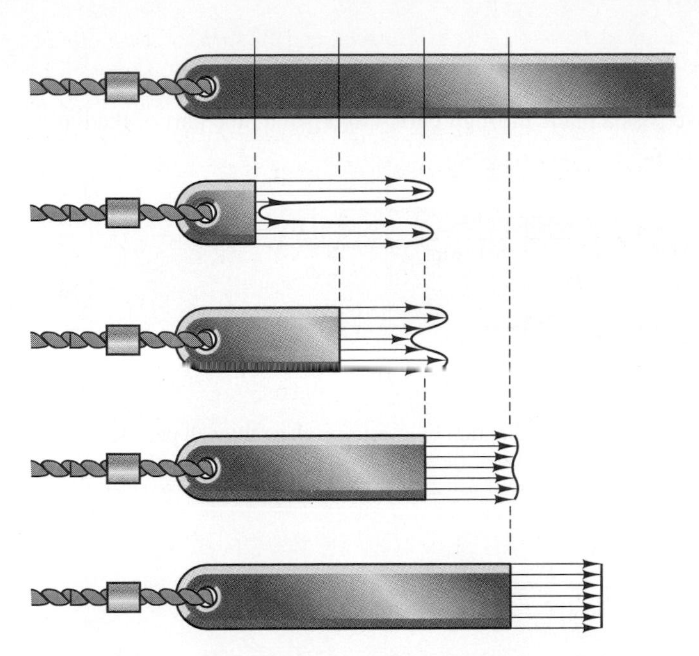

FIGURE 3-6 Stress distributions at increasing distances from the end of the bar.

problems that the stress distribution on a plane perpendicular to the axis of an axially loaded bar consists of uniform normal stress. But keep in mind that this assumption is invalid near the ends of the bar. A separate analysis, again beyond the scope of our discussion, is necessary to determine the distribution of stress near the ends.

EXAMPLE 3-1

Part A of the bar in Fig. 3-7 has a 2-in.-diameter circular cross section and part B has a 4-in.-diameter circular cross section. Determine the normal stress on a plane perpendicular to the bar's axis (**a**) in part A; (**b**) in part B.

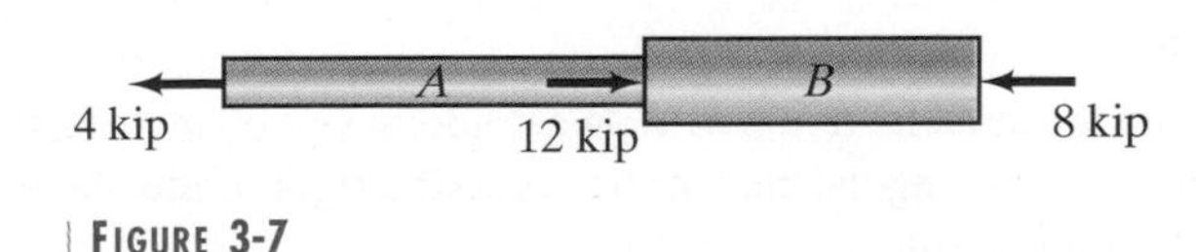

FIGURE 3-7

Strategy

By passing a plane through part A of the bar and isolating the part of the bar on one side of the plane, we can determine the axial load in part A, then use Eq. (3-1) to determine the normal stress. We can determine the normal stress in part B in the same way.

Solution

(**a**) In Fig. (a) we pass a plane through part A and isolate the part of the bar to the left of the plane.

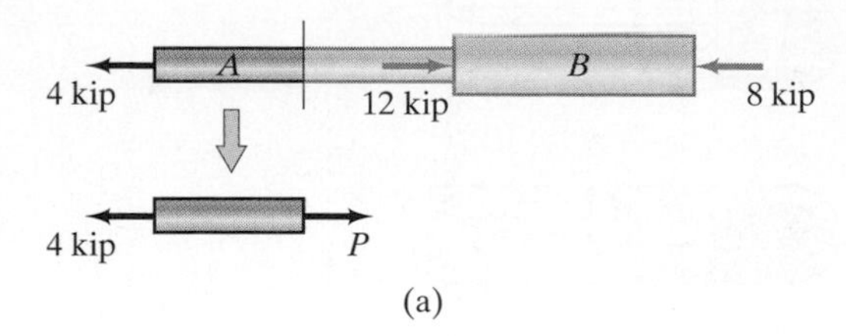

(a) Obtaining a free-body diagram by passing a plane through part A.

From the resulting free-body diagram, the axial load in part A is $P = 4$ kip. The normal stress in part A is

$$\sigma = \frac{P}{A} = \frac{4000}{\pi(2)^2/4} = 1273 \text{ psi.}$$

(**b**) In Fig. (b) we pass a plane through part B and isolate the part of the bar to the left of the plane.

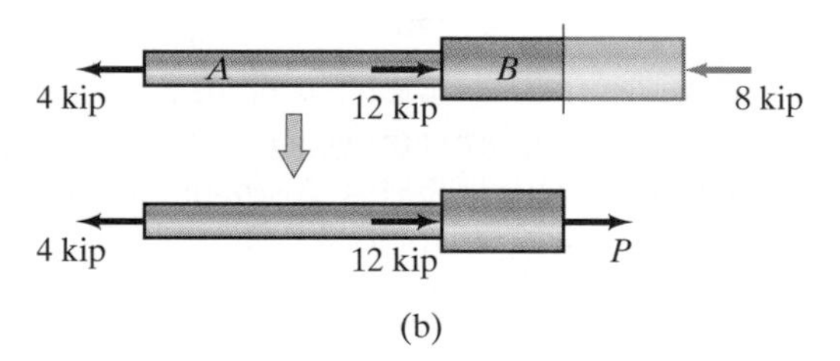

(b) Obtaining a free-body diagram by passing a plane through part B.

From the equilibrium equation $P + 12 - 4 = 0$, the axial load in part B is $P = -8$ kip. The normal stress in part B is

$$\sigma = \frac{P}{A} = \frac{-8000}{\pi(4)^2/4} = -637 \text{ psi.}$$

Discussion

In determining the axial load in part B, we could have obtained a simpler free-body diagram by isolating the part of the bar to the right of the plane [Fig. (c)]. We obtain the same result, $P = -8$ kip.

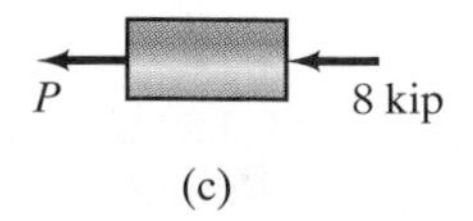

(c)

(c) Free-body diagram of the right part of the bar.

EXAMPLE 3-2

The members of the truss in Fig. 3-8 have equal cross-sectional areas $A = 400\ \text{mm}^2$. The suspended mass is $m = 3400$ kg. What are the normal stresses in the members?

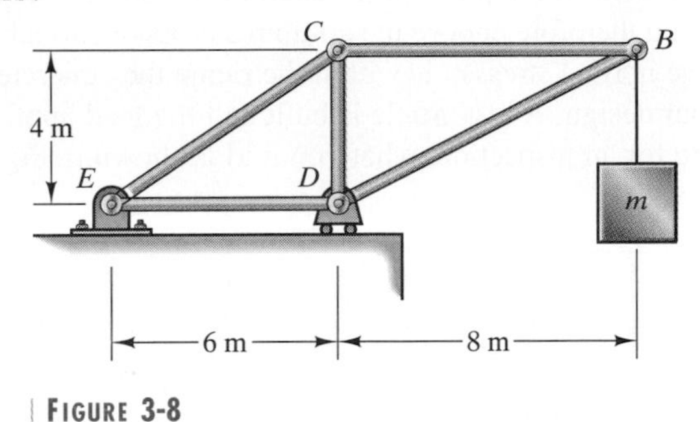

FIGURE 3-8

Strategy

We can use the method of joints to determine the axial force in each member and divide by A to determine the normal stress.

Solution

In Fig. (a) we draw the free-body diagram of joint B of the truss. The angle $\theta = \arctan(4/8) = 26.6°$.

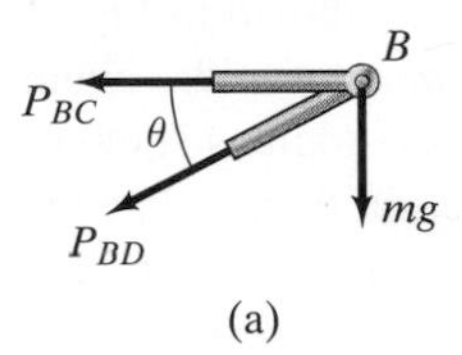

(a) Joint B.

From the equilibrium equations

$$\Sigma F_x = -P_{BC} - P_{BD}\cos\theta = 0,$$
$$\Sigma F_y = -P_{BD}\sin\theta - mg = 0,$$

we obtain $P_{BC} = 2mg$, $P_{BD} = -2.24mg$. Continuing in this way, we obtain the axial forces:

Member:	BC	BD	CD	CE	DE
Axial force:	$2mg$	$-2.24mg$	$-1.33mg$	$2.40mg$	$-2mg$

Substituting the values $m = 3400$ kg and $g = 9.81\ \text{m/s}^2$ and dividing by $A = 400 \times 10^{-6}\ \text{m}^2$, the stresses are

Member:	BC	BD	CD	CE	DE
Normal stress (MPa):	167	−186	−111	200	−167

Stresses on Oblique Planes

Suppose that you have just been hired as a structural engineer, and your first assignment is to design a cylindrical concrete column to support a given weight (Fig. 3-9). You draw a free-body diagram by passing a plane through the column (Fig. 3-10) and write the equilibrium equation, obtaining the normal stress $\sigma = -W/A$. You therefore choose the column's cross-sectional area A so that the compressive normal stress σ is within the range the concrete will support, and submit your design. A test article is built and the load applied. When you drive to the site for an inspection, what you find is shown in Fig. 3-11. Where did you go wrong?

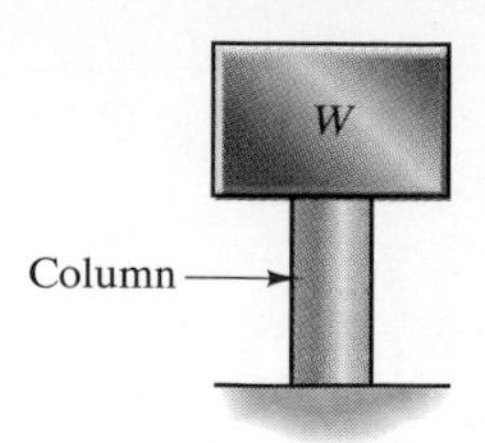

FIGURE 3-9 Column supporting a weight W.

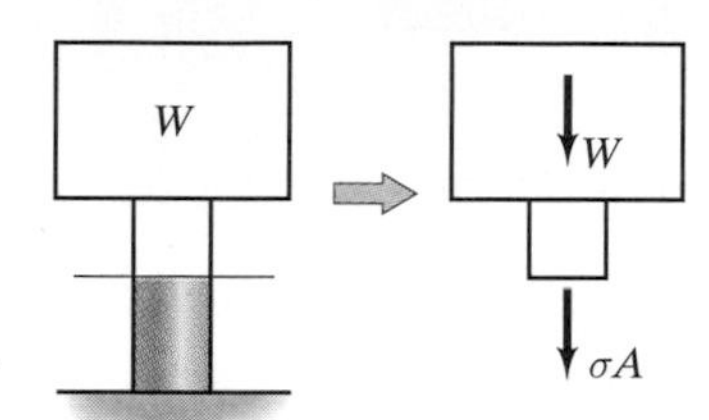

FIGURE 3-10 Obtaining a free-body diagram by passing a plane through the column perpendicular to its axis.

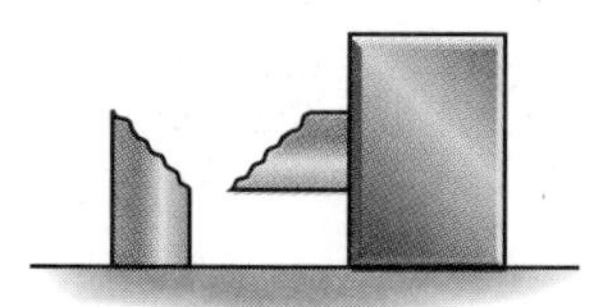

FIGURE 3-11 Why did the column fail?

In design you cannot consider the stress distribution on only one particular plane or subset of planes. We can demonstrate this for an axially loaded bar and explain why the column failed. Let's determine the normal and shear stresses acting on a plane $\mathcal{P}$ oriented as shown in Fig. 3-12. The angle θ is measured from the bar's axis to a line *normal* to the plane $\mathcal{P}$. To determine the stresses on $\mathcal{P}$, we define a free-body diagram as shown in Fig. 3-13: We first isolate a rectangular element from the bar, then isolate the part of the element to the left of plane $\mathcal{P}$. Our objective is to obtain a geometrically simple free-body diagram subjected to the normal and shear stresses on $\mathcal{P}$. We denote the normal stress by σ_θ and the shear stress by τ_θ, indicating that they act upon the plane defined by the angle θ. We have arbitrarily chosen the direction of τ_θ.

We will determine σ_θ and τ_θ by writing equilibrium equations for the triangular free-body diagram we cut from the element. Letting A_θ be the area of the slanted face, we can express the areas of the horizontal and vertical faces in terms of A_θ and θ (Fig. 3-14a). Then the forces on the free-body diagram are the products of the areas and the uniform stresses (Fig. 3-14b). Summing forces in the σ_θ direction,

$$\sigma_\theta A_\theta - (\sigma A_\theta \cos\theta)\cos\theta = 0,$$

we determine the normal stress on $\mathcal{P}$:

$$\sigma_\theta = \sigma \cos^2\theta. \qquad \textbf{(3-2)}$$

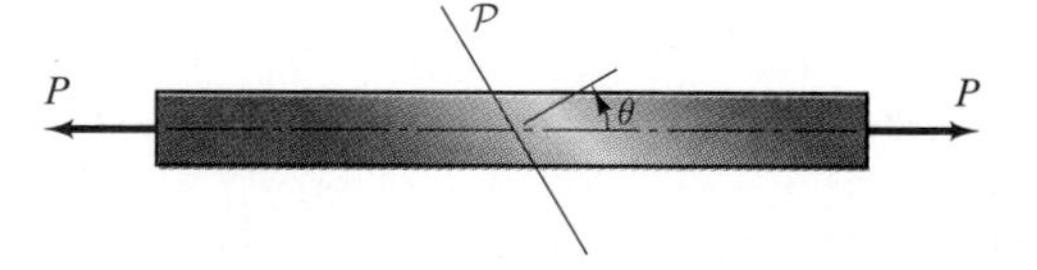

FIGURE 3-12 Passing a plane $\mathcal{P}$ through a bar at an arbitrary angle relative to its axis.

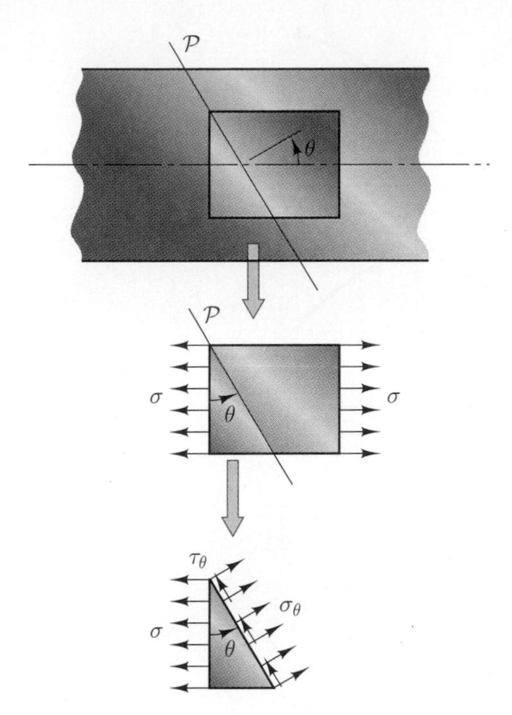

FIGURE 3-13 Obtaining a free-body diagram for determining the stresses on $\mathcal{P}$.

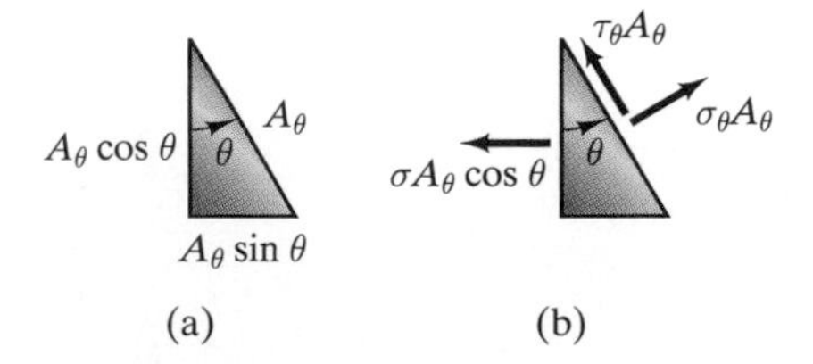

FIGURE 3-14 (a) Areas of the faces. (b) Forces on the free-body diagram.

Summing forces in the τ_θ direction,

$$\tau_\theta A_\theta + (\sigma A_\theta \cos\theta)\sin\theta = 0,$$

we determine the shear stress on $\mathcal{P}$:

$$\tau_\theta = -\sigma \sin\theta \cos\theta. \quad \textbf{(3-3)}$$

Equations (3-2) and (3-3) give the normal and shear stresses acting on a plane oriented at an arbitrary angle relative to the axis of a prismatic bar. In Fig. 3-15 we plot the ratios σ_θ/σ and τ_θ/σ from $\theta = 0$ to $\theta = 180°$. We also show the elements and the values of σ_θ and τ_θ corresponding to particular values of θ.

Figure 3-15 permits us to examine the normal and shear stresses acting on all planes through an axially loaded prismatic bar. Notice that there is no plane for which the normal stress is larger in magnitude than the normal stress on the plane perpendicular to the bar's axis ($\theta = 0$). When you use the equation $\sigma = P/A$ to determine the normal stress on the plane perpendicular to a bar's axis, you are determining the largest tensile or compressive normal stress acting on any plane within the bar. On the other hand, there is no shear stress on the plane perpendicular to the bar's axis, but there are shear stresses on oblique planes. On planes oriented at 45° relative to the bar's axis, the magnitude of the shear stress is one-half the magnitude of the maximum normal stress: $|\tau_\theta| = |\sigma/2|$. We can confirm that this is the maximum magnitude of the shear stress from

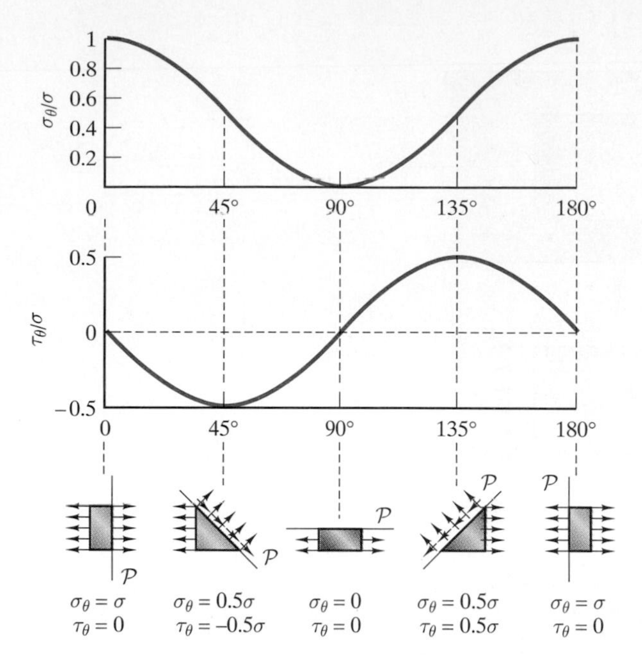

FIGURE 3-15 Normal and shear stresses as functions of θ.

Eq. (3-3). The derivative of τ_θ with respect to θ is

$$\frac{d\tau_\theta}{d\theta} = \sigma(\sin^2\theta - \cos^2\theta).$$

Equating this expression to zero to determine the angles at which τ_θ has a maximum or minimum, we obtain $\theta = 45^\circ$ and $\theta = 135^\circ$. At $\theta = 45^\circ$, $\tau_\theta = -\sigma/2$, and at $\theta = 135^\circ$, $\tau_\theta = \sigma/2$.

We have shown that the maximum tensile or compressive normal stress on any plane through an axially loaded prismatic bar is $\sigma = P/A$, and the magnitude of the maximum shear stress is $|\sigma/2|$.

We can now return to the example of the design of the concrete column in Fig. 3-9 and explain why it failed. You chose the column's cross-sectional area so that the concrete would support the compressive normal stress acting on the plane perpendicular to the column's axis. Concrete—and masonry materials in general—will support large compressive stresses but will fail under substantially smaller shear stresses. Because you designed the column based only upon the maximum normal stress, it failed due to the maximum shear stress acting on planes oriented at 45° relative to the axis (Fig. 3-11). In design, you must consider the stresses acting on all planes and must be particularly concerned with the maximum normal and shear stresses.

When we refer to the normal stress in an axially loaded bar without specifying the plane, we will mean the normal stress on a plane perpendicular to the bar's axis.

EXAMPLE 3-3

Worksheet 1

To test a glue, two plates are glued together as shown in Fig. 3-16. The bar formed by the joined plates is then subjected to a tensile axial load of 200 N. What normal and shear stresses act on the plane where the plates are glued together? (In other words, what stresses must the glue support?)

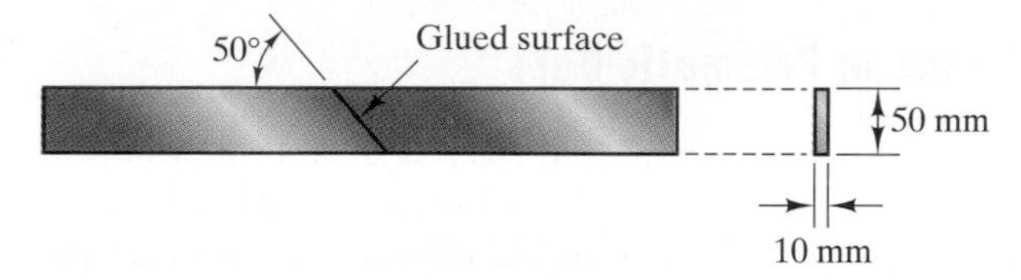

FIGURE 3-16

Strategy

We know the axial load and the bar's cross-sectional area, so we can determine the normal stress $\sigma = P/A$ on a plane perpendicular to the bar's axis. Then we can use Eqs. (3-2) and (3-3) to determine the normal and shear stresses on the glued plane.

Solution

The normal stress on a plane perpendicular to the bar's axis is

$$\sigma = \frac{P}{A} = \frac{200}{(0.01)(0.05)} = 400{,}000 \text{ Pa}.$$

Our objective is to determine the normal stress σ_θ and shear stress τ_θ on the plane where the plates are glued [Fig. (a)].

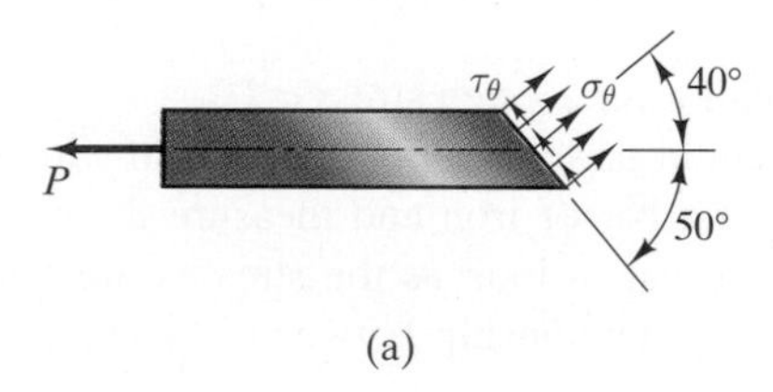

(a)

(a) Stresses on the glued plane.

The angle between the normal to the glued plane and the bar's axis is 40°. From Eq. (3-2),

$$\sigma_\theta = \sigma \cos^2\theta = (400{,}000)\cos^2 40^\circ = 235{,}000 \text{ Pa},$$

and from Eq. (3-3),

$$\tau_\theta = -\sigma \sin\theta \cos\theta = -(400{,}000)\sin 40^\circ \cos 40^\circ = -197{,}000 \text{ Pa}.$$

The glued surface is subjected to a tensile normal stress of 235 kPa and a shear stress of magnitude 197 kPa. The negative value of the shear stress indicates that the shear stress acts in the direction opposite to the defined direction of τ_θ.

3-2 | Strains in Prismatic Bars

Pulling on a rubber band stretches it. If you were to pull on a bar of tungsten steel, it would also stretch. You wouldn't be able to see it stretch—in fact, the deformations of many structural elements under their operating loads are so small they can't be seen—but we show in this section that you can calculate how much the bar would stretch. We also show how you can determine the change in the bar's dimensions in the directions perpendicular to its axis.

Axial Strain and Modulus of Elasticity

Our objective is to determine how much a bar stretches or contracts due to an axial load. Let's return for a moment to our hypothetical example of a prismatic bar loaded at its ends by uniform normal stresses σ (Fig. 3-17). Every plane perpendicular to the bar's axis is subjected to exactly the same normal stress, so it is reasonable to assume that the extensional strain ϵ in the direction parallel to the bar's axis, the axial strain, is the same at each point along the bar's axis. (We verify this assumption in Chapter 6.) We can therefore use Eq. (2-6) to determine the bar's change in length in terms of its reference length L:

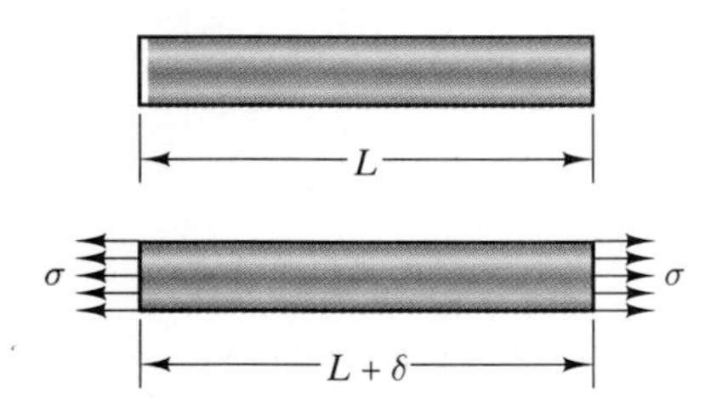

FIGURE 3-17 Loading a bar by normal stresses at its ends.

$$\delta = \epsilon L. \qquad \textbf{(3-4)}$$

But what is the value of ϵ for a given stress σ?

Suppose that we could perform an experiment in which we apply uniform stresses to the ends of a bar of iron and measure the resulting change in its length. We would find that as long as the stresses are not too large, there is an approximately linear relationship between σ and the axial strain ϵ. This relationship is written as

$$\sigma = E\epsilon, \qquad \textbf{(3-5)}$$

where E is a constant called the *modulus of elasticity* or *Young's modulus*. Since ϵ is dimensionless, the dimensions of E are the same as the dimensions of σ, force/area. The value of E we would measure for a steel bar would be different from the value we would measure for a bar of a different material. For example, a given strain ϵ would cause a much larger stress σ in a bar of steel than in a bar of rubber. The modulus of elasticity E is a property of solid materials. Typical values are given in Appendix B.

We have pointed out that if we apply loads to the ends of a prismatic bar of cross-sectional area A that are equivalent to axial forces P applied at the centroid of its cross section, the stress distribution on planes perpendicular to its axis approximates a uniform normal stress $\sigma = P/A$ except near the ends of the bar. As a result, Eq. (3-5) relates the normal stress and axial strain in a prismatic bar except near the ends, and we can use Eq. (3-4) to approximate the change in the bar's length:

$$\delta = \frac{PL}{EA}. \qquad \textbf{(3-6)}$$

If you know a prismatic bar's length and cross-sectional area and the elastic modulus of the material comprising it, you can use this equation to determine the change in the bar's length resulting from a given axial load. This result is based upon the assumption that the material obeys Eq. (3-5). Many solid materials used in engineering applications do satisfy this relation within some range of values of ϵ and some range of temperatures. We discuss relationships between stress and strain further in Section 3-6 and in Chapter 6.

In many engineering applications the strains that occur in axially loaded bars are small, and as a result their changes in length are small in comparison to their lengths. For example, the modulus of elasticity of pure aluminum is 70 GPa, and the largest normal stress a bar of pure aluminum can support without breaking is approximately 70 MPa. If we subject a 1-m-long bar to this normal stress, the resulting change in length is $\delta = (70 \times 10^6)(1)/(70 \times 10^9) = 0.001$ m, or 1 mm. This often simplifies the analysis of structures. For example, when you draw free-body diagrams of a truss to determine the axial forces in its members, you can often ignore changes in the geometry of the truss due to the changes in length of the members. In Example 3-6 we demonstrate this and also show how small changes in the lengths of its members simplify determination of the deformation of a truss.

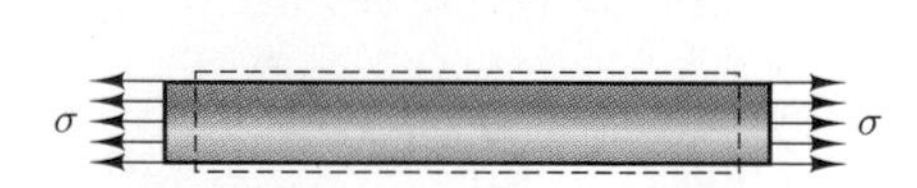

FIGURE 3-18 A bar lengthens in the axial direction and contracts in the lateral direction when subjected to tensile stress.

Lateral Strain and Poisson's Ratio

When you stretch a rubber band, you can see that it becomes thinner. A bar subjected to tensile axial stresses contracts in the lateral direction (Fig. 3-18). Imagine drawing an infinitesimal line dL on an arbitrary cross section of the bar in its undeformed state (Fig. 3-19a). After tensile axial forces are applied to the bar, the line contracts to a length dL' (Fig. 3-19b). The extensional strain of the material in the lateral direction,

$$\epsilon_{\text{lat}} = \frac{dL' - dL}{dL},$$

is negative. Of course, if the bar is subjected to compressive axial forces, it expands in the lateral direction and ϵ_{lat} is positive. The strain ϵ_{lat} has the same value in every direction perpendicular to the bar's cross section and is uniform over the cross section. (Here we assume that the material has the property of

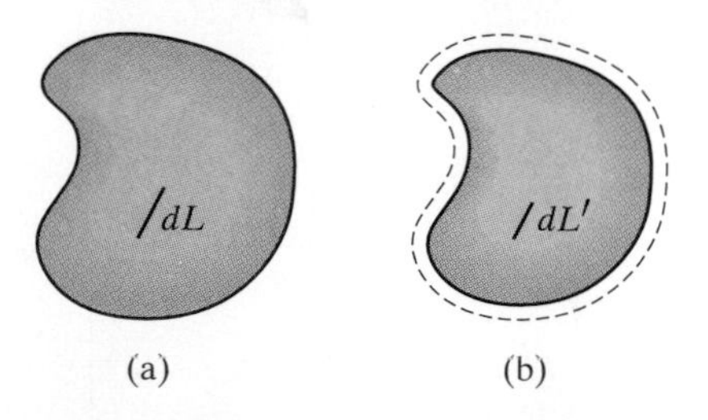

FIGURE 3-19 (a) Infinitesimal line on the unloaded bar's cross section. (b) Cross section and line after the bar is subjected to tensile loads.

isotropy, which we discuss in Chapter 6.) Therefore, if we know the lateral strain we can calculate the length of a finite lateral line L in the deformed state,

$$L' = (1 + \epsilon_{\text{lat}})L,$$

and its change in length,

$$\delta_{\text{lat}} = L' - L = \epsilon_{\text{lat}} L.$$

For example, if a bar has a circular cross section of diameter D in the unloaded state, its diameter in the deformed state is $D' = (1 + \epsilon_{\text{lat}})D$ (Fig. 3-20). But how can we determine the lateral strain?

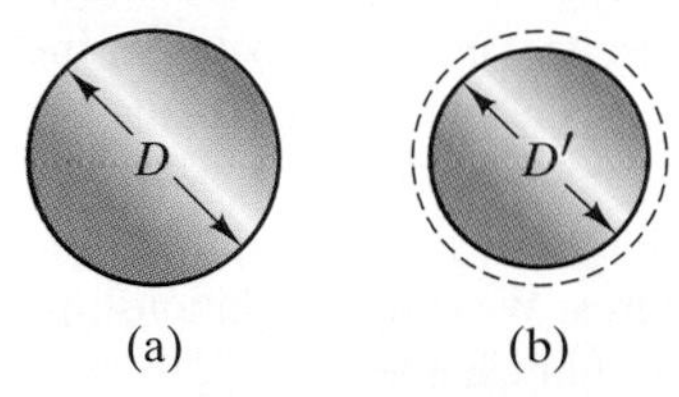

FIGURE 3-20 Circular cross section: (a) in the undeformed state; (b) after the bar is subjected to tensile forces.

The negative of the ratio of the lateral strain of a prismatic bar to its axial strain when it is subjected to axial forces is called *Poisson's ratio:*

$$\nu = -\frac{\epsilon_{\text{lat}}}{\epsilon}. \tag{3-7}$$

Since ϵ_{lat} is negative if the bar is in tension, the minus sign is included in the definition so that the ratio is a positive number. Like the modulus of elasticity, Poisson's ratio is a property of solid materials, and typical values are given in Appendix B. Being a ratio, it is dimensionless.

If we know the modulus of elasticity and Poisson's ratio of a material, we can determine the lateral strain in a prismatic bar of cross-sectional area A subjected to an axial load P. The normal stress is $\sigma = P/A$, and the axial strain is $\epsilon = \sigma/E = P/EA$. Then from Eq. (3-7), the lateral strain is

$$\epsilon_{\text{lat}} = -\frac{\nu P}{EA}. \tag{3-8}$$

EXAMPLE 3-4

A prismatic bar with length $L = 200$ mm and a circular cross section with diameter $D = 10$ mm is subjected to a tensile load $P = 16$ kN (Fig. 3-21). The length and diameter of the deformed bar are measured and determined to be $L' = 200.60$ mm and $D' = 9.99$ mm. What are the modulus of elasticity and Poisson's ratio of the material?

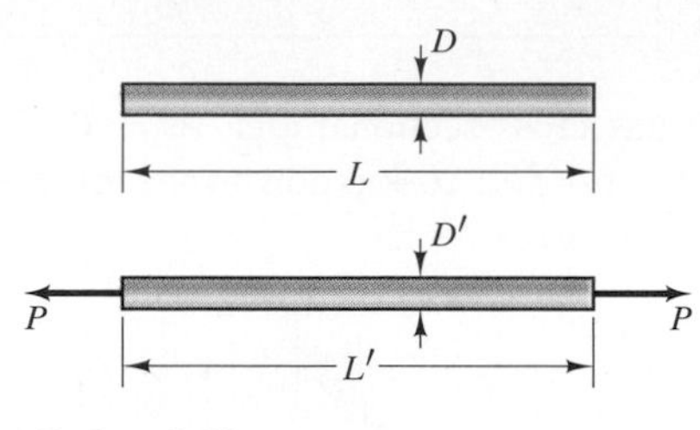

FIGURE 3-21

Stratogy

We can calculate the normal stress $\sigma = P/A$. Since we know L, L', D, and D', we can also calculate the axial and lateral strains. Then we can use Eqs. (3-5) and (3-7) to determine E and ν.

Solution

The normal stress is

$$\sigma = \frac{P}{A} = \frac{16000}{\pi(0.01)^2/4} = 203.7 \text{ MPa}.$$

The axial strain is

$$\epsilon = \frac{L' - L}{L} = \frac{200.60 - 200}{200} = 0.003,$$

and the lateral strain is

$$\epsilon_{\text{lat}} = \frac{D' - D}{D} = \frac{9.99 - 10}{10} = -0.001.$$

The modulus of elasticity is the ratio of the normal stress to the axial strain,

$$E = \frac{\sigma}{\epsilon} = \frac{203.7 \times 10^6}{0.003} = 67.9 \text{ GPa},$$

and Poisson's ratio is

$$\nu = -\frac{\epsilon_{\text{lat}}}{\epsilon} = -\frac{-0.001}{0.003} = 0.333.$$

Discussion

In Chapter 6 we show that the elastic properties of an important class of materials called isotropic linearly elastic materials are completely characterized by two parameters, the modulus of elasticity and Poisson's ratio. In this example we demonstrate that a simple test, subjecting a bar to axial load and measuring its axial and lateral strains, is sufficient to determine both of these parameters. We discuss this test further in Section 3-6.

EXAMPLE 3-5

The bar in Fig. 3-22 has cross-sectional area $A = 0.4\ \text{in}^2$ and modulus of elasticity $E = 12 \times 10^6$ psi. If a 10-kip downward force is applied at B, how far down does point B move?

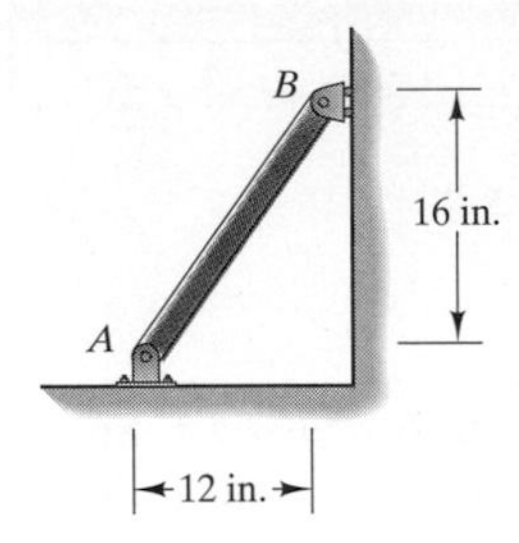

FIGURE 3-22

Strategy

By drawing the free-body diagram of joint B, we can determine the axial force in the bar. We can then use Eq. (3-6) to determine the bar's change in length. We must then use geometry to determine how far down point B moves. To simplify the analysis we will take advantage of the fact that the change in length of the bar is small in comparison to its length.

Solution

We draw the free-body diagram of joint B showing the 10-kip force in Fig. (a). The angle $\theta = \arctan(16/12) = 53.1°$.

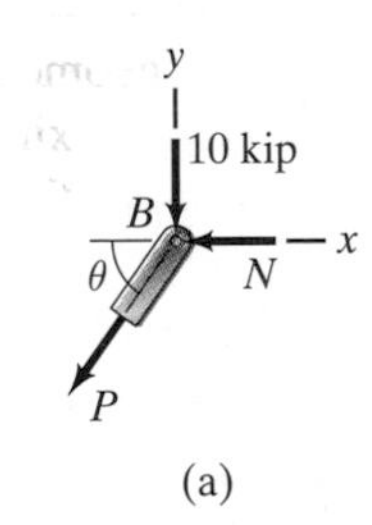

(a) Joint B.

From the equilibrium equation in the vertical direction,

$$\Sigma F_y = -10 - P\sin\theta = 0,$$

we obtain $P = -12.5$ kip. (The bar is in compression.) The bar's length is $L = 20$ in., so from Eq. (3-6), its change in length is

$$\delta = \frac{PL}{EA} = \frac{(-12{,}500)(20)}{(12 \times 10^6)(0.4)} = -0.0521 \text{ in.}$$

Let v be the distance point B moves downward when the force is applied [Fig. (b)]. The bar's deformed length is $(20 + \delta)$ in. Applying the Pythagorean theorem to the dashed triangle,

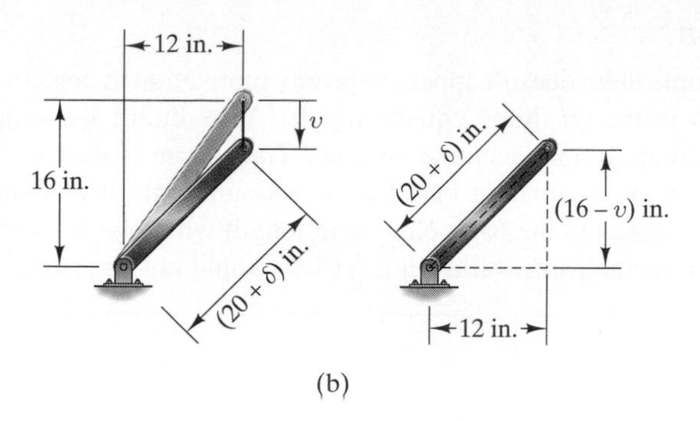

(b) Analyzing the bar's geometry.

$$(12)^2 + (16 - v)^2 = (20 + \delta)^2,$$

we obtain an equation relating v to δ:

$$-32v + v^2 = 40\delta + \delta^2. \qquad \textbf{(3-9)}$$

Since the change in length of the bar is small in comparison to its length, we can neglect the second-order terms v^2 and δ^2, obtaining the linear equation

$$-32v = 40\delta. \qquad \textbf{(3-10)}$$

Solving, we obtain $v = (-40/32)(-0.0521) = 0.0651$ in.

We can also derive Eq. (3-10) using a geometric approach that is more intuitive and direct but also obscures the approximation being made. Consider the right triangle in Fig. (c). Since v is small, the angle of the triangle labeled θ is approximately equal to θ and the side of the triangle opposite θ is approximately equal to $|\delta|$. From this triangle we obtain the relation

$$\sin\theta = \frac{|\delta|}{v}.$$

Since $\sin\theta = 16/20 = 32/40$ and δ is negative, we obtain Eq. (3-10).

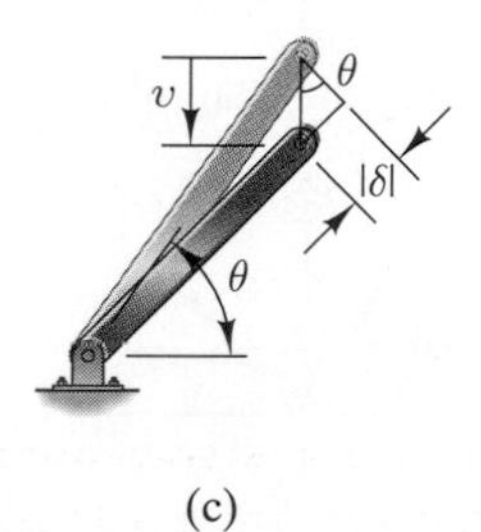

(c) Triangle for obtaining an approximate relation between v and δ.

Discussion

In this example there doesn't appear to be any motivation to neglect the second-order terms in the quadratic equation (3-9). Why didn't we simply solve it for v and obtain a more accurate answer? The reason is that we had already introduced an approximation based on the assumption of a small change in length—we neglected the bar's change in length when we determined P—so retaining the second-order terms in Eq. (3-9) would not be justified.

EXAMPLE 3-6

Bars AB and AC in Fig. 3-23 each have cross-sectional area $A = 60 \text{ mm}^2$ and modulus of elasticity $E = 200$ GPa. The dimension $h = 200$ mm. If a downward force $F = 40$ kN is applied at A, what are the resulting horizontal and vertical displacements of point A?

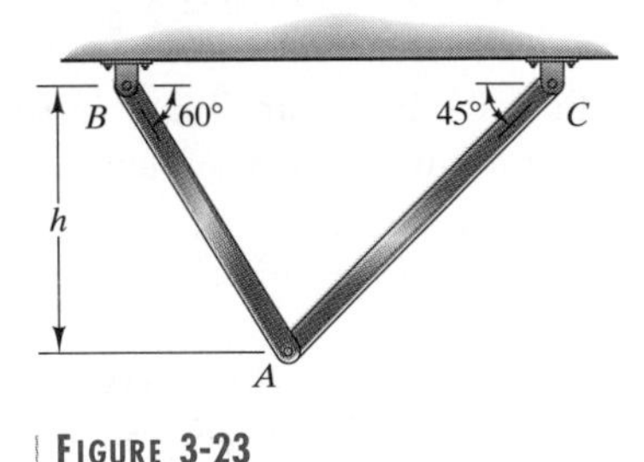

FIGURE 3-23

Strategy

By drawing the free-body diagram of joint A, we can determine the axial forces in the two bars. Knowing the axial forces, we can use Eq. (3-6) to determine the change in length of each bar. We must then use geometry to determine the horizontal and vertical displacements of point A. We simplify the analysis by taking advantage of the fact that the changes in length of the bars are small in comparison to their lengths. (See Example 3-5.)

Solution

In Fig. (a) we draw the free-body diagram of joint A showing the force F.

(a)

(a) Joint A.

The equilibrium equations are

$$\Sigma F_x = -P_{AB}\cos 60° + P_{AC}\cos 45° = 0,$$
$$\Sigma F_y = P_{AB}\sin 60° + P_{AC}\sin 45° - F = 0.$$

Solving these equations, the axial loads in the bars are

$$P_{AB} = \frac{F\cos 45°}{D}, \qquad P_{AC} = \frac{F\cos 60°}{D}, \tag{3-11}$$

where $D = \sin 45° \cos 60° + \cos 45° \sin 60°$.

The lengths of the bars are $L_{AB} = h/\sin 60°$ and $L_{AC} = h/\sin 45°$. Using Eqs. (3-6) and (3-11), the changes in length of the bars are

$$\delta_{AB} = \frac{P_{AB}L_{AB}}{EA} = \frac{FL_{AB}\cos 45°}{EAD},$$
$$\delta_{AC} = \frac{P_{AC}L_{AC}}{EA} = \frac{FL_{AC}\cos 60°}{EAD}. \tag{3-12}$$

To determine the displacement of point A, we first consider bar AB, denoting the horizontal and vertical displacements of point A by u and v [Fig. (b)]. The direction of u is chosen arbitrarily; we don't know beforehand whether point A will move to the left or right.

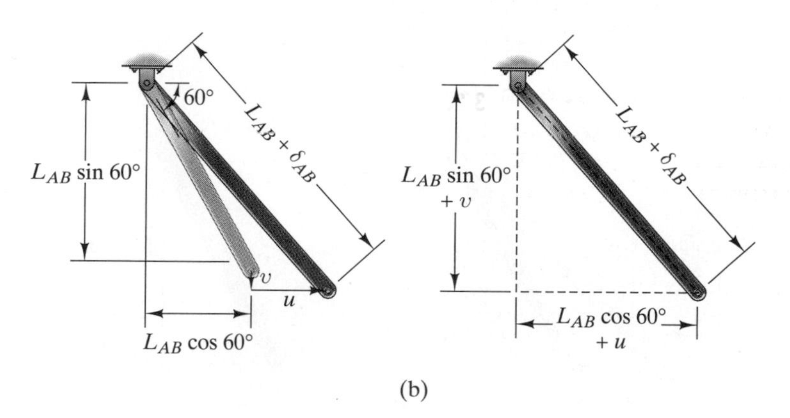

(b) Analyzing the geometry of bar AB.

The deformed length of the bar is $L_{AB} + \delta_{AB}$. From the dashed right triangle in Fig. (b), we obtain an equation relating δ_{AB}, u, and v:

$$(L_{AB}\sin 60° + v)^2 + (L_{AB}\cos 60° + u)^2 = (L_{AB} + \delta_{AB})^2.$$

Squaring these expressions and using the identity $\sin^2 60° + \cos^2 60° = 1$, we obtain

$$2vL_{AB}\sin 60° + v^2 + 2uL_{AB}\cos 60° + u^2 = 2\delta_{AB}L_{AB} + \delta_{AB}^2.$$

Because u, v, and δ_{AB} are small in comparison to L_{AB}, we can neglect the second-order terms v^2, u^2, and δ_{AB}^2, obtaining a linear equation relating δ_{AB}, u, and v:

$$v \sin 60^\circ + u \cos 60^\circ = \delta_{AB}. \tag{3-13}$$

We can also derive this result geometrically as shown in Fig. (c). Because u and v are small, the angles of the right triangles labeled 60° are approximately 60°, and the distance labeled δ_{AB} is approximately δ_{AB}. From these triangles, we obtain Eq. (3-13).

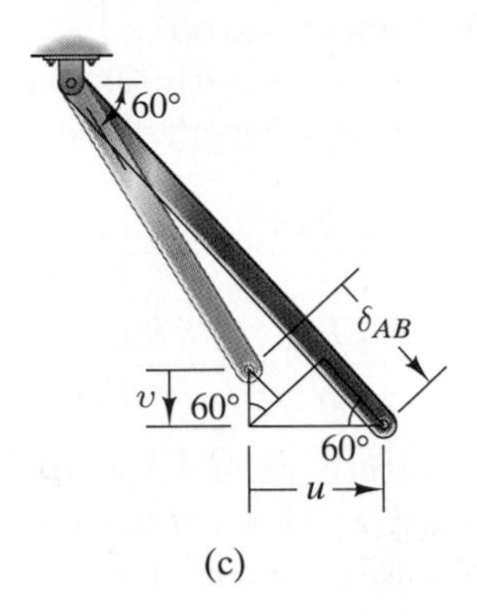

(c) Obtaining the linear equation relating u, v, and δ_{AB}.

We next consider bar AC [Fig. (d)].

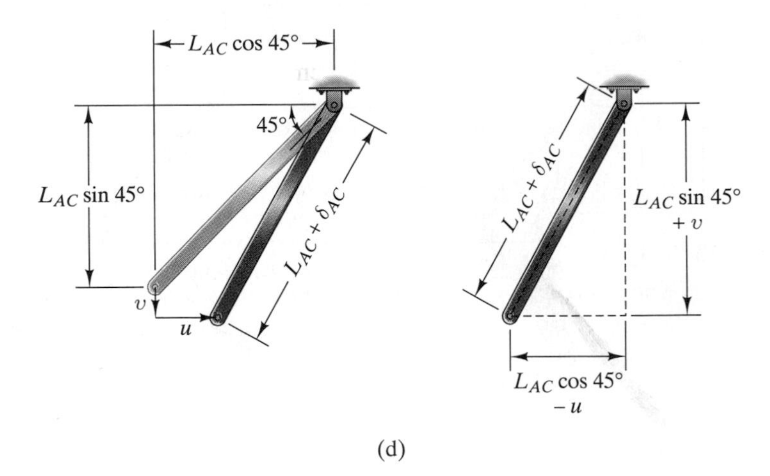

(d) Analyzing the geometry of bar AC.

From the dashed right triangle, we obtain an equation relating δ_{AC}, u, and v:

$$(L_{AC} \sin 45^\circ + v)^2 + (L_{AC} \cos 45^\circ - u)^2 = (L_{AC} + \delta_{AC})^2.$$

Squaring these expressions, we obtain

$$2vL_{AC} \sin 45^\circ + v^2 - 2uL_{AC} \cos 45^\circ + u^2 = 2\delta_{AC} L_{AC} + \delta_{AC}^2.$$

Neglecting the second-order terms v^2, u^2, and δ_{AC}^2 gives a linear equation relating δ_{AC}, u, and v:

$$v \sin 45^\circ - u \cos 45^\circ = \delta_{AC}. \quad \textbf{(3-14)}$$

The geometrical derivation of this result is shown in Fig. (e).

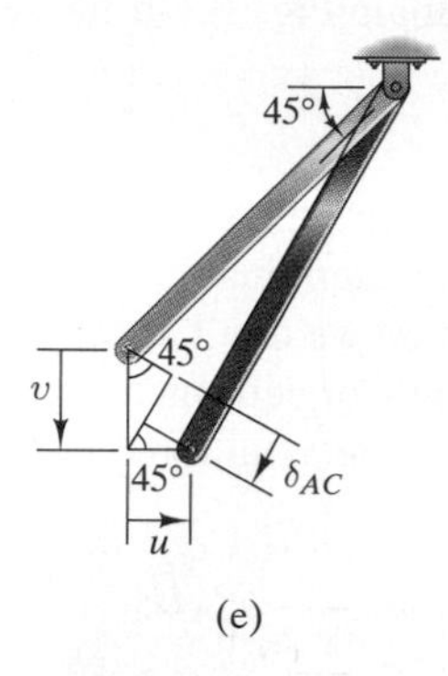

(e) Obtaining the linear equation relating u, v, and δ_{AC}.

Solving Eqs. (3-13) and (3-14) for u and v and substituting Eqs. (3-12), we obtain

$$u = \frac{F(L_{AB} \sin 45^\circ \cos 45^\circ - L_{AC} \sin 60^\circ \cos 60^\circ)}{EAD^2},$$

$$v = \frac{F(L_{AB} \cos^2 45^\circ + L_{AC} \cos^2 60^\circ)}{EAD^2}.$$

Substituting the numerical values, the displacements are

$$u = -0.0250 \text{ mm}, \qquad v = 0.6652 \text{ mm}.$$

Point A moves 0.0250 mm to the left and 0.6652 mm downward.

Discussion

Notice that in drawing the free-body diagram of joint A in Fig. (a), we disregarded the changes in the 45° and 60° angles due to the changes in length of the bars. This approximation requires that the changes in length of the bars be small compared to their lengths.

3-3 | Statically Indeterminate Problems

You have already encountered statics problems in which the number of unknown reactions exceeded the number of independent equilibrium equations. Such problems are said to be *statically indeterminate*. They are common in engineering, because safe and conservative design frequently requires the use of redundant supports—supports in addition to the minimum number necessary for equilibrium. We can now solve statically indeterminate problems involving axially loaded bars by supplementing the equilibium equations with the relationships between the axial loads in bars and their changes in length.

Example

The bar in Fig. 3-24a consists of two segments A and B with different lengths and cross-sectional areas. It is fixed at both ends and subjected to an axial force. We draw its free-body diagram in Fig. 3-24b. The equilibrium equation is

$$F - P_A + P_B = 0. \tag{3-15}$$

We cannot determine the two reactions P_A and P_B from this equation. This simple problem is statically indeterminate.

Let us disregard the fact that we don't know the reactions P_A and P_B, and determine the changes in length, or deformations, of the two parts of the bar as if we did know the reactions. The axial force in part A is P_A (Fig. 3-25a). Its change in length is therefore

$$\delta_A = \frac{P_A L_A}{E A_A}. \tag{3-16}$$

The axial force in part B is P_B (Fig. 3-25b), so its change in length is

$$\delta_B = \frac{P_B L_B}{E A_B}. \tag{3-17}$$

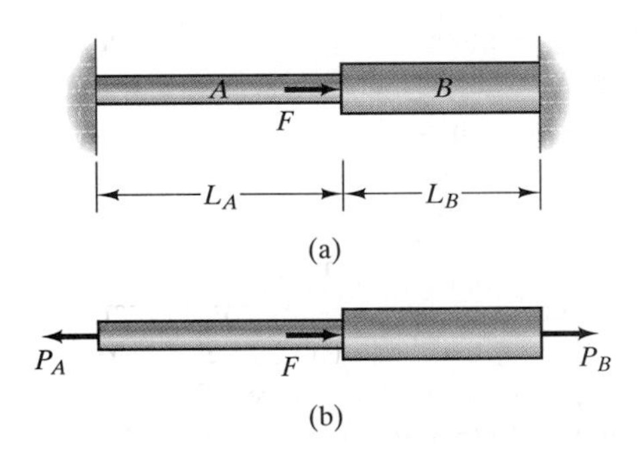

FIGURE 3-24 (a) Axially loaded bar fixed at both ends. (b) Free-body diagram of the bar.

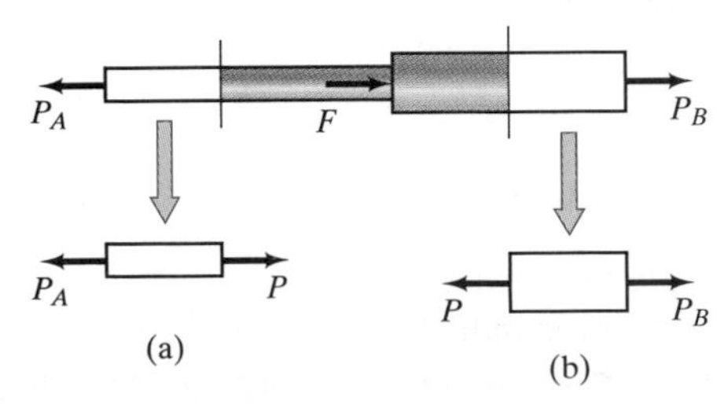

FIGURE 3-25 Determining the internal axial forces in parts A and B.

But we know that the change in length of the entire bar is zero, because it is fixed at both ends. Therefore, the changes in length of the two parts must satisfy the equation

$$\delta_A + \delta_B = 0. \quad \textbf{(3-18)}$$

This is called a *compatibility condition*. It ensures that the changes in length of the two parts are compatible with the constraint that the bar's total length cannot change. We substitute Eqs. (3-16) and (3-17) into this equation, obtaining

$$\frac{P_A L_A}{E A_A} + \frac{P_B L_B}{E A_B} = 0. \quad \textbf{(3-19)}$$

We can solve this equation together with Eq. (3-15) to determine P_A and P_B, obtaining

$$P_A = \frac{F}{1 + L_A A_B / L_B A_A}, \qquad P_B = \frac{-F}{1 + L_B A_A / L_A A_B}.$$

Now that we know the reactions, we can determine the change in length of each part of the bar from Eqs. (3-16) and (3-17). We can also determine the normal stress in each part of the bar: $\sigma_A = P_A/A_A$ and $\sigma_B = P_B/A_B$.

Notice that our solution was based on three elements: (1) relations between the axial forces in the parts of the bar and their changes in length, or deformations; (2) compatibility; and (3) equilibrium. Other statically indeterminate problems involving axially loaded bars, as well as more elaborate problems in structural analysis, can be solved using these three elements.

Flexibility and Stiffness Methods

In this section we describe two approaches that are commonly used to analyze statically indeterminate problems. By applying them to the example discussed in the preceding section, we compare the two methods in a simple context and explain their terminology. They are both based on the same three elements that we applied in the preceding section: relations between forces and deformations, compatibility, and equilibrium. For the bar in Fig. 3-24, the equations expressing these elements are:

$$\left.\begin{aligned} \delta_A &= \frac{P_A L_A}{E A_A} \\ \delta_B &= \frac{P_B L_B}{E A_B} \end{aligned}\right\} \quad \textbf{Force–Deformation Relations}$$

$$\delta_A + \delta_B = 0 \qquad \textbf{Compatibility}$$

$$F - P_A + P_B = 0 \qquad \textbf{Equilibrium}$$

Flexibility Method

We write the force–deformation relations as

$$\delta_A = \frac{L_A}{EA_A} P_A = f_A P_A,$$

$$\delta_B = \frac{L_B}{EA_B} P_B = f_B P_B,$$

where the constants $f_A = L_A/EA_A$ and $f_B = L_B/EA_B$ are called the *flexibilities* of parts A and B of the bar. We substitute these equations into the compatibility equation, obtaining

$$f_A P_A + f_B P_B = 0.$$

This equation can be solved together with the equilibrium equation for the reactions P_A and P_B. Once P_A and P_B are known, the changes in length can be determined from the force–deformation relations.

Except for some added terminology, this is the method of solution we used in the preceding section. It is also called the *force method* because it results in a system of equations in terms of forces.

Stiffness Method

In this method we write the force–deformation relations as

$$P_A = \frac{EA_A}{L_A} \delta_A = k_A \delta_A,$$

$$P_B = \frac{EA_B}{L_B} \delta_B = k_B \delta_B,$$

where the constants $k_A = EA_A/L_A$ and $k_B = EA_B/L_B$ are called the *stiffnesses* of parts A and B of the bar. We substitute these equations into the equilibrium equation, obtaining

$$F - k_A \delta_A + k_B \delta_B = 0.$$

This equation can be solved together with the compatibility equation for the changes in length δ_A and δ_B. Once they are known, the forces P_A and P_B can be determined from the force–deformation relations. This approach is also called the *displacement method* because it results in a system of equations in terms of deformations, which can be expressed in terms of displacements.

From these examples you may wonder why the flexibility and stiffness methods are given distinguishing names, since they apparently differ only in terminology and the order in which equations are solved. The reason is that the differences between the methods become significant when they are applied to more elaborate problems in structural analysis. For the statically indeterminate problems in this book it is not necessary (unless you are asked to do so) to apply one of these specific approaches and use its associated terminology. You need only apply their essential elements—force–deformation relations, compatibility, and equilibrium—and you can choose how and in what order they are applied.

EXAMPLE 3-7

The bar in Fig. 3-26 has modulus of elasticity $E = 12 \times 10^6$ psi. Part A has length $L_A = 10$ in. and a 2-in.-diameter circular cross section. Part B has length $L_B = 8$ in. and a 4-in.-diameter circular cross section. Part A is fixed at its left end, and there is a gap $b = 0.02$ in. between the right end of part B and the rigid wall. If a 160-kip axial force pointing to the right is applied at the joint between parts A and B, what are the normal stresses in parts A and B?

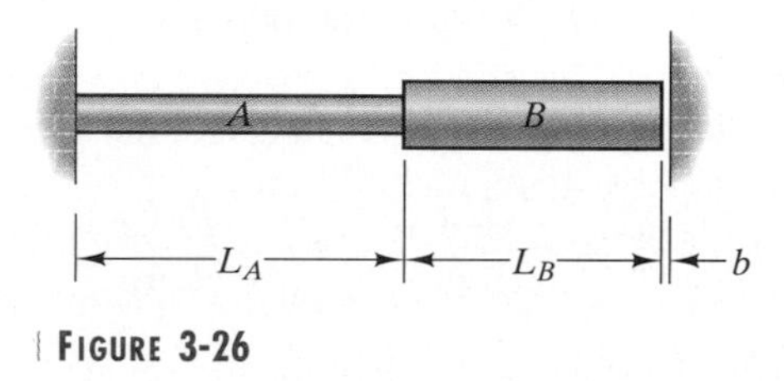

FIGURE 3-26

Strategy

We will first consider the possibility that the 160-kip force doesn't cause the right end of the bar to come into contact with the wall. If that is the case, the normal stress in part B is zero and the normal stress in part A is $160{,}000/A_A$. If the bar does contact the right wall, the problem is statically indeterminate and we can apply force–deformation relations, compatibility, and equilibrium to determine the axial forces in parts A and B. Here the compatibility condition is that the total change in length of the two parts of the bar must equal b.

Solution

In Fig. (a) we draw the free-body diagram of the bar under the assumption that it doesn't contact the right wall.

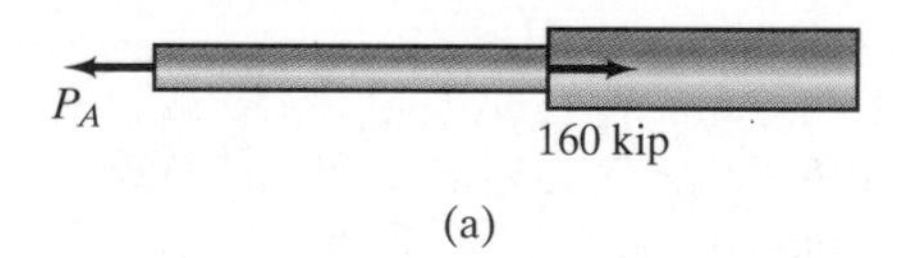

(a)

(a) Assuming that the bar doesn't contact the wall.

The axial force in part A is $P_A = 160{,}000$ lb and there is no axial force in part B. The change in length of part A is

$$\delta_A = \frac{P_A L_A}{E A_A} = \frac{(160{,}000)(10)}{(12 \times 10^6)[\pi(2)^2/4]} = 0.0424 \text{ in.}$$

Since $\delta_A > b$, the bar does come into contact with the right wall and the problem is statically indeterminate.

In Fig. (b) we draw the free-body diagram of the bar, including the reaction exerted on part B by the wall.

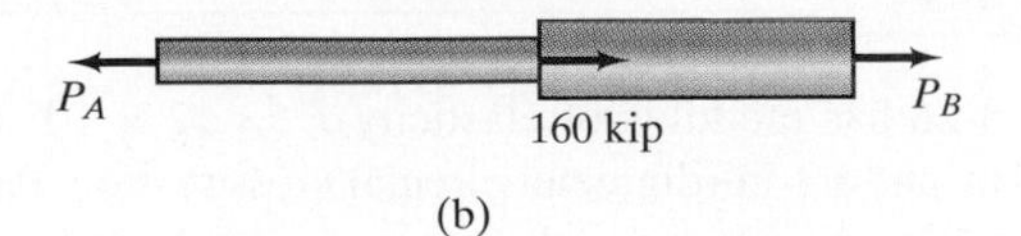

(b)

(b) Free-body diagram showing the reactions at both walls.

Equilibrium The equilibrium equation is

$$-P_A + P_B + 160{,}000 = 0.$$

Force–deformation relations The changes in length of the two parts of the bar are

$$\delta_A = \frac{P_A L_A}{E A_A}, \qquad \delta_B = \frac{P_B L_B}{E A_B}.$$

Compatibility The total change in length of the bar equals b:

$$\delta_A + \delta_B = b.$$

We substitute the force–deformation relations into this equation, obtaining

$$\frac{P_A L_A}{E A_A} + \frac{P_B L_B}{E A_B} = b:$$

$$\frac{P_A(10)}{(12 \times 10^6)[\pi(2)^2/4]} + \frac{P_B(8)}{(12 \times 10^6)[\pi(4)^2/4]} = 0.02.$$

We can solve this equation together with the equilibrium equation for the axial forces P_A and P_B. The results are $P_A = 89{,}500$ lb, $P_B = -70{,}500$ lb. The normal stresses are

$$\sigma_A = \frac{P_A}{A_A} = \frac{89{,}500}{\pi(2)^2/4} = 28{,}490 \text{ psi},$$

$$\sigma_B = \frac{P_B}{A_B} = \frac{-70{,}500}{\pi(4)^2/4} = -5610 \text{ psi}.$$

EXAMPLE 3-8

In Fig. 3-27 two aluminum bars ($E_{\text{Al}} = 10.0 \times 10^6$ psi) are attached to a rigid support at the left and a cross-bar at the right. An iron bar ($E_{\text{Fe}} = 28.5 \times 10^6$ psi) is attached to the rigid support at the left and there is a gap $b = 0.02$ in. between the right end of the iron bar and the cross-bar. The cross-sectional area of each bar is $A = 0.5\ \text{in}^2$ and $L = 10$ in. If the iron bar is stretched until it contacts the cross-bar and welded to it, what are the normal stresses in the bars afterward?

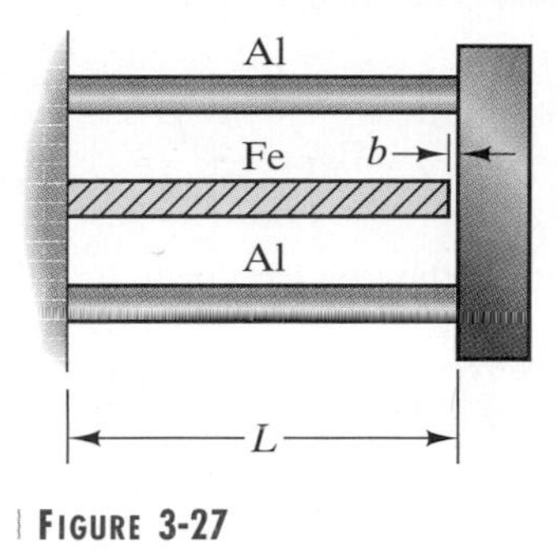

FIGURE 3-27

Strategy

By drawing a free-body diagram of the cross-bar we can obtain one equilibrium equation, but the problem is statically indeterminate because the axial forces in the aluminum and iron bars may be different. (Notice, however, that the axial forces in the two aluminum bars are equal because of the symmetry of the system.) The compatibility condition is that the change in length of the iron bar must equal the change in length of the aluminum bars plus the gap b.

Solution

Equilibrium In Fig. (a) we assume the iron bar has been welded to the cross-bar and obtain a free-body diagram of the cross-bar by passing a plane through the right ends of the three bars. The equilibrium equation for the cross-bar is

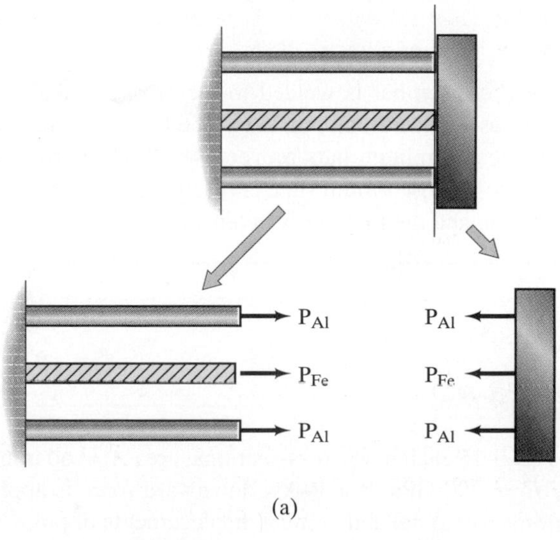

(a) Isolating the cross-bar.

$$2P_{\mathrm{Al}} + P_{\mathrm{Fe}} = 0.$$

Force–deformation relations The change in length of one of the aluminum bars due to its axial force is

$$\delta_{\mathrm{Al}} = \frac{P_{\mathrm{Al}} L}{E_{\mathrm{Al}} A},$$

and the change in length of the iron bar due to its axial force is

$$\delta_{\mathrm{Fe}} = \frac{P_{\mathrm{Fe}}(L - b)}{E_{\mathrm{Fe}} A}.$$

Compatibility The compatibility condition is

$$\delta_{\mathrm{Fe}} = \delta_{\mathrm{Al}} + b.$$

We substitute the force–deformation relations into this equation, obtaining

$$\frac{P_{\mathrm{Fe}}(L - b)}{E_{\mathrm{Fe}} A} = \frac{P_{\mathrm{Al}} L}{E_{\mathrm{Al}} A} + b.$$

We can solve this equation together with the equilibrium equation for the axial forces P_{Al} and P_{Fe}. The results are $P_{\mathrm{Al}} = -5880$ lb, $P_{\mathrm{Fe}} = 11{,}760$ lb. The normal stresses in the bars are

$$\sigma_{\mathrm{Al}} = \frac{P_{\mathrm{Al}}}{A} = \frac{-5880}{0.5} = -11{,}760 \text{ psi},$$

$$\sigma_{\mathrm{Fe}} = \frac{P_{\mathrm{Fe}}}{A} = \frac{11{,}760}{0.5} = 23{,}520 \text{ psi}.$$

Discussion

When the stretched iron bar is welded to the cross-bar and then released, it contracts, compressing the aluminum bars. The forces exerted on the cross-bar by the compressed aluminum bars prevent the iron bar from returning to its original length, so an equilibrium state is reached in which the aluminum bars are in compression and the iron bar is in tension.

EXAMPLE 3-9

M

Worksheet 2

The bars in Fig. 3-28 each have cross-sectional area $A = 60$ mm^2 and modulus of elasticity $E = 200$ GPa. If a 40-kN downward force is applied at A, what are the resulting horizontal and vertical displacements of point A?

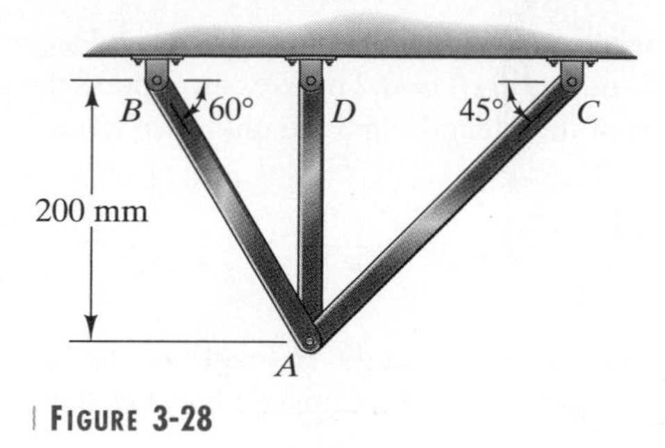

FIGURE 3-28

Strategy

In Example 3-6 we solved this same problem except that the vertical bar AD was absent. (You should review that solution.) In this case the problem is statically indeterminate because we cannot determine the axial forces in the bars from equilibrium alone. We can obtain two equilibrium equations from the free-body diagram of joint A, but there are three unknown axial forces. However, we can approach the problem in exactly the same way that we approached Example 3-6. We can express the change in length of each bar in terms of its unknown axial force. Compatibility conditions arise from the fact that the bars are pinned together: The horizontal and vertical displacements of point A must be the same for each bar. The equilibrium equations, force–deformation equations, and compatibility conditions will provide a complete system of equations for the axial forces, the changes in length of the bars, and the horizontal and vertical displacements of point A.

Solution

Equilibrium In Fig. (a) we draw the free-body diagram of joint A showing the 40-kN force.

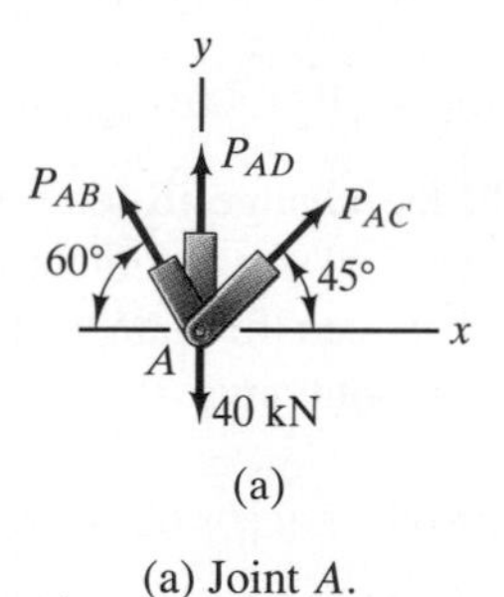

(a) Joint A.

The equilibrium equations are

$$\Sigma F_x = -P_{AB}\cos 60^\circ + P_{AC}\cos 45^\circ = 0,$$

$$\Sigma F_y = P_{AB}\sin 60^\circ + P_{AD} + P_{AC}\sin 45^\circ - 40{,}000 = 0.$$

Force–deformation relations The lengths of the bars are $L_{AB} = 0.2/\sin 60^\circ$ m, $L_{AC} = 0.2/\sin 45^\circ$ m, and $L_{AD} = 0.2$ m. We can express the changes in length of the bars in terms of their lengths and the unknown axial forces:

$$\delta_{AB} = \frac{P_{AB}L_{AB}}{EA}, \qquad \delta_{AC} = \frac{P_{AC}L_{AC}}{EA}, \qquad \delta_{AD} = \frac{P_{AD}L_{AD}}{EA}. \qquad \textbf{(3-20)}$$

Compatibility We now relate the changes in length of the bars to the horizontal and vertical displacements of point A. Consider bar AD, denoting the horizontal and vertical displacements of point A by u and v [Fig. (b)].

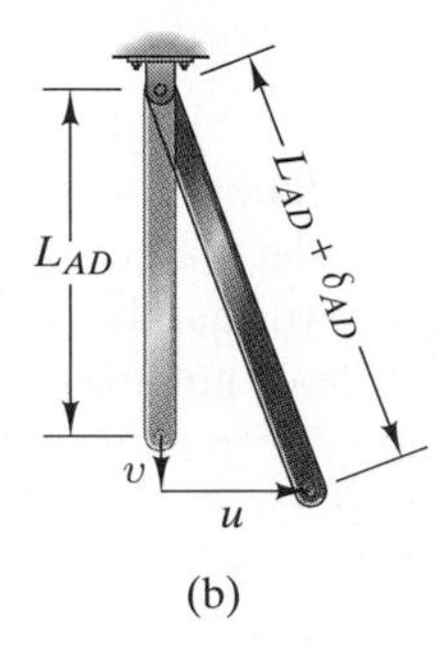

(b) Analyzing the geometry of bar AD.

From the figure we obtain the relationship

$$(L_{AD} + v)^2 + u^2 = (L_{AD} + \delta_{AD})^2,$$

which we can write as

$$2vL_{AD} + v^2 + u^2 = 2\delta_{AD}L_{AD} + \delta_{AD}^2.$$

Neglecting the second-order terms v^2, u^2, and δ_{AD}^2 yields the equation

$$v = \delta_{AD}. \qquad \textbf{(3-21)}$$

Because u and v are small, the change in length of the vertical bar AD is approximately equal to the vertical displacement of point A. In Example 3-6 we derived the relationships between the changes in length of bars AB and AC and the displacements u and v, obtaining

$$v \sin 60^\circ + u \cos 60^\circ = \delta_{AB}, \qquad \textbf{(3-22)}$$

$$v \sin 45^\circ - u \cos 45^\circ = \delta_{AC}. \qquad \textbf{(3-23)}$$

Equations (3-21)–(3-23) are the compatibility conditions for this problem. The changes in length of the bars are constrained because they are pinned together at A. These constraints are enforced by these equations, which require the end of each bar to undergo the same horizontal and vertical displacement.

The equilibrium equations, Eqs. (3-20), and Eqs. (3-21)–(3-23) provide eight equations in the eight unknowns P_{AB}, P_{AC}, P_{AD}, δ_{AB}, δ_{AC}, δ_{AD}, u, and v. Solving them, we obtain

$$u = -0.013 \text{ mm}, \qquad v = 0.333 \text{ mm}.$$

3-4 | Nonprismatic Bars and Distributed Loads

In this section we extend our analysis of prismatic bars to additional important applications by considering bars whose cross sections vary with distance along their axes and also bars subjected to axial loads that are distributed along their axes.

Bars with Gradually Varying Cross Sections

Let us consider a bar whose cross-sectional area varies with distance along its axis as shown in Fig. 3-29a. The cross-sectional area is a function of x, which we indicate by writing it as $A(x)$. If we subject the bar to axial loads and the change in $A(x)$ with x is gradual [$dA(x)/dx$ is small], the stress distribution on a plane perpendicular to the bar's axis can be approximated by a uniform normal stress (Fig. 3-29b):

$$\sigma = \frac{P}{A(x)}. \tag{3-24}$$

In other words, for a given perpendicular plane the stress distribution is approximately the same as for the case of a prismatic bar. But since $A(x)$ varies with distance along the bar's axis, the normal stress does also.

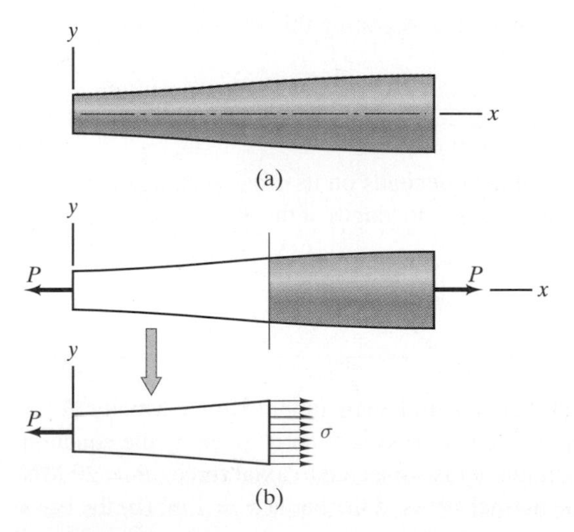

FIGURE 3-29 (a) Bar with a varying cross-sectional area. (b) Approximate stress distribution.

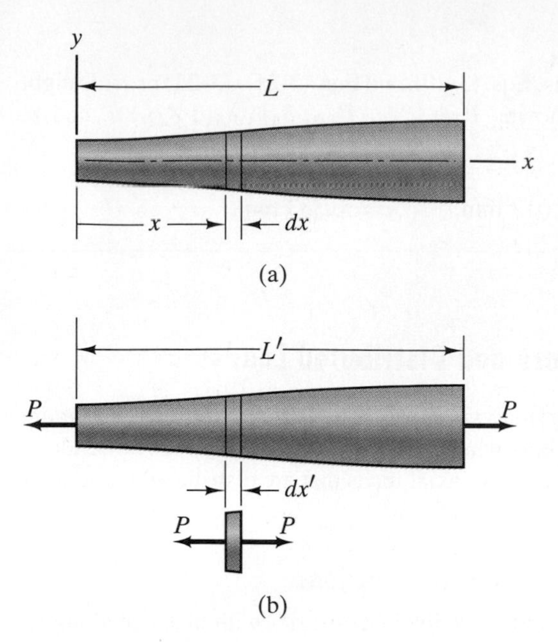

FIGURE 3-30 (a) Element of the unloaded bar. (b) Length of the element in the loaded state.

Since the cross-sectional area and normal stress vary along the length of the bar, how can we determine the change in the bar's length? We begin by considering an infinitesimal element of the bar whose length is dx in the unloaded state (Fig. 3-30a). We can use Eq. (3-6) to determine the element's change in length (Fig. 3-30b):

$$dx' - dx = \frac{P\,dx}{EA(x)}.$$

We obtain the bar's change in length by integrating this expression from $x = 0$ to $x = L$:

$$\delta = L' - L = \int_0^L \frac{P\,dx}{EA(x)}. \qquad \textbf{(3-25)}$$

The change in length of each element depends on its cross-sectional area, and we must add up, or integrate, the changes in length of the elements to determine the total change in length.

EXAMPLE 3-10

The bar in Fig. 3-31 consists of material with modulus of elasticity $E = 120$ GPa. The area of the bar's circular cross section is given by the equation $A(x) = 0.03 + 0.008x^2$ m^2. If the bar is subjected to axial forces $P = 20$ MN at the ends, determine **(a)** the normal stress in the bar at $x = 1$ m; **(b)** the bar's change in length.

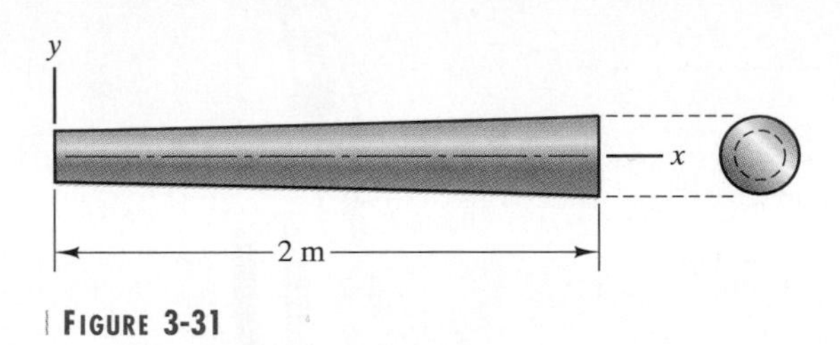

FIGURE 3-31

Strategy

(a) From the given equation for $A(x)$ we can determine the cross-sectional area at $x = 1$ m. Then the normal stress is $P/A(x)$. **(b)** Since the bar's cross-sectional area is a function of x, we must determine the bar's change in length from Eq. (3-25).

Solution

(a) The cross-sectional area at $x = 1$ m is

$$A(1) = 0.03 + 0.008(1)^2 = 0.038 \text{ m}^2,$$

so the normal stress at $x = 1$ m is

$$\sigma = \frac{P}{A(1)} = \frac{20 \times 10^6}{0.038} = 526 \text{ MN}.$$

(b) From Eq. (3-25), the change in length of the bar is

$$\delta = \int_0^L \frac{P\,dx}{EA(x)} = \int_0^2 \frac{(20 \times 10^6)\,dx}{(120 \times 10^9)(0.03 + 0.008x^2)} = 8.62 \text{ mm}.$$

Distributed Axial Loads

In some situations bars are subjected to axial forces that are continuously distributed along some part of the bar's axis, or to axial forces that can be modeled as continuous distributions. For example, a pile driven into the ground is subjected to a resisting friction force that is distributed along the pile's length (Fig. 3-32).

To describe a distributed axial force, we introduce a function q defined such that the force on each element dx of the bar is $q\,dx$ (Fig. 3-33). Since the product of q and dx is a force, the dimensions of q are force/length.

For example, the bar in Fig. 3-34a is fixed at the left end and subjected to a distributed axial force throughout its length. In Fig. 3-34b we pass a plane through the bar at an arbitrary position x and draw the free-body diagram of the part of the bar to the right of the plane. From the equilibrium equation

$$-P + \int_x^L q_0 \left(\frac{x}{L}\right)^2 dx = 0,$$

FIGURE 3-32 A driven pile is subjected to a distributed axial force.

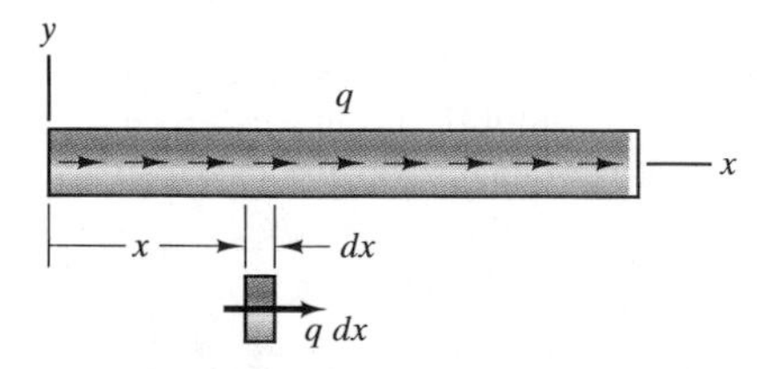

FIGURE 3-33 Describing a distributed axial force by a function. The force on an element of length dx is $q\,dx$.

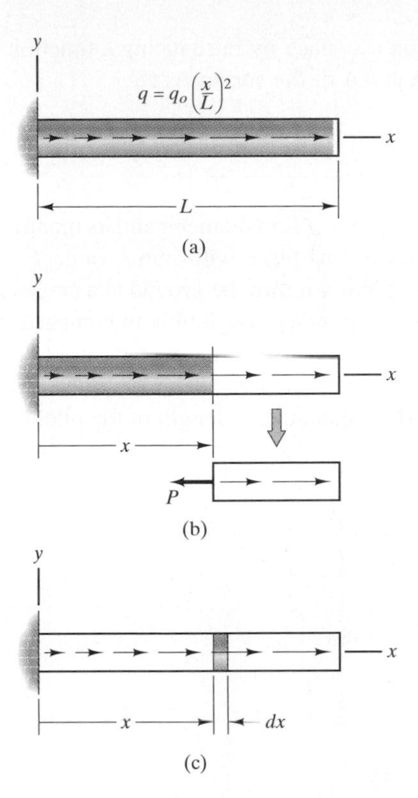

FIGURE 3-34 (a) Bar subjected to a distributed load. (b) Determining the internal axial force at x. (c) Element of length dx.

we determine the axial load in the bar at the position x:

$$P = \frac{q_0}{3}\left(L - \frac{x^3}{L^2}\right).$$

The distributed axial force causes the internal axial force in the bar to vary with axial position. To determine the bar's change in length, we consider an element of the bar of length dx (Fig. 3-34c). From Eq. (3-6), its change in length is

$$\delta_{\text{element}} = \frac{P\,dx}{EA} = \frac{q_0}{3EA}\left(L - \frac{x^3}{L^2}\right)dx.$$

Integrating this result from $x = 0$ to $x = L$, we obtain the change in length of the bar:

$$\delta = \int_0^L \frac{q_0}{3EA}\left(L - \frac{x^3}{L^2}\right)dx = \frac{q_0 L^2}{4EA}.$$

Notice the parallel between the definition of the distributed axial force q and the definitions of other distributed loads with which you are familiar: We have represented distributed lateral loads on beams by a function w defined such that the lateral force on an element dx of the beam's axis is $w\,dx$, and we represented the stress distribution on a surface by introducing a function **t** defined such that the force on an element dA of the surface is $\mathbf{t}\,dA$.

EXAMPLE 3-11

The pile in Fig. 3-35 has a circular cross section 1 ft in diameter and its modulus of elasticity is $E = 2 \times 10^6$ psi. A downward force with initial value $F = 480$ kip is applied to the top of the pile, driving it into the ground at a constant rate. The resisting force on the bottom of the pile is negligible in comparison to the axial frictional force on its lateral surface. Assume that the frictional force is uniformly distributed. As the driving process begins, determine **(a)** the maximum compressive normal stress; **(b)** the change in length of the pile.

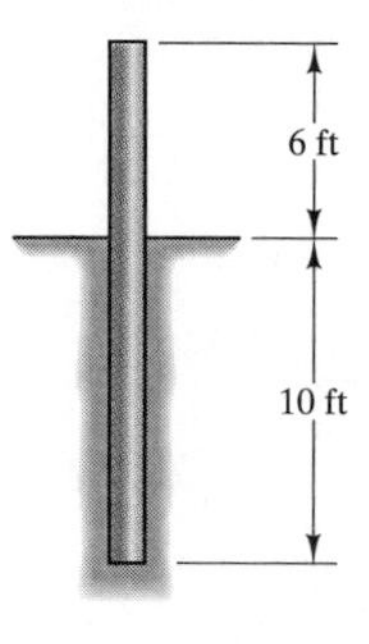

FIGURE 3-35

Strategy

The part of the pile below the ground is subjected to a uniform distributed force q [Fig. (a)]. From the equilibrium equation for the entire pile we can determine the value of q. Then, by passing a plane through the pile at an arbitrary position x, we can determine the internal axial force in the pile [Fig. (b)]. Once we know the axial force, we can determine the maximum compressive stress and the change in the pile's length.

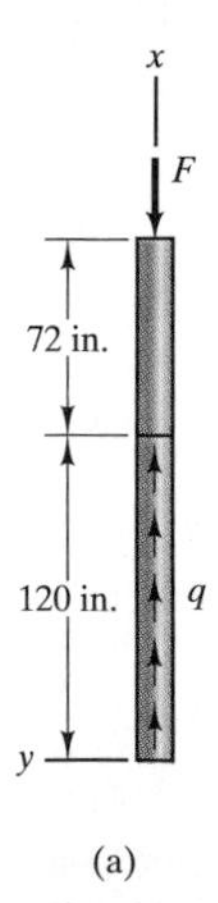

(a)

(a) Free-body diagram of the entire pile.

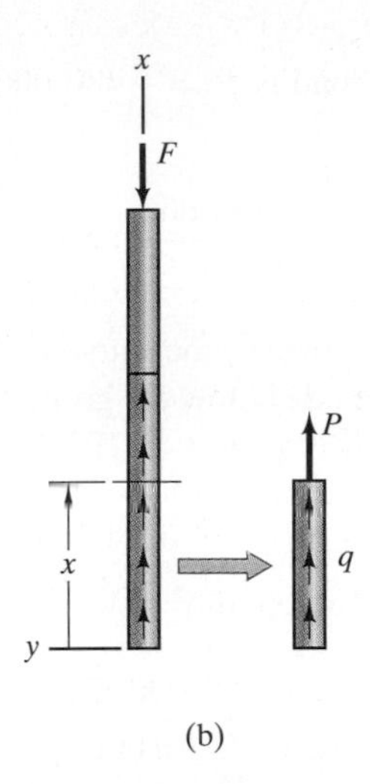

(b) Passing a plane at an arbitrary position x.

Solution

(a) From the equilibrium equation for the entire pile [Fig. (a)],

$$\int_0^{120} q\,dx - F = 0,$$

we determine that $q = F/120 = 4000$ lb/in. Then from the free-body diagram in Fig. (b), we obtain an equation for the internal axial force at an arbitrary position x:

$$\int_0^{x} q\,dx + P = 0.$$

Solving this equation, the internal axial load from $x = 0$ to $x = 120$ in. is $P = -qx = -4000x$ lb. We show the distribution of the internal axial load in the pile in Fig. (c).

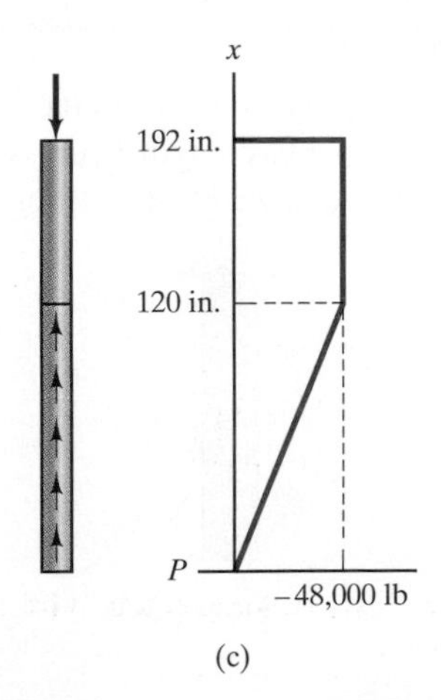

(c) Axial load P as a function of x.

The maximum compressive load is $P = -480{,}000$ lb, so the maximum compressive stress in the pile is

$$\sigma = \frac{P}{A} = \frac{-480{,}000}{\pi(12)^2/4} = -4240 \text{ psi}.$$

(b) The internal axial force is given by the equation $P = -4000x$ lb from $x = 0$ to $x = 120$ in. (the part of the pile below the ground) and has the constant value $P = -480{,}000$ lb from $x = 120$ in. to $x = 192$ in. (the part above the ground). To determine the pile's change in length, we need to analyze the parts above and below the ground separately.

Above the ground The change in length is

$$\delta_{\text{above}} = \frac{PL}{EA} = \frac{(-480{,}000)(72)}{(2 \times 10^6)[\pi(12)^2/4]} = -0.153 \text{ in.}$$

Below the ground Below the ground the internal axial load varies with x. The change in length of an element of the beam of length dx is

$$\delta_{\text{element}} = \frac{P\,dx}{EA} = \frac{-4000x\,dx}{EA}.$$

Integrating this expression from $x = 0$ to $x = 120$ in., the change in length of the part of the pile below the ground is

$$\delta_{\text{below}} = \int_0^{120} \frac{-4000x\,dx}{EA} = \frac{(-4000)(120)^2}{(2)(2 \times 10^6)[\pi(12)^2/4]} = -0.127 \text{ in.}$$

The change in length of the pile is $\delta_{\text{above}} + \delta_{\text{below}} = -0.280$ in.

EXAMPLE 3-12

The prismatic bar in Fig. 3-36 is suspended from the ceiling and loaded only by its own weight. If the unloaded bar has length L, cross-sectional area A, weight density γ, and modulus of elasticity E, what is its length when suspended?

FIGURE 3-36

Strategy

We can treat the weight as an axial force distributed along the bar's axis. The force exerted on an element of the bar of length dx by its weight is equal to the product of the weight density and the volume of the element [Fig. (a)].

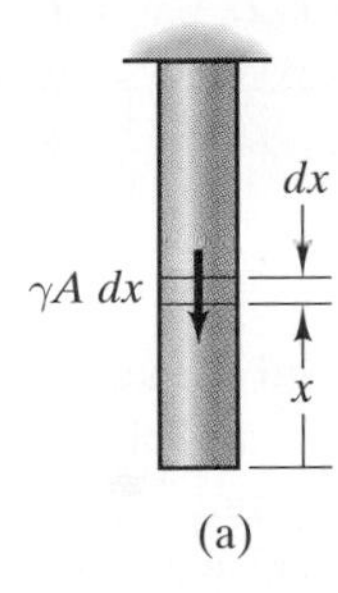

(a) Axial force on an element of length dx.

Solution

To determine the internal axial force, we obtain a free-body diagram by passing a plane through the bar at an arbitrary distance x from the bottom [Fig. (b)].

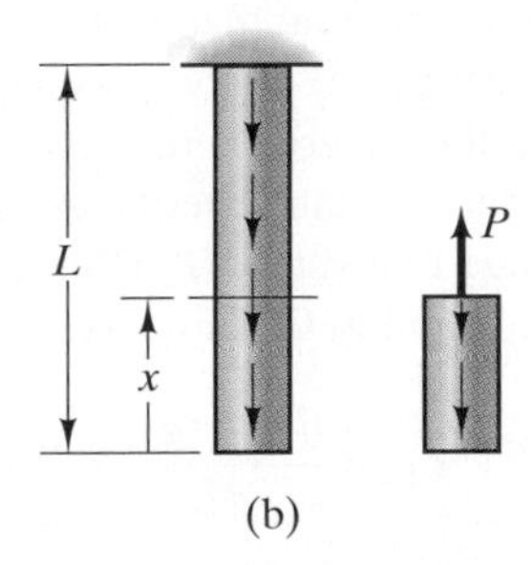

(b) Determining the internal axial force.

From the equilibrium equation

$$P - \int_0^x \gamma A \, dx = 0,$$

we determine that $P = \gamma A x$. The internal axial force increases linearly from zero at the bottom of the bar to $\gamma A L = W$, the bar's weight, at the top. The change in length of the element dx in Fig. (a) is

$$\delta_{\text{element}} = \frac{P \, dx}{EA} = \frac{\gamma A x \, dx}{EA}.$$

We integrate this expression from $x = 0$ to $x = L$ to determine the bar's change in length:

$$\delta = \int_0^L \frac{\gamma A x \, dx}{EA} = \frac{\gamma A L^2}{2EA}.$$

In terms of the bar's weight $W = \gamma AL$, the change in length is

$$\delta = \frac{WL}{2EA}.$$

The length of the suspended bar is $L + \delta = L + WL/2EA$.

3-5 | Thermal Strains

All structures that are not provided with a controlled environment are subject to changes in temperature. Most materials tend to expand when their temperatures rise and contract when they fall, and this can have important effects on their use in structures and mechanisms. You are familiar with such dramatic effects as heated glass dishes breaking when suddenly cooled and concrete sidewalks developing cracks due to environmental temperature changes. Changes in temperature can also affect the mechanical properties of materials. Structural engineers must be aware of thermal effects and account for them in design. In this section we introduce the concept of thermal strain and demonstrate its effects in axially loaded bars.

Consider a sample of solid material that is at a uniform temperature T and is unconstrained—not subject to any forces or couples. Imagine drawing an infinitesimal line dL within the material (Fig. 3-37a). If the temperature of the sample changes by an amount ΔT, the material will expand or contract and the length of the line will change to a value dL' (Fig. 3-37b). The *thermal strain* of the material at the location and in the direction of the line dL is

$$\epsilon_T = \frac{dL' - dL}{dL}.$$

If ΔT is positive, most materials will expand and ϵ_T will be positive, and if ΔT is negative, the material will contract and ϵ_T will be negative. For many materials used in engineering, the relationship between the thermal strain and the change in temperature can be approximated by a linear equation over some

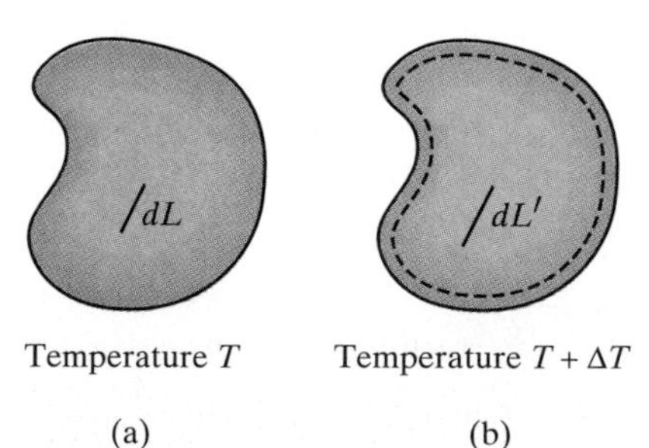

FIGURE 3-37 (a) Infinitesimal line within a sample of material. (b) Sample and line after a temperature change.

range of temperature:

$$\epsilon_T = \alpha \, \Delta T. \tag{3-26}$$

This equation gives the thermal strain relative to the unconstrained material at temperature T. The constant α is a material property called the *coefficient of thermal expansion*. Since strain is dimensionless, the dimensions of α are $(\text{temperature})^{-1}$. Typical values are given in Appendix B. Some materials exhibit different thermal strains in different directions (the value of α depends on direction), but we consider only materials in which the thermal strain in an unconstrained sample is the same in every direction.

The bar in Fig. 3-38a is fixed at both ends. What happens if we raise its temperature by an amount ΔT? To answer this question we first imagine that the right support is removed and raise the temperature by the amount ΔT. The temperature increase causes the bar's length to increase an amount $\epsilon_T L = \alpha \, \Delta T \, L$ (Fig. 3-38b). But the length of the actual bar cannot increase, so the right support must exert a compressive force sufficient to compress the released bar by the amount the temperature increase caused it to expand (Fig. 3-38c):

$$\frac{CL}{EA} = \alpha \, \Delta T \, L.$$

(Notice that we calculated the change in length of the bar due to the compressive force C in terms of the original length L of the bar, neglecting the increase in the bar's length due to the change in temperature. We can do so because the change in length is small compared to L.) From this expression we see that the supports subject the bar to compressive axial forces $C = \alpha \, \Delta T \, EA$, and the resulting compressive normal stress in the bar is $\sigma = -C/A = -\alpha \Delta T \, E$.

This example explains one of the reasons that concrete sidewalks sometimes develop cracks. Concrete is weak in tension, and if the sidewalk is not designed so that it can contract freely when its temperature decreases, it is subjected to tensile stresses and cracks form.

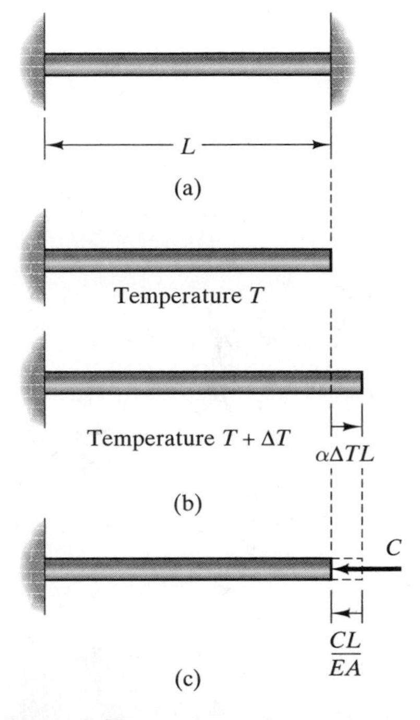

FIGURE 3-38 (a) Bar fixed at both ends. (b) "Releasing" the right end and increasing the temperature. (c) Compressive force exerted by the support.

EXAMPLE 3-13

In Fig. 3-39 two aluminum bars ($E_{\text{Al}} = 10.0 \times 10^6$ psi, $\alpha_{\text{Al}} = 13.3 \times 10^{-6}\ {}^\circ\text{F}^{-1}$) and an iron bar ($E_{\text{Fe}} = 28.5 \times 10^6$ psi, $\alpha_{\text{Fe}} = 6.5 \times 10^{-6}\ {}^\circ\text{F}^{-1}$) are attached to a rigid support at the left and a cross-bar at the right. The cross-sectional area of each bar is $A = 0.5\ \text{in}^2$ and $L = 10$ in. The normal stresses in the bars are initially zero. If the temperature is increased by $\Delta T = 100^\circ$F, what are the resulting normal stresses in the three bars? The deformation of the cross-bar can be neglected.

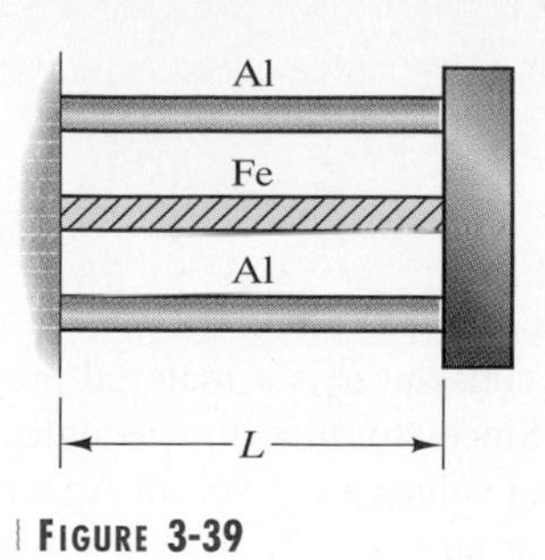

FIGURE 3-39

Strategy

Because the two materials have different coefficients of thermal expansion, the aluminum and iron bars tend to lengthen by different amounts. As a result, they will develop internal axial forces when the temperature increases. By drawing a free-body diagram of the cross-bar we can obtain one equilibrium equation, but the problem is statically indeterminate because the axial forces in the aluminum and iron bars may be different. (Notice, however, that the axial forces in the two aluminum bars are equal because of the symmetry of the system.) We can express the changes in length of the bars as the sum of their changes in length due to their internal axial forces and their changes in length due to thermal strain. We can then apply the compatibility condition that the changes in length of the bars must be equal because they are attached to the rigid cross-bar.

Solution

Equilibrium In Fig. (a) we obtain a free-body diagram of the cross-bar by passing a plane through the right ends of the three bars.

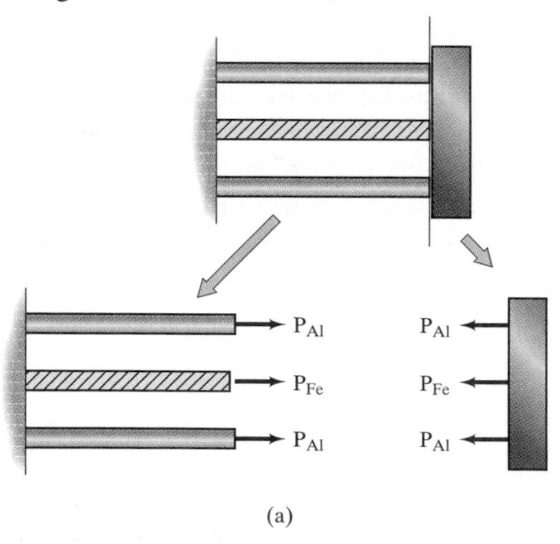

(a) Isolating the cross-bar.

The equilibrium equation for the cross-bar is

$$2P_{\mathrm{Al}} + P_{\mathrm{Fe}} = 0.$$

Deformation relations The change in length of one of the aluminum bars due to its axial force and the change in temperature is

$$\delta_{\mathrm{Al}} = \frac{P_{\mathrm{Al}} L}{E_{\mathrm{Al}} A} + \alpha_{\mathrm{Al}}\, \Delta T\, L,$$

and the change in length of the iron bar is

$$\delta_{\mathrm{Fe}} = \frac{P_{\mathrm{Fe}} L}{E_{\mathrm{Fe}} A} + \alpha_{\mathrm{Fe}}\, \Delta T\, L.$$

Compatibility The compatibility condition is that the changes in length of the bars are equal:

$$\delta_{\mathrm{Al}} = \delta_{\mathrm{Fe}}.$$

Substituting the deformation relations into this equation gives

$$\frac{P_{\mathrm{Al}} L}{E_{\mathrm{Al}} A} + \alpha_{\mathrm{Al}}\, \Delta T\, L = \frac{P_{\mathrm{Fe}} L}{E_{\mathrm{Fe}} A} + \alpha_{\mathrm{Fe}}\, \Delta T\, L.$$

This equation can be solved together with the equilibrium equation for the axial forces P_{Al} and P_{Fe}. Solving them, we obtain $P_{\mathrm{Al}} = -2000$ lb, $P_{\mathrm{Fe}} = 4000$ lb. The normal stresses in the bars are

$$\sigma_{\mathrm{Al}} = \frac{P_{\mathrm{Al}}}{A} = \frac{-2000}{0.5} = -4000 \text{ psi},$$

$$\sigma_{\mathrm{Fe}} = \frac{P_{\mathrm{Fe}}}{A} = \frac{4000}{0.5} = 8000 \text{ psi}.$$

Discussion

Beyond being an exercise to illustrate thermal strains, this example demonstrates an important problem in design. The elements of structures and machines constructed of various materials are subject to different thermal strains due to their different coefficients of thermal expansion. As a result, temperature changes can cause both stresses and deformations that must be considered in design.

EXAMPLE 3-14

The bars in Fig. 3-40 each have cross-sectional area $A = 60 \text{ mm}^2$, modulus of elasticity $E = 200$ GPa, and coefficient of thermal expansion $\alpha = 12 \times 10^{-6}\ {}^\circ\mathrm{C}^{-1}$. If a 40-kN downward force is applied at A and the temperature of the bars is raised 30°C, what are the resulting horizontal and vertical displacements of point A?

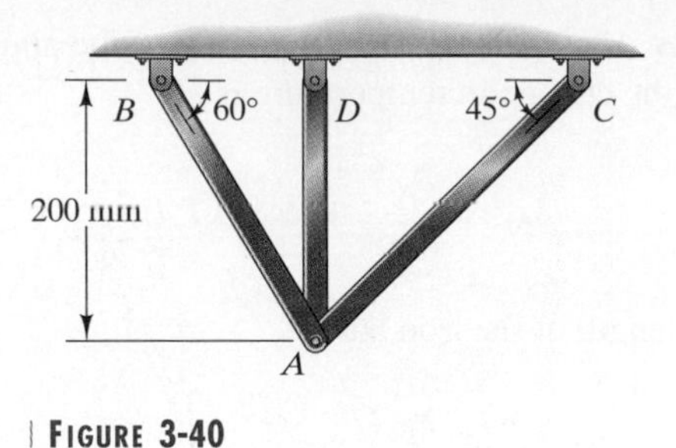

| FIGURE 3-40

Strategy

In this problem we extend Example 3-9 by introducing a change in temperature. The only change in the method of solution we used in that example is that we must express the change in length of each bar as a sum of the change due to its unknown axial force and the change due to its thermal strain.

Solution

Equilibrium In Fig. (a) we draw the free-body diagram of joint A showing the 40-kN force.

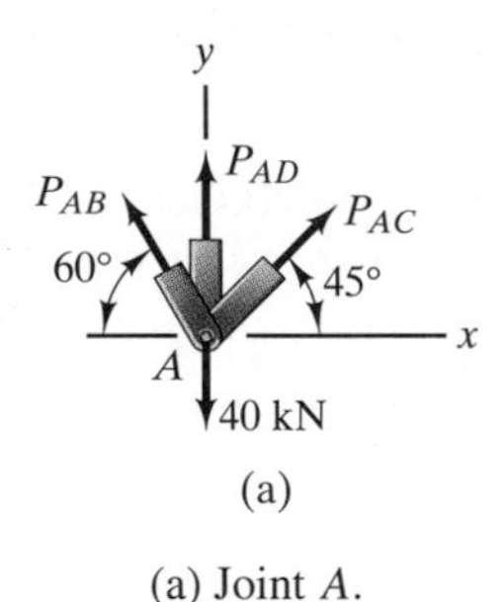

(a) Joint A.

The equilibrium equations are

$$\Sigma F_x = -P_{AB}\cos 60^\circ + P_{AC}\cos 45^\circ = 0,$$
$$\Sigma F_y = P_{AB}\sin 60^\circ + P_{AD} + P_{AC}\sin 45^\circ - 40{,}000 = 0.$$

Deformation relations The lengths of the bars are $L_{AB} = 0.2/\sin 60^\circ$ m, $L_{AC} = 0.2/\sin 45^\circ$ m, and $L_{AD} = 0.2$ m. The change in length of each bar will be the sum of its change in length due to its axial force and the change in length due to thermal strain:

$$\delta_{AB} = \frac{P_{AB}L_{AB}}{EA} + \alpha\,\Delta T L_{AB},$$
$$\delta_{AC} = \frac{P_{AC}L_{AC}}{EA} + \alpha\,\Delta T L_{AC}, \qquad \textbf{(3-27)}$$
$$\delta_{AD} = \frac{P_{AD}L_{AD}}{EA} + \alpha\,\Delta T L_{AD}.$$

Compatibility We denote the horizontal and vertical displacements of joint A by u and v [Fig. (b)].

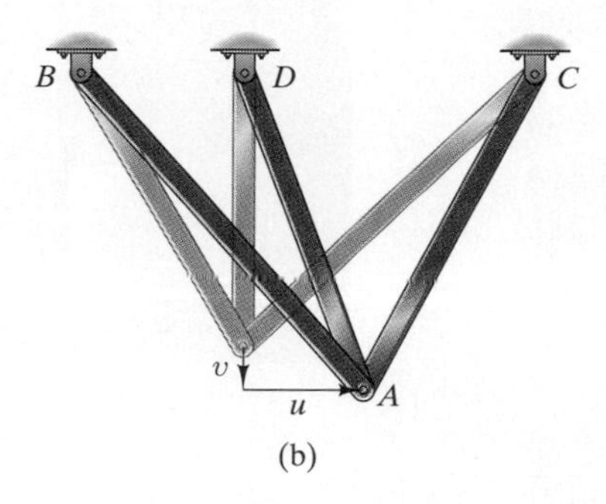

(b) Displacements of joint A.

From Example 3-9, the relationships between the changes in length of the bars and the displacements u and v are

$$\begin{aligned} v\sin 60^\circ + u\cos 60^\circ &= \delta_{AB}, \\ v\sin 45^\circ - u\cos 45^\circ &= \delta_{AC}, \\ v &= \delta_{AD}. \end{aligned} \qquad \textbf{(3-28)}$$

These equations are the compatibility conditions. The equilibrium equations, Eqs. (3-27), and Eqs. (3-28) provide eight equations in the unknowns P_{AB}, P_{AC}, P_{AD}, δ_{AB}, δ_{AC}, δ_{AD}, u, and v. Solving them, we obtain

$$u = -0.042 \text{ mm}, \qquad v = 0.426 \text{ mm}.$$

3-6 | Material Behavior

Until now we have assumed that the normal stress and axial strain in an axially loaded bar are related by the linear equation $\sigma = E\epsilon$, where E is the modulus of elasticity of the material. In this section we discuss this assumption and introduce the richly varied subject of material behavior.

Axial Force Tests

Leonardo da Vinci (1452–1519) and Galileo (1564–1642) investigated the strengths of wires and bars by subjecting them to axial tension. In a modern *tension test,* still the most common test of the mechanical behavior of materials, a bar of material is mounted in a machine that subjects it to a tensile axial force (Fig. 3-41). For a given value of the tensile force P, the normal stress $\sigma = P/A$ can be determined. (The stress is normally defined in terms of the cross-sectional area A of the undeformed bar. The term *true stress* denotes the stress calculated using the cross-sectional area of the deformed bar. In most

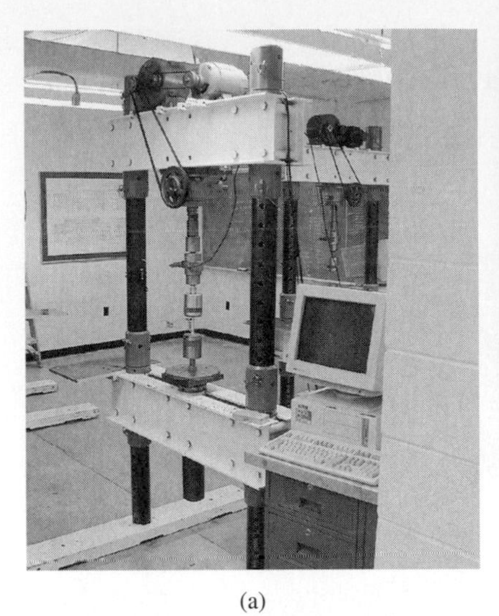
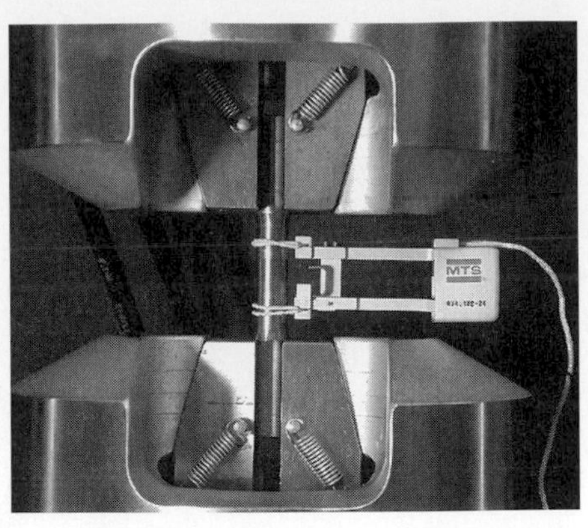

(a) (b)

FIGURE 3-41 (a) Machine for subjecting a bar of material to axial force. (b) Aluminum specimen mounted in the machine. The extensometer measures the specimen's change in length.

engineering applications the difference between these definitions of stress is negligible because the change in the cross-sectional area is small.) The bar's axial strain ϵ is determined by measuring the change in the distance between two axial positions relative to the distance in the unloaded state, which is called the *gauge length*. The axial positions are not chosen near the ends of the bar so that ϵ will be approximately uniform throughout the gauge length. When the test proceeds until the bar fractures, or breaks, the *elongation,* the final change in the gauge length expressed as a percentage of the gauge length, can also be measured. Because the axial strain becomes increasingly nonuniform as fracture approaches, the gauge length used in determining the elongation must be specified.

A *compression test* is carried out in much the same way as a tension test, applying compressive axial force to a specimen and measuring the strain. A much shorter square or cylindrical specimen is used and the axial forces are usually applied by compressing the material between flat platens (Fig. 3-42).

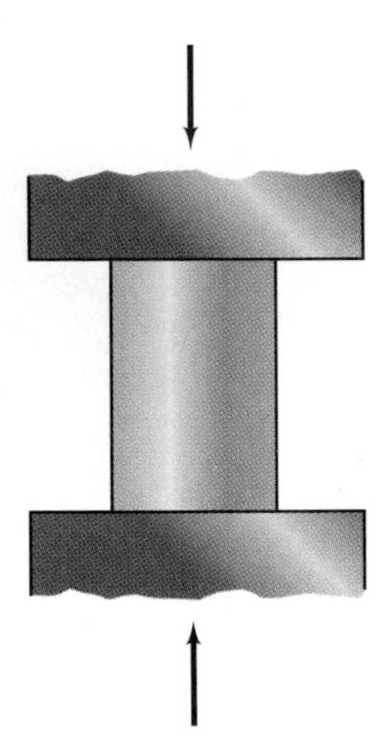

FIGURE 3-42 Typical compression specimen. Compressive axial load is applied to the ends by flat platens.

DUCTILE MATERIALS

We first describe some phenomena that occur in a tension test of a low-carbon steel. Progressively increasing the axial strain and recording the corresponding stress, the graph of the stress as a function of the strain, or *stress-strain diagram,* appears qualitatively as shown in Fig. 3-43. For small values of the strain, the relationship between the stress and strain is linear. The slope of this linear portion of the graph is the modulus of elasticity E. At a point called the *proportional limit,* the graph deviates from a straight line. (The stress is no longer proportional to the strain.) At a point called the *yield point,* the strain begins increasing with no change in stress. The corresponding stress is called the *yield stress* σ_Y. Eventually, the stress again begins increasing with increasing

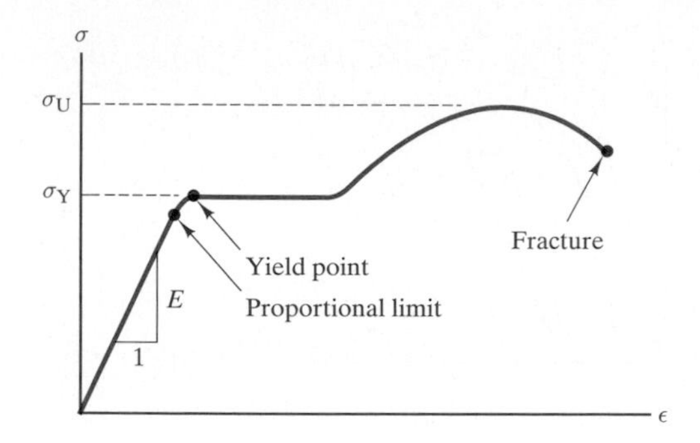

FIGURE 3-43 Stress-strain diagram for a tension test of low-carbon steel.

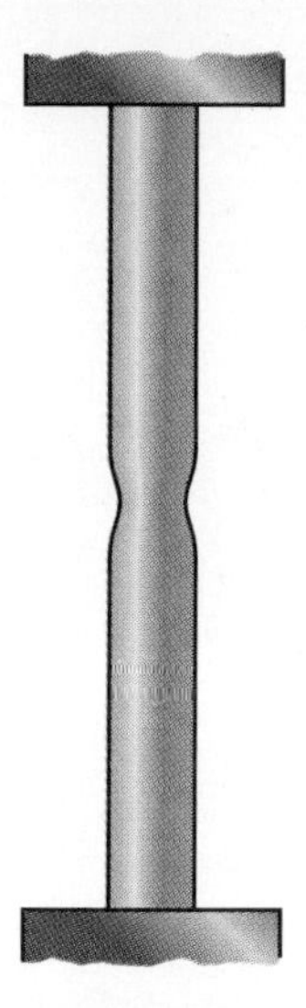

FIGURE 3-44 Necking in a tension test specimen.

strain, a phenomenon called *strain hardening,* and reaches a maximum value, the *ultimate stress* σ_U.

As the strain continues to increase beyond the occurrence of the ultimate stress, the stress decreases until the bar fractures, or breaks. The decreasing stress is associated with the formation of a region of decreased cross-sectional area in the bar, a phenomenon called *necking* (Fig. 3-44). The apparently decreasing stress is an artifact caused by defining σ in terms of the cross-sectional area of the undeformed bar. The true stress in the necked portion of the bar continues to increase with strain until fracture occurs.

If the strain to which the steel is subjected remains below the proportional limit, so that $\sigma = E\epsilon$, we say that its stress-strain relationship is *linearly elastic,* and if the stress is removed, the bar returns to its original length. If the bar is strained beyond the yield point and the stress is then decreased, the resulting path in the σ–ϵ plane does not return to the origin along the original curve. Instead, it returns to zero stress along a straight line with slope E, and a residual strain remains (Fig. 3-45a). If the bar is then reloaded, it follows the new straight path until it reaches the stress-strain curve obtained with monotonically increasing strain (Fig. 3-45b). Strain beyond the value of strain corresponding to the yield point, which remains as residual strain if the bar is unloaded, is called *plastic strain.* Thus the yield stress σ_Y is the stress at which plastic strain begins.

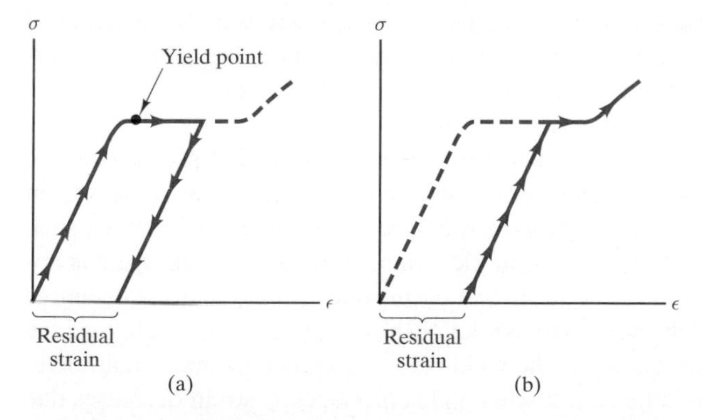

FIGURE 3-45 (a) Straining the bar beyond the yield point and then unloading it. (b) Reloading the bar.

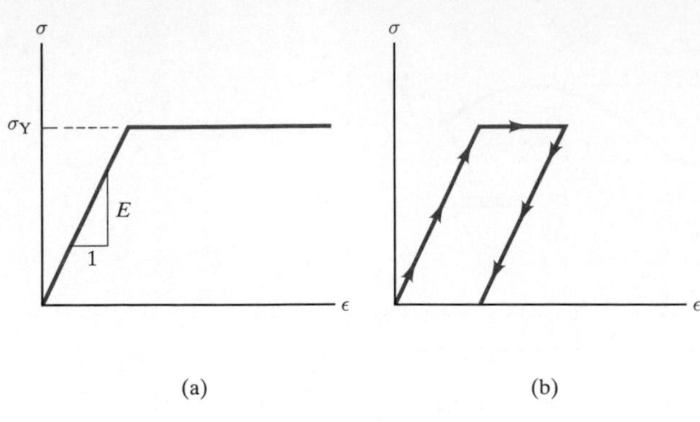

FIGURE 3-46 (a) Model of an elastic–perfectly plastic material. (b) Loading–unloading behavior.

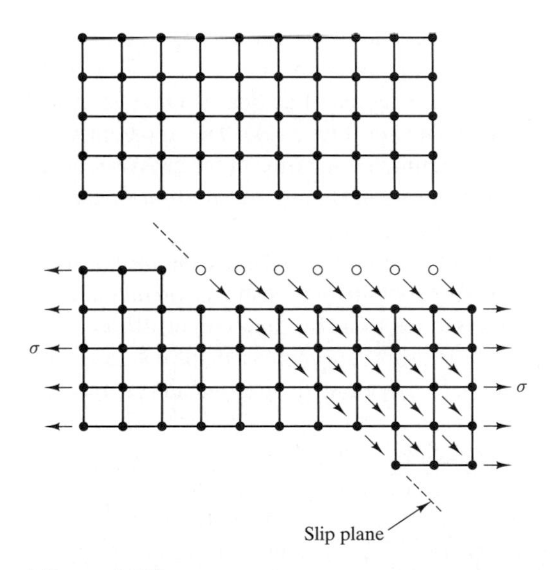

FIGURE 3-47 Plastic strain of a crystal lattice by the mechanism of slip.

Materials with stress-strain relationships of this type are sometimes modeled by representing their stress-strain relationship as shown in Fig. 3-46, which is referred to as an *elastic–perfectly plastic material.* ("Perfectly" plastic means it is assumed that there is no strain hardening.)

The explanation for this somewhat bizarre behavior is that plastic strain in a crystalline material like steel occurs in part by mechanisms that result in rearrangements of its lattices of atoms. Figure 3-47 illustrates how one such mechanism, called *slip,* results in plastic strain. When increasing stress is applied to the unloaded material, strain first occurs due to changes in the distances between atoms, and the stress-strain relationship is linearly elastic. Beyond the value of strain corresponding to the yield stress, strain occurs due to rearrangements of the crystal lattices. But when unloading occurs, strain decreases due to changes in the distances between atoms, and the stress-strain relationship is again linearly elastic.

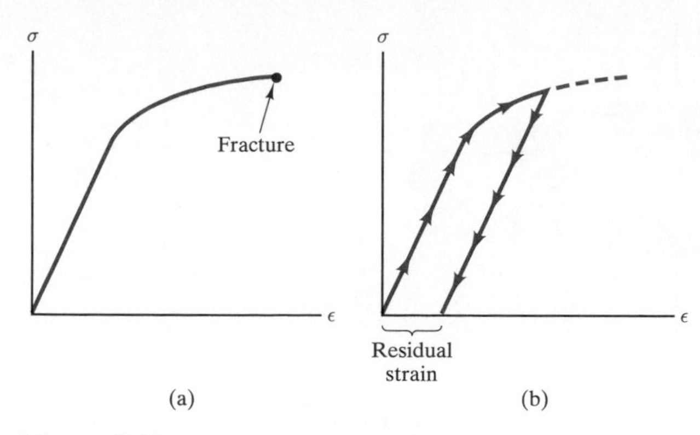

FIGURE 3-48 (a) Stress-strain relationship for a ductile material that does not exhibit an identifiable yield stress. (b) Loading–unloading behavior.

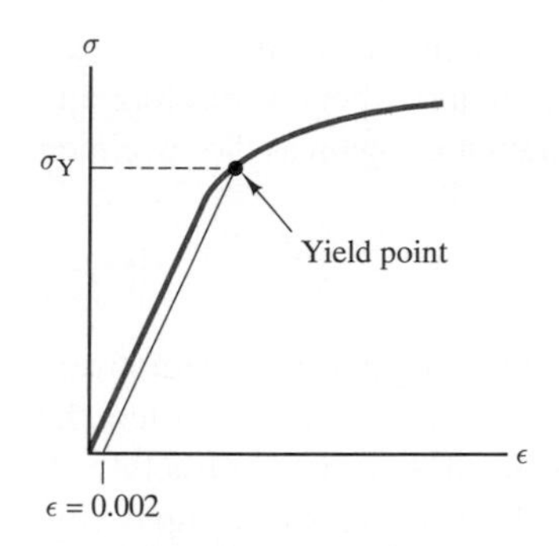

FIGURE 3-49 Designating a yield point and yield stress by the offset method.

Materials that undergo significant plastic strain before fracture are said to be *ductile*. Our discussion of tensile behavior has focused on low-carbon steel in part due to its importance in structural applications, but also because describing the stress-strain behavior of steel permitted us to introduce concepts and terminology—linearly elastic, yield stress, ultimate stress, plastic strain, fracture—that apply to many other ductile materials. Some ductile materials, such as aluminum and copper, do not exhibit an easily identifiable yield stress but undergo a smooth transition from linearly elastic to plastic behavior with strain hardening (Fig. 3-48a). However, the yield stress is a very useful parameter in design, so it has become traditional to artificially designate a yield point and yield stress for such materials. This is done by drawing a line parallel to the linear part of the stress-strain diagram which intersects the strain axis at some arbitrarily chosen value, often $\epsilon = 0.002$. The point at which this line intersects the stress-strain diagram is defined to be the yield point, and the corresponding stress is defined to be the yield stress (Fig. 3-49). The arbitrary strain chosen is called the *offset,* and this technique is called the offset method of determining the yield point.

When a typical ductile material is subjected to a compression test (Fig. 3-50), it initially exhibits linearly elastic behavior governed by $\sigma = E\epsilon$. After yielding occurs, the stress-strain behavior is different in character from

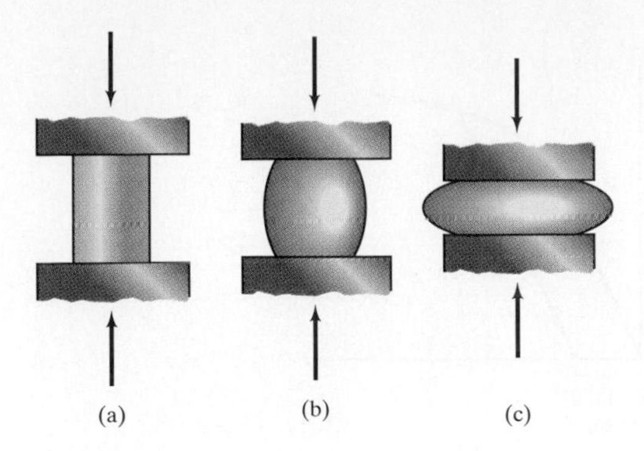

FIGURE 3-50 (a) Beginning a compression test of a ductile material. (b) Soon after yielding occurs, the specimen becomes barrel shaped. (c) The specimen eventually becomes flattened.

that observed in the tension test (Fig. 3-51). Although the slope of the stress as a function of strain initially decreases as the material begins yielding and the specimen becomes barrel shaped, the slope increases again as the specimen begins to flatten out.

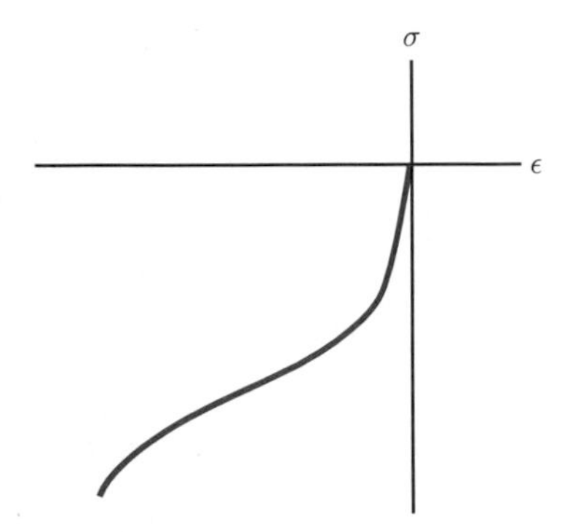

FIGURE 3-51 Stress as a function of strain for a compression test of a typical ductile material.

BRITTLE MATERIALS

Materials such as cast iron, high-carbon steel, and masonry that exhibit relatively little plastic strain prior to fracture are said to be *brittle*. In a tension test the stress increases monotonically, so that the ultimate stress occurs at fracture. A notable feature of brittle materials is that their ultimate stress in compression is considerably greater in magnitude than their ultimate stress in tension. This is illustrated in Fig. 3-52, in which we show the stress as a function of strain for a typical brittle material in both tension and compression. Unlike ductile materials, brittle materials subjected to a compression test undergo fracture. In different materials the sample may crush or may fracture along a plane of maximum shear stress as shown in Fig. 3-11.

Other Aspects of Material Behavior

The ordinary tension test we have described measures the strain induced in a bar of material by a static normal stress. This information alone is not usually adequate even for the design of a static structure composed of conventional materials and exposed to normal environmental conditions, much less the rapidly varying loads, unconventional materials, and extreme environments associated with many current structures. Modern structural engineers must have a broad understanding of material behavior, which is beyond the scope of a first course. The terms we define in this section, together with our previous discussion of axial load testing, simply provide an introduction and an initial working vocabulary on material behavior.

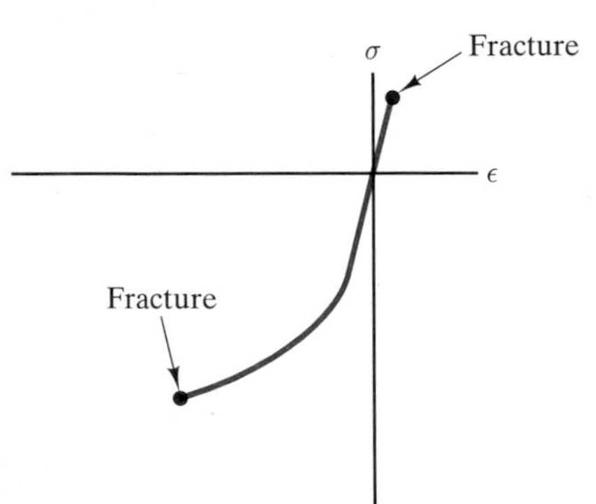

FIGURE 3-52 Stress as a function of strain for tension and compression tests of a typical brittle material.

Aging *Aging* refers to gradual changes in material properties that can occur with time. These changes may occur more rapidly at elevated temperatures. Aging can be part of the normal processing of a material, as in the curing of concrete.

Alloy An *alloy* is a material, usually a metal, consisting of a combination of different elements. Steel is an alloy of iron, manganese, carbon, and sometimes other elements, and bronze is an alloy of copper, tin, and sometimes other elements. The behavior of an alloy can depend strongly on the percentages of its elements, and tabulated lists of properties often specify them.

Annealing Heating a material to a high temperature and then cooling it in a controlled way is called *annealing*. It is often carried out to achieve a specific granular structure in a metal and is also used to relieve internal stresses resulting from uneven cooling during previous processing. Annealing substantially affects a material's mechanical properties, and tabulated lists of properties often specify whether materials are in the annealed state.

Brittleness A material is said to be *brittle* if it fractures before undergoing significant plastic deformation. Chalk is a familiar brittle material. Masonry—brick, concrete, and stone—are also brittle. (*See* Ductility.)

Creep *Creep* refers to a gradual increase in strain under constant applied stress. It is an example of viscoelastic behavior (*see* Viscoelasticity) and occurs in metals at elevated temperatures. The creep strength of a material is defined to be the value of stress that will cause a specified amount of creep at a given temperature.

Ductility A material capable of significant plastic deformation before fracturing is said to be *ductile*. A material that is not ductile is called *brittle*.

Elasticity *Elasticity* is a model of material behavior based upon the assumption that the stress is a single-valued function of the strain. If the stress is assumed to be a linear function of the strain, the model is called *linear elasticity*. Otherwise, it is called *nonlinear elasticity*.

Fatigue Some materials subjected to repeated stresses above a particular magnitude will eventually fail. In metals, failure can occur due to the gradual growth and coalescence of small cracks. This phenomenon, called *fatigue,* is observed in rotating machinery and in the wings of airplanes, which undergo repeated flexure due to gusts. The fatigue life of a material is defined to be the number of cycles of specified stress prior to failure, and the fatigue limit is a specified stress below which a material does not exhibit fatigue. We analyze fatigue in Chapter 12.

Hardness *Hardness* is a rather general term referring to a material's resistance to plastic deformation. It is sometimes evaluated in a comparative way by abrading or cutting a material. Standard comparative tests have also been established for which tabulated results are available. For example, in the Brinell hardness test a steel or carbide ball of specified diameter is pressed into a test material by a specified force. The force in kilograms divided by the surface area of the impression made in the test material in square millimeters is called the *Brinell hardness number* of the material.

Heat Treatment *Heat treatment* refers to heating and cooling a material to alter its mechanical properties. Tabulated lists of mechanical properties often indicate what kind of heat treatment a material has received.

Impact Test An *impact test* is a comparative test of the response of materials to rapid loading. In the Charpy and Izod tests, a material specimen is fractured by a swinging pendulum. The energy absorbed, determined by measuring the swing of the pendulum following the impact, is used as an indication of impact strength.

Phase *Phase* refers to the solid, liquid, and gaseous states of a material, but is also applied to different crystalline forms of a material. For example, steel can exist in crystalline phases called austenite, ferrite, and pearlite, and its mechanical properties will depend on the percentages of these phases present.

Plasticity *Plasticity* is a model of material behavior based on the assumption that there is a yield stress and permanent or plastic strain can develop when the yield stress is reached. The relationship between the stress and the plastic strain is called a *flow rule*.

Thermoelasticity *Thermoelasticity* is a model of material behavior that accounts for temperature effects as well as deformation. In Section 3-5 we applied a simple form of this model in which the relationship between the stress and strain is independent of the temperature. A model in which the mechanical and thermal behaviors of the material are assumed to be independent is referred to as *uncoupled thermoelasticity*. In *coupled thermoelasticity* the deformation and temperature fields must be determined simultaneously. Whether a material must be modeled as thermoelastic in a given situation depends on the range of temperatures involved.

Viscoelasticity If you apply a constant axial load to a bar of certain materials, the bar will elongate, then slowly increase in length and approach a different equilibrium length. This behavior is called *viscoelastic;* it combines aspects of an elastic solid and a viscous fluid. *Viscoelasticity* is a model of material behavior in which the relationship between the stress and strain involves the time and must be expressed as a differential or integral equation. Polymeric materials must usually be modeled as viscoelastic materials. Metals can also exhibit viscoelastic behavior at elevated temperatures or if they are subjected to rapidly varying loads. In some circumstances it is necessary to assume that the stress-strain relationship of a material depends on both time and temperature, a model of material behavior called *thermoviscoelasticity*.

3-7 Design Issues

In this chapter we have provided some of the essential background for designing bars to support constant axial loads and structures made up of such bars. If there is a dictum for the structural designer equivalent to the physician's "First, do no harm," it is "First, prevent failure." Although *failure* must be defined in different ways for different applications, at a minimum the normal and shear stresses in a structural member must not be allowed to exceed the values the material will support.

Allowable Stress

In a prismatic bar subjected to a specified axial load P, we have shown in Section 3-1 that there is no plane on which the normal stress is larger in magnitude than the stress $\sigma = P/A$ on a plane perpendicular to the bar's axis. The magnitude of the maximum shear stress, which occurs on planes oriented at 45° relative to the bar's axis, is $|\sigma/2| = |P/2A|$. A criterion frequently used in design is to try to ensure that σ does not exceed the yield stress σ_Y of the material, or a specified fraction of the yield stress called the *allowable stress* σ_{allow}. The ratio of the yield stress to the allowable stress is called the *factor of safety:*

$$S = \frac{\sigma_Y}{\sigma_{\text{allow}}}.$$

In choosing the factor of safety, the design engineer must consider how accurately the loads to which the bar will be subjected in normal use can be estimated. Whenever possible, a conservatively large factor of safety would be desirable, but compromises are usually necessary. For example, cost may require compromise in the properties of the materials used. Excessive structural weight can discourage the use of high factors of safety, especially in vehicle applications. Contingencies beyond normal use must also be considered, such as potential earthquake loads in the structure of a building, or loads exerted on the members of a car's suspension during emergency maneuvers. If the item being designed will be mass produced, the factor of safety must be chosen to account for anticipated variations in dimensions and material properties. The engineer must balance these considerations within the essential constraint of arriving at a reliable and safe design.

Other Design Considerations

A bar subjected to compression can fail by buckling (geometric instability) at a much smaller axial load than is necessary to cause yielding of the material. We analyze buckling in Chapter 10. Furthermore, even when the stresses to which a structural member is subjected are small compared to the yield stress, failure can occur if they are applied repeatedly. We discuss failure due to cyclic loads in Chapter 12. Also, our analysis of the stress distribution in an axially loaded bar does not apply near the ends where the loads are applied. Those regions normally require a detailed stress analysis that is beyond our scope.

Our examples and problems are limited primarily to designing axially loaded members and structures to meet the objective of supporting given loads. But in addition to the overriding concern of preventing failure, the structural designer is usually confronted with a broad array of decisions relevant to a particular application, and the finest designs are achieved by successfully meeting a spectrum of requirements. Material cost and availability, cost and feasibility of processing and manufacture, resistance to corrosion in the expected environment, compatibility with other materials, and the effect of aging on the material's properties may be important considerations. Decisions on a given design can also be influenced to a greater or lesser extent by concern for safety, ease of maintenance, and aesthetics.

EXAMPLE 3-15

Worksheet 3

The truss in Fig. 3-53 is to be constructed of members with yield stress $\sigma_Y =$ 700 MPa and equal cross-sectional areas. If the structure must support a mass m as large as 3400 kg with factor of safety $S = 3$, what should the cross-sectional area of the members be?

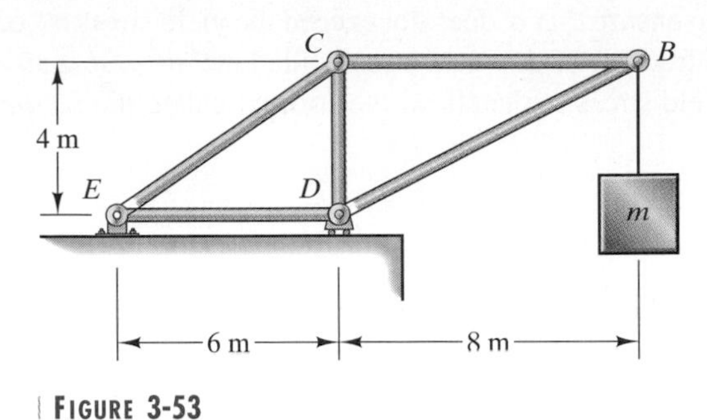

FIGURE 3-53

Strategy

We can use the method of joints to determine the axial forces in the members when $m = 3400$ kg. We must then choose the cross-sectional area so that the member subjected to the largest axial load has a factor of safety of 3.

Solution

With $m = 3400$ kg, the axial forces in the members are (see Example 3-2):

Member:	BC	BD	CD	CE	DE
Axial force (N):	66,700	−74,600	−44,500	80,200	−66,700

The largest axial force, 80,200 N, occurs in member CE. With a factor of safety of 3, the allowable stress is

$$\sigma_{\text{allow}} = \frac{\sigma_Y}{S} = 233 \text{ MPa}.$$

Equating the normal stress in member CE to the allowable stress,

$$\frac{80,200}{A} = 233 \times 10^6,$$

we obtain $A = 0.000344 \text{ m}^2$.

Chapter Summary

Stresses in Prismatic Bars

If axial forces P are applied at the centroid of the cross section of a prismatic bar, the stress distribution on any plane perpendicular to the bar's axis that is not near the ends of the bar [Fig. (a)] can be approximated by a uniform normal stress

$$\sigma = \frac{P}{A}. \qquad \text{Eq. (3-1)}$$

(a)

The normal and shear stresses on a plane $\mathcal{P}$ oriented as shown in Fig. (b) are

$$\sigma_\theta = \sigma \cos^2\theta, \qquad \text{Eq. (3-2)}$$

$$\tau_\theta = -\sigma \sin\theta \cos\theta. \qquad \text{Eq. (3-3)}$$

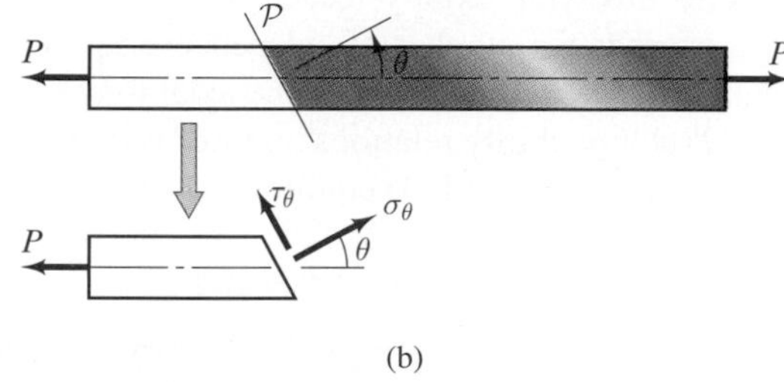

(b)

There is no plane on which the normal stress is larger in magnitude than the stress $\sigma = P/A$ on the plane perpendicular to the bar's axis. The magnitude of the maximum shear stress, which occurs on planes oriented at 45° relative to the bar's axis, is $|\sigma/2|$.

Strains in Prismatic Bars

The change in length δ of a prismatic bar subjected to axial load [Fig. (c)] is related to the axial strain ϵ by

$$\delta = \epsilon L, \qquad \text{Eq. (3-4)}$$

(c)

where L is the length of the undeformed bar. The normal stress is related to the axial strain by

$$\sigma = E\epsilon, \qquad \text{Eq. (3-5)}$$

where E is the *modulus of elasticity*. Typical values of E are given in Appendix B. The change in the bar's length resulting from a given axial load P

is

$$\delta = \frac{PL}{EA}, \qquad \text{Eq. (3-6)}$$

where A is the cross-sectional area.

The negative of the ratio of the lateral strain of a prismatic bar to its axial strain when it is subjected to axial forces is called *Poisson's ratio:*

$$\nu = -\frac{\epsilon_{\text{lat}}}{\epsilon}. \qquad \text{Eq. (3-7)}$$

Typical values are given in Appendix B. The lateral strain resulting from an axial load P is

$$\epsilon_{\text{lat}} = -\frac{\nu P}{EA}. \qquad \text{Eq. (3-8)}$$

Statically Indeterminate Problems

Solutions to problems involving axially loaded bars in which the number of unknown reactions exceeds the number of independent equilibrium equations involve three elements: (1) relations between the axial forces in bars and their changes in length; (2) compatibility relations imposed on the changes in length by the geometry of the problem; and (3) equilibrium.

Bars with Gradually Varying Cross Sections

If the cross-sectional area of a bar is a gradually varying function of axial position $A(x)$ [Fig. (d)], the stress distribution on a plane perpendicular to the bar's axis can be approximated by a uniform normal stress:

$$\sigma = \frac{P}{A(x)}. \qquad \text{Eq. (3-24)}$$

The bar's change in length is

$$\delta = \int_0^L \frac{P\,dx}{EA(x)}. \qquad \text{Eq. (3-25)}$$

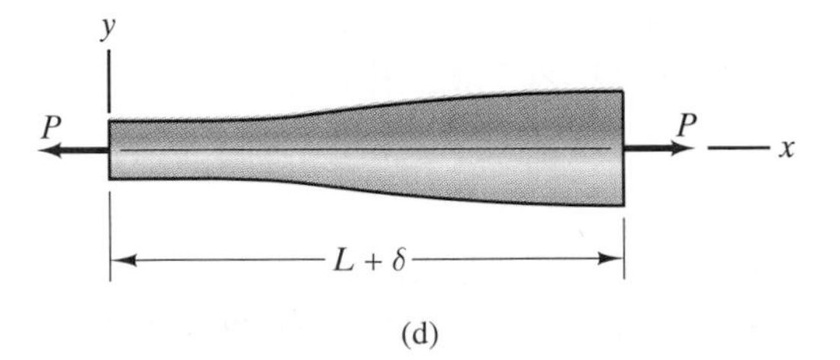

(d)

Distributed Axial Loads

A distributed axial force on a bar can be described by a function q defined such that the force on each element dx of the bar is $q\,dx$ [Fig. (e)].

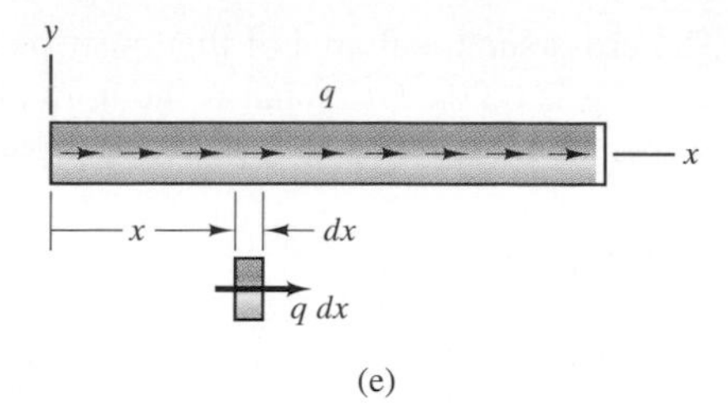

(e)

Thermal Strains

Consider an infinitesimal line dL within an unconstrained material at a uniform temperature T. If the temperature of the sample changes by an amount ΔT, the material will expand or contract and the length of the line will change to a value dL' [Fig. (f)]. The *thermal strain* of the material at the location and in the direction of the line dL is

$$\epsilon_T = \frac{dL' - dL}{dL}.$$

Temperature T Temperature $T + \Delta T$

(f)

The thermal strain is related to the change in temperature by

$$\epsilon_T = \alpha\,\Delta T, \qquad \text{Eq. (3-26)}$$

where α is the *coefficient of thermal expansion*. Typical values of α are given in Appendix B.

PROBLEMS

Worksheet 1 can be used to solve Problems 3-1.20 through 3-1.22.

3-1.1. A prismatic bar with cross-sectional area $A = 0.1\ \text{m}^2$ is loaded at the ends in two ways: **(a)** by 100-Pa uniform normal tractions; **(b)** by 10-N axial forces acting at the centroid of the bar's cross section. What are the normal and shear stress distributions at the plane $\mathcal{P}$ in the two cases?

3-1.2. A prismatic bar with cross-sectional area $A = 4\ \text{in}^2$ is subjected to tensile axial loads P. It consists of a material

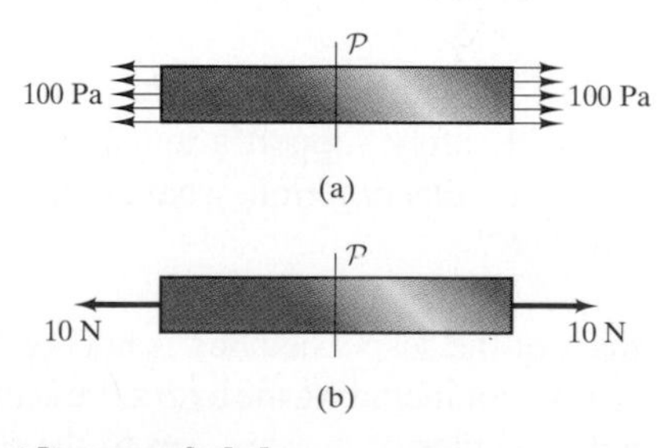

Problem 3-1.1

that will safely support a tensile normal stress of 60 ksi. Based on this criterion, what is the largest safe value of P?

3-1.3. A prismatic bar has a solid circular cross section with 20-mm diameter. It consists of a material that will safely support a tensile normal stress of 300 MPa. Based on this criterion, what is the largest tensile load P to which the bar can be subjected?

3-1.4. The cross-sectional area of bar AB is 0.5 in^2. If the force $F = 3$ kip, what is the normal stress on a plane perpendicular to the axis of bar AB? (*Strategy:* You can determine the axial force in bar AB by drawing a free-body diagram of the horizontal bar and summing moments about point C.)

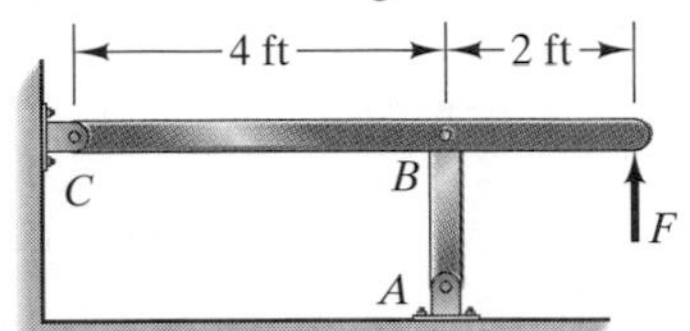

| PROBLEM 3-1.4

3-1.5. Bar AB of the frame in Problem 3-1.4 consists of a material that will safely support a tensile normal stress of 20 ksi. If you want to design the frame to support forces F as large as 8 kip, what is the minimum required cross-sectional area of bar AB?

3-1.6. The cross-sectional area of bar AB is 0.015 m^2. If the force $F = 20$ kN, what is the normal stress on a plane perpendicular to the axis of bar AB?

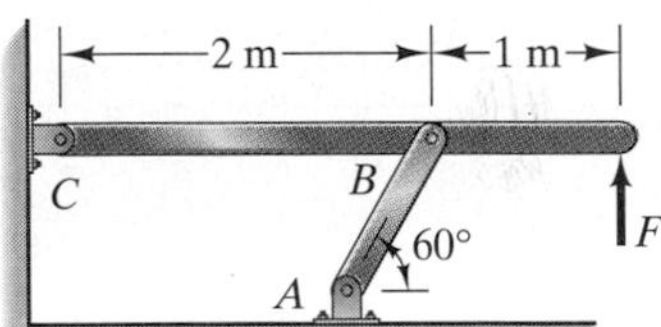

| PROBLEM 3-1.6

3-1.7. Bar AB of the frame in Problem 3-1.6 consists of a material that will safely support a tensile normal stress of 20 MPa. Based on this criterion, what is the largest safe value of the force F?

3-1.8. The mass of the suspended box is 800 kg. The mass of the crane's arm (not including the hydraulic actuator BC) is 200 kg, and its center of mass is 2 m to the right of A. The cross-sectional area of the upper part of the hydraulic actuator is 0.004 m^2. What is the normal stress on a plane perpendicular to the axis of the upper part of the actuator?

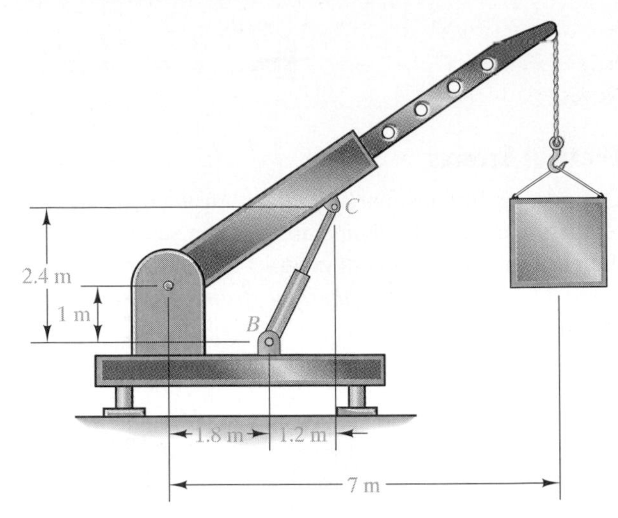

| PROBLEM 3-1.8

3-1.9. The cross-sectional area of the lower part of the hydraulic actuator BC in Problem 3-1.8 is 0.010 m^2. What is the normal stress on a plane perpendicular to the axis of the lower part of the actuator?

3-1.10. The cross-sectional area of each bar is A. What is the normal stress on a plane perpendicular to the axis of one of the bars?

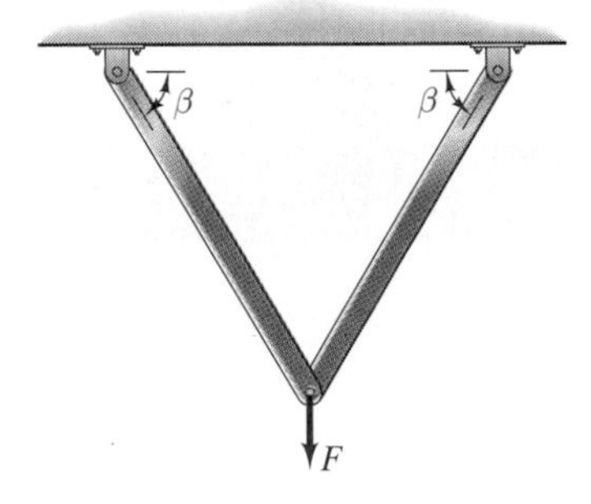

| PROBLEM 3-1.10

3-1.11. The angle β of the system in Problem 3-1.10 is 60°. The bars are made of a material that will safely support a tensile normal stress of 8 ksi. Based on this criterion, if you want to design the system so that it will support a force $F = 3$ kip, what is the minimum necessary value of the cross-sectional area A?

3-1.12. Suppose that the horizontal distance between the supports of the system in Problem 3-1.10 and the load F are specified, and the prismatic bars are made of a material that will safely support a tensile normal stress σ_0. You want to choose the angle β and the cross-sectional area A of the bars so that the total volume of material used is a minimum. What are β and A?

3-1.13. The cross-sectional area of each bar is 60 mm^2. If $F = 40$ kN, what are the normal stresses on planes perpendicular to the axes of the bars?

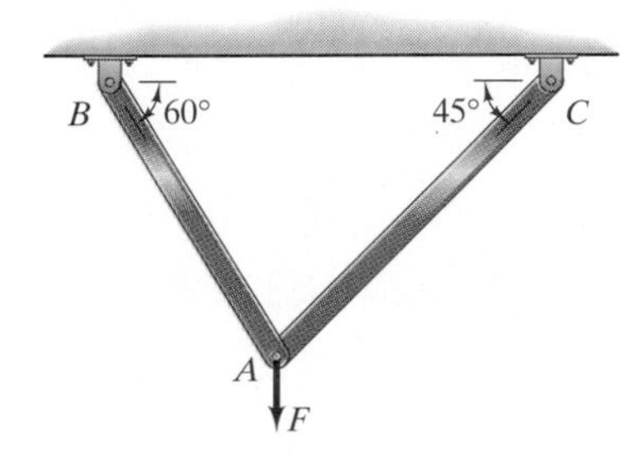

PROBLEM 3-1.13

3-1.14. The bars of the truss in Problem 3-1.13 are made of material that will safely support a tensile normal stress of 600 MPa. Based on this criterion, what is the largest safe value of the force F?

3-1.15. The cross-sectional area of each bar of the truss is 400 mm^2. If $F = 30$ kN, what is the normal stress on a plane perpendicular to the axis of member BE?

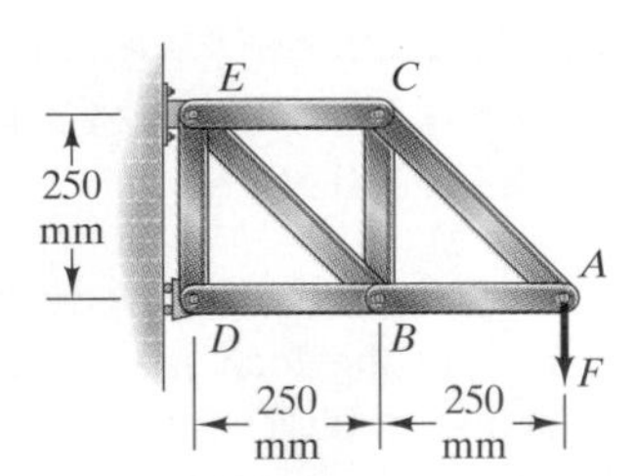

PROBLEM 3-1.15

3-1.16. In Problem 3-1.15, what is the normal stress on a plane perpendicular to the axis of member BC?

3-1.17. The truss in Problem 3-1.15 is made of a material that will safely support a normal stress (tension or compression) of 340 MPa. Based on this criterion, what is the largest safe value of the force F?

3-1.18. The system shown supports half of the weight of the 680-kg excavator. The cross-sectional area of member AB is 0.0012 m^2. If the system is stationary, what normal stress acts on a plane perpendicular to the axis of member AB?

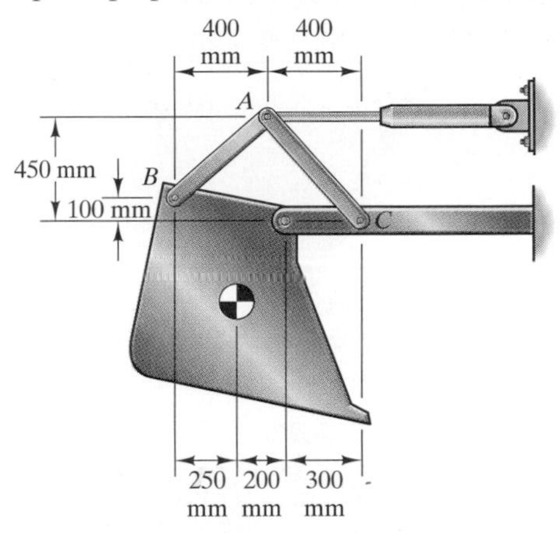

PROBLEM 3-1.18

3-1.19. Member AC in Problem 3-1.18 has a cross-sectional area of 0.0014 m^2. If the system is stationary, what normal stress acts on a plane perpendicular to the axis of member AC?

3-1.20. The cross-sectional area of the prismatic bar is $A = 2$ in^2 and the axial force $P = 20$ kip. Determine the normal and shear stresses on the plane $\mathcal{P}$. Draw a diagram isolating the part of the bar to the left of plane $\mathcal{P}$ and show the stresses.

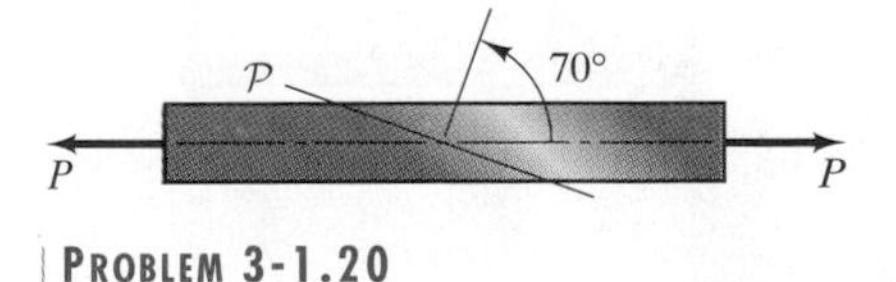

PROBLEM 3-1.20

3-1.21. If the normal stress on the plane $\mathcal{P}$ in Problem 3-1.20 is 6000 psi, what is the axial force P?

3-1.22. The cross-sectional area of the prismatic bar is 0.02 m^2. If the normal and shear stresses on the plane $\mathcal{P}$ are $\sigma_\theta = 1.25$ MPa and $\tau_\theta = -1.50$ MPa, what are the angle θ and the axial force P?

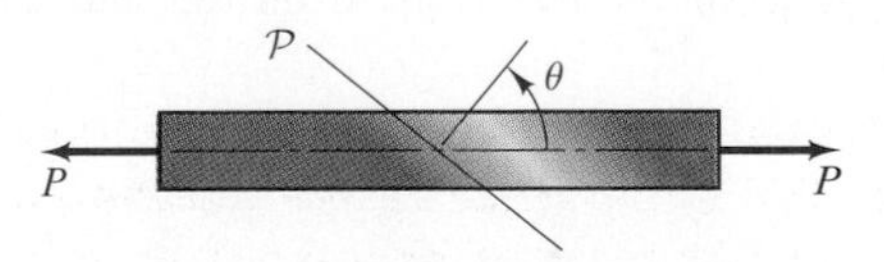

PROBLEM 3-1.22

3-1.23. The cross-sectional area of the bars is $A = 0.5\ \text{in}^2$ and the force $F = 3000$ lb. Determine the normal stresses and the magnitudes of the shear stresses on the planes (a) and (b).

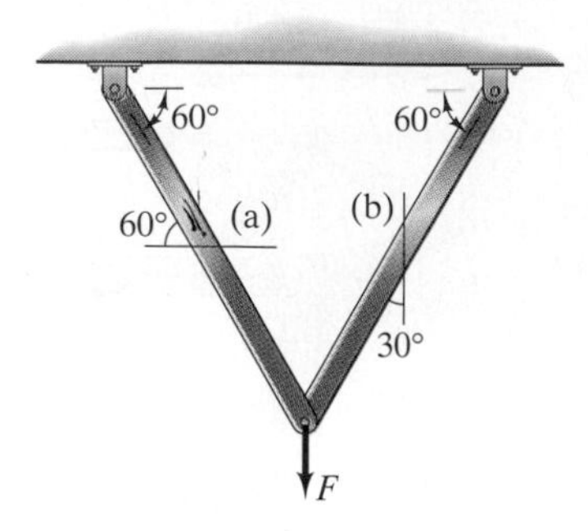

PROBLEM 3-1.23

3-1.24. The truss in Problem 3-1.23 is constructed of a material that will safely support a normal stress of 8 ksi and a shear tress of 3 ksi. Based on these criteria, what is the largest force F that can safely be applied?

3-1.25. In Problem 3-1.18, what are the normal stress and the magnitude of the shear stress in member AB on a plane oriented at 45° relative to the member's axis?

3-1.26. The cross-sectional area of bar BE is $3\ \text{in}^2$. Determine the normal stress and the magnitude of the shear stress on the indicated plane through BE.

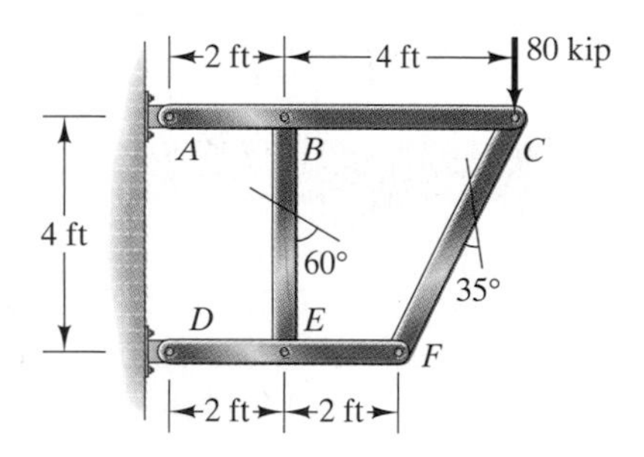

PROBLEM 3-1.26

3-1.27. The cross-sectional area of bar CF of the frame in Problem 3-1.26 is $4\ \text{in}^2$. Determine the normal stress and the magnitude of the shear stress on the indicated plane through CF.

3-1.28. The space truss supports a vertical 800-lb load. The cross-sectional area of the bars is $0.2\ \text{in}^2$. Determine the normal stress in member AB. (*Strategy:* Draw a free-body diagram of joint A.)

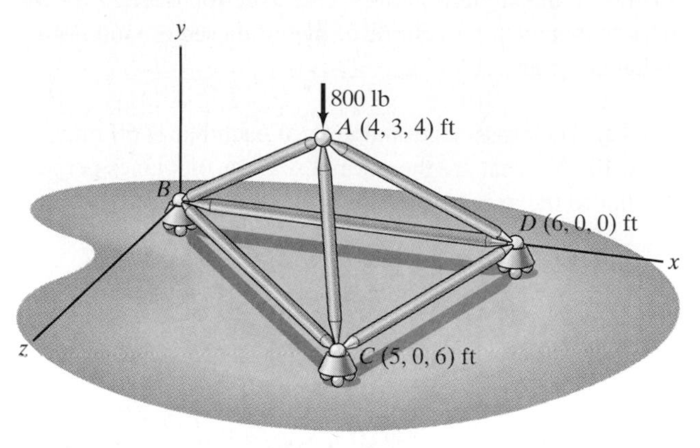

PROBLEM 3-1.28

3-1.29. The space truss in Problem 3-1.28 has roller supports at B, C, and D. Determine the normal stress in member BC.

3-1.30. The free-body diagram of the part of the construction crane to the left of the plane is shown. The coordinates (in meters) of the joints A, B, and C are (1.5, 1.5, 0), (0, 0, 1), and (0, 0, −1), respectively. The axial forces P_1, P_2, and P_3 are parallel to the x axis. The axial forces P_4, P_5, and P_6 point in the directions of the unit vectors

$$\mathbf{e}_4 = 0.640\,\mathbf{i} - 0.640\,\mathbf{j} - 0.426\,\mathbf{k},$$
$$\mathbf{e}_5 = 0.640\,\mathbf{i} - 0.640\,\mathbf{j} + 0.426\,\mathbf{k},$$
$$\mathbf{e}_6 = 0.832\,\mathbf{i} - 0.555\,\mathbf{k}.$$

The total force exerted on the free-body diagram by the weight of the crane and the load it supports is $-F\mathbf{j} = -44\mathbf{j}$ kN acting at the point (−20, 0, 0) m. The cross-sectional area of members 1, 2, and 3 is $5000\ \text{mm}^2$, and the cross-sectional area of members 4, 5 and 6 is $1600\ \text{mm}^2$. What is the normal stress in member 3? (*Strategy:* Calculate the moment about the line that passes through joints A and B.)

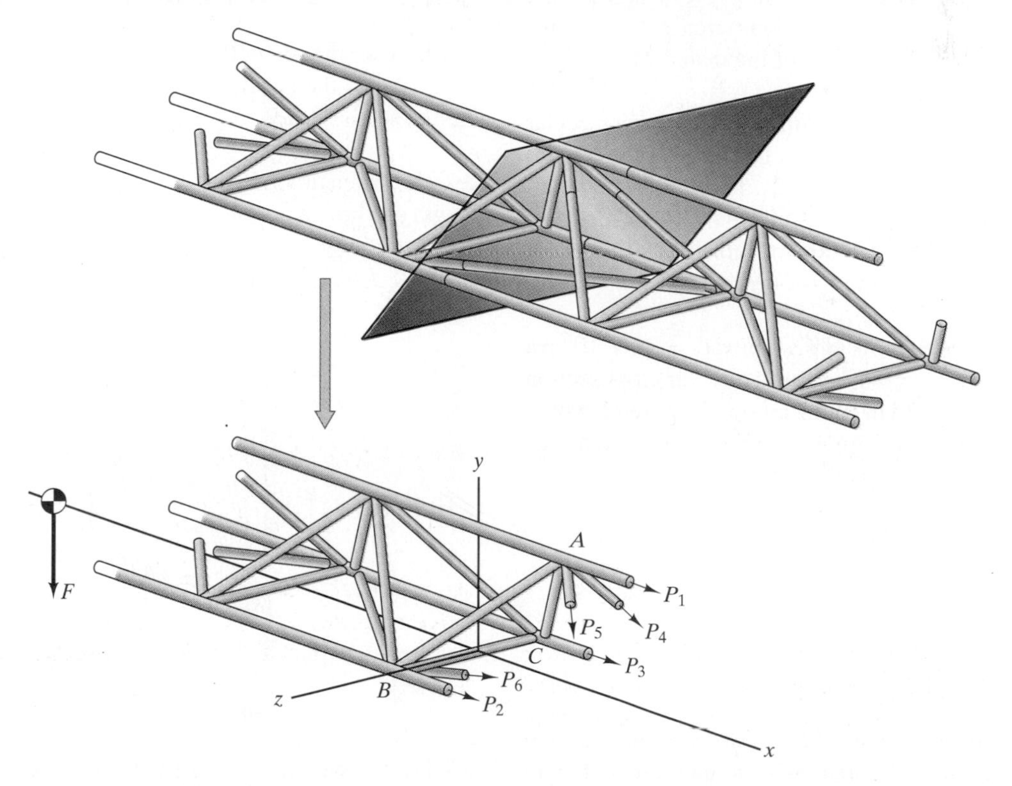

PROBLEM 3-1.30

3-1.31. In Problem 3-1.30, determine the normal stresses in members 1, 4, and 5.

3-2.1. Two marks are made 2 in. apart on an unloaded bar. When the bar is subjected to axial forces P, the marks are 2.004 in. apart. What is the axial strain of the loaded bar?

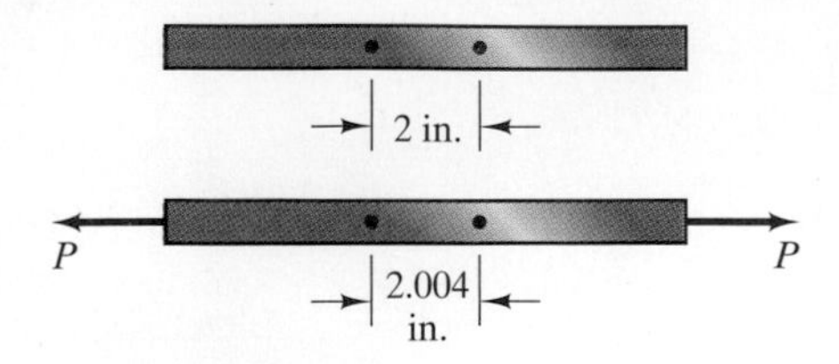

PROBLEM 3-2.1

3-2.2. The total length of the unloaded bar in Problem 3-2.1 is 10 in. Use the result of Problem 3-2.1 to determine the total length of the loaded bar. What assumption are you making when you do so?

3-2.3. If the forces exerted on the bar in Problem 3-2.1 are $P = 20$ kip and the bar's cross-sectional area is $A = 1.5\text{ in}^2$, what is the modulus of elasticity of the material?

3-2.4. A prismatic bar with length $L = 6$ m and a circular cross section with diameter $D = 0.02$ m is subjected to 20-kN compressive forces at its ends. The length and diameter of the deformed bar are measured and determined to be $L' = 5.940$ m and $D' = 0.02006$ m. What are the modulus of elasticity and Poisson's ratio of the material?

3-2.5. The bar has modulus of elasticity $E = 30 \times 10^6$ psi and Poisson's ratio $\nu = 0.32$. It has a circular cross section with diameter $D = 0.75$ in. What compressive force would have to be exerted on the right end of the bar to increase its diameter to 0.752 in.?

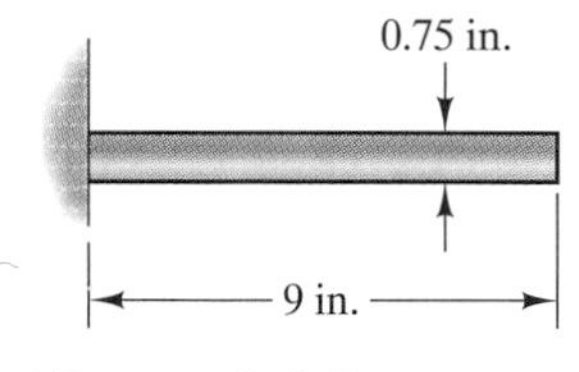

PROBLEM 3-2.5

3-2.6. What tensile force would have to be exerted on the right end of the bar in Problem 3-2.5 to increase its length to 9.02 in.? What is the bar's diameter after this load is applied?

3-2.7. A prismatic bar is 300 mm long and has a circular cross section with 20-mm diameter. Its modulus of elasticity is 120 GPa and its Poisson's ratio is 0.33. Axial forces P are applied to the ends of the bar that cause its diameter to decrease to 19.948 mm. **(a)** What is the length of the loaded bar? **(b)** What is the value of P?

3-2.8. The bar has modulus of elasticity $E = 30 \times 10^6$ psi, Poisson's ratio $\nu = 0.32$, and a circular cross section with diameter $D = 0.75$ in. There is a gap $b = 0.02$ in. between the right end of the bar and the rigid wall. If the bar is stretched so that it contacts the rigid wall and is welded to it, what is the bar's diameter afterward?

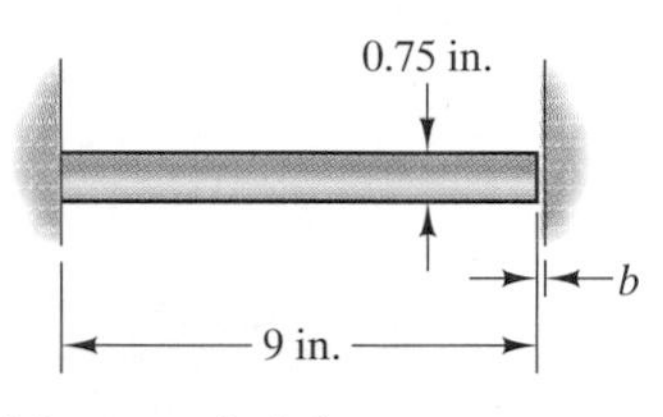

PROBLEM 3-2.8

3-2.9. After the bar in Problem 3-2.8 is welded to the rigid wall, what is the normal stress on a plane perpendicular to the bar's axis?

3-2.10. When unloaded, bars AB and AC are each 36 in. in length and have a cross-sectional area of 2 in^2. Their modulus of elasticity is $E = 1.6 \times 10^6$ psi. When the weight W is suspended at A, bar AB increases in length by 0.1 in. What is the change in length of bar AC?

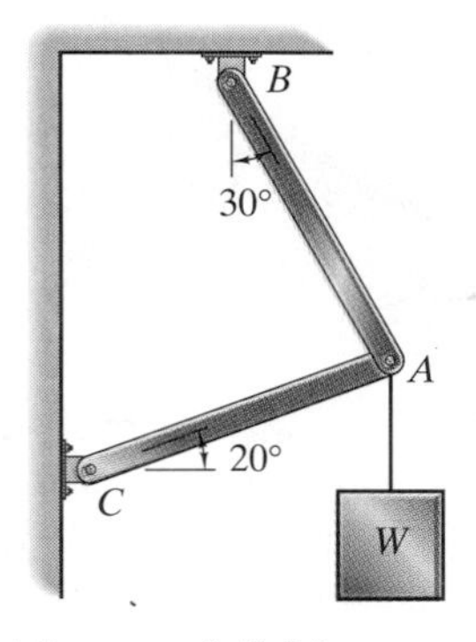

PROBLEM 3-2.10

3-2.11. If a weight $W = 12{,}000$ lb is suspended from the truss in Problem 3-2.10, what are the changes in length of the two bars?

3-2.12. Bars AB and AC are each 300 mm in length, have a cross-sectional area of 500 mm^2, and have modulus of elasticity $E = 72$ GPa. If a 24-kN downward force is applied at A, what is the resulting displacement of point A?

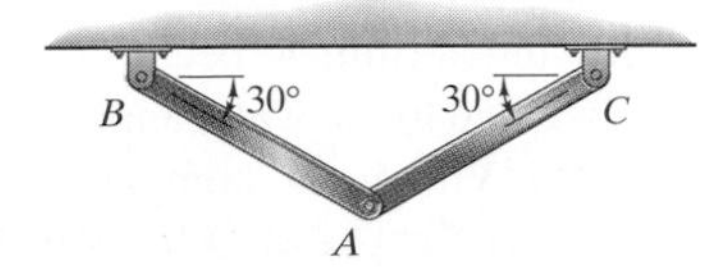

PROBLEM 3-2.12

3-2.13. Bars AB and AC of the truss shown in Problem 3-2.12 are each 300 mm in length, have a cross-sectional area of 500 mm^2, and are made of the same material. When a 30-kN downward force is applied at point A, it deflects downward 0.4 mm. What is the modulus of elasticity of the material?

3-2.14. Bar AB has cross-sectional area $A = 100$ mm^2 and modulus of elasticity $E = 102$ GPa. The distance $H = 400$ mm. If a 200-kN downward force is applied to bar CD at D, through what angle in degrees does bar CD rotate? (You can neglect the deformation of bar CD.) [*Strategy:* Because the change in length of bar AB is small, you can assume that the downward displacement v of point B is vertical and that the angle (in radians) through which bar CD rotates is v/H.]

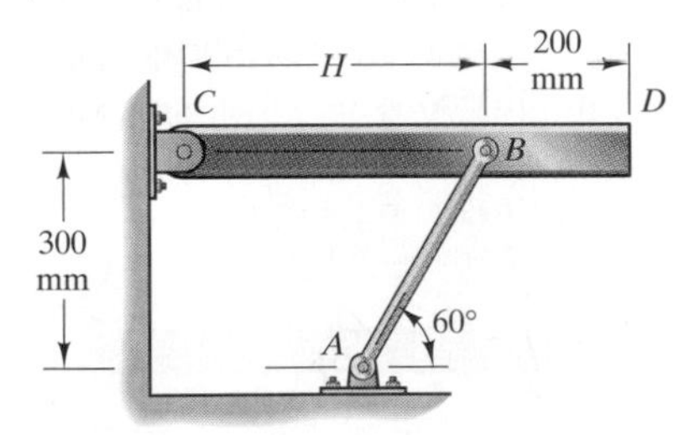

PROBLEM 3-2.14

3-2.15. Bar AB in Problem 3-2.14 is made of a material that will safely support a normal stress (in tension or compression) of 5 GPa. Based on this criterion, through what angle in degrees can bar CD safely be rotated relative to the position shown? (You can neglect the deformation of bar CD.)

3-2.16. If an upward force is applied at H that causes bar GH to rotate 0.02° in the counterclockwise direction, what are the axial strains in bars AB, CD, and EF? (You can neglect the deformation of bar GH.)

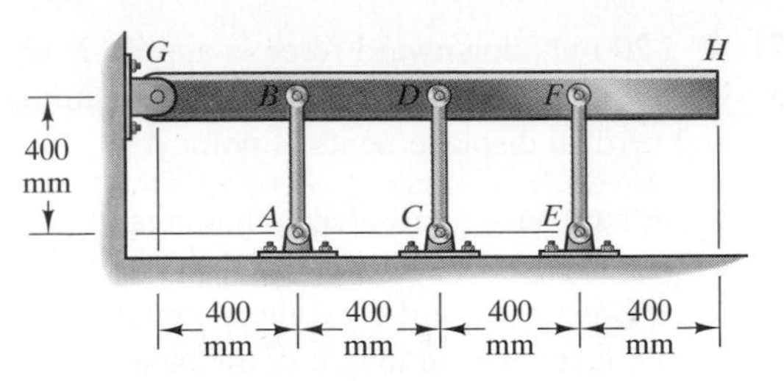

PROBLEM 3-2.16

3-2.17. Bars AB, CD, and EF in Problem 3-2.16 each has a cross-sectional area of 25 mm^2 and modulus of elasticity $E = 200$ GPa. What downward force applied at H would cause bar GH to rotate 0.02° in the clockwise direction?

3-2.18. Bar BE of the frame has cross-sectional area $A = 0.5$ in^2 and modulus of elasticity $E = 14 \times 10^6$ psi. If a 1000-lb downward force is applied at C, what is the resulting change in the length of bar BE?

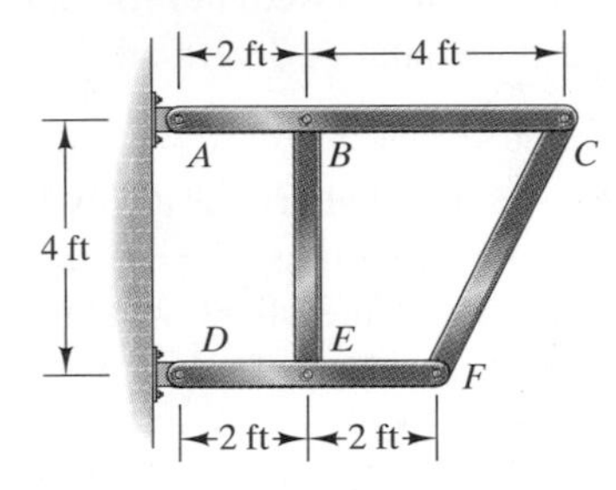

PROBLEM 3-2.18

3-2.19. Bar CF of the frame in Problem 3-2.18 has cross-sectional area $A = 0.5$ in^2 and modulus of elasticity $E = 14 \times 10^6$ psi. After a downward force is applied at C, its length is measured and determined to have decreased by 0.125 in. What force was applied at C?

3-2.20. Both bars have a cross-sectional area of 0.002 m^2 and modulus of elasticity $E = 70$ GPa. If an 80-kN downward force is applied at A, what are the resulting changes in length of the bars?

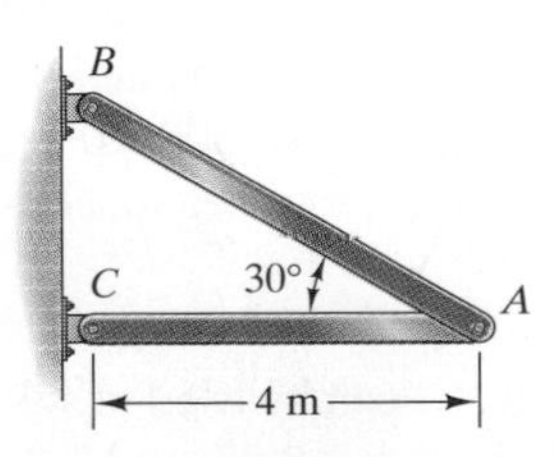

PROBLEM 3-2.20

3-2.21. If a 200-kN downward force is applied at point A of the system in Problem 3-2.20, what are the resulting horizontal and vertical displacements of point A?

3-2.22. Both bars have a cross-sectional area of 3 in^2 and modulus of elasticity $E = 12 \times 10^6$ lb/in^2. If a 40-kip horizontal force directed toward the right is applied at A, what are the resulting changes in length of the bars?

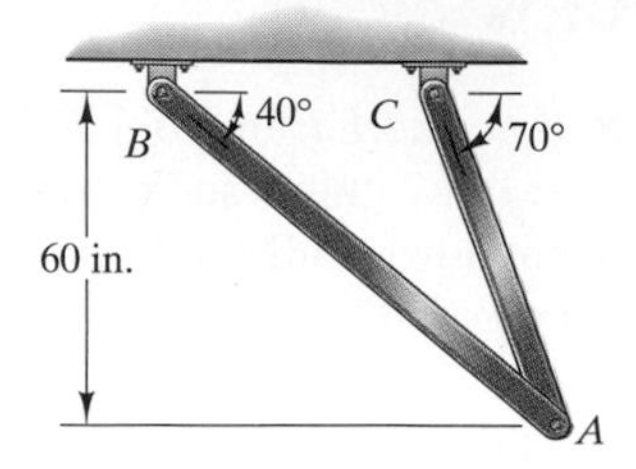

PROBLEM 3-2.22

3-2.23. If a 40-kip horizontal force directed toward the right is applied at point A of the system in Problem 3-2.22, what are the resulting horizontal and vertical displacements of point A?

3-2.24. The link AB of the pliers has a cross-sectional area of 40 mm^2 and elastic modulus $E = 210$ GPa. If forces $F = 150$ N are applied to the pliers, what is the change in length of link AB?

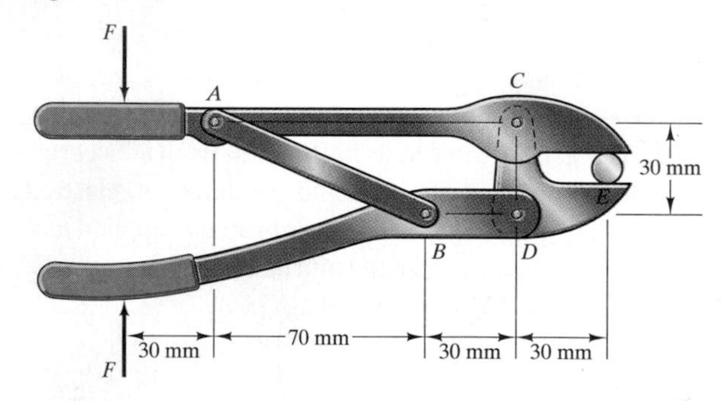

PROBLEM 3-2.24

3-2.25. Suppose that you want to design the pliers in Problem 3-2.24 so that forces F as large as 450 N can be applied. The link AB is to be made of a material that will support a compressive normal stress of 200 MPa. Based on this criterion, what minimum cross-sectional area must link AB have?

Worksheet 2 can be used to solve Problems 3-3.13, 3-3.14, 3-3.22, and 3-3.23.

3-3.1. The bar has cross-sectional area A and modulus of elasticity E. The left end of the bar is fixed. There is initially a gap b between the right end of the bar and the rigid wall (Fig. 1). The bar is stretched until it comes into contact with the rigid wall and is welded to it (Fig. 2). Notice that this problem is statically indeterminate because the axial force in the bar after it is welded to the wall cannot be determined from statics alone. **(a)** What is the compatibility condition in this problem? **(b)** What is the axial force in the bar after it is welded to the wall?

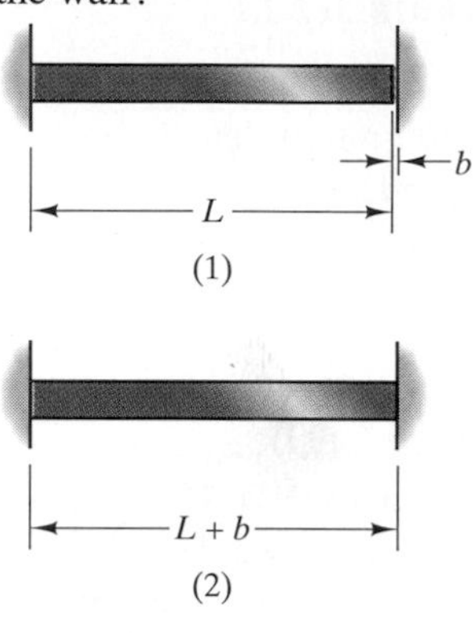

PROBLEM 3-3.1

3-3.2. The bar has cross-sectional area A and modulus of elasticity E. If an axial force F directed toward the right is applied at C, what is the normal stress in the part of the bar to the left of C? (*Strategy:* Draw the free-body diagram of the entire bar and write the equilibrium equation. Then apply the compatibility condition that the increase in length of the part of the bar to the left of C must equal the decrease in length of the part to the right of C.)

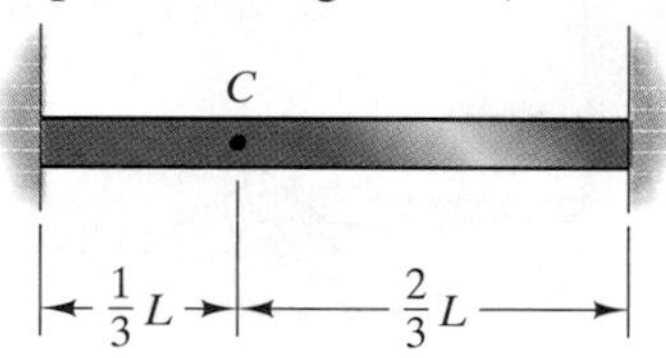

PROBLEM 3-3.2

3-3.3. In Problem 3-3.2, what is the resulting displacement of point C?

3-3.4. The bar in Problem 3-3.2 has cross-sectional area $A = 0.005$ m^2, modulus of elasticity $E = 72$ GPa, and $L = 1$ m. It is made of a material that will safely support a normal stress (in tension and compression) of 120 MPa. Based on this criterion, what is the largest axial force that can be applied at C?

3-3.5. In Example 3-7, determine the normal stresses in parts A and B of the bar if the force applied at the joint between parts A and B is **(a)** 40 kip; **(b)** 200 kip.

3-3.6. The bar has a circular cross section and modulus of elasticity $E = 70$ GPa. Parts A and C are 40 mm in diameter and part B is 80 mm in diameter. If $F_1 = 60$ kN and $F_2 = 30$ kN, what is the normal stress in part B?

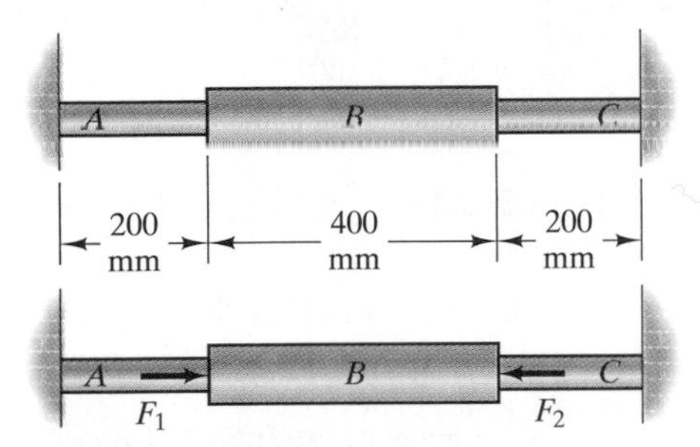

PROBLEM 3-3.6

3-3.7. In Problem 3-3.6, if $F_1 = 60$ kN, what force F_2 will cause the normal stress in part C to be zero?

3-3.8. The bar in Problem 3-3.6 consists of a material that will safely support a normal stress of 40 MPa. If $F_2 = 20$ kN, what is the largest safe value of F_1?

3-3.9. Two aluminum bars ($E_{Al} = 10.0 \times 10^6$ psi) are attached to a rigid support at the left and a cross-bar at the right. An iron bar ($E_{Fe} = 28.5 \times 10^6$ psi) is attached to the rigid support at the left and there is a gap b between the right end of the iron bar and the cross-bar. The cross-sectional area of each bar is $A = 0.5$ in^2 and $L = 10$ in. The iron bar is stretched until it contacts the cross-bar and welded to it. Afterward, the axial strain of the iron bar is measured and determined to be $\epsilon_{Fe} = 0.002$. What was the size of the gap b?

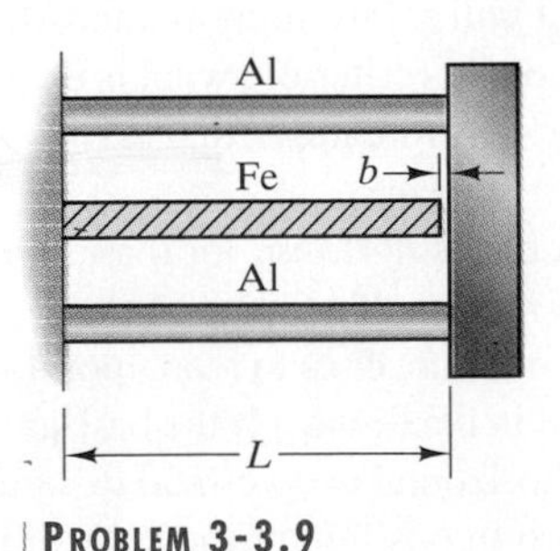

PROBLEM 3-3.9

3-3.10. In Problem 3-3.9, the iron will safely support a tensile stress of 100 ksi and the aluminum will safely support a compressive stress of 40 ksi. What is the largest safe value of the gap b?

3-3.11. Bars AB and AC each have cross-sectional area A and modulus of elasticity E. If a downward force F is applied at A, show that the resulting downward displacement of point A is

$$\frac{Fh}{EA}\left(\frac{1}{1+\cos^3\theta}\right)$$

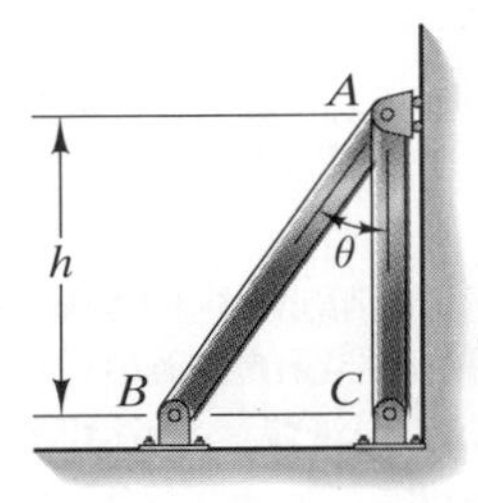

PROBLEM 3-3.11

3-3.12. If a downward force F is applied at point A of the system shown in Problem 3-3.11, what are the resulting normal stresses in bars AB and AC?

3-3.13. Each bar has a 500-mm^2 cross-sectional area and modulus of elasticity $E = 72$ GPa. If a 160-kN downward force is applied at A, what is the resulting displacement of point A?

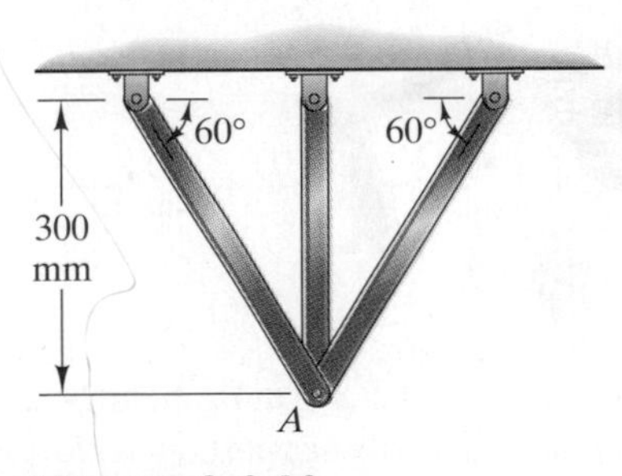

PROBLEM 3-3.13

3-3.14. The bars in Problem 3-3.13 are made of material that will safely support a tensile stress of 270 MPa. Based on this criterion, what is the largest downward force that can safely be applied at A?

3-3.15. Each bar has a 500-mm^2 cross-sectional area and modulus of elasticity $E = 72$ GPa. If there is a gap $h = 2$ mm between the hole in the vertical bar and the pin A connecting bars AB and AD, what are the normal stresses in the three bars after the vertical bar is connected to the pin at A?

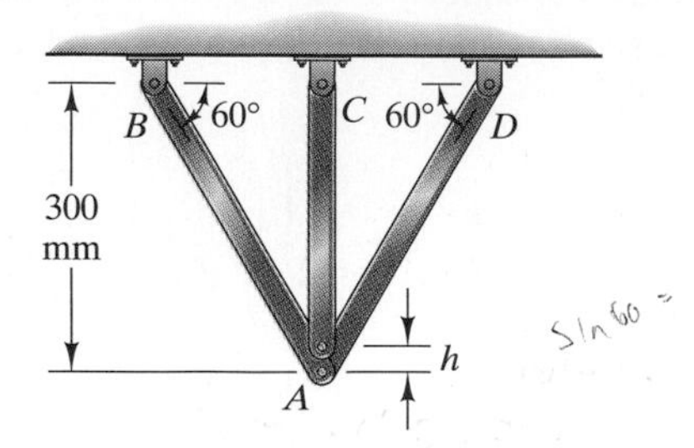

PROBLEM 3-3.15

3-3.16. The bars in Problem 3-3.15 are made of material that will safely support a normal stress (tension or compression) of 400 MPa. Based on this criterion, what is the largest safe value of the gap h?

3-3.17. Bars AB, CD, and EF each have a cross-sectional area of 25 mm^2 and modulus of elasticity $E = 200$ GPa. If a 5-kN upward force is applied at H, what are the normal stresses in the three bars? (You can neglect the deformation of bar GH.)

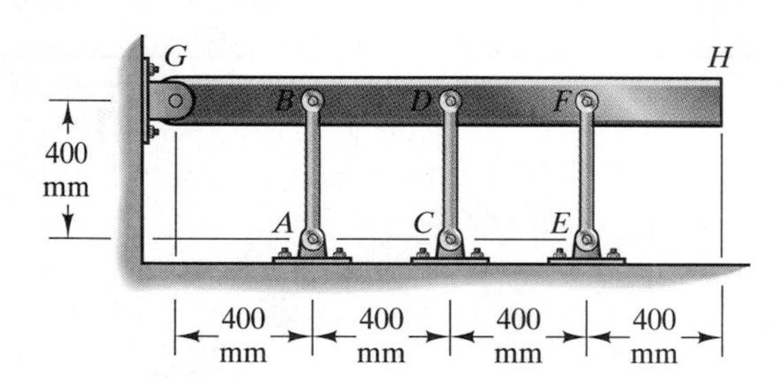

PROBLEM 3-3.17

3-3.18. Bars AB, CD, and EF in Problem 3-3.17 are made of a material that will safely support a tensile normal stress of 340 MPa. Based on this criterion, what is the largest upward force that can safely be applied at H?

3-3.19. Bars AB and AC have cross-sectional area $A = 100$ mm^2, modulus of elasticity $E = 102$ GPa, and are pinned at A. If a 200-kN downward force is applied to bar DE at E, through what angle in degrees does bar DE rotate? (You can neglect the deformation of bar DE.)

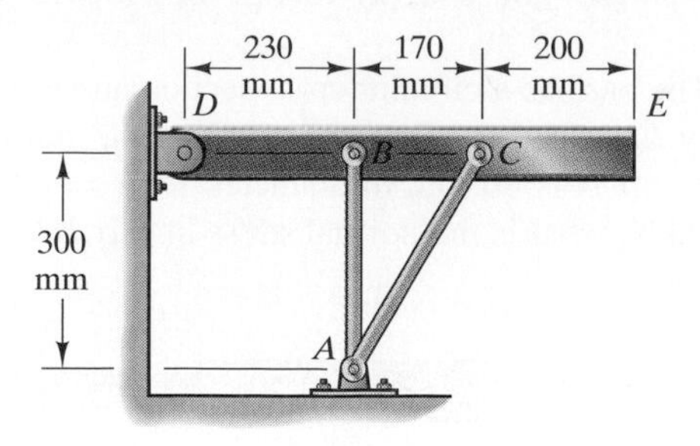

PROBLEM 3-3.19

3-3.20. In Problem 3-3.19, what are the normal stresses in the bars AB and AC when the 200-kN downward force is applied to bar DE at E?

3-3.21. In Example 3-9, what are the normal stresses in the three bars?

3-3.22. Each bar has a cross-sectional area of 3 in^2 and modulus of elasticity $E = 12 \times 10^6$ lb/in^2. If a 40-kip horizontal force directed toward the right is applied at A, what are the normal stresses in the bars?

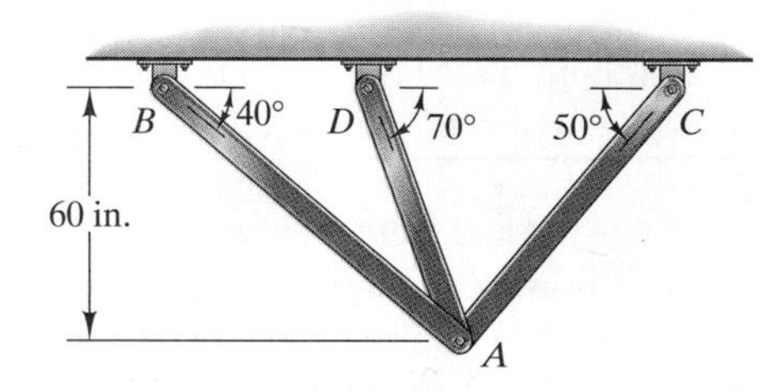

PROBLEM 3-3.22

3-3.23. The bars of the system in Problem 3-3.22 consist of a material that will safely support a tensile normal stress of 20 ksi. Based on this criterion, what is the largest *downward* force that can safely be applied at A?

3-3.24. Each bar is 400 mm long and has cross-sectional area $A = 100$ mm^2 and modulus of elasticity $E = 102$ GPa. Each bar is 2 mm too short to reach point G. (This distance is exaggerated in the figure.) If the bars are pinned together, what are the horizontal and vertical distances from point G to the equilibrium position of the ends of the bars?

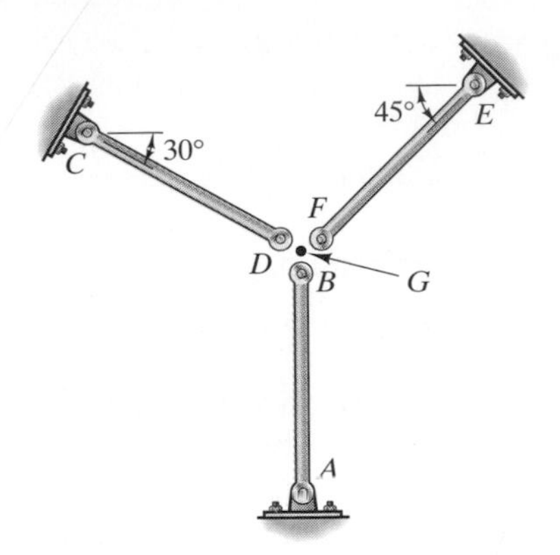

PROBLEM 3-3.24

3-3.25. In Problem 3-3.24, what are the normal stresses in the three bars after they are pinned together?

3-4.1. The bar's cross-sectional area is $A = (1 + 0.1x)\ \text{in}^2$ and the modulus of elasticity of the material is $E = 12 \times 10^6$ psi. If the bar is subjected to tensile axial forces $P = 20$ kip at its ends, what is the normal stress at $x = 6$ in.?

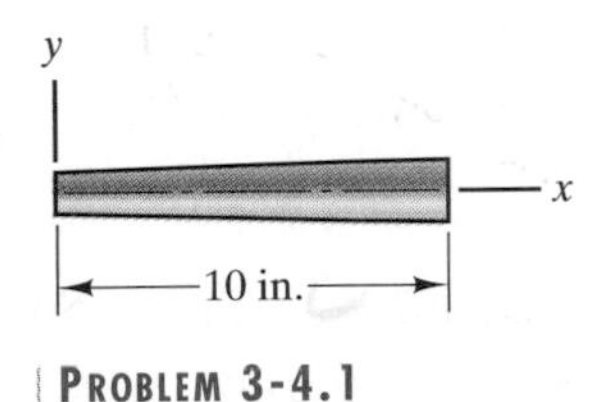

PROBLEM 3-4.1

3-4.2. What is the change in length of the bar in Problem 3-4.1?

3-4.3. The cross-sectional area of the bar in Problem 3-4.1 is $A = (1 + ax)\ \text{in}^2$, where a is a constant, and the modulus of elasticity of the material is $E = 8 \times 10^6$ psi. When the bar is subjected to tensile axial forces $P = 14$ kip at its ends, its change in length is $\delta = 0.01$ in. What is the value of the constant a? (*Strategy:* Estimate the value of a by drawing a graph of δ as a function of a.)

3-4.4. From $x = 0$ to $x = 100$ mm, the bar's height is 20 mm. From $x = 100$ mm to $x = 200$ mm, its height varies linearly from 20 to 40 mm. From $x = 200$ mm to $x = 300$ mm, its height is 40 mm. The flat bar's thickness is 20 mm. The modulus of elasticity of the material is $E = 70$ GPa. If the bar is subjected to tensile axial forces $P = 50$ kN at its ends, what is its change in length?

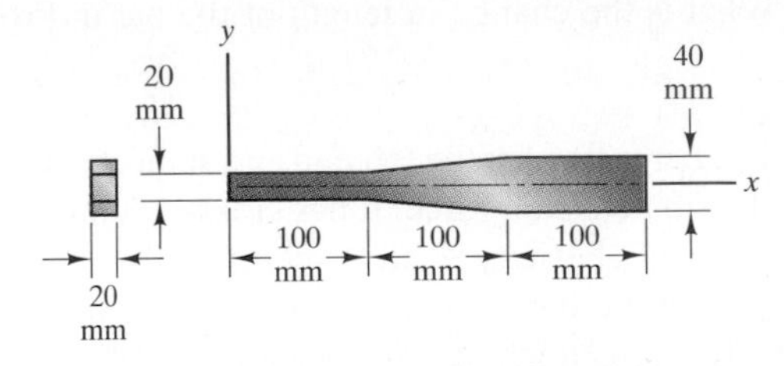

PROBLEM 3-4.4

3-4.5. From $x = 0$ to $x = 10$ in., the bar's cross-sectional area is $A = 1\ \text{in}^2$. From $x = 10$ in. to $x = 20$ in., $A = (0.1x)\ \text{in}^2$. The modulus of elasticity of the material is $E = 12 \times 10^6$ psi. There is a gap $b = 0.02$ in. between the right end of the bar and the rigid wall. If the bar is stretched so that it contacts the rigid wall and is welded to it, what is the axial force in the bar afterward?

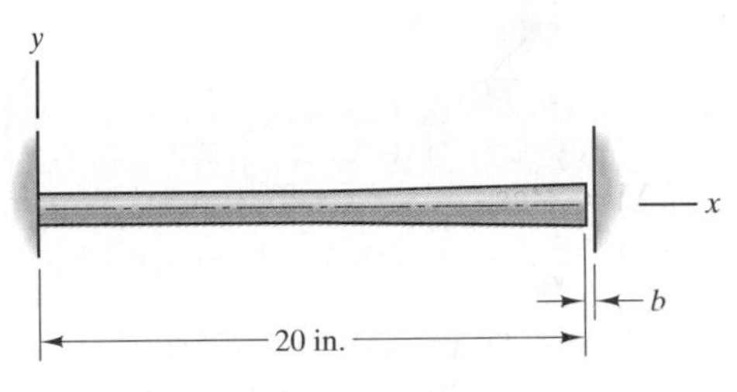

PROBLEM 3-4.5

3-4.6. From $x = 0$ to $x = 10$ in., the cross-sectional area of the bar shown in Problem 3-4.5 is $A = 1\ \text{in}^2$. From $x = 10$ in. to $x = 20$ in., $A = (0.1x)\ \text{in}^2$. The modulus of elasticity of the material is $E = 12 \times 10^6$ psi. There is a gap $b = 0.02$ in. between the right end of the bar and the rigid wall. If a 40-kip axial force toward the right is applied to the bar at $x = 10$ in., what is the resulting normal stress in the left half of the bar?

3-4.7. The diameter of the bar's circular cross section varies linearly from 10 mm at its left end to 20 mm at its right end. The modulus of elasticity of the material is $E = 45$ GPa. If the bar is subjected to tensile axial forces $P = 6$ kN at its ends, what is the normal stress at $x = 80$ mm?

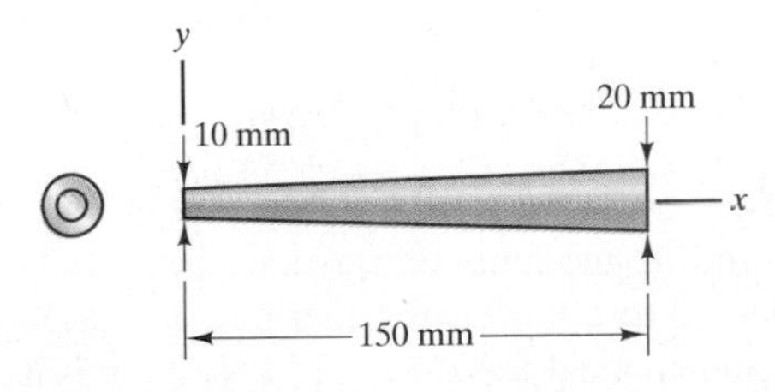

PROBLEM 3-4.7

3-4.8. What is the change in length of the bar in Problem 3-4.7?

3-4.9. The bar is fixed at the left end and subjected to a uniformly distributed axial force. It has cross-sectional area A and modulus of elasticity E. **(a)** Determine the internal axial force P in the bar as a function of x. **(b)** What is the bar's change in length?

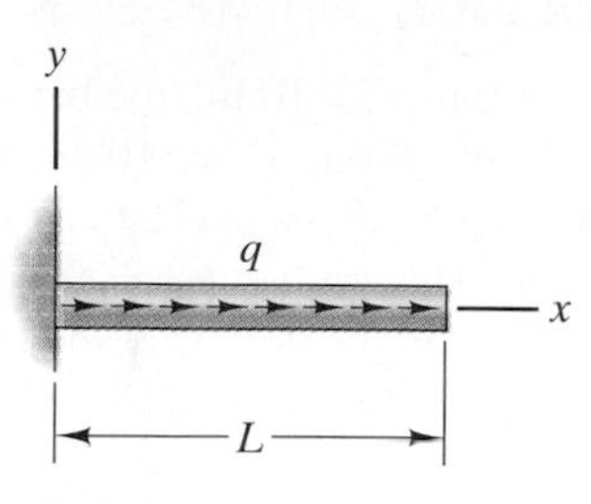

| **PROBLEM 3-4.9**

3-4.10. The bar shown in Problem 3-4.9 has length $L = 2$ m, cross-sectional area $A = 0.03\ \text{m}^2$, and modulus of elasticity $E = 200$ GPa. It is subjected to a distributed axial force $q = 12(1 + 0.4x)$ MN/m. What is the bar's change in length?

3-4.11. A cylindrical bar with 1-in. diameter fits tightly in a circular hole in a 5-in. thick plate. The modulus of elasticity of the material is $E = 14 \times 10^6$ psi. A 1000-lb tensile force is applied at the left end of the bar, causing it to begin slipping out of the hole. At the instant slipping begins, determine **(a)** the magnitude of the uniformly distributed axial force exerted on the bar by the plate; **(b)** the total change in the bar's length.

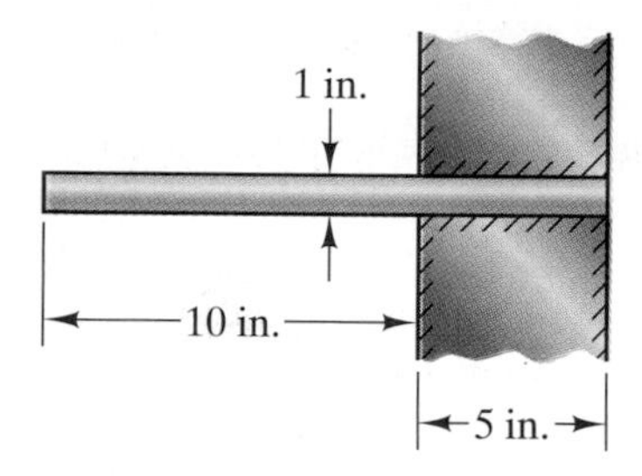

| **PROBLEM 3-4.11**

3-4.12. The bar has a circular cross section with 0.002-m diameter and its modulus of elasticity is $E = 86.6$ GPa. It is subjected to a uniformly distributed axial force $q = 75$ kN/m and an axial force $F = 15$ kN. What is its change in length?

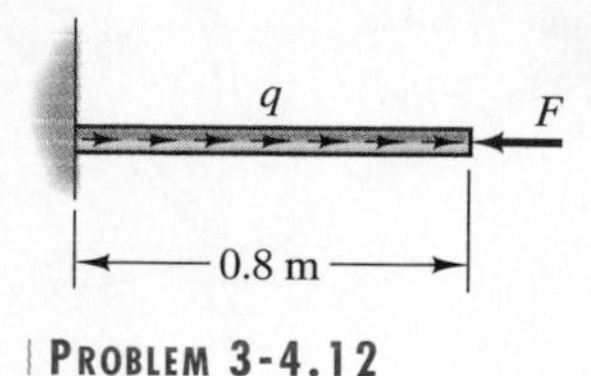

| **PROBLEM 3-4.12**

3-4.13. In Problem 3-4.12, what axial force F would cause the bar's change in length to be zero?

3-4.14. If the bar in Problem 3-4.12 is subjected to a distributed force $q = 75(1 + 0.2x)$ kN/m and an axial force $F = 15$ kN, what is its change in length?

3-4.15. The bar is fixed at A and B and is subjected to a uniformly distributed axial force. It has cross-sectional area A and modulus of elasticity E. What are the reactions at A and B?

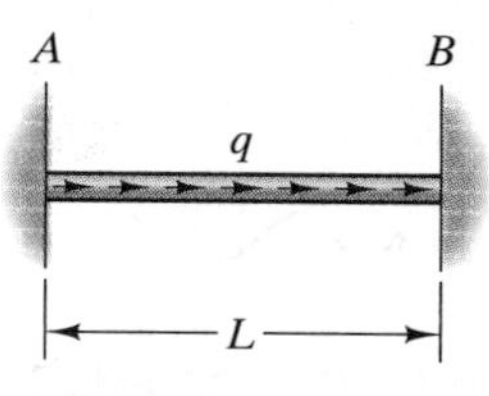

| **PROBLEM 3-4.15**

3-4.16. What point of the bar in Problem 3-4.15 undergoes the largest displacement, and what is the displacement?

3-4.17. The bar has a cross-sectional area of $0.0025\ \text{m}^2$ and modulus of elasticity $E = 200$ GPa. The bar is fixed at both ends and is subjected to a distributed axial force $q = 80x^2$ kN/m. What are the maximum tensile and compressive stresses in the bar?

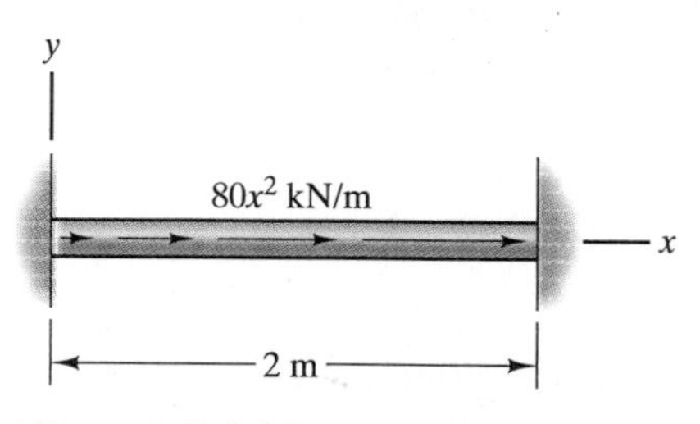

| **PROBLEM 3-4.17**

3-4.18. What point of the bar in Problem 3-4.17 undergoes the largest displacement, and what is the displacement?

3-4.19. The bar has cross-sectional area A and modulus of elasticity E. It is fixed at A and there is a gap b between the right end and the rigid wall B. If the bar is subjected to a uniformly distributed axial force q directed toward the right, what force is exerted on the bar by the wall B?

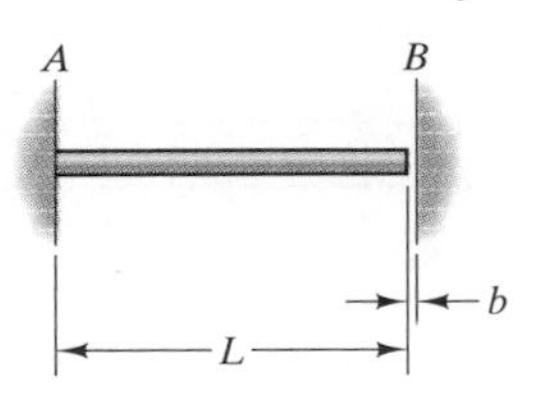

PROBLEM 3-4.19

3-4.20. Bar A has a cross-sectional area of 2 in^2 and its modulus of elasticity is 10×10^6 psi. Bar B has a cross-sectional area of 4 in^2 and its modulus of elasticity is 16×10^6 psi. There is a gap $b = 0.01$ in. between the bars. A uniformly distributed load of 20 kip/in. directed to the right is applied to bar A, and a uniformly distributed load of 20 kip/in. directed to the left is applied to bar B. What are the maximum tensile stresses in the two bars?

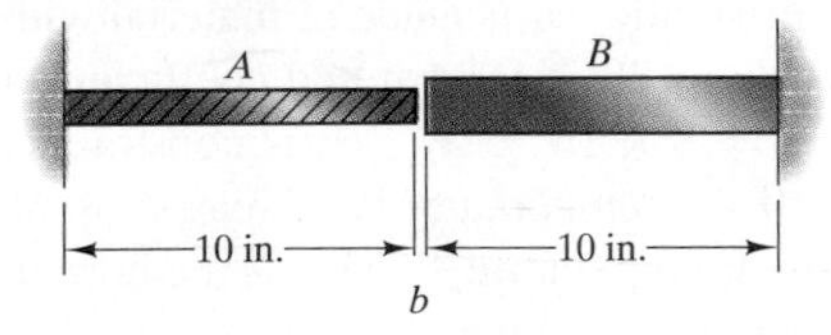

PROBLEM 3-4.20

3-4.21. A conical indentor is used to test samples of soil. A force pushes the indentor into the soil at a constant rate. The soil exerts a distributed axial force on the indentor that is proportional to the circumference of the indentor, which means that the distributed force is given by a linear equation $q = cx$. **(a)** Determine the value of c in terms of F and d. **(b)** Determine the average normal stress in the indentor as a function of x for $0 < x < d$.

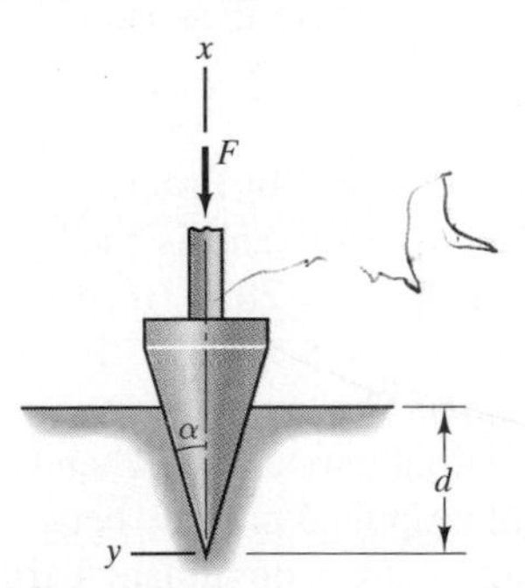

PROBLEM 3-4.21

3-4.22. Determine the change in length of the part of the indentor in Problem 3-4.21 that is below the surface of the soil.

3-4.23. The column is suspended and loaded only by its own weight. The mass density of the material is $\rho = 2300$ kg/m^3. (The weight density is $g\rho = 9.81\rho$ N/m^3.) The column's width varies linearly from 1 m at the bottom to 2 m at the top. What is the normal stress at $x = 3$ m?

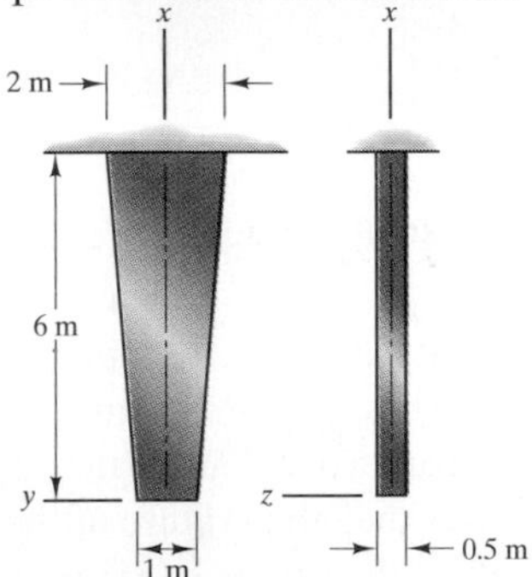

PROBLEM 3-4.23

3-4.24. What is the change in length of the column in Problem 3-4.23 if $E = 25$ GPa?

3-4.25. Suppose that you want to design a column to support a uniform compressive stress σ_0. If the column's cross-sectional area were uniform, the compressive stress at increasing distances x from the top would increase due to the column's weight. You can optimize your use of material by varying the column's cross-sectional area in such a way that the compressive stress at *every* position x equals σ_0. Prove that this is accomplished if the column's cross-sectional area is

$$A = A_0 e^{(\gamma/\sigma_0)x},$$

where A_0 is the cross-sectional area at $x = 0$ and γ is the weight density of the material.

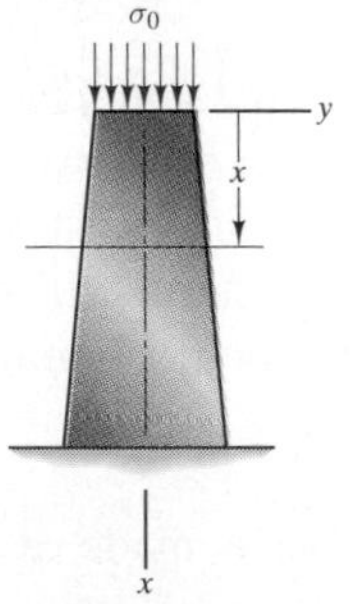

PROBLEM 3-4.25

3-5.1. A line L within an unconstrained sample of material is 200 mm long. The coefficient of thermal expansion of the material is $\alpha = 22 \times 10^{-6}\ {}^\circ\mathrm{C}^{-1}$. If the temperature of the material is increased by 30°C, what is the length of the line?

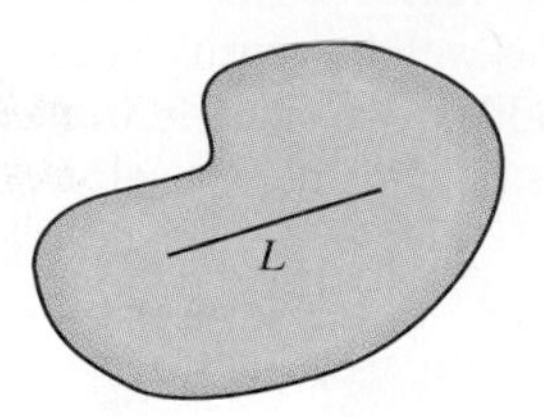

PROBLEM 3-5.1

3-5.2. The length of the line L within the unconstrained sample of material shown in Problem 3-5.1 is 2 in. The coefficient of thermal expansion of the material is $\alpha = 8 \times 10^{-6}\ {}^\circ\mathrm{F}^{-1}$. After the temperature of the material is increased, the length of the line is 2.002 in. How much was the temperature increased?

3-5.3. Consider a 1 in. × 1 in. × 1 in. cube within an unconstrained sample of material. The coefficient of thermal expansion of the material is $\alpha = 14 \times 10^{-6}\ {}^\circ\mathrm{F}^{-1}$. If the temperature of the material is decreased by 40°F, what is the volume of the cube?

3-5.4. A prismatic bar is 200 mm long and has a circular cross section with 30-mm diameter. After the temperature of the unconstrained bar is increased, its length is measured and determined to be 200.160 mm. What is the bar's diameter after the increase in temperature?

3-5.5. If the increase in temperature in Problem 3-5.4 is 20°C, what is the coefficient of thermal expansion of the bar?

3-5.6. If the increase in temperature in Problem 3-5.4 is 20°C and the modulus of elasticity of the material is $E = 72$ GPa, what is the normal stress on a plane perpendicular to the bar's axis after the increase in temperature? (*Strategy:* Obtain a free-body diagram by passing a plane through the bar.)

3-5.7. The prismatic bar is made of material with modulus of elasticity $E = 28 \times 10^6$ psi and coefficient of thermal expansion $\alpha = 8 \times 10^{-6}\ {}^\circ\mathrm{F}^{-1}$. The temperature of the unconstrained bar is increased by 50°F above its initial temperature T. **(a)** What is the change in the bar's length? **(b)** What is the change in the bar's diameter? **(c)** What is the normal stress on a plane perpendicular to the bar's axis after the increase in temperature?

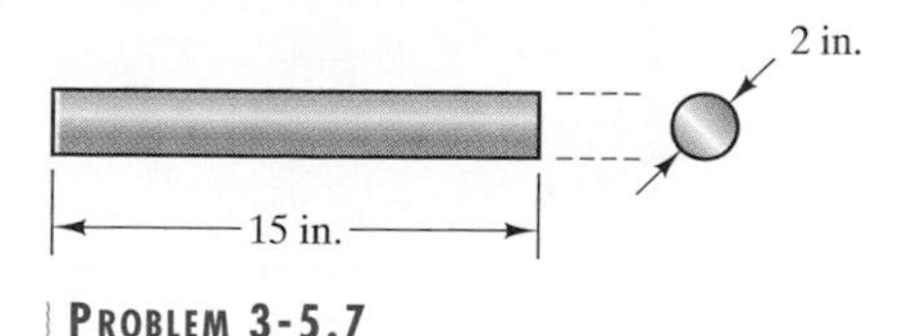

PROBLEM 3-5.7

3-5.8. Suppose that the temperature of the unconstrained bar in Problem 3-5.7 is increased by 50°F above its initial temperature T and the bar is also subjected to 30,000-lb tensile axial forces at the ends. What is the resulting change in the bar's length? Determine the change in length assuming that **(a)** the temperature is first increased and then the axial forces are applied; **(b)** the axial forces are first applied and then the temperature is increased.

3-5.9. The prismatic bar is made of material with modulus of elasticity $E = 28 \times 10^6$ psi and coefficient of thermal expansion $\alpha = 8 \times 10^{-6}\ {}^\circ\mathrm{F}^{-1}$. It is constrained between rigid walls. If the temperature is increased by 50°F above the bar's initial temperature T, what is the normal stress on a plane perpendicular to the bar's axis?

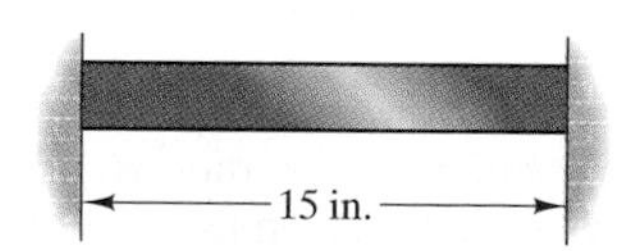

PROBLEM 3-5.9

3-5.10. The walls between which the prismatic bar in Problem 3-5.9 is constrained will safely support a compressive normal stress of 30,000 psi. Based on this criterion, what is the largest safe temperature increase to which the bar can be subjected?

3-5.11. The prismatic bar in Problem 3-5.9 has a cross-sectional area $A = 3\ \mathrm{in}^2$ and is made of material with modulus of elasticity $E = 28 \times 10^6$ psi and coefficient of thermal expansion $\alpha = 8 \times 10^{-6}\ {}^\circ\mathrm{F}^{-1}$. It is constrained between rigid walls. The temperature is increased by 50°F above the bar's initial temperature T and a 20,000-lb axial force to the right is applied midway between the two walls. What is the normal stress on a plane perpendicular to the bar's axis to the right of the point where the force is applied?

3-5.12. The prismatic bar is made of material with modulus of elasticity $E = 28 \times 10^6$ psi and coefficient of thermal expansion $\alpha = 8 \times 10^{-6}$ °F^{-1}. It is fixed to a rigid wall at the left. There is a gap $b = 0.002$ in. between the bar's right end and the rigid wall. If the temperature is increased by 50°F above the bar's initial temperature T, what is the normal stress on a plane perpendicular to the bar's axis?

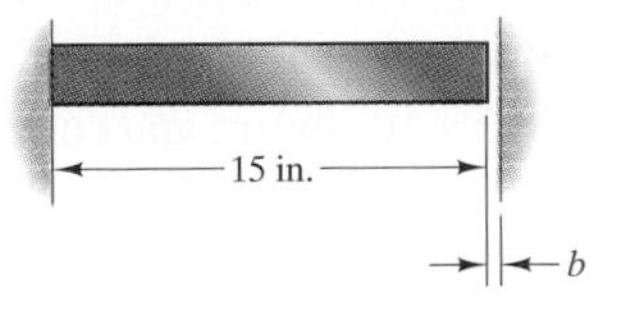

PROBLEM 3-5.12

3-5.13. Bar A has a cross-sectional area of 0.04 m^2, modulus of elasticity $E = 70$ GPa, and coefficient of thermal expansion $\alpha = 14 \times 10^{-6}$ °C^{-1}. Bar B has a cross-sectional area of 0.01 m^2, modulus of elasticity $E = 120$ GPa, and coefficient of thermal expansion $\alpha = 16 \times 10^{-6}$ °C^{-1}. There is a gap $b = 0.4$ mm between the ends of the bars. What minimum increase in the temperature of the bars above their initial temperature T is necessary to cause them to come into contact?

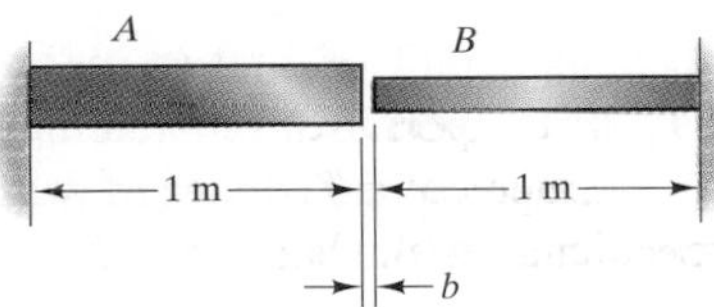

PROBLEM 3-5.13

3-5.14. If the temperature of the bars in Problem 3-5.13 is increased by 40°C above their initial temperature T, what are the normal stresses in the bars?

3-5.15. Each bar has a 2-in^2 cross-sectional area, modulus of elasticity $E = 14 \times 10^6$ psi, and coefficient of thermal expansion $\alpha = 11 \times 10^{-6}$ °F^{-1}. If their temperature is increased by 40°F from their initial temperature T, what is the resulting displacement of point A?

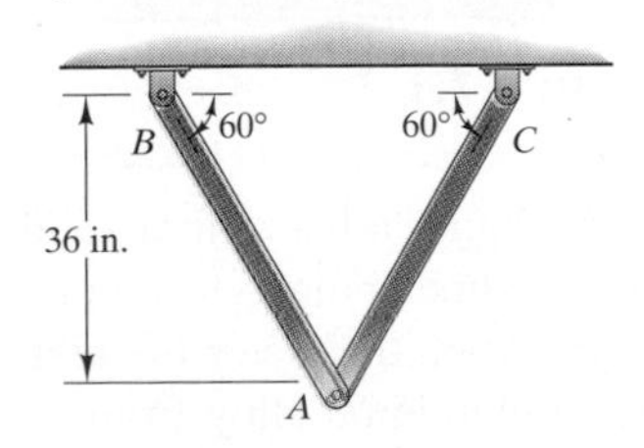

PROBLEM 3-5.15

3-5.16. If the temperature of the bars in Problem 3-5.15 is decreased by 30°F from their initial temperature T, what force would need to be applied at A so that the total displacement of point A caused by the temperature change and the force is zero?

3-5.17. Both bars have cross-sectional area 3 in^2, modulus of elasticity $E = 12 \times 10^6$ lb/in^2, and coefficient of thermal expansion $\alpha = 6.6 \times 10^{-6}$ °F^{-1}. If a 40-kip horizontal force directed toward the right is applied at A and their temperature is increased by 30°F from their initial temperature T, what are the resulting changes in length of the bars?

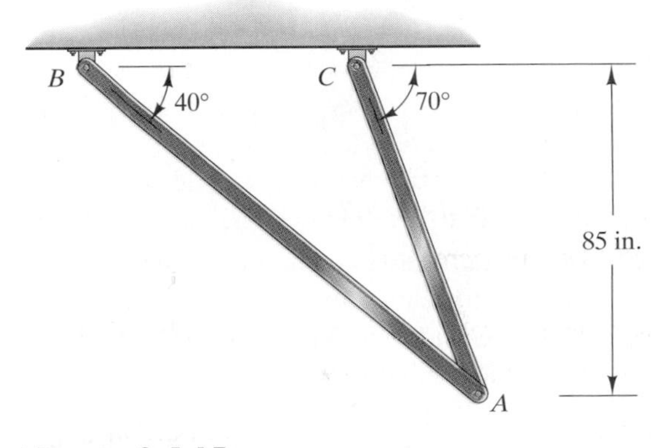

PROBLEM 3-5.17

3-5.18. In Problem 3-5.17, what are the resulting horizontal and vertical displacements of point A?

3-5.19. Both bars have cross-sectional area 0.002 m^2, modulus of elasticity $E = 70$ GPa, and coefficient of thermal expansion $\alpha = 23 \times 10^{-6}$ °C^{-1}. If the temperature of the bars is increased by 30°C from their initial temperature T, what are the resulting horizontal and vertical displacements of joint A?

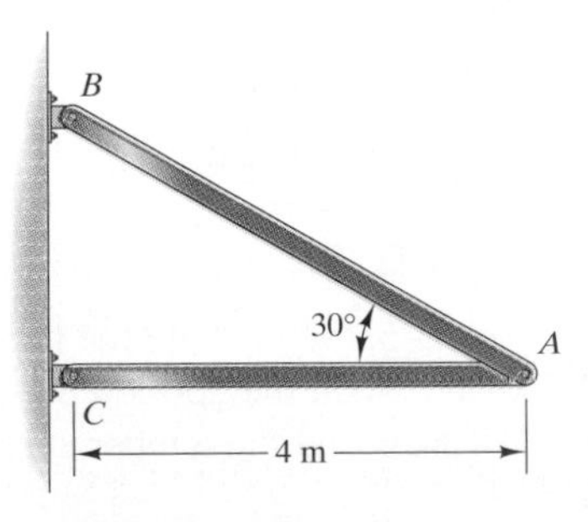

PROBLEM 3-5.19

3-5.20. If the temperature of the bars in Problem 3-5.19 is increased by 30°C from their initial temperature T and an 80-kN downward force is applied at joint A, what are the resulting horizontal and vertical displacements of joint A?

3-5.21. Each bar has a 500-mm^2 cross-sectional area, modulus of elasticity $E = 72$ GPa, and coefficient of thermal expansion $\alpha = 25 \times 10^{-6}\ ^\circ\text{C}^{-1}$. The normal stresses in the bars are initially zero. If their temperature is increased by 20°C, what are the normal stresses in the bars?

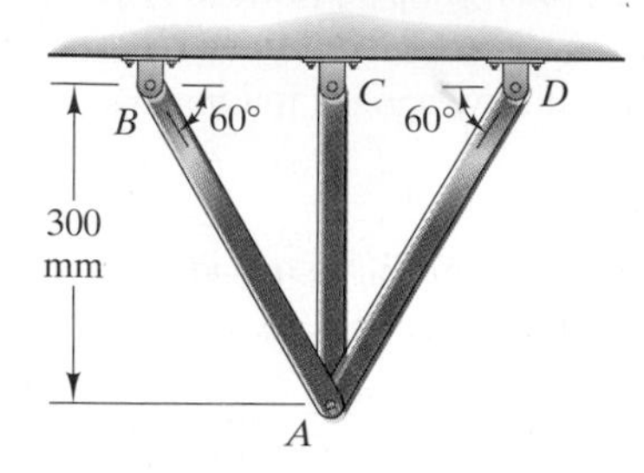

| **PROBLEM 3-5.21**

3-5.22. The normal stresses in the bars of the system shown in Problem 3-5.21 are initially zero. If their temperature is increased by 30°C and an 80-kN downward force is applied at A, what are the normal stresses in the bars?

3-5.23. Bars AB and AC have cross-sectional area $A = 100$ mm^2, modulus of elasticity $E = 102$ GPa, and coefficient of thermal expansion $\alpha = 18 \times 10^{-6}\ ^\circ\text{C}^{-1}$. The two bars are pinned at A. If a 200-kN downward force is applied to bar DE at E and the temperature is lowered by 40°C, through what angle in degrees does bar DE rotate? (You can neglect the deformation of bar DE.)

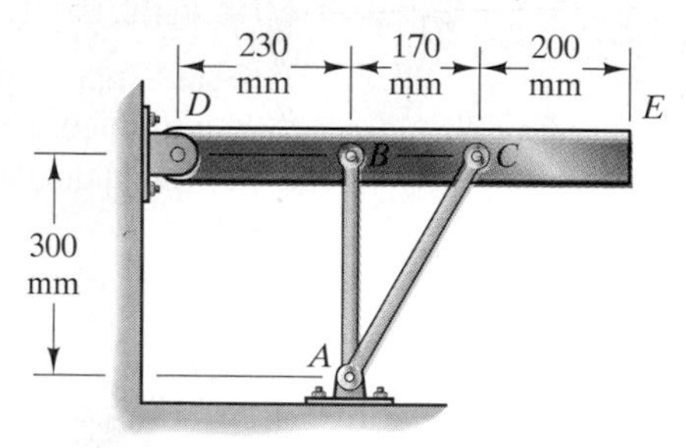

| **PROBLEM 3-5.23**

3-5.24. In Problem 3-5.23, what are the normal stresses in the bars AB and AC when the 200-kN downward force is applied to bar DE at E and the temperature is lowered by 40°C?

3-5.25. If a 12-kN downward force is applied to bar DE of the system shown in Problem 3-5.23 at E, what increase in temperature would be necessary for bar DE to be horizontal?

3-5.26. In Example 3-14, what are the normal stresses in the three bars?

3-5.27. Each bar has a cross-sectional area of 3 in^2, modulus of elasticity $E = 12 \times 10^6$ lb/in^2, and coefficient of thermal expansion $\alpha = 6 \times 10^{-6}\ ^\circ\text{F}^{-1}$. The normal stresses in the bars are initially zero. If their temperature is increased by 80°F above their initial temperature T, what are the normal stresses in the bars?

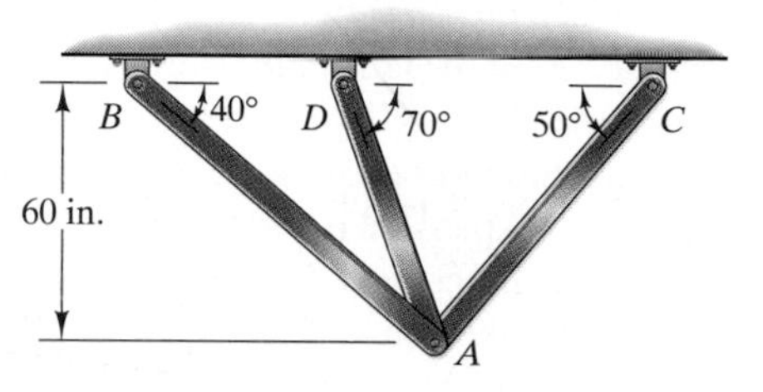

| **PROBLEM 3-5.27**

3-5.28. If the temperature of the bars in Problem 3-5.27 is increased by 120°F above their initial temperature T and a 20-kip horizontal force directed toward the right is applied at A, what are the normal stresses in the bars?

Worksheet 3 can be used to solve Problems 3-7.5, 3-7.6, 3-7.9, and 3-7.10.

3-7.1. You are designing a bar with a solid circular cross section that is to support a 4-kN tensile axial load. You have decided to use 6061-T6 aluminum alloy (see Appendix B), and you want the factor of safety to be $S = 2$. Based on this criterion, what should the bar's diameter be?

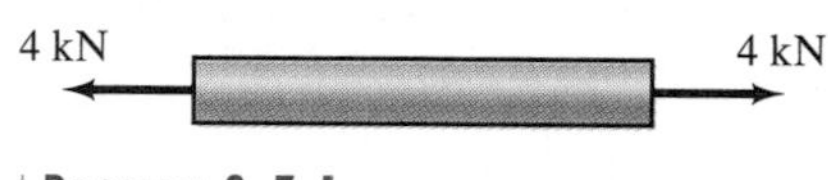

| **PROBLEM 3-7.1**

3-7.2. You are designing a bar with a solid circular cross section with 5-mm diameter that is to support a 4-kN tensile axial load, and you want the factor of safety to be at least $S = 2$. Choose an aluminum alloy from Appendix B that satisfies this requirement.

3-7.3. You are designing a bar with a solid circular cross section that is to support a 4000-lb tensile axial load. You have decided to use ASTM-A572 structural steel (see Appendix B), and you want the factor of safety to be $S = 1.5$. Based on this criterion, what should the bar's diameter be?

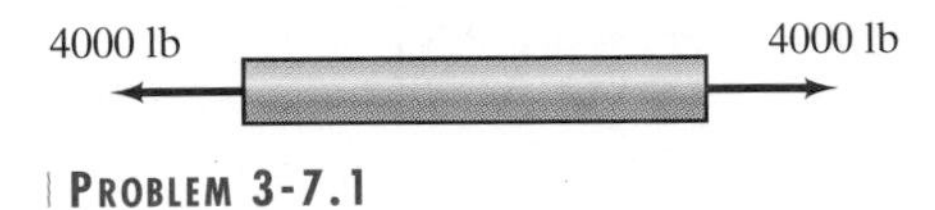

PROBLEM 3-7.1

3-7.4. You are designing a bar with a solid circular cross section with $\frac{1}{2}$-in. diameter that is to support a 4000-lb tensile axial load, and you want the factor of safety to be at least $S = 3$. Choose a structural steel from Appendix B that satisfies this requirement.

3-7.5. The horizontal beam of length $L = 2$ m supports a load $F = 30$ kN. The beam is supported by a pin support and the brace BC. The dimension $h = 0.54$ m. Suppose that you want to make the brace out of existing stock that has cross-sectional area $A = 0.0016\ \text{m}^2$ and yield stress $\sigma_Y = 400$ MPa. If you want the brace to have a factor of safety $S = 1.5$, what should the angle θ be?

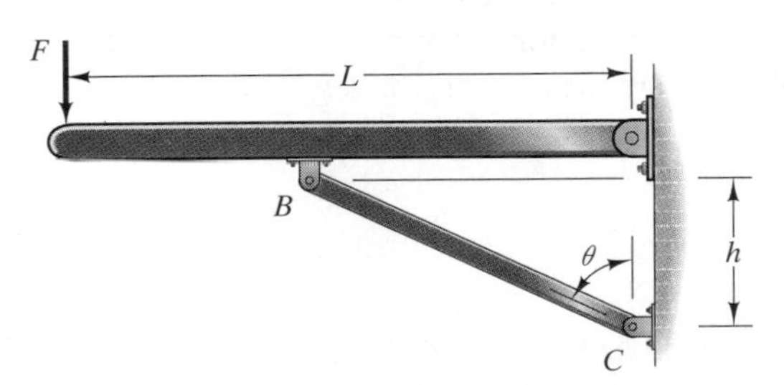

PROBLEM 3-7.5

3-7.6. Consider the system shown in Problem 3-7.5. The horizontal beam of length $L = 4$ ft supports a load $F = 20$ kip. The beam is supported by a pin support and the brace BC. The dimension $h = 1$ ft and the angle $\theta = 60°$. Suppose that you want to make the brace out of existing stock that has yield stress $\sigma_Y = 50$ ksi. If you want to design the brace BC to have a factor of safety $S = 2$, what should its cross-sectional area be?

3-7.7. The horizontal beam shown in Problem 3-7.5 is of length L and supports a load F. The beam is supported by a pin support and the brace BC. Suppose that the brace is to consist of a specified material for which you have chosen an allowable stress σ_{allow}, and you want to design the brace so that its weight is a minimum. You can do this by assuming that the brace is subjected to the allowable stress and choosing the angle θ so that the volume of the brace is a minimum. What is the necessary angle θ?

3-7.8. In Problem 3-7.7, draw a graph showing the dependence of the volume of the brace on the angle θ for $5° \leq \theta \leq 85°$. Notice that the graph is relatively flat near the optimum angle, meaning that the designer can choose θ within a range of angles near the optimum value and still obtain a nearly optimum design.

3-7.9. The truss is a preliminary design for a structure to attach one end of a stretcher to a rescue helicopter. Based on dynamic simulations, the design engineer estimates that the downward forces the stretcher will exert will be no greater than 360 lb at A and at B. Assume that the members of the truss have the same cross-sectional area. Choose a material from Appendix B and determine the cross-sectional area so that the structure has a factor of safety $S = 2.5$.

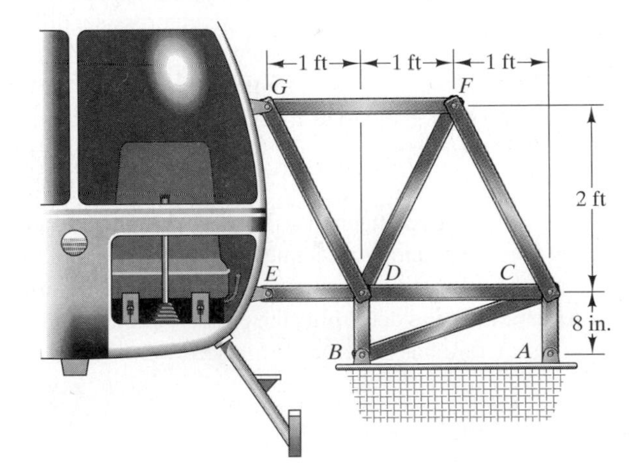

PROBLEM 3-7.9

3-7.10. Upon learning of an upgrade in the helicopter's engine, the engineer designing the truss shown in Problem 3-7.9 does new simulations and concludes that the downward forces the stretcher will exert at A and at B may be as large as 400 lb. He also decides the truss will be made of existing stock with cross-sectional area $A = 0.1\ \text{in}^2$. Choose an aluminum alloy from Appendix B so that the structure will have a factor of safety of at least $S = 5$.

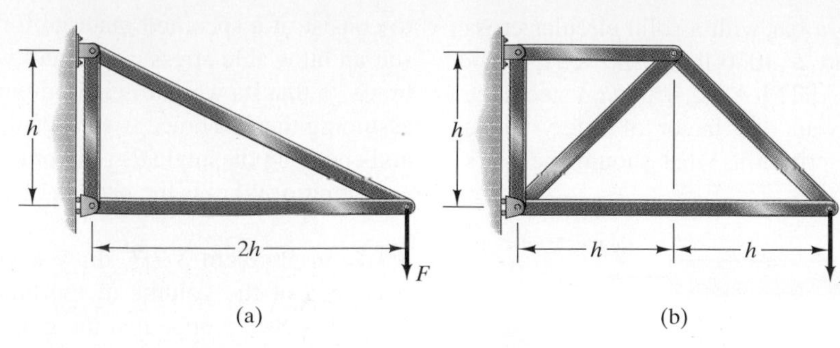

PROBLEM 3-7.11

3-7.11. Two candidate truss designs to support the load F are shown. Members of a given cross-sectional area A and yield stress σ_Y are to be used. Compare the factors of safety and weights of the two designs and discuss reasons that might lead you to choose one design over the other. (The weights can be compared by calculating the total lengths of their members.)

3-7.12. In Problem 3-7.11, two trusses that support a load F that is a horizontal distance $2h$ from the roller support are shown. They consist of members of a given cross-sectional area A. Design (b) has a higher factor of safety than design (a). Try to design a truss with the same pin and roller supports that supports the load F and has a higher factor of safety than design (b).

3-7.13. Design a truss attached at A and B that will support the 2-kN loads at C and D. Assume that the members of the truss have the same cross-sectional area. Choose a material from Appendix B and determine the cross-sectional area so that your structure has a factor of safety $S = 2$.

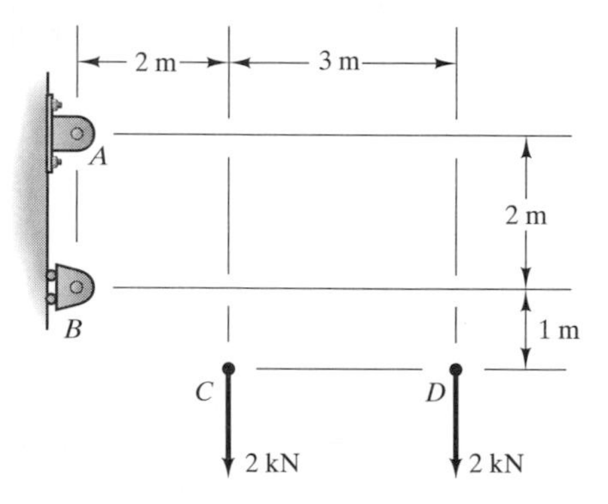

PROBLEM 3-7.13

3-7.14. Design a truss attached at A and B that will support the loads at C and D. Assume that the members of the truss have the same cross-sectional area. Choose a material from Appendix B and determine the cross-sectional area so that your structure has a factor of safety $S = 3$.

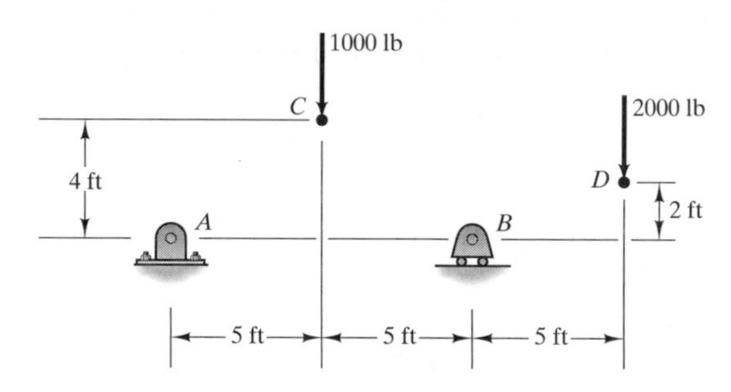

PROBLEM 3-7.14

3-7.15. Design a truss attached at A and B that will support the 2-kN load at C and clear the obstacle. Assume that the members of the truss have the same cross-sectional area. Choose a material from Appendix B and determine the cross-sectional area so that your structure has a factor of safety $S = 2$.

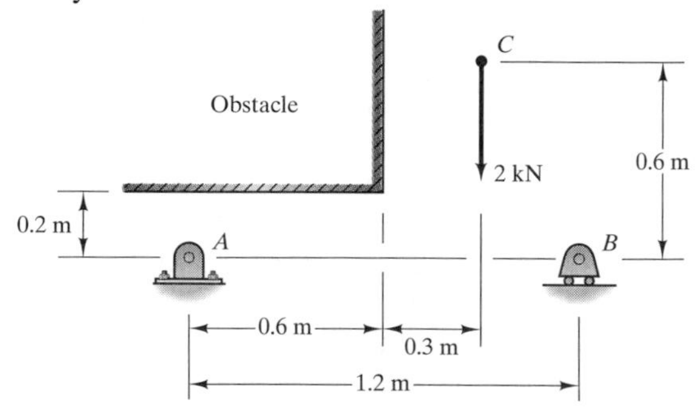

PROBLEM 3-7.15

3-7.16. The space truss is going to be constructed of 7075-T6 aluminum alloy to support the 800-lb vertical load. If you want to design member AD to have a factor of safety $S = 4$, what should its cross-sectional area be?

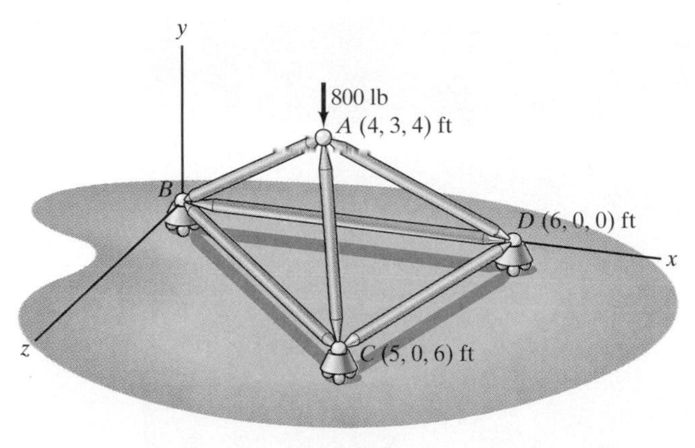

Problem 3-7.16

3-7.17. The space truss in Problem 3-7.16 has roller supports at B, C, and D. It is going to be constructed of 7075-T6 aluminum alloy to support the 800-lb vertical load and each member is to have the same cross-sectional area. If you want the structure to have a factor of safety $S = 4$, what should the cross-sectional area of the bars be?

3-7.18. The construction crane in Problem 3-1.30 is to be made of ASTM-A572 structural steel. If you want to design member 3 to have a factor of safety $S = 2$, what should its cross-sectional area be? (*Strategy:* To determine the axial force in member 3, calculate the moment about the line that passes through joints A and B.)

3-7.19. The construction crane in Problem 3-1.30 is to be made of ASTM-A572 structural steel. Members 1, 2, and 3 are to have the same cross-sectional area, and members 4, 5, and 6 are to have the same cross-sectional area. If you want to choose the cross-sectional areas of these members to obtain a factor of safety $S = 2$, what should they be?

This vintage Corvette has many structural elements designed to support and transmit torsional loads, including the drive shaft, axles, and crankshaft.

CHAPTER 4

Torsion

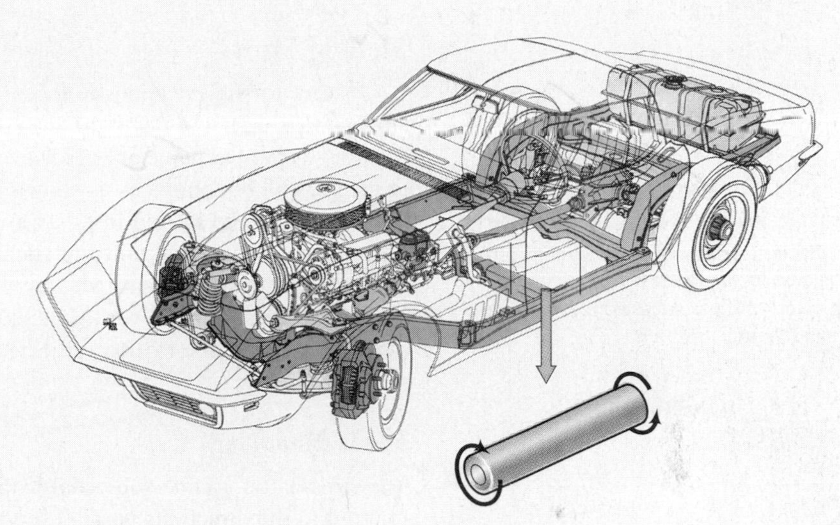

The figure shows the couples, or torques, acting on a segment of a car's drive shaft. The drive shaft transmits power from the car's engine to its rear wheels. The power produced by turbine engines and electrical generators is also transmitted by shafts subjected to torques about their longitudinal axes. Bars are subjected to axial torques in many engineering applications, and we analyze the resulting stresses and deformations in this chapter. To do so, we must first introduce the concept of a state of pure shear stress and the shear modulus.

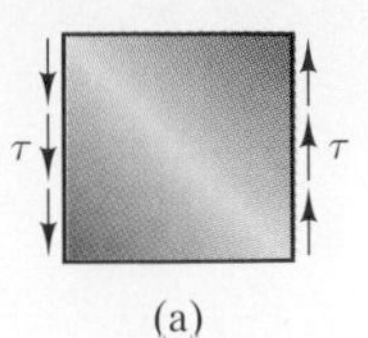

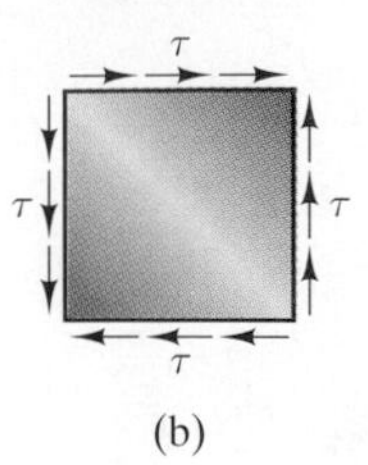

FIGURE 4-1 (a) A cube is not in equilibrium if only two faces are subjected to shear stress. (b) A cube subjected to shear stresses in this way is in equilibrium.

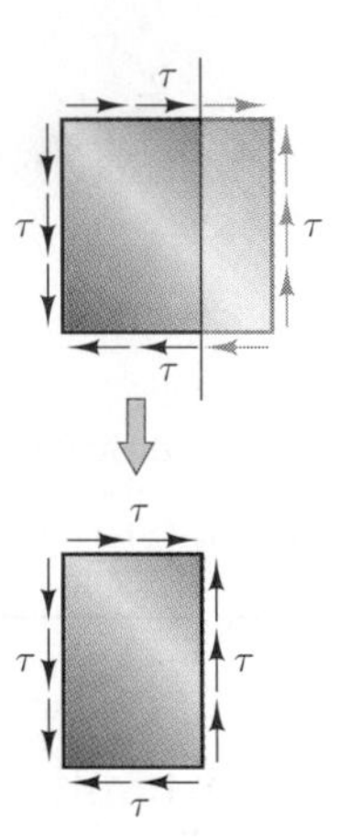

FIGURE 4-2 Planes parallel to the faces of the cube are subjected to the same uniform shear stress.

4-1 | Pure Shear Stress

You have seen in Chapter 3 that subjecting a bar to axial forces results in a state of uniform normal stress and no shear stress on planes perpendicular to the bar's axis. We begin this chapter by describing how, at least in principle, a uniform distribution of shear stress and no normal stress can be achieved on particular planes within a material.

State of Stress

If we were to subject opposite faces of a cube of material to uniform shear stresses τ as shown in Fig. 4-1a, the cube would not be in equilibrium because the stresses exert a couple on it. However, if we also apply uniform shear stresses to the top and bottom faces as shown in Fig. 4-1b, the cube is in equilibrium. If a cube could be loaded in this way, the stress on any plane parallel to the loaded faces consists of uniform shear stress and no normal stress, as shown in Fig. 4-2. For this reason, we say that the cube is in a state of *pure shear stress*. An element from within the cube having rectangular faces parallel to the faces of the cube is also subjected to the same state of pure shear stress (Fig. 4-3).

Shear Modulus

The stresses on a cube subjected to pure shear stress cause a shear strain γ referred to the directions parallel to the loaded faces as shown in Fig. 4-4. If

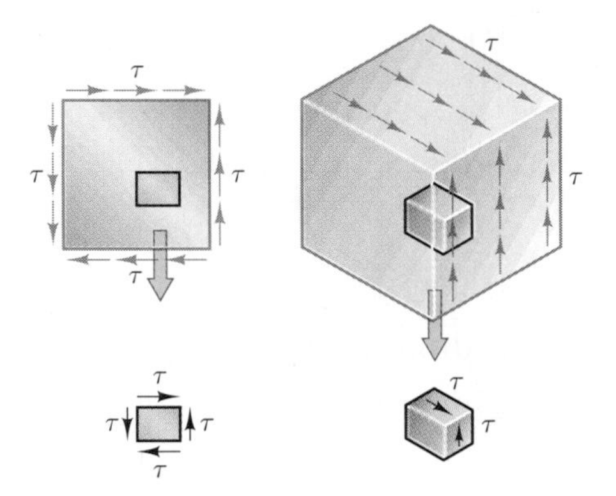

FIGURE 4-3 An element with faces parallel to the faces of the cube is subjected to the same state of pure shear stress.

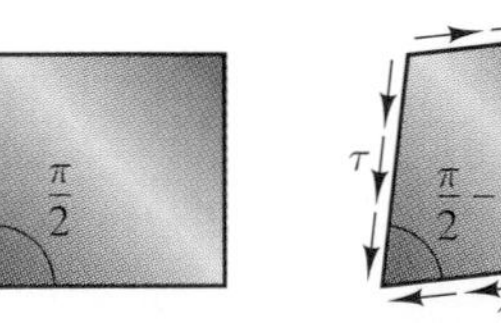

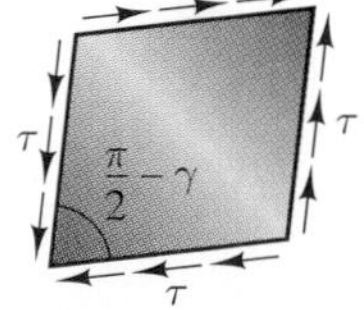

FIGURE 4-4 Shear strain resulting from the shear stresses on the cube.

we were to perform an experiment in which we apply a pure shear stress τ to a cube and measure the resulting shear strain γ, for many materials we would find that as long as the stress is not too large, there is an approximately linear relationship between τ and γ. This relationship is written

$$\tau = G\gamma, \tag{4-1}$$

where the constant G is called the *shear modulus* or *modulus of rigidity*. Because γ is dimensionless, the dimensions of G are the same as the dimensions of τ, force/area. Typical values are given in Appendix B. In Chapter 6 we show that the shear modulus is not independent of the modulus of elasticity E and Poisson's ratio ν, but is related to them by the equation

$$G = \frac{E}{2(1+\nu)}. \tag{4-2}$$

Stresses on Oblique Planes

We have described the stresses acting on the faces of a cube subjected to pure shear stress, but in design you cannot consider only the stresses on particular planes. In our discussion of bars subjected to axial forces in Chapter 3, we found that the values of the normal and shear stresses were different for planes having different orientations. Let us consider a cube of material subjected to pure shear stress and obtain a free-body diagram as shown in Fig. 4-5. The normal and shear stresses on the plane $\mathcal{P}$ are denoted by σ_θ and τ_θ. Letting A_θ be the area of the slanted face of the free-body diagram, the areas of the horizontal and vertical faces in terms of A_θ and θ are shown in Fig. 4-6a. The forces on the free-body diagram are the products of the areas and the uniform stresses (Fig. 4-6b). Summing forces in the σ_θ direction,

$$\sigma_\theta A_\theta - (\tau A_\theta \cos\theta)\sin\theta - (\tau A_\theta \sin\theta)\cos\theta = 0,$$

we determine the normal stress on $\mathcal{P}$:

$$\sigma_\theta = 2\tau \sin\theta \cos\theta. \tag{4-3}$$

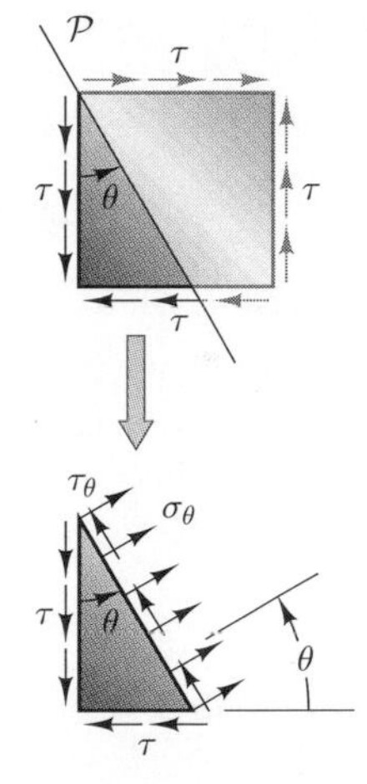

FIGURE 4-5 Obtaining a free-body diagram by passing a plane $\mathcal{P}$ through the cube.

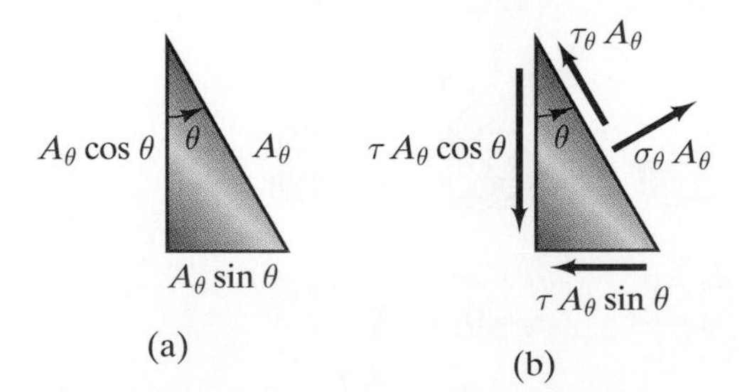

FIGURE 4-6 (a) Areas of the faces. (b) Forces on the free-body diagram.

Then by summing forces in the τ_θ direction,

$$\tau_\theta A_\theta - (\tau A_\theta \cos\theta)\cos\theta + (\tau A_\theta \sin\theta)\sin\theta = 0,$$

we determine the shear stress on $\mathcal{P}$:

$$\tau_\theta = \tau(\cos^2\theta - \sin^2\theta). \tag{4-4}$$

In Fig. 4-7 we plot the ratios σ_θ/τ and τ_θ/τ obtained from Eqs. (4-3) and (4-4). We also show the free-body diagrams and values of σ_θ and τ_θ corresponding to particular values of θ. There is no value of θ for which the shear stress is larger in magnitude than τ. Notice that although there are no normal stresses on the faces of the element in Fig. 4-3, normal stresses do occur on oblique planes. On the plane oriented at $\theta = 45°$, $\sigma_\theta = \tau$, and at $\theta = 135°$, $\sigma_\theta = -\tau$. These are the maximum tensile and compressive stresses for any value of θ. Although we have considered only a subset of planes, it can be shown that these are the maximum shear, tensile, and compressive stresses.

In summary, we have shown that if a material is subjected to a state of pure shear stress of magnitude τ, the maximum shear stress, maximum tensile stress, and maximum compressive stress each equal τ.

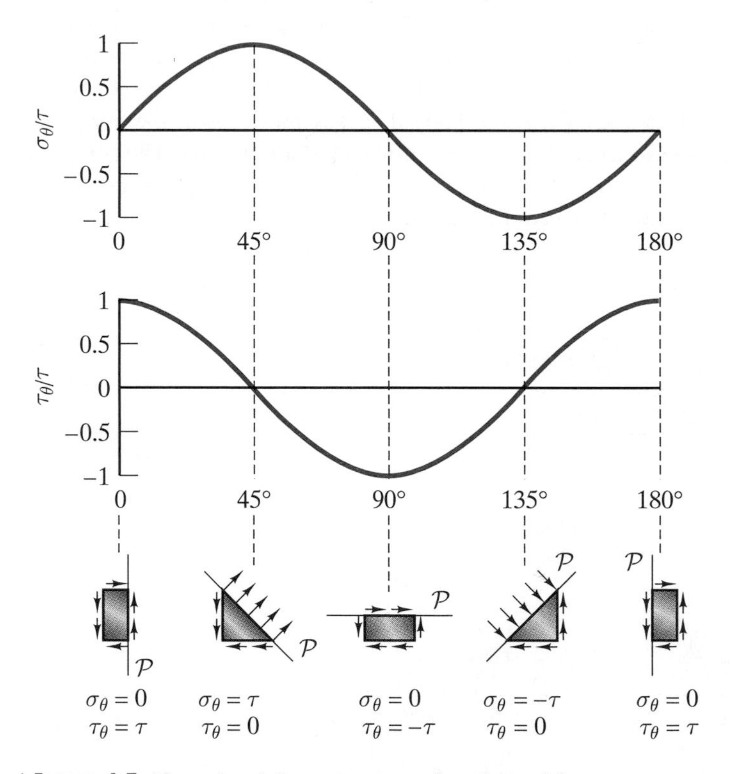

FIGURE 4-7 Normal and shear stresses as functions of θ.

4-2 | Torsion of Prismatic Circular Bars

Our analysis of prismatic bars subjected to axial forces in Chapter 3 applied to bars of arbitrary cross section. In contrast, simple analytical solutions for the deformation and stresses in a bar subjected to axial torsion exist only for bars with circular cross sections. We analyze such bars in this section.

Stresses and Strains

Our objective is to describe the deformation and stresses that result when a cylindrical bar fixed at one end is subjected to an axial torque T at the free end (Fig. 4-8). To describe the bar's deformation, we begin with a cylindrical shell within the bar that has inner radius r and infinitesimal thickness dr (Fig. 4-9a). We draw a radial line from the center of the end of the bar to the cylindrical shell and extend it along the length of the shell parallel to the bar's axis. When the torque is applied (Fig.4-9b), *we assume that each cross section of the bar undergoes a rigid rotation*. The angle ϕ through which the end of the bar rotates is the *angle of twist*. If ϕ is expressed in radians, the circumferential distance $c = r\phi$.

Now imagine slicing the deformed cylindrical shell along its length and laying it out flat (Fig. 4-10). We assume that the longitudinal line viewed in this way remains straight after the torque is applied. If the angle γ expressed in radians is small, we can approximate it by $\gamma = c/L$. Using the relation $c = r\phi$, we obtain

$$\gamma = \frac{r\phi}{L}. \tag{4-5}$$

Let us consider an infinitesimal rectangular element of the shell with sides parallel and perpendicular to the bar's axis before the torque is applied

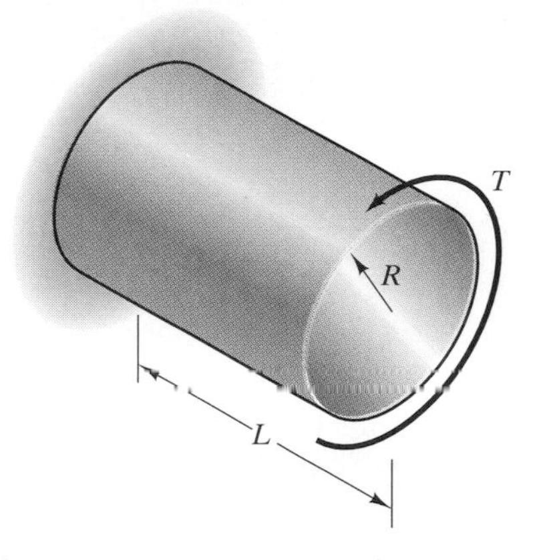

FIGURE 4-8 Subjecting a bar to an axial torque.

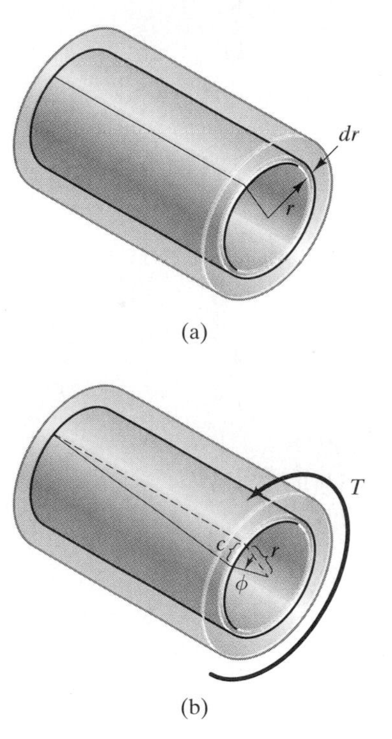

FIGURE 4-9 (a) Cylindrical shell within the bar. (b) Rotations of the radial and longitudinal lines when torque is applied.

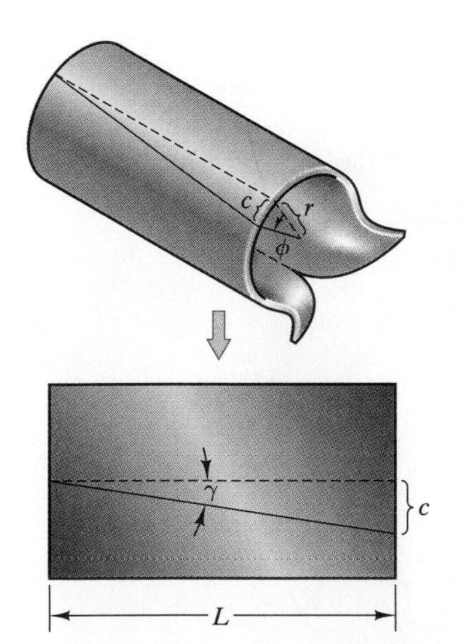

FIGURE 4-10 Laying the shell out flat to visualize its deformation.

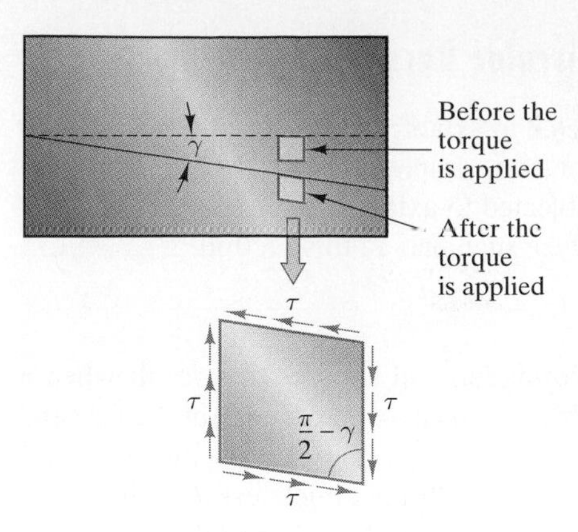

FIGURE 4-11 Element of the shell before and after the torque is applied.

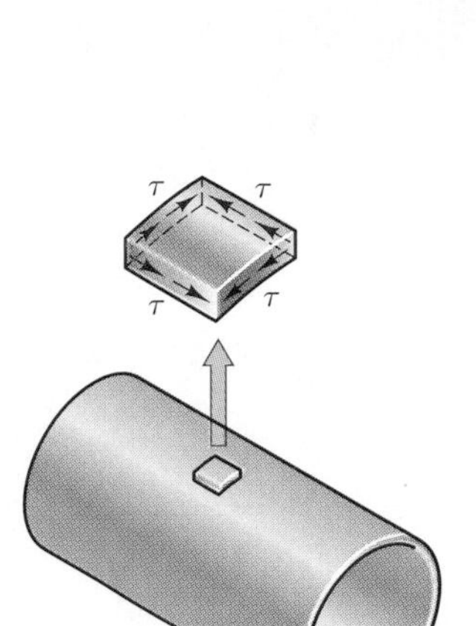

FIGURE 4-12 Stress distribution on an element of the shell.

(Fig. 4-11). If we assume there are no extensional strains in the axial and circumferential directions, the torque subjects the element to a shear strain γ, which implies that the element is in a state of pure shear stress

$$\tau = G\gamma = \frac{Gr\phi}{L}. \tag{4-6}$$

We have given a suggestive argument (not a proof) that the applied torque subjects infinitesimal rectangular elements of the cylindrical shell oriented as shown in Fig. 4-12 to pure shear stress given by Eq. (4-6). If we pass a plane perpendicular to the shell's axis (Fig. 4-13), every element around the

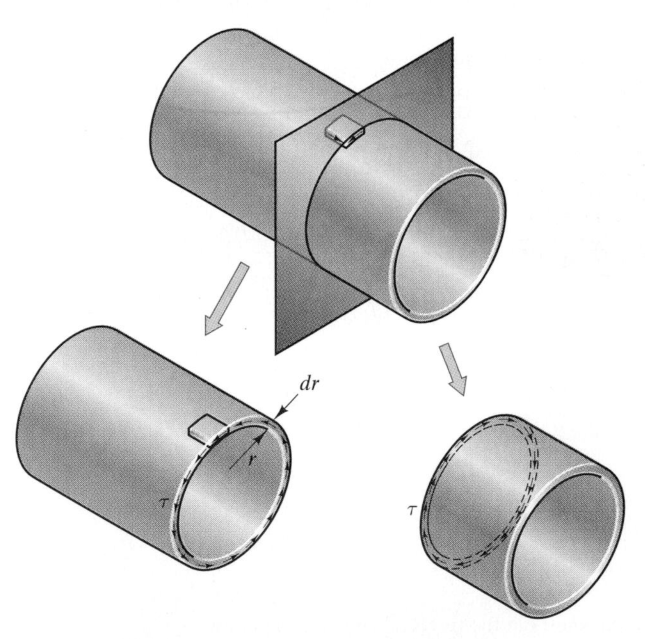

FIGURE 4-13 Passing a plane perpendicular to the shell's axis.

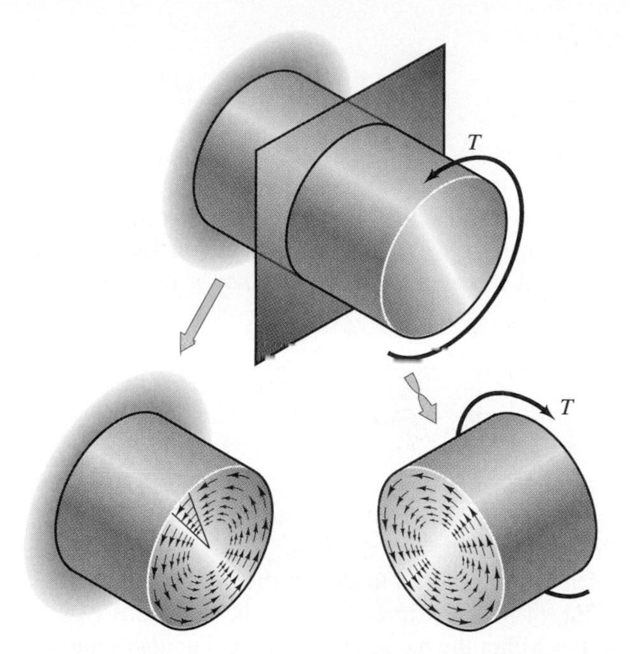

FIGURE 4-14 Stress distribution on a plane perpendicular to the axis of the bar.

circumference is acted upon by the same state of shear stress, so the exposed surfaces are subjected to a uniform circumferential stress τ. By using Fig. 4-13 and Eq. (4-6), we can describe the stress distribution on a plane perpendicular to the axis of the cylindrical bar. It is subjected to a circumferential shear stress distribution τ whose magnitude is proportional to the distance from the center of the bar (Fig. 4-14).

We have now described the stress distribution in the bar, but Eq. (4-6) gives the shear stress in terms of the bar's angle of twist. Our next objective is to determine the shear stress in terms of the applied torque T. From the free-body diagram in Fig. 4-14, equilibrium requires that the moment about the bar's axis due to the shear stress distribution must equal T. Let dA be an infinitesimal element of the bar's cross-sectional area at a distance r from the center. The moment about the bar's axis due to the shear stress acting on dA is $r\tau\, dA$ (Fig. 4-15). Integrating to determine the total moment due to the shear stress, we obtain the equilibrium equation

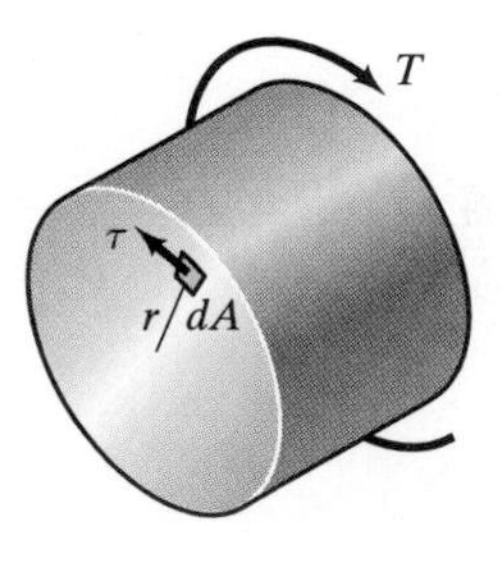

FIGURE 4-15 Calculating the moment about the bar's axis due to the shear stress distribution.

$$\int_A r\tau\, dA = T.$$

Substituting Eq. (4-6) into this equation, we can solve for the bar's angle of twist ϕ. The result is

$$\phi = \frac{TL}{GJ}, \qquad \textbf{(4-7)}$$

where

$$J = \int_A r^2 \, dA \tag{4-8}$$

is the polar moment of inertia of the bar's cross-sectional area about its axis. [Notice the similarity between Eq. (4-7) and the equation $\delta = PL/EA$ for the change in length of a bar subjected to axial forces.] We can now substitute Eq. (4-7) back into Eq. (4-6), obtaining an expression for the shear stress in terms of the applied torque T and the distance r from the bar's axis:

$$\tau = \frac{Tr}{J}. \tag{4-9}$$

We have attempted to make Eqs. (4-7) and (4-9) plausible, but we have not proven that they determine the angle of twist and distribution of stress in a solid or hollow circular bar. Although the proof is beyond our scope, it can be shown that if such a bar could be loaded at the ends by shear stress distributions satisfying Eq. (4-9), our expressions for the angle of twist and the shear stress distribution within the bar are the exact and unique solutions. Furthermore, no matter how the torques are applied, the stress distribution at axial distances from the ends greater than a few times the bar's width is given approximately by Eq. (4-9). This is another example of Saint-Venant's principle (see Section 3-1). As a consequence, Eq. (4-7) approximates the angle of twist of a slender bar no matter how the torques are applied at the ends. These results require that the material have the property of isotropy, which we discuss in Chapter 6.

You can perform a simple experiment that dramatically demonstrates the effect of the state of stress in a bar subjected to torsion. Apply torsion to a piece of blackboard chalk as shown in Fig. 4-16a. The maximum tensile stress resulting from the state of pure shear stress you apply occurs on planes at 45° relative to the chalk's axis (see Figs. 4-7 and 4-16b). Brittle materials such as chalk are weak in tension, and you will observe that the chalk fails along a clearly defined 45° spiral line (Fig. 4-16c).

Polar Moment of Inertia

To evaluate J, we can use an annular element of area with radius r and thickness dr (Fig. 4-17). The area of the element is the product of its circumference and thickness, $dA = 2\pi r \, dr$, so the polar moment of intertia is

$$J = \int_A r^2 \, dA = \int_0^R r^2 (2\pi r \, dr).$$

Integrating, we obtain

$$J = \frac{\pi}{2} R^4. \qquad \text{solid circular cross section} \tag{4-10}$$

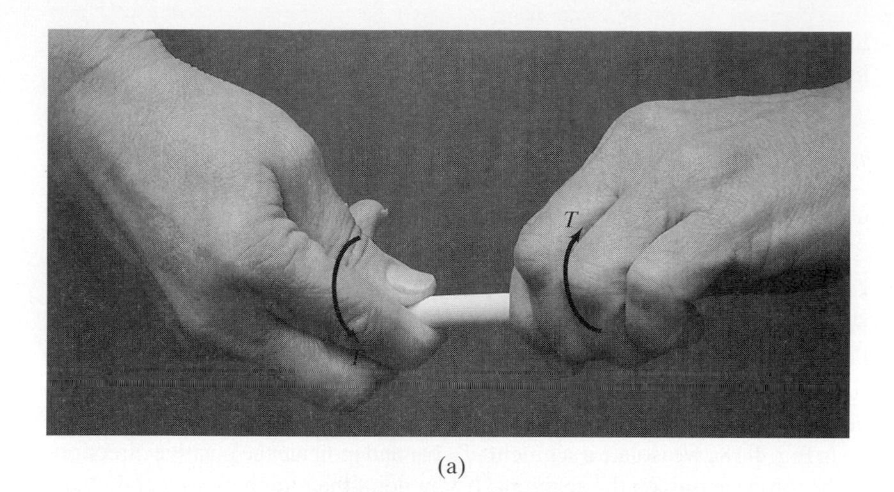

(a)

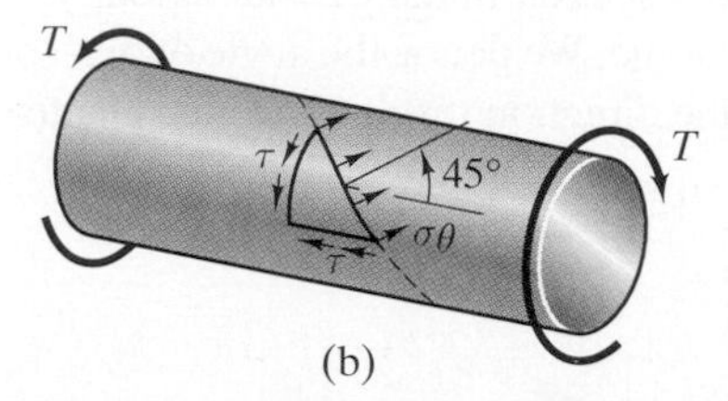

(b)

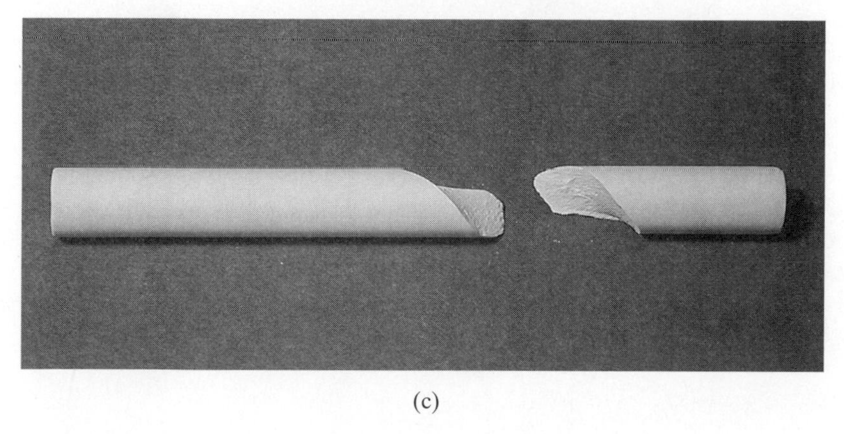

(c)

FIGURE 4-16 (a) Applying axial torques to a piece of chalk. (b) The maximum tensile stress occurs at 45° relative to the longitudinal axis. (c) Failure occurs along the plane subjected to the maximum tensile stress.

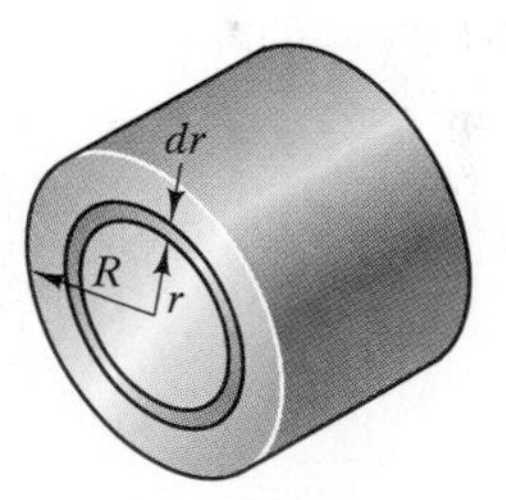

FIGURE 4-17 Element of area for calculating J.

Equations (4-7) and (4-9) also apply to a bar with a hollow circular cross section. In that case the polar moment of inertia is given by

$$J = \frac{\pi}{2}(R_o^4 - R_i^4), \qquad \text{hollow circular cross section} \qquad \textbf{(4-11)}$$

where R_i and R_o are the bar's inner and outer radii.

Positive Directions of the Torque and Angle of Twist

Equations (4-7) and (4-9), with the polar moment of inertia J evaluated using Eq. (4-10) or (4-11), determine the angle of twist and distribution of shear stress for a circular bar subjected to a torque T. For some applications it will be convenient to have sign conventions for the torque and angle of twist.

In Fig. 4-18a we isolate a segment of a bar and indicate the positive directions of the torque acting on the segment: If you point the thumb of your right hand outward from the cross section under consideration, your fingers point in the direction of positive torque. We define the angle of twist of a segment of a bar to be positive if it is in the direction resulting from a positive torque (Fig. 4-18b).

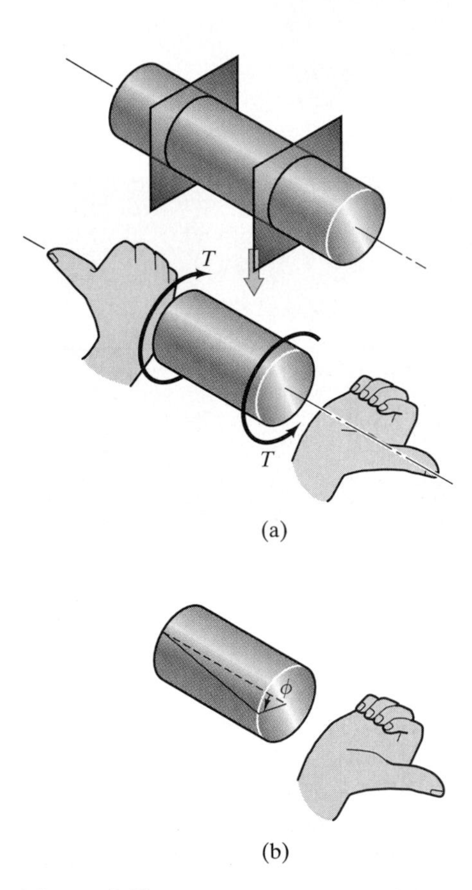

FIGURE 4-18 (a) Positive directions of the torque. (b) Segment with a positive angle of twist ϕ.

EXAMPLE 4-1

Worksheet 4

The bar in Fig. 4-19 consists of material with shear modulus $G = 28$ GPa and has a solid circular cross section. Part A is 40 mm in diameter and part B is 20 mm in diameter. **(a)** Determine the magnitudes of the maximum shear stresses in parts A and B. **(b)** Determine the angle of twist of the right end of the bar relative to the wall.

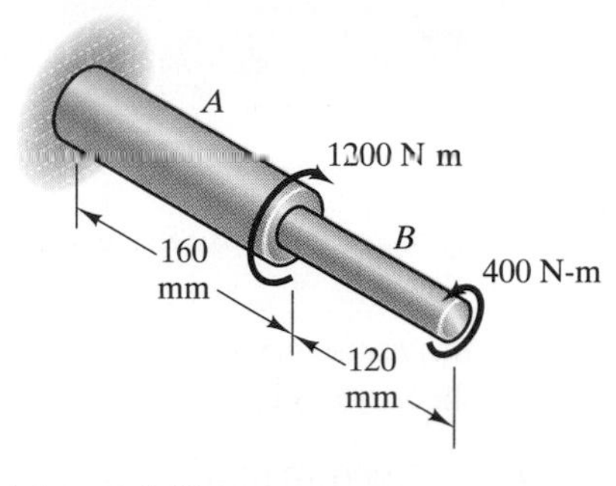

FIGURE 4-19

Strategy

By passing a plane through part A of the bar and isolating the part of the bar on one side of the plane, we can determine the internal torque in part A. We can then use Eq. (4-9) to determine the maximum shear stress in part A and use Eq. (4-7) to determine the angle of twist of part A. We can determine the maximum shear stress in part B and the angle of twist of part B in the same way.

Solution

(a) In Fig. (a) we draw the free-body diagram of the entire bar.

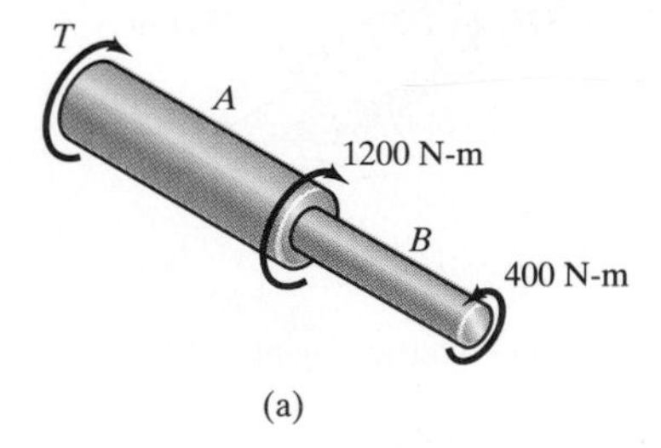

(a)

(a) Isolating the bar from the wall.

From the equilibrium equation $400 - 1200 - T = 0$, we find that the torque exerted on the bar by the wall is $T = -800$ N-m. In Fig. (b) we pass a plane through part A and isolate the part of the bar to the left of the plane. From the equilibrium equation $T + 800 = 0$, the torque in part A of the bar is $T = -800$ N-m. The polar moment of inertia of the cross section in part A is

$$J = \frac{\pi}{2}R^4 = \frac{\pi}{2}(0.02)^4 = 2.51 \times 10^{-7}\ \text{m}^4.$$

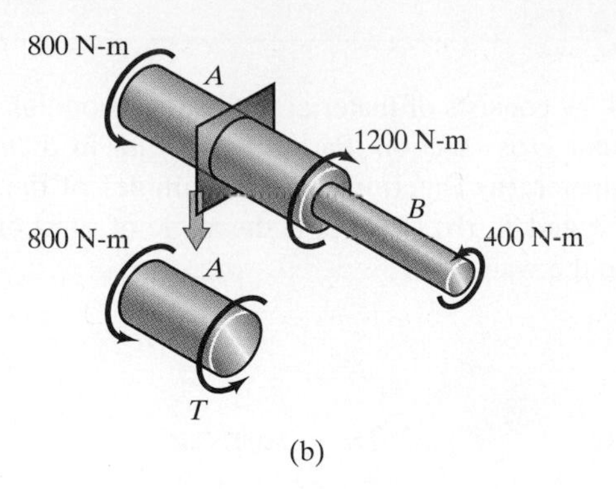

(b) Obtaining a free-body diagram by passing a plane through part *A*.

The maximum shear stress in part *A* occurs at $r = 0.02$ m. Therefore, the magnitude of the maximum shear stress is

$$\tau = \frac{|T|r}{J} = \frac{(800)(0.02)}{2.51 \times 10^{-7}} = 63.7 \text{ MPa}.$$

In Fig. (c) we pass a plane through part *B* and isolate the part of the bar to the left of the plane.

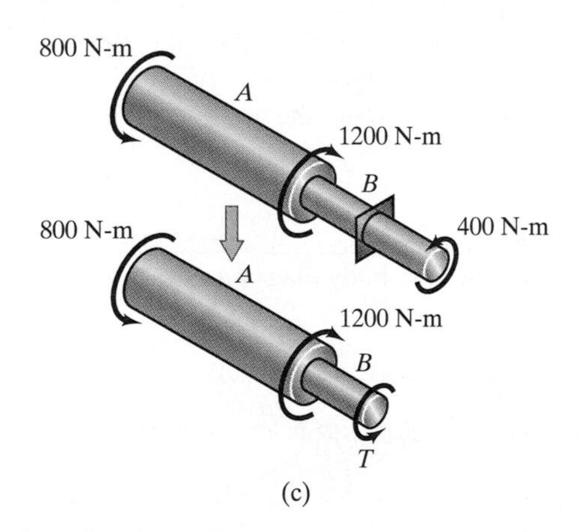

(c) Obtaining a free-body diagram by passing a plane through part *B*.

From the equilibrium equation $T - 1200 + 800 = 0$, the torsion in part *B* is $T = 400$ N-m. The polar moment of inertia of the cross section in part *B* is

$$J = \frac{\pi}{2}R^4 = \frac{\pi}{2}(0.01)^4 = 1.57 \times 10^{-8} \text{ m}^4.$$

The maximum shear stress in part *B* occurs at $r = 0.01$ m. The magnitude of the maximum shear stress in part *B* is

$$\tau = \frac{|T|r}{J} = \frac{(400)(0.01)}{1.57 \times 10^{-8}} = 255 \text{ MPa}.$$

(b) The torque in part A of the bar is $T = -800$ N-m, so the angle of twist of part A is

$$\phi_{\text{part } A} = \frac{TL}{GJ} = \frac{(-800)(0.16)}{(28 \times 10^9)(2.51 \times 10^{-7})} = -0.0182 \text{ rad} = -1.04^\circ.$$

This is the angle of twist of the right end of part A relative to the wall [Fig. (d)]. The torque in part B is $T = 400$ N-m, so the angle of twist of part B is

$$\phi_{\text{part } B} = \frac{TL}{GJ} = \frac{(400)(0.12)}{(28 \times 10^9)(1.57 \times 10^{-8})} = 0.1091 \text{ rad} = 6.25^\circ.$$

This is the angle of twist of the right end of part B relative to the end attached to part A [Fig. (e)]. The angle of twist of the right end of part B relative to the wall is

$$\phi = \phi_{\text{part } A} + \phi_{\text{part } B} = -1.04 + 6.25 = 5.21^\circ.$$

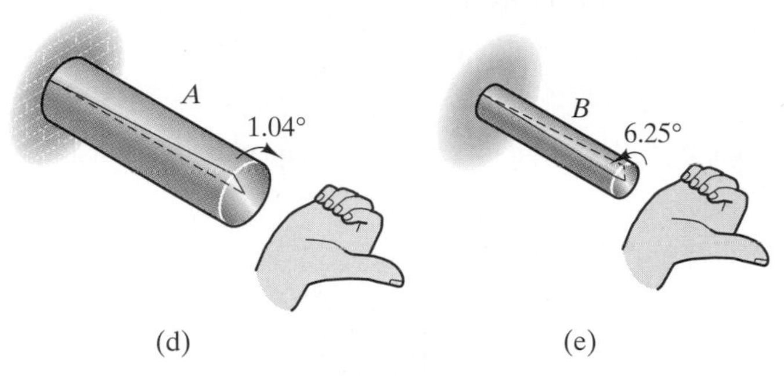

(d) Angle of twist of part A. (e) Angle of twist of part B.

4-3 | Statically Indeterminate Problems

The approach you learned in Chapter 3 for solving statically indeterminate problems involving axially loaded bars—applying equilibrium, force–deformation relations, and compatibility—applies to virtually all statically indeterminate problems. In this section we demonstrate the solution of statically indeterminate problems involving bars subjected to torsion. The force–deformation relations will now be relations between torques and angles of twist, and the compatibility conditions will be constraints imposed on the angles of twist of torsionally loaded bars.

To emphasize that you can use the same procedure you applied to axially loaded bars, we present an example equivalent to the one discussed in Section 3-3 and solve it by the same steps. (You should compare the two solutions.) The bar in Fig. 4-20a consists of two segments A and B with different lengths and diameters. It is fixed at both ends and subjected to an axial torsion. We draw

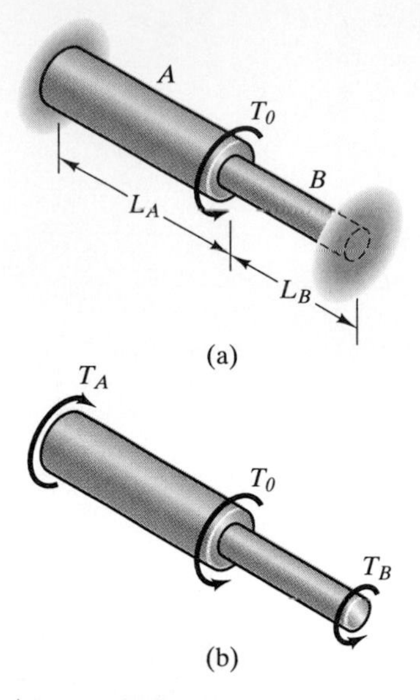

FIGURE 4-20 (a) Torsionally loaded bar fixed at both ends. (b) Free-body diagram of the bar.

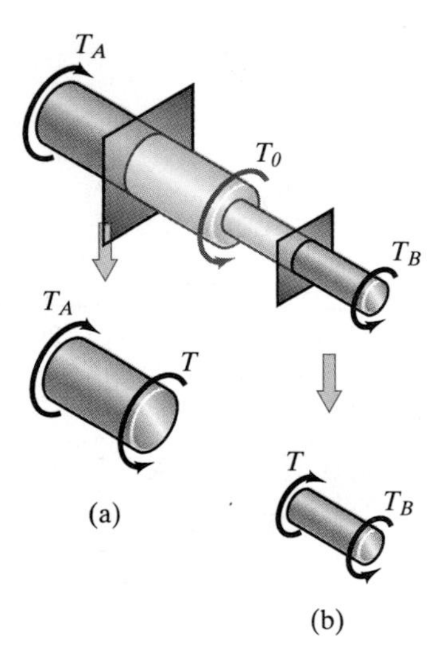

FIGURE 4-21 Determining the internal torques in parts A and B.

its free-body diagram in Fig. 4-20b. The equilibrium equation is

$$T_0 - T_A + T_B = 0. \tag{4-12}$$

We cannot determine the two reactions T_A and T_B from this equation. The problem is statically indeterminate.

The torque in part A of the bar is T_A (Fig. 4-21a), so the angle of twist of part A is

$$\phi_A = \frac{T_A L_A}{G J_A}. \tag{4-13}$$

The torque in part B is T_B (Fig. 4-21b). Its angle of twist is

$$\phi_B = \frac{T_B L_B}{G J_B}. \tag{4-14}$$

Since the bar is fixed at both ends, the compatibility condition is that the angle of twist of the entire bar equals zero:

$$\phi_A + \phi_B = 0. \tag{4-15}$$

We substitute Eqs. (4-13) and (4-14) into this equation, obtaining

$$\frac{T_A L_A}{G J_A} + \frac{T_B L_B}{G J_B} = 0. \tag{4-16}$$

We can solve this equation simultaneously with the equilibrium equation to determine the reactions T_A and T_B. The results are

$$T_A = \frac{T_0}{1 + L_A J_B / L_B J_A}, \qquad T_B = \frac{-T_0}{1 + L_B J_A / L_A J_B}.$$

Now that we know the reactions, we can determine the angle of twist of each part of the bar from Eqs. (4-13) and (4-14). We can also determine the shear stress distribution in each part of the bar: $\tau_A = T_A r / J_A$ and $\tau_B = T_B r / J_B$.

4-4 | Nonprismatic Bars and Distributed Loads

In this section we discuss torsion of circular bars whose diameters vary with distance along their axes and also bars subjected to torsional loads that are distributed along their axes. The derivations closely follow our treatment in Section 3-4 of analogous applications involving axially loaded bars.

Bars with Gradually Varying Cross Sections

Figure 4-22a shows a bar with a hollow circular cross section whose inner and outer radii vary with distance along the bar's axis. The polar moment of inertia of the cross section depends upon x, which we indicate by writing it as $J(x)$.

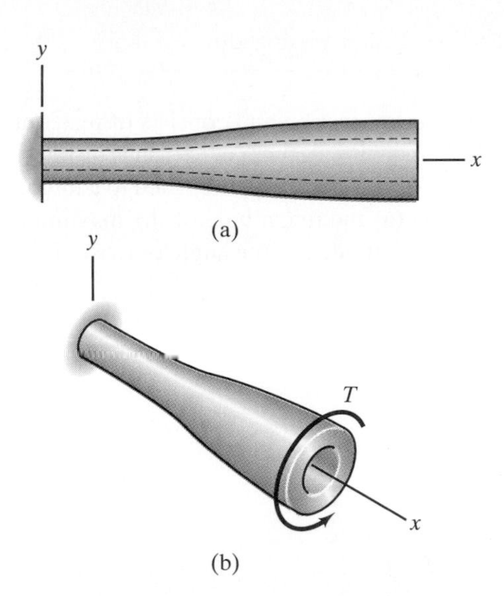

FIGURE 4-22 (a) Bar with a varying cross section. (b) Subjecting the bar to a torque T.

If we subject the bar to a torsional load (Fig. 4-22b) and the change in $J(x)$ with x is gradual, we can approximate the stress distribution on a plane at axial position x by using Eq. (4-9):

$$\tau = \frac{Tr}{J(x)}. \tag{4-17}$$

For a given perpendicular plane, the stress distribution is approximately the same as for the case of a prismatic bar. But since $J(x)$ varies with distance along the bar's axis, the shear stress distribution does also.

To determine the bar's angle of twist, we begin by considering an infinitesimal element of the bar of length dx (Fig. 4-23). We can use Eq. (4-7) to determine the element's angle of twist:

$$\phi_{\text{element}} = \frac{T\,dx}{GJ(x)}.$$

We obtain the angle of twist of the entire bar by integrating this expression from $x = 0$ to $x = L$:

$$\phi = \int_0^L \frac{T\,dx}{GJ(x)}. \tag{4-18}$$

The angle of twist of each element depends on its polar moment of inertia, and we must add up the angles of twist of the elements to determine the angle of twist of the bar.

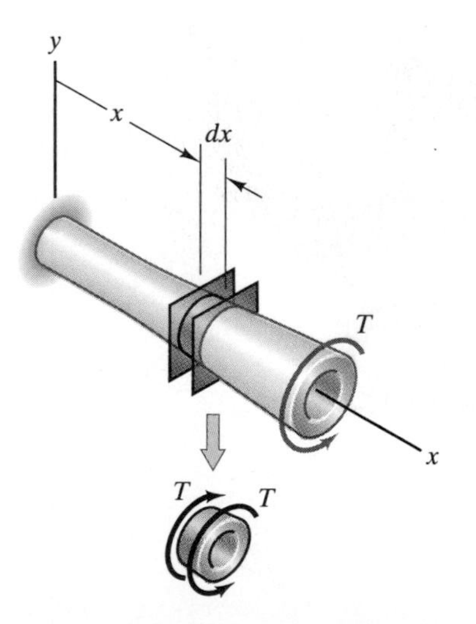

FIGURE 4-23 Determining the angle of twist of an infinitesimal element of the bar.

EXAMPLE 4-2

The bar in Fig. 4-24 has a solid circular cross section and consists of material with shear modulus $G = 47$ GPa. The bar's polar moment of inertia is given by the equation $J(x) = 0.00016 + 0.0006x^2$ m^4. If the bar is subjected to a torque $T = 200$ kN-m at its free end, determine **(a)** the magnitude of the maximum shear stress in the bar at $x = 1$ m; **(b)** the magnitude of the angle of twist of the entire bar in degrees.

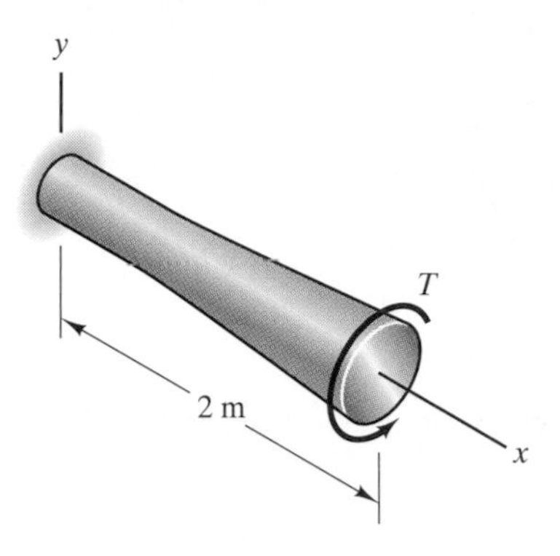

FIGURE 4-24

Strategy

(a) From the given equation for $J(x)$ we can determine the polar moment of inertia and the bar's radius R at $x = 1$ m. Then the maximum shear stress is $TR/J(x)$. **(b)** Since the bar's polar moment of inertia is a function of x, we must determine the bar's angle of twist from Eq. (4-18).

Solution

(a) The polar moment of inertia at $x = 1$ m is

$$J(1) = 0.00016 + 0.0006(1)^2 = 0.00076 \text{ m}^4.$$

Solving the equation $J = (\pi/2)R^4$ for the bar's radius at $x = 1$ m, we obtain $R = 0.148$ m. Therefore, the magnitude of the maximum shear stress at $x = 1$ m is

$$|\tau| = \frac{TR}{J(1)} = \frac{(200{,}000)(0.148)}{0.00076} = 39.0 \text{ MPa}.$$

(b) From Eq. (4-18), the magnitude of the bar's angle of twist is

$$|\phi| = \int_0^L \frac{T\,dx}{GJ(x)} = \int_0^2 \frac{(200{,}000)\,dx}{(47 \times 10^9)(0.00016 + 0.0006x^2)} = 0.0181 \text{ rad},$$

which is 1.04°.

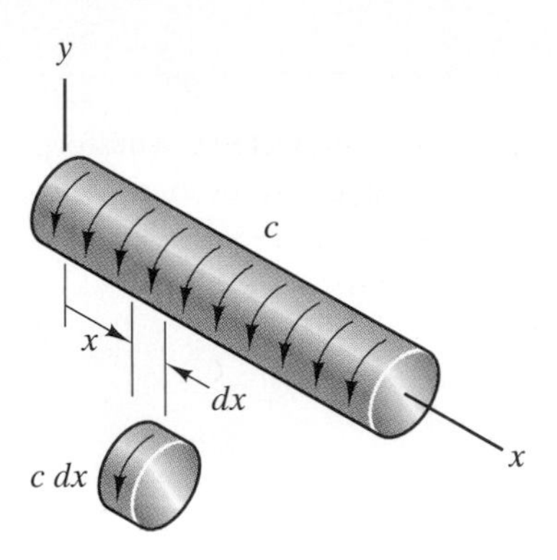

FIGURE 4-25 Describing a distributed torque by a function. The torque on an element of length dx is $c\,dx$.

Distributed Torsional Loads

We can describe a distributed torsional load on a bar by introducing a function c defined such that the axial torque on each element of the bar is $c\,dx$ (Fig. 4-25). Since the product of c and dx is a moment, the dimensions of c are moment/length. For example, the bar in Fig. 4-26a is fixed at the left end and subjected to a distributed axial torque throughout its length. In Fig. 4-26b we pass a plane through the bar at an arbitrary position x and draw the free-body diagram of the part of the bar to the right of the plane. From the equilibrium equation

$$-T + \int_x^L c_0 \left(\frac{x}{L}\right)^2 dx = 0,$$

we determine the internal torque in the bar at the position x:

$$T = \frac{c_0}{3}\left(L - \frac{x^3}{L^2}\right).$$

The distributed torque causes the internal torque in the bar to vary with axial position. To determine the angle of twist of the right end of the bar relative to the wall, we consider an element of the bar of length dx (Fig. 4-26c). From Eq. (4-7), its angle of twist is

$$\phi_{\text{element}} = \frac{T\,dx}{GJ} = \frac{c_0}{3GJ}\left(L - \frac{x^3}{L^2}\right) dx.$$

Integrating this result from $x = 0$ to $x = L$, we obtain the angle of twist of the right end:

$$\phi = \int_0^L \frac{c_0}{3GJ}\left(L - \frac{x^3}{L^2}\right) dx = \frac{c_0 L^2}{4GJ}.$$

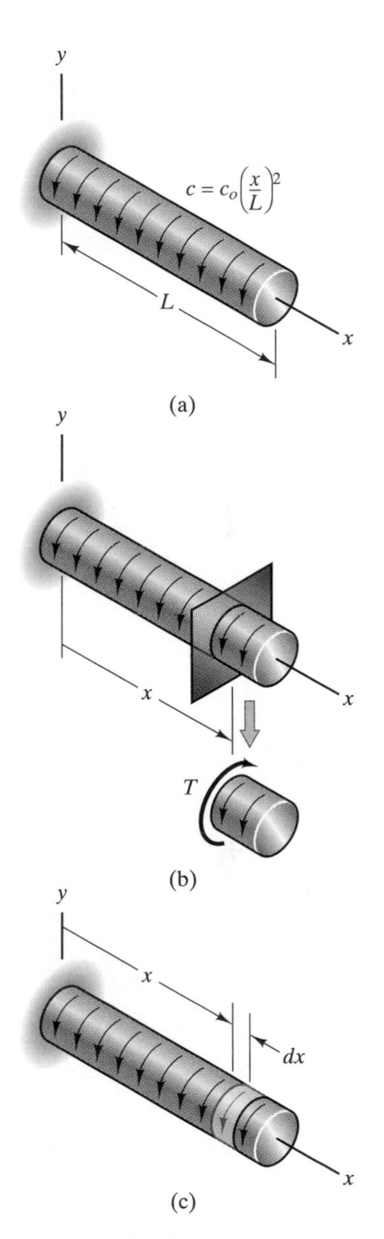

FIGURE 4-26 (a) Bar subjected to a distributed torque. (b) Determining the internal torque at x. (c) Element of length dx.

EXAMPLE 4-3

The cylindrical bar in Fig. 4-27 is fixed at both ends and is subjected to a uniform distributed torque c from $x = 0$ to $x = L_A$. What torques are exerted on the bar by the walls?

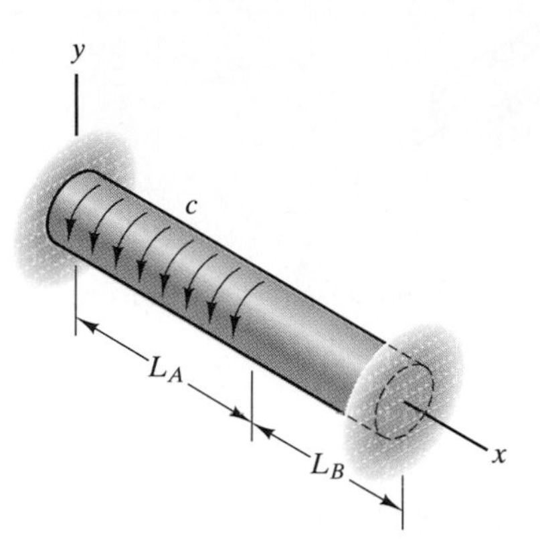

FIGURE 4-27

Strategy

By using the equilibrium equation for the entire bar and the compatibility condition that the angle of twist of the right end of the bar relative to the left end is zero, we can solve for the two torques exerted on the bar by the walls.

Solution

In Fig. (a) we draw the free-body diagram of the entire bar.

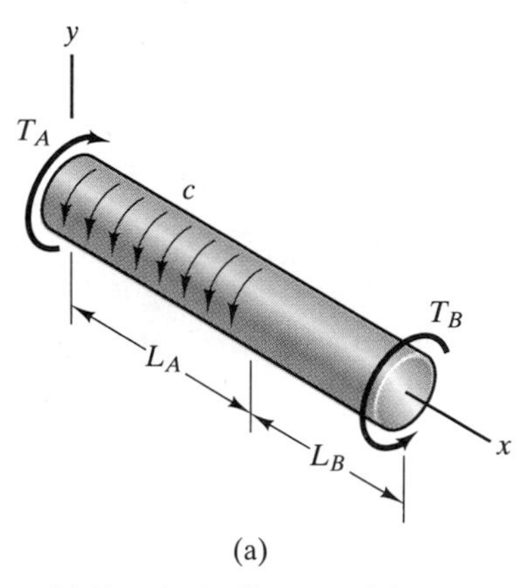

(a)

(a) Free-body diagram of the bar.

Because c is constant, the total moment exerted by the distributed torque is the product of c and L_A. The equilibrium equation is

$$-T_A + T_B + cL_A = 0.$$

We cannot solve this equation for T_A and T_B; the problem is statically indeterminate. In Fig. (b) we pass a plane through part A of the bar at an arbitrary position x. From the equilibrium equation $-T_A + T + cx = 0$, the internal torque in part A of the bar is

$$T = T_A - cx.$$

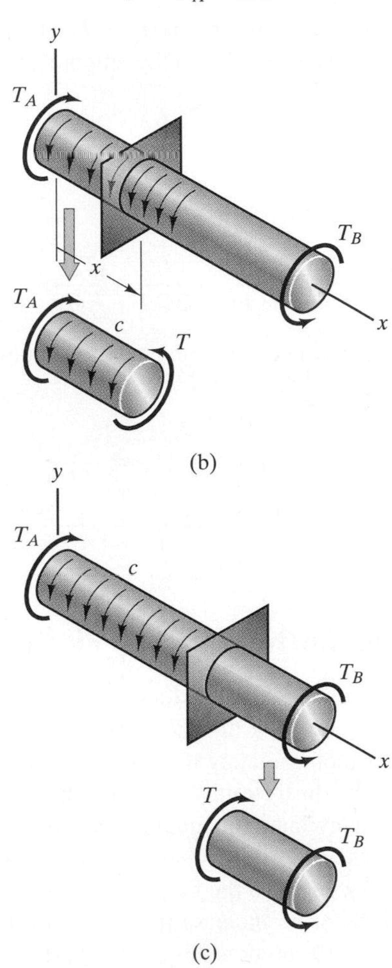

(b) Determining the internal torque in part A. (c) Determining the internal torque in part B.

Consider an element of part A of the bar that has length dx and is located at the position x. The angle of twist of this element is

$$d\phi_A = \frac{T\,dx}{GJ} = \frac{(T_A - cx)\,dx}{GJ}.$$

Integrating this expression to determine the angle of twist of the right end of part A relative to the left end, we obtain

$$\phi_A = \int_0^{L_A} \frac{T\,dx}{GJ} = \int_0^{L_A} \frac{(T_A - cx)\,dx}{GJ} = \frac{T_A L_A}{GJ} - \frac{cL_A^2}{2GJ}.$$

The internal torque in part B of the bar is $T = T_B$ [Fig. (c)], so the angle of twist of the right end of part B relative to the left end of part B is

$$\phi_B = \frac{T_B L_B}{GJ}.$$

The compatibility condition is

$$\phi_A + \phi_B = \frac{T_A L_A}{GJ} - \frac{cL_A^2}{2GJ} + \frac{T_B L_B}{GJ} = 0.$$

Solving this equation simultaneously with the equilibrium equation, we obtain the torques exerted by the walls:

$$T_A = \frac{1/2 + L_B/L_A}{1 + L_B/L_A} cL_A, \qquad T_B = -\frac{1/2}{1 + L_B/L_A} cL_A.$$

4-5 | Torsion of an Elastic–Perfectly Plastic Circular Bar

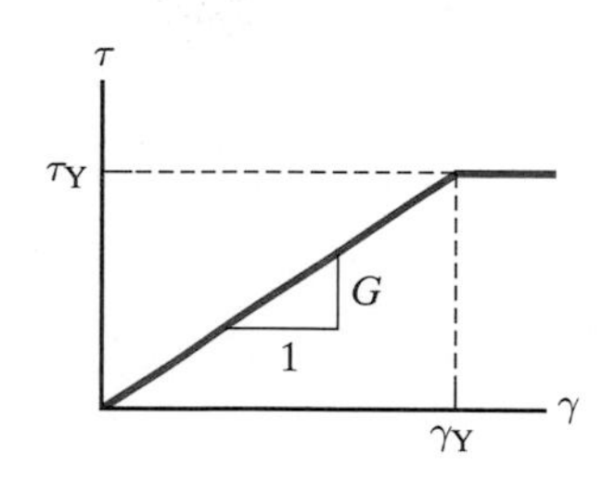

FIGURE 4-28 Elastic–perfectly plastic model of shear stress–shear strain behavior.

In our discussion of bars subjected to torsional loads, we have assumed that the shear stress is a linear function of the shear strain: $\tau = G\gamma$. Many materials satisfy this relationship approximately if the shear strain does not become too large. At large values of the shear strain, the relationship between the shear stress and shear strain is for many materials qualitatively similar to the relationship between the normal stress and extensional strain (see Section 3-6). In ductile materials, a yield stress occurs and the material enters a plastic range in which strain increases with little or no increase in stress until the material fractures. The simplest model of this phenomenon is elastic–perfectly plastic behavior, in which it is assumed that plastic strain occurs with no increase in stress once the yield stress is reached (Fig. 4-28). Although more realistic models of plastic behavior must usually be used in actual design, you can gain understanding from analyses based on the elastic–perfectly plastic model. In this section we apply this model to torsion of a circular bar.

As in our analysis of the torsion of an elastic bar in Section 4-2, we begin with the assumption that each cross section of the bar undergoes a rigid rotation. This leads to Eq. (4-5), which states that the shear strain is proportional to the radial distance r from the bar's axis:

$$\gamma = \frac{r\phi}{L}, \qquad \textbf{(4-19)}$$

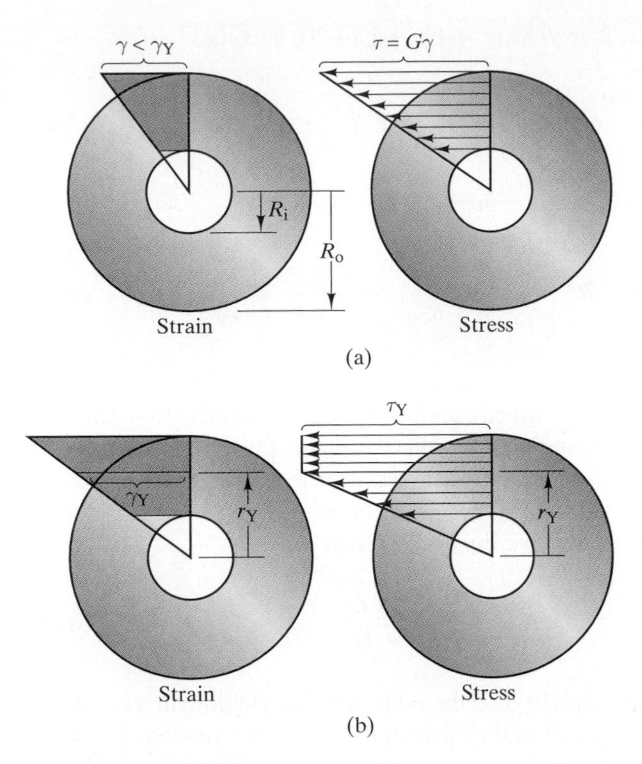

FIGURE 4-29 (a) Strain and stress distributions when the maximum strain is less than the strain corresponding to the yield stress. (b) Strain and stress distributions when the maximum strain is greater than the strain corresponding to the yield stress.

where ϕ is the angle of twist of the bar and L is its length. If the maximum shear strain in the bar (the shear strain at the bar's outer surface) does not exceed the value γ_Y corresponding to the yield stress, $\tau = G\gamma$ and the shear stress is also proportional to r (Fig. 4-29a). This is the elastic solution: The shear stress distribution is given by Eq. (4-9) and the angle of twist by Eq. (4-7).

If the shear strain does exceed the value corresponding to the yield stress at some radial position r_Y, the shear stress is proportional to r until it reaches the yield stress at $r = r_Y$, then remains constant (Fig. 4-29b). Equilibrium requires that the total moment about the axis of the bar due to the stress distribution be equal to T. We can use this condition to determine r_Y. The force due to the shear stress τ acting on an element dA at a radial distance r from the axis is $\tau\, dA$. The moment due to this force about the axis is $r(\tau\, dA)$, so

$$T = \int_A r\tau\, dA. \tag{4-20}$$

Let us apply this condition to the shear stress distribution shown in Fig. 4-29b. For radial distances in the range $R_i \leq r \leq r_Y$, the shear stress $\tau = (\tau_Y/r_Y)r$. For radial distances greater than r_Y, the shear stress $\tau = \tau_Y$. If we use an annular element of area $dA = 2\pi r\,dr$ (Fig. 4-17), Eq. (4-20) becomes

$$T = \int_{R_i}^{r_Y} r\left(\frac{\tau_Y r}{r_Y}\right)(2\pi r\,dr) + \int_{r_Y}^{R_o} r\tau_Y(2\pi r\,dr). \quad \textbf{(4-21)}$$

Evaluating the integrals in this expression, we obtain

$$T = \pi\tau_Y R_o^3\left[\frac{2}{3} - \frac{1}{6}\left(\frac{r_Y}{R_o}\right)^3 - \frac{1}{2}\left(\frac{R_i}{R_o}\right)^4\frac{R_o}{r_Y}\right]. \quad \textbf{(4-22)}$$

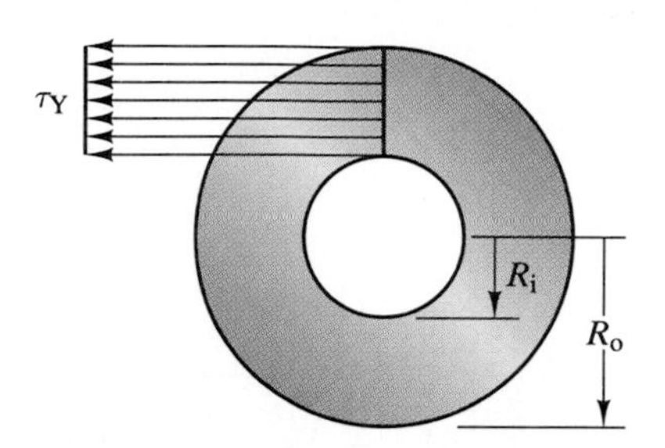

FIGURE 4-30 Stress distribution when the entire cross section has yielded. This condition defines the upper bound of the torque an elastic–perfectly plastic bar can support.

For a given value of the external torque T, this nonlinear algebraic equation can be solved for the value of r_Y, which establishes the stress distribution. Once the value of r_Y is known, we can also determine the angle of twist. Equation (4-19), which gives the shear strain in the bar in terms of the radial position and the bar's angle of twist, applies if $r \leq r_Y$. At the radial position $r = r_Y$, the strain $\gamma = \gamma_Y$. Therefore, we can solve Eq. (4-19) for ϕ:

$$\phi = \frac{\gamma_Y L}{r_Y} = \frac{\tau_Y L}{r_Y G}. \quad \textbf{(4-23)}$$

As the torque T increases beyond the value at which yielding first occurs, r_Y decreases. When $r_Y = R_i$, all of the material of the bar has yielded (Fig. 4-30) and the bar will twist with no further increase in T. Therefore, the upper bound of the torque the bar will support is obtained by setting $r_Y = R_i$ in Eq. (4-22).

EXAMPLE 4-4

A manganese bronze bar 2 m in length has the cross section shown in Fig. 4-31. The shear modulus $G = 39$ GPa and the yield stress $\tau_Y = 450$ MPa. The bar is fixed at one end and subjected to an axial torque T at the other end. Model the material as elastic–perfectly plastic. **(a)** If $T = 3$ kN-m, what is the bar's angle of twist? **(b)** What is the upper bound of the torque T the bar will support?

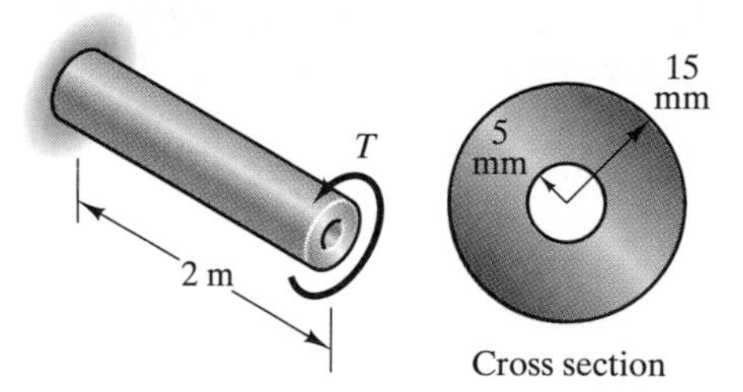

FIGURE 4-31

Strategy

(a) We will first determine whether we even need the elastic–perfectly plastic solution for the angle of twist. By substituting the yield stress $\tau_Y = 450$ MPa and $r = R_o = 0.015$ m into the elastic equation $\tau = Tr/J$, we will determine the maximum value of T for which the elastic solution applies. If the applied torque does not exceed that value, we can use the elastic solution to determine the bar's angle of twist. If it does exceed it, we must determine the value of r_Y from Eq. (4-22), then determine the angle of twist from Eq. (4-23). **(b)** The upper bound of the torque the bar will support is given by Eq. (4-22) with $r_Y = R_i$.

Solution

(a) To determine the largest torque for which the elastic solution would apply, we substitute the yield stress and the bar's outer radius into the elastic equation for the stress distribution:

$$\tau_Y = \frac{TR_o}{J}:$$

$$450 \times 10^6 = \frac{T(0.015)}{(\pi/2)[(0.015)^4 - (0.005)^4]}.$$

Solving, we obtain $T = 2.36$ kN-m. The applied torque of 3 kN-m exceeds this value, so we must use the elastic–perfectly plastic solution. From Eq. (4-22), the relation between the applied torque T and r_Y is

$$T = \pi(450 \times 10^6)(0.015)^3 \left[\frac{2}{3} - \frac{1}{6}\left(\frac{r_Y}{0.015}\right)^3 - \frac{1}{2}\left(\frac{0.005}{0.015}\right)^4 \frac{0.015}{r_Y}\right]. \tag{4-24}$$

Figure (a) shows the graph of T as a function of r_Y for $R_i \le r_Y \le R_o$. From the graph we estimate that $T = 3000$ N-m corresponds to a value of r_Y of 8.1 mm. By using software designed to solve nonlinear algebraic equations, we obtain $r_Y = 8.128$ mm.

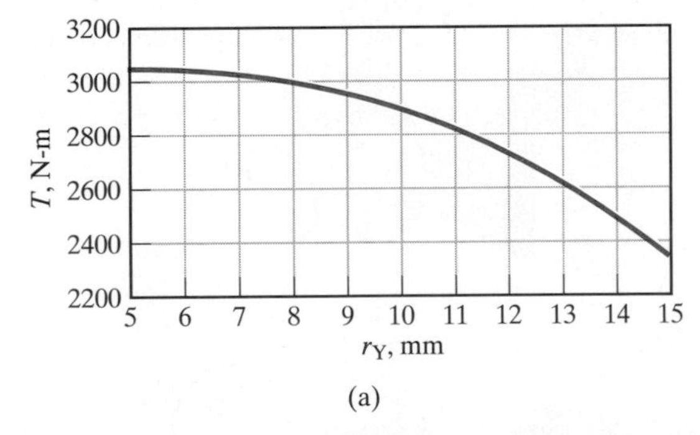

(a)

(a) Graph of T as a function of the radial position at which the yield stress is reached.

From Eq. (4-23), the bar's angle of twist is

$$\phi = \frac{\tau_Y L}{r_Y G} = \frac{(450 \times 10^6)(2)}{(0.008128)(39 \times 10^9)} = 2.84 \text{ rad},$$

which is 163°.

(b) To determine the upper bound of the torque T the bar will support, we set $r_Y = R_i = 0.005$ m in Eq. (4-24), obtaining $T = 3060$ N-m.

4-6 | Torsion of Thin-Walled Tubes

The results we have presented for the deformations and states of stress of bars subjected to torsional loads apply only to bars with circular cross sections. Extending these results to an arbitrary cross-sectional shape generally requires the use of advanced analytical techniques or a numerical approach such as the finite element method. In this section we describe a clever approximate analysis of a limited class of bars with noncircular cross sections that was presented by R. Bredt, a German engineer, in 1896.

Stress

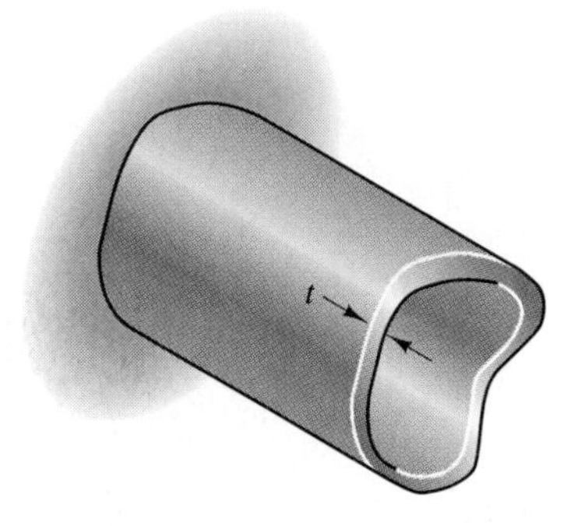

FIGURE 4-32 Thin-walled prismatic tube.

We consider a prismatic tube whose wall thickness t is small compared to the lateral dimensions of the tube (Fig. 4-32). As we indicate in the figure, the wall thickness need not be uniform. In Fig. 4-33 we subject the tube to a torque T and obtain a free-body diagram by passing a plane perpendicular to its axis. The stress distribution for a circular bar suggests the assumption that the tube is subjected to a shear stress parallel to the tube wall. Here we interpret τ as the average value of the shear stress across the wall's thickness. We do not assume that τ is uniform around the tube's circumference; in fact, we will

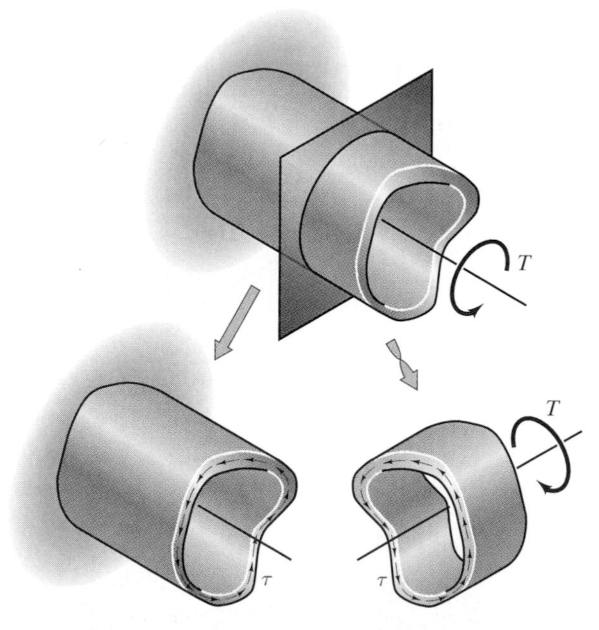

FIGURE 4-33 Assumed shear stress distribution on a plane perpendicular to the bar's axis.

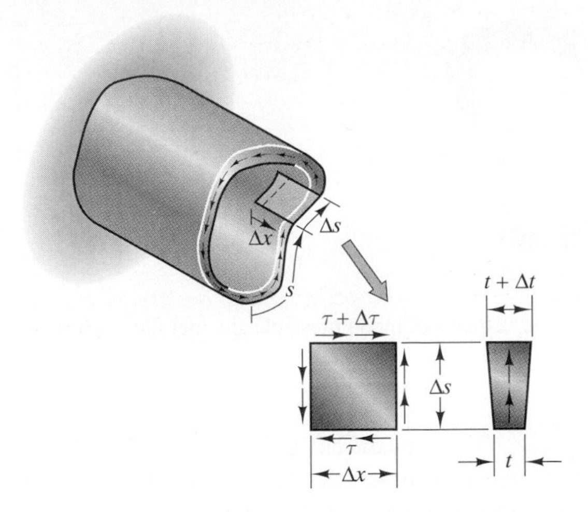

FIGURE 4-34 Element of the tube wall. The shear stress τ may vary in the circumferential direction.

show that its value must vary if the wall thickness t varies around the circumference.

Our first step in determining τ is to consider the element in Fig. 4-34. The coordinate s measures distance along the circumference of the wall relative to some reference point. As s increases by the amount Δs, the wall thickness may change by an amount Δt and the shear stress may change by an amount $\Delta\tau$. The sum of the forces on the element in the direction parallel to the bar's axis (the x direction) is

$$(\tau + \Delta\tau)(\Delta x)(t + \Delta t) - \tau \Delta x t = 0.$$

We divide this equation by $\Delta x \; \Delta s$ and write it as

$$\tau \frac{\Delta t}{\Delta s} + \frac{\Delta\tau}{\Delta s} t + \frac{\Delta\tau}{\Delta s}\frac{\Delta t}{\Delta s} \Delta s = 0.$$

Taking the limit of this expression as $\Delta s \to 0$, we obtain

$$\tau \frac{dt}{ds} + \frac{d\tau}{ds} t = \frac{d}{ds}(\tau t) = 0.$$

This result indicates that the product of the shear stress and the wall thickness, which we denote by

$$f = \tau t, \tag{4-25}$$

does not depend on s. That is, it is constant in the circumferential direction.

The quantity f is called the *shear flow*. This term originates from an interesting analogy with fluid flow. If you visualize the bar's cross section as a channel of steadily flowing incompressible fluid of uniform depth and width t,

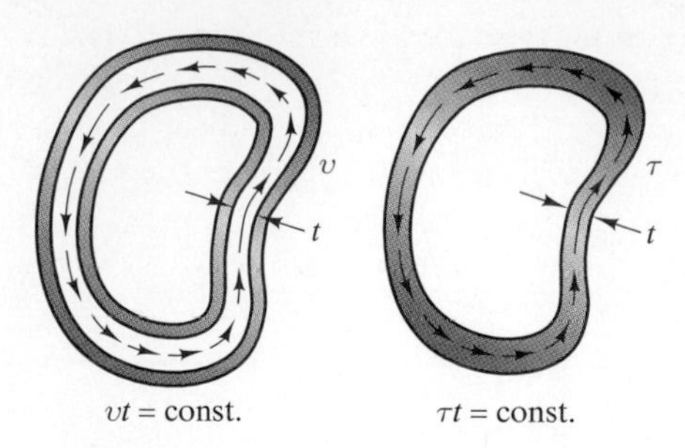

FIGURE 4-35 Analogy between the velocity of incompressible channel flow and the shear stress.

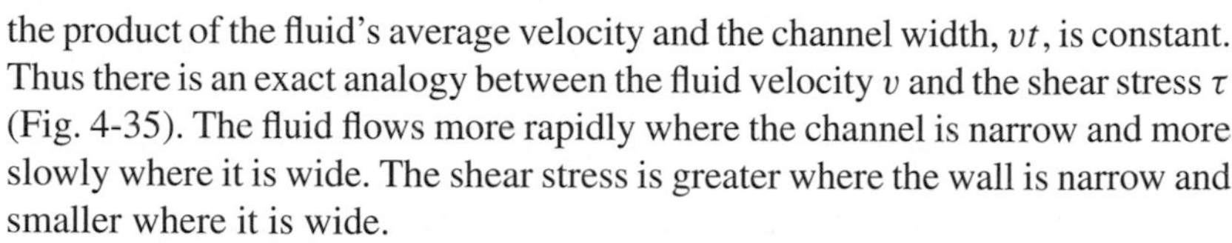

the product of the fluid's average velocity and the channel width, vt, is constant. Thus there is an exact analogy between the fluid velocity v and the shear stress τ (Fig. 4-35). The fluid flows more rapidly where the channel is narrow and more slowly where it is wide. The shear stress is greater where the wall is narrow and smaller where it is wide.

We now know that the shear flow $f = \tau t$ is constant around the tube's circumference, but we don't know its value. We must determine it by equating the couple exerted by the shear stress distribution to the external torque T (Fig. 4-33). Remarkably, we can determine the couple exerted by the shear stress distribution without specifying the shape of the tube's cross section. The force exerted by the shear stress on an element of the cross section of length ds is $\tau t\, ds = f\, ds$ (Fig. 4-36a). The moment due to this force about the tube's axis is $hf\, ds$, where h is the perpendicular distance to the line of action of the force. (Since the moment due to a couple is the same about any point, our choice of the location of the axis is arbitrary.) The integral of this expression over the entire circumference must equal T:

$$T = f \int_s h\, ds.$$

At this point things don't look promising, because h depends on the shape of the cross section. However, notice that the area of the triangle in Fig. 4-36b is $dA = h\, ds/2$. Therefore, we can write the equilibrium equation as

$$T = 2f \int_A dA = 2fA,$$

where A is the cross-sectional area of the tube (Fig. 4-36c). From this result we obtain the value of the shear flow in terms of the external torque:

$$f = \tau t = \frac{T}{2A}. \qquad \text{(4-26)}$$

Maximizing the tube's cross-sectional area minimizes the shear flow, which explains the popularity of circular tubes for supporting torsional loads.

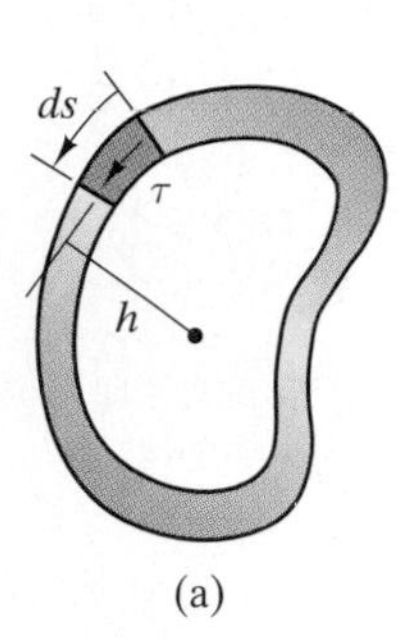

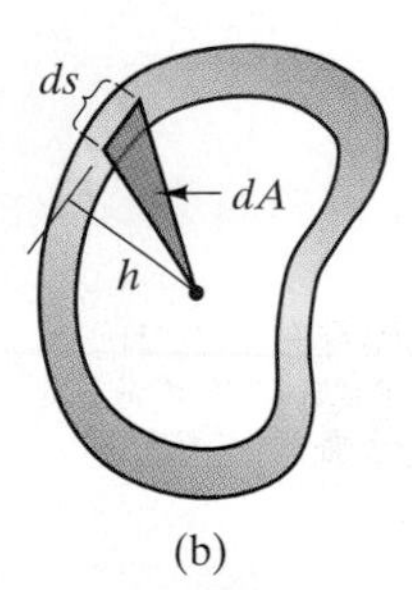

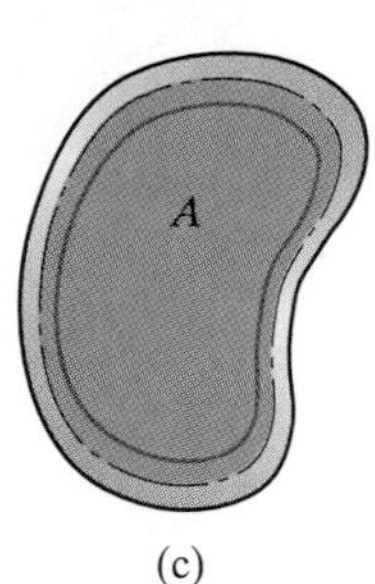

FIGURE 4-36 (a) Determining the moment due to the shear stress on an element ds. (b) The moment can be expressed in terms of the area dA. (c) A is the area within the midline of the tube wall.

We now know the stress distribution in the tube in terms of the external torque T. We can determine the shear flow from Eq. (4-26). Then for a given circumferential position on the tube wall with thickness t, we can evaluate the average shear stress τ from Eq. (4-25). (See Problem 4-6.17, in which the shear stress obtained in this way is compared to the exact shear stress for the case of a circular thin-walled tube.) Our next objective is to determine the tube's angle of twist in terms of T.

Angle of Twist

When the torque T is applied to the tube, twisting it through an angle ϕ (Fig. 4-37), work is done on the tube. The work done by a torque acting through an angle ϕ is

$$\text{work} = \int_0^{\phi} T\, d\phi.$$

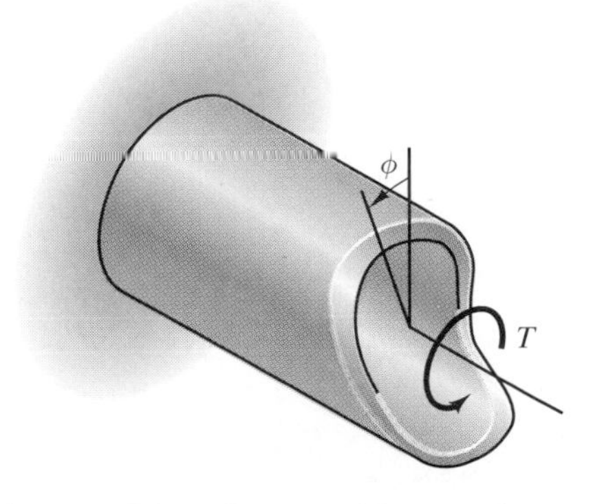

FIGURE 4-37 The torque T rotates the end of the tube through an angle ϕ.

Because the angle through which the elastic tube rotates is a linear function of the torque (Fig. 4-38), the work equals one-half the product of the torque and the angle of twist:

$$\text{work} = \tfrac{1}{2} T\phi. \qquad \textbf{(4-27)}$$

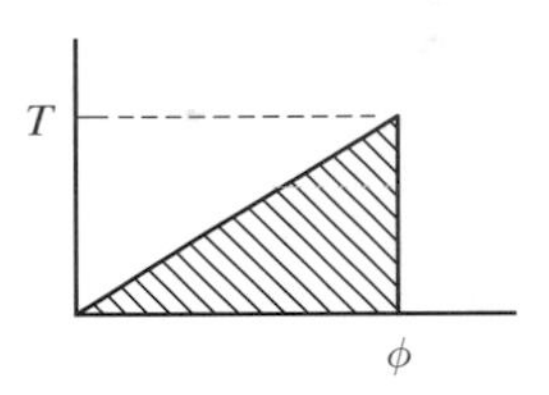

FIGURE 4-38 The angle of twist of the tube is a linear function of the applied torque. The triangular area equals the work done.

The work done in stretching a linear spring is stored in the spring as an equal amount of potential energy. In the same way, the work done in twisting an elastic tube is stored within the material of the deformed tube as *strain energy*, which we discuss in Chapter 11. By equating the work done to the total strain energy, we can determine the tube's angle of twist.

Consider the infinitesimal element of the tube shown in Fig. 4-39a. In Section 11.1 we show that the strain energy per unit volume of an element subjected to a pure shear stress τ is

$$u = \tfrac{1}{2}\tau\gamma,$$

where γ is the shear strain. The volume of the element in Fig. 4-39a is $t\, ds\, dx$ and the shear strain $\gamma = \tau/G$, so the strain energy of the element is

$$U_{\text{element}} = \frac{\tau^2 t\, ds\, dx}{2G}.$$

From Eq. (4-26), the shear stress $\tau = T/2At$. Using this expression, the strain energy of the element is

$$U_{\text{element}} = \frac{T^2\, ds\, dx}{8A^2 G t}.$$

By integrating this result with respect to x over the length of the tube, we obtain the strain energy of a strip element with circumferential dimension ds

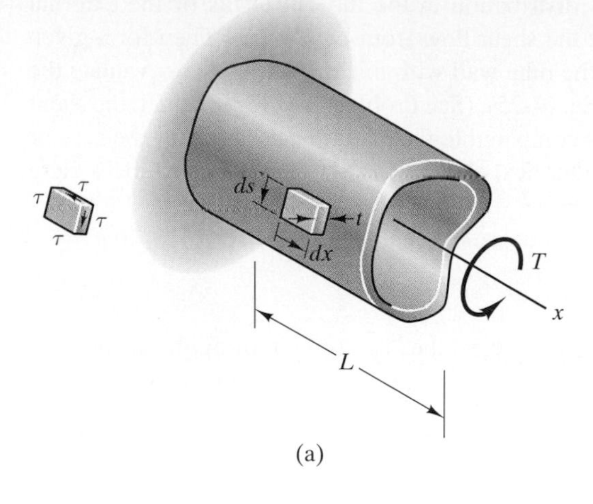

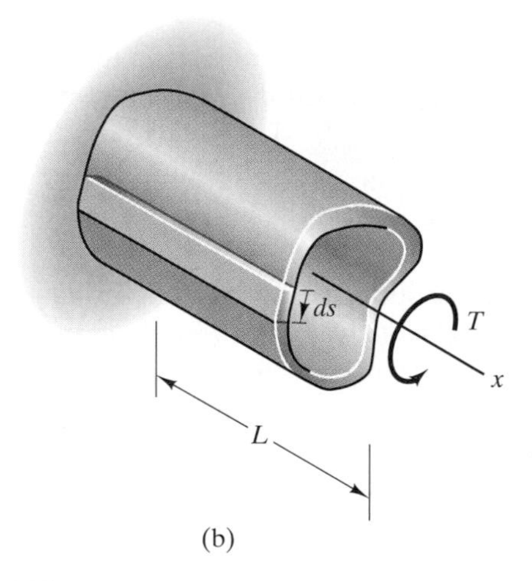

| **FIGURE 4-39** Integrating to determine the strain energy in the tube.

(Fig. 4-39b):

$$U_{\text{strip}} = \int_0^L \frac{T^2\, ds\, dx}{8A^2Gt} = \frac{T^2L\, ds}{8A^2Gt}.$$

To obtain the total strain energy of the tube, we must integrate this expression with respect to s over the tube's circumference:

$$U = \frac{T^2L}{8A^2G}\int_s \frac{ds}{t}.$$

We can now equate the strain energy stored within the material of the tube to the work done on the bar in twisting it, given by Eq. (4-27), and solve for the angle of twist:

$$\phi = \frac{TL}{4A^2G} \int_s \frac{ds}{t}. \tag{4-28}$$

To evaluate the integral in this expression, the dependence of the tube's wall thickness t on the circumferential coordinate s must be specified.

EXAMPLE 4-5

The prismatic tube in Fig. 4-40 consists of material with shear modulus $G =$ 28 GPa and has the cross section shown. The straight part of the tube wall has a thickness of 4 mm, and the semicircular part has a thickness of 2 mm. The tube is subjected to a torque $T = 800$ N-m at the end. Determine **(a)** the magnitude of the maximum shear stress in the tube; **(b)** the angle of twist.

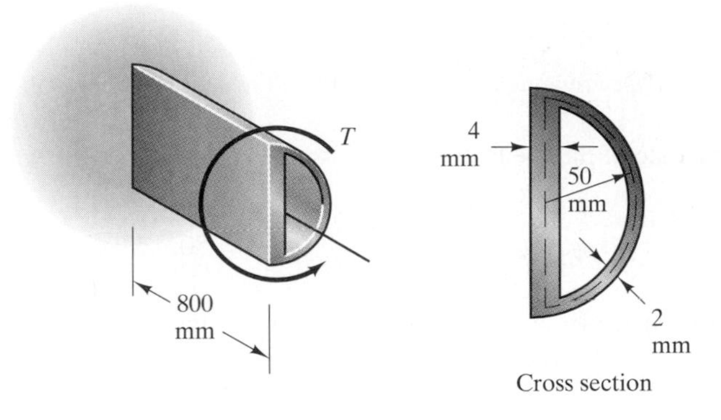

FIGURE 4-40

Strategy

(a) To determine the shear flow from Eq. (4-26), we must calculate the area A, which is the area within the dashed centerline of the tube wall. Once we know the shear flow, we can use Eq. (4-25) to determine the values of the shear stress in the straight and semicircular parts of the wall. Since the shear stress is inversely proportional to the wall thickness, clearly the maximum shear stress occurs in the semicircular part of the wall. **(b)** The angle of twist is given by Eq. (4-28). Since the wall thickness is constant in the straight and semicircular parts of the wall, the integral in Eq. (4-28) can easily be evaluated if we express it as the sum of integrals over the two parts.

Solution

(a) The area A [Fig. (a)] is $\pi(0.05)^2/2 = 0.00393$ m^2, so the shear flow is

$$f = \frac{T}{2A} = \frac{800}{(2)(0.00393)} = 101.9 \text{ kN/m}.$$

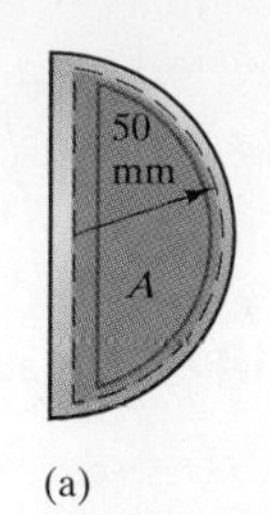

(a)

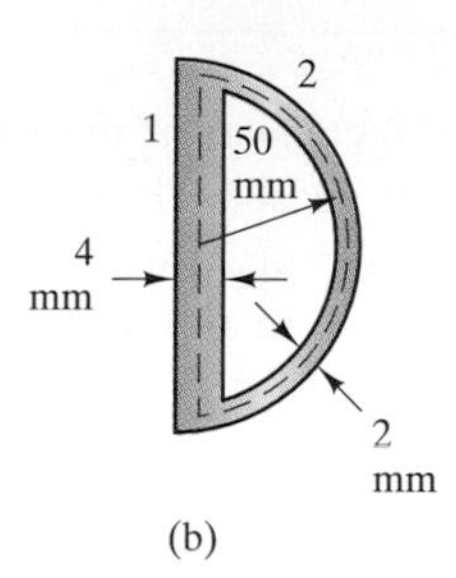

(b)

(a) Determining the area A. (b) Parts 1 and 2 of the tube wall.

The magnitude of the shear stress in the semicircular part of the wall is

$$\tau_{\text{semicircle}} = \frac{f}{t} = \frac{101.9 \times 10^3}{0.002} = 50.9 \text{ MPa}.$$

The straight part of the wall is twice as thick, so the magnitude of the shear stress in the straight part is one-half this value.

(b) Denote the straight and semicircular parts of the tube wall as parts 1 and 2 [Fig. (b)]. The integral in Eq. (4-28) is

$$\begin{aligned}\int_s \frac{ds}{t} &= \int_{s_1} \frac{ds}{t} + \int_{s_2} \frac{ds}{t} \\ &= \frac{1}{t_1}\int_{s_1} ds + \frac{1}{t_2}\int_{s_2} ds \\ &= \frac{s_1}{t_1} + \frac{s_2}{t_2} \\ &= \frac{0.1}{0.004} + \frac{\pi(0.05)}{0.002} \\ &= 103.5.\end{aligned}$$

The angle of twist of the bar is

$$\phi = \frac{TL}{4A^2G}\int_s \frac{ds}{t} = \frac{(800)(0.8)(103.5)}{(4)(0.00393)^2(28 \times 10^9)} = 0.0384 \text{ rad},$$

which is 2.20°.

4-7 Design Issues

The results we have presented in this chapter can be applied to the design of bars to support torsional loads.

Cross Sections

Except for our analysis of thin-walled cross sections in Section 4-6, we have restricted our study of torsion to bars with circular cross sections. Our primary reason for doing so is that consideration of other cross-sectional shapes requires advanced analytical methods, numerical solutions, or the use of empirical information. But this is not as serious a limitation for a discussion of design as it might appear. The reason is that in normal circumstances a circular cross section is optimal for a bar supporting an axial torsional load. Figure 4-41 compares the ratios of the maximum shear stresses resulting from a given torque applied to bars with circular, elliptical, and square cross sections of equal area.

Furthermore, a hollow circular cross section is usually preferable to a solid one. In Fig. 4-42 a torque T is applied to bars with solid and hollow cross sections. The maximum shear stress in the solid bar is

$$\tau_{\text{solid}} = \frac{TR}{(\pi/2)R^4}, \tag{4-29}$$

and the maximum shear stress in the hollow bar is

$$\tau_{\text{hollow}} = \frac{TR_{\text{o}}}{(\pi/2)\left(R_{\text{o}}^4 - R_{\text{i}}^4\right)}. \tag{4-30}$$

If the bars have equal cross-sectional areas, $\pi R^2 = \pi R_{\text{o}}^2 - \pi R_{\text{i}}^2$, and we denote the wall thickness of the hollow bar by $t = R_{\text{o}} - R_{\text{i}}$, we can write the ratio of

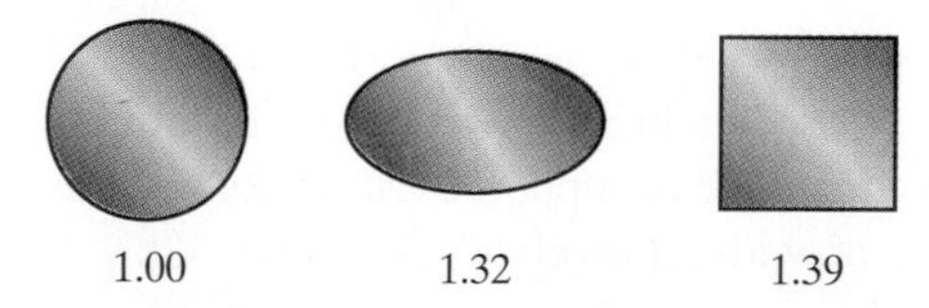

FIGURE 4-41 Ratio of the maximum shear stress to the value for a circular cross section of equal area.

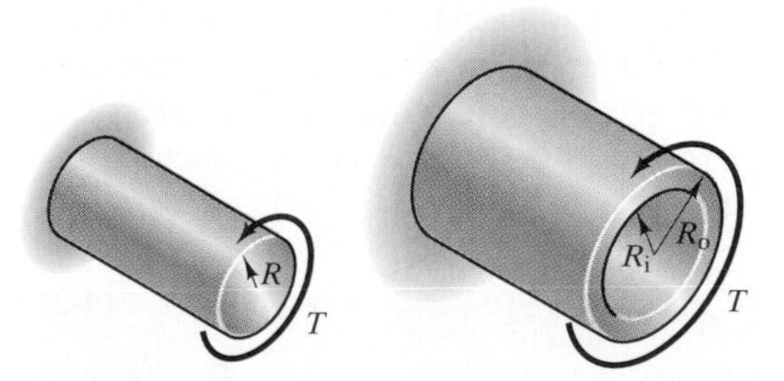

FIGURE 4-42 Applying the same torque T to bars with solid and hollow circular cross sections.

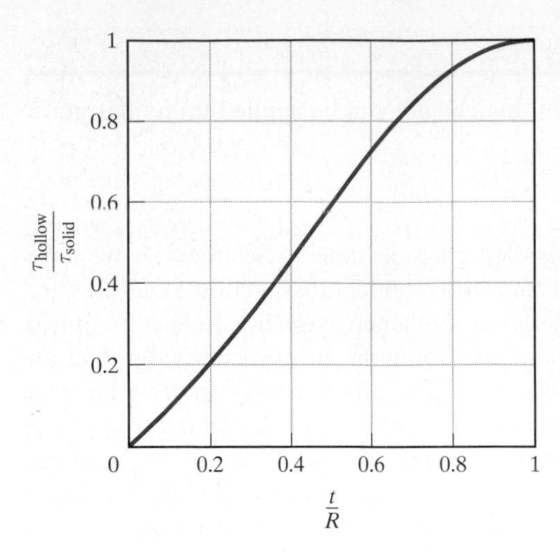

FIGURE 4-43 Ratio of the maximum stress in a hollow circular bar to that in a solid circular bar.

the maximum shear stresses as

$$\frac{\tau_{\text{hollow}}}{\tau_{\text{solid}}} = \frac{(t/R)\left[1 + (t/R)^2\right]}{1 + (t/R)^4}.$$

From this relationship, plotted in Fig. 4-43, the case for using hollow bars to support torsional loads is clear: For a given torque, the maximum shear stress in a hollow bar is smaller than in a solid bar of the same cross-sectional area (which means a bar of the same weight). The angle of twist for a given torque is also smaller for the hollow bar.

The maximum stress in the hollow bar continues to decrease as the wall thickness decreases. But the wall must not be made too thin, as can readily be illustrated by applying torque to an aluminum soft drink can (Fig. 4-44). The wall of the can forms wrinkles and fails by buckling, or geometric instability. Because of these considerations, bars designed to support axial torsional loads often have hollow circular cross sections with relatively thick walls.

Allowable Stress

In a circular bar subjected to torsion there is no plane on which the shear stress is larger in magnitude than the values given by Eq. (4-29) or (4-30), and the maximum tensile and compressive stresses are equal in magnitude to the maximum shear stress. As a result, the allowable maximum shear stress τ_{allow} for a bar subjected to torsion can be defined to be some specified fraction of the yield stress σ_{Y}, and the factor of safety defined by

$$S = \frac{\sigma_{\text{Y}}}{\tau_{\text{allow}}}. \qquad \textbf{(4-31)}$$

Some of the constraints that the design engineer must typically balance against the desire for a conservatively large factor of safety are discussed in Section 3-7.

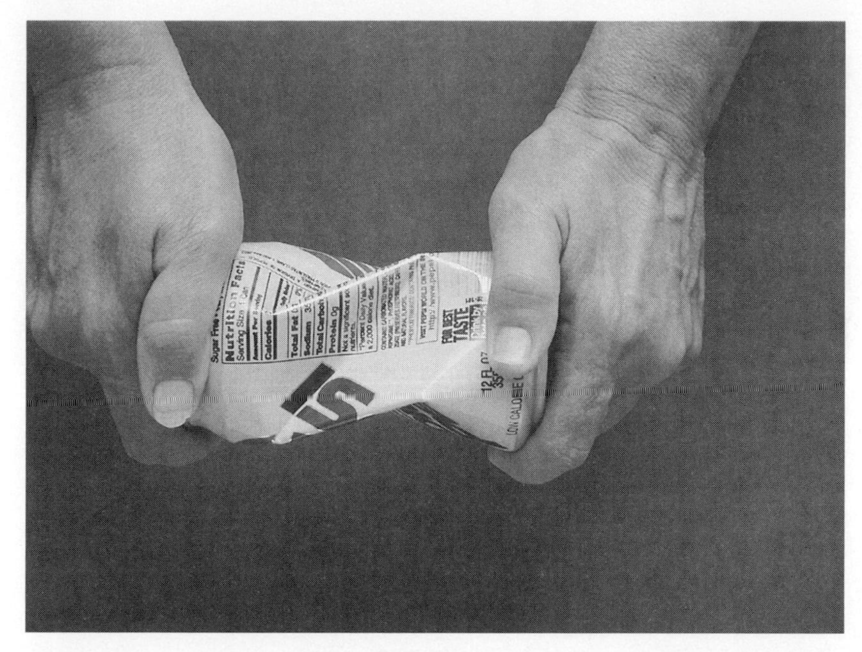

FIGURE 4-44 Applying torque to a bar with a thin wall (here an aluminum drink can) can cause it to fail by buckling.

EXAMPLE 4-6

The power (work per unit time) transmitted by a rotating shaft is $P = T\omega$, where T is the axial torque and ω is the shaft's angular velocity in radians per second. The maximum torque produced by the engine of the car in Fig. 4-45 occurs at 4750 rpm, when the engine is generating 286 horsepower (hp). Design a drive shaft for the car that is constructed of steel with yield stress $\sigma_Y = 80{,}000$ psi and has a factor of safety $S = 4$. (Although the shaft is rotating, assume that the maximum shear stress can be adequately approximated by assuming that the material is in equilibrium.)

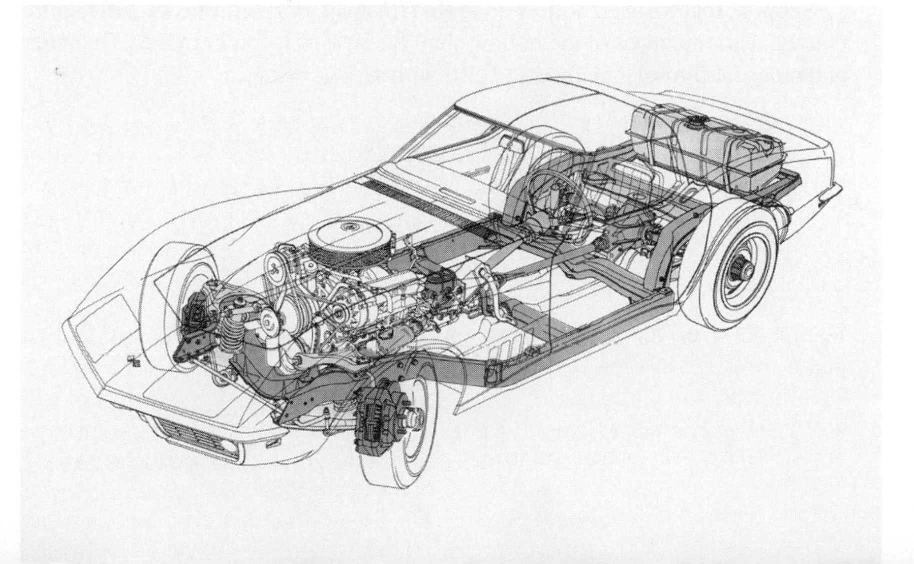

Strategy

Knowing the power transmitted by the shaft and its angular velocity, we can determine the torque. We can use Eq. (4-31) to determine the allowable maximum shear stress, then use Eq. (4-29) or (4-30) to determine the dimensions of the cross section.

Solution

One horsepower is 550 ft-lb/s. Determining the torque from the expression

$$P = T\omega:$$
$$(286)(550) = T\left[\frac{(4750)(2\pi)}{60}\right],$$

we obtain $T = 316$ ft-lb $= 3790$ in-lb. The allowable maximum shear stress is

$$\tau_{\text{allow}} = \frac{\sigma_Y}{S} = \frac{80{,}000}{4} = 20{,}000 \text{ psi.}$$

If we use a solid drive shaft, the radius of the shaft is determined from Eq. (4-29) with $\tau_{\text{solid}} = \tau_{\text{allow}}$:

$$\tau_{\text{allow}} = \frac{T}{(\pi/2)R^3}:$$
$$20{,}000 = \frac{3790}{(\pi/2)R^3}.$$

Solving, we obtain $R = 0.494$ in.

Suppose that we use a hollow drive shaft instead. For stability, we will require that the wall thickness of the hollow shaft be 30% of its outer radius. The inner and outer radii must satisfy Eq. (4-30) with $\tau_{\text{hollow}} = \tau_{\text{allow}}$:

$$\tau_{\text{allow}} = \frac{TR_o}{(\pi/2)\left(R_o^4 - R_i^4\right)}:$$
$$20{,}000 = \frac{3790R_o}{(\pi/2)\left(R_o^4 - R_i^4\right)}.$$

Setting $R_i = 0.7R_o$ in this expression and solving, we obtain $R_o = 0.542$ in. and $R_i = 0.379$ in.

Discussion

Using 15 slug/ft^3 as the density of steel, the solid drive shaft weighs 2.57 lb/ft while the hollow drive shaft weighs 1.58 lb/ft. The hollow shaft also has superior dynamic (vibrational) properties.

Chapter Summary

Pure Shear Stress

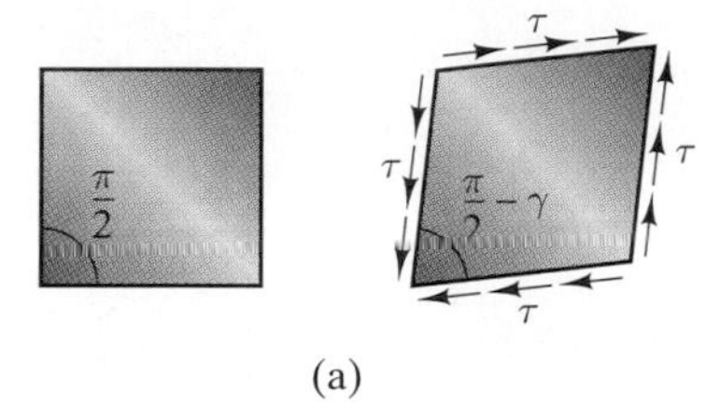

(a)

In Fig. (a) a cube is subjected to a state of pure shear stress. If the stress is not too large, for many materials it is related to the shear strain by

$$\tau = G\gamma, \qquad \text{Eq. (4-1)}$$

where the constant G is the *shear modulus*. Typical values of G are given in Appendix B. The shear modulus is related to the modulus of elasticity and Poisson's ratio by

$$G = \frac{E}{2(1+\nu)}. \qquad \text{Eq. (4-2)}$$

Stresses on Oblique Planes

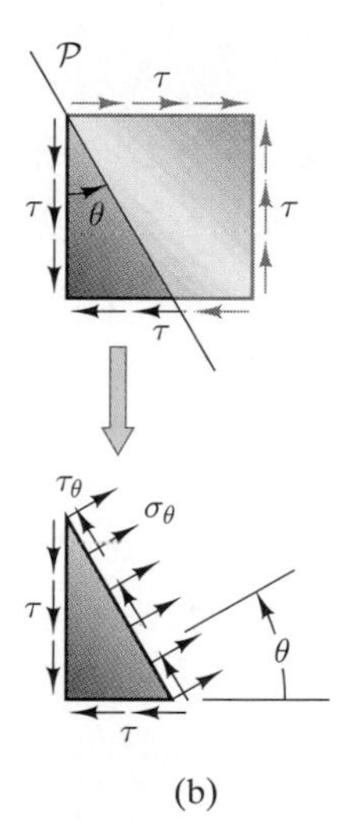

(b)

The normal and shear stresses on a plane $\mathcal{P}$ oriented as shown in Fig. (b) are

$$\sigma_\theta = 2\tau \sin\theta \cos\theta, \qquad \text{Eq. (4-3)}$$

$$\tau_\theta = \tau(\cos^2\theta - \sin^2\theta). \qquad \text{Eq. (4-4)}$$

There is no value of θ for which the shear stress is larger in magnitude than τ. The maximum tensile and compressive stresses occur at $\theta = 45^\circ$ where $\sigma_\theta = \tau$ and at $\theta = 135^\circ$ where $\sigma_\theta = -\tau$.

Torsion of Prismatic Circular Bars

Consider a cylindrical bar of length L and radius R subjected to an axial torque T [Fig. (c)]. The resulting angle of twist of the end of the bar is

$$\phi = \frac{TL}{GJ}, \qquad \text{Eq. (4-7)}$$

where

$$J = \frac{\pi}{2}R^4 \qquad \text{Eq. (4-10)}$$

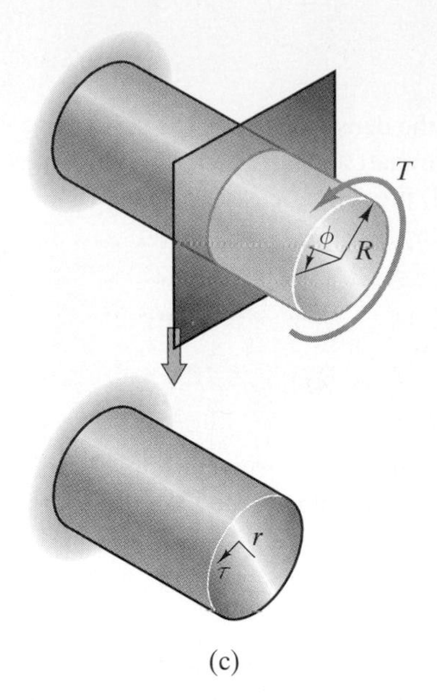

(c)

is the polar moment of inertia of the bar's cross-sectional area about its axis. The shear stress on the plane shown at a distance r from the bar's axis is

$$\tau = \frac{Tr}{J}. \qquad \text{Eq. (4-9)}$$

Equations (4-7) and (4-9) also apply to a bar with a hollow circular cross section. In that case the polar moment of inertia is given by

$$J = \frac{\pi}{2}\left(R_o^4 - R_i^4\right), \qquad \text{Eq. (4-11)}$$

where R_i and R_o are the bar's inner and outer radii.

Statically Indeterminate Problems

Solutions to problems involving torsionally loaded bars in which the number of unknown reactions exceeds the number of independent equilibrium equations involve three elements: (1) relations between the torques in bars and their angles of twist; (2) compatibility relations imposed on the angles of twist by the geometry of the problem; and (3) equilibrium.

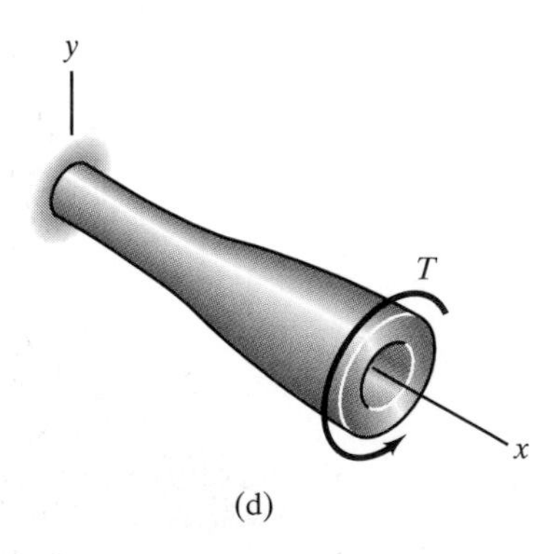

(d)

Bars with Gradually Varying Cross Sections

If the polar moment of inertia is a gradually varying function of axial position $J(x)$ [Fig. (d)], the stress distribution on a plane perpendicular to the bar's axis at axial position x can be approximated by

$$\tau = \frac{Tr}{J(x)}. \qquad \text{Eq. (4-17)}$$

The angle of twist of the bar is

$$\phi = \int_0^L \frac{T\,dx}{GJ(x)}. \qquad \text{Eq. (4-18)}$$

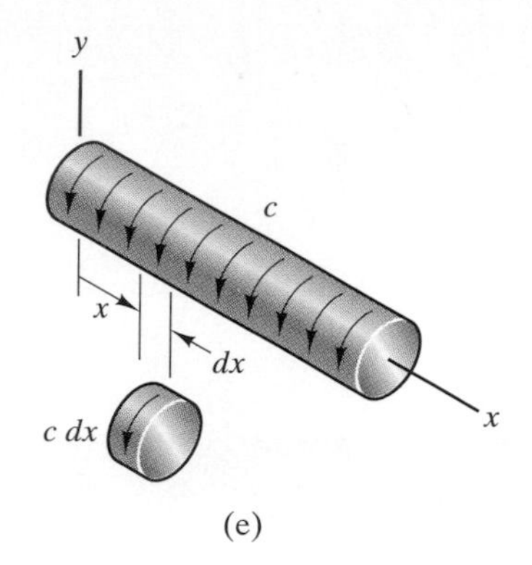

(e)

Distributed Torsional Loads

A distributed torsional load on a bar can be described by a function c defined such that the axial torque on each element dx of the bar is $c\,dx$ [Fig. (e)].

Torsion of an Elastic–Perfectly Plastic Circular Bar

The elastic–perfectly plastic model of shear stress-shear strain behavior is shown in Fig. (f). If the shear strain at the bar's outer surface does not exceed the value γ_Y corresponding to the yield stress, the shear stress distribution is given by Eq. (4-9) and the angle of twist by Eq. (4-7). If the shear strain does exceed the value corresponding to the yield stress at some radial position r_Y, the shear stress distribution is as shown in Fig. (g). The torque T and radial distance r_Y satisfy the equilibrium equation

$$T = \pi\tau_Y R_o^3 \left[\frac{2}{3} - \frac{1}{6}\left(\frac{r_Y}{R_o}\right)^3 - \frac{1}{2}\left(\frac{R_i}{R_o}\right)^4 \frac{R_o}{r_Y}\right]. \qquad \text{Eq. (4-22)}$$

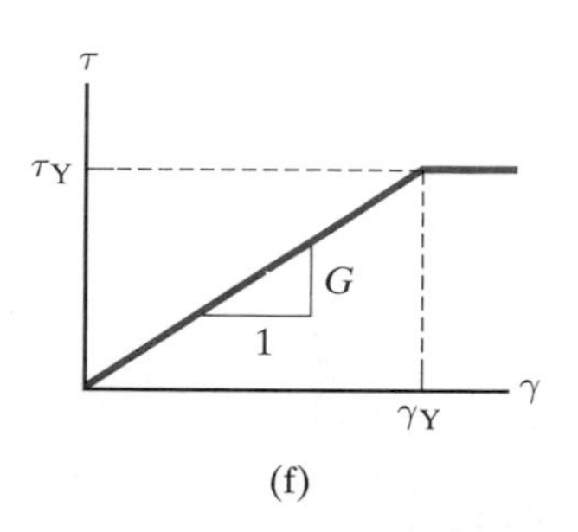

(f)

The upper bound of the torque the bar will support is obtained by setting $r_Y = R_i$ in Eq. (4-22). The bar's angle of twist is

$$\phi = \frac{\gamma_Y L}{r_Y} = \frac{\tau_Y L}{r_Y G} \qquad \text{Eq. (4-23)}$$

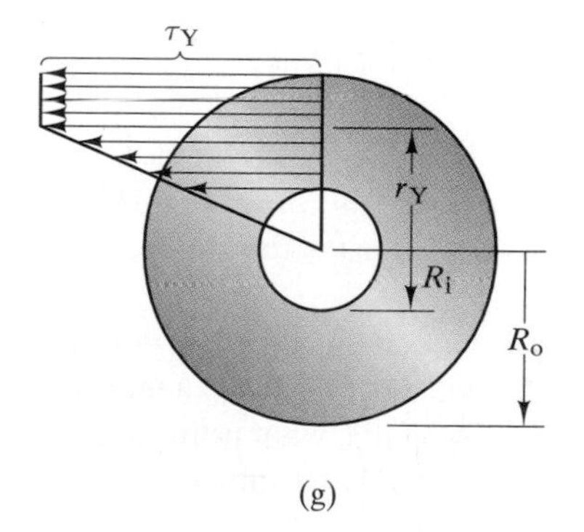

(g)

Torsion of Thin-Walled Tubes

Consider a prismatic tube whose wall thickness t is small compared to the lateral dimensions of the tube [Fig. (h)]. The *shear flow* f is given by

$$f = \tau t = \frac{T}{2A}, \qquad \text{Eq. (4-26)}$$

where A is the cross-sectional area of the tube [Fig. (i)]. The angle of twist of

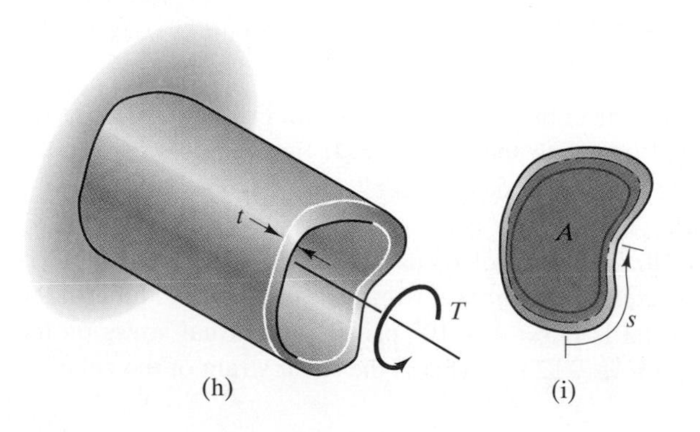

(h) (i)

the tube is

$$\phi = \frac{TL}{4A^2G}\int_s \frac{ds}{t}, \qquad \text{Eq. (4-28)}$$

where s measures distance along the circumference of the wall relative to some reference point.

PROBLEMS

4-1.1. The cube of material is subjected to a pure shear stress $\tau = 9$ MPa. The angle β is measured and determined to be 89.98°. What is the shear modulus G of the material?

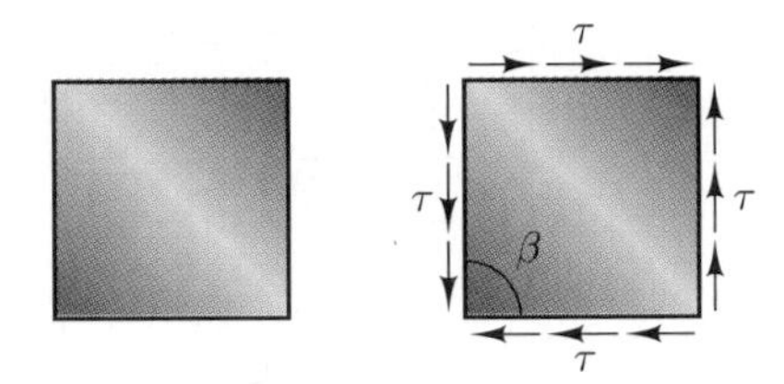

PROBLEM 4-1.1

4-1.2. If the cube in Problem 4-1.1 consists of material with shear modulus $G = 4.6 \times 10^6$ psi and the shear stress $\tau = 8000$ psi, what is the angle β in degrees?

4-1.3. If the cube in Problem 4-1.1 consists of aluminum alloy that will safely support a pure shear stress of 270 MPa and $G = 26.3$ GPa, what is the largest shear strain to which the cube can safely be subjected?

4-1.4. The cube of material is subjected to a pure shear stress $\tau = 12$ MPa. What are the normal stress and the magnitude of the shear stress on the plane $\mathcal{P}$?

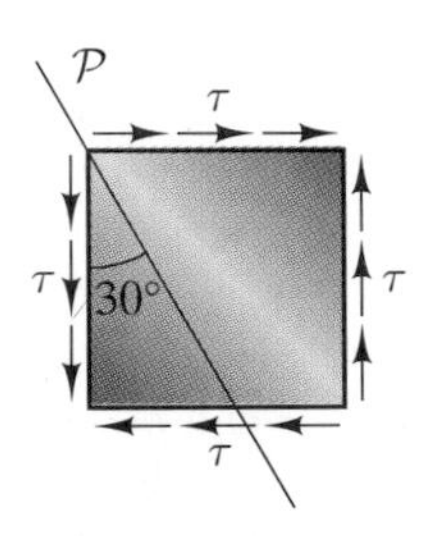

PROBLEM 4-1.4

4-1.5. In Problem 4-1.4, what are the magnitudes of the maximum tensile, compressive, and shear stresses to which the material is subjected?

4-1.6. The cube of material shown in Problem 4-1.4 is subjected to a pure shear stress τ. If the normal stress on the plane $\mathcal{P}$ is 14 MPa, what is τ?

4-1.7. The cube of material shown in Problem 4-1.4 is subjected to a pure shear stress τ. The shear modulus of the material is $G = 28$ GPa. If the normal stress on the plane $\mathcal{P}$ is 80 MPa, what is the shear strain of the cube?

4-1.8. The cube of material is subjected to a pure shear stress $\tau = 20$ ksi. **(a)** What are the normal stress and the magnitude of the shear stress on the plane $\mathcal{P}$? **(b)** What are the magnitudes of the maximum tensile, compressive, and shear stresses to which the material is subjected?

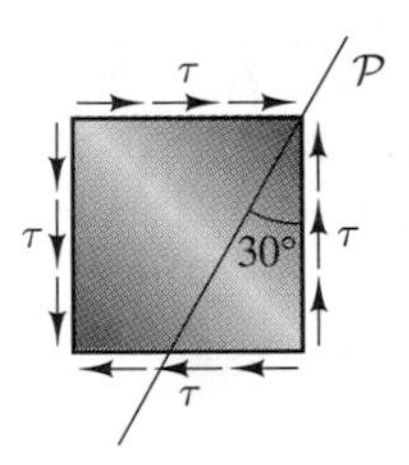

PROBLEM 4-1.8

4-1.9. The cube of material shown in Problem 4-1.8 is subjected to a pure shear stress τ. If the normal stress on the plane $\mathcal{P}$ is -20 ksi, what is τ?

4-1.10. The cube of material shown in Problem 4-1.8 is subjected to a pure shear stress τ. The shear modulus of the material is $G = 4 \times 10^6$ psi. If the normal stress on the plane $\mathcal{P}$ is -12 ksi, what is the shear strain of the cube?

Worksheet 4 can be used to solve Problems 4-2.14, 4-2.15, 4-2.19, and 4-2.20.

4-2.1. If a bar has a solid circular cross section with 15-mm diameter, what is the polar moment of inertia of its cross section in m^4?

4-2.2. If a bar has a hollow circular cross section with 2-in. outer radius and 1-in. inner radius, what is the polar moment of inertia of its cross section?

4-2.3. The bar has a circular cross section with 15-mm diameter and the shear modulus of the material is G = 26 GPa. If the torque T = 10 N-m, determine **(a)** the magnitude of the maximum shear stress in the bar; **(b)** the angle of twist of the end of the bar in degrees.

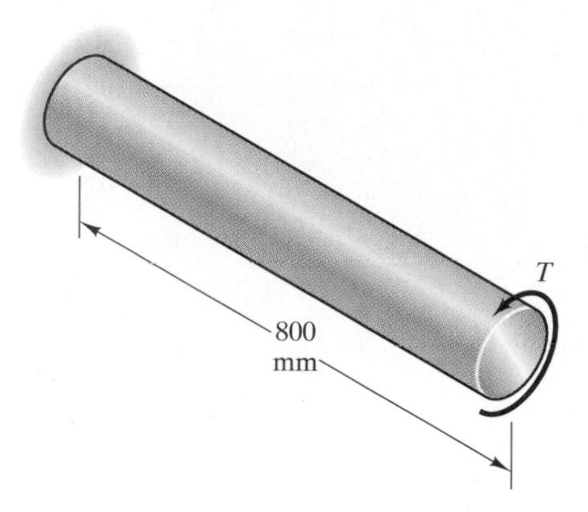

PROBLEM 4-2.3

4-2.4. If the bar in Problem 4-2.3 is subjected to a torque T that causes the end of the bar to rotate 4°, what is the magnitude of the maximum shear stress in the bar?

4-2.5. The bar in Problem 4-2.3 is to be used in an application that requires that it be subjected to an angle of twist no greater than 1°. What is the maximum allowable value of the torque T?

4-2.6. The solid circular shaft that connects the turbine blades of the hydroelectric power unit to the generator has a 0.4-m radius and supports a torque T = 2 MN-m. What is the maximum shear stress in the shaft?

4-2.7. Consider the solid circular shaft in Problem 4-2.6. The shear modulus of the material is G = 80 GPa. What angle of twist per unit meter of length is caused by the 2-MN-m torque?

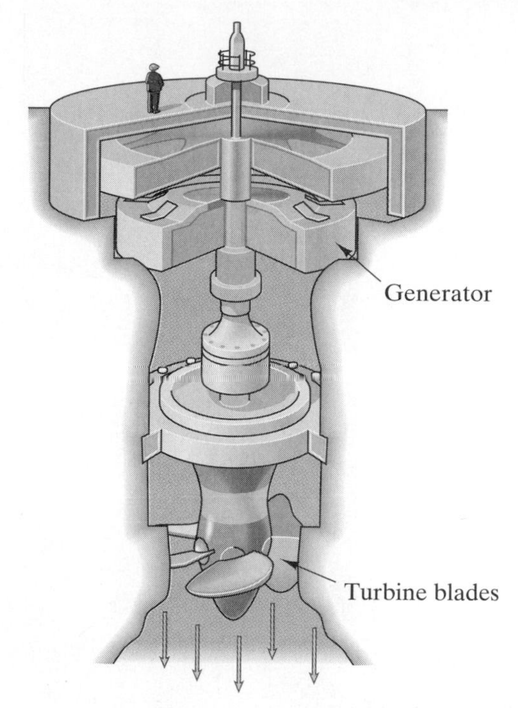

PROBLEM 4-2.6

4-2.8. If the shaft in Problem 4-2.6 has a hollow circular cross section with 0.5-m outer radius and 0.3-m inner radius, what is the maximum shear stress?

4-2.9. One type of high-strength steel drill pipe used in drilling oil wells has a 5-in. outside diameter and 4.28-in. inside diameter. If the steel will safely support a shear stress of 95 ksi, what is the largest torque to which the pipe can safely be subjected?

4-2.10. The drill pipe described in Problem 4-2.9 has a shear modulus $G = 12 \times 10^6$ psi. If it is used to drill an oil well 20,000 ft deep and the drilling operation subjects the bottom of the pipe to a torque T = 7500 in-lb, what is the resulting angle of twist (in degrees) of the 20,000-ft pipe?

4-2.11. The propeller of a wind generator is supported by a hollow circular shaft with 0.4-m outer radius and 0.3-m inner radius. The shear modulus of the material is

| **PROBLEM 4-2.11**

$G = 80$ GPa. If the propeller exerts an 840 kN-m torque on the shaft, what is the resulting maximum shear stress?

4-2.12. In Problem 4-2.11, what is the angle of twist of the propeller shaft per meter of length?

4-2.13. In designing a new shaft for the wind generator in Problem 4-2.11, the engineer wants to limit the maximum shear stress in the shaft to 10 MPa, but design constraints require retaining the 0.4-m outer radius. What new inner radius should she use?

4-2.14. The bar has a circular cross section with 1-in. diameter and the shear modulus of the material is $G = 5.8 \times 10^6$ psi. If the torque $T = 1000$ in-lb, determine **(a)** the magnitude of the maximum shear stress in the bar; **(b)** the magnitude of the angle of twist of the right end of the bar relative to the wall in degrees.

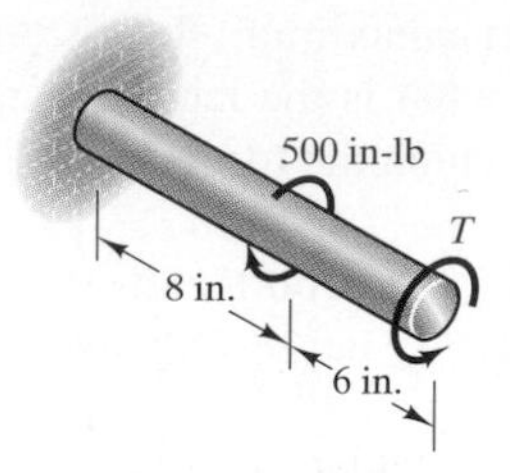

PROBLEM 4-2.14

4-2.15. For the bar in Problem 4-2.14, what value of the torque T would cause the angle of twist of the end of the bar to be zero?

4-2.16. Part A of the bar has a solid circular cross section and part B has a hollow circular cross section. The shear modulus of the material is $G = 3.8 \times 10^6$ psi. Determine the magnitudes of the maximum shear stresses in parts A and B of the bar.

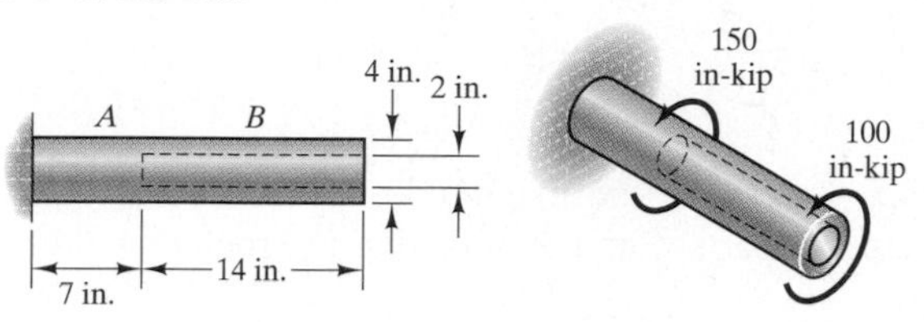

PROBLEM 4-2.16

4-2.17. For the bar in Problem 4-2.16, determine the magnitude of the angle of twist of the end of the bar relative to the wall in degrees.

4-2.18. For the bar in Problem 4-2.16, determine the magnitudes of the maximum shear stresses in parts A and B of the bar and the magnitude of the angle of twist of the end of the bar in degrees if the 150 in-kip couple acts in the opposite direction.

4-2.19. The lengths $L_A = L_B = 200$ mm and L_C = 240 mm. The diameter of parts A and C of the bar is 25 mm and the diameter of part B is 50 mm. The shear modulus of the material is $G = 80$ GPa. If the torque $T = 2.2$ kN-m, determine the magnitude of the angle of twist of the right end of the bar relative to the wall in degrees.

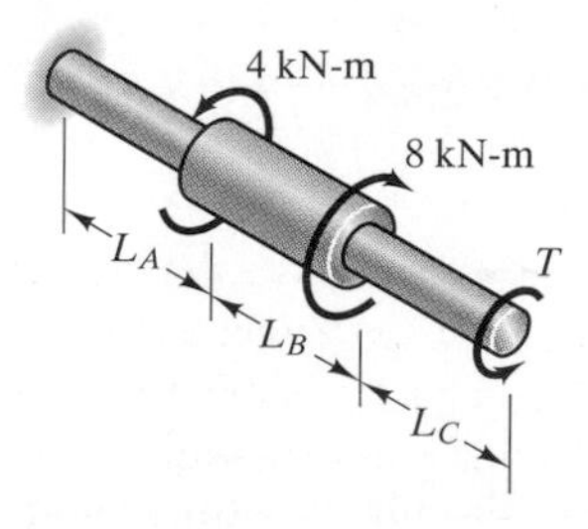

PROBLEM 4-2.19

4-2.20. For the bar in Problem 4-2.19, what value of the torque T would cause the angle of twist of the right end of the bar relative to the wall to be zero?

4-2.21. The bar in Problem 4-2.19 is made of a material that can safely support a pure shear stress of 1.1 GPa. Based on this criterion, what is the range of positive values of the torque T that can safely be applied?

4-2.22. The bars AB and CD each have a solid circular cross section with 30-mm diameter, are each 1 m in length, and consist of a material with shear modulus $G = 28$ GPa. The radii of the gears are $r_B = 120$ mm and $r_C = 90$ mm. If the torque $T_A = 200$ N-m, what are the maximum shear stresses in the bars?

PROBLEM 4-2.22

4-2.23. In Problem 4-2.22, what is the angle of rotation at A? (Assume that the deformations of the gears are negligible.)

4-2.24. Consider the system shown in Problem 4-2.22. The bars AB and CD each have a solid circular cross section with 30-mm diameter. The radii of the gears must satisfy the relation $r_B + r_C = 210$ mm. If the torque $T_A = 200$ N-m and the bars are made of a material that will safely support a pure shear stress of 40 MPa, what is the largest safe value of the radius r_C?

4-2.25. The radius $R = 200$ mm. The infinitesimal element is at the surface of the bar. What are the normal stress and the magnitude of the shear stress on the plane $\mathcal{P}$?

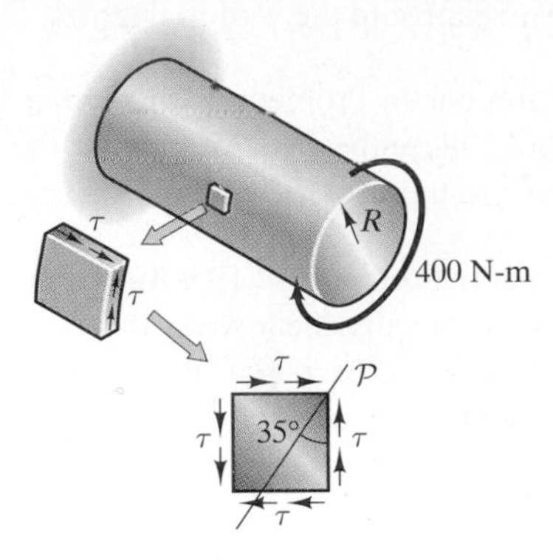

PROBLEM 4-2.25

4-2.26. For the element in Problem 4-2.25, determine the normal stress and the magnitude of the shear stress on the plane $\mathcal{P}$ shown.

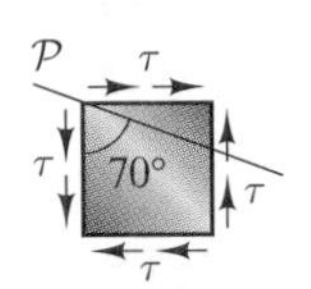

PROBLEM 4-2.26

4-3.1. The bar has a circular cross section with 1-in. diameter. If the torque $T_O = 1000$ in-lb, determine the magnitudes of the maximum shear stresses in parts A and B of the bar.

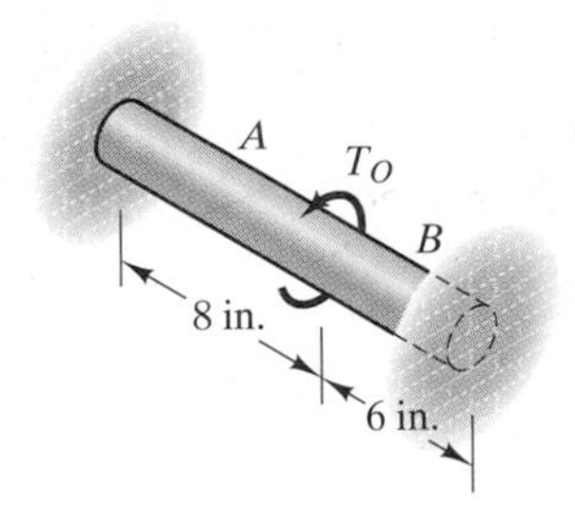

PROBLEM 4-3.1

4-3.2. Suppose that the bar in Problem 4-3.1 consists of a material that will safely support a maximum shear stress of 40 ksi. Based on this criterion, what is the maximum safe magnitude of the torque T_O?

4-3.3. Suppose that the bar in Problem 4-3.1 is subjected to a torque $T_O = 10{,}000$ in-lb and consists of a material that will safely support a maximum shear stress of 40 ksi. Based on this criterion, what is the largest distance from the left end of the bar at which the torque can safely be applied?

4-3.4. The bar is fixed at both ends. It consists of material with shear modulus $G = 28$ GPa and has a solid circular cross section. Part A is 40 mm in diameter and part B is 20 mm in diameter. Determine the torques exerted on the bar by the walls.

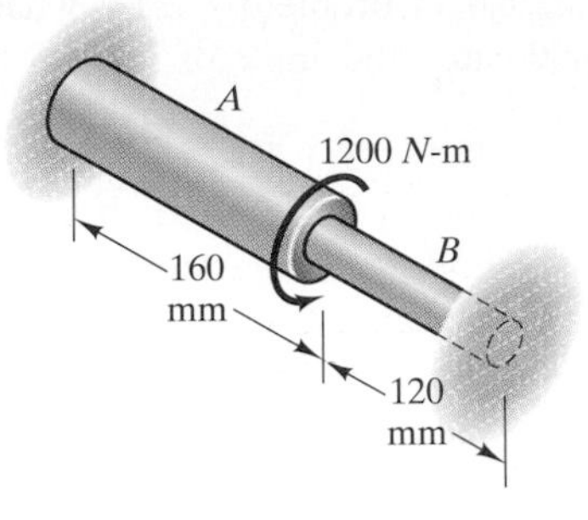

PROBLEM 4-3.4

4-3.5. Determine the magnitudes of the maximum shear stresses in parts A and B of the bar in Problem 4-3.4.

4-3.6. In Problem 4-3.4, the total length of the bar is constrained to be 280 mm. Determine the necessary lengths of parts A and B so that the maximum shear stresses in the two parts are equal. What is the magnitude of the resulting maximum shear stress?

4-3.7. Part A of the bar has a solid circular cross section and part B has a hollow circular cross section. The bar is fixed at both ends and the shear modulus of the material is $G = 3.8 \times 10^6$ psi. Determine the torques exerted on the bar by the walls.

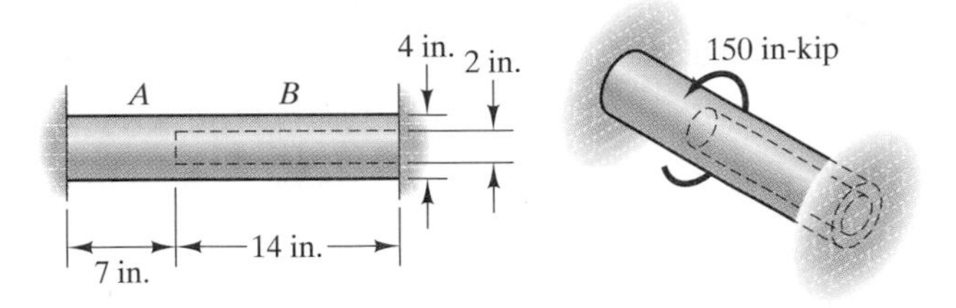

PROBLEM 4-3.7

4-3.8. Determine the magnitudes of the maximum shear stresses in parts A and B of the bar in Problem 4-3.7.

4-3.9. Suppose that you want to decrease the weight of the bar in Problem 4-3.7 by increasing the inside diameter of part B. The bar is made of material that will safely support a pure shear stress of 10 ksi. Based on this criterion, what is the largest safe value of the inside diameter?

4-3.10. Each bar is 10 in. long and has a solid circular cross section. Bar A has a diameter of 1 in. and its shear modulus is 6×10^6 psi. Bar B has a diameter of 2 in. and its shear modulus is 3.8×10^6 psi. The ends of the bars are separated by a small gap. The free end of bar A is rotated 2° about the bar's axis and the bars are welded together. What are the magnitudes of the angles of twist (in degrees) of the two bars afterward?

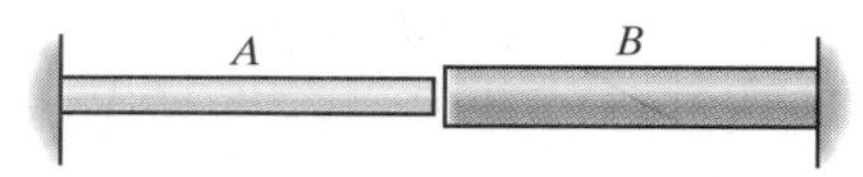

PROBLEM 4-3.10

4-3.11. In Problem 4-3.10, the ends of the bars are separated by a small gap. Suppose that the free end of bar A is rotated 2° about its axis, the free end of bar B is rotated 2° about its axis in the opposite direction, and the bars are welded together. What are the magnitudes of the maximum shear stresses in the two bars afterward?

4-3.12. The lengths $L_A = L_B = 200$ mm and $L_C =$ 240 mm. The diameter of parts A and C of the bar is 25 mm and the diameter of part B is 50 mm. The shear modulus of the material is $G = 80$ GPa. What is the magnitude of the maximum shear stress in the bar?

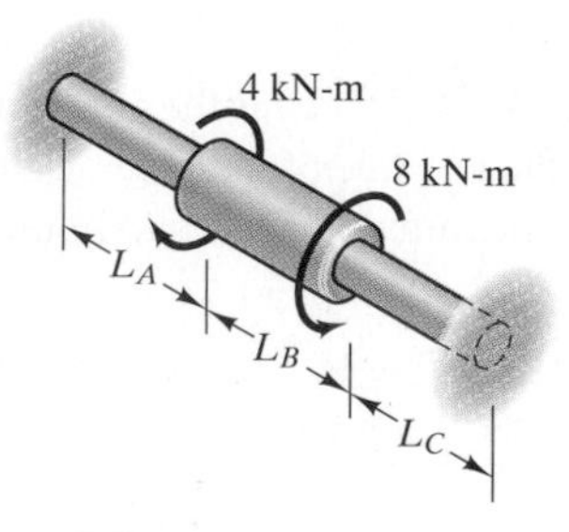

PROBLEM 4-3.12

4-3.13. In Problem 4-3.12, through what angle does the bar rotate at the position where the 8-kN-m couple is applied?

4-3.14. A collar is rigidly attached to bar A. The cylindrical bar A is 80 mm in diameter and its shear modulus is $G = 66$ GPa. There are gaps $b = 2$ mm between the arms of the collar and the ends of the identical bars B and C. Bars B and C are 30 mm in diameter and their modulus of elasticity is $E = 170$ GPa. If the bars B and C are stretched until they come into contact with the arms of the collar and are welded to them, what is the magnitude of the maximum shear stress in bar A afterward?

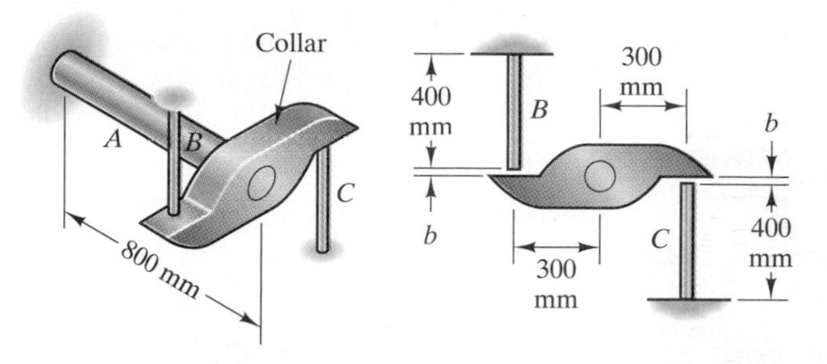

PROBLEM 4-3.14

4-3.15. Bar A in Problem 4-3.14 is made of a material that will safely support a tensile or compressive normal stress of 170 MPa. Based on this criterion, what is the largest safe value of the gap b?

4-4.1. In Example 4-2, what is the magnitude of the maximum shear stress in the bar?

4-4.2. In Example 4-2, suppose that the torque T is applied to the bar at $x = 1$ m. What is the magnitude of the angle of twist of the entire bar?

4-4.3. The bar has a solid circular cross section. Its polar moment of inertia is given by $J = (0.1 + 0.15x)$ in^4, where x is the axial position in inches, and the shear modulus of the material is $G = 4.6 \times 10^6$ psi. If the bar is subjected to an axial torque $T = 20$ in-kip, what is the magnitude of the maximum shear stress at $x = 6$ in.?

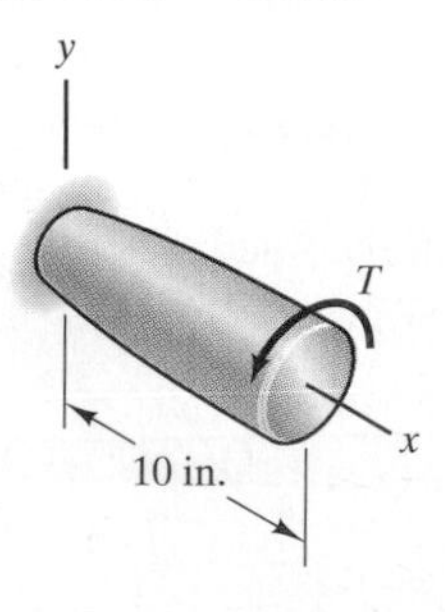

PROBLEM 4-4.3

4-4.4. What is the angle of twist (in degrees) of the entire bar in Problem 4-4.3?

4-4.5. Suppose that an axial hole is drilled through the bar in Problem 4-4.3 so that it has a hollow circular cross section with inner radius $r_i = 0.3$ in. What is the angle of twist (in degrees) of the entire bar due to the 20-in-kip torque?

4-4.6. The radius of the bar's circular cross section varies linearly from 10 mm at $x = 0$ to 5 mm at $x = 150$ mm. The shear modulus of the material is $G = 17$ GPa. What torque T would cause a maximum shear stress of 10 MPa at $x = 80$ mm?

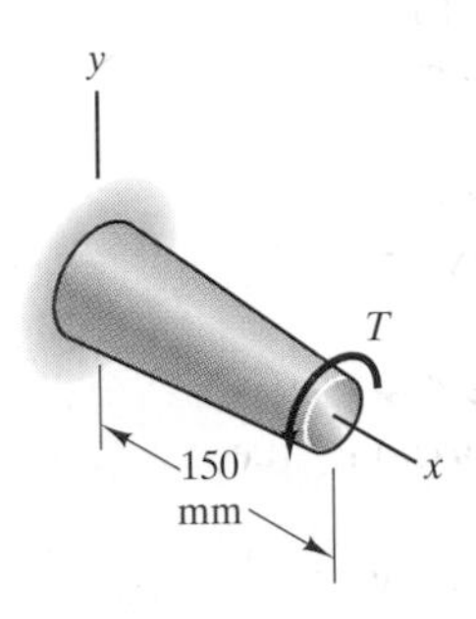

PROBLEM 4-4.6

4-4.7. In Problem 4-4.6, what torque T would cause the end of the bar to rotate 1°?

4-4.8. In Problem 4-4.6, suppose that the torque T at the end of the bar is 20 N-m and you want to apply a torque in the opposite direction at $x = 75$ mm so that the angle through which the end of the bar rotates is zero. What is the magnitude of the torque you must apply?

4-4.9. Bars A and B have solid circular cross sections and consist of material with shear modulus $G = 17$ GPa. Bar A is 150 mm long and its radius varies linearly from 10 mm at its left end to 5 mm at its right end. The prismatic bar B is 100 mm long and its radius is 5 mm. There is a small gap between the bars. The end of bar A is given an axial rotation of 1° and the bars are welded together. What is the torque in the bars afterward?

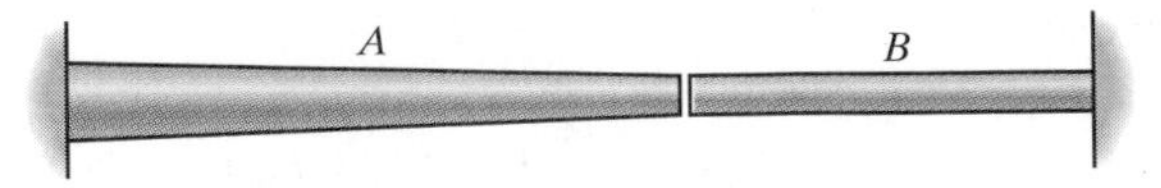

PROBLEM 4-4.9

4-4.10. The aluminum alloy bar has a circular cross section with 20-mm diameter, length $L = 120$ mm, and a shear modulus of 28 GPa. If the distributed torque is uniform and causes the end of the bar to rotate 0.5°, what is the magnitude of the maximum shear stress in the bar?

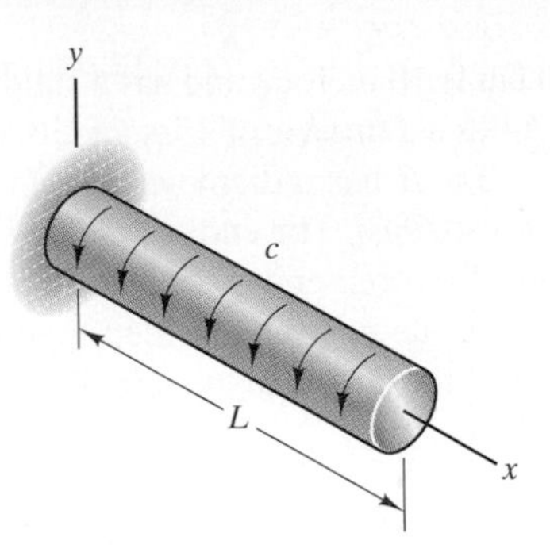

PROBLEM 4-4.10

4-4.11. If the distributed torque in Problem 4-4.10 is given by the equation $c = c_0(x/L)^3$ and causes the end of the bar to rotate 0.5°, what is the magnitude of the maximum shear stress in the bar?

4-4.12. A cylindrical bar with 1-in. diameter fits tightly in a circular hole in a 5-in.-thick plate. The shear modulus of the material is $G = 5.6 \times 10^6$ psi. A 12,000-in-lb axial torque is applied at the left end of the bar. The distributed torque exerted on the bar by the plate is given by the equation

$$c = c_0\left[1 - \left(\frac{x}{5}\right)^{1/2}\right] \text{ in-lb/in.,}$$

where c_0 is a constant and x is the axial position in inches measured from the left side of the plate. Determine the constant c_0 and the magnitude of the maximum shear stress in the bar at $x = 2$ in.

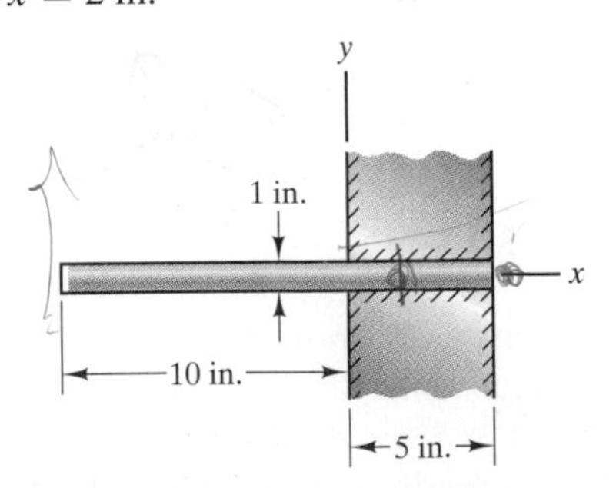

PROBLEM 4-4.12

4-4.13. In Problem 4-4.12, what is the magnitude of the angle of twist of the left end of the bar relative to its right end?

4-4.14. The aluminum alloy bar has a circular cross section with 20-mm diameter and a shear modulus of 28 GPa. What is the magnitude of the maximum shear stress in the bar due to the uniformly distributed torque?

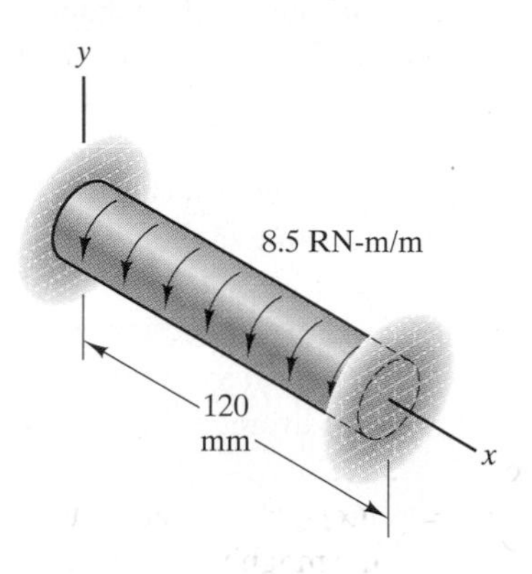

PROBLEM 4-4.14

4-4.15. In Problem 4-4.14, what is the bar's angle of twist (in degrees) at $x = 60$ mm?

4-4.16. The bar has a circular cross section with polar moment of inertia J and shear modulus G. The distributed torque $c = c_0(x/L)^2$. What are the magnitudes of the torques exerted on the bar by the left and right walls?

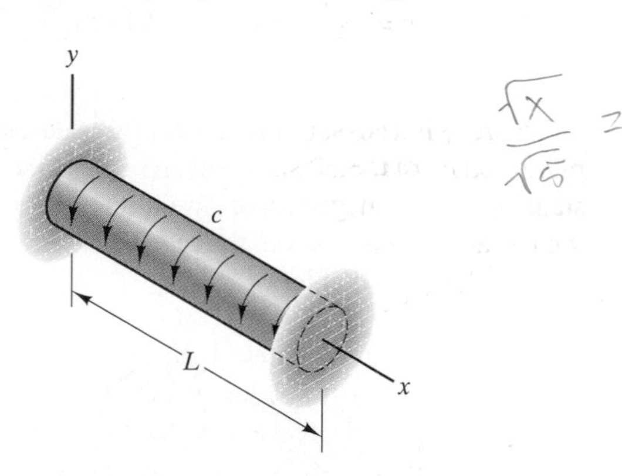

PROBLEM 4-4.16

4-4.17. In Problem 4-4.16, at what axial position x is the bar's angle of twist the greatest, and what is its magnitude?

4-4.18. If the bar in Problem 4-4.16 is acted upon by the distributed load $c = c_0(x/L)^2$ from $x = 0$ to $x = L/2$ and is free of external torque from $x = L/2$ to $x = L$, what are the magnitudes of the torques exerted on the bar by the left and right walls?

4-5.1. A bar has a circular cross section with radius $R =$ 20 mm. If the bar is steel and is modeled as an elastic–perfectly plastic material with yield stress $\tau_Y = 400$ MPa, what is the upper bound of the torque the bar will support?

4-5.2. Suppose that the bar described in Problem 4-5.1 is subjected to an increasing torque T. For what value of T is $r_Y = 10$ mm? (That is, the shear stress is equal to the yield stress from $r = 10$ mm to $r = 20$ mm.)

4-5.3. In Example 4-4, what are the external torque T and the bar's angle of twist (in degrees) if $r_Y = 12$ mm?

4-5.4. In Example 4-4, determine the bar's angle of twist (in degrees) if the axial torque $T = 2.8$ kN-m.

4-5.5. The 48-in. bar has a solid circular cross section with 4-in. diameter. The shear modulus $G = 4 \times 10^6$ psi and the yield stress $\tau_Y = 60{,}000$ psi. The bar is fixed at one end and subjected to an axial torque T at the other end. Model the material as elastic–perfectly plastic. What are the external torque T and the bar's angle of twist (in degrees) if $r_Y = 1$ in.?

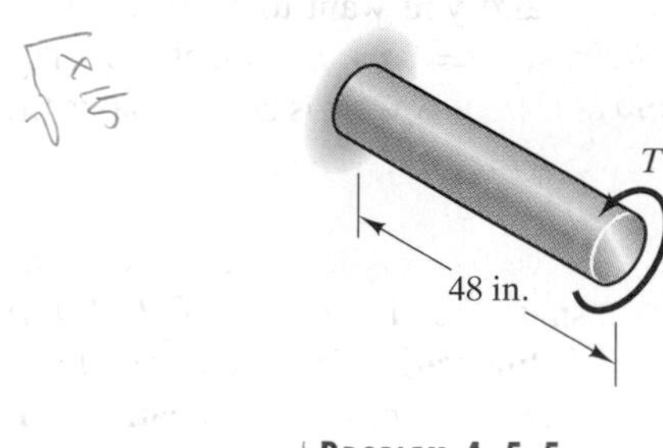

PROBLEM 4-5.5

4-5.6. If the bar described in Problem 4-5.5 is subjected to a torque $T = 1 \times 10^6$ in-lb, what are r_Y and the bar's angle of twist in degrees?

4-5.7. What is the upper bound of the torque T the bar described in Problem 4-5.5 will support?

4-5.8. The aluminum alloy bar, 4 m in length, has the cross section shown. The shear modulus $G = 28$ GPa and the yield stress $\tau_Y = 410$ MPa. The bar is fixed at one end and subjected to an axial torque T at the other end. Model the material as elastic–perfectly plastic. If the bar's angle of twist is $\phi = 130°$, what is the torque T?

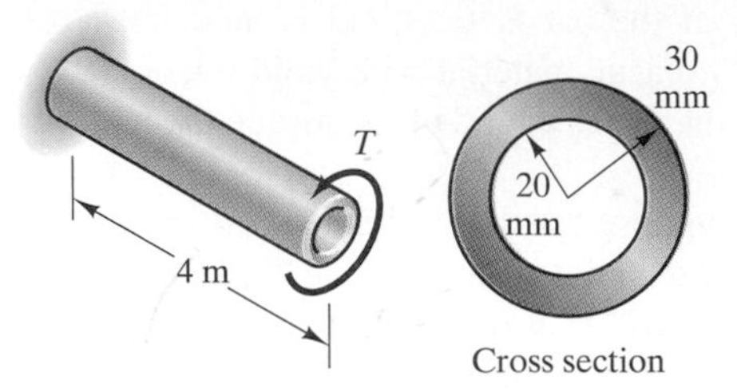

| PROBLEM 4-5.8

4-5.9. What is the upper bound of the torque T the aluminum alloy bar described in Problem 4-5.8 will support?

4-5.10. If the aluminum alloy bar described in Problem 4-5.8 is subjected to a torque $T = 16$ kN-m, what is the bar's angle of twist in degrees?

4-5.11. The aluminum alloy bar described in Problem 4-5.8 is fixed at both ends and subjected to an axial torque T 1 m from the left end. If the bar's angle of twist where the torque is applied is 30°, what is the torque T?

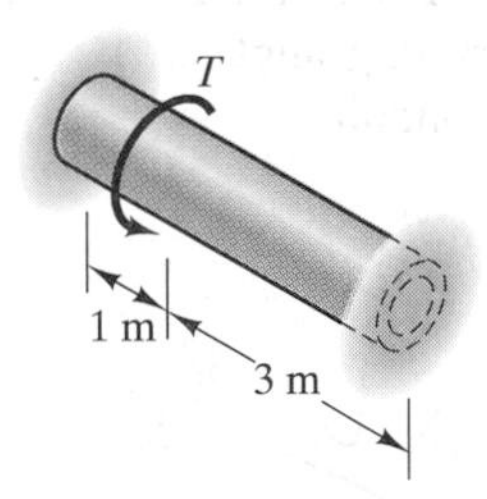

| PROBLEM 4-5.11

4-5.12. In Problem 4-5.11, what is the upper bound of the torque T the bar will support?

4-5.13. By evaluating the integrals in Eq. (4-21), derive the relation (4-22) between the torque T and the radial distance r_Y at which the shear stress becomes equal to the yield stress.

4-6.1. In Example 4-5, what is the magnitude of the maximum shear stress in the tube if the radius of the semicircular part of the cross section is 80 mm?

4-6.2. In Example 4-5, what is the angle of twist of the tube if the radius of the semicircular part of the cross section is 80 mm?

4-6.3. A thin-walled tube has wall thickness $t = \frac{1}{16}$ in. and is subjected to a torque $T = 12$ in-kip. Determine the average shear stress in the tube if its cross-sectional shape is **(a)** circular with radius $R = 2$ in.; **(b)** square with the same wall length as the circular cross section. (Notice that the weight of the tube is the same in the two cases.)

4-6.4. A steel tube has the cross section shown. If it is fixed at one end and subjected to an axial torque $T = 2$ kN-m at the other end, what is the average shear stress in the tube?

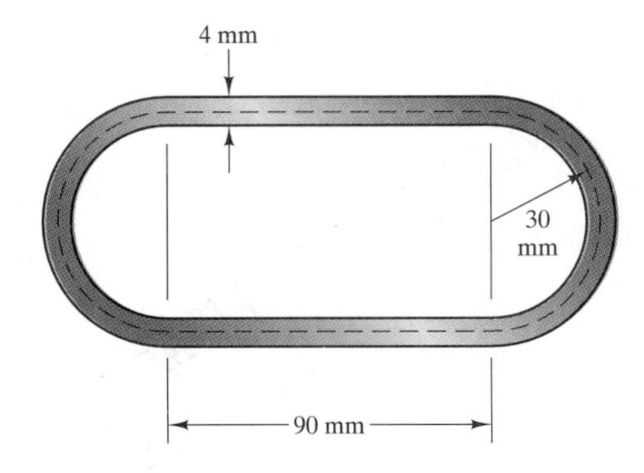

| PROBLEM 4-6.4

4-6.5. If the cross section of the steel tube described in Problem 4-6.4 is circular instead of the shape shown, with the same thickness and the same circumferential length, what is the average shear stress in the tube? Compare your answer to the answer to Problem 4-6.4.

4-6.6. If the steel tube described in Problem 4-6.4 is 2 m long and the shear modulus of the steel is 80 GPa, what angle of twist (in degrees) is caused by the 2-kN-m torque?

4-6.7. The steel tube described in Problem 4-6.4 is 2 m long and the shear modulus of the steel is 80 GPa. If the cross section is circular instead of the shape shown, with the same thickness and the same circumferential length, what angle of twist (in degrees) is caused by the 2-kN-m torque? Compare your answer to the answer to Problem 4-6.6.

4-6.8. The midline of the tube's cross section is an equilateral triangle with 2-in. sides. If the thickness $t = \frac{1}{16}$ in. and the axial torque $T = 1200$ in-lb, what is the average shear stress in the $\frac{1}{16}$-in. walls?

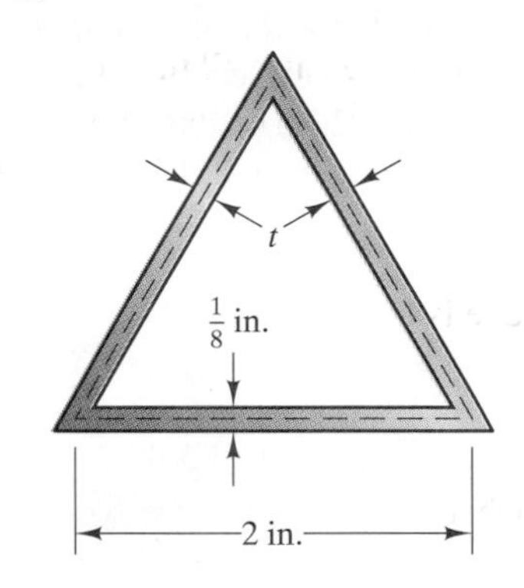

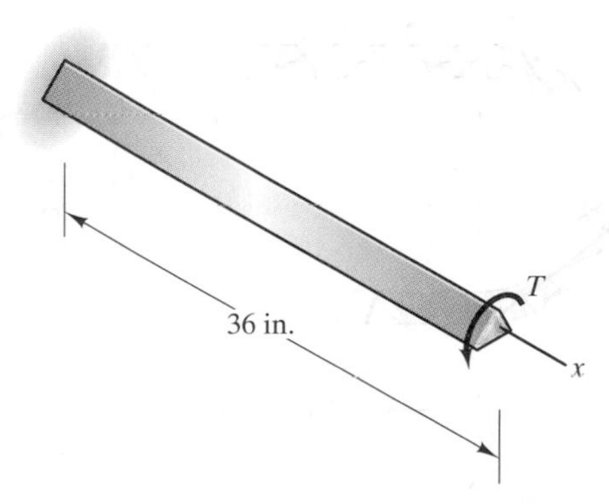

PROBLEM 4-6.8

4-6.9. If the tube described in Problem 4-6.8 has a shear modulus of 5.8×10^6 psi, what angle of twist (in degrees) is caused by the 1200-in-lb torque?

4-6.10. If the tube described in Problem 4-6.8 is subjected to an axial torque $T = 1000$ in-lb and a design requirement is that the average shear stress in each wall must not exceed 3600 in-lb, what is the minimum required value of the thickness t? If t has the minimum value, what angle of twist (in degrees) is caused by the 1000 in-lb torque?

4-6.11. The tube is subjected to a distributed torque $c = 200(x/36)^2$ in-lb/in. Its cross section is shown in Problem 4-6.8, where the thickness $t = \frac{1}{16}$ in. What is the average shear stress in the $\frac{1}{16}$-in. walls at $x = 12$ in?

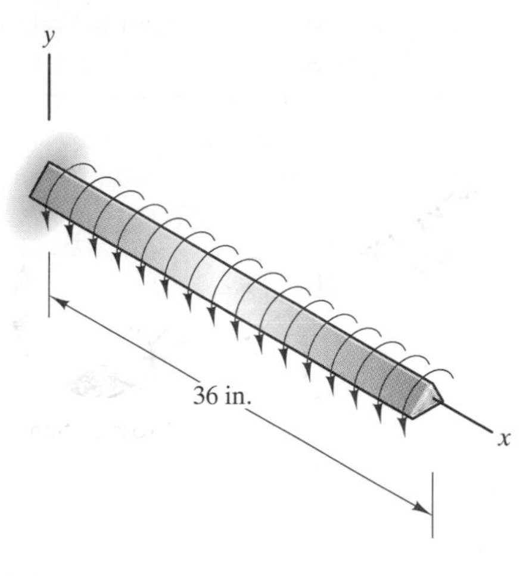

PROBLEM 4-6.11

4-6.12. In Problem 4-6.11, what is the tube's angle of twist in degrees if $G = 5.8x10^6$ psi?

4-6.13. An engineering trainee is assigned to measure the cross-sectional area A of a prismatic conduit made of 2014-T6 aluminum. She measures the uniform wall thickness and determines it to be $t = 2$ mm, and measures the circumferential length of the wall and determines it to be 260 mm. She then takes a 1-m-long section of the conduit, fixes one end, and applies a 1-kN-m axial torque to the other end. Measuring the resulting angle of twist, she finds it to be 5.2°. What is the conduit's cross-sectional area?

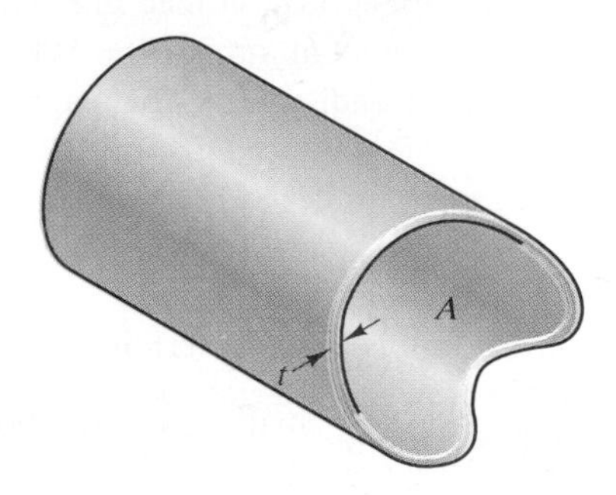

PROBLEM 4-6.13

4-6.14. To what average shear stress is the material subjected as a result of the test described in Problem 4-6.13?

4-6.15. The conduit described in Problem 4-6.13 will safely support an average shear stress in the walls of 100 MPa. Based on this criterion, what is the largest axial torque T that can safely be applied to a section of the conduit?

4-6.16. A tube has the circular cross section shown with $R = 2$ in. and $t = 0.1$ in. The tube is fixed at one end and subjected to an axial torque $T = 4000$ in-lb at the other end. **(a)** Use Eqs. (4-25) and (4-26) to determine the average shear stress in the tube. **(b)** Use Eq. (4-9) to determine the shear stress in the tube at $r = R$.

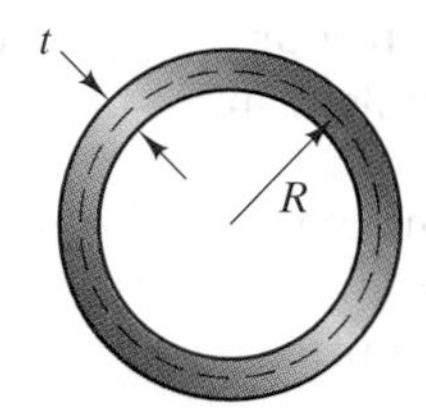

| **PROBLEM 4-6.16**

4-6.17. Consider a tube with the circular cross section shown in Problem 4-6.16 that is subjected to a torque T. If you use Eqs. (4-25) and (4-26) to determine the average shear stress τ_{av} in the tube and use Eq. (4-9) to determine the exact shear stress τ_{ex} in the tube at $r = R$, show that

$$\frac{\tau_{av}}{\tau_{ex}} = 1 + \frac{1}{4}\left(\frac{t}{R}\right)^2.$$

4-6.18. Suppose that the tube described in Example 4-5 is fixed at both ends and an axial torque $T = 800$ N-m is applied 300 mm from one end. What is the magnitude of the maximum shear stress in the tube?

4-6.19. Suppose that the tube described in Example 4-5 is fixed at both ends and an axial torque $T = 800$ N-m is applied 300 mm from one end. Determine the magnitude of the tube's angle of twist (in degrees) at the axial position where the torque is applied.

4-7.1. Suppose that you are designing a bar with a solid circular cross section that is to support an 800-N-m torsional load. The bar is to be made of 6061-T6 aluminum alloy (see Appendix B), and you want the factor of safety to be $S = 2$. Based on these criteria, what should the bar's diameter be?

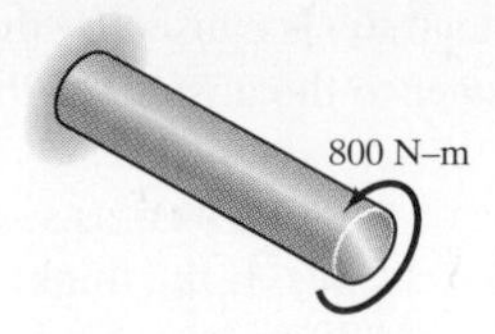

| **PROBLEM 4-7.1**

4-7.2. A hollow circular tube that is to support an airplane's aileron may be subjected to service torques as large as 260 N-m. It is to have a 20-mm outside diameter and 14-mm inside diameter. Choose an aluminum alloy from Appendix B so that the tube has a factor of safety of at least 2.

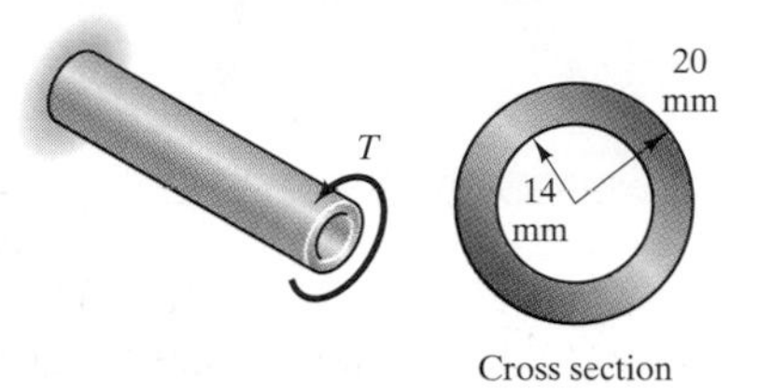

| **PROBLEM 4-7.2**

4-7.3. Suppose that you are designing a bar with a solid circular cross section that is to support a 6500-in-lb torsional load. The bar is to be made of ASTM-A572 structural steel (see Appendix B), and you want the factor of safety to be $S = 1.5$. Based on these criteria, what should the bar's diameter be?

4-7.4. A bar with a solid circular cross section is to support a 1200-N-m torsional load. Choose a material from Appendix B and determine the bar's diameter so that the factor of safety is $S = 3$.

4-7.5. A bar with a hollow circular cross section is to support a 1200-N-m torsional load. Choose a material from Appendix B and determine the bar's inner and outer radii so that the inner radius is one-half of the outer radius and the factor of safety is $S = 3$.

4-7.6. A bar with a solid circular cross section is to support an 8000-in-lb torsional load. Choose a material from Appendix B and determine the bar's diameter so that the factor of safety is $S = 2.5$.

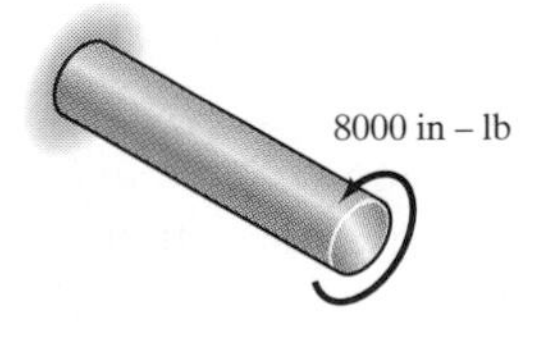

| **PROBLEM 4-7.6**

4-7.7. A bar with a hollow circular cross section is to support an 8000-in-lb torsional load. Choose a material from Appendix B and determine the bar's inner and outer radii so that the inner radius is 60% of the outer radius and the factor of safety is $S = 3$.

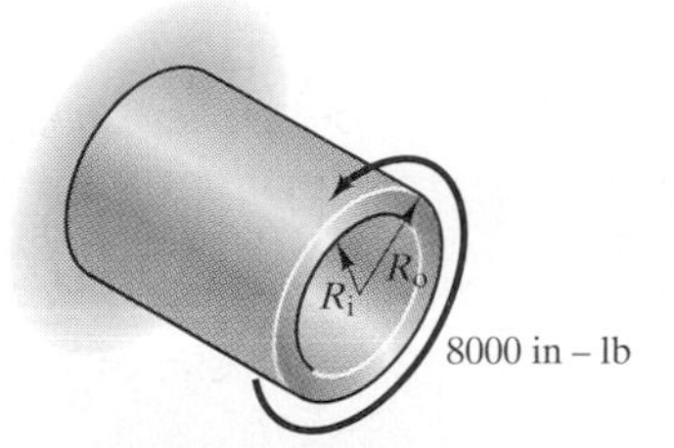

PROBLEM 4-7.7

4-7.8. After being equipped with a turbocharger, the engine of the car in Example 4-6 produces its maximum torque at 5000 rpm when the engine is generating 320 horsepower. Choose a material from Appendix B and design a drive shaft for the car that has a solid circular cross section and a factor of safety $S = 3$.

PROBLEM 4-7.8

4-7.9. A Catalina 30 sailboat has a Universal 5411 diesel engine with 11 hp. Assume that the propeller shaft transmits 11 hp at 1200 rpm. If the shaft is to have a solid circular cross section, be made of soft manganese bronze, and you want it to have a factor of safety $S = 3$, what should its diameter be? (See Example 4-6.)

4-7.10. The shaft that connects the turbine blades of the hydroelectric power unit shown in Problem 4-2.6 to the generator transmits 160 MW of power at 150 rpm. Choose a material from Appendix B and design a shaft with a solid circular cross section that has a factor of safety $S = 2$. (See Example 4-6.)

4-7.11. At peak power, the propeller of the wind generator shown in Problem 4-2.11 produces 3 MW of power at 34 rpm. Based on this criterion, choose a material from Appendix B and design a propeller shaft that has a solid circular cross section and a factor of safety $S = 2$ (see Example 4-6). Notice that this analysis does not account for the load exerted on the shaft by the propeller's weight.

4-7.12. Design a hollow shaft for the propeller in Problem 4-7.11 in which the inner radius is 75% of the outer radius and the factor of safety is $S = 2$.

4-7.13. Suppose that you want to design the system shown so that the torque exerted on the fixed support at D by the bar CD is 600 N-m when a torque $T_A = 200$ N-m is applied at A. Determine the radii of the gears and design the shafts AB and CD so that the system has a factor of safety $S = 2.5$. Determine the angle of twist at A that results from your design when the torque $T_A = 200$ N-m is applied.

PROBLEM 4-7.13

The hull of the deep submersible vehicle must be designed to support the large stresses resulting from the pressure of the surrounding water.

CHAPTER 5

States of Stress

$$\begin{bmatrix} \sigma_x & \tau_{xy} & \tau_{xz} \\ \tau_{yx} & \sigma_y & \tau_{yz} \\ \tau_{zx} & \tau_{zy} & \sigma_z \end{bmatrix} = \begin{bmatrix} \sigma & 0 & 0 \\ 0 & \sigma_{\mathrm{h}} & 0 \\ 0 & 0 & -p_{\mathrm{o}} \end{bmatrix}$$

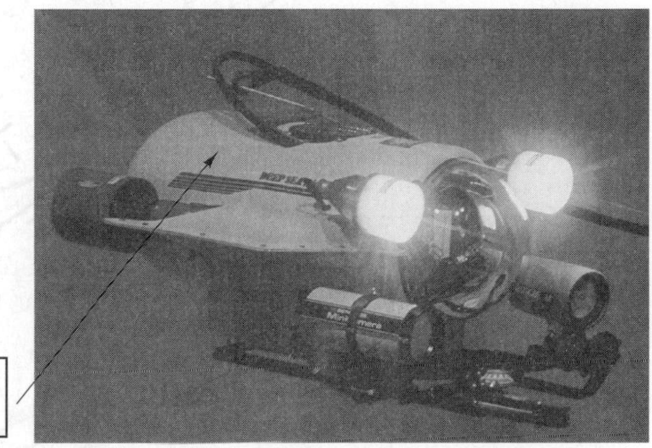

In this chapter we explain how to completely describe the state of stress at a point of an object such as the submersible vehicle's hull, and we show that the state of stress can be used to determine the normal and shear stresses acting on an arbitrary plane through that point. Because of the importance of the maximum values of the normal and shear stresses in design, we describe how to determine maximum stresses and the orientations of the planes on which they act.

5-1 | Components of Stress

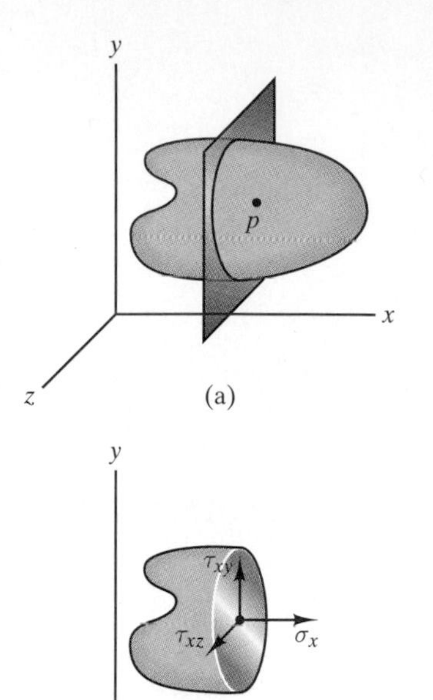

FIGURE 5-1 (a) Passing a plane perpendicular to the x axis. (b) Normal stress and components of the shear stress at p.

Suppose that we are interested in the stresses acting at a point p. Let us introduce a coordinate system and pass a plane through p that is perpendicular to the x axis (Fig. 5-1a). The normal stress acting on this plane at p is denoted by σ_x (Fig. 5-1b). The shear stress may act at any direction parallel to the y–z plane and therefore may have components in both the y and z directions. These components are denoted by τ_{xy} (the shear stress on the plane perpendicular to the x axis that acts in the y direction) and τ_{xz} (the shear stress on the plane perpendicular to the x axis that acts in the z direction). Next, we pass a plane through p that is perpendicular to the y axis (Fig. 5-2a). The normal stress on this plane is σ_y and the components of the shear stress are τ_{yx} and τ_{yz} (Fig. 5-2b). Finally, we pass a plane through p perpendicular to the z axis (Fig. 5-3a). The normal stress is σ_z and the components of the shear stress are τ_{zx} and τ_{zy} (Fig. 5-3b).

At this point you may object that we have not done anything new. We have simply passed three different planes through a point and named the normal and shear stresses on those planes. If a different plane is passed through the point, the normal and shear stresses will generally be different from those acting on the planes perpendicular to the coordinate axes, so what have we achieved? The answer is that if the normal and shear stresses on these three planes are known at a point, the normal and shear stresses on any plane through the point can be determined. For this reason, the stresses on these planes are called the *state of stress* at the point. We can represent the state of stress compactly as a matrix in terms of the *components of stress*:

$$\begin{bmatrix} \sigma_x & \tau_{xy} & \tau_{xz} \\ \tau_{yx} & \sigma_y & \tau_{yz} \\ \tau_{zx} & \tau_{zy} & \sigma_z \end{bmatrix}. \qquad \textbf{(5-1)}$$

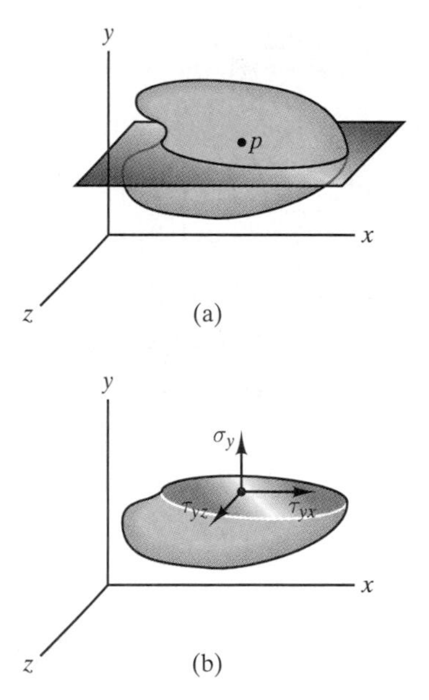

FIGURE 5-2 (a) Passing a plane perpendicular to the y axis. (b) Normal stress and components of the shear stress at p.

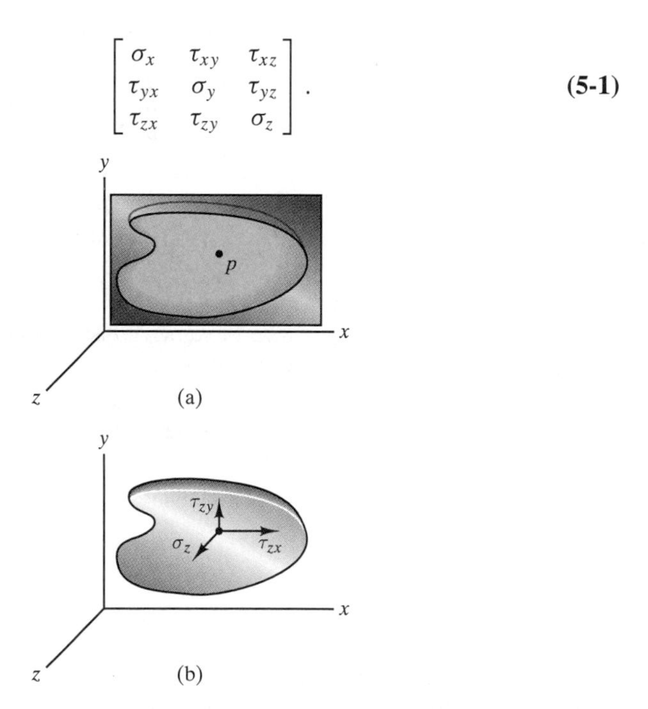

FIGURE 5-3 (a) Passing a plane perpendicular to the z axis. (b) Normal stress and components of the shear stress at p.

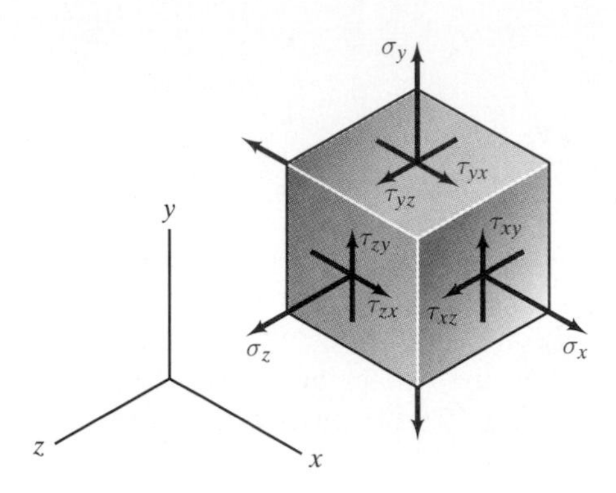

FIGURE 5-4 Components of stress on an element containing p.

Figure 5-4 shows the components of stress on an element whose faces are perpendicular to the coordinate axes. If the state of stress is uniform, or *homogeneous*, in a finite neighborhood surrounding the point, you can regard the stresses in Fig. 5-4 as the stresses on an element of finite size containing p. Otherwise, you must interpret them as the average values of the stress components on an element containing p. These average values approach the state of stress at p in the limit as the element shrinks.

The shear stresses $\tau_{xy} = \tau_{yx}$, $\tau_{yz} = \tau_{zy}$, and $\tau_{xz} = \tau_{zx}$, so that the stress matrix (5-1) is symmetric. To show this, let the element in Fig. 5-5 be a cube with dimension b. The sum of the moments about the z axis is

$$\begin{aligned}(\sigma_x b^2)(b/2) &- (\sigma_x b^2)(b/2) + (\sigma_y b^2)(b/2) - (\sigma_y b^2)(b/2) + (\tau_{xy} b^2)(b) \\ &- (\tau_{yx} b^2)(b) + (\tau_{zy} b^2)(b/2) - (\tau_{zy} b^2)(b/2) \\ &- (\tau_{zx} b^2)(b/2) + (\tau_{zx} b^2)(b/2) = 0.\end{aligned}$$

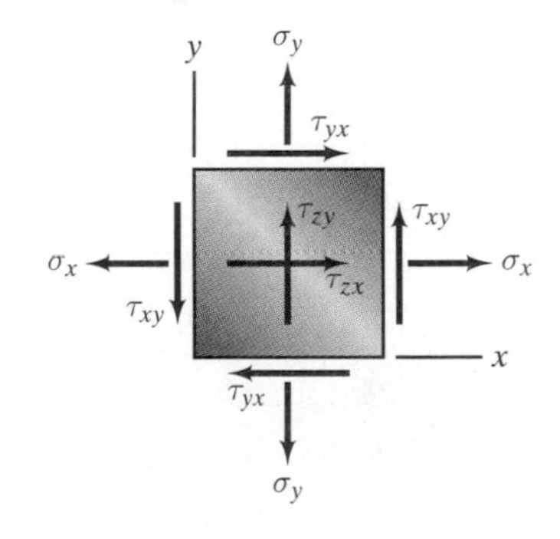

FIGURE 5-5 Summing moments about the z axis to show that $\tau_{xy} = \tau_{yx}$.

Therefore, $\tau_{xy} = \tau_{yx}$, and by summing moments about the x and y axes it can be shown that $\tau_{yz} = \tau_{zy}$ and $\tau_{xz} = \tau_{zx}$. Although we have assumed the material to be in equilibrium, these results hold even when the material is not in equilibrium.

As a simple example of a state of stress, consider the pressure in a fluid (a liquid or gas) at rest. The stresses exerted on the faces of an element are the pressures exerted by the surrounding fluid (Fig. 5-6). The normal stresses are $\sigma_x = \sigma_y = \sigma_z = -p$ and the shear stresses are zero. (The shear stresses on an element of a flowing fluid are not generally zero, although in some situations they can be neglected.) Therefore, the state of stress at a point in a fluid at rest is

$$\begin{bmatrix} -p & 0 & 0 \\ 0 & -p & 0 \\ 0 & 0 & -p \end{bmatrix}. \tag{5-2}$$

Applying axial loads to the ends of a prismatic bar results in another simple state of stress. If we orient the coordinate system with its x axis parallel to the axis of the bar (Fig. 5-7), the only nonzero stress component is $\sigma_x = P/A$,

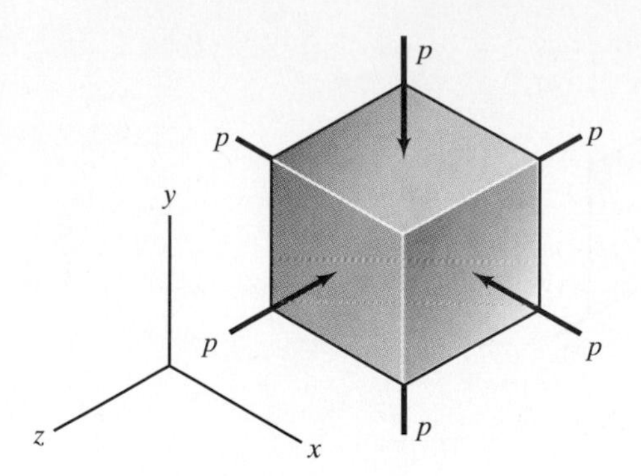

FIGURE 5-6 Stresses on an element of a fluid at rest.

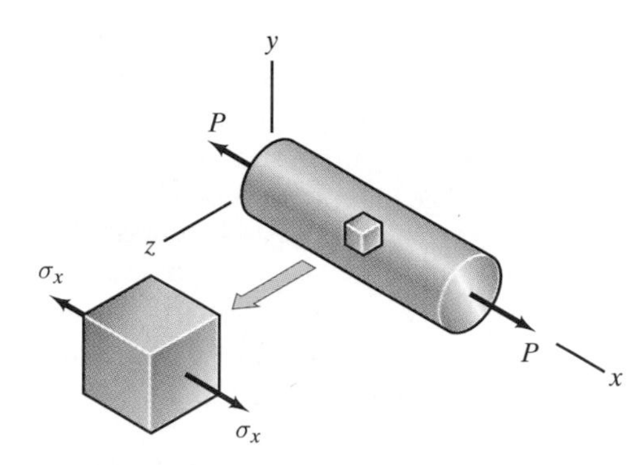

FIGURE 5-7 Stresses on an element of a bar subjected to axial forces.

where A is the bar's cross-sectional area. The state of stress is

$$\begin{bmatrix} \sigma_x & 0 & 0 \\ 0 & 0 & 0 \\ 0 & 0 & 0 \end{bmatrix}.$$

In general, this state of stress applies only to elements that are not near the ends of the bar where the forces are applied.

In Fig. 5-8 we isolate an infinitesimal element of a cylindrical bar subjected to axial torsion. If we orient the coordinate system so that its x axis is parallel to the axis of the bar and the element lies in the $x - y$ plane, the element is in a state of pure shear stress:

$$\begin{bmatrix} 0 & \tau_{xy} & 0 \\ \tau_{yx} & 0 & 0 \\ 0 & 0 & 0 \end{bmatrix}.$$

This state of stress also generally applies only to elements that are not near the ends of the bar.

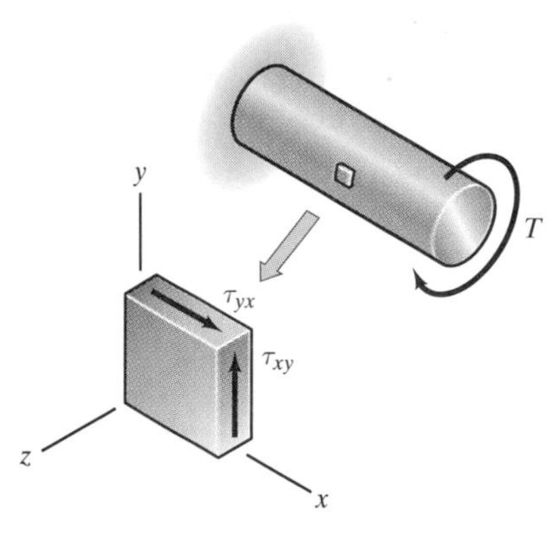

FIGURE 5-8 Stresses on an element of a bar subjected to torsion.

In citing these examples, we have emphasized that *the values of the components of normal and shear stress that define the state of stress at a point depend on the orientation of the coordinate system.* If the coordinate system is rotated, the orientations of the planes perpendicular to the axes change, and so in general the normal and shear stresses acting on them change. By doing so, we can determine the normal and shear stresses acting on different planes through the point p. We undertake this task in the following section.

5-2 | Transformations of Plane Stress

The stress at a point is said to be a state of *plane stress* if it is of the form

$$\begin{bmatrix} \sigma_x & \tau_{xy} & 0 \\ \tau_{yx} & \sigma_y & 0 \\ 0 & 0 & 0 \end{bmatrix}. \tag{5-3}$$

That is, the stress components σ_z, τ_{xz}, and τ_{yz} are zero (Fig. 5-9). This does not imply that the three stress components σ_x, σ_y, and τ_{xy} are each necessarily nonzero, but they are the only stress components that may be nonzero in plane stress. The state of stress shown in Fig. 5-7 for a bar subjected to axial forces is plane stress, and the state of stress shown in Fig. 5-8 for a bar subjected to axial torsion is plane stress. But the state of stress shown in Fig. 5-6 for a fluid at rest is not plane stress, because $\sigma_z \neq 0$.

Many important applications in stress analysis besides bars subjected to axial and torsional loads result in states of plane stress. In this section we assume that the state of plane stress at a point of a material is known and we wish to know the states of stress on planes other than the three planes perpendicular to the coordinate axes. We also address the most crucial questions from the standpoint of design: What are the maximum normal and shear stresses, and what are the orientations of the planes on which they act?

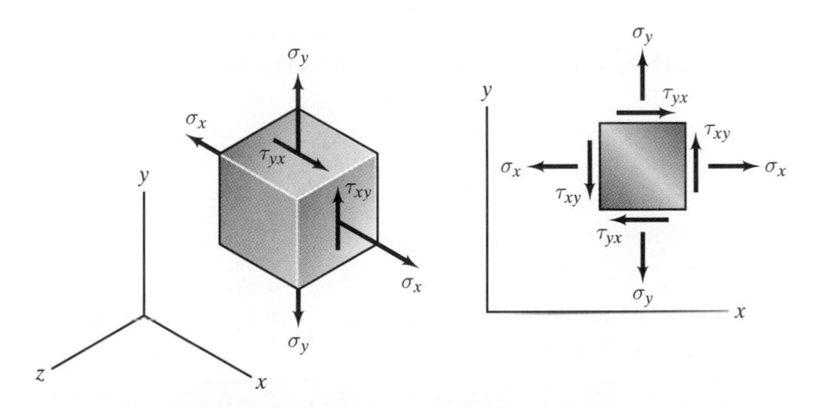

| **FIGURE 5-9** Plane stress.

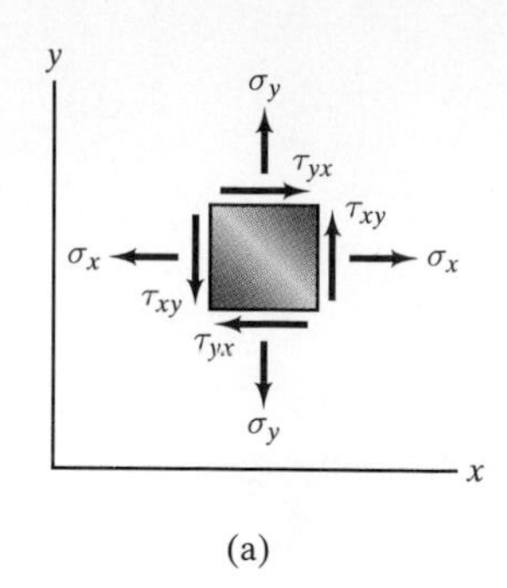

(a)

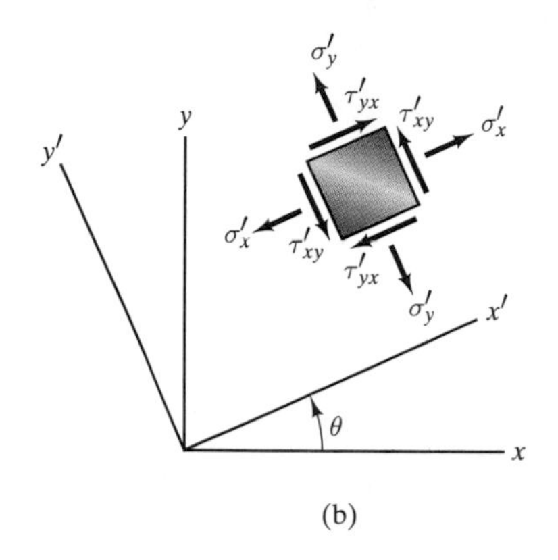

(b)

FIGURE 5-10 Stress components at p: (a) in terms of the xyz coordinate system; (b) in terms of the $x'y'z'$ coordinate system.

Coordinate Transformations

Suppose that we know the state of plane stress at a point p of a material in terms of the coordinate system shown in Fig. 5-10a, and we want to know the state of stress at p in terms of a coordinate system $x'y'z'$ oriented as shown in Fig. 5-10b. (The z and z' axes are coincident.) We begin with the element in Fig. 5-10a and pass a plane through it as shown in Fig. 5 11. The oblique surface of the resulting free-body diagram is perpendicular to the x' axis, so the normal and shear stresses acting on it are σ'_x and τ'_{xy}. We can determine σ'_x and τ'_{xy} by writing the equilibrium equations for this free-body diagram.

Let the area of the oblique surface of the free-body diagram be ΔA. The sum of the forces in the x' direction is

$$\sigma'_x\,\Delta A - (\sigma_x\,\Delta A\cos\theta)\cos\theta - (\sigma_y\,\Delta A\sin\theta)\sin\theta$$
$$- (\tau_{xy}\,\Delta A\cos\theta)\sin\theta - (\tau_{yx}\,\Delta A\sin\theta)\cos\theta = 0.$$

Solving for σ'_x, we obtain

$$\sigma'_x = \sigma_x\cos^2\theta + \sigma_y\sin^2\theta + 2\tau_{xy}\sin\theta\cos\theta. \quad \textbf{(5-4)}$$

The sum of the forces in the y' direction is

$$\tau'_{xy}\Delta A + (\sigma_x\,\Delta A\cos\theta)\sin\theta - (\sigma_y\,\Delta A\sin\theta)\cos\theta$$
$$- (\tau_{xy}\,\Delta A\cos\theta)\cos\theta + (\tau_{yx}\,\Delta A\sin\theta)\sin\theta = 0.$$

The solution for τ'_{xy} is

$$\tau'_{xy} = -(\sigma_x - \sigma_y)\sin\theta\cos\theta + \tau_{xy}(\cos^2\theta - \sin^2\theta). \quad \textbf{(5-5)}$$

By using the trigonometric identities

$$\begin{aligned} 2\cos^2\theta &= 1 + \cos 2\theta, \\ 2\sin^2\theta &= 1 - \cos 2\theta, \\ 2\sin\theta\cos\theta &= \sin 2\theta, \\ \cos^2\theta - \sin^2\theta &= \cos 2\theta, \end{aligned} \quad \textbf{(5-6)}$$

we can write Eqs. (5-4) and (5-5) in alternative forms that will be useful:

$$\sigma'_x = \frac{\sigma_x + \sigma_y}{2} + \frac{\sigma_x - \sigma_y}{2}\cos 2\theta + \tau_{xy}\sin 2\theta, \quad \textbf{(5-7)}$$

$$\tau'_{xy} = -\frac{\sigma_x - \sigma_y}{2}\sin 2\theta + \tau_{xy}\cos 2\theta. \quad \textbf{(5-8)}$$

We can obtain an equation for σ'_y by setting θ equal to $\theta + 90°$ in the expression for σ'_x. The result is

$$\sigma'_y = \frac{\sigma_x + \sigma_y}{2} - \frac{\sigma_x - \sigma_y}{2}\cos 2\theta - \tau_{xy}\sin 2\theta. \quad \textbf{(5-9)}$$

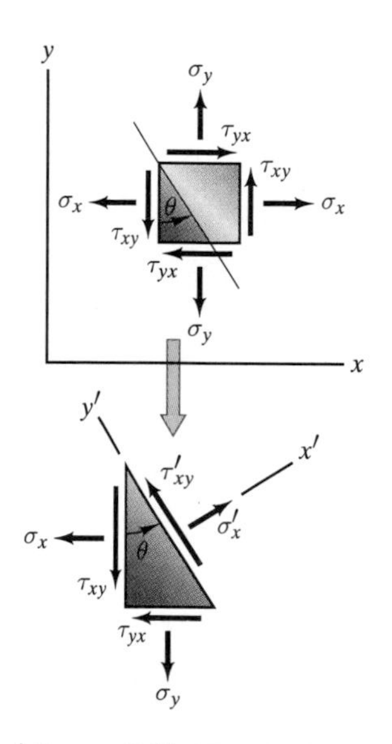

FIGURE 5-11 Free-body diagram for determining σ'_x and τ'_{xy}.

Given the state of plane stress shown in Fig. 5-10a, Eqs. (5-7)–(5-9) determine the state of plane stress shown in Fig. 5-10b for any value of θ. This means that when we know a state of plane stress at a point p, we can determine the normal and shear stresses on planes through p other than the three planes perpendicular to the coordinate axes. Notice, however, that we can do so only for planes that are parallel to the z axis.

EXAMPLE 5-1

The components of plane stress at point p of the material in Fig. 5-12 are $\sigma_x = 4$ ksi, $\sigma_y = -2$ ksi, and $\tau_{xy} = 2$ ksi. What are the normal stress and the magnitude of the shear stress on the plane $\mathcal{P}$ at point p?

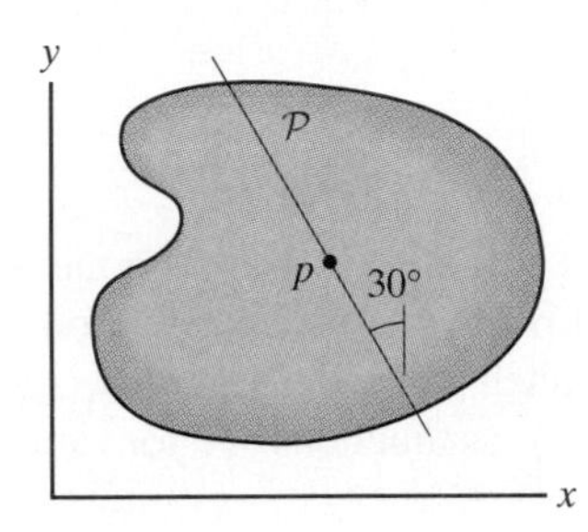

FIGURE 5-12

Strategy

If we orient the $x'y'$ coordinate system so that the x' axis is perpendicular to the plane $\mathcal{P}$, then σ'_x and τ'_{xy} are the normal and shear stresses on $\mathcal{P}$ (Fig. 5-11). We can use Eqs. (5-7) and (5-8) to determine σ'_x and τ'_{xy}.

Solution

The x' axis is perpendicular to the plane $\mathcal{P}$ if $\theta = 30°$ [Fig. (a)].

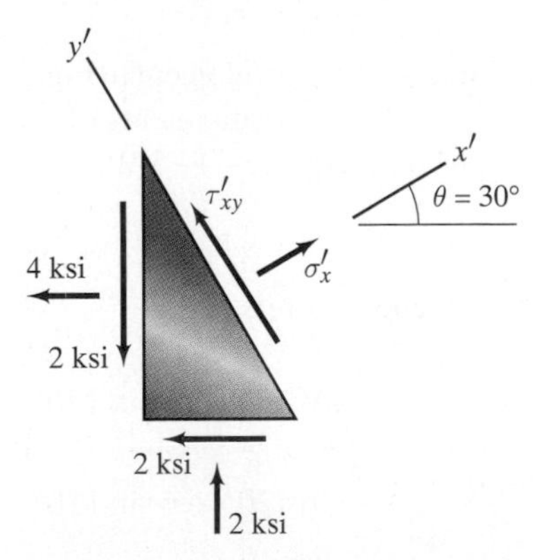

(a) Orienting the $x'y'$ coordinate system so that the x' axis is perpendicular to $\mathcal{P}$.

From Eq. (5-7),

$$\sigma'_x = \frac{4 + (-2)}{2} + \frac{4 - (-2)}{2} \cos 60^\circ + 2 \sin 60^\circ = 4.23 \text{ ksi},$$

and from Eq. (5-8),

$$\tau'_{xy} = -\frac{4 - (-2)}{2} \sin 60^\circ + 2 \cos 60^\circ = -1.60 \text{ ksi}.$$

The normal stress on $\mathcal{P}$ at point p is 4.23 ksi and the magnitude of the shear stress is 1.60 ksi.

EXAMPLE 5-2

The state of plane stress at a point p is shown on the left element in Fig. 5-13. **(a)** Determine the state of plane stress at p acting on the right element in Fig. 5-13 if $\theta = 60^\circ$. Draw a sketch of the element showing the stresses acting on it. **(b)** Draw graphs of σ'_x and τ'_{xy} as functions of θ for values of θ from zero to 360°.

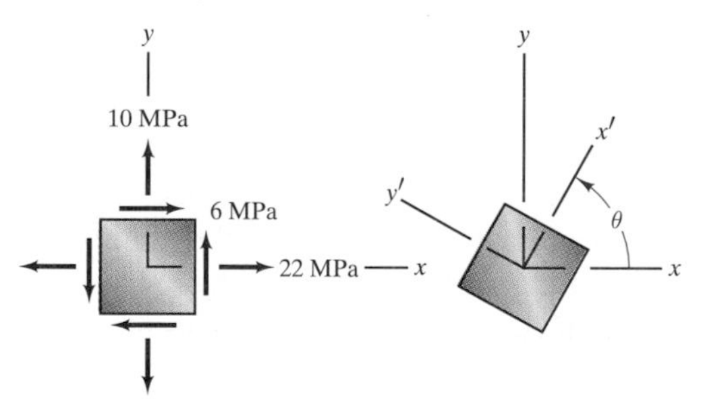

FIGURE 5-13

Strategy

The components of plane stress on the left element in Fig. 5-13 are $\sigma_x = 22$ MPa, $\sigma_y = 10$ MPa, and $\tau_{xy} = 6$ MPa. The components of plane stress on the right element in Fig. 5-13 are given by Eqs. (5-7)–(5-9).

Solution

(a) For $\theta = 60^\circ$, the components of stress are

$$\sigma'_x = \frac{22 + 10}{2} + \frac{22 - 10}{2} \cos 120^\circ + 6 \sin 120^\circ = 18.20 \text{ MPa},$$

$$\sigma'_y = \frac{22 + 10}{2} - \frac{22 - 10}{2} \cos 120^\circ - 6 \sin 120^\circ = 13.80 \text{ MPa},$$

$$\tau'_{xy} = -\frac{22 - 10}{2} \sin 120^\circ + 6 \cos 120^\circ = -8.20 \text{ MPa}.$$

We show the components of stress acting on the element in Fig. (a). Notice the directions of the shear stresses due to the negative value of τ'_{xy}.

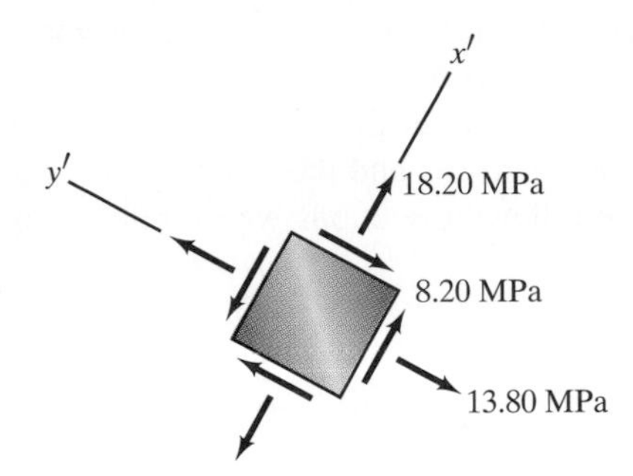

(a) State of stress for $\theta = 60°$.

(b) The stresses σ'_x and τ'_{xy} as functions of θ are

$$\begin{aligned}\sigma'_x &= \frac{22+10}{2} + \frac{22-10}{2}\cos 2\theta + 6\sin 2\theta \\ &= 16 + 6\cos 2\theta + 6\sin 2\theta, \\ \tau'_{xy} &= -\frac{22-10}{2}\sin 2\theta + 6\cos 2\theta \\ &= -6\sin 2\theta + 6\cos 2\theta.\end{aligned}$$

The graphs of these expressions are shown in Fig. (b).

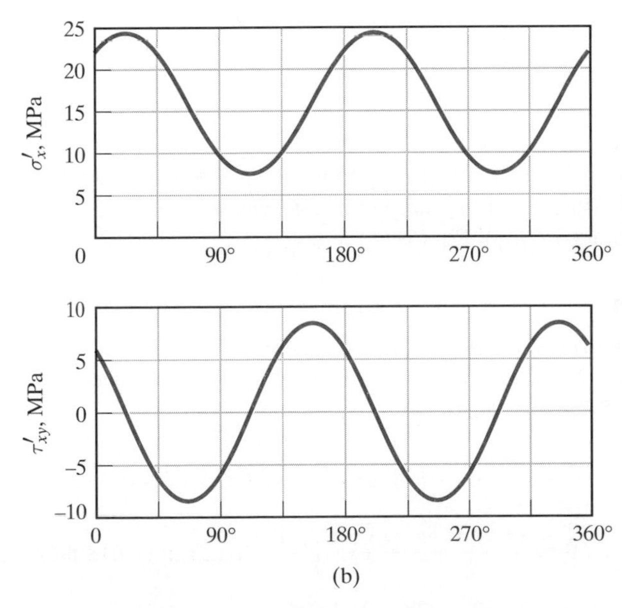

(b) Graphs of σ'_x and τ'_{xy} as functions of θ.

Discussion

The graphs of σ'_x and τ'_{xy} show that the normal stress and shear stress attain maximum and minimum values at particular values of θ. (Not, however, at the same values of θ. Notice that at angles for which the normal stress is a maximum or minimum, the shear stress is zero.) Since the maximum values of the stresses are so important with regard to design, we could use graphs such as these to determine the maximum stresses and the orientations of the planes on which they occur. But in the following sections we introduce more efficient ways to obtain this information.

EXAMPLE 5-3

The state of plane stress at a point p is shown on the left element in Fig. 5-14, and the values of the stresses σ'_x and τ'_{xy} are shown on a rotated element. What are the normal stress σ'_y and the angle θ?

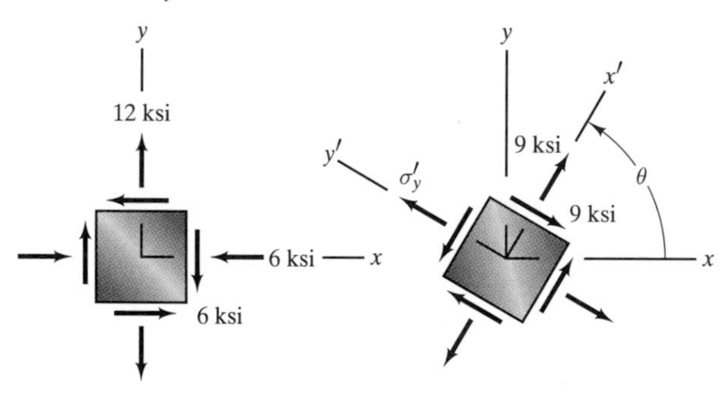

FIGURE 5-14

Strategy

Equations (5-7)–(5-9) give the components of stress on the rotated element in terms of θ. Since σ'_x and τ'_{xy} are known, we can solve Eqs. (5-7) and (5-8) for θ, then determine σ'_y from Eq. (5-9).

Solution

The components of stress on the left element are $\sigma_x = -6$ ksi, $\tau_{xy} = -6$ ksi, and $\sigma_y = 12$ ksi. On the rotated element, $\sigma'_x = 9$ ksi and $\tau'_{xy} = -9$ ksi. Equation (5-7) is

$$\sigma'_x = \frac{\sigma_x + \sigma_y}{2} + \frac{\sigma_x - \sigma_y}{2} \cos 2\theta + \tau_{xy} \sin 2\theta :$$

$$9 = \frac{(-6) + 12}{2} + \frac{(-6) - 12}{2} \cos 2\theta + (-6) \sin 2\theta$$

$$= 3 - 9 \cos 2\theta - 6 \sin 2\theta,$$

and Eq. (5-8) is

$$\tau'_{xy} = -\frac{\sigma_x - \sigma_y}{2}\sin 2\theta + \tau_{xy}\cos 2\theta :$$

$$-9 = -\frac{(-6) - 12}{2}\sin 2\theta + (-6)\cos 2\theta$$

$$= 9\sin 2\theta - 6\cos 2\theta.$$

We can solve these two equations for $\sin 2\theta$ and $\cos 2\theta$. The results are $\sin 2\theta = -1$ and $\cos 2\theta = 0$, from which we obtain $\theta = 135°$. Substituting this result into Eq. (5-9), the stress σ'_y is

$$\sigma'_y = \frac{\sigma_x + \sigma_y}{2} - \frac{\sigma_x - \sigma_y}{2}\cos 2\theta - \tau_{xy}\sin 2\theta$$

$$= \frac{(-6) + 12}{2} - \frac{(-6) - 12}{2}\cos 2(135°) - (-6)\sin 2(135°)$$

$$= -3 \text{ ksi}.$$

The stresses are shown on the rotated element in Fig. (a).

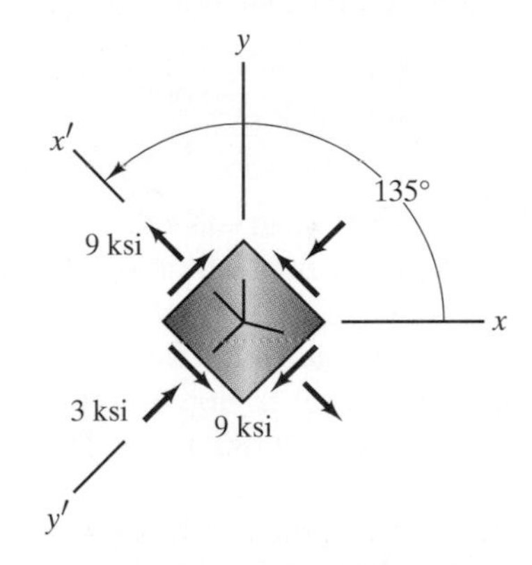

(a) Stresses on the properly oriented element.

Maximum and Minimum Stresses

Given the state of plane stress at a point p, you have seen that the normal and shear stresses on the plane shown in Fig. 5-15 are given by Eqs. (5-7) and (5-8):

$$\sigma'_x = \frac{\sigma_x + \sigma_y}{2} + \frac{\sigma_x - \sigma_y}{2}\cos 2\theta + \tau_{xy}\sin 2\theta, \quad \textbf{(5-10)}$$

$$\tau'_{xy} = -\frac{\sigma_x - \sigma_y}{2}\sin 2\theta + \tau_{xy}\cos 2\theta. \quad \textbf{(5-11)}$$

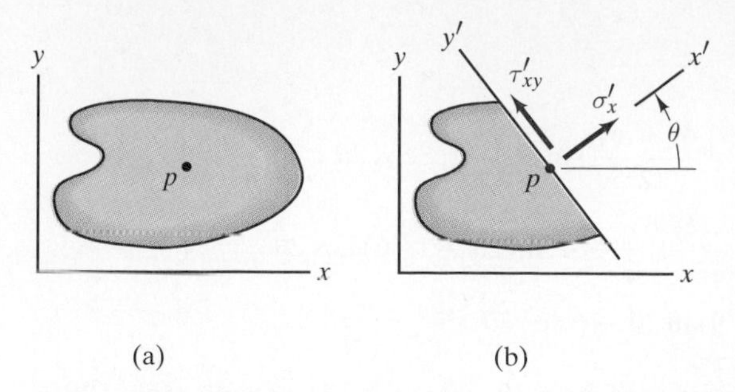

FIGURE 5-15 (a) Point p of a material. (b) Normal and shear stresses at p.

Thus you can determine the normal and shear stresses at p for any value of θ. But it is the maximum values of the stresses that determine whether a material will fail. How can you determine them and the orientations of the planes through the point p on which they act?

PRINCIPAL STRESSES

Let a value of θ for which the normal stress σ_x^1 is a maximum or minimum be denoted by θ_p. By evaluating the derivative of Eq. (5-10) with respect to 2θ and setting it equal to zero, we obtain the equation

$$\tan 2\theta_p = \frac{2\tau_{xy}}{\sigma_x - \sigma_y}. \tag{5-12}$$

When σ_x, σ_y, and τ_{xy} are known, we can solve this equation for $\tan 2\theta_p$, which allows us to determine θ_p. Then we can substitute θ_p into Eq. (5-10) to determine the maximum or minimum value of the normal stress.

Equation (5-12) yields more than one solution for θ_p, because of the periodic nature of the tangent. Observe in Fig. 5-16 that if $2\theta_p$ is a solution of Eq. (5-12), so are $2\theta_p + 180°$, $2\theta_p + 2(180°)$, ... This means that the normal

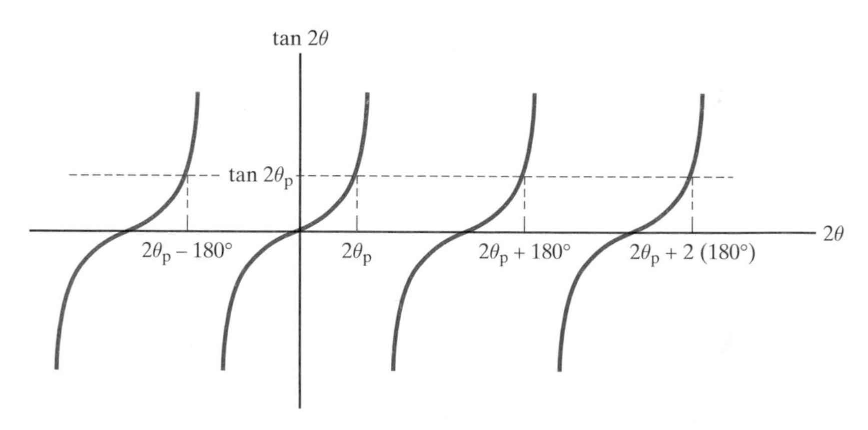

FIGURE 5-16 The periodic nature of the tangent gives rise to multiple roots.

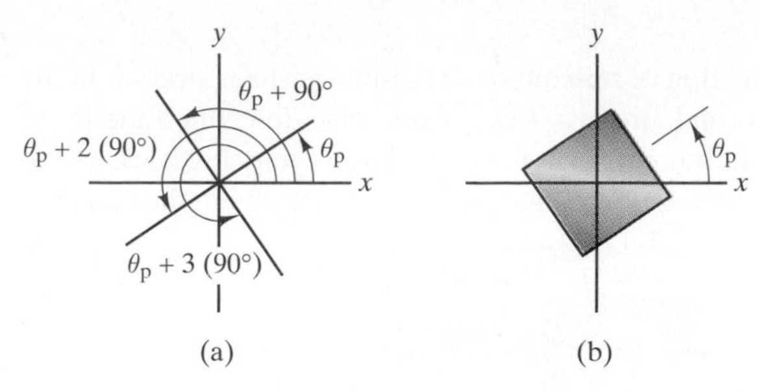

FIGURE 5-17 (a) Angles at which maximum or minimum normal stresses occur. (b) The planes correspond to the faces of a rectangular element.

stress is a maximum or minimum at θ_p, $\theta_p + 90°$, $\theta_p + 2(90°)$, ... (Fig. 5-17a). *The planes on which the maximum and minimum normal stresses act correspond to the faces of a rectangular element* (Fig. 5-17b).

Recognizing that the maximum and minimum normal stresses act on the faces of a particular rectangular element makes it easy to determine and visualize the planes on which these stresses act. We simply determine one value of θ_p from Eq. (5-12), establishing the orientation of the element. We can then determine the values of the maximum and minimum stresses, called the principal stresses and denoted σ_1 and σ_2 (Fig. 5-18), from Eq. (5-10).

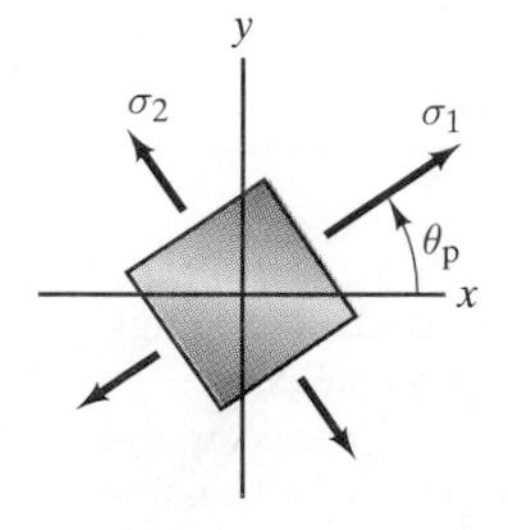

FIGURE 5-18 Principal stresses and planes on which they act.

We can obtain analytical expressions for the values of the principal stresses that are useful when you are not concerned with the planes on which they act. By solving the equations

$$\frac{\sin 2\theta_p}{\cos 2\theta_p} = \tan 2\theta_p = \frac{2\tau_{xy}}{\sigma_x - \sigma_y} \tag{5-13}$$

and

$$\sin^2 2\theta_p + \cos^2 2\theta_p = 1 \tag{5-14}$$

for $\sin 2\theta_p$ and $\cos 2\theta_p$ and substituting the results into Eq. (5-10), we obtain

$$\sigma_1, \sigma_2 = \frac{\sigma_x + \sigma_y}{2} \pm \sqrt{\left(\frac{\sigma_x - \sigma_y}{2}\right)^2 + \tau_{xy}^2}. \tag{5-15}$$

By substituting the expressions for $\sin 2\theta_p$ and $\cos 2\theta_p$ into Eq. (5-11), we obtain $\tau'_{xy} = 0$. *On the planes on which the principal stresses act, the shear stresses are zero.*

The principal stresses are the maximum and minimum normal stresses acting on planes through point p that are parallel to the z axis. There are no normal stresses of greater magnitude on any plane through p. We will see that the situation is more complicated in the case of the maximum shear stress.

Maximum Shear Stresses

We approach the determination of maximum or minimum shear stresses in the same way that we did normal stresses. Let a value of θ for which the shear stress is a maximum or minimum be denoted by θ_s. Evaluating the derivative of Eq. (5-11) with respect to 2θ and setting it equal to zero, we obtain the equation

$$\tan 2\theta_s = -\frac{\sigma_x - \sigma_y}{2\tau_{xy}}. \tag{5-16}$$

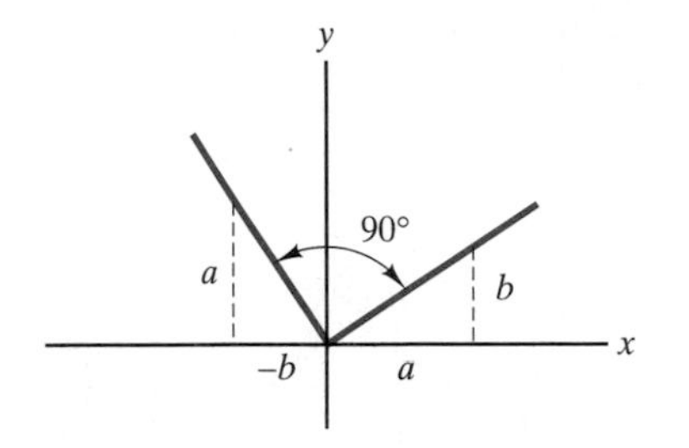

Figure 5-19 The tangents of the angles defining these two perpendicular directions relative to the x axis are b/a and $-a/b$. One tangent is the negative inverse of the other.

With this equation we can determine θ_s and substitute it into Eq. (5-11) to determine the maximum or minimum value of the shear stress.

As in the case of the normal stress, if the shear stress is a maximum or minimum at θ_s, it is also a maximum or minimum at $\theta_s + 90°$, $\theta_s + (2)(90°)$, ... *The planes on which the maximum and minimum shear stresses act also correspond to the faces of a rectangular element.* Since the shear stresses on the faces of a rectangular element are equal in magnitude, the magnitudes of the maximum and minimum shear stresses are equal. Furthermore, the orientation of this element is related in a simple way to the orientation of the element on which the principal stresses act. Notice from Eqs. (5-12) and (5-16) that $\tan 2\theta_s$ is the negative inverse of $\tan 2\theta_p$. This implies that the directions defined by the angles $2\theta_s$ and $2\theta_p$ are perpendicular (Fig. 5-19), which means that *the element on which the maximum and minimum shear stresses act is rotated* 45° *relative to the element on which the principal stresses act.* Once we have determined the orientation of the element on which the principal stresses act, we also know the orientation of the element on which the maximum and minimum shear stresses act (Fig. 5-20).

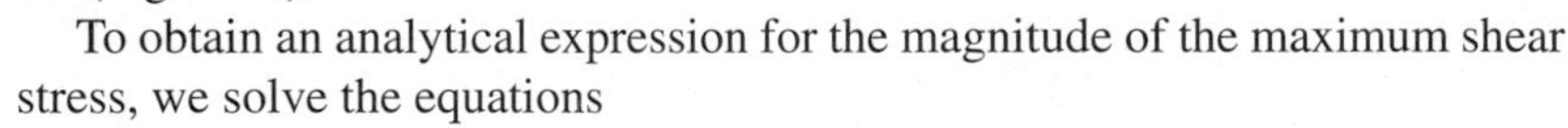

To obtain an analytical expression for the magnitude of the maximum shear stress, we solve the equations

$$\frac{\sin 2\theta_s}{\cos 2\theta_s} = \tan 2\theta_s = -\frac{\sigma_x - \sigma_y}{2\tau_{xy}} \tag{5-17}$$

and

$$\sin^2 2\theta_s + \cos^2 2\theta_s = 1 \tag{5-18}$$

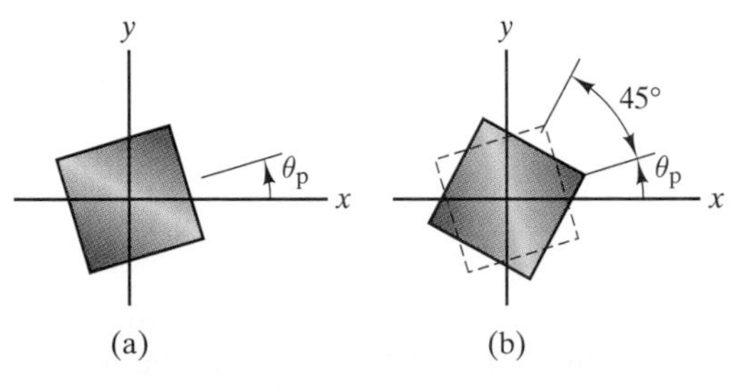

Figure 5-20 Relationship between the orientation of the elements (a) on which the principal stresses act and (b) on which the maximum and minimum shear stresses act.

for $\sin 2\theta_s$ and $\cos 2\theta_s$ and substitute the results into Eq. (5-11). The result is

$$\tau_{max} = \sqrt{\left(\frac{\sigma_x - \sigma_y}{2}\right)^2 + \tau_{xy}^2}. \tag{5-19}$$

This stress is called the *maximum in-plane shear stress,* because it is the greatest shear stress that occurs on any plane parallel to the z axis. However, we will see that greater shear stresses may occur on other planes through p.

To determine the complete state of plane stress on the element on which the maximum in-plane shear stresses act, we must evaluate the normal stresses. Substituting our results for $\sin 2\theta_s$ and $\cos 2\theta_s$ into Eqs. (5-7) and (5-9), we obtain

$$\sigma_x' = \sigma_y' = \frac{\sigma_x + \sigma_y}{2}.$$

The normal stresses on the element on which the maximum in-plane shear stresses act are equal. We denote this normal stress by σ_s:

$$\sigma_s = \frac{\sigma_x + \sigma_y}{2}. \tag{5-20}$$

Figure 5-21 shows the complete state of stress.

Now let us consider whether shear stresses greater in magnitude than the value given by Eq. (5-19) occur on other planes through p. We begin with the element on which the principal stresses occur, and realign the coordinate system with the faces of the element (Fig. 5-22a). In terms of this new coordinate system, the components of plane stress are $\sigma_x = \sigma_1$, $\sigma_y = \sigma_2$, and $\tau_{xy} = 0$. Substituting these components into Eq. (5-19), we obtain a different expression for the magnitude of the maximum in-plane shear stress:

$$\sqrt{\left(\frac{\sigma_x - \sigma_y}{2}\right)^2 + \tau_{xy}^2} = \sqrt{\left(\frac{\sigma_1 - \sigma_2}{2}\right)^2 + 0} = \left|\frac{\sigma_1 - \sigma_2}{2}\right|. \tag{5-21}$$

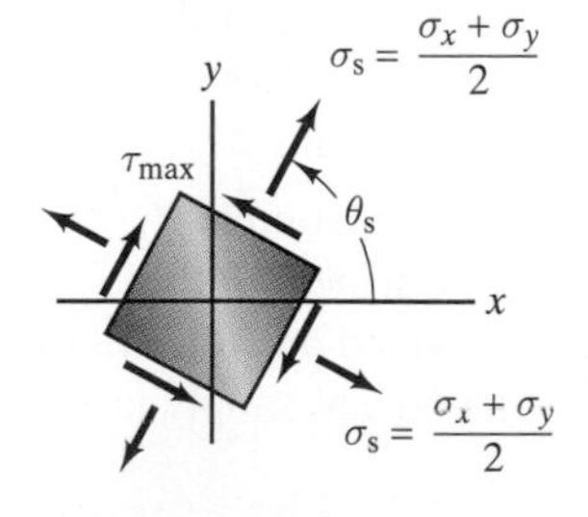

FIGURE 5-21 Element on which the maximum and minimum in-plane shear stresses act.

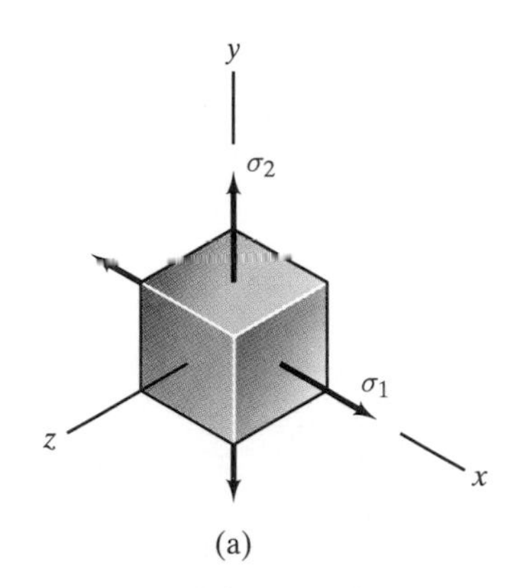

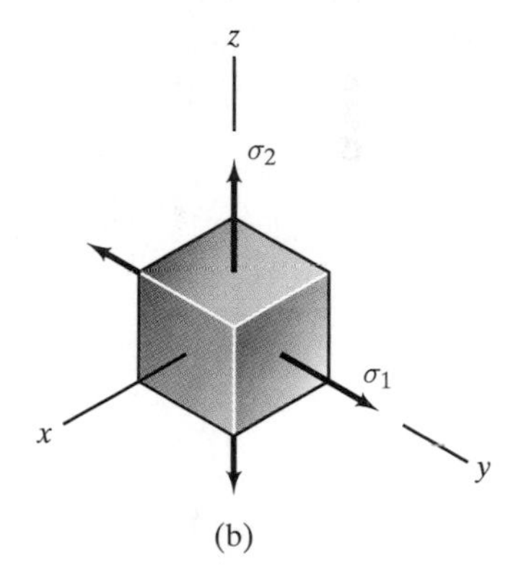

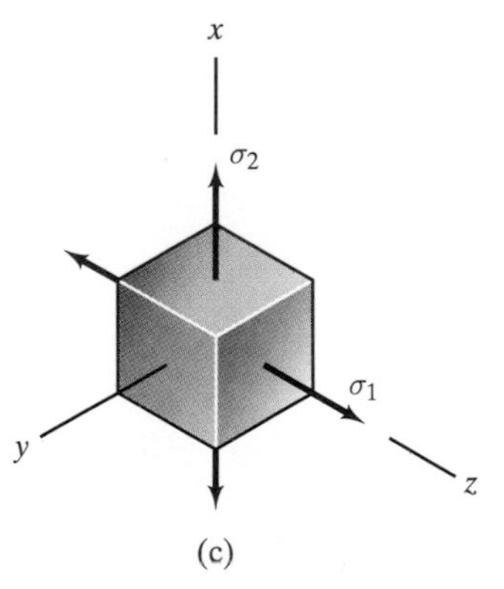

FIGURE 5-22 Element on which the principal stresses act. (a) Realigned coordinate system. (b), (c) Other orientations of the coordinate system.

This equation is expressed in terms of the principal stresses, but it gives the same result as Eq. (5-19). We have still considered only planes parallel to the original z axis. Now, however, let's consider the element on which the principal stresses occur and reorient the coordinate system as shown in Fig. 5-22b. In terms of this coordinate system, $\sigma_x = 0$, $\sigma_y = \sigma_1$, and $\tau_{xy} = 0$. Substituting these components into Eq. (5-19), we obtain the maximum shear stress on planes parallel to the new z axis:

$$\sqrt{\left(\frac{\sigma_x - \sigma_y}{2}\right)^2 + \tau_{xy}^2} = \sqrt{\left(\frac{0 - \sigma_1}{2}\right)^2 + 0} = \left|\frac{\sigma_1}{2}\right|. \qquad \textbf{(5-22)}$$

Next, we reorient the coordinate system as shown in Fig. 5-22c. In terms of this coordinate system, $\sigma_x = \sigma_2$, $\sigma_y = 0$, and $\tau_{xy} = 0$. Substituting these components into Eq. (5-19), we obtain the maximum shear stress on planes parallel to this z axis:

$$\sqrt{\left(\frac{\sigma_x - \sigma_y}{2}\right)^2 + \tau_{xy}^2} = \sqrt{\left(\frac{\sigma_2 - 0}{2}\right)^2 + 0} = \left|\frac{\sigma_2}{2}\right|. \qquad \textbf{(5-23)}$$

Depending on the values of the principal stresses, Eq. (5-22) and/or Eq. (5-23) can result in larger values of the magnitude of the maximum shear stress than the maximum in-plane shear stress. Although we have still considered only a subset of the possible planes through point p, there are no shear stresses of greater magnitude on any plane through p than the largest value given by Eq. (5-21), (5-22), or (5-23), which is called the *absolute maximum shear stress.*

SUMMARY: DETERMINING THE PRINCIPAL STRESSES AND THE MAXIMUM SHEAR STRESS

Here we give a sequence of steps you can use to determine the maximum and minimum stresses at a point p subjected to a known state of plane stress and the orientations of the planes on which they act.

1. Use Eq. (5-12) to determine θ_p, establishing the orientation of the element on which the principal stresses act.
2. Determine the two principal stresses by substituting first θ_p and then $\theta_p + 90^\circ$ into Eq. (5-10). Since no shear stresses act on the element on which the principal stresses act, this determines the complete state of stress on the element. Notice that you can determine the values of the principal stresses from Eq. (5-15), but this does not tell you which principal stress acts on which faces of the element.
3. Use Eq. (5-16) to determine θ_s, establishing the orientation of the element on which the maximum and minimum in-plane shear stresses act. Alternatively, since this element is rotated 45° relative to the element on which the principal stresses act, you can simply use $\theta_s = \theta_p + 45^\circ$.
4. Determine the shear stress on the element by substituting θ_s into Eq. (5-11). The magnitude of this shear stress is the maximum

in-plane shear stress. Once you have determined the shear stress on one face of the element, you know the shear stresses on all the faces. The normal stress on each face of this element is $(\sigma_x + \sigma_y)/2$, which completes the determination of the state of stress on the element.

5. The absolute maximum shear stress (the maximum shear stress on any plane through p) is the largest of the three values

$$\left|\frac{\sigma_1 - \sigma_2}{2}\right|, \quad \left|\frac{\sigma_1}{2}\right|, \quad \left|\frac{\sigma_2}{2}\right|. \tag{5-24}$$

EXAMPLE 5-4

Worksheet 5

The state of plane stress at a point p is shown on the element in Fig. 5-23. Determine the principal stresses and the maximum in-plane shear stress and show them acting on properly oriented elements. Also determine the absolute maximum shear stress.

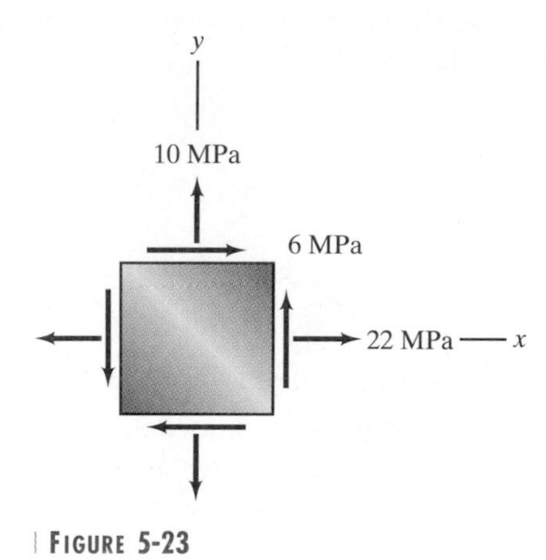

FIGURE 5-23

Strategy

The components of plane stress on the element are $\sigma_x = 22$ MPa, $\sigma_y = 10$ MPa, and $\tau_{xy} = 6$ MPa. We can follow the steps given in the preceding summary for determining the principal stresses and the maximum shear stress.

Solution

Step 1 From Eq. (5-12),

$$\tan 2\theta_p = \frac{2\tau_{xy}}{\sigma_x - \sigma_y} = \frac{2(6)}{22 - 10} = 1.$$

Solving this equation, we obtain $\theta_p = 22.5°$. This angle tells us the orientation of the element on which the principal stresses act.

Step 2 We substitute θ_p into Eq. (5-10) to determine the first principal stress.

$$\begin{aligned}\sigma_1 &= \frac{\sigma_x + \sigma_y}{2} + \frac{\sigma_x - \sigma_y}{2}\cos 2\theta_p + \tau_{xy}\sin 2\theta_p \\ &= \frac{22+10}{2} + \frac{22-10}{2}\cos 45^\circ + 6\sin 45^\circ \\ &= 24.49 \text{ MPa.}\end{aligned}$$

We then substitute $\theta_p + 90^\circ$ into Eq. (5-10) to determine the second principal stress.

$$\begin{aligned}\sigma_2 &= \frac{\sigma_x + \sigma_y}{2} + \frac{\sigma_x - \sigma_y}{2}\cos 2(\theta_p + 90^\circ) + \tau_{xy}\sin 2(\theta_p + 90^\circ) \\ &= \frac{22+10}{2} + \frac{22-10}{2}\cos 225^\circ + 6\sin 225^\circ \\ &= 7.51 \text{ MPa.}\end{aligned}$$

The principal stresses are shown on the properly oriented element in Fig. (a). No shear stresses act on this element.

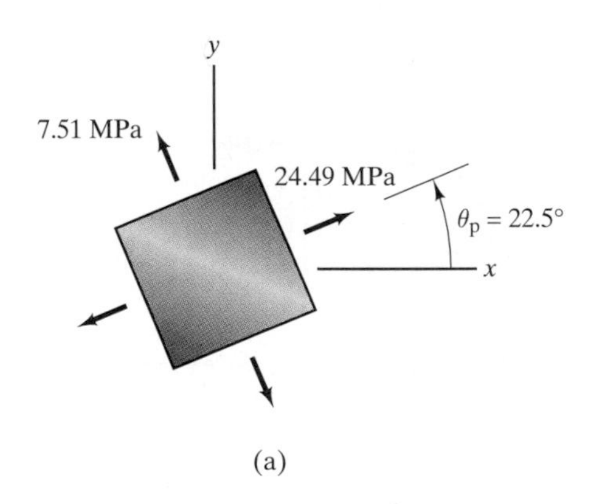

(a) The principal stresses.

We can also determine the values of the principal stresses from Eq. (5-15),

$$\begin{aligned}\sigma_1, \sigma_2 &= \frac{\sigma_x + \sigma_y}{2} \pm \sqrt{\left(\frac{\sigma_x - \sigma_y}{2}\right)^2 + \tau_{xy}^2} \\ &= \frac{22+10}{2} \pm \sqrt{\left(\frac{22-10}{2}\right)^2 + (6)^2} \\ &= 24.49, 7.51 \text{ MPa,}\end{aligned}$$

but this procedure does not tell us which faces of the element the stresses act on.

Step 3 From Eq. (5-16),

$$\tan 2\theta_s = -\frac{\sigma_x - \sigma_y}{2\tau_{xy}} = -\frac{22 - 10}{2(6)} = -1,$$

from which we obtain $\theta_s = -22.5°$. [We could choose instead to determine θ_s by using the fact that the element on which the maximum in-plane shear stresses act is rotated 45° relative to the element on which the principal stresses act. In this way we obtain $\theta_s = \theta_p + 45° = 67.5°$. This angle differs from our previous result by 90°, so the resulting orientation of the element is the same. This emphasizes that the angle θ_s can only be determined from Eq. (5-16) within a multiple of 90°.]

Step 4 We substitute θ_s into Eq. (5-11) to determine the maximum in-plane shear stress.

$$\begin{aligned}\tau_{max} &= -\frac{\sigma_x - \sigma_y}{2}\sin 2\theta_s + \tau_{xy}\cos 2\theta_s \\ &= -\frac{22 - 10}{2}\sin(-45°) + 6\cos(-45°) \\ &= 8.49 \text{ MPa}.\end{aligned}$$

The normal stress on each face of this element is

$$\sigma_s = \frac{\sigma_x + \sigma_y}{2} = \frac{22 + 10}{2} = 16 \text{ MPa}.$$

The maximum in-plane shear stresses and associated normal stresses are shown on the properly oriented element in Fig. (b).

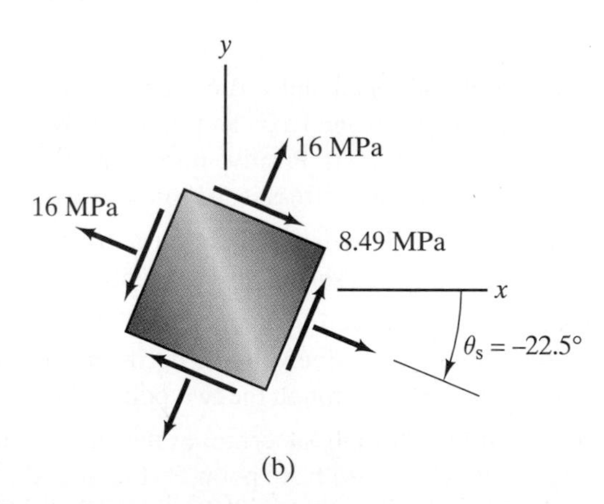

(b) Maximum in-plane shear stresses and associated normal stresses.

Step 5 The absolute maximum shear stress is given by the largest of the three values

$$\left|\frac{\sigma_1 - \sigma_2}{2}\right| = \left|\frac{24.49 - 7.51}{2}\right| = 8.49 \text{ MPa},$$

$$\left|\frac{\sigma_1}{2}\right| = \left|\frac{24.49}{2}\right| = 12.24 \text{ MPa},$$

$$\left|\frac{\sigma_2}{2}\right| = \left|\frac{7.51}{2}\right| = 3.76 \text{ MPa}.$$

The absolute maximum shear stress is 12.24 MPa.

5-3 | Mohr's Circle for Plane Stress

Mohr's circle is a graphical method for solving Eqs. (5-7), (5-8), and (5-9). Given a state of plane stress (Fig. 5-24a), Mohr's circle allows you to determine the components of stress in terms of a coordinate system rotated through a specified angle θ (Fig. 5-24b). You may wonder why we discuss this method in an age when computers have made most graphical methods obsolete. The reason is that Mohr's circle allows you to visualize the solutions to Eqs. (5-7)–(5-9), and understand their properties, to an extent not possible with other approaches. We first explain how to apply Mohr's circle and then show why it works.

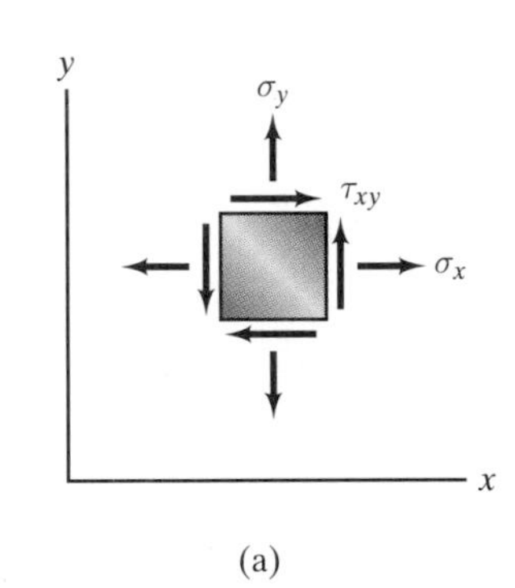

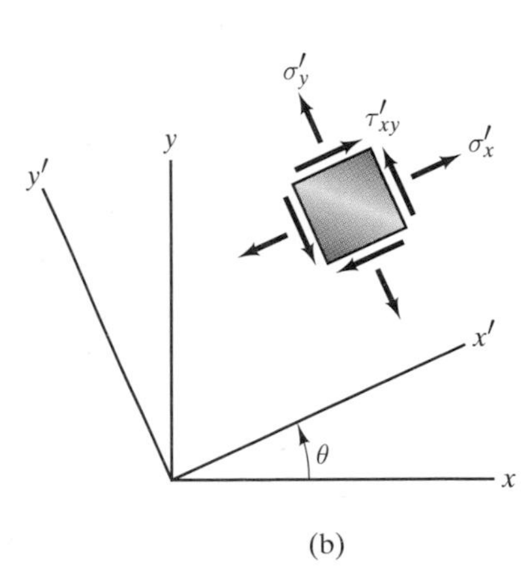

FIGURE 5-24 (a) State of plane stress. (b) Components in terms of a rotated coordinate system.

Constructing the Circle

Suppose that we know the components σ_x, τ_{xy}, and σ_y, and we want to determine the components σ'_x, τ'_{xy}, and σ'_y for a given angle θ. Determining this information with Mohr's circle involves four steps:

1. Establish a set of horizontal and vertical axes with normal stress measured along the horizontal axis and shear stress measured along the vertical axis (Fig. 5-25a). Positive normal stress is measured to the right and positive shear stress is measured *downward*.
2. Plot two points, point P with coordinates (σ_x, τ_{xy}) and point Q with coordinates $(\sigma_y, -\tau_{xy})$, as shown in Fig. 5-25b.
3. Draw a straight line connecting points P and Q. Using the intersection of the straight line with the horizontal axis as the center, draw a circle that passes through the two points (Fig. 5-25c).
4. Draw a straight line through the center of the circle at an angle 2θ measured counterclockwise from point P (Fig. 5-25d). The point P' at which this line intersects the circle has coordinates (σ'_x, τ'_{xy}), and the point Q' has coordinates $(\sigma'_y, -\tau'_{xy})$.

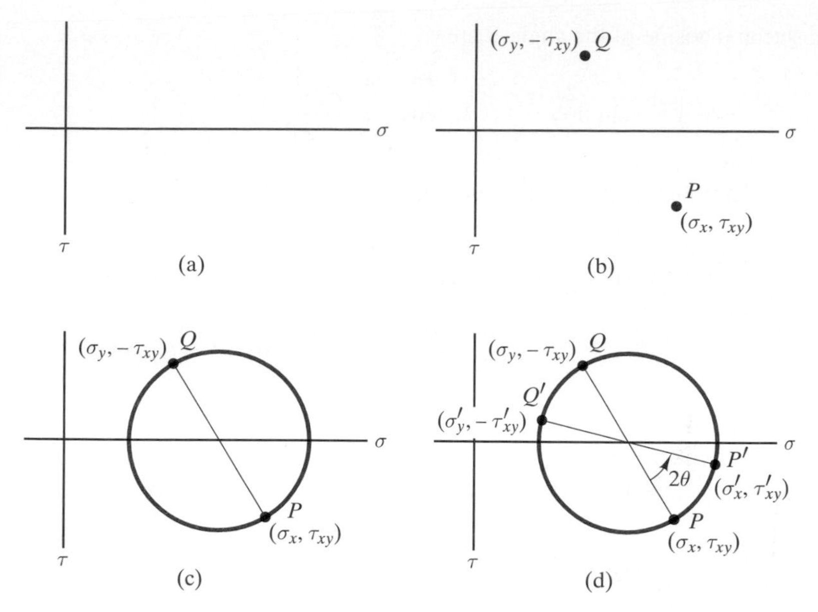

FIGURE 5-25 (a) Establishing the axes. Shear stress is positive downward. (b) Plotting points P and Q. (c) Drawing Mohr's circle. The center of the circle is the intersection of the line between points P and Q with the horizontal axis. (d) Determining the stresses.

This construction indicates that for any value of the angle θ, you can determine the stress components σ'_x, τ'_{xy}, and σ'_y from the coordinates of two points on Mohr's circle. But we must prove this result.

Why Mohr's Circle Works

We will now prove that Mohr's circle solves Eqs. (5-7)–(5-9):

$$\sigma'_x = \frac{\sigma_x + \sigma_y}{2} + \frac{\sigma_x - \sigma_y}{2}\cos 2\theta + \tau_{xy}\sin 2\theta,$$

$$\tau'_{xy} = -\frac{\sigma_x - \sigma_y}{2}\sin 2\theta + \tau_{xy}\cos 2\theta,$$

$$\sigma'_y = \frac{\sigma_x + \sigma_y}{2} - \frac{\sigma_x - \sigma_y}{2}\cos 2\theta - \tau_{xy}\sin 2\theta.$$

In Fig. 5-26a we show the points P and Q and Mohr's circle. Notice that the horizontal coordinate of the center of the circle is $(\sigma_x + \sigma_y)/2$, and R, the radius of the circle, is given by

$$R = \sqrt{\left(\frac{\sigma_x - \sigma_y}{2}\right)^2 + (\tau_{xy})^2}.$$

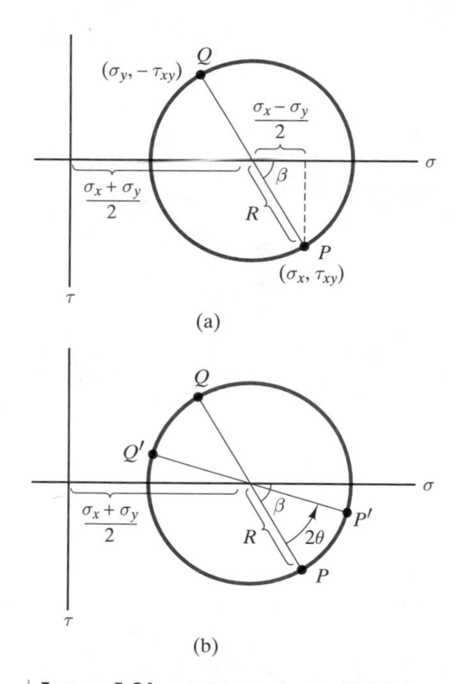

FIGURE 5-26 (a) Dimensions of Mohr's circle. (b) Dimensions including the points P' and Q'.

The sine and cosine of the angle β are

$$\sin\beta = \frac{\tau_{xy}}{R}, \qquad \cos\beta = \frac{\sigma_x - \sigma_y}{2R}.$$

From Fig. 5-26b, the horizontal coordinate of point P' is

$$\begin{aligned}\frac{\sigma_x + \sigma_y}{2} + R\cos(\beta - 2\theta) &= \frac{\sigma_x + \sigma_y}{2} + R(\cos\beta\cos 2\theta + \sin\beta\sin 2\theta)\\ &= \frac{\sigma_x + \sigma_y}{2} + \frac{\sigma_x - \sigma_y}{2}\cos 2\theta + \tau_{xy}\sin 2\theta\\ &= \sigma'_x,\end{aligned}$$

and the horizontal coordinate of point Q' is

$$\begin{aligned}\frac{\sigma_x + \sigma_y}{2} - R\cos(\beta - 2\theta) &= \frac{\sigma_x + \sigma_y}{2} - R(\cos\beta\cos 2\theta + \sin\beta\sin 2\theta)\\ &= \frac{\sigma_x + \sigma_y}{2} - \frac{\sigma_x - \sigma_y}{2}\cos 2\theta - \tau_{xy}\sin 2\theta\\ &= \sigma'_y.\end{aligned}$$

The vertical coordinate of point P' is

$$\begin{aligned}R\sin(\beta - 2\theta) &= R(-\cos\beta\sin 2\theta + \sin\beta\cos 2\theta)\\ &= -\frac{\sigma_x - \sigma_y}{2}\sin 2\theta + \tau_{xy}\cos 2\theta\\ &= \tau'_{xy},\end{aligned}$$

which implies that the vertical coordinate of point Q' is $-\tau'_{xy}$. Thus we have shown that the coordinates of point P' are (σ'_x, τ'_{xy}) and the coordinates of point Q' are $(\sigma'_y, -\tau'_{xy})$.

Determining Principal Stresses and the Maximum In-Plane Shear Stress

Mohr's circle is a map of the stresses σ'_x, τ'_{xy}, and σ'_y for all values of θ, so once we have constructed the circle we can immediately see the values of the maximum and minimum normal stresses and the magnitude of the maximum in-plane shear stress. The coordinates of the points where the circle intersects the horizontal axis determine the two principal stresses (Fig. 5-27). The coordinates of the points at the bottom and top of the circle determine the maximum and minimum in-plane shear stresses, so the radius of the circle equals the magnitude of the maximum in-plane shear stress. Notice that Mohr's circle demonstrates very clearly that the shear stresses are zero on the planes on which the principal

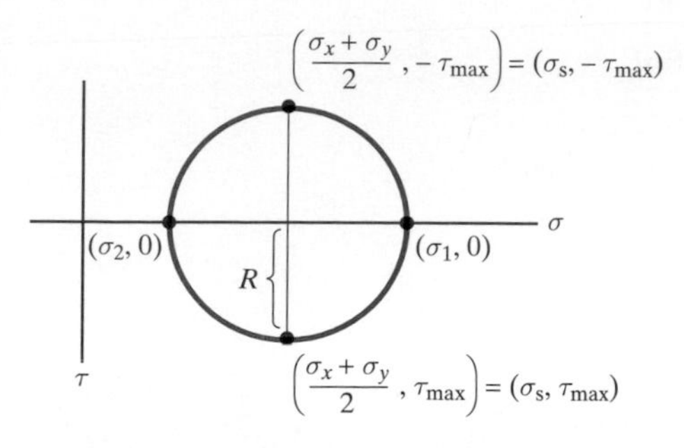

FIGURE 5-27 Mohr's circle indicates the values of the principal stresses and the magnitude of the maximum in-plane shear stress.

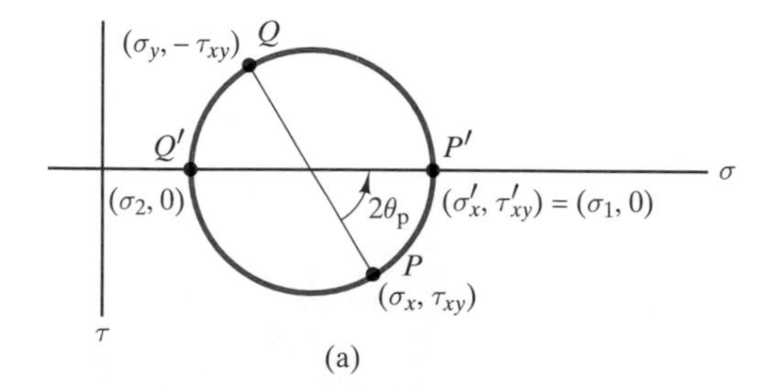

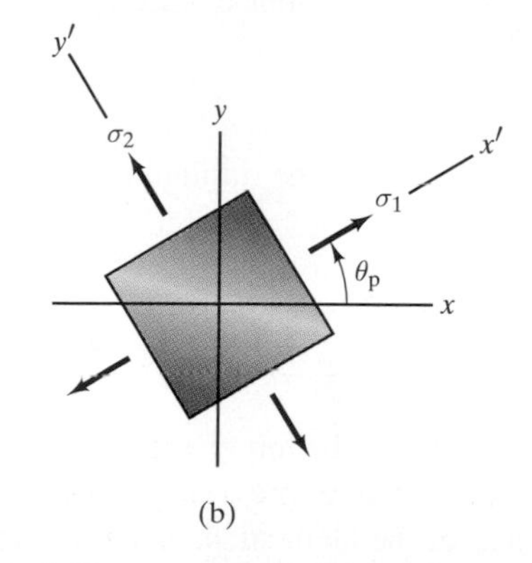

FIGURE 5-28 Using Mohr's circle to determine the orientation of the element on which the principal stresses act.

stresses act, and also that the normal stresses are equal on the planes on which the maximum and minimum in-plane shear stresses act.

We can also use Mohr's circle to determine the orientations of the elements on which the principal stresses and maximum in-plane shear stresses act. If we let the point P' coincide with either principal stress (Fig. 5-28a), we can measure the angle $2\theta_p$ and thereby determine the orientation of the plane on which that principal stress acts (Fig. 5-28b).

We can then let P' coincide with either the maximum or minimum shear stress (Fig. 5-29a) and measure the angle $2\theta_s$, determining the orientation of the plane on which that shear stress acts (Fig. 5-29b). Notice in the example

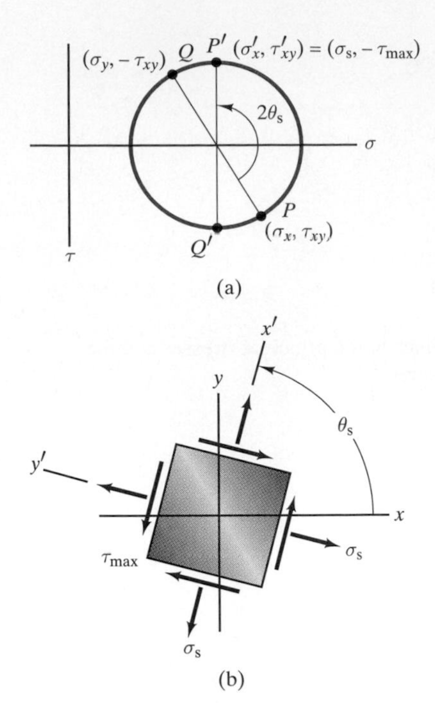

FIGURE 5-29 Using Mohr's circle to determine the orientation of the element on which the maximum in-plane shear stresses act.

illustrated in Fig. 5-29 that point P' coincides with the minimum shear stress, so $\tau'_{xy} = -\tau_{max}$.

EXAMPLE 5-5

The state of plane stress at a point p is shown on the left element in Fig. 5-30. Use Mohr's circle to determine the state of plane stress at p acting on the right element in Fig. 5-30. Draw a sketch of the element showing the stresses acting on it.

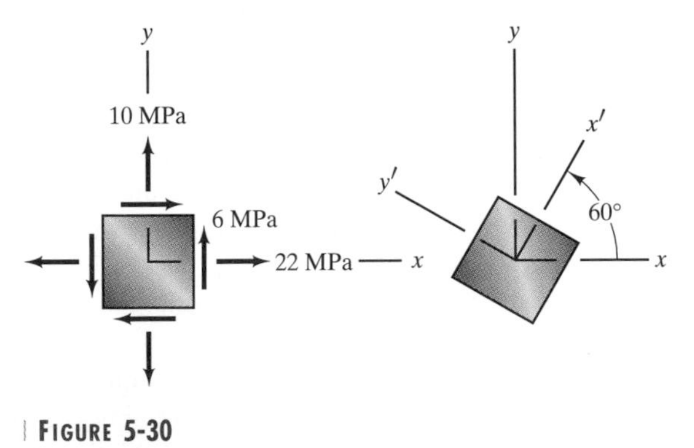

FIGURE 5-30

Strategy

The components of plane stress on the left element in Fig. 5-30 are $\sigma_x = 22$ MPa, $\sigma_y = 10$ MPa, and $\tau_{xy} = 6$ MPa. We can follow the steps given earlier for constructing Mohr's circle and determining the stress components σ'_x, τ'_{xy}, and σ'_y.

Solution

Step 1 Establish a set of horizontal and vertical axes with normal stress measured along the horizontal axis and shear stress measured along the vertical axis [Fig. (a)]. Positive shear stress is measured downward. The scales of normal stress and shear stress must be equal and chosen so that the circle will fit on the page but be large enough for reasonable accuracy. (This may require some trial and error.)

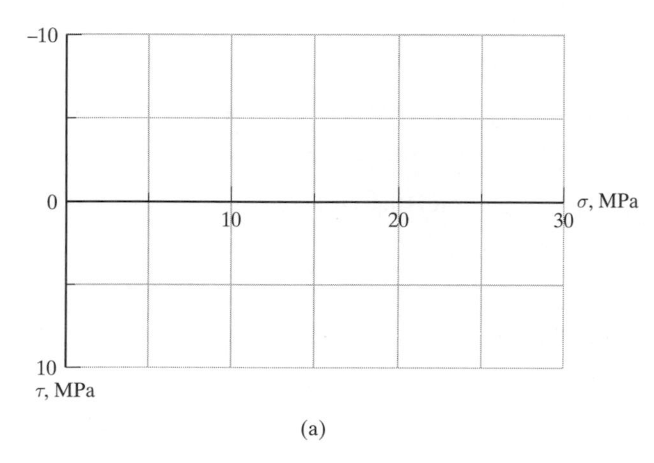

(a) Establishing the axes.

Step 2 Plot point P with coordinates $(\sigma_x, \tau_{xy}) = (22, 6)$ and point Q with coordinates $(\sigma_y, -\tau_{xy}) = (10, -6)$, as shown in Fig. (b).

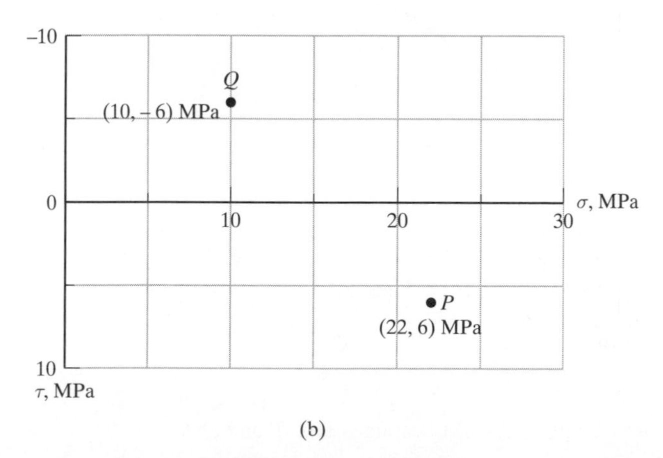

(b) Plotting points P and Q.

Step 3 Draw a straight line connecting points P and Q. Using the intersection of the straight line with the horizontal axis as the center, draw a circle that passes through the two points [Fig. (c)].

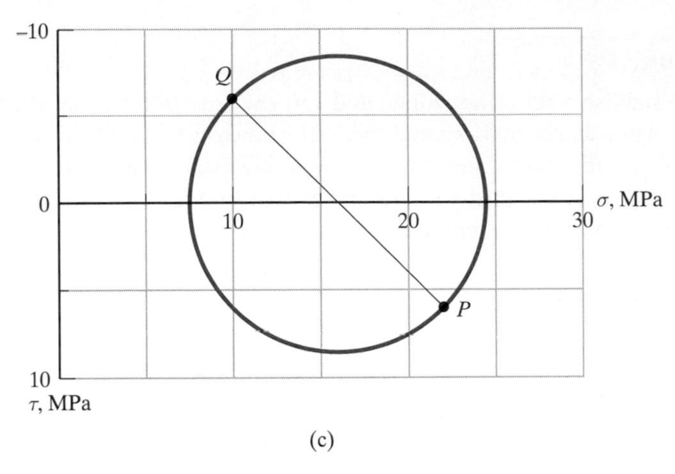

(c) Drawing the circle.

Step 4 Draw a straight line through the center of the circle at an angle $2\theta = 120°$ measured counterclockwise from point P [Fig. (d)]. The point P' at which this line intersects the circle has coordinates (σ'_x, τ'_{xy}). The values we estimate from the graph are $\sigma'_x = 18.3$ MPa and $\tau'_{xy} = -8.3$ MPa. The point Q' has coordinates $(\sigma'_y, -\tau'_{xy})$, from which we estimate that $\sigma'_y = 13.8$ MPa. The stresses are shown in Fig. (e).

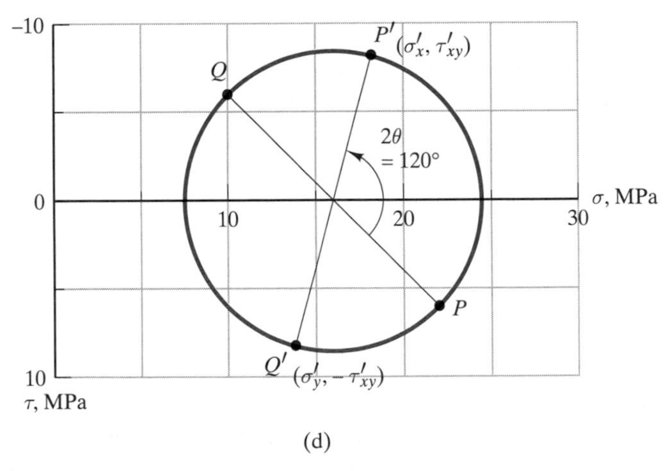

(d) Locating points P' and Q'.

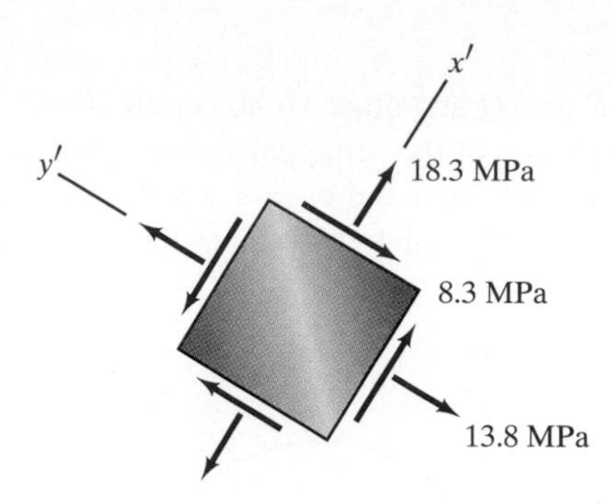

(c) Stresses on the rotated element.

Discussion

Compare this application of Mohr's circle to the analytical solution of the same problem in Example 5-2.

EXAMPLE 5-6

The state of plane stress at a point p is shown on the element in Fig. 5-31. Use Mohr's circle to determine the principal stresses and the maximum in-plane shear stress and show them acting on properly oriented elements.

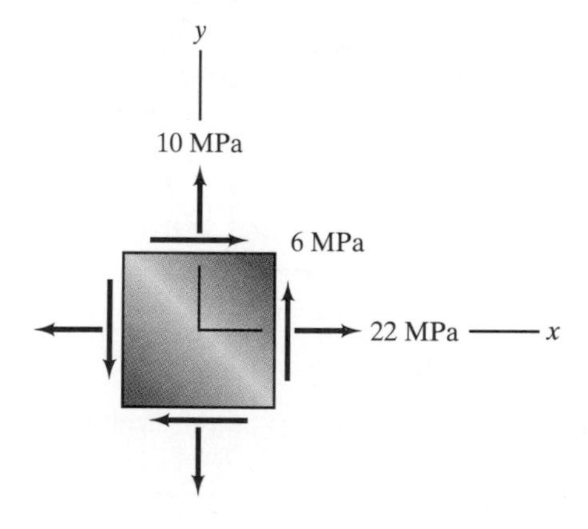

FIGURE 5-31

Strategy

The components of plane stress on the element are $\sigma_x = 22$ MPa, $\sigma_y = 10$ MPa, and $\tau_{xy} = 6$ MPa. Mohr's circle determines the principal stresses and the maximum and minimum in-plane shear stresses. By letting the point P' coincide first with one of the principal stresses and then with the maximum or minimum in-plane shear stress, we can determine the orientations of the elements on which these stresses act.

Solution

We first plot points P and Q and draw Mohr's circle [Fig. (a)]. Then we let the point P' coincide with one of the principal stresses [Fig. (b)]. From the circle we estimate that $\sigma_1 = 24.5$ MPa and $\sigma_2 = 7.5$ MPa. Measuring the angle $2\theta_p$, we estimate that $\theta_p = 22.5°$, which determines the orientation of the plane on which the principal stress σ_1 acts [Fig. (c)].

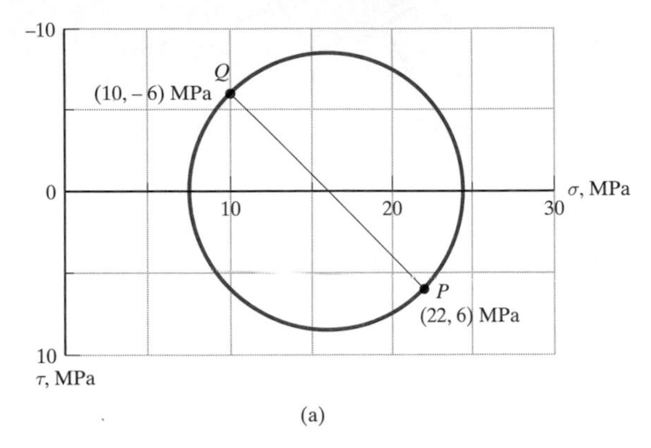

(a) Mohr's circle.

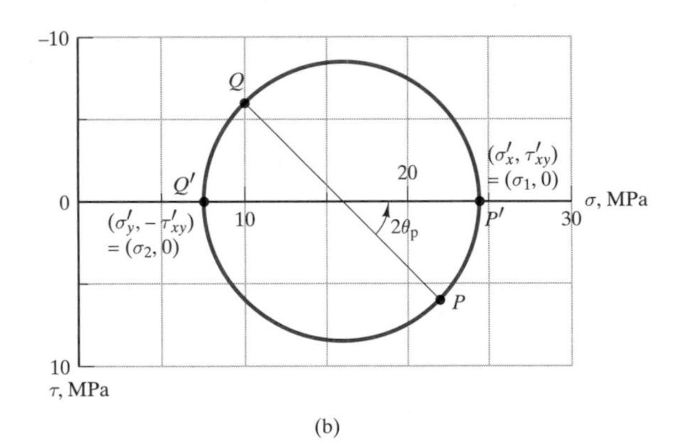

(b) Letting P' coincide with the principal stress σ_1.

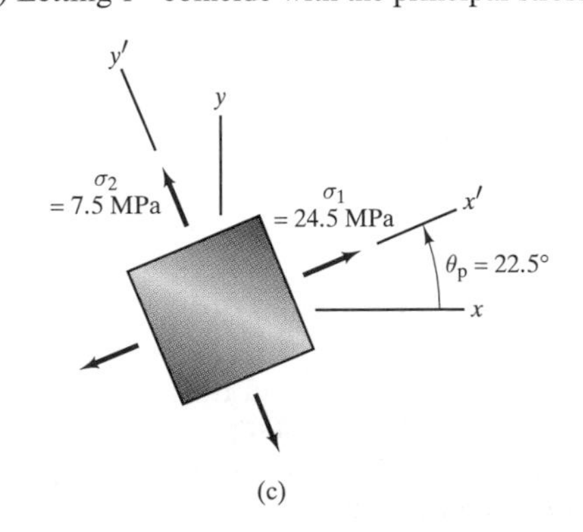

(c) Element on which the principal stresses act.

We next let the point P' coincide with either the minimum or maximum in-plane shear stress. In this case we choose the minimum stress [Fig. (d)]. We estimate that $\tau_{\max} = 8.5$ MPa and the normal stress $\sigma_s = 16$ MPa. Measuring the angle $2\theta_s$, we estimate that $\theta_s = 67.5°$, which determines the orientation of the plane on which the minimum in-plane shear stress acts [Fig. (e)].

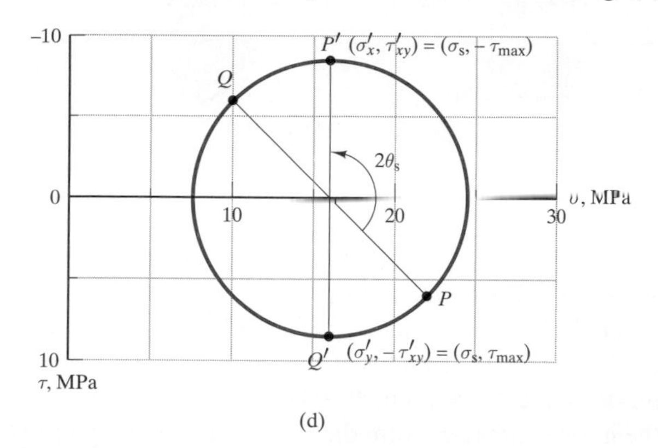

(d)

(d) Letting P' coincide with the minimum in-plane shear stress.

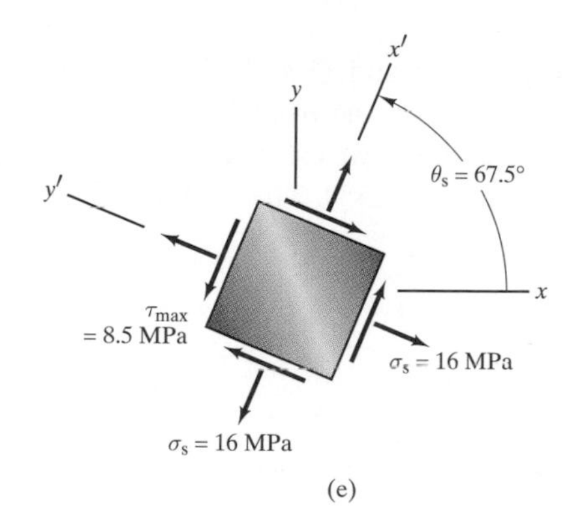

(e)

(e) Element on which the maximum and minimum in-plane shear stresses act. Notice that $\tau'_{xy} = -\tau_{\max}$.

Discussion

Compare this solution to Example 5-4, in which we begin with the same state of stress and analytically determine the principal stresses and maximum in-plane shear stress.

EXAMPLE 5-7

The state of plane stress at a point p is shown on the left element in Fig. 5-32, and the values of the stresses σ'_x and τ'_{xy} are shown on the right element. What are the normal stress σ'_y and the angle θ?

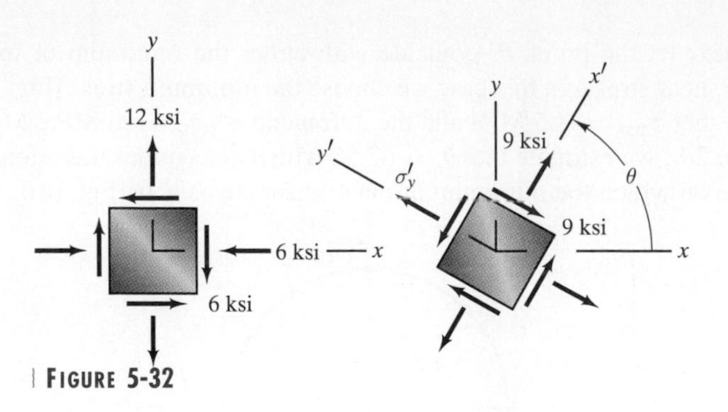

| FIGURE 5-32

Strategy

The components of stress on the left element are $\sigma_x = -6$ ksi, $\tau_{xy} = -6$ ksi, and $\sigma_y = 12$ ksi. With this information we can plot the points P and Q and draw Mohr's circle. On the rotated element, the stresses $\sigma'_x = 9$ ksi and $\tau'_{xy} = -9$ ksi, which permits us to locate the point P'. Then we can measure the angle 2θ and determine the normal stress σ'_y from the coordinates of the point Q'.

Solution

Figure (a) shows the points P and Q and Mohr's circle. In Fig. (b) we plot the point P' and draw a straight line from P' through the center of the circle. From the coordinates of point Q', we estimate that $\sigma'_y = -3$ ksi. Measuring the angle 2θ, we estimate that $\theta = 135°$. The stresses are shown on the properly oriented element in Fig. (c).

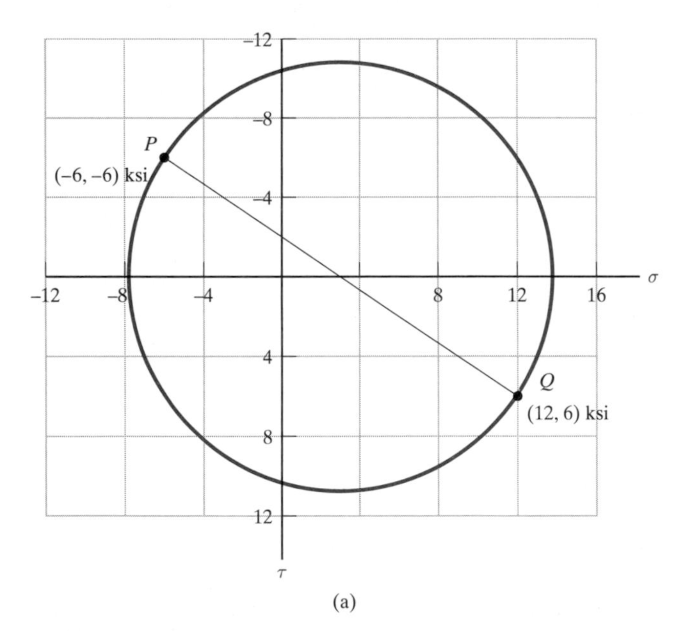

(a) Plotting the points P and Q and drawing Mohr's circle.

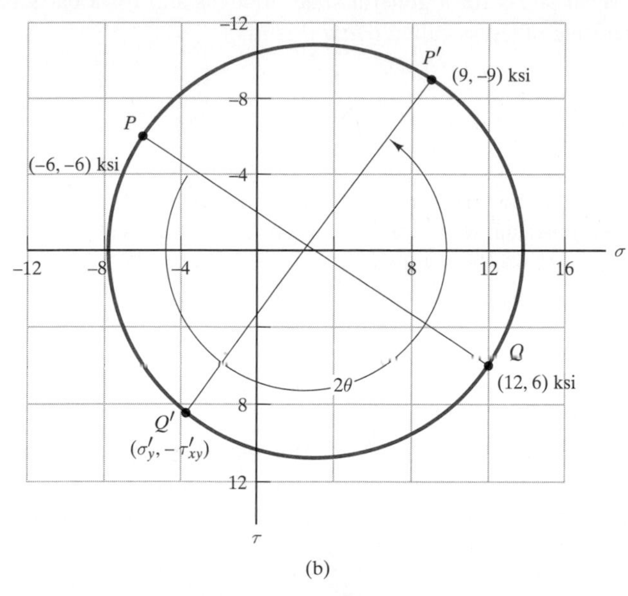

(b)

(b) Plotting P'.

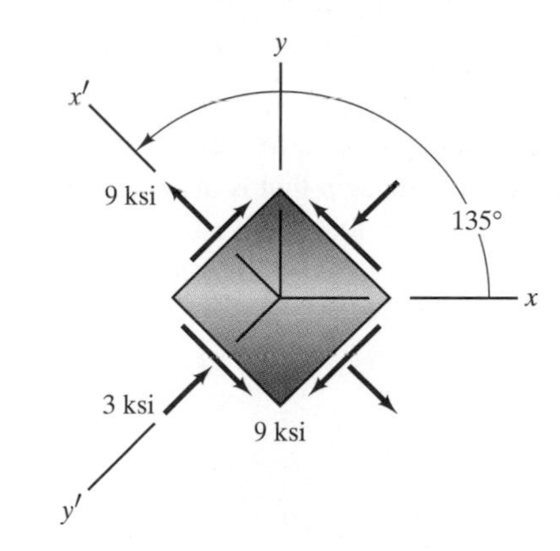

(c) Stresses on the properly oriented element.

Discussion

Compare this solution with Example 5-3, in which we solve the same problem analytically.

5-4 | Principal Stresses in Three Dimensions

Many important applications involve states of stress more general than plane stress. Because of the crucial importance of maximum stresses in design, in this section we explain how to determine the principal stresses and absolute maximum shear stress for a general state of stress and for a particular three-dimensional state of stress called *triaxial stress.*

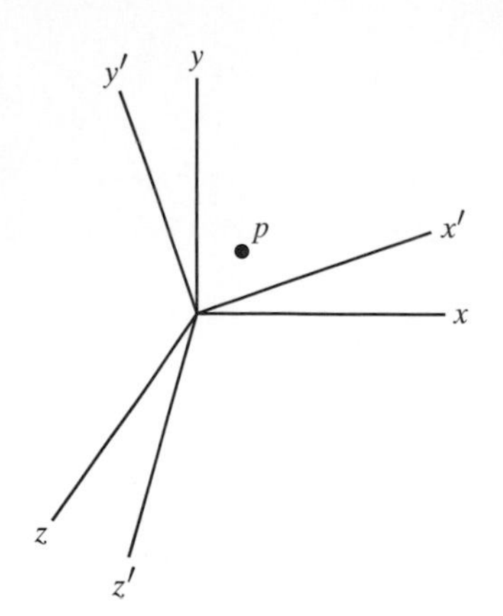

FIGURE 5-33 The xyz coordinate system and a system $x'y'z'$ with a different orientation.

General State of Stress

We have seen in our discussion of plane stress that the components of the state of stress depend on the orientation of the coordinate system in which they are expressed. Suppose that we know the components of stress at a point p in terms of a particular coordinate system xyz:

$$[T] = \begin{bmatrix} \sigma_x & \tau_{xy} & \tau_{xz} \\ \tau_{yx} & \sigma_y & \tau_{yz} \\ \tau_{zx} & \tau_{zy} & \sigma_z \end{bmatrix}.$$

These components will generally have different values when expressed in terms of a coordinate system $x'y'z'$ having a different orientation (Fig. 5-33). For *any* state of stress, at least one coordinate system $x'y'z'$ exists for which the state of stress is of the form

$$\begin{bmatrix} \sigma'_x & \tau'_{xy} & \tau'_{xz} \\ \tau'_{yx} & \sigma'_y & \tau'_{yz} \\ \tau'_{zx} & \tau'_{zy} & \sigma'_z \end{bmatrix} = \begin{bmatrix} \sigma_1 & 0 & 0 \\ 0 & \sigma_2 & 0 \\ 0 & 0 & \sigma_3 \end{bmatrix}.$$

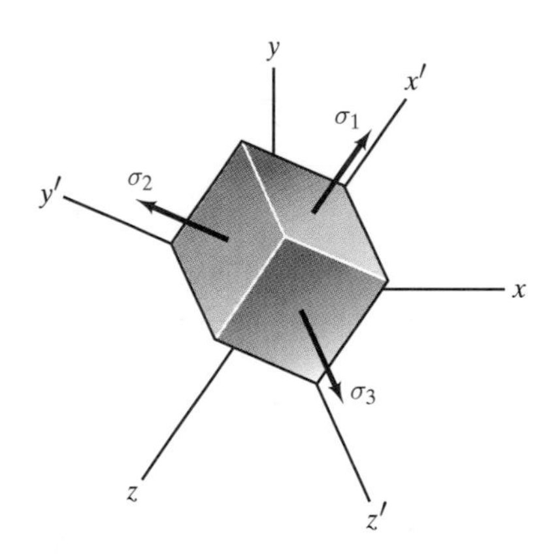

FIGURE 5-34 Stresses on an element oriented with the principal axes.

The axes x', y', z' are called *principal axes* and σ_1, σ_2, and σ_3 are the principal stresses. An infinitesimal element at p that is oriented with the principal axes is subject to the principal stresses and no shear stress (Fig. 5-34). It can be shown that the principal stresses are the roots of the cubic equation

$$\sigma^3 - I_1\sigma^2 + I_2\sigma - I_3 = 0, \qquad \textbf{(5-25)}$$

where

$$\begin{aligned} I_1 &= \sigma_x + \sigma_y + \sigma_z, \\ I_2 &= \sigma_x\sigma_y + \sigma_y\sigma_z + \sigma_z\sigma_x - \tau_{xy}^2 - \tau_{yz}^2 - \tau_{zx}^2, \\ I_3 &= \sigma_x\sigma_y\sigma_z - \sigma_x\tau_{yz}^2 - \sigma_y\tau_{xz}^2 - \sigma_z\tau_{xy}^2 + 2\tau_{xy}\tau_{yz}\tau_{zx}. \end{aligned} \qquad \textbf{(5-26)}$$

[Although the components of the state of stress depend on the orientation of the coordinate system in which they are expressed, the values of these three coefficients do not. This can be deduced from the fact that the principal stresses, the roots of Eq. (5-25), cannot depend on the orientation of the coordinate system used to evaluate them. For this reason I_1, I_2, and I_3 are called *stress invariants.*] Thus for a given state of stress, the principal stresses can be determined by evaluating the coefficients I_1, I_2, and I_3 and solving Eq. (5-25).

The absolute maximum shear stress can be determined by the same approach as that applied to plane stress in Section 5-2. We begin with the element on which the principal stresses occur, and align the coordinate system with the faces of the element in the three ways shown in Fig. 5-35. Applying Eq. (5-19) to each

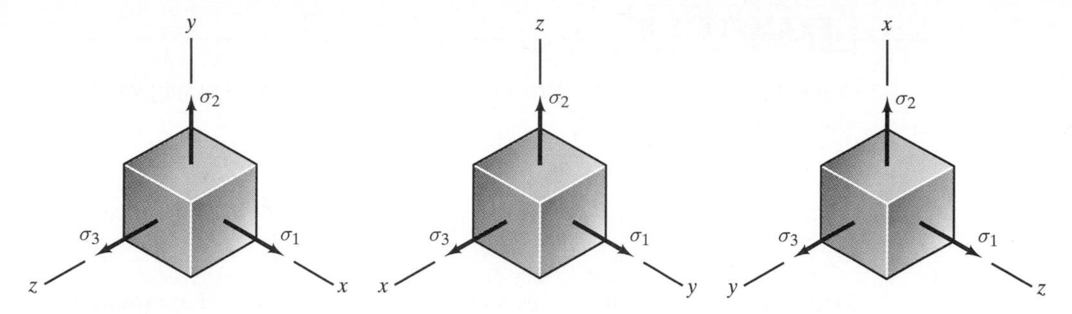

FIGURE 5-35 Different orientations of the coordinate system relative to the element on which the principal stresses act.

of these orientations, we find that the absolute maximum shear stress is the largest of the three values

$$\left|\frac{\sigma_1 - \sigma_2}{2}\right|, \quad \left|\frac{\sigma_1 - \sigma_3}{2}\right|, \quad \left|\frac{\sigma_2 - \sigma_3}{2}\right|. \tag{5-27}$$

Although this analysis considers only a subset of the possible planes through point p, no shear stresses of greater magnitude act on any plane.

In Fig. 5-36 we show that the absolute maximum shear stress determined in this way can be visualized very clearly by superimposing the Mohr's circles obtained from the three orientations of the coordinate system in Fig. 5-35. Notice that if $\sigma_1 > \sigma_2 > \sigma_3$, the absolute maximum shear stress is $(\sigma_1 - \sigma_3)/2$.

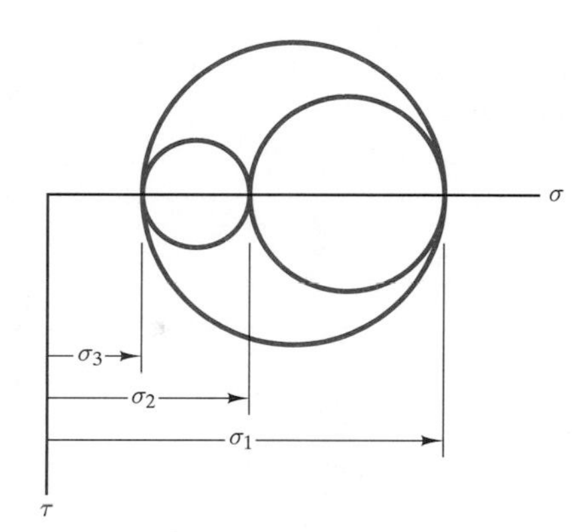

FIGURE 5-36 Superimposing the Mohr's circles graphically demonstrates the absolute maximum shear stress.

Triaxial Stress

The state of stress at a point is said to be *triaxial* if it is of the form

$$\begin{bmatrix} \sigma_x & 0 & 0 \\ 0 & \sigma_y & 0 \\ 0 & 0 & \sigma_z \end{bmatrix}. \tag{5-28}$$

The shear stress components τ_{xy}, τ_{xz}, and τ_{yz} are zero (Fig. 5-37). For example, the state of stress at a point in a fluid at rest [Eq. 5-2] is triaxial. In triaxial stress, x, y, and z are principal axes and σ_x, σ_y, and σ_z are the principal stresses. From Eq. (5-27), the absolute maximum shear stress in triaxial stress is the largest of the three values

$$\left|\frac{\sigma_x - \sigma_y}{2}\right|, \quad \left|\frac{\sigma_x - \sigma_z}{2}\right|, \quad \left|\frac{\sigma_y - \sigma_z}{2}\right|. \tag{5-29}$$

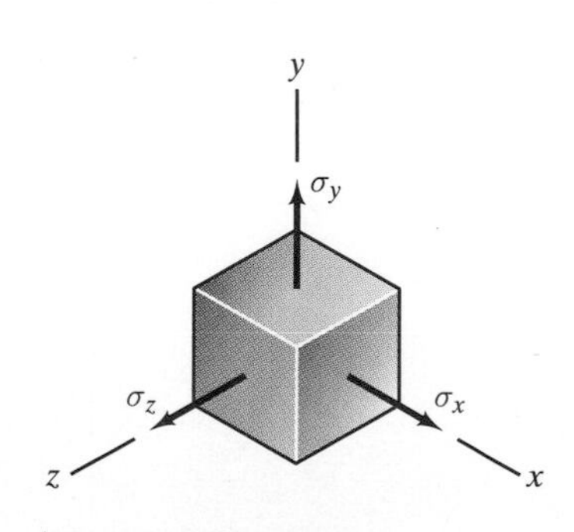

FIGURE 5-37 Triaxial stress.

EXAMPLE 5-8

A point p in the frame of the racing motorcycle in Fig. 5-38 is subjected to the state of stress (in MPa)

$$\begin{bmatrix} \sigma_x & \tau_{xy} & \tau_{xz} \\ \tau_{yx} & \sigma_y & \tau_{yz} \\ \tau_{zx} & \tau_{zy} & \sigma_z \end{bmatrix} = \begin{bmatrix} 4 & 2 & 1 \\ 2 & 2 & 1 \\ 1 & 1 & 3 \end{bmatrix}.$$

Determine the principal stresses and the absolute maximum shear stress at p.

FIGURE 5-38

Strategy

Since we know the state of stress, we can use Eqs. (5-26) to evaluate the coefficients I_1, I_2, and I_3, then solve Eq. (5-25) to determine the principal stresses. The absolute maximum shear stress is the largest of the three values in Eq. (5-27).

Solution

Principal stresses Substituting the components of stress in MPa into Eqs. (5-26), we obtain $I_1 = 9$, $I_2 = 20$, and $I_3 = 10$, so Eq. (5-25) is

$$\sigma^3 - 9\sigma^2 + 20\sigma - 10 = 0.$$

We can estimate the roots of this cubic equation by drawing a graph of the left side as a function of σ [Fig. (a)]. By using software designed to obtain roots of nonlinear algebraic equations, we obtain $\sigma_1 = 5.895$ MPa, $\sigma_2 = 2.397$ MPa, and $\sigma_3 = 0.708$ MPa.

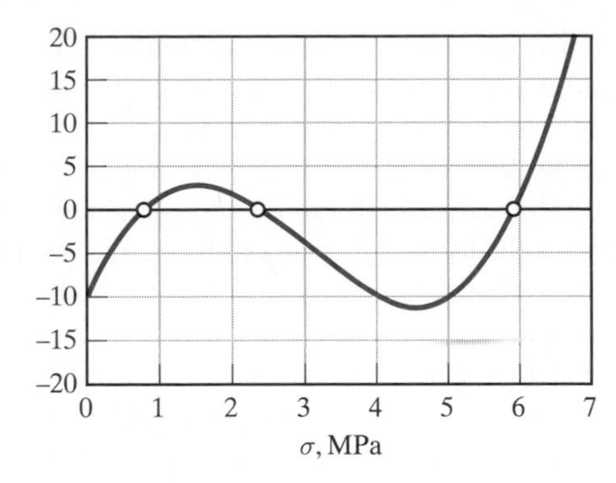

(a) Graph of $\sigma^3 - 9\sigma^2 + 20\sigma - 10$.

Absolute maximum shear stress The three values in Eq. (5-27) are

$$\left|\frac{\sigma_1 - \sigma_2}{2}\right| = \left|\frac{5.895 - 2.397}{2}\right| = 1.749 \text{ MPa},$$

$$\left|\frac{\sigma_1 - \sigma_3}{2}\right| = \left|\frac{5.895 - 0.708}{2}\right| = 2.594 \text{ MPa},$$

$$\left|\frac{\sigma_2 - \sigma_3}{2}\right| = \left|\frac{2.397 - 0.708}{2}\right| = 0.845 \text{ MPa}.$$

The absolute maximum shear stress is 2.594 MPa.

5-5 Design Issues: Pressure Vessels

A faculty member in our college once proposed that the first course in mechanics of materials no longer be part of the mechanical engineering curriculum, to which a colleague responded "Well, if that happens I'm going to stop standing near boilers!" Underlying his facetious remark on the need for mechanical engineers to understand solid mechanics (the proposal was soundly rejected) was a serious concern: the proper design of pressure vessels. The critical importance of this problem in engineering has been emphasized by pressure vessel accidents from the early days of steam power to *Apollo 13*. We discuss this subject in this chapter because pressure vessels provide interesting examples of triaxial states of stress (see Section 5-4).

Spherical Vessels

Consider a spherical pressure vessel with radius R and wall thickness t, where $t \ll R$ (Fig. 5-39). We assume that the vessel contains a gas with uniform

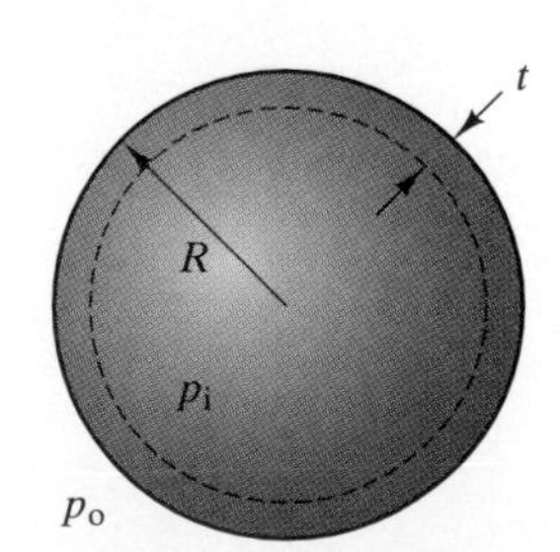

FIGURE 5-39 Spherical pressure vessel. The wall thickness is exaggerated.

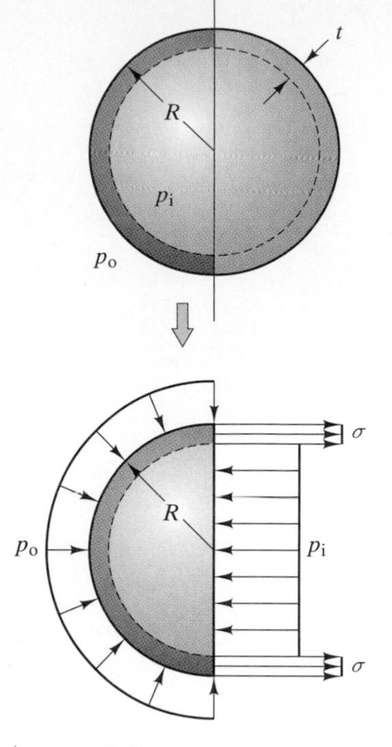

FIGURE 5-40 Free-body diagram of half of the pressure vessel, including the enclosed gas.

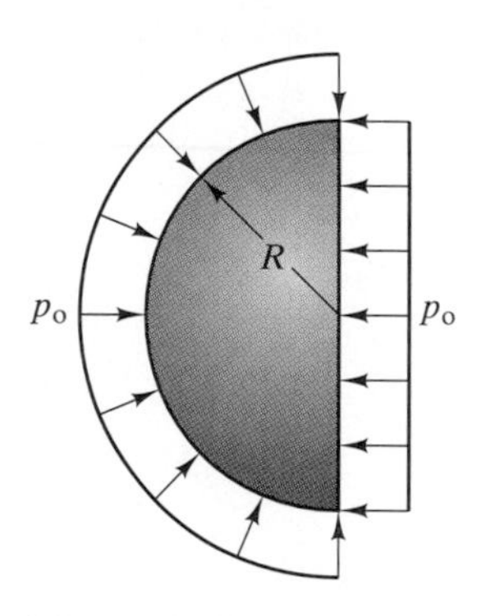

FIGURE 5-41 Hemisphere subjected to uniform pressure p_o.

pressure p_i and that the outer wall is subjected to a uniform pressure p_o, and we denote the difference between the inner and outer pressures by $p = p_i - p_o$. In applications, p_o is often atmospheric pressure (approximately 10^5 Pa or 14.7 psi).

We can approximate the state of stress in a thin-walled spherical pressure vessel if we assume that the effects of loads other than those exerted by the internal and external pressures are negligible. In Fig. 5-40 we bisect the vessel by a plane and draw a free-body diagram of one half, *including the gas it contains*. The normal stress in the wall, which can be approximated as being uniformly distributed for a thin-walled vessel, is denoted by σ. We want to determine the stress σ by summing the horizontal forces on the free-body diagram, but what horizontal force is exerted by the pressure p_o? Suppose that a solid hemisphere of radius R is subjected to a uniform pressure p_o (Fig. 5-41). No net force is exerted on an object by a uniform distribution of pressure, so the horizontal force to the right exerted on the hemispherical surface must equal the force to the left exerted on the plane circular face: $p_o(\pi R^2)$. Therefore, the sum of the horizontal forces on the free-body diagram in Fig. 5-40 is

$$p_o(\pi R^2) + \sigma(2\pi R t) - p_i(\pi R^2) = 0.$$

Notice that the assumption $t \ll R$ is used in writing this equation. From this equation we obtain the normal stress σ in terms of the dimensions of the vessel and the pressure difference p:

$$\sigma = \frac{(p_i - p_o)R}{2t} = \frac{pR}{2t}. \tag{5-30}$$

An element isolated from the vessel's outer surface is subjected to the stresses shown in Fig. 5-42. In terms of the coordinate system shown (the z axis is

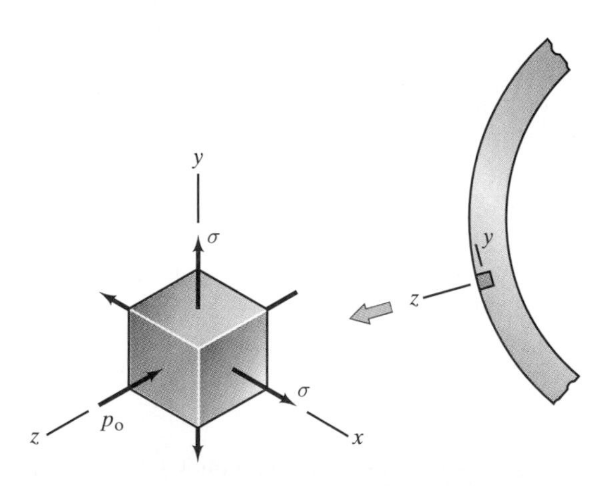

FIGURE 5-42 Stresses on an element at the outer surface.

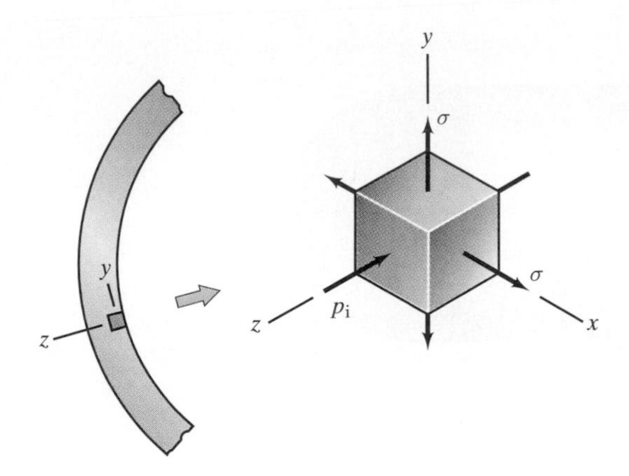

FIGURE 5-43 Stresses on an element at the inner surface.

perpendicular to the wall), the triaxial state of stress is

$$\begin{bmatrix} \sigma_x & \tau_{xy} & \tau_{xz} \\ \tau_{yx} & \sigma_y & \tau_{yz} \\ \tau_{zx} & \tau_{zy} & \sigma_z \end{bmatrix} = \begin{bmatrix} \sigma & 0 & 0 \\ 0 & \sigma & 0 \\ 0 & 0 & -p_o \end{bmatrix}. \quad \textbf{(5-31)}$$

As we discussed in Section 5-4, the principal stresses for this state of stress are $\sigma, \sigma, -p_o$, and from Eq. (5-29) the absolute maximum shear stress is $|(\sigma + p_o)/2|$. The triaxial state of stress on an element isolated from the vessel's inner surface (Fig. 5-43) is

$$\begin{bmatrix} \sigma_x & \tau_{xy} & \tau_{xz} \\ \tau_{yx} & \sigma_y & \tau_{yz} \\ \tau_{zx} & \tau_{zy} & \sigma_z \end{bmatrix} = \begin{bmatrix} \sigma & 0 & 0 \\ 0 & \sigma & 0 \\ 0 & 0 & -p_i \end{bmatrix}. \quad \textbf{(5-32)}$$

In this case the principal stresses are $\sigma, \sigma, -p_i$ and the absolute maximum shear stress is $|(\sigma + p_i)/2|$.

Cylindrical Vessels

We now consider a cylindrical vessel with radius R and wall thickness t (Fig. 5-44). The vessel shown has hemispherical ends, but the results we will obtain for the state of stress in the cylindrical wall do not depend on the shapes of the ends. We again assume that the vessel contains a gas with uniform pressure p_i and that the outer wall is subjected to a uniform pressure p_o.

In Fig. 5-45 we obtain a free-body diagram by passing a plane through the cylindrical wall perpendicular to its axis. The normal stress in the wall is denoted by σ. The horizontal forces on this free-body diagram are identical to those on the free-body diagram of the spherical vessel in Fig. 5-40, so the normal stress σ is given by Eq. (5-30):

$$\sigma = \frac{(p_i - p_o)R}{2t} = \frac{pR}{2t}. \quad \textbf{(5-33)}$$

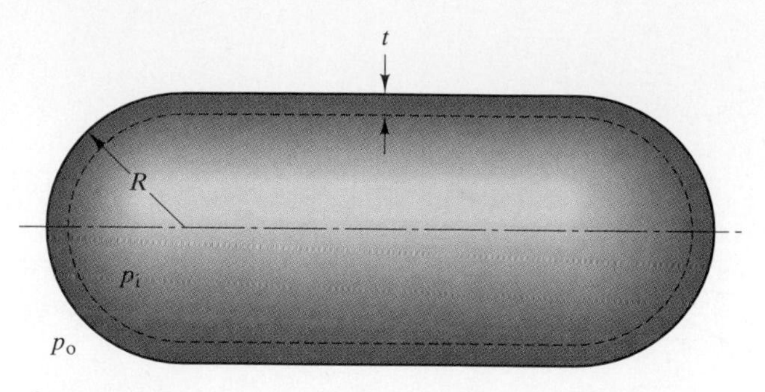

FIGURE 5-44 Cylindrical pressure vessel.

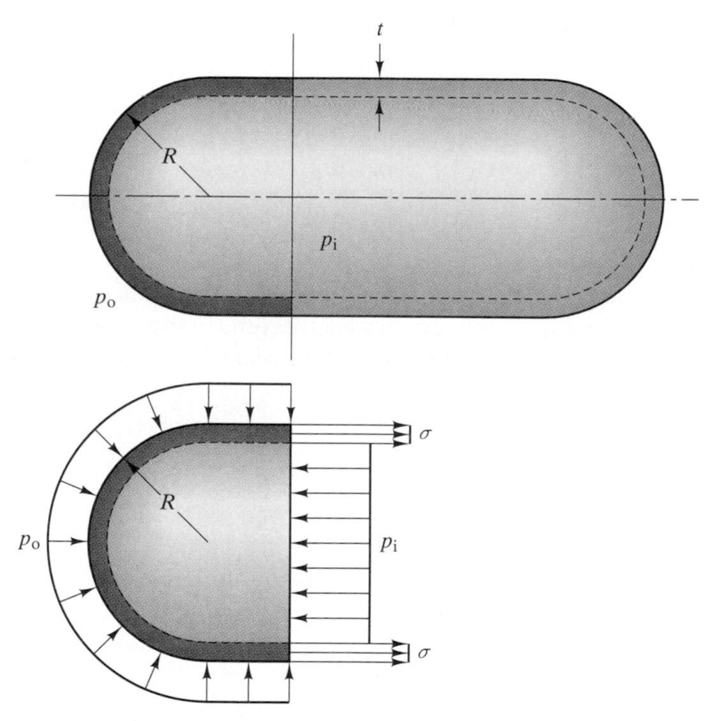

FIGURE 5-45 Free-body diagram obtained by passing a plane perpendicular to the cylinder axis.

But the state of stress in a cylindrical vessel is slightly more complicated than that in a spherical vessel. In Fig. 5-46 we first isolate a "slice" of the cylindrical wall of length Δx, including the gas it contains, then obtain a free-body diagram by bisecting the slice by a plane parallel to the cylinder axis. The normal stress σ_h on the resulting free-body diagram is called the *hoop stress* (so named because it plays the same role as the tensile stresses in the metal hoops used to reinforce wooden barrels). Planes on which σ_h acts are perpendicular to planes on which the normal stress σ acts (Fig. 5-47).

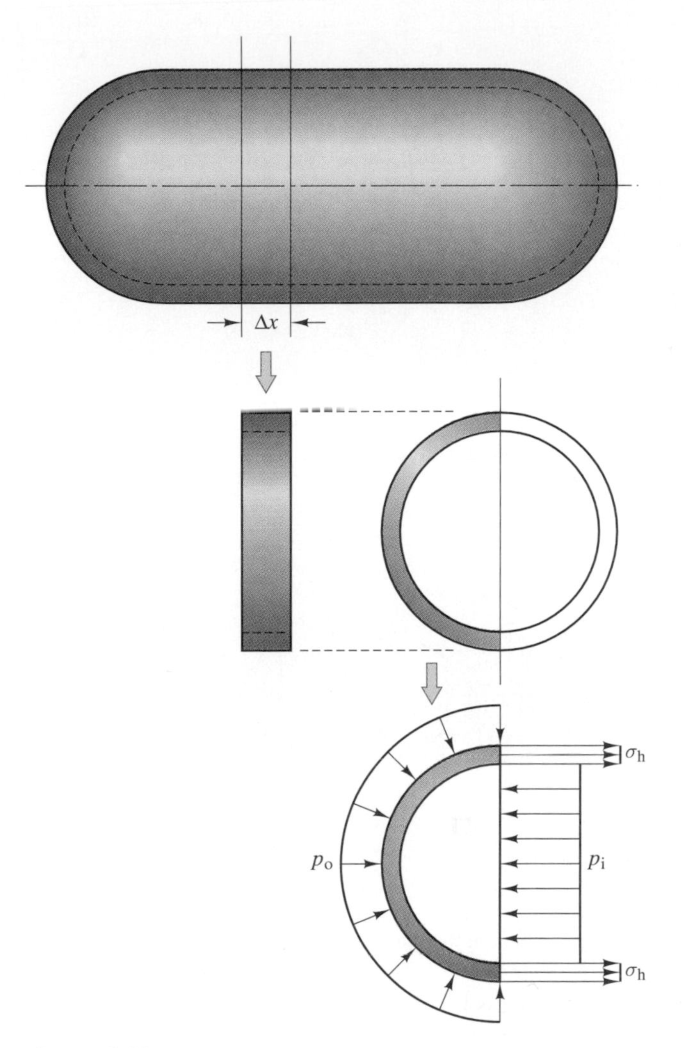

FIGURE 5-46 Free-body diagram for determining the hoop stress σ_h.

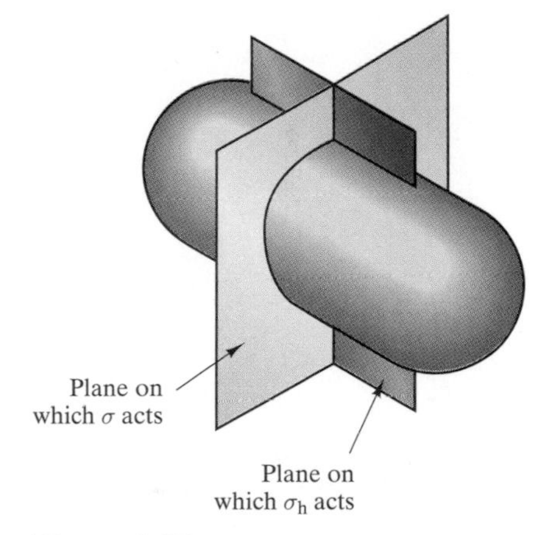

FIGURE 5-47 Planes on which the normal stresses σ and σ_h act.

The sum of the horizontal forces on the free-body diagram in Fig. 5-46 is

$$p_o(2R\,\Delta x) + \sigma_h(2t\,\Delta x) - p_i(2R\,\Delta x) = 0.$$

Solving for the hoop stress, we obtain

$$\sigma_h = \frac{(p_i - p_o)R}{t} = \frac{pR}{t}. \tag{5-34}$$

Observe that $\sigma_h = 2\sigma$, which helps explain the popularity of spherical pressure vessels.

An element isolated from the outer surface of the cylindrical wall is subjected to the stresses shown in Fig. 5-48. In terms of the coordinate system shown (the z axis is perpendicular to the wall and the x axis is parallel to the axis of the

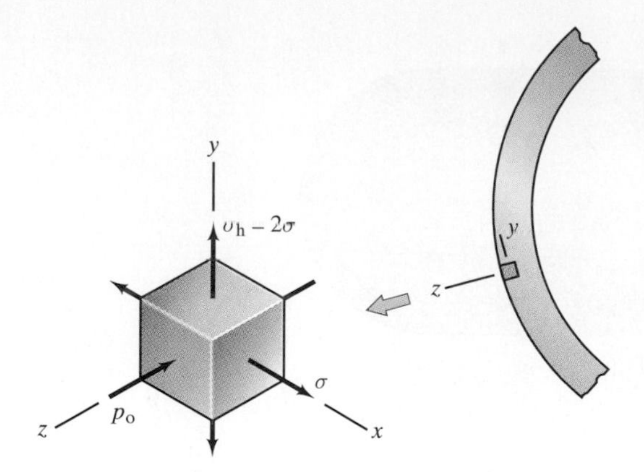

FIGURE 5-48 Stresses on an element at the outer surface.

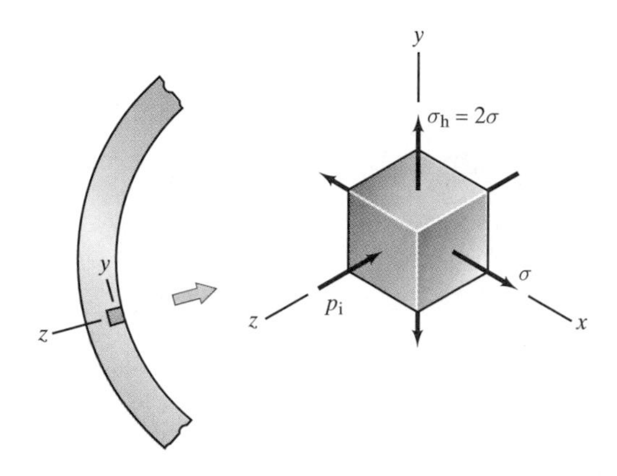

FIGURE 5-49 Stresses on an element at the inner surface.

cylinder), the triaxial state of stress is

$$\begin{bmatrix} \sigma_x & \tau_{xy} & \tau_{xz} \\ \tau_{yx} & \sigma_y & \tau_{yz} \\ \tau_{zx} & \tau_{zy} & \sigma_z \end{bmatrix} = \begin{bmatrix} \sigma & 0 & 0 \\ 0 & \sigma_h & 0 \\ 0 & 0 & -p_o \end{bmatrix}. \tag{5-35}$$

The principal stresses are σ_h, σ, $-p_o$ and from Eq. (5-29) the absolute maximum shear stress is the largest of the three values

$$\left|\frac{\sigma}{2}\right|, \qquad \left|\frac{\sigma + p_o}{2}\right|, \qquad \left|\frac{\sigma_h + p_o}{2}\right|. \tag{5-36}$$

The triaxial state of stress on an element isolated from the vessel's inner surface (Fig. 5-49) is

$$\begin{bmatrix} \sigma_x & \tau_{xy} & \tau_{xz} \\ \tau_{yx} & \sigma_y & \tau_{yz} \\ \tau_{zx} & \tau_{zy} & \sigma_z \end{bmatrix} = \begin{bmatrix} \sigma & 0 & 0 \\ 0 & \sigma_h & 0 \\ 0 & 0 & -p_i \end{bmatrix}. \tag{5-37}$$

In this case the principal stresses are σ_h, σ, $-p_i$ and the absolute maximum shear stress is the largest of the three values

$$\left|\frac{\sigma}{2}\right|, \qquad \left|\frac{\sigma + p_i}{2}\right|, \qquad \left|\frac{\sigma_h + p_i}{2}\right|. \tag{5-38}$$

Allowable Stress

The one aspect of pressure vessel design we consider is the most important: making sure that the stresses in the vessel walls do not exceed those the material will support. We have determined the principal stresses and absolute maximum shear stresses in the walls of spherical and cylindrical vessels. The criterion for design we use in our example and problems is to ensure that the absolute maximum shear stress in the material does not exceed a shear yield stress τ_Y, or some specified allowable shear stress τ_{allow}. This is called the *Tresca criterion*, and is discussed further in Chapter 12. The factor of safety is now defined by

$$S = \frac{\tau_Y}{\tau_{allow}}. \tag{5-39}$$

Since the maximum shear stress in a tensile test is one-half the applied normal stress, we will assume that the shear yield stress τ_Y is given in terms of the normal yield stress σ_Y tabulated in Appendix B by the relation

$$\tau_Y = \tfrac{1}{2}\sigma_Y. \tag{5-40}$$

Of course, the values in Appendix B are merely representative, and in actual design the yield stress must be determined for the specific material being used. We also emphasize that our analyses apply only to thin-walled vessels made of homogeneous and isotropic materials.

EXAMPLE 5-9

A cylindrical pressure vessel with 2-m radius and hemispherical ends is to be designed to support an internal pressure as large as $p_i = 8 \times 10^5$ Pa (8 atmospheres) with the outer pressure equal to atmospheric pressure $p_o = 1 \times 10^5$ Pa. It is to be constructed of ASTM-A514 steel. Determine the vessel's wall thickness so that it has a factor of safety $S = 4$.

Strategy

We can determine the normal yield stress σ_Y for ASTM-A514 steel from Appendix B, then use Eqs. (5-40) and (5-39) to determine the allowable value of the absolute maximum shear stress τ_{allow}. Comparing the terms (5-36) and (5-38), and remembering that $\sigma_h = 2\sigma$, it is clear that the absolute maximum shear stress is $(\sigma_h + p_i)/2$. (The absolute value signs are unnecessary because σ_h is positive.) By equating this expression to τ_{allow} and using Eq. (5-34), we can determine the necessary wall thickness.

Solution

From Appendix B the normal yield stress σ_Y for ASTM-A514 steel is 700 MPa, so from Eq. (5-40), $\tau_Y = 350$ MPa. Then from Eq. (5-39), the allowable value of the absolute maximum shear stress is

$$\tau_{allow} = \frac{\tau_Y}{S} = \frac{350}{4} = 87.5 \text{ MPa}.$$

We equate the absolute maximum shear stress to the allowable value:

$$\frac{\sigma_h + p_i}{2} = \tau_{allow}.$$

Substituting Eq. (5-34) into this expression and solving for the wall thickness, we obtain

$$\begin{aligned} t &= \frac{(p_i - p_o)R}{2\tau_{allow} - p_i} \\ &= \frac{(7 \times 10^5)(2)}{(2)(87.5 \times 10^6) - 8 \times 10^5} \\ &= 0.00804 \text{ m}. \end{aligned}$$

The necessary wall thickness is 8.04 mm.

5-6 Tetrahedron Argument

We have seen that the normal and shear stresses acting on a plane through a point p of a material depend in general on the orientation of the plane. We have derived equations and also a graphical solution, Mohr's circle, which allow determination of the normal and shear stresses, but only for a subset of planes through p and for the special case of plane stress. A derivation called the *tetrahedron argument* leads to equations that allow us to determine the normal and shear stresses on any plane through p and for a general state of stress.

Determining the Traction

We consider a point p of a sample of material and a tetrahedral element containing p (Fig. 5-50). Three faces of the tetrahedron are perpendicular to the coordinate axes. The fourth face is perpendicular to a unit vector $\mathbf{n}$ which specifies its orientation. Let the area of the face perpendicular to $\mathbf{n}$ be ΔA, and let the areas of the faces perpendicular to the x, y, and z axes be ΔA_x, ΔA_y, and ΔA_z (Fig. 5-51). Denoting the angle between the vector $\mathbf{n}$ and the x axis by θ_x, the x component of $\mathbf{n}$ is $n_x = \cos\theta_x$. Since $\mathbf{n}$ is perpendicular to the surface with area ΔA and the x axis is perpendicular to the surface with area ΔA_x, the angle between these two surfaces is θ_x. The area ΔA_x is the projection onto

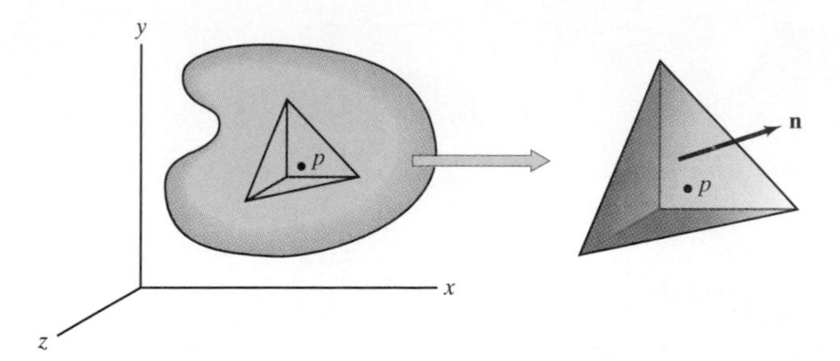

FIGURE 5-50 Point p of a material and a tetrahedron containing p.

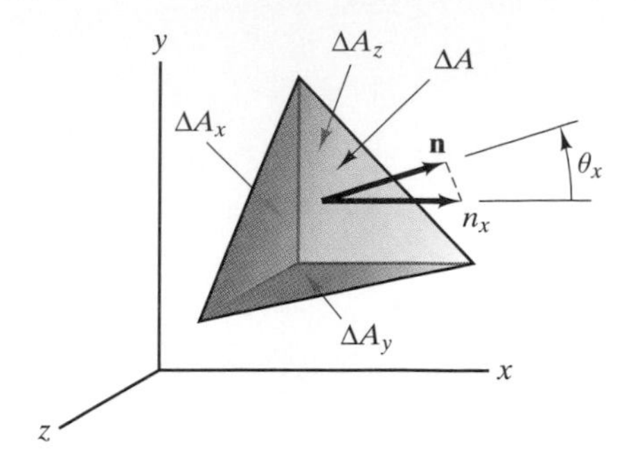

FIGURE 5-51 Areas of the faces of the tetrahedron. The angle between the vector **n** and its component n_x is θ_x.

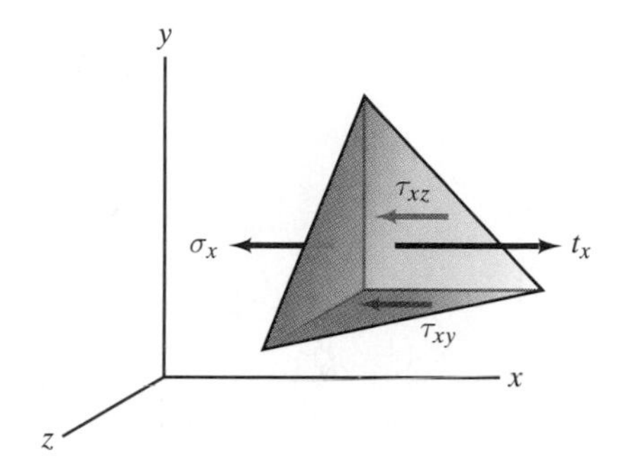

FIGURE 5-52 Component of the average traction and the average stresses on the tetrahedron that exert forces in the x direction.

the y–z plane of the area ΔA:

$$\begin{aligned} \Delta A_x &= \Delta A \cos\theta_x \\ &= \Delta A\, n_x. \end{aligned} \tag{5-41}$$

Applying the same argument to the surface A_y perpendicular to the y axis and the surface A_z perpendicular to the z axis yields the relations

$$\Delta A_y = \Delta A\, n_y, \qquad \Delta A_z = \Delta A\, n_z. \tag{5-42}$$

Let **t** be the average value of the traction acting on the surface with area ΔA. Figure 5-52 shows the x component of **t** and also the average values of those stress components on the surfaces ΔA_x, ΔA_y, and ΔA_z which exert forces in the x direction. The sum of the forces on the tetrahedron in the x direction is

$$t_x\,\Delta A - \sigma_x\,\Delta A_x - \tau_{xy}\,\Delta A_y - \tau_{xz}\,\Delta A_z = 0.$$

Dividing this equation by ΔA and using Eqs. (5-41) and (5-42), we obtain

$$t_x - \sigma_x n_x - \tau_{xy} n_y - \tau_{xz} n_z = 0.$$

As the size of the tetrahedron decreases, the average traction and average stress components in this equation approach their values at point p. Corresponding expressions can be obtained by summing the forces on the tetrahedron in the y and z directions. The resulting equations for the components of the traction **t** are

$$\begin{aligned} t_x &= \sigma_x n_x + \tau_{xy} n_y + \tau_{xz} n_z, \\ t_y &= \tau_{yx} n_x + \sigma_y n_y + \tau_{yz} n_z, \\ t_z &= \tau_{zx} n_x + \tau_{zy} n_y + \sigma_z n_z, \end{aligned} \tag{5-43}$$

which we can write as the matrix equation

$$\begin{bmatrix} t_x \\ t_y \\ t_z \end{bmatrix} = \begin{bmatrix} \sigma_x & \tau_{xy} & \tau_{xz} \\ \tau_{yx} & \sigma_y & \tau_{yz} \\ \tau_{zx} & \tau_{zy} & \sigma_z \end{bmatrix} \begin{bmatrix} n_x \\ n_y \\ n_z \end{bmatrix}. \tag{5-44}$$

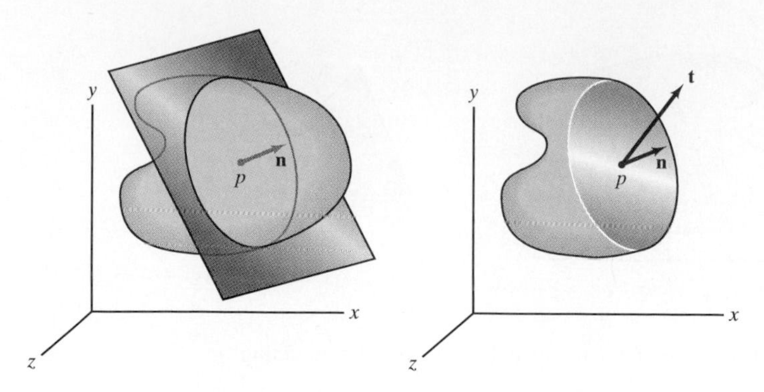

FIGURE 5-53 Traction **t** on a plane whose orientation is specified by a unit vector **n**.

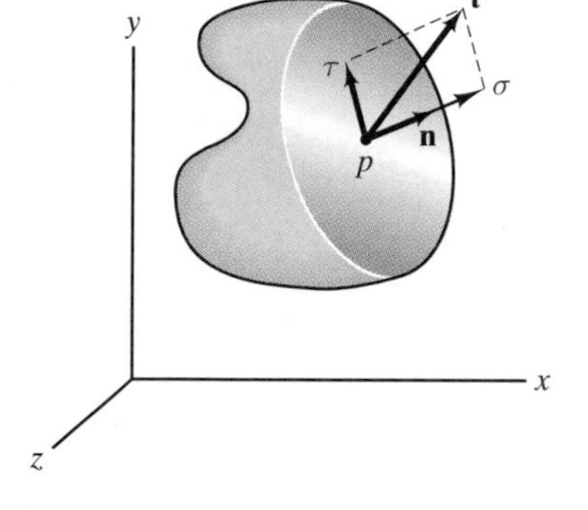

FIGURE 5-54 Normal and shear stresses.

Although we have assumed the material to be in equilibrium in deriving these equations, they hold even when the material is not in equilibrium. If you know the state of stress at a point p, Eqs. (5-43) or (5-44) determine the components of the traction **t** at p acting on the plane through p whose orientation is defined by the unit vector **n** (Fig. 5-53).

Determining the Normal and Shear Stresses

Once the components of the traction **t** are known, we can easily determine the normal stress and the magnitude of the shear stress. The normal and shear stresses at p are the components of the traction **t** normal and parallel to the given plane (Fig. 5-54). The normal stress can be determined from the relation

$$\sigma = \mathbf{t} \cdot \mathbf{n} = t_x n_x + t_y n_y + t_z n_z. \tag{5-45}$$

The vector **t** is the sum of its vector component $\sigma\mathbf{n}$ normal to the plane and its vector component parallel to the plane, so the shear stress can be determined by evaluating the magnitude of the parallel vector component $\mathbf{t} - \sigma\mathbf{n}$:

$$|\tau| = |\mathbf{t} - \sigma\mathbf{n}|. \tag{5-46}$$

These results confirm the statement we made when introducing the state of stress: If the state of stress is known at a point, the normal and shear stresses on any plane through the point can be determined.

EXAMPLE 5-10

The state of stress at point p of the material in Fig. 5-55a is

$$\begin{bmatrix} 4 & -2 & 3 \\ -2 & 2 & 0 \\ 3 & 0 & -2 \end{bmatrix} \text{MPa}.$$

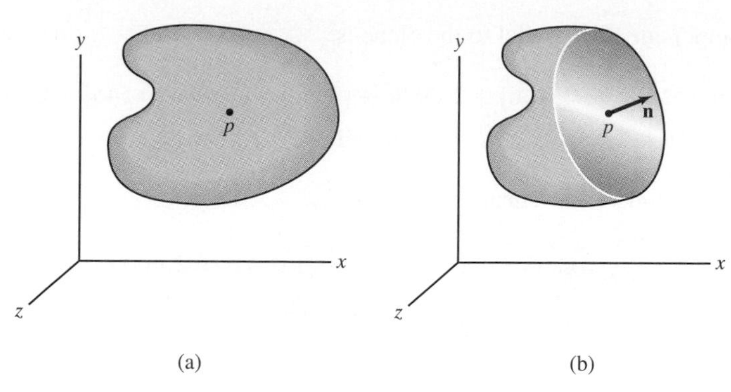

FIGURE 5-55

(a) Determine the traction **t** at p acting on the plane in Fig. 5-55b if $\mathbf{n} = 0.818\mathbf{i} + 0.545\mathbf{j} + 0.181\mathbf{k}$. **(b)** Determine the normal stress and the magnitude of the shear stress at p acting on the plane in Fig. 5-55b.

Strategy

(a) The components of the traction are given in terms of the components of **n** by Eqs. (5-43) or (5-44). **(b)** The normal stress and the magnitude of the shear stress are given by Eqs. (5-45) and (5-46).

Solution

(a) The stress components are $\sigma_x = 4$ MPa, $\sigma_y = 2$ MPa, $\sigma_z = -2$ MPa, $\tau_{xy} = -2$ MPa, $\tau_{xz} = 3$ MPa, and $\tau_{yz} = 0$. From Eqs. (5-43), the components of the traction are

$$\begin{aligned} t_x &= \sigma_x n_x + \tau_{xy} n_y + \tau_{xz} n_z \\ &= (4)(0.818) + (-2)(0.545) + (3)(0.181) \\ &= 2.725 \text{ MPa}, \end{aligned}$$

$$\begin{aligned} t_y &= \tau_{yx} n_x + \sigma_y n_y + \tau_{yz} n_z \\ &= (-2)(0.818) + (2)(0.545) + (0)(0.181) \\ &= -0.546 \text{ MPa}, \end{aligned}$$

$$\begin{aligned} t_z &= \tau_{zx} n_x + \tau_{zy} n_y + \sigma_z n_z \\ &= (3)(0.818) + (0)(0.545) + (-2)(0.181) \\ &= 2.092 \text{ MPa}. \end{aligned}$$

The traction is $\mathbf{t} = 2.725\mathbf{i} - 0.546\mathbf{j} + 2.092\mathbf{k}$ MPa.

(b) From Eq. (5-45), the normal stress is

$$\begin{aligned} \sigma = \mathbf{t} \cdot \mathbf{n} &= t_x n_x + t_y n_y + t_z n_z \\ &= (2.725)(0.818) + (-0.546)(0.545) + (2.092)(0.181) \\ &= 2.310 \text{ MPa}. \end{aligned}$$

The component of **t** parallel to the plane is

$$\mathbf{t} - \sigma\mathbf{n} = 2.725\mathbf{i} - 0.546\mathbf{j} + 2.092\mathbf{k} - (2.310)(0.818\mathbf{i} + 0.545\mathbf{j} + 0.181\mathbf{k})$$
$$= 0.835\mathbf{i} - 1.805\mathbf{j} + 1.674\mathbf{k} \text{ (MPa)},$$

so the magnitude of the shear stress is

$$|\tau| = \sqrt{(0.835)^2 + (-1.805)^2 + (1.674)^2} = 2.600 \text{ MPa}.$$

EXAMPLE 5-11

If the material in Fig. 5-56 is subjected to the state of plane stress

$$\begin{bmatrix} \sigma_x & \tau_{xy} & 0 \\ \tau_{yx} & \sigma_y & 0 \\ 0 & 0 & 0 \end{bmatrix}$$

at point p, the shear stress τ'_{xy} is given by Eq. (5-5):

$$\tau'_{xy} = -(\sigma_x - \sigma_y)\sin\theta\cos\theta + \tau_{xy}(\cos^2\theta - \sin^2\theta).$$

Derive this equation by using Eq. (5-44).

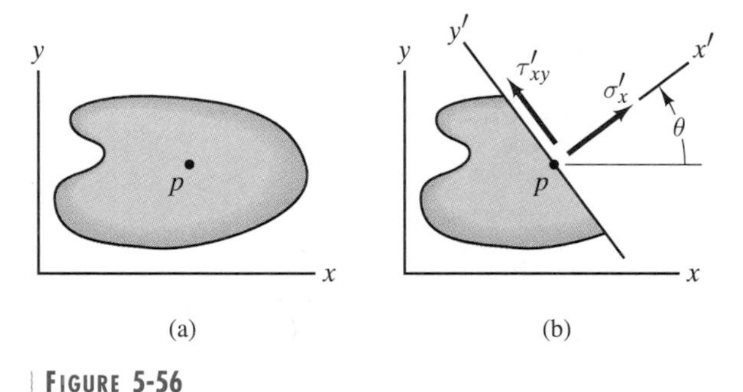

FIGURE 5-56

Strategy

With Eq. (5-5) we can obtain the components of the traction on the plane shown in Fig. 5-56b. Then we can determine the stress τ'_{xy} by calculating the component of the traction parallel to the plane.

Solution

The unit vector normal to the plane [Fig. (a)] has components $\mathbf{n} = \cos\theta\mathbf{i} + \sin\theta\mathbf{j}$.

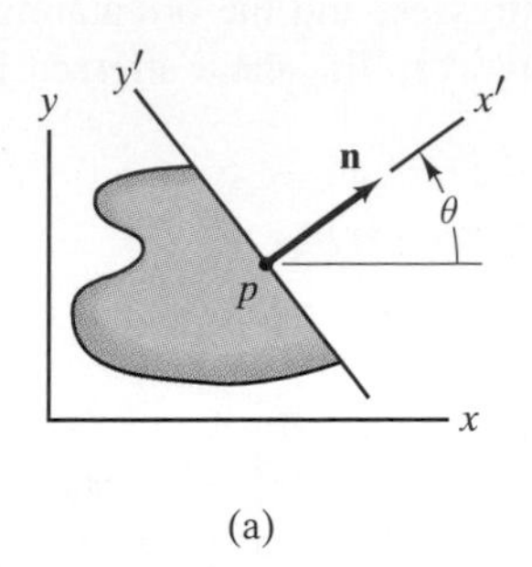

(a)

(a) Unit vector **n** normal to the plane.

The components of the traction are

$$\begin{bmatrix} t_x \\ t_y \\ t_z \end{bmatrix} = \begin{bmatrix} \sigma_x & \tau_{xy} & 0 \\ \tau_{yx} & \sigma_y & 0 \\ 0 & 0 & 0 \end{bmatrix} \begin{bmatrix} \cos\theta \\ \sin\theta \\ 0 \end{bmatrix} = \begin{bmatrix} \sigma_x \cos\theta + \tau_{xy} \sin\theta \\ \tau_{xy} \cos\theta + \sigma_y \sin\theta \\ 0 \end{bmatrix}.$$

The components of the traction are shown in Fig. (b).

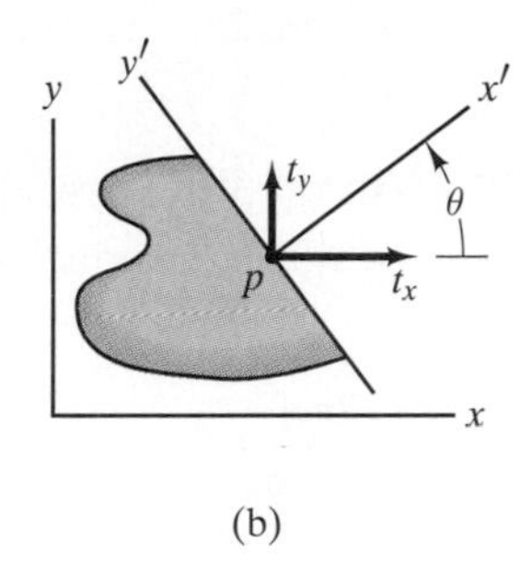

(b)

(b) The x and y components of the traction.

Summing the components of t_x and t_y parallel to the plane, we obtain the shear stress:

$$\begin{aligned} \tau'_{xy} &= t_y \cos\theta - t_x \sin\theta \\ &= (\tau_{xy} \cos\theta + \sigma_y \sin\theta) \cos\theta - (\sigma_x \cos\theta + \tau_{xy} \sin\theta) \sin\theta \\ &= -(\sigma_x - \sigma_y) \sin\theta \cos\theta + \tau_{xy}(\cos^2\theta - \sin^2\theta). \end{aligned}$$

Chapter Summary

Components of Stress

In terms of a given coordinate system, the *state of stress* at a point p of a material is defined by the *components of stress*

$$\begin{bmatrix} \sigma_x & \tau_{xy} & \tau_{xz} \\ \tau_{yx} & \sigma_y & \tau_{yz} \\ \tau_{zx} & \tau_{zy} & \sigma_z \end{bmatrix}. \qquad \text{Eq. (5-1)}$$

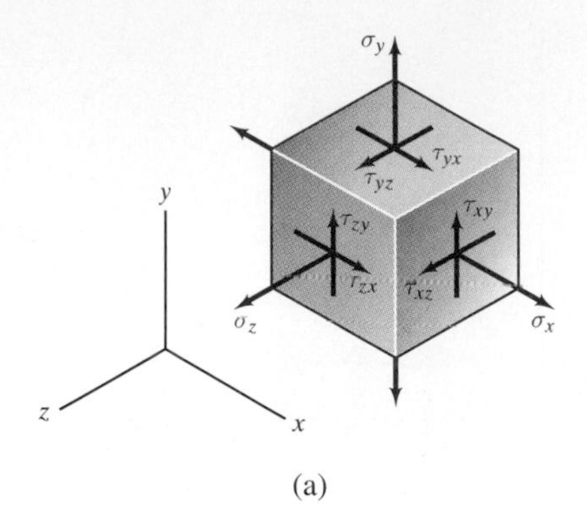

(a)

The directions of these stresses, and the orientations of the planes on which they act, are shown in Fig. (a). The shear stresses $\tau_{xy} = \tau_{yx}$, $\tau_{yz} = \tau_{zy}$, and $\tau_{xz} = \tau_{zx}$.

Transformations of Plane Stress

The stress at a point is said to be a state of *plane stress* if it is of the form

$$\begin{bmatrix} \sigma_x & \tau_{xy} & 0 \\ \tau_{yx} & \sigma_y & 0 \\ 0 & 0 & 0 \end{bmatrix}. \qquad \text{Eq. (5-3)}$$

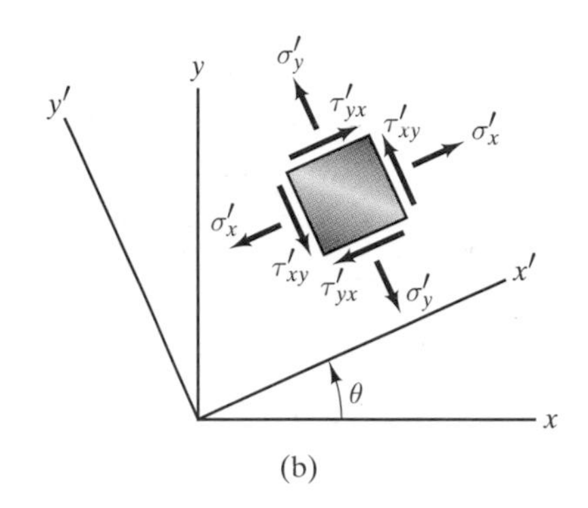

(b)

In terms of a coordinate system $x'y'z'$ oriented as shown in Fig. (b), the components of stress are

$$\sigma'_x = \frac{\sigma_x + \sigma_y}{2} + \frac{\sigma_x - \sigma_y}{2}\cos 2\theta + \tau_{xy}\sin 2\theta, \qquad \text{Eq. (5-7)}$$

$$\tau'_{xy} = -\frac{\sigma_x - \sigma_y}{2}\sin 2\theta + \tau_{xy}\cos 2\theta, \qquad \text{Eq. (5-8)}$$

$$\sigma'_y = \frac{\sigma_x + \sigma_y}{2} - \frac{\sigma_x - \sigma_y}{2}\cos 2\theta - \tau_{xy}\sin 2\theta. \qquad \text{Eq. (5-9)}$$

Maximum and Minimum Stresses in Plane Stress

The orientation of the element on which the principal stresses act [Fig. (c)] is determined from the equation

$$\tan 2\theta_p = \frac{2\tau_{xy}}{\sigma_x - \sigma_y}. \qquad \text{Eq. (5-12)}$$

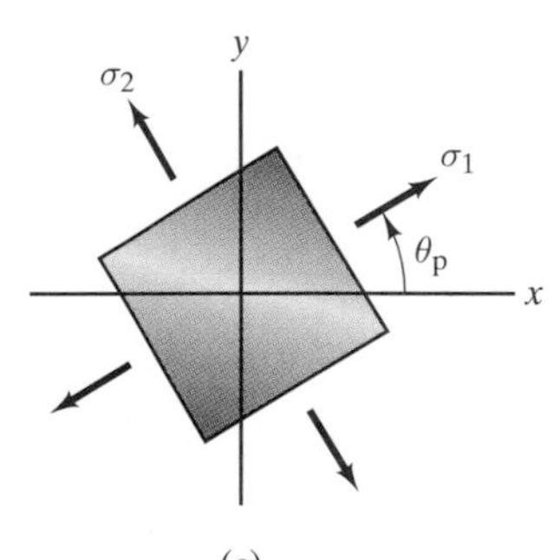

(c)

The values of the principal stresses can be obtained by substituting θ_p into Eqs. (5-7) and (5-9). Their values can also be determined from the equation

$$\sigma_1, \sigma_2 = \frac{\sigma_x + \sigma_y}{2} \pm \sqrt{\left(\frac{\sigma_x - \sigma_y}{2}\right)^2 + \tau_{xy}^2}, \qquad \text{Eq. (5-15)}$$

although this equation does not indicate the planes on which they act.

The orientation of the element on which the maximum in-plane shear stresses act [Fig. (d)] can be determined from the equation

$$\tan 2\theta_s = -\frac{\sigma_x - \sigma_y}{2\tau_{xy}}, \qquad \text{Eq. (5-16)}$$

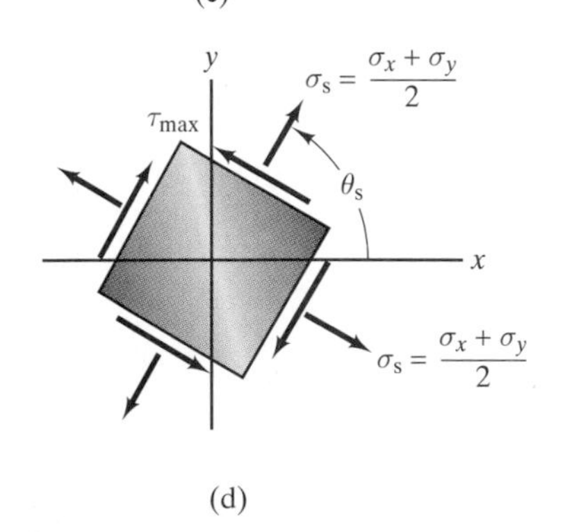

(d)

or by using the relation $\theta_s = \theta_p + 45°$. The normal stresses σ_s acting on this element are shown. The value of the maximum shear stress can be obtained by substituting θ_s into Eq. (5-8). Its value can also be determined from the equation

$$\tau_{max} = \sqrt{\left(\frac{\sigma_x - \sigma_y}{2}\right)^2 + \tau_{xy}^2}, \qquad \text{Eq. (5-19)}$$

although this equation does not indicate the direction of the stress on the element.

The *absolute maximum shear stress* (the maximum shear stress on any plane through p) is the largest of the three values

$$\left|\frac{\sigma_1 - \sigma_2}{2}\right|, \quad \left|\frac{\sigma_1}{2}\right|, \quad \left|\frac{\sigma_2}{2}\right|. \qquad \text{Eq. (5-24)}$$

Mohr's Circle for Plane Stress

Given a state of plane stress σ_x, τ_{xy}, and σ_y, establish a set of horizontal and vertical axes with normal stress measured to the right along the horizontal axis and shear stress measured downward along the vertical axis. Plot two points, point P with coordinates (σ_x, τ_{xy}) and point Q with coordinates $(\sigma_y, -\tau_{xy})$. Draw a straight line connecting points P and Q. Using the intersection of the straight line with the horizontal axis as the center, draw a circle that passes through the two points [Fig. (e)]. Draw a straight line through the center of the circle at an angle 2θ measured counterclockwise from point P. The point P' at which this line intersects the circle has coordinates (σ'_x, τ'_{xy}), and the point Q' has coordinates $(\sigma'_y, -\tau'_{xy})$, as shown in Fig. (f).

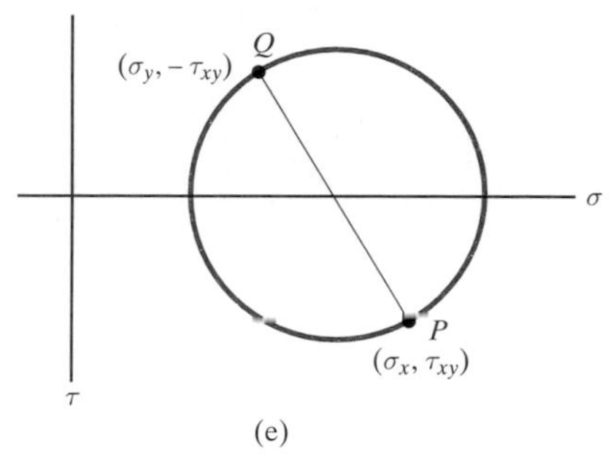

(e)

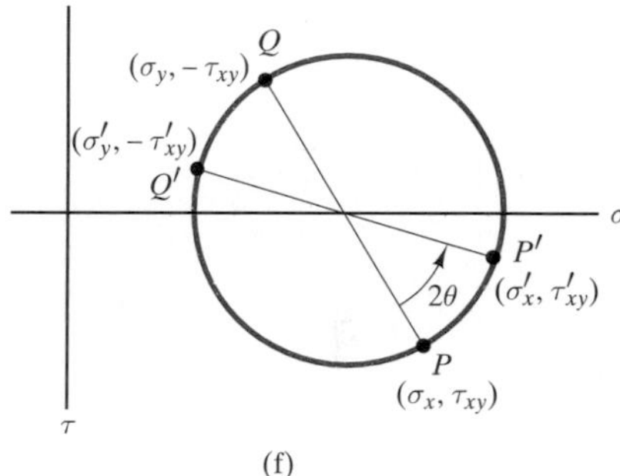

(f)

Principal Stresses in Three Dimensions

The principal stresses for a general state of stress are the roots of the cubic equation

$$\sigma^3 - I_1\sigma^2 + I_2\sigma - I_3 = 0, \qquad \text{Eq. (5-25)}$$

where

$$\begin{aligned} I_1 &= \sigma_x + \sigma_y + \sigma_z, \\ I_2 &= \sigma_x\sigma_y + \sigma_y\sigma_z + \sigma_z\sigma_x - \tau_{xy}^2 - \tau_{yz}^2 - \tau_{zx}^2, \qquad \text{Eq. (5-26)} \\ I_3 &= \sigma_x\sigma_y\sigma_z - \sigma_x\tau_{yz}^2 - \sigma_y\tau_{xz}^2 - \sigma_z\tau_{xy}^2 + 2\tau_{xy}\tau_{yz}\tau_{zx}. \end{aligned}$$

The absolute maximum shear stress is the largest of the three values

$$\left|\frac{\sigma_1 - \sigma_2}{2}\right|, \quad \left|\frac{\sigma_1 - \sigma_3}{2}\right|, \quad \left|\frac{\sigma_2 - \sigma_3}{2}\right|. \qquad \text{Eq. (5-27)}$$

The state of stress at a point is said to be *triaxial* if it is of the form

$$\begin{bmatrix} \sigma_x & 0 & 0 \\ 0 & \sigma_y & 0 \\ 0 & 0 & \sigma_z \end{bmatrix}. \qquad \text{Eq. (5-28)}$$

In triaxial stress, x, y, and z are principal axes and σ_x, σ_y, and σ_z are the principal stresses. The absolute maximum shear stress is the largest of the three values

$$\left|\frac{\sigma_x - \sigma_y}{2}\right|, \quad \left|\frac{\sigma_x - \sigma_z}{2}\right|, \quad \left|\frac{\sigma_y - \sigma_z}{2}\right|. \qquad \text{Eq. (5-29)}$$

Tetrahedron Argument

Given the state of stress at a point p, the components of the traction $\mathbf{t}$ at p acting on a plane through p whose orientation is defined by a perpendicular unit vector $\mathbf{n}$ [Fig. (g)] are

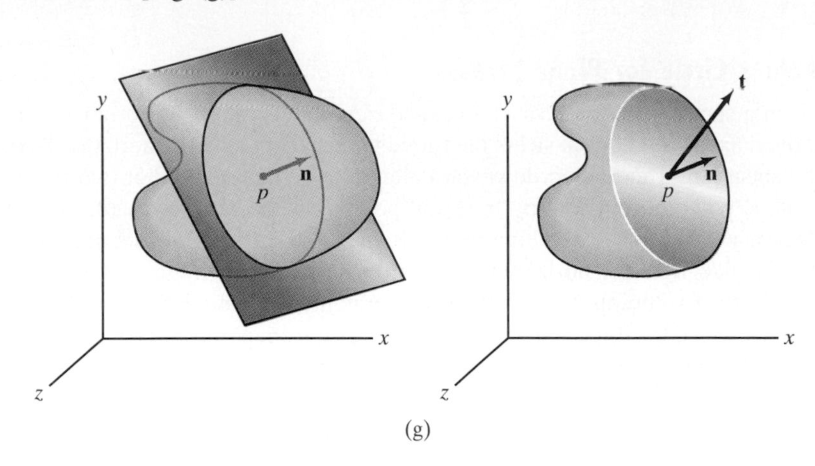

(g)

$$t_x = \sigma_x n_x + \tau_{xy} n_y + \tau_{xz} n_z,$$
$$t_y = \tau_{yx} n_x + \sigma_y n_y + \tau_{yz} n_z, \qquad \text{Eq. (5-43)}$$
$$t_z = \tau_{zx} n_x + \tau_{zy} n_y + \sigma_z n_z.$$

The normal stress at p is

$$\sigma = \mathbf{t} \cdot \mathbf{n} = t_x n_x + t_y n_y + t_z n_z, \qquad \text{Eq. (5-45)}$$

and the magnitude of the shear stress is

$$|\tau| = |\mathbf{t} - \sigma \mathbf{n}|. \qquad \text{Eq. (5-46)}$$

PROBLEMS

5-2.1. The components of plane stress at a point p of a material are $\sigma_x = 20$ MPa, $\sigma_y = 0$, and $\tau_{xy} = 0$. If $\theta = 45°$, what are the stresses σ'_x, σ'_y, and τ'_{xy} at point p?

5-2.2. The components of plane stress at a point p of a material are $\sigma_x = 0$, $\sigma_y = 0$, and $\tau_{xy} = 25$ ksi. If $\theta = 45°$, what are the stresses σ'_x, σ'_y, and τ'_{xy} at point p?

5-2.3. The components of plane stress at a point p of a material are $\sigma_x = -8$ ksi, $\sigma_y = 6$ ksi, and $\tau_{xy} = -6$ ksi. If $\theta = 30°$, what are the stresses σ'_x, σ'_y, and τ'_{xy} at point p?

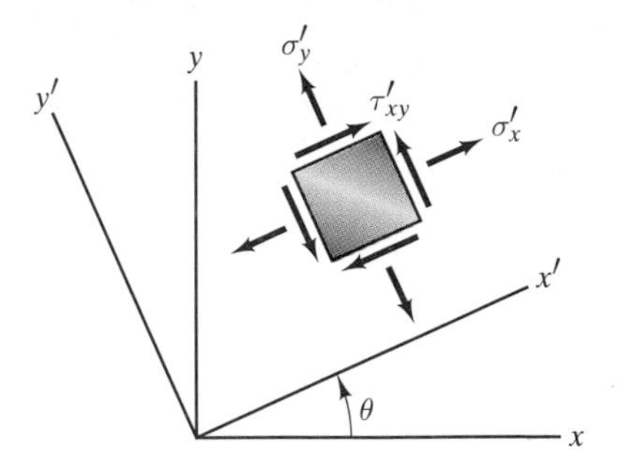

PROBLEMS 5-2.1–5-2.7

5-2.4. During liftoff, strain gauges attached to one of the Space Shuttle main engine nozzles determine that the components of plane stress $\sigma'_x = 66.46$ MPa, $\sigma'_y = 82.54$ MPa, and $\tau'_{xy} = 6.75$ MPa at $\theta = 20°$. What are the stresses σ_x, σ_y, and τ_{xy} at that point?

5-2.5. The components of plane stress at a point p of a material are $\sigma_x = 240$ MPa, $\sigma_y = -120$ MPa, and $\tau_{xy} = 240$ MPa, and the components referred to the $x'y'z'$ coordinate system are $\sigma'_x = 347$ MPa, $\sigma'_y = -227$ MPa, and $\tau'_{xy} = -87$ MPa. What is the angle θ? [*Strategy:* Solve Eqs. (5-7) and (5-8) for $\sin 2\theta$ and $\cos 2\theta$. Knowing these two quantities, you can determine θ. (Why are the values of both $\sin 2\theta$ and $\cos 2\theta$ needed to uniquely determine θ?)]

5-2.6. The components of plane stress at a point p of a bit during a drilling operation are $\sigma_x = 40$ ksi, $\sigma_y = -30$ ksi, and $\tau_{xy} = 30$ ksi, and the components referred to the $x'y'z'$ coordinate system are $\sigma'_x = 12.5$ ksi, $\sigma'_y = -2.5$ ksi, and $\tau'_{xy} = 45.5$ ksi. What is the angle θ?

5-2.7. The components of plane stress at a point p of a material referred to the $x'y'z'$ coordinate system are $\sigma'_x = -8$ MPa, $\sigma'_y = 6$ MPa, and $\tau'_{xy} = -16$ MPa. If $\theta = 20°$, what are the stresses σ_x, σ_y, and τ_{xy} at point p?

5-2.8. A point p of the car's frame is subjected to the components of plane stress $\sigma'_x = 32$ MPa, $\sigma'_y = -16$ MPa, and $\tau'_{xy} = -24$ MPa. If $\theta = 20°$, what are the stresses σ_x, σ_y, and τ_{xy} at p?

5-2.9. In Problem 5-2.8, what are the stresses σ_x, σ_y, and τ_{xy} at point p if $\theta = -40°$?

5-2.10. The components of plane stress at point p of the material shown are $\sigma_x = 4$ ksi, $\sigma_y = -2$ ksi, and $\tau_{xy} =$

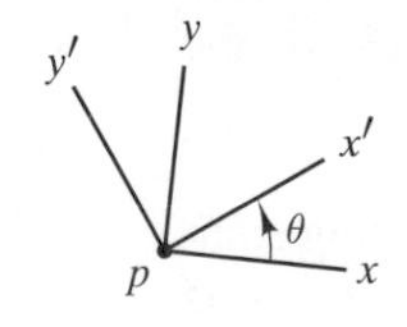

PROBLEM 5-2.8

2 ksi. What are the normal stress and the magnitude of the shear stress on the plane $\mathcal{P}$ at point p?

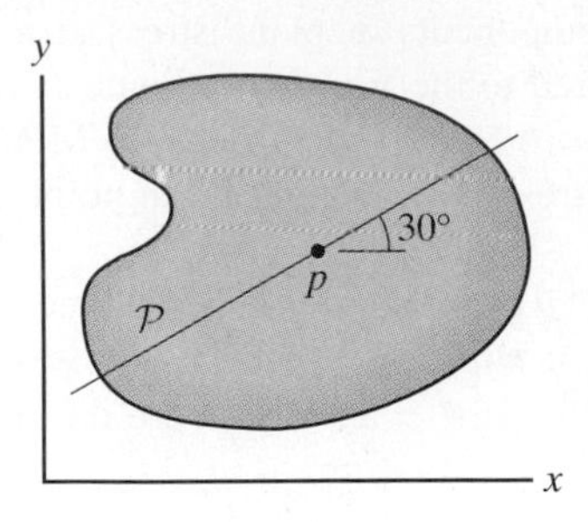

PROBLEM 5-2.10

5-2.11. The components of plane stress at point p of the material shown in Problem 5-2.10 are $\sigma_x = -10.5$ MPa, $\sigma_y = 6.0$ MPa, and $\tau_{xy} = -4.5$ MPa. What are the normal stress and the magnitude of the shear stress on the plane $\mathcal{P}$ at point p?

5-2.12. Determine the stresses σ and τ **(a)** by writing equilibrium equations for the element shown; **(b)** by using Eqs. (5-7) and (5-8).

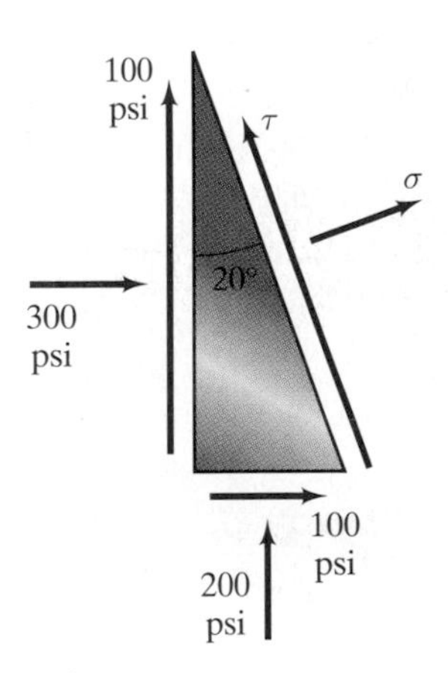

PROBLEM 5-2.12

5-2.13. Solve Problem 5-2.12 if the 300-psi stress on the element is in tension instead of compression.

5-2.14. Determine the stresses σ and τ **(a)** by writing equilibrium equations for the element shown; **(b)** by using Eqs. (5-7) and (5-8).

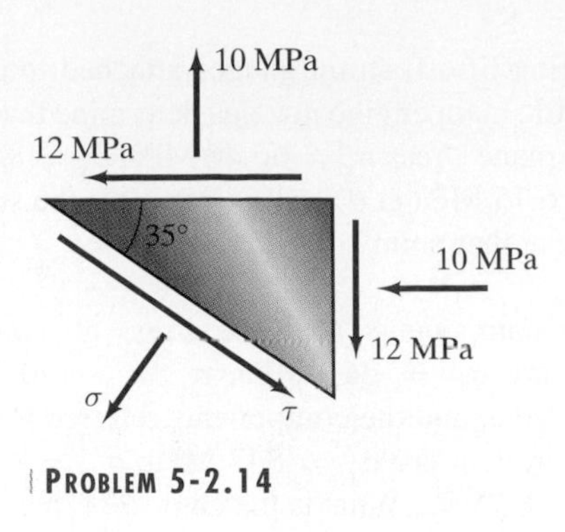

PROBLEM 5-2.14

5-2.15. The stress $\tau_{xy} = 14$ MPa and the angle $\theta = 25°$. Determine the components of stress on the right element.

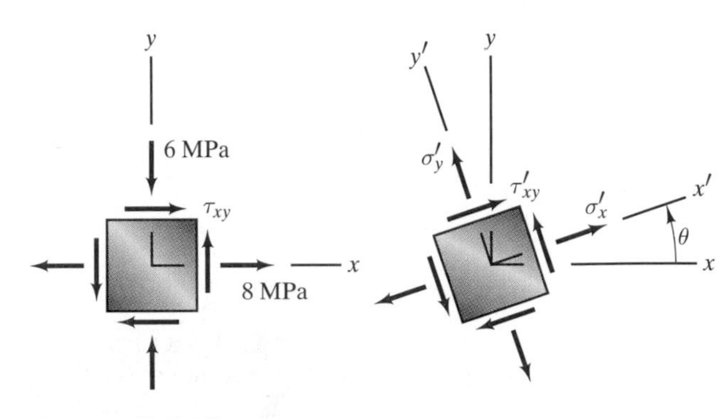

PROBLEM 5-2.15

5-2.16. On the elements shown in Problem 5-2.15, the stresses $\tau_{xy} = 12$ MPa, $\sigma'_x = 14$ MPa, and $\sigma'_y = -12$ MPa. Determine the stress τ'_{xy} and the angle θ.

5-2.17. On the elements shown in Problem 5-2.15, the stress $\tau'_{xy} = 12$ MPa and the angle $\theta = 35°$. Determine σ'_x, σ'_y, and τ_{xy}.

5-2.18. A point p of the airplane's wing is subjected to plane stress. When $\theta = 55°$, $\sigma_x = 100$ psi, $\sigma_y = -200$ psi, and $\sigma'_x = -175$ psi. Determine the stresses τ_{xy} and τ'_{xy} at p.

5-2.19. Under a different flight condition of the airplane in Problem 5-2.18, the stress components $\sigma_x = 80$ psi, $\sigma_y = -120$ psi, $\tau_{xy} = -100$ psi, $\sigma'_x = -80$ psi, and $\sigma'_y = 40$ psi. Determine the stress τ'_{xy} and the angle θ.

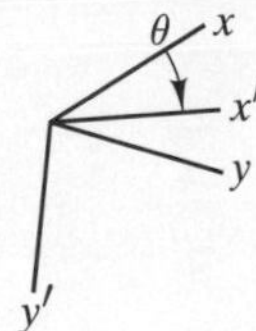

PROBLEM 5-2.18

5-2.20. Equations (5-7)–(5-9) apply to plane stress, but they also apply to states of stress of the form

$$\begin{bmatrix} \sigma_x & \tau_{xy} & 0 \\ \tau_{yx} & \sigma_y & 0 \\ 0 & 0 & \sigma_z \end{bmatrix}.$$

Show that for a fluid at rest, $\sigma'_x = -p$ and $\tau'_{xy} = 0$ for any value of θ (see Eq. 5-2). That is, the normal stress at a point is the negative of the pressure and the shear stress is zero for any plane through the point.

5-2.21. By substituting the trigonometric identities (5-6) into Eqs. (5-4) and (5-5), derive Eqs. (5-7) and (5-8).

5-2.22. The components of plane stress acting on an element of a bar subjected to axial loads are shown in Fig. 5-7. Assuming the stress σ_x to be known, determine the principal stresses and the maximum in-plane shear stress and show them acting on properly oriented elements.

5-2.23. The components of plane stress acting on an element of a bar subjected to torsion are shown in Fig. 5-8. Assuming the stress τ_{xy} to be known, determine the principal stresses and the maximum in-plane shear stress and show them acting on properly oriented elements.

For the states of plane stress given in Problems 5-2.24–5-2.27, determine the principal stresses and the maximum in-plane shear stress and show them acting on properly oriented elements.

5-2.24. $\sigma_x = 20$ MPa, $\sigma_y = 10$ MPa, and $\tau_{xy} = 0$.

5-2.25. $\sigma_x = 25$ ksi, $\sigma_y = 0$, and $\tau_{xy} = -25$ ksi.

5-2.26. $\sigma_x = -8$ ksi, $\sigma_y = 6$ ksi, and $\tau_{xy} = -6$ ksi.

5-2.27. $\sigma_x = 240$ MPa, $\sigma_y = -120$ MPa, and $\tau_{xy} = 240$ MPa.

5-2.28. For the state of plane stress $\sigma_x = 20$ MPa, $\sigma_y = 10$ MPa, and $\tau_{xy} = 0$, what is the absolute maximum shear stress?

5-2.29. For the state of plane stress $\sigma_x = 25$ ksi, $\sigma_y = 0$, and $\tau_{xy} = -25$ ksi, what is the absolute maximum shear stress?

5-2.30. For the state of plane stress $\sigma_x = 8$ ksi, $\sigma_y = 6$ ksi, and $\tau_{xy} = -6$ ksi, what is the absolute maximum shear stress?

5-2.31. For the state of plane stress $\sigma_x = 240$ MPa, $\sigma_y = -120$ MPa, and $\tau_{xy} = 240$ MPa, what is the absolute maximum shear stress?

5-2.32. A point p of the antenna's supporting structure is subjected to the state of plane stress $\sigma_x = 40$ MPa, $\sigma_y = -20$ MPa, and $\tau_{xy} = 30$ MPa. Determine the principal stresses and the absolute maximum shear stress.

5-2.33. A point p of the supporting structure of the antenna shown in Problem 5-2.32 is subjected to plane stress. The components $\sigma_y = -20$ MPa and $\tau_{xy} = 30$ MPa. The allowable normal stress of the material (in tension and compression) is 80 MPa. Based on this criterion, what is the allowable range of values of the stress component σ_x?

5-2.34. The element in Fig. 5-4 is a cube with dimension b. By summing moments about the x and y axes, show that $\tau_{yz} = \tau_{zy}$ and $\tau_{xz} = \tau_{zx}$ if the material is in equilibrium.

| **PROBLEM 5-2.32**

(In fact, these results hold even if the material is not in equilibrium.)

5-2.35. By setting θ equal to $\theta + 90^\circ$ in Eq. (5-7), derive Eq. (5-9).

5-2.36. By solving Eqs. (5-13) and (5-14) for $\sin 2\theta_p$ and $\cos 2\theta_p$ and substituting the results into Eq. (5-10), derive Eq. (5-15) for the principal stresses.

5-2.37. By solving Eqs. (5-17) and (5-18) for $\sin 2\theta_s$ and $\cos 2\theta_s$ and substituting the results into Eq. (5-11), derive Eq. (5-19) for the maximum in-plane shear stress.

5-3.1. The components of plane stress at a point p of a material are $\sigma_x = 20$ MPa, $\sigma_y = 0$, and $\tau_{xy} = 0$, and the angle $\theta = 45^\circ$. Use Mohr's circle to determine the stresses σ'_x, σ'_y, and τ'_{xy} at point p.

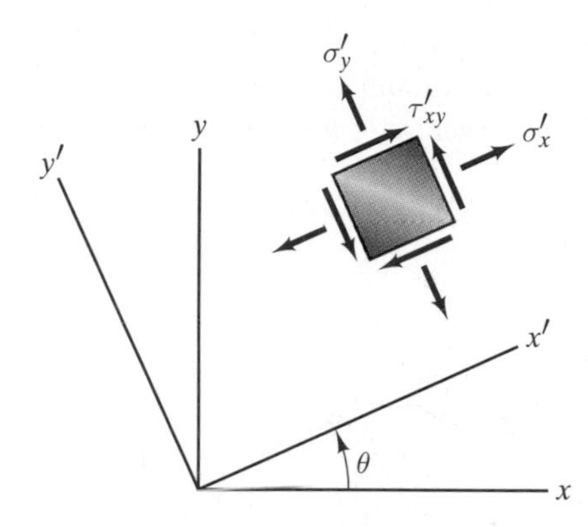

| PROBLEMS 5-3.1–5-3.6

5-3.2. The components of plane stress at a point p of a material are $\sigma_x = 0$, $\sigma_y = 0$, and $\tau_{xy} = 25$ ksi, and the angle $\theta = 45^\circ$. Use Mohr's circle to determine the stresses σ'_x, σ'_y, and τ'_{xy} at point p.

5-3.3. The components of plane stress at a point p of a material are $\sigma_x = 240$ MPa, $\sigma_y = -120$ MPa, and $\tau_{xy} = 240$ MPa, and the components referred to the $x'y'z'$ coordinate system are $\sigma'_x = 347$ MPa, $\sigma'_y = -227$ MPa, and $\tau'_{xy} = -87$ MPa. Use Mohr's circle to determine the angle θ.

5-3.4. Use Mohr's circle to determine the components of stress at a point of one of the Space Shuttle's main engine nozzles in Problem 5-2.4.

5-3.5. The components of plane stress at a point p of a material referred to the $x'y'z'$ coordinate system are $\sigma'_x = -8$ MPa, $\sigma'_y = 6$ MPa, and $\tau'_{xy} = -16$ MPa, and the angle $\theta = 20^\circ$. Use Mohr's circle to determine the stresses σ_x, σ_y, and τ_{xy} at point p.

5-3.6. The components of plane stress at a point p of a bit during a drilling operation are $\sigma_x = 40$ ksi, $\sigma_y = -30$ ksi, and $\tau_{xy} = 30$ ksi, and the components referred to the $x'y'z'$ coordinate system are $\sigma'_x = 12.5$ ksi, $\sigma'_y = -2.5$ ksi, and $\tau'_{xy} = 45.5$ ksi. Use Mohr's circle to estimate the angle θ.

5-3.7. The components of plane stress at point p of the material shown are $\sigma_x = 4$ ksi, $\sigma_y = -2$ ksi, and $\tau_{xy} = 2$ ksi. Use Mohr's circle to determine the normal stress and the magnitude of the shear stress on the plane $\mathcal{P}$ at point p.

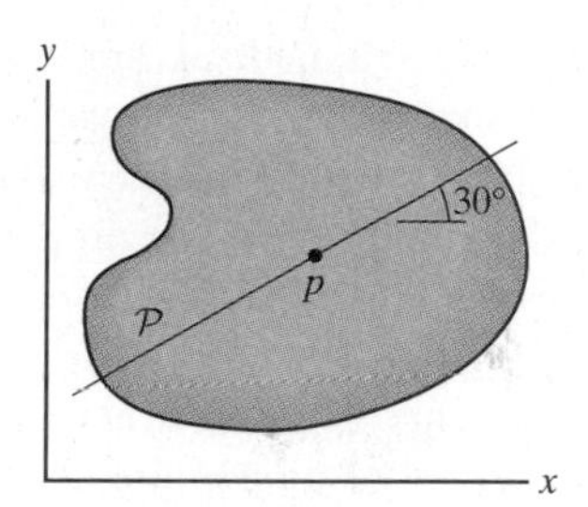

| PROBLEM 5-3.7

5-3.8. The components of plane stress at point p of the material shown in Problem 5-3.7 are $\sigma_x = -10.5$ MPa, $\sigma_y = 6.0$ MPa, and $\tau_{xy} = -4.5$ MPa. Use Mohr's circle to determine the normal stress and the magnitude of the shear stress on the plane $\mathcal{P}$ at point p.

5-3.9. Determine the stresses σ and τ **(a)** by using Mohr's circle; **(b)** by using Eqs. (5-7) and (5-8).

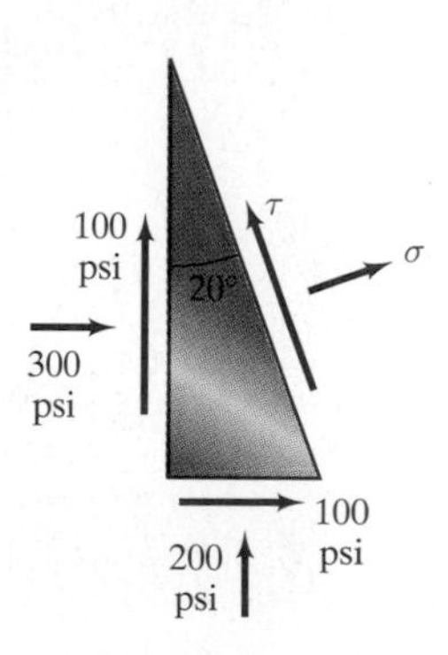

| PROBLEM 5-3.9

5-3.10. Solve Problem 5-3.9 if the 300-psi stress on the element is in tension instead of compression.

5-3.11. Determine the stresses σ and τ **(a)** by using Mohr's circle; **(b)** by using Eqs. (5-7) and (5-8).

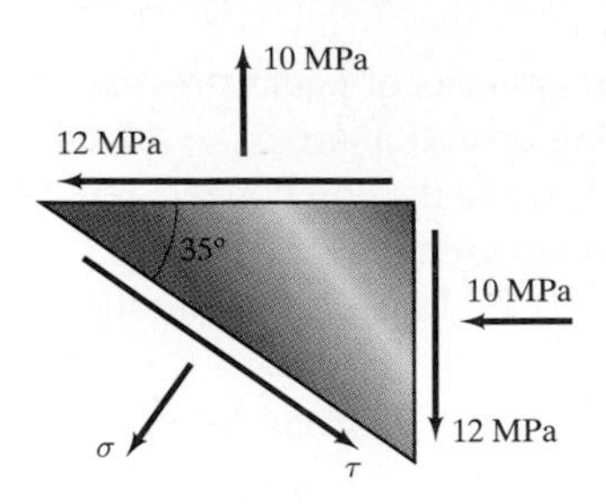

PROBLEM 5-3.11

5-3.12. The components of plane stress acting on an element of a bar subjected to axial loads are shown in Fig. 5-7. Assuming the stress σ_x to be known, use Mohr's circle to determine the principal stresses and the maximum in-plane shear stress and show them acting on properly oriented elements.

5-3.13. The components of plane stress acting on an element of a bar subjected to torsion are shown in Fig. 5-8. Assuming the stress τ_{xy} to be known, use Mohr's circle to determine the principal stresses and the maximum in-plane shear stress and show them acting on properly oriented elements.

For the states of plane stress given in Problems 5-3.14–5-3.17, use Mohr's circle to determine the principal stresses and the maximum in-plane shear stress and show them acting on properly oriented elements.

5-3.14. $\sigma_x = 20$ MPa, $\sigma_y = 10$ MPa, and $\tau_{xy} = 0$.

5-3.15. $\sigma_x = 25$ ksi, $\sigma_y = 0$, and $\tau_{xy} = -25$ ksi.

5-3.16. $\sigma_x = -8$ ksi, $\sigma_y = 6$ ksi, and $\tau_{xy} = -6$ ksi.

5-3.17. $\sigma_x = 240$ MPa, $\sigma_y = -120$ MPa, and $\tau_{xy} = 240$ MPa.

5-3.18. Use Mohr's circle to determine the principal stresses and the maximum in-plane shear stress at point p of the supporting structure of the antenna in Problem 5-2.32.

5-3.19. At touchdown, a point p of the Space Shuttle's landing gear is subjected to the state of plane stress

PROBLEM 5-3.19

$\sigma_x = -120$ MPa, $\sigma_y = 80$ MPa, and $\tau_{xy} = -50$ MPa. Use Mohr's circle to determine the principal stresses and the maximum in-plane shear stress.

5-3.20. For the state of plane stress $\sigma_x = 8$ ksi, $\sigma_y = 6$ ksi, and $\tau_{xy} = -6$ ksi, use Mohr's circle to determine the absolute maximum shear stress. [*Strategy:* Use Mohr's circle to determine the principal stresses and then determine the the absolute maximum shear stress from the expressions (5-24).]

5-3.21. For the state of plane stress $\sigma_x = 240$ MPa, $\sigma_y = -120$ MPa, and $\tau_{xy} = 240$ MPa, use Mohr's circle to determine the absolute maximum shear stress.

5-4.1. At a point p a material is subjected to the state of *plane* stress $\sigma_x = 20$ MPa, $\sigma_y = 10$ MPa, $\tau_{xy} = 0$. Use Eq. (5-25) to determine the principal stresses and use Eq. (5-27) to determine the absolute maximum shear stress. Confirm the absolute maximum shear stress by drawing the superimposed Mohr's circle as shown in Fig. 5-36.

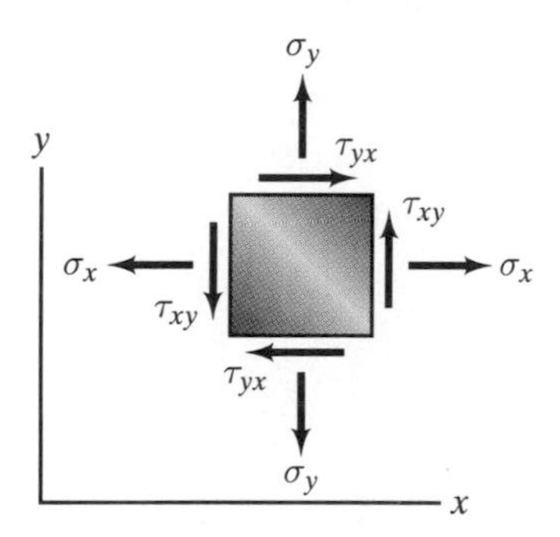

5-4.2. At a point p a material is subjected to the state of *plane* stress $\sigma_x = 25$ ksi, $\sigma_y = 0$, $\tau_{xy} = -25$ ksi. Use Eq. (5-25) to determine the principal stresses and use Eq. (5-27) to determine the absolute maximum shear stress.

5-4.3. At a point p a material is subjected to the state of *plane* stress $\sigma_x = 240$ MPa, $\sigma_y = -120$ MPa, $\tau_{xy} = 240$ MPa. Use Eq. (5-25) to determine the principal stresses and use Eq. (5-27) to determine the absolute maximum shear stress. Confirm the absolute maximum shear stress by drawing the superimposed Mohr's circle as shown in Fig. 5-36.

5-4.4. Strain gauges attached to one of the Space Shuttle main engine nozzles determine that the components of *plane* stress are $\sigma_x = 67.34$ MPa, $\sigma_y = 82.66$ MPa, and $\tau_{xy} = 6.43$ MPa. Use Eq. (5-25) to determine the principal stresses and use Eq. (5-27) to determine the absolute maximum shear stress.

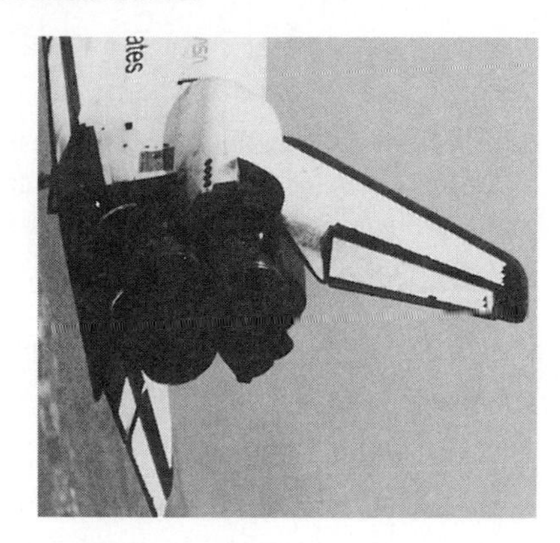

5-4.5. Use Eq. (5-25) to determine the principal stresses for an arbitrary state of plane stress σ_x, σ_y, τ_{xy} and confirm Eq. (5-15).

5-4.6. At a point p a material is subjected to the state of *triaxial* stress $\sigma_x = 240$ MPa, $\sigma_y = -120$ MPa, $\sigma_z = 240$ MPa. Determine the principal stresses and the absolute maximum shear stress.

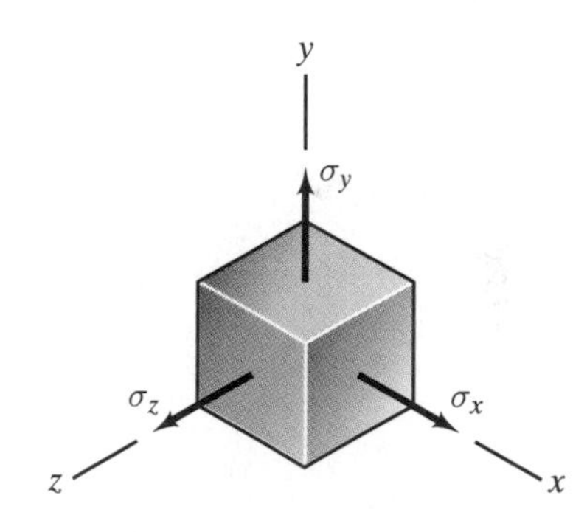

5-4.7. At a point p a material is subjected to the state of *triaxial* stress $\sigma_x = 40$ ksi, $\sigma_y = 80$ ksi, $\sigma_z = -20$ ksi. Determine the principal stresses and the absolute maximum shear stress.

5-4.8. A machine element is subjected to the state of *triaxial* stress $\sigma_x = 300$ MPa, $\sigma_y = -200$ MPa, σ_z. If the material is not to be subjected to a shear stress greater than 400 MPa, what is the acceptable range of the stress σ_z?

5-4.9. At a point p a material is subjected to the state of stress (in MPa)

$$\begin{bmatrix} \sigma_x & \tau_{xy} & \tau_{xz} \\ \tau_{yx} & \sigma_y & \tau_{yz} \\ \tau_{zx} & \tau_{zy} & \sigma_z \end{bmatrix} = \begin{bmatrix} 4 & 2 & 1 \\ 2 & -2 & 1 \\ 1 & 1 & -3 \end{bmatrix}.$$

Determine the principal stresses and the absolute maximum shear stress. Confirm the absolute maximum shear stress by drawing the superimposed Mohr's circle as shown in Fig. 5-36.

5-4.10. At a point p a material is subjected to the state of stress (in ksi)

$$\begin{bmatrix} \sigma_x & \tau_{xy} & \tau_{xz} \\ \tau_{yx} & \sigma_y & \tau_{yz} \\ \tau_{zx} & \tau_{zy} & \sigma_z \end{bmatrix} = \begin{bmatrix} 300 & 150 & -100 \\ 150 & 200 & 100 \\ -100 & 100 & -200 \end{bmatrix}.$$

Determine the principal stresses and the absolute maximum shear stress. Confirm the absolute maximum shear stress by drawing the superimposed Mohr's circle as shown in Fig. 5-36.

5-4.11. A finite element analysis of a bearing housing indicates that at a point p the material is subjected to the state of stress (in MPa)

$$\begin{bmatrix} \sigma_x & \tau_{xy} & \tau_{xz} \\ \tau_{yx} & \sigma_y & \tau_{yz} \\ \tau_{zx} & \tau_{zy} & \sigma_z \end{bmatrix} = \begin{bmatrix} 20 & 20 & 0 \\ 20 & -30 & -10 \\ 0 & -10 & 40 \end{bmatrix}.$$

Determine the principal stresses and the absolute maximum shear stress.

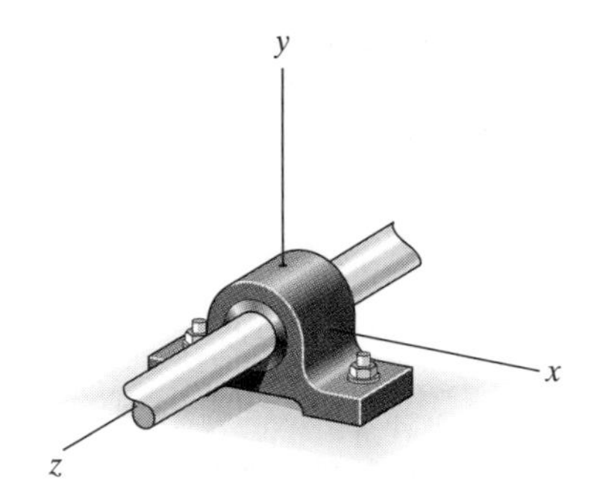

5-4.12. The components of *plane* stress at a point p of a material are $\sigma_x = -8$ ksi, $\sigma_y = 6$ ksi, and $\tau_{xy} = -6$ ksi. Use Eqs. (5-7)–(5-9) to determine the components of stress σ'_x, σ'_y, and τ'_{xy} corresponding to a coordinate system $x'y'z'$ oriented at $\theta = 30°$. (a) Determine the principal stresses from Eq. (5-25) using the components of stress σ_x, σ_y, and τ_{xy}. (b) Determine the principal stresses from Eq. (5-25) using the components of stress σ'_x, σ'_y, and τ'_{xy}.

5-4.13. By using results from Section 5-2, prove for a state of *plane* stress that the coefficient I_1 in Eq. (5-25) does not depend on the orientation of the coordinate system used to evaluate it.

5-4.14. By using results from Section 5-2, prove for a state of *plane* stress that the coefficient I_2 in Eq. (5-25) does not depend on the orientation of the coordinate system used to evaluate it.

5-5.1. A spherical pressure vessel has a 2.5-m radius and a 5-mm wall thickness. It contains a gas with pressure $p_i = 6 \times 10^5$ Pa and the outer wall is subjected to atmospheric pressure $p_o = 1 \times 10^5$ Pa. Determine the maximum normal stress in the vessel wall.

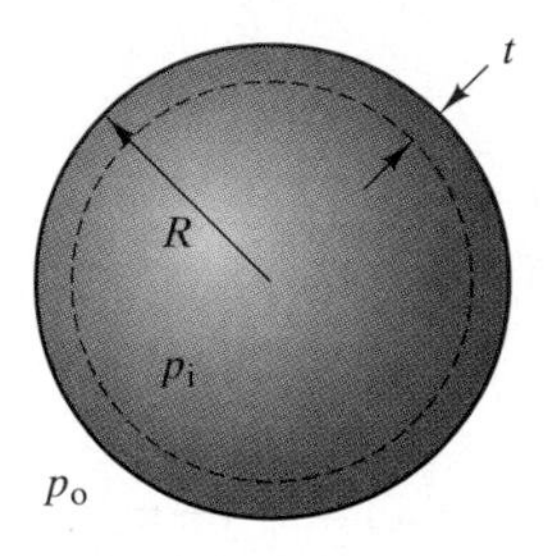

5-5.2. A spherical pressure vessel has a 1-m radius and a 0.002-m wall thickness. It contains a gas with pressure $p_i = 1.8 \times 10^5$ Pa, and the outer wall is subjected to atmospheric pressure $p_o = 1 \times 10^5$ Pa. Determine the maximum normal stress and the absolute maximum shear stress at the vessel's inner surface.

5-5.3. For the spherical pressure vessel described in Problem 5-5.2, suppose that the allowable value of the absolute maximum shear stress is $\tau_{allow} = 14$ MPa. If the outer wall is subjected to atmospheric pressure $p_o = 1 \times 10^5$ Pa, what is the maximum allowable internal pressure?

5-5.4. A spherical pressure vessel has a 24-in. radius and a $\frac{1}{64}$-in. wall thickness. It contains a gas with pressure $p_i = 200$ psi and the outer wall is subjected to atmospheric pressure $p_o = 14.7$ psi. Determine the maximum normal stress and the absolute maximum shear stress at the vessel's inner surface.

5-5.5. Suppose that the spherical pressure vessel described in Problem 5-5.4 is made of material with a yield shear stress $\tau_Y = 100$ ksi. If the vessel is designed to contain gas with a maximum pressure $p_i = 150$ psi and the outer wall is subjected to atmospheric pressure $p_o = 14.7$ psi, what is the factor of safety?

5-5.6. A cylindrical pressure vessel with hemispherical ends has a 2.5-m radius and a 5-mm wall thickness. It contains a gas with pressure $p_i = 6 \times 10^5$ Pa, and the outer wall is subjected to atmospheric pressure $p_o = 1 \times 10^5$ Pa. Determine the maximum normal stress in the vessel wall. Compare your answer to the answer to Problem 5-5.1.

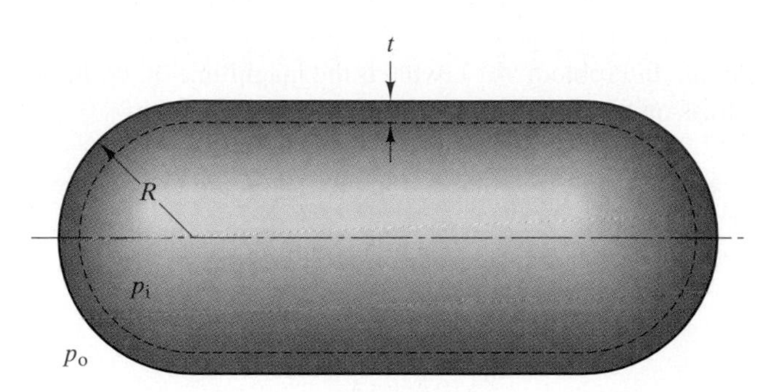

5-5.7. In Example 5-9, the wall thickness of the cylindrical vessel is determined to be 8.04 mm. What is the resulting maximum normal stress in the vessel wall?

5-5.8. A cylindrical pressure vessel has a 600-mm radius and an 8-mm wall thickness. It contains a gas with pressure $p_i = 3 \times 10^5$ Pa and the outer wall is subjected to atmospheric pressure $p_o = 1 \times 10^5$ Pa. Determine the maximum normal stress and the absolute maximum shear stress at the inner surface of the vessel's cylindrical wall.

5-5.9. A cylindrical pressure vessel used for natural gas storage has a 6-ft radius and a $\frac{1}{2}$-in. wall thickness. It contains gas with pressure $p_i = 80$ psi and the outer wall is subjected to atmospheric pressure $p_o = 14.7$ psi. Determine the maximum normal stress and the absolute maximum shear stress at the inner surface of the vessel's cylindrical wall.

5-5.10. In Example 5-9, determine the vessel's wall thickness so that the factor of safety is $S = 3$.

5-5.11. Suppose that you are designing a spherical pressure vessel with a 200-mm radius to be used in a fuel cell to provide power in a satellite. The maximum internal pressure will be 16 MPa and the external pressure will be negligible. Choose an aluminum alloy from Appendix B and determine the wall thickness to obtain a safety factor $S = 1.5$.

5-5.12. If design constraints require that the spherical pressure vessel in Problem 5-5.11 be a cylindrical vessel with 150-mm radius, hemispherical ends, and the same internal volume as the spherical vessel, determine the wallthickness needed to obtain a safety factor $S = 1.5$. Compare the weight of the resulting vessel to the weight of the spherical vessel.

5-5.13. Suppose that you are making preliminary design calculations for a deep submersible vehicle that is to have a cylindrical hull with hemispherical ends and a 1.5-m radius. At its operating depth, the internal pressure will be 1×10^5 Pa and the external pressure will be 330×10^5 Pa. Choose a steel from Appendix B and determine the necessary thickness of the hull so that it has a safety factor $S = 2$.

5-5.14. Choose an aluminum alloy from Appendix B and determine the necessary thickness of the hull described in Problem 5-5.13 so that it has a safety factor $S = 2$. Compare the weight of your aluminum hull design with that of the steel hull designed in Problem 5-5.13 if the length of the cylinder is 6 m.

5-6.1. The state of stress at a point p of a material is

$$\begin{bmatrix} 8 & 0 & 0 \\ 0 & 4 & 0 \\ 0 & 0 & 6 \end{bmatrix} \text{MPa.}$$

Determine the traction $\mathbf{t}$ acting on the plane whose orientation is specified by the unit vector $\mathbf{n} = -0.857\,\mathbf{i} + 0.429\,\mathbf{j} + 0.286\,\mathbf{k}$. [*Strategy:* The components of the traction are given in terms of the components of $\mathbf{n}$ by Eqs. (5-43) or (5-44).]

5-6.2. In Problem 5-6.1, determine the normal stress σ and the magnitude of the shear stress $|\tau|$ acting on the plane specified.

5-6.3. The state of stress at a point p of a material is

$$\begin{bmatrix} -30 & 20 & 0 \\ 20 & 45 & -20 \\ 0 & -20 & 60 \end{bmatrix} \text{ksi.}$$

Determine the normal stress σ and the magnitude of the shear stress $|\tau|$ acting on the plane whose orientation is specified by the unit vector $\mathbf{n} = 0.381\mathbf{i} - 0.889\mathbf{j} - 0.254\mathbf{k}$.

5-6.4. The state of stress at point p of the material shown is

$$\begin{bmatrix} 2 & 2 & 1 \\ 2 & -3 & -1 \\ 1 & -1 & 4 \end{bmatrix} \text{ksi,}$$

and the unit vector $\mathbf{n} = 0.667\mathbf{i} + 0.333\mathbf{j} + 0.667\mathbf{k}$. Determine the traction $\mathbf{t}$ on the plane shown.

5-6.5. In Problem 5-6.4, determine the normal stress and the magnitude of the shear stress on the plane shown.

5-6.6. In Problem 5-6.4, suppose that the state of stress at point p is

$$\begin{bmatrix} 6 & -2 & 3 \\ -2 & 2 & 1 \\ 3 & 1 & 4 \end{bmatrix} \text{GPa,}$$

and that the plane shown passes through the three points (3, 0, 0) m, (0, 6, 0) m, and (0, 0, 2) m. Determine the normal stress and the magnitude of the shear stress acting on the plane. (You must determine the unit vector **n**.)

5-6.7. The state of stress at a point p of a material is

$$\begin{bmatrix} 20 & -15 & 30 \\ -15 & -40 & 25 \\ 30 & 25 & \sigma_z \end{bmatrix} \text{MPa.}$$

The normal stress on the plane whose orientation is specified by the unit vector $\mathbf{n} = -0.857\mathbf{i} + 0.429\mathbf{j} + 0.286\mathbf{k}$ is $\sigma = 12.6$ MPa. What is the stress component σ_z?

5-6.8. In Problem 5-6.7, what is the magnitude of the shear stress on the specified plane?

5-6.9. The state of stress at a point in a liquid or gas at rest is

$$\begin{bmatrix} -p & 0 & 0 \\ 0 & -p & 0 \\ 0 & 0 & -p \end{bmatrix},$$

where p is the pressure. Show that the normal stress is $\sigma = -p$ and the shear stress is zero on any plane through the point.

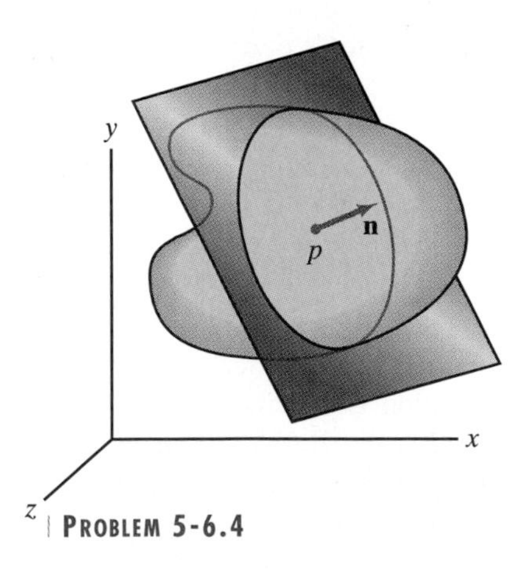

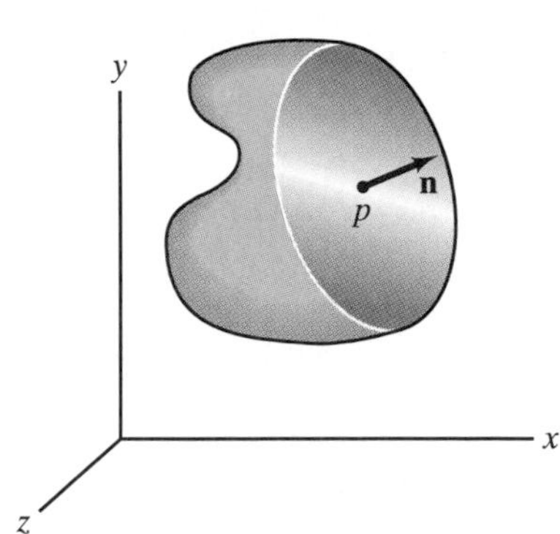

| PROBLEM 5-6.4

5-6.10. For *any* state of stress at a point p, at least one coordinate system xyz exists for which the state of stress is of the form

$$\begin{bmatrix} \sigma_x & \tau_{xy} & \tau_{xz} \\ \tau_{yx} & \sigma_y & \tau_{yz} \\ \tau_{zx} & \tau_{zy} & \sigma_z \end{bmatrix} = \begin{bmatrix} \sigma_1 & 0 & 0 \\ 0 & \sigma_2 & 0 \\ 0 & 0 & \sigma_3 \end{bmatrix},$$

where σ_1, σ_2, and σ_3 are the principal stresses (see Section 5-4). Use Eqs. (5-43), (5-45), and (5-46) to determine the normal stress and the magnitude of the shear stress on a plane through p that is perpendicular to the x axis. (Your answers will be in terms of the principal stresses.)

5-6.11. In Problem 5-6.10, determine the normal stress on a plane through p whose orientation is specified by the unit vector $(1/\sqrt{3})(\mathbf{i}+\mathbf{j}+\mathbf{k})$.

5-6.12. For the state of plane stress shown, use Eqs. (5-45) and (5-46) to determine the normal stress σ and the magnitude of the shear stress τ.

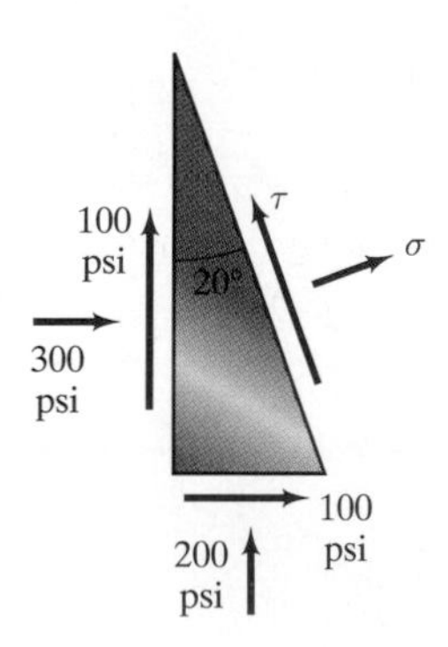

PROBLEM 5-6.12

5-6.13. For the state of plane stress shown, use Eqs. (5-45) and (5-46) to determine the normal stress σ and the magnitude of the shear stress τ.

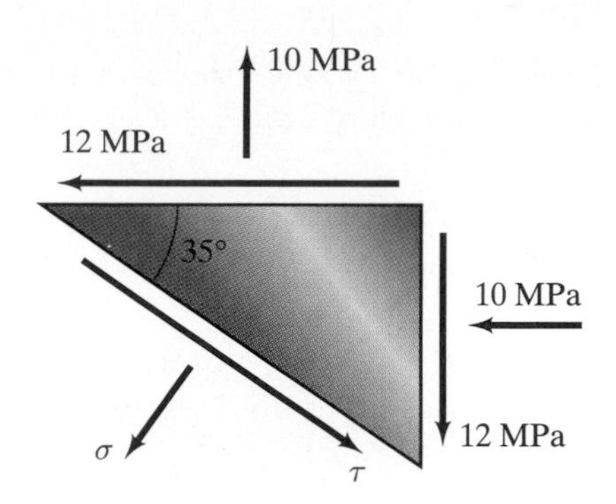

PROBLEM 5-6.13

5-6.14. If the material in Fig. (a) is subjected to the state of plane stress

$$\begin{bmatrix} \sigma_x & \tau_{xy} & 0 \\ \tau_{yx} & \sigma_y & 0 \\ 0 & 0 & 0 \end{bmatrix}$$

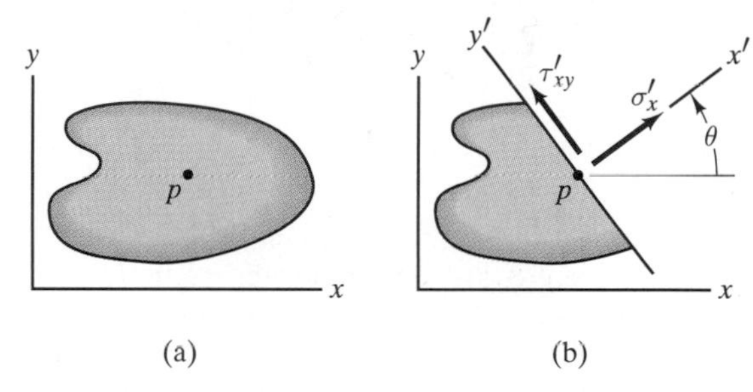

PROBLEM 5-6.14

at point p, the normal stress σ'_x in Fig. (b) is given by Eq. (5-4):

$$\sigma'_x = \sigma_x \cos^2\theta + \sigma_y \sin^2\theta + 2\tau_{xy} \sin\theta \cos\theta.$$

Derive this equation by using Eq. (5-44) (see Example 5-11).

5-6.15. By drawing appropriate free-body diagrams of the tetrahedron in Fig. 5-50 and summing forces in the x, y, and z directions, derive Eqs. (5-43).

Aerodynamic stresses cause the wings of the Concorde to bend and warp, subjecting the material to extensional and shear strains.

CHAPTER 6

States of Strain

In this chapter we define the state of strain at a point of an object such as the Concorde's wing, and show that it specifies the deformation of the material in the neighborhood of the point. We then develop the relationship between the state of stress and the state of strain for an elastic material subject to small strains.

6-1 | Components of Strain

We have introduced two types of strain, extensional strain and shear strain. The extensional strain ϵ is a measure of the change in length of a material line element whose length is dL in a reference state. If the length of the line element in a deformed state is dL' (Fig. 6-1), the extensional strain is defined to be the change in length divided by its original length:

$$\epsilon = \frac{dL' - dL}{dL}.$$

The value of the extensional strain at a point depends in general on the direction of the line element. We say that ϵ is the extensional strain in the direction of dL.

The shear strain γ is a measure of the change in the angle between two line elements dL_1 and dL_2 that are perpendicular in a reference state. The angle between the elements in a deformed state is defined to be $(\pi/2) - \gamma$ (Fig. 6-2). The value of γ at a point depends in general on the directions of the two elements. We say that γ is the shear strain referred to the directions of the elements dL_1 and dL_2.

To define the state of strain at a point p of a material, we introduce a coordinate system (Fig. 6-3a) and introduce six components of strain:

- ϵ_x The extensional strain determined with the element dL parallel to the x axis (Fig. 6-3b)
- ϵ_y The extensional strain determined with the element dL parallel to the y axis
- ϵ_z The extensional strain determined with the element dL parallel to the z axis

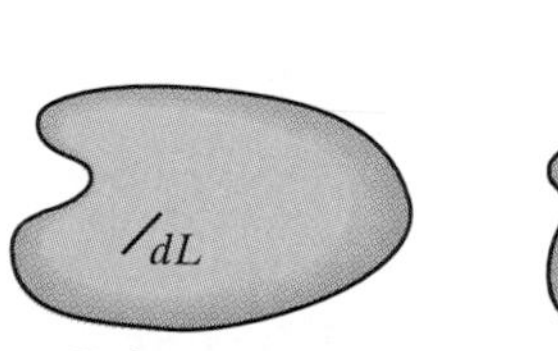

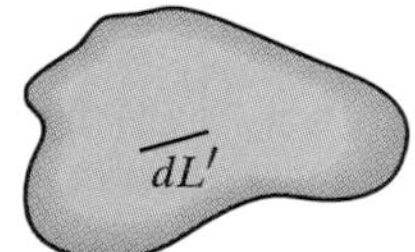

FIGURE 6-1 Material line element in the reference and deformed states.

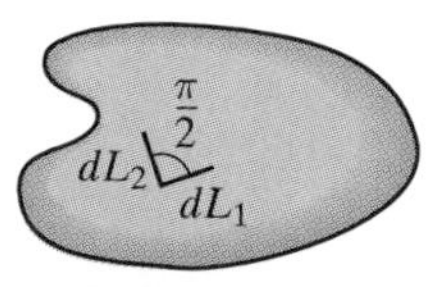

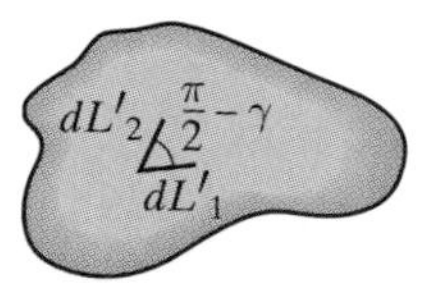

FIGURE 6-2 Line elements that are perpendicular in the reference state. The decrease in the angle between them in the deformed state is the shear strain.

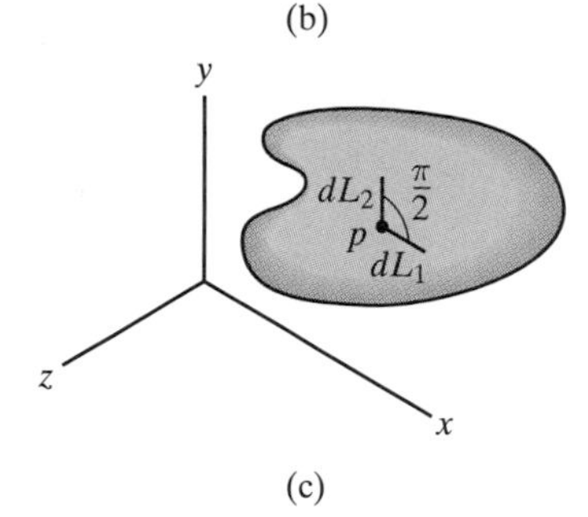

FIGURE 6-3 (a) Introducing a coordinate system. (b) Line element for determining the value of ϵ_x at p. (c) Line elements for determining the value of γ_{xy} at p.

γ_{xy} The shear strain determined with the elements dL_1 and dL_2 in the positive x and y directions (Fig. 6-3c)

γ_{yz} The shear strain determined with the elements dL_1 and dL_2 in the positive y and z directions

γ_{xz} The shear strain determined with the elements dL_1 and dL_2 in the positive x and z directions

If these six components of strain are known at a point, the extensional strain in any direction and the shear strain referred to any two perpendicular directions can be determined. (These results require that the components of strain be sufficiently small that products of the components are negligible in comparison to the components themselves. In this introductory treatment we consider only small strains.) For this reason these components are called the *state of strain* at the point. We say that the strain is *homogeneous* in a region if the state of strain is the same at each point. In analogy to the state of stress, we can represent the state of strain as the matrix

$$\begin{bmatrix} \epsilon_x & \gamma_{xy} & \gamma_{xz} \\ \gamma_{yx} & \epsilon_y & \gamma_{yz} \\ \gamma_{zx} & \gamma_{zy} & \epsilon_z \end{bmatrix}, \tag{6-1}$$

where $\gamma_{yx} = \gamma_{xy}$, $\gamma_{zx} = \gamma_{xz}$, and $\gamma_{zy} = \gamma_{yz}$.

The state of strain determines the change in volume of a material due to a deformation. Consider an element of material in the reference state with dimensions dx_0, dy_0, dz_0 (Fig. 6-4). The volume of the element is $dV_0 = dx_0\,dy_0\,dz_0$. In the deformed state, the lengths of the edges of the element are $dx = (1 + \epsilon_x)\,dx_0$, $dy = (1 + \epsilon_y)\,dy_0$, and $dz = (1 + \epsilon_z)\,dz_0$ (Fig. 6-4). Its volume in the deformed state is

$$dV = dx\,dy\,dz = (1 + \epsilon_x)(1 + \epsilon_y)(1 + \epsilon_z)\,dx_0\,dy_0\,dz_0.$$

Neglecting products of the strains, we can write this result as

$$dV = (1 + \epsilon_x + \epsilon_y + \epsilon_z)\,dV_0. \tag{6-2}$$

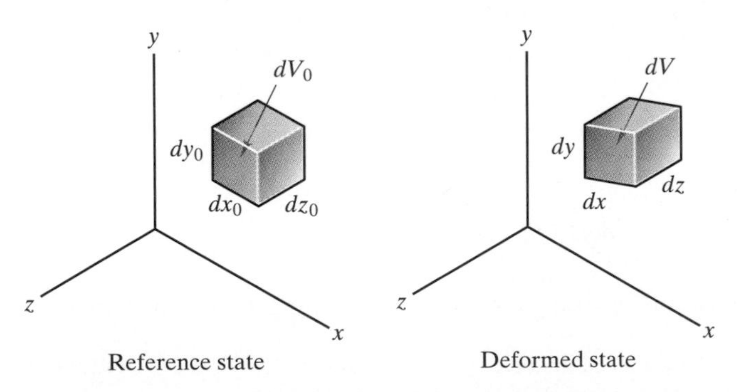

FIGURE 6-4 Element of material in the reference and deformed states.

(When products of strains are negligible, the components of shear strain do not affect the change in volume of the material.) The change in volume of the material per unit volume, denoted by e, is called the *dilatation:*

$$e = \frac{dV - dV_0}{dV_0} = \epsilon_x + \epsilon_y + \epsilon_z. \qquad \textbf{(6-3)}$$

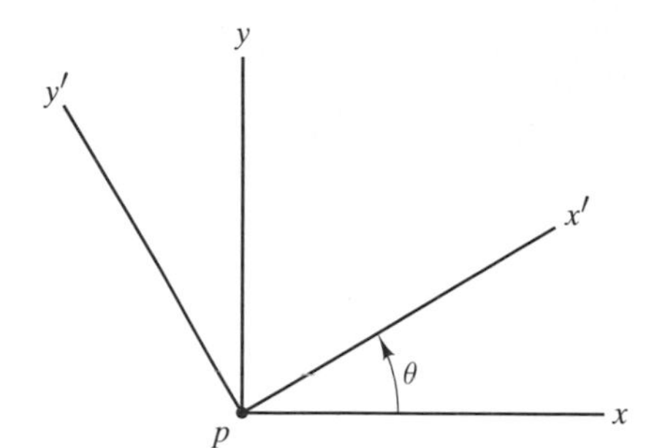

FIGURE 6-5 Coordinate systems xyz and $x'y'z'$. The z and z' axes are coincident.

6-2 | Transformations of Plane Strain

For a state of *plane strain,* defined by

$$\begin{bmatrix} \epsilon_x & \gamma_{xy} & 0 \\ \gamma_{yx} & \epsilon_y & 0 \\ 0 & 0 & 0 \end{bmatrix}, \qquad \textbf{(6-4)}$$

we can derive transformation equations equivalent to the equations for plane stress developed in Section 5-2. Suppose that we know the state of plane strain at a point p of a material in terms of the xyz coordinate system shown in Fig. 6-5 and we want to know the components of strain ϵ'_x, γ'_{xy}, and ϵ'_y.

To determine the extensional strain ϵ'_x, we begin with an infinitesimal element of material at p which in the reference state has the triangular shape shown in Fig. 6-6a. We denote the infinitesimal length of the hypotenuse by dL_0. In the deformed state (Fig. 6-6b), we know the lengths of the sides in terms of the extensional strains ϵ_x, ϵ_y, and ϵ'_x, and we can express the angle between the sides that were perpendicular in the reference state in terms of γ_{xy}. By analyzing this triangle we can determine ϵ'_x in terms of the strains ϵ_x, ϵ_y, and γ_{xy}.

Applying the law of cosines to the deformed element gives

$$dL_0^2(1+\epsilon'_x)^2 = dL_0^2 \sin^2\theta(1+\epsilon_y)^2 + dL_0^2\cos^2\theta(1+\epsilon_x)^2$$
$$- 2dL_0^2 \sin\theta\cos\theta(1+\epsilon_x)(1+\epsilon_y)\cos(\pi/2+\gamma_{xy}). \qquad \textbf{(6-5)}$$

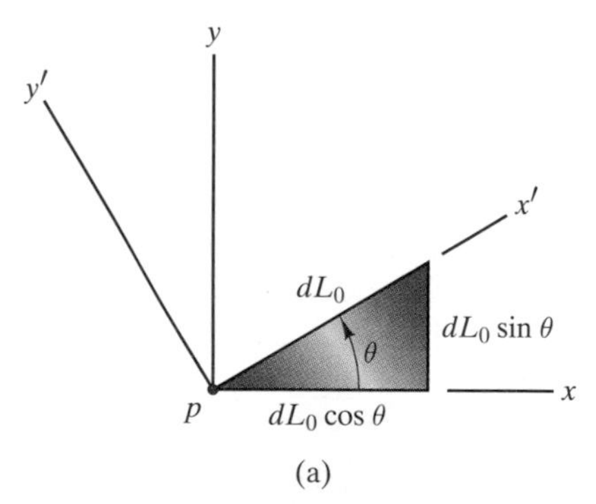

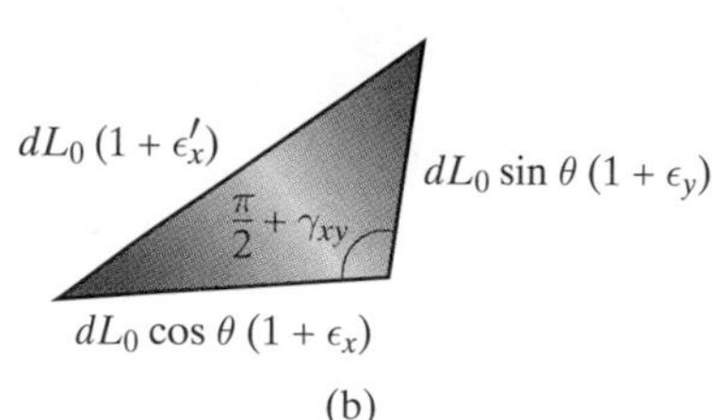

FIGURE 6-6 (a) Triangular element in the reference state. (b) Element in the deformed state.

We apply the identity

$$\cos(\pi/2+\gamma_{xy}) = \cos(\pi/2)\cos\gamma_{xy} - \sin(\pi/2)\sin\gamma_{xy}$$
$$= -\sin\gamma_{xy},$$

and since strains are assumed to be small, we can approximate $\sin\gamma_{xy}$ by γ_{xy}, obtaining

$$\cos(\pi/2+\gamma_{xy}) = -\gamma_{xy}.$$

Substituting this result into Eq. (6-5) and neglecting products of strains, it becomes

$$\epsilon'_x = \epsilon_x\cos^2\theta + \epsilon_y\sin^2\theta + \gamma_{xy}\sin\theta\cos\theta. \qquad \textbf{(6-6)}$$

By using the trigonometric identities

$$2\cos^2\theta = 1 + \cos 2\theta,$$
$$2\sin^2\theta = 1 - \cos 2\theta,$$
$$2\sin\theta\cos\theta = \sin 2\theta,$$

we can write Eq. (6-6) in the form

$$\epsilon'_x = \frac{\epsilon_x + \epsilon_y}{2} + \frac{\epsilon_x - \epsilon_y}{2}\cos 2\theta + \frac{\gamma_{xy}}{2}\sin 2\theta. \quad \textbf{(6-7)}$$

We can obtain an equation for ϵ'_y by setting θ equal to $\theta + 90°$ in this expression. The result is

$$\epsilon'_y = \frac{\epsilon_x + \epsilon_y}{2} - \frac{\epsilon_x - \epsilon_y}{2}\cos 2\theta - \frac{\gamma_{xy}}{2}\sin 2\theta. \quad \textbf{(6-8)}$$

We determine the shear strain γ'_{xy} by using Eq. (6-7). Instead of using this equation to express the extensional strain ϵ'_x in terms of the strains ϵ_x, ϵ_y, and γ_{xy}, we can reverse its role and use it to express the extensional strain ϵ_x in terms of the strains ϵ'_x, ϵ'_y, and γ'_{xy} simply by replacing θ by $-\theta$. The result is

$$\epsilon_x = \frac{\epsilon'_x + \epsilon'_y}{2} + \frac{\epsilon'_x - \epsilon'_y}{2}\cos 2\theta - \frac{\gamma'_{xy}}{2}\sin 2\theta.$$

Substituting Eqs. (6-7) and (6-8) into this equation, we can solve for γ'_{xy} in terms of ϵ_x, ϵ_y, and γ_{xy}:

$$\frac{\gamma'_{xy}}{2} = -\frac{\epsilon_x - \epsilon_y}{2}\sin 2\theta + \frac{\gamma_{xy}}{2}\cos 2\theta. \quad \textbf{(6-9)}$$

Compare Eqs. (6-7), (6-8), and (6-9) to the transformation equations for plane stress, Eqs. (5-7), (5-9), and (5-8). They are identical in form, with the normal stress replaced by the extensional strain and the shear stress replaced by one-half the shear strain. The state of stress and the state of strain, with the shear strains γ_{xy}, γ_{yz}, and γ_{xz} replaced by $\gamma_{xy}/2$, $\gamma_{yz}/2$, and $\gamma_{xz}/2$, are both quantities called *tensors*. Although a complete discussion of tensors is beyond our scope, these examples demonstrate how components of tensors transform between coordinate systems. From our present point of view, the similarities of these equations means that the analysis of strains follows the same path used for stresses, and you will therefore find the results quite familiar.

Although we have derived the strain transformation equations under the assumption of a state of plane strain, we will show in Section 6-4 that for many materials they also apply to the components of strain resulting from a state of plane stress. The strain component ϵ_z is not generally zero in plane stress, but that does not affect the derivations of Eqs. (6-7), (6-8), and (6-9).

EXAMPLE 6-1

The components of plane strain at point p of the material shown in Fig. 6-7 are $\epsilon_x = 0.003$, $\epsilon_y = 0.001$, and $\gamma_{xy} = -0.006$. Determine the components of plane strain in terms of the $x'y'$ coordinate system.

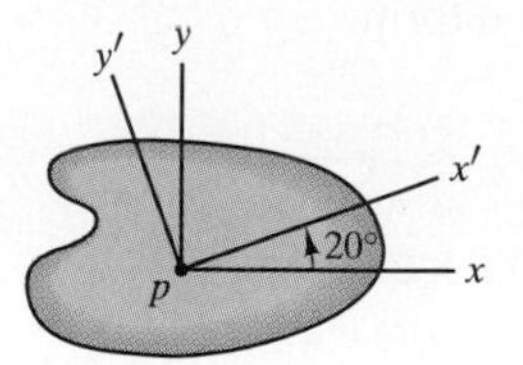

FIGURE 6-7

Strategy

We can use Eqs. (6-7)–(6-9) to determine the strain components ϵ_x', ϵ_y', and γ_{xy}'.

Solution

The components of plane strain in terms of the $x'y'$ coordinate system are

$$\begin{aligned}
\epsilon_x' &= \frac{\epsilon_x + \epsilon_y}{2} + \frac{\epsilon_x - \epsilon_y}{2}\cos 2\theta + \frac{\gamma_{xy}}{2}\sin 2\theta \\
&= \frac{0.003 + 0.001}{2} + \frac{0.003 - 0.001}{2}\cos 2(20^\circ) + \frac{-0.006}{2}\sin 2(20^\circ) \\
&= 0.00084,
\end{aligned}$$

$$\begin{aligned}
\epsilon_y' &= \frac{\epsilon_x + \epsilon_y}{2} - \frac{\epsilon_x - \epsilon_y}{2}\cos 2\theta - \frac{\gamma_{xy}}{2}\sin 2\theta \\
&= \frac{0.003 + 0.001}{2} - \frac{0.003 - 0.001}{2}\cos 2(20^\circ) - \frac{-0.006}{2}\sin 2(20^\circ) \\
&= 0.00316,
\end{aligned}$$

$$\begin{aligned}
\gamma_{xy}' &= 2\left(-\frac{\epsilon_x - \epsilon_y}{2}\sin 2\theta + \frac{\gamma_{xy}}{2}\cos 2\theta\right) \\
&= -(0.003 - 0.001)\sin 2(20^\circ) + (-0.006)\cos 2(20^\circ) \\
&= -0.00588.
\end{aligned}$$

Strain Gauge Rosette

Before continuing our discussion of the analysis of strains, we will describe an interesting and important application of the strain transformation equations. The term *strain gauge* refers to an instrument for measuring strains. The type we are concerned with here, called a *resistance strain gauge,* is based on the observation that the electrical resistance of a wire varies when the wire is subjected to axial strain (Fig. 6-8). Once the relationship between the strain of a given wire and its electrical resistance has been established experimentally (a procedure called *calibration*), the axial strain of the wire can be determined by measuring its resistance, which can be done very accurately.

To use the calibrated wire as a strain gauge, it is bonded to the surface of an unloaded specimen. When the specimen is loaded, the strain of the specimen in the direction of the wire is determined by measuring the wire's resistance. (The wire must be sufficiently thin that the force the strained wire exerts on the specimen is negligible.) The wire is typically arranged in the pattern shown in Fig. 6-9 to minimize the size of the gauge and so measure the strain within a relatively small neighborhood of a point.

Extensional strain can be measured by the type of strain gauge we have described, but how can shear strain be measured? This can be done in a clever way by a *strain gauge rosette,* which consists of three superimposed strain gauges measuring extensional strain in three directions, as shown in Fig. 6-11a. [The term *rosette* is said to derive from the resemblance between strain gauge rosettes mounted on colored felt and a small cloth ornament called a rosette that was worn on hats during the French Revolution (Fig. 6-10).] We introduce a coordinate system and denote the directions of the strain gauges by θ_a, θ_b, and θ_c (Fig. 6-11b). Using Eq. (6-6) to express the extensional strains in the directions of the three strain gauges in terms of the strain components ϵ_x, ϵ_y,

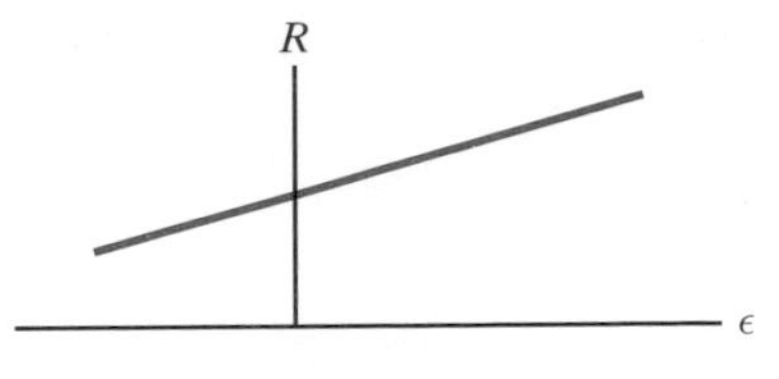

FIGURE 6-8 The electrical resistance R of a wire is a function of its axial strain ϵ.

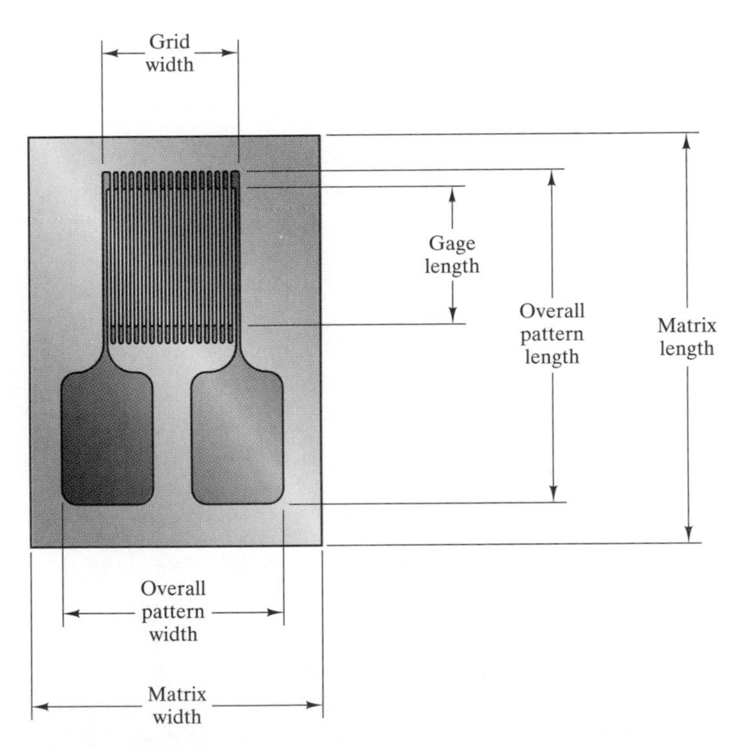

FIGURE 6-9 Typical resistance strain gauge.

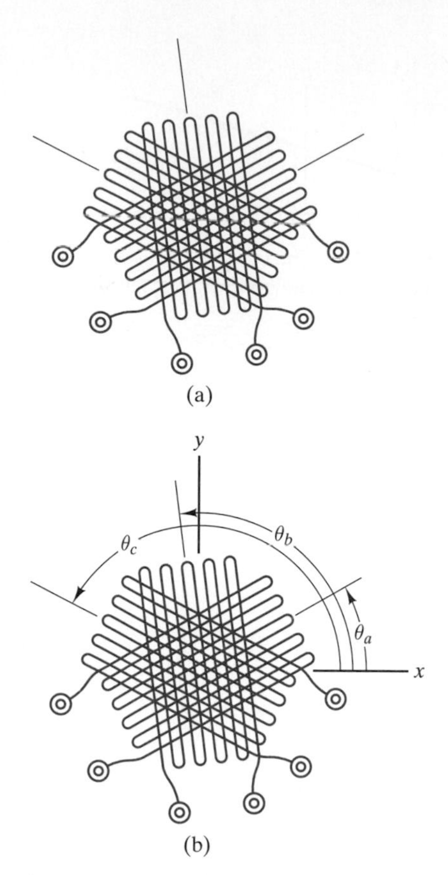

FIGURE 6-11 (a) Strain gauge rosette. (b) Introducing a coordinate system. The angles θ_a, θ_b, and θ_c specify the directions of the three strain gauges.

FIGURE 6-10 The original rosette.

and γ_{xy}, we obtain

$$\begin{aligned}\epsilon_a &= \epsilon_x \cos^2\theta_a + \epsilon_y \sin^2\theta_a + \gamma_{xy}\sin\theta_a\cos\theta_a,\\ \epsilon_b &= \epsilon_x \cos^2\theta_b + \epsilon_y \sin^2\theta_b + \gamma_{xy}\sin\theta_b\cos\theta_b,\\ \epsilon_c &= \epsilon_x \cos^2\theta_c + \epsilon_y \sin^2\theta_c + \gamma_{xy}\sin\theta_c\cos\theta_c.\end{aligned} \quad \textbf{(6-10)}$$

By measuring the extensional strains ϵ_a, ϵ_b, and ϵ_c, this system of equations can be solved for the strain components ϵ_x, ϵ_y, and γ_{xy}, thus determining the shear strain.

EXAMPLE 6-2

The strains measured by a strain gauge rosette oriented as shown in Fig. 6-12 are $\epsilon_a = 0.004$, $\epsilon_b = -0.003$, and $\epsilon_c = 0.002$. Determine the components of strain ϵ_x, ϵ_y, and γ_{xy}.

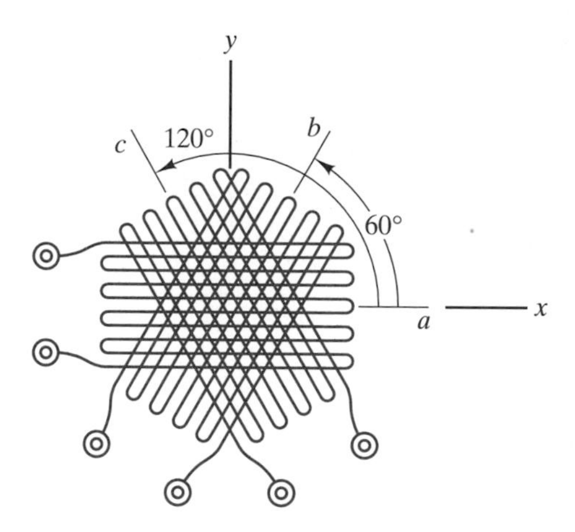

FIGURE 6-12

Strategy

The angles specifying the directions of the strain gauges relative to the x axis are $\theta_a = 0$, $\theta_b = 60^\circ$, and $\theta_c = 120^\circ$. By substituting these values and the values of the strains ϵ_a, ϵ_b, and ϵ_c into Eqs. (6-10), we can solve for ϵ_x, ϵ_y, and γ_{xy}.

Solution

Equations (6-10) are

$$0.004 = \epsilon_x,$$

$$-0.003 = \epsilon_x \cos^2 60^\circ + \epsilon_y \sin^2 60^\circ + \gamma_{xy} \sin 60^\circ \cos 60^\circ,$$

$$0.002 = \epsilon_x \cos^2 120^\circ + \epsilon_y \sin^2 120^\circ + \gamma_{xy} \sin 120^\circ \cos 120^\circ.$$

Solving, we obtain $\epsilon_x = 0.00400$, $\epsilon_y = -0.00200$, and $\gamma_{xy} = -0.00577$.

Maximum and Minimum Strains

Given a state of plane strain at a point p, Eqs. (6-7), (6-8), and (6-9) determine the strain components ϵ'_x, ϵ'_y, and γ'_{xy} corresponding to the coordinate system shown in Fig. 6-5 for any value of θ. In this section we consider the following questions: For what values of θ are the extensional strain ϵ'_x and shear strain γ'_{xy} a maximum or minimum, and what are their values? In other words, what are the maximum and minimum extensional and shear strains in the x–y plane?

Principal Strains

Let a value of θ for which ϵ'_x is a maximum or minimum be denoted by θ_p. By evaluating the derivative of Eq. (6-7) with respect to 2θ and setting it equal to zero, we obtain the equation

$$\tan 2\theta_p = \frac{\gamma_{xy}}{\epsilon_x - \epsilon_y}. \qquad \textbf{(6-11)}$$

When ϵ_x, ϵ_y, and γ_{xy} are known, you can solve this equation for θ_p and substitute it into Eq. (6-7) to determine the maximum or minimum value of ϵ'_x. Just as in the case of the maximum and minimum normal stresses, if $2\theta_p$ is a solution of Eq. (6-11), then so is $2\theta_p + 180^\circ$. This means that the extensional strain is a maximum or minimum in the direction θ_p and also in the direction $\theta_p + 90^\circ$, which means that *the maximum and minimum extensional strains occur in the x' and y' axis directions.* Once you have determined θ_p, you can determine the maximum and minimum extensional strains in the x–y plane, called the *principal strains* and denoted ϵ_1 and ϵ_2, from Eqs. (6-7) and (6-8). There are no extensional strains of greater magnitude in any direction at the point p.

To obtain analytical expressions for the values of the principal strains, we solve the equations

$$\frac{\sin 2\theta_p}{\cos 2\theta_p} = \tan 2\theta_p = \frac{\gamma_{xy}}{\epsilon_x - \epsilon_y} \qquad \textbf{(6-12)}$$

and

$$\sin^2 2\theta_p + \cos^2 2\theta_p = 1 \tag{6-13}$$

for $\sin 2\theta_p$ and $\cos 2\theta_p$ and substitute the results into Eq. (6-7), obtaining

$$\epsilon_1, \epsilon_2 = \frac{\epsilon_x + \epsilon_y}{2} \pm \sqrt{\left(\frac{\epsilon_x - \epsilon_y}{2}\right)^2 + \left(\frac{\gamma_{xy}}{2}\right)^2}. \tag{6-14}$$

By substituting the expressions for $\sin 2\theta_p$ and $\cos 2\theta_p$ into Eq. (6-9), we find that the value of the shear strain γ'_{xy} at $\theta = \theta_p$ is zero. The physical interpretation of this result is interesting: An infinitesimal square element oriented as shown in Fig. 6-13a is subjected to the principal strains in the x' and y' directions and undergoes no shear strain. The element is rectangular in the deformed state (Fig. 6-13b).

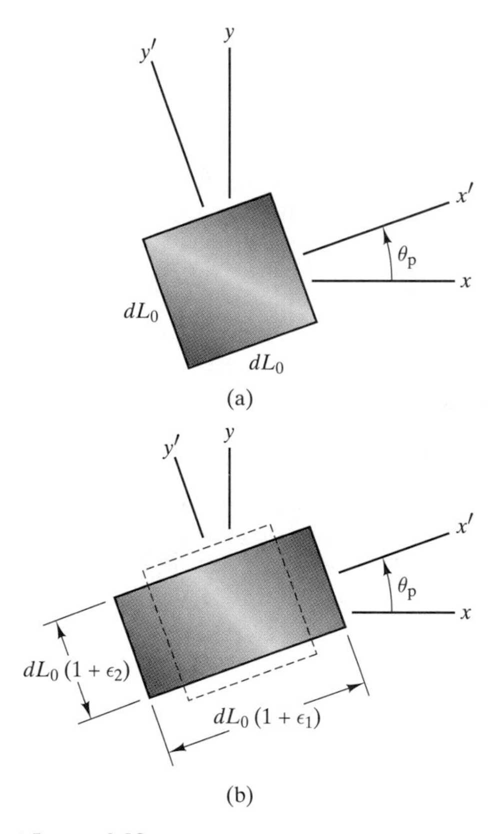

FIGURE 6-13 (a) Reference state of a square element aligned with the directions of the principal strains. (b) The deformed element undergoes no shear strain. (The strains are exaggerated. The largest principal strain may occur in either the x' or the y' direction.)

Maximum Shear Strains

Let a value of θ for which γ'_{xy} is a maximum or minimum be denoted by θ_s. Evaluating the derivative of Eq. (6-9) with respect to 2θ and setting it equal to zero, we obtain the equation

$$\tan 2\theta_s = -\frac{\epsilon_x - \epsilon_y}{\gamma_{xy}}. \qquad \textbf{(6-15)}$$

With this equation we can determine θ_s and substitute it into Eq. (6-9) to obtain the maximum or minimum value of γ'_{xy}. If θ_s is a solution of Eq. (6-15), then so are $\theta_s + 90°$, $\theta_s + 2(90°)$, ... As we illustrate in Fig. 6-14, *the maximum and minimum shear strains describe the shear strain of a rectangular element.* Furthermore, because $\tan 2\theta_s$ is the negative inverse of $\tan 2\theta_p$, this element is rotated 45° relative to the element which is subjected to the maximum and minimum extensional strains. We can demonstrate why this is the case by superimposing these two elements (Fig. 6-15).

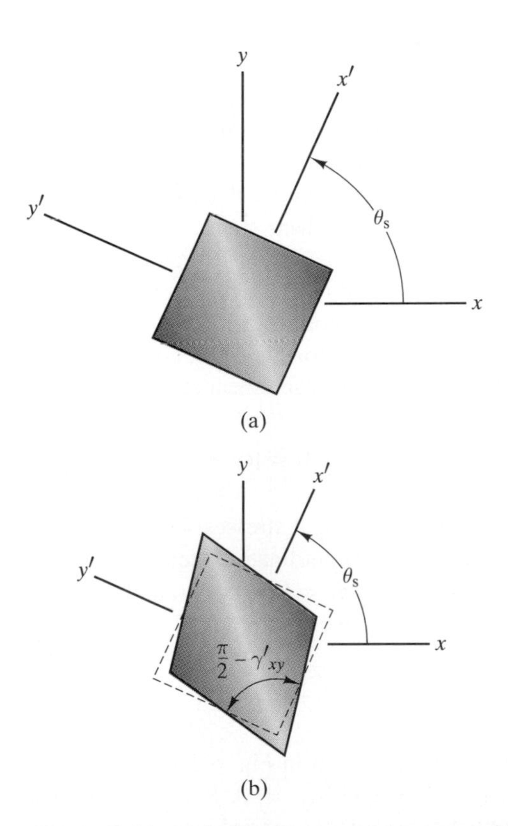

FIGURE 6-14 (a) Reference state of the square element that is subjected to the maximum and minimum shear strains. (b) Deformed element. (The shear strain is exaggerated. The strain γ'_{xy} may be either positive or negative.)

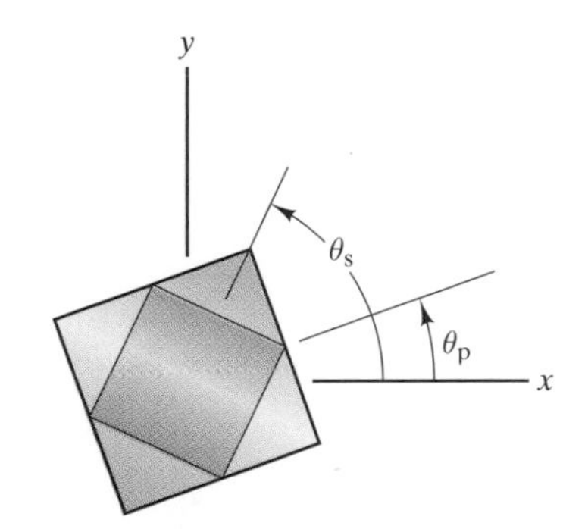

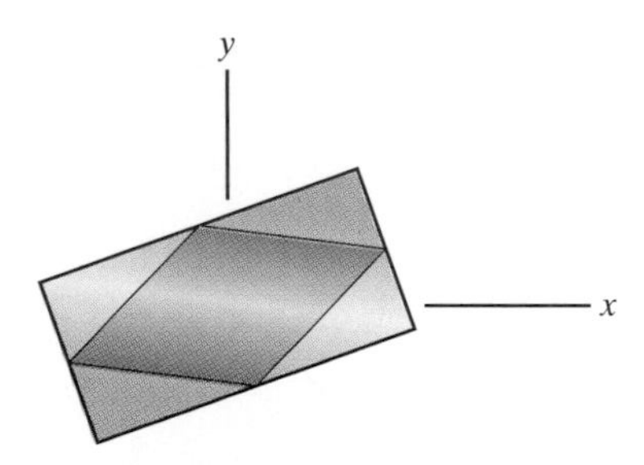

FIGURE 6-15 Element subjected to the greatest shear strain superimposed onto the element subjected to the principal strains.

To obtain an analytical expression for the maximum magnitude of the shear strain, we solve the equations

$$\frac{\sin 2\theta_s}{\cos 2\theta_s} = \tan 2\theta_s = -\frac{\epsilon_x - \epsilon_y}{\gamma_{xy}} \tag{6-16}$$

and

$$\sin^2 2\theta_s + \cos^2 2\theta_s = 1 \tag{6-17}$$

for $\sin 2\theta_s$ and $\cos 2\theta_s$ and substitute the results into Eq. (6-9). The result is

$$\gamma_{max} = \sqrt{(\epsilon_x - \epsilon_y)^2 + \gamma_{xy}^2}. \tag{6-18}$$

This is called the *maximum in-plane shear strain,* because it is the maximum shear strain in the x–y plane.

By substituting our results for $\sin 2\theta_s$ and $\cos 2\theta_s$ into Eqs. (6-7) and (6-8), we obtain

$$\epsilon'_x = \epsilon'_y = \frac{\epsilon_x + \epsilon_y}{2}.$$

The element that is subjected to the maximum and minimum shear strains (Fig. 6-14) is subjected to equal extensional strains in the x' and y' directions.

We can obtain expressions for the absolute maximum shear strain the same way that we determined the absolute maximum shear stress. We begin with the element that is subjected to the principal strains (Fig. 6-13) and realign the coordinate system with the faces of the element (Fig. 6-16a). In terms of this new coordinate system, the components of plane strain are $\epsilon_x = \epsilon_1, \epsilon_y = \epsilon_2$, and $\gamma_{xy} = 0$. Substituting these components into Eq. (6-18), we obtain a different expression for the magnitude of the maximum in-plane shear strain:

$$\sqrt{(\epsilon_x - \epsilon_y)^2 + \gamma_{xy}^2} = \sqrt{(\epsilon_1 - \epsilon_2)^2 + 0} = |\epsilon_1 - \epsilon_2|. \tag{6-19}$$

This equation is expressed in terms of the principal stresses, but it gives the same result as Eq. (6-18). Now we reorient the coordinate system as shown in Fig. 6-16b. In terms of this coordinate system, $\epsilon_x = 0$, $\epsilon_y = \epsilon_1$, and $\gamma_{xy} = 0$. Substituting these components into Eq. (6-18), we obtain

$$\sqrt{(\epsilon_x - \epsilon_y)^2 + \gamma_{xy}^2} = \sqrt{(0 - \epsilon_1)^2 + 0} = |\epsilon_1|. \tag{6-20}$$

Next, we reorient the coordinate system as shown in Fig. 6-16c. In terms of this coordinate system, $\epsilon_x = 0$, $\epsilon_y = \epsilon_2$, and $\gamma_{xy} = 0$. Substituting these components into Eq. (6-18), we obtain

$$\sqrt{(\epsilon_x - \epsilon_y)^2 + \gamma_{xy}^2} = \sqrt{(0 - \epsilon_2)^2 + 0} = |\epsilon_2|. \tag{6-21}$$

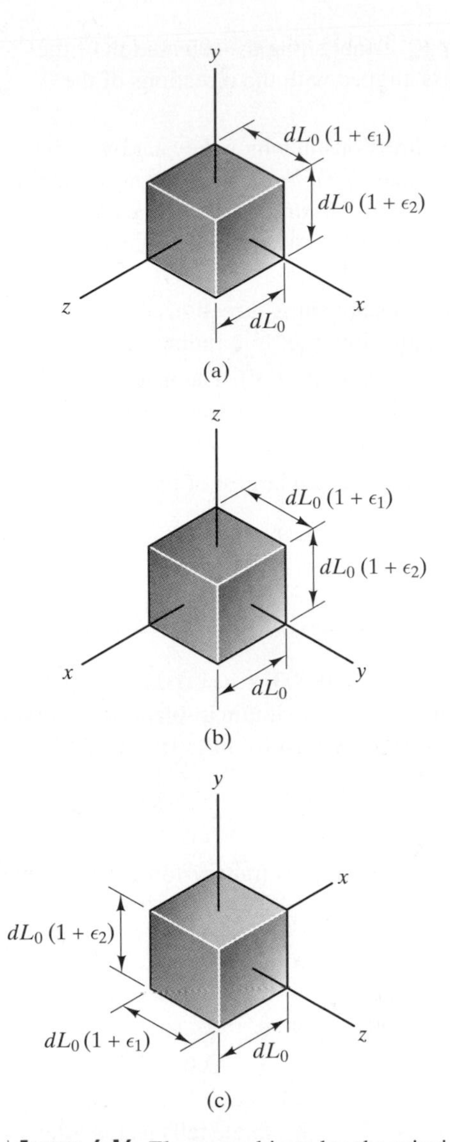

FIGURE 6-16 Element subjected to the principal strains. (a) Realigned coordinate system. (b), (c) Other orientations of the coordinate system.

Depending on the values of the principal strains, Eq. (6-20) and/or Eq. (6-21) can result in larger values of the magnitude of the maximum shear strain than the maximum in-plane shear strain. The absolute maximum shear strain is the largest value given by Eq. (6-19), (6-20), or (6-21). No greater shear strain occurs for any orientation of the coordinate system.

SUMMARY: DETERMINING THE PRINCIPAL STRAINS AND THE MAXIMUM SHEAR STRAIN

Here we give a sequence of steps you can use to determine the principal strains and the maximum shear strain for a given state of plane strain.

1. Use Eq. (6-11) to determine θ_p, establishing the orientation of the $x'y'$ coordinate system that is aligned with the directions of the principal strains.
2. Determine ϵ_1 and ϵ_2 and the directions in which they act by substituting θ_p into Eqs. (6-7) and (6-8). You can also determine the values of ϵ_1 and ϵ_2 from Eq. (6-14), but this equation does not tell you which principal strain acts in the x' direction and which one acts in the y' direction.
3. Use Eq. (6-15) to determine θ_s, establishing directions of the x' and y' axes for which γ'_{xy} is a maximum or minimum.
4. Determine γ'_{xy} by substituting θ_s into Eq. (6-9). The magnitude of this shear strain is the maximum in-plane shear strain. You can also determine the maximum in-plane shear strain from Eq. (6-18).
5. The absolute maximum shear strain is the largest of the three values

$$|\epsilon_1 - \epsilon_2|, \qquad |\epsilon_1|, \qquad |\epsilon_2|. \tag{6-22}$$

EXAMPLE 6-3

The state of plane strain at a point p is $\epsilon_x = 0.003$, $\epsilon_y = 0.001$, and $\gamma_{xy} = -0.006$. Determine the principal strains and the maximum in-plane shear strain and show the orientations of the elements subjected to these strains. Also determine the absolute maximum shear strain.

Strategy

We can follow the steps given in the preceeding summary to determine the principal strains and the maximum shear strain.

Solution

Step 1 From Eq. (6-11),

$$\tan 2\theta_p = \frac{\gamma_{xy}}{\epsilon_x - \epsilon_y} = \frac{-0.006}{0.003 - 0.001} = -3.0.$$

Solving this equation, we obtain $\theta_p = -35.78°$. This angle tells us the orientation of the $x'y'$ coordinate system aligned with the principal strains.

Step 2 We substitute θ_p into Eqs. (6-7) and (6-8) to determine the principal strains.

$$\begin{aligned}
\epsilon'_x &= \frac{\epsilon_x + \epsilon_y}{2} + \frac{\epsilon_x - \epsilon_y}{2}\cos 2\theta_p + \frac{\gamma_{xy}}{2}\sin 2\theta_p \\
&= \frac{0.003 + 0.001}{2} + \frac{0.003 - 0.001}{2}\cos 2(-35.78°) \\
&\quad + \frac{-0.006}{2}\sin 2(-35.78°) \\
&= 0.00516,
\end{aligned}$$

$$\epsilon'_y = \frac{\epsilon_x + \epsilon_y}{2} - \frac{\epsilon_x - \epsilon_y}{2}\cos 2\theta_p - \frac{\gamma_{xy}}{2}\sin 2\theta_p$$

$$= \frac{0.003 + 0.001}{2} - \frac{0.003 - 0.001}{2}\cos 2(-35.78^\circ)$$

$$- \frac{-0.006}{2}\sin 2(-35.78^\circ)$$

$$= -\,0.00116.$$

The principal strains are $\epsilon_1 = 0.00516$ and $\epsilon_2 = -0.00116$. They are shown on the properly oriented element in Fig. (a). This element is subjected to no shear strain.

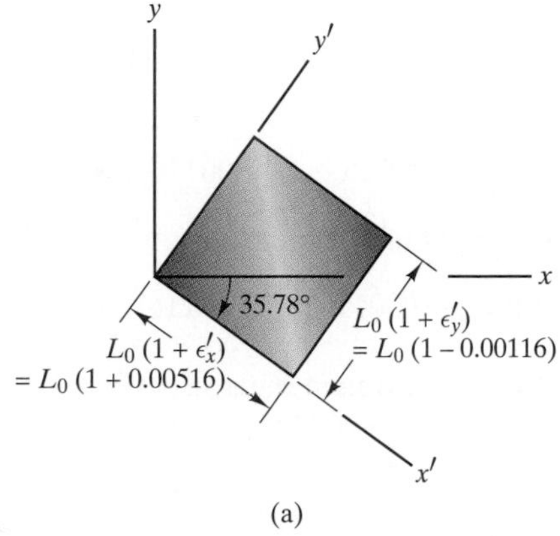

(a)

(a) The principal strains. L_0 is the dimension of the square element in the reference state.

Step 3 From Eq. (6-15),

$$\tan 2\theta_s = -\frac{\epsilon_x - \epsilon_y}{\gamma_{xy}} = -\frac{0.003 - 0.001}{-0.006} = 0.333,$$

from which we obtain $\theta_s = 9.22^\circ$.

Step 4 We substitute θ_s into Eq. (6-9) to determine the maximum in-plane shear strain.

$$\frac{\gamma'_{xy}}{2} = -\frac{\epsilon_x - \epsilon_y}{2}\sin 2\theta_s + \frac{\gamma_{xy}}{2}\cos 2\theta_s$$

$$= -\frac{0.003 - 0.001}{2}\sin 2(9.22^\circ) + \frac{-0.006}{2}\cos 2(9.22^\circ)$$

$$= -0.00316.$$

We see that $\gamma'_{xy} = -0.00632$. The maximum in-plane shear strain is shown on the properly oriented element in Fig. (b).

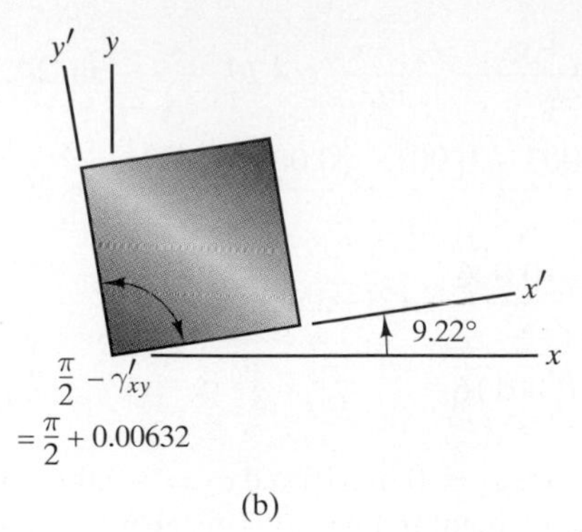

(b)

(b) Maximum in-plane shear strain. Notice that γ'_{xy} is negative.

Step 5 The absolute maximum shear strain is given by the largest of the three values

$$|\epsilon_1 - \epsilon_2| = |0.00516 - (-0.00116)| = 0.00632,$$

$$|\epsilon_1| = |0.00516| = 0.00516,$$

$$|\epsilon_2| = |-0.00116| = 0.00116.$$

In this example the absolute maximum shear strain equals the magnitude of the maximum in-plane shear strain, 0.00632.

Discussion

Observe that the element subjected to the maximum in-plain shear strain, Fig. (b), is rotated 45° relative to the element that is subjected to the principal strains, Fig. (a).

6-3 | Mohr's Circle for Plane Strain

Because the equations for transforming strains between coordinate systems are so similar to the corresponding equations for stresses, we can apply Mohr's circle to strains in the same way we applied it to stresses. Recall that Eqs. (6-7), (6-8), and (6-9),

$$\begin{aligned} \epsilon'_x &= \frac{\epsilon_x + \epsilon_y}{2} + \frac{\epsilon_x - \epsilon_y}{2}\cos 2\theta + \frac{\gamma_{xy}}{2}\sin 2\theta, \\ \epsilon'_y &= \frac{\epsilon_x + \epsilon_y}{2} - \frac{\epsilon_x - \epsilon_y}{2}\cos 2\theta - \frac{\gamma_{xy}}{2}\sin 2\theta, \\ \frac{\gamma'_{xy}}{2} &= -\frac{\epsilon_x - \epsilon_y}{2}\sin 2\theta + \frac{\gamma_{xy}}{2}\cos 2\theta, \end{aligned} \qquad \textbf{(6-23)}$$

are identical to Eqs. (5-7), (5-9), and (5-8) for plane stress, with the normal stresses replaced by the extensional strains and the shear stress replaced by one-half the shear strain. Accounting for the factor of $\frac{1}{2}$ multiplying the shear strain, we can construct Mohr's circle for strain in exactly the same way that we did for stresses.

Constructing the Circle

Suppose that we know the components of plane strain ϵ_x, ϵ_y, and γ_{xy} at a point p. Mohr's circle allows us to solve graphically for the components ϵ'_x, ϵ'_y, and γ'_{xy} for a given angle θ. This involves four steps:

1. Establish a set of horizontal and vertical axes with extensional strain measured along the horizontal axis and one-half the shear strain measured along the vertical axis (Fig. 6-17a). Positive extensional strain is measured to the right and positive shear strain is measured *downward.*
2. Plot two points, point P with coordinates $(\epsilon_x, \gamma_{xy}/2)$ and point Q with coordinates $(\epsilon_y, -\gamma_{xy}/2)$, as shown in Fig. 6-17b.
3. Draw a straight line connecting points P and Q. Using the intersection of the straight line with the horizontal axis as the center, draw a circle that passes through the two points (Fig. 6-17c).

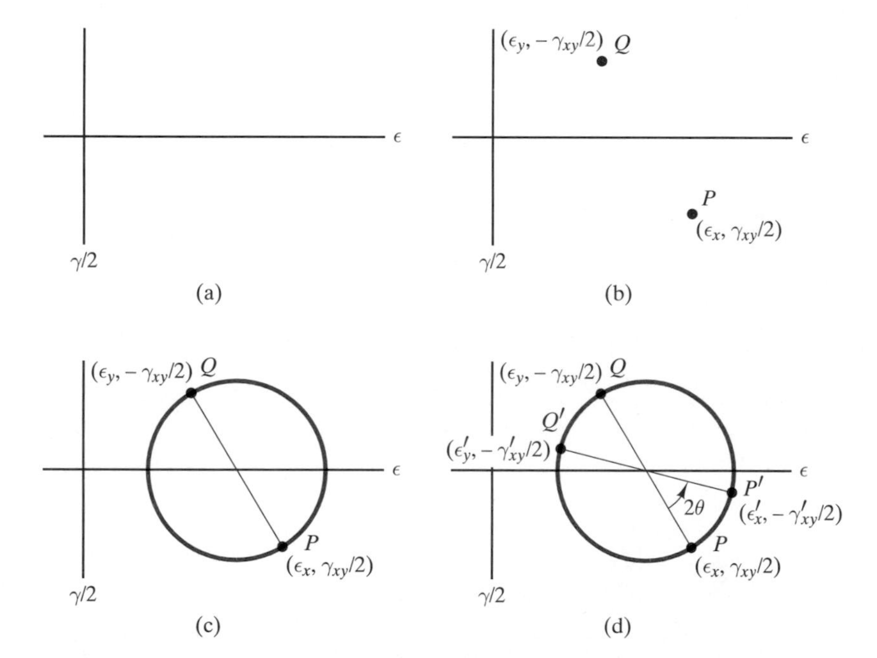

FIGURE 6-17 (a) Establishing the axes. Shear strain is positive downward. (b) Plotting points P and Q. (c) Drawing Mohr's circle. The center of the circle is the intersection of the line between points P and Q with the horizontal axis. (d) Determining the strains.

4. Draw a straight line through the center of the circle at an angle 2θ measured counterclockwise from point P (Fig. 6-17d). The point P' at which this line intersects the circle has coordinates $(\epsilon'_x, \gamma'_{xy}/2)$, and the point Q' has coordinates $(\epsilon'_y, -\gamma'_{xy}/2)$.

Determining Principal Strains and the Maximum In-Plane Shear Strain

Once we have constructed Mohr's circle, we can immediately see the values of the maximum and minimum extensional strains and the magnitude of the maximum in-plane shear strain. The coordinates of the points where the circle intersects the horizontal axis determine the two principal strains (Fig. 6-18). The coordinates of the points at the bottom and top of the circle determine the maximum and minimum in-plane shear strains, so the radius of the circle equals one-half the magnitude of the maximum in-plane shear strain.

We can also use Mohr's circle to determine the orientations of the elements subjected to the principal strains and the maximum in-plane shear strain. If we let the point P' coincide with either principal strain (Fig. 6-19a), we can measure the angle $2\theta_p$ and thereby determine the orientation of the x' and y' directions in which the principal strains occur (Fig. 6-19b).

We can then let P' coincide with either the maximum or minimum shear strain (Fig. 6-20a) and measure the angle $2\theta_s$, determining the orientation of the x'–y' axes associated with that shear strain (Fig. 6-20b). Notice that in the example illustrated in Fig. 6-20, point P' coincides with the minimum shear strain, so $\gamma'_{xy} = -\gamma_{max}$.

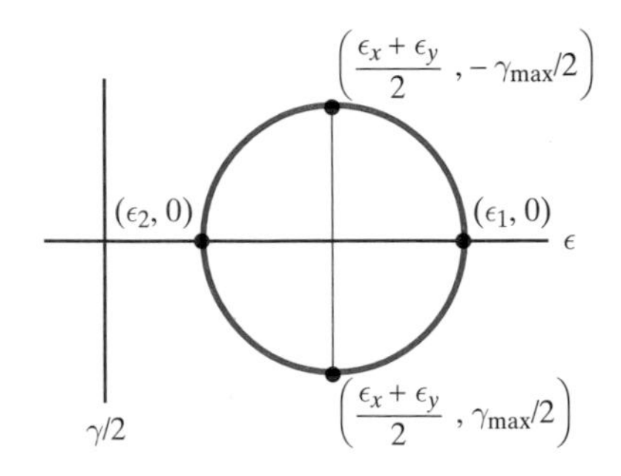

FIGURE 6-18 Mohr's circle indicates the values of the principal strains and the magnitude of the maximum in-plane shear strain.

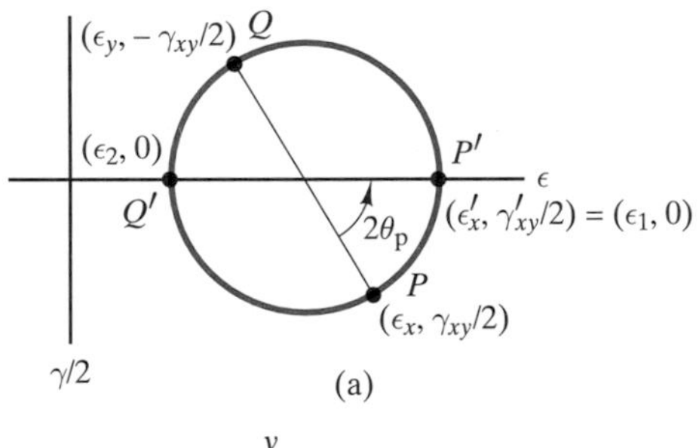

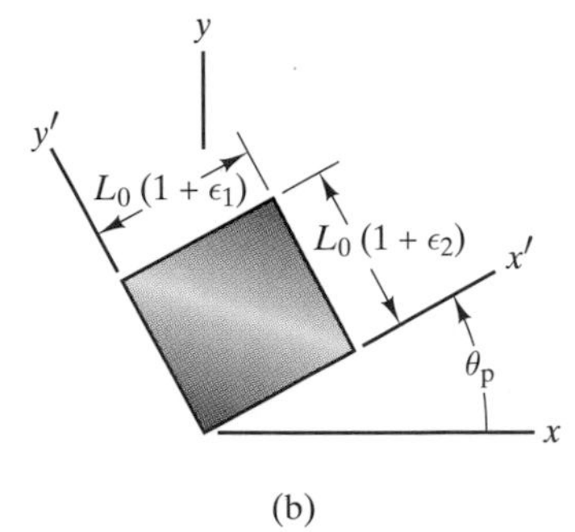

FIGURE 6-19 Using Mohr's circle to determine the orientation of the element subjected to the principal strains.

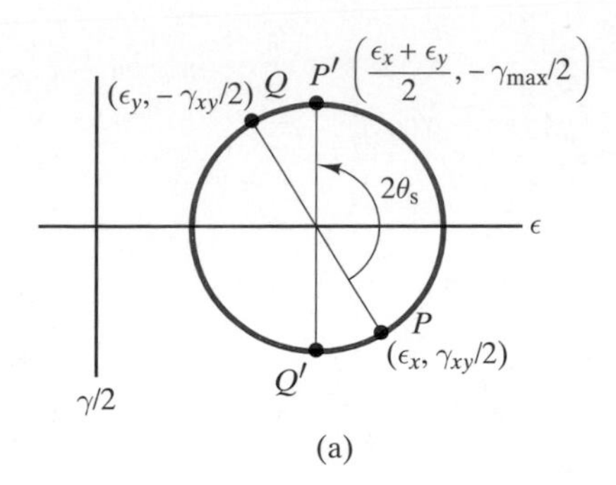

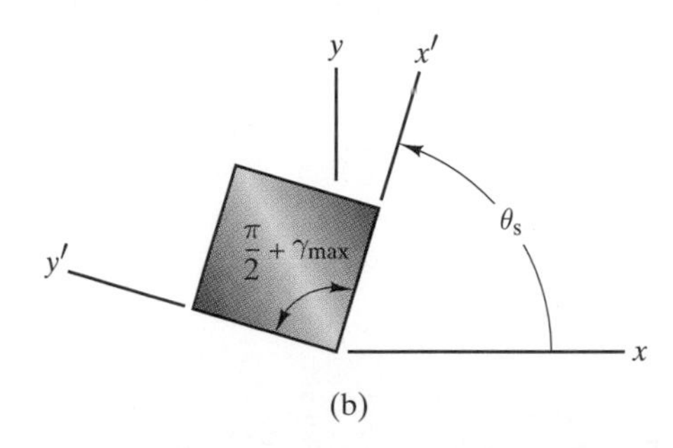

FIGURE 6-20 Using Mohr's circle to determine the orientation of the element subjected to the maximum in-plane shear strain.

EXAMPLE 6-4

The components of plane strain at a point p are $\epsilon_x = 0.003$, $\epsilon_y = 0.001$, and $\gamma_{xy} = -0.006$. Use Mohr's circle to determine the principal strains and the maximum in-plane shear strain and show the orientations of the elements subjected to these strains.

Strategy

By letting the point P' of Mohr's circle coincide first with one of the principal strains and then with the maximum or minimum in-plane shear strain, we can determine the values of these strains and the orientations of the elements subjected to them.

Solution

We first plot points P and Q and draw Mohr's circle [Fig. (a)]. Then we let the point P' coincide with one of the principal strains. In this case we choose the minimum principal strain ϵ_2 [Fig. (b)]. From the circle we estimate that $\epsilon_1 = 0.0052$ and $\epsilon_2 = -0.0012$. Measuring the angle $2\theta_p$, we estimate that $\theta_p = 54.5°$, which determines the orientation of the element subjected to the principal strains [Fig. (c)].

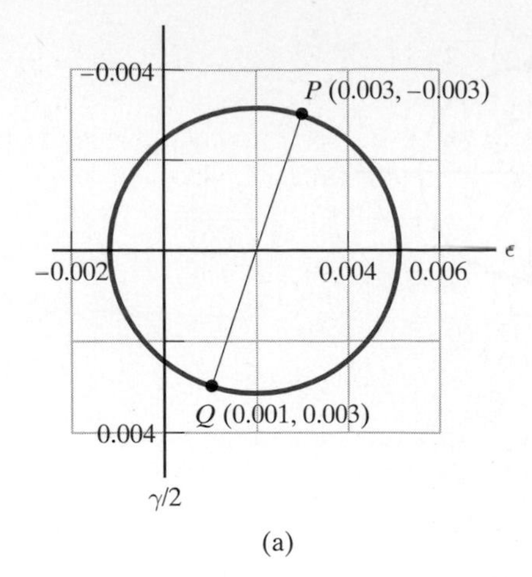

(a) Mohr's circle.

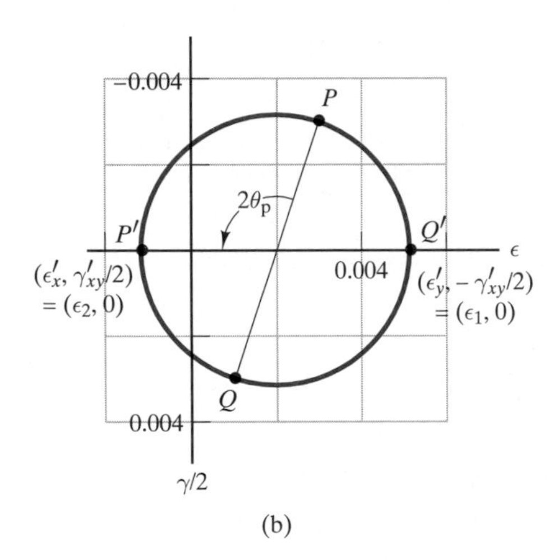

(b) Letting P' coincide with the principal strain ϵ_2.

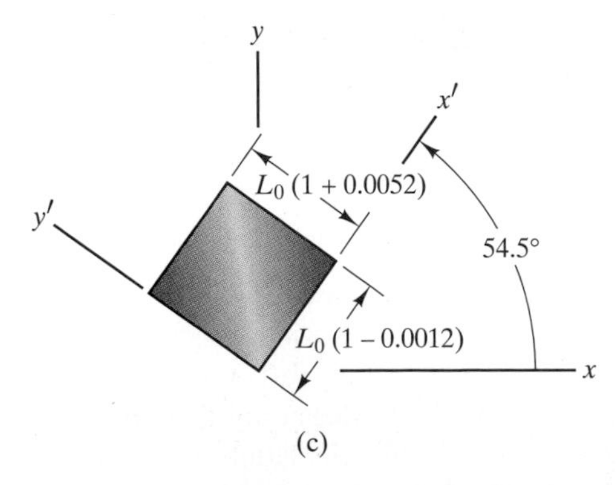

(c) Element subjected to the principal strains.

Letting the point P' coincide with the minimum in-plane shear strain [Fig. (d)], we estimate that $\gamma_{max} = 0.0062$. Measuring the angle $2\theta_s$, we estimate that $\theta_s = 9.5°$, which determines the orientation of the element subjected to the maximum and minimum in-plane shear strains [Fig. (e)].

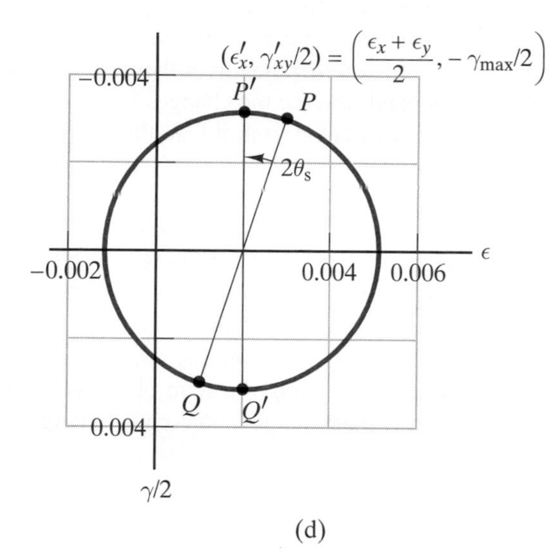

(d)

(d) Letting P' coincide with the minimum in-plane shear strain.

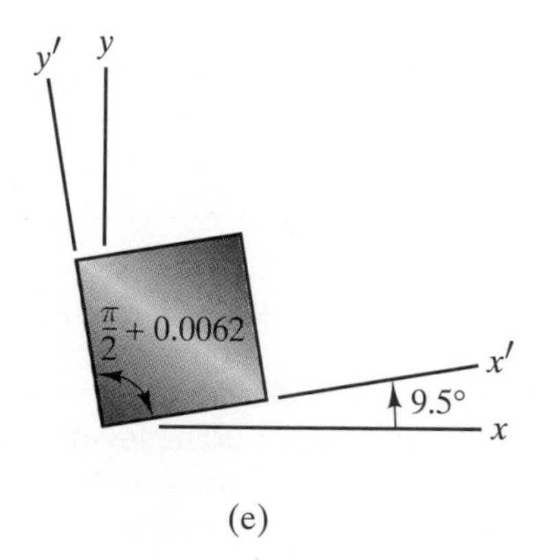

(e)

(e) Element subjected to the maximum and minimum in-plane shear strains. Notice that $\gamma_{xy}' = -\gamma_{max}$.

Discussion

Compare this solution to Example 6-3, in which we begin with the same state of strain and analytically determine the principal strains and maximum in-plane shear strain.

6-4 Stress-Strain Relations

We have discussed the state of stress, which is related to the internal forces at a material point, and the state of strain, related to the deformation in the neighborhood of a material point. Is there a relationship between them? Consider the example of a crystalline material, such as iron, that can be modeled as a lattice of atoms connected by bonds that behave like springs. When such a material is deformed, the distances between atoms change and the "springs" become stretched or compressed, altering the forces they exert. Since the internal forces near a point will depend only on the changes in the distances between atoms near that point, it is reasonable to postulate that the state of stress depends only on the state of strain. This is a very simple conceptual model of material behavior, and in general the state of stress at a point of a material can depend on the temperature as well as the history of the state of strain. A material for which the state of stress at a point is a single-valued function of the current state of strain at that point is said to be *elastic*. Some materials, including most metals subjected to limited ranges of temperature and stress, can be modeled adequately by assuming them to be elastic. In this section we derive the equations relating the state of stress in an elastic material to its state of strain for linearly elastic materials having the property of isotropy.

Linearly Elastic Materials

An elastic material is one for which the state of stress at a point is a function only of the current state of strain at that point. We can express each component of stress as a function of the components of strain:

$$\begin{aligned}
\sigma_x &= \sigma_x(\epsilon_x, \epsilon_y, \epsilon_z, \gamma_{xy}, \gamma_{yz}, \gamma_{xz}),\\
\sigma_y &= \sigma_y(\epsilon_x, \epsilon_y, \epsilon_z, \gamma_{xy}, \gamma_{yz}, \gamma_{xz}),\\
\sigma_z &= \sigma_z(\epsilon_x, \epsilon_y, \epsilon_z, \gamma_{xy}, \gamma_{yz}, \gamma_{xz}),\\
\tau_{xy} &= \tau_{xy}(\epsilon_x, \epsilon_y, \epsilon_z, \gamma_{xy}, \gamma_{yz}, \gamma_{xz}),\\
\tau_{yz} &= \tau_{yz}(\epsilon_x, \epsilon_y, \epsilon_z, \gamma_{xy}, \gamma_{yz}, \gamma_{xz}),\\
\tau_{xz} &= \tau_{xz}(\epsilon_x, \epsilon_y, \epsilon_z, \gamma_{xy}, \gamma_{yz}, \gamma_{xz}).
\end{aligned}$$

These *stress-strain relations,* which determine the state of stress in a given material in terms of its state of strain, are examples of what are called *constitutive equations*. They are functions that depend on the constitution, or physical structure, of a material. Let us express the equation for σ_x as a power series in terms of the components of strain,

$$\begin{aligned}
\sigma_x = a_{10} &+ a_{11}\epsilon_x + a_{12}\epsilon_y + a_{13}\epsilon_z + a_{14}\gamma_{xy} + a_{15}\gamma_{yz} + a_{16}\gamma_{xz}\\
&+ a_{17}\epsilon_x^2 + a_{18}\epsilon_x\epsilon_y + \cdots,
\end{aligned}$$

where the coefficients $a_{10}, a_{11}, \ldots$ are constants. If we assume that the stress σ_x is zero when the components of strain are zero, the coefficient $a_{10} = 0$. If we also assume that the components of strain are sufficiently small that products of the components are negligible in comparison to the components themselves, we obtain

$$\sigma_x = a_{11}\epsilon_x + a_{12}\epsilon_y + a_{13}\epsilon_z + a_{14}\gamma_{xy} + a_{15}\gamma_{yz} + a_{16}\gamma_{xz}.$$

Expressing each component of stress in this way, we obtain the equations

$$\begin{aligned}
\sigma_x &= a_{11}\epsilon_x + a_{12}\epsilon_y + a_{13}\epsilon_z + a_{14}\gamma_{xy} + a_{15}\gamma_{yz} + a_{16}\gamma_{xz}, \\
\sigma_y &= a_{21}\epsilon_x + a_{22}\epsilon_y + a_{23}\epsilon_z + a_{24}\gamma_{xy} + a_{25}\gamma_{yz} + a_{26}\gamma_{xz}, \\
\sigma_z &= a_{31}\epsilon_x + a_{32}\epsilon_y + a_{33}\epsilon_z + a_{34}\gamma_{xy} + a_{35}\gamma_{yz} + a_{36}\gamma_{xz}, \\
\tau_{xy} &= a_{41}\epsilon_x + a_{42}\epsilon_y + a_{43}\epsilon_z + a_{44}\gamma_{xy} + a_{45}\gamma_{yz} + a_{46}\gamma_{xz}, \\
\tau_{yz} &= a_{51}\epsilon_x + a_{52}\epsilon_y + a_{53}\epsilon_z + a_{54}\gamma_{xy} + a_{55}\gamma_{yz} + a_{56}\gamma_{xz}, \\
\tau_{xz} &= a_{61}\epsilon_x + a_{62}\epsilon_y + a_{63}\epsilon_z + a_{64}\gamma_{xy} + a_{65}\gamma_{yz} + a_{66}\gamma_{xz}.
\end{aligned} \qquad \textbf{(6-24)}$$

An elastic material that can be modeled by these stress-strain relations is said to be *linearly elastic*. The components of stress are linear functions of the components of strain. To model a given material, the 36 constants $a_{11}, a_{12}, \ldots$ must be known. At this juncture we are apparently faced with the daunting prospect of performing 36 independent experiments to determine the stress-strain relations of a single material. But we will show in the following section that far fewer than 36 constants need to be determined to characterize most linearly elastic materials.

Isotropic Materials

The simplest way to explain what is meant by an isotropic material is to consider a familiar material that is not isotropic—wood. If we apply a normal stress σ to opposite faces of a cube of wood and measure the resulting extensional strain ϵ, it is clear that we obtain one result if the grain of the wood is parallel to the direction of the strain ϵ (Fig. 6-21a) and a different result if the grain of the wood is perpendicular to the direction of the strain ϵ (Fig. 6-21b). The behavior of the wood depends on the direction of its grain. The behavior of a material that is not isotropic, or *anisotropic*, depends on the orientation of the material. The behavior of a material that is *isotropic* (which, roughly translated from

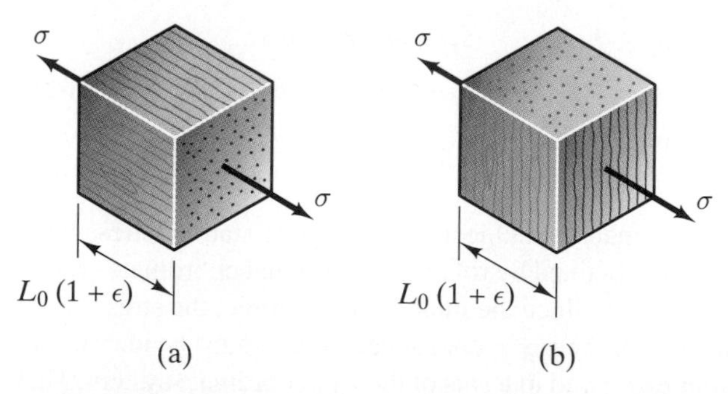

FIGURE 6-21 Stretching a cube of wood: (a) parallel to the grain; (b) perpendicular to the grain.

Latin, means "the same in all directions") does not depend on the orientation of the material. Its stress-strain relations are the same for any orientation of the material. Many materials used in engineering are approximately isotropic, although the use of intentionally created anisotropic materials, such as fiber-reinforced and layered composite materials, is increasing.

Another way to state the definition of an isotropic material is that *the stress-strain relations are the same for any orientation of the coordinate system relative to the material.* In other words, instead of requiring the material properties to be the same for different orientations of the material, we require the material properties to be the same for different orientations of the frame of reference relative to the material. Using this definition, we now investigate how material isotropy affects the stress-strain relations of a linearly elastic material.

Isotropic Stress-Strain Relations

If we regard the stress-strain equations (6-24) as six linear algebraic equations for the components of strain in terms of the components of stress, we can in principle invert them to obtain linear equations for the components of strain in terms of the components of stress. We write the resulting equations as

$$
\begin{aligned}
\epsilon_x &= b_{11}\sigma_x + b_{12}\sigma_y + b_{13}\sigma_z + b_{14}\tau_{xy} + b_{15}\tau_{yz} + b_{16}\tau_{xz}, \\
\epsilon_y &= b_{21}\sigma_x + b_{22}\sigma_y + b_{23}\sigma_z + b_{24}\tau_{xy} + b_{25}\tau_{yz} + b_{26}\tau_{xz}, \\
\epsilon_z &= b_{31}\sigma_x + b_{32}\sigma_y + b_{33}\sigma_z + b_{34}\tau_{xy} + b_{35}\tau_{yz} + b_{36}\tau_{xz}, \\
\gamma_{xy} &= b_{41}\sigma_x + b_{42}\sigma_y + b_{43}\sigma_z + b_{44}\tau_{xy} + b_{45}\tau_{yz} + b_{46}\tau_{xz}, \\
\gamma_{yz} &= b_{51}\sigma_x + b_{52}\sigma_y + b_{53}\sigma_z + b_{54}\tau_{xy} + b_{55}\tau_{yz} + b_{56}\tau_{xz}, \\
\gamma_{xz} &= b_{61}\sigma_x + b_{62}\sigma_y + b_{63}\sigma_z + b_{64}\tau_{xy} + b_{65}\tau_{yz} + b_{66}\tau_{xz}.
\end{aligned}
\tag{6-25}
$$

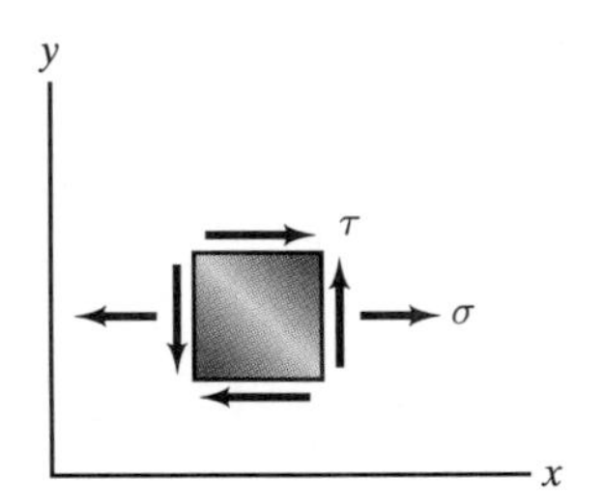

FIGURE 6-22 Applying normal and shear stresses to an isotropic material.

Isotropy places severe restrictions on the possible values of the constants b_{11}, $b_{12}, \ldots$. Suppose that we subject an isotropic material to a normal stress σ and shear stress τ (Fig. 6-22). In terms of the coordinate system shown, the only nonzero stress components are $\sigma_x = \sigma$ and $\tau_{xy} = \tau$. Therefore, Eqs. (6-25) are

$$
\begin{aligned}
\epsilon_x &= b_{11}\sigma + b_{14}\tau, & \gamma_{xy} &= b_{41}\sigma + b_{44}\tau, \\
\epsilon_y &= b_{21}\sigma + b_{24}\tau, & \gamma_{yz} &= b_{51}\sigma + b_{54}\tau, \\
\epsilon_z &= b_{31}\sigma + b_{34}\tau, & \gamma_{xz} &= b_{61}\sigma + b_{64}\tau.
\end{aligned}
\tag{6-26}
$$

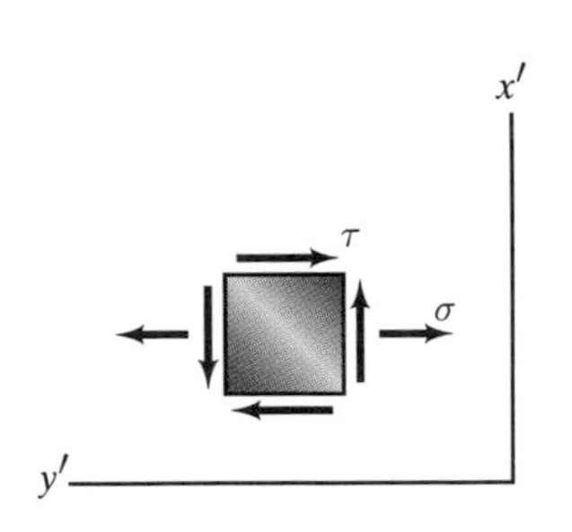

FIGURE 6-23 Rotating the xyz coordinate system 90° about the z axis.

We now consider the same material subjected to the same state of stress but a different coordinate system obtained by rotating the original coordinate system 90° about its z axis (Fig. 6-23). Since the material is isotropic, the stress-strain relations expressed in terms of the $x'y'z'$ coordinate system must be identical to the stress-strain relations expressed in terms of the xyz coordinate system. That is, the coefficients $b_{11}, b_{12}, \ldots$ must be the same. In terms of the $x'y'z'$ coordinate system, the only nonzero stress components are $\sigma'_y = \sigma$ and $\tau'_{xy} = -\tau$, so

FIGURE 6-24 Demonstrating that $\gamma_{xy} = -\gamma'_{xy}$.

from Eqs. (6-25) we obtain

$$\begin{aligned} \epsilon'_x &= b_{12}\sigma - b_{14}\tau, & \gamma'_{xy} &= b_{42}\sigma - b_{44}\tau, \\ \epsilon'_y &= b_{22}\sigma - b_{24}\tau, & \gamma'_{yz} &= b_{52}\sigma - b_{54}\tau, \\ \epsilon'_z &= b_{32}\sigma - b_{34}\tau, & \gamma'_{xz} &= b_{62}\sigma - b_{64}\tau. \end{aligned} \qquad \textbf{(6-27)}$$

Because the x and y' axes are parallel, the extensional strains ϵ_x and ϵ'_y are equal.

$$\epsilon_x = \epsilon'_y :$$
$$b_{11}\sigma + b_{14}\tau = b_{22}\sigma - b_{24}\tau.$$

From this equation we conclude that the constants $b_{11} = b_{22}$ and $b_{14} = -b_{24}$.

Next, by sketching the element subjected to a positive shear strain γ_{xy} (Fig. 6-24), we see that the shear strains $\gamma_{xy} = -\gamma'_{xy}$,

$$\gamma_{xy} = -\gamma'_{xy} :$$
$$b_{41}\sigma + b_{44}\tau = -b_{42}\sigma + b_{44}\tau.$$

We conclude that the constants $b_{41} = -b_{42}$. The complete set of strain correspondences and the resulting conclusions are

$$\begin{aligned} \epsilon_x &= \epsilon'_y & &\Rightarrow & b_{11} &= b_{22}, & b_{14} &= -b_{24}, \\ \epsilon_y &= \epsilon'_x & &\Rightarrow & b_{21} &= b_{12}, & b_{24} &= -b_{14}, \\ \epsilon_z &= \epsilon'_z & &\Rightarrow & b_{31} &= b_{32}, & b_{34} &= -b_{34}, \\ \gamma_{xy} &= -\gamma'_{xy} & &\Rightarrow & b_{41} &= -b_{42}, & b_{44} &= b_{44}, \\ \gamma_{yz} &= \gamma'_{xz} & &\Rightarrow & b_{51} &= b_{62}, & b_{54} &= -b_{64}, \\ \gamma_{xz} &= -\gamma'_{yz} & &\Rightarrow & b_{61} &= -b_{52}, & b_{64} &= b_{54}. \end{aligned}$$

Isotropy—the requirement that the stress-strain relations must be the same for any orientation of the coordinate system—has forced us to conclude that the coefficients in Eqs. (6-25) cannot have any values, but must satisfy the restrictions $b_{11} = b_{22}$, $b_{12} = b_{21}$, $b_{14} = -b_{24}$, $b_{31} = b_{32}$, $b_{34} = 0$, $b_{41} = -b_{42}$,

$b_{51} = b_{62}$, $b_{52} = -b_{61}$, $b_{54} = 0$, and $b_{64} = 0$. As a result, the 36 independent constants in Eqs. (6-25) are reduced to 26.

By considering additional orientations of the coordinate system (described in Appendix F), we obtain additional restrictions which reduce the stress-strain relations for an isotropic linearly elastic material to the forms

$$\epsilon_x = b_{11}\sigma_x + b_{12}\sigma_y + b_{12}\sigma_z, \tag{6-28}$$

$$\epsilon_y = b_{12}\sigma_x + b_{11}\sigma_y + b_{12}\sigma_z, \tag{6-29}$$

$$\epsilon_z = b_{12}\sigma_x + b_{12}\sigma_y + b_{11}\sigma_z, \tag{6-30}$$

$$\gamma_{xy} = b_{44}\tau_{xy}, \tag{6-31}$$

$$\gamma_{yz} = b_{44}\tau_{yz}, \tag{6-32}$$

$$\gamma_{xz} = b_{44}\tau_{xz}. \tag{6-33}$$

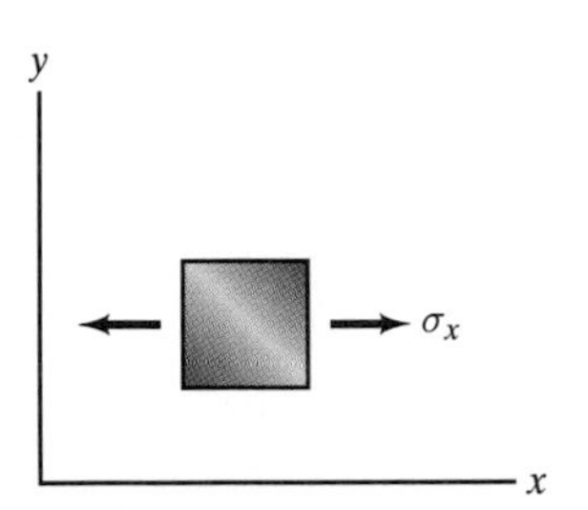

FIGURE 6-25 Applying a normal stress σ_x.

The number of coefficients is reduced from 36 to 3. Furthermore, we can express these coefficients in familiar terms. If we subject an isotropic material to a normal stress σ_x and the other components of stress are zero (Fig. 6-25), the ratio of σ_x to the resulting extensional strain ϵ_x is the modulus of elasticity E of the material:

$$\frac{\sigma_x}{\epsilon_x} = E. \tag{6-34}$$

Appying Eq. (6-28) to this state of stress, we obtain

$$\epsilon_x = b_{11}\sigma_x. \tag{6-35}$$

By comparing Eqs. (6-34) and (6-35), we determine the constant b_{11} in terms of E:

$$b_{11} = \frac{1}{E}. \tag{6-36}$$

For the state of stress shown in Fig. 6-25, the negative of the ratio of the lateral strain (either ϵ_y or ϵ_z) to the axial strain ϵ_x is Poisson's ratio ν of the material:

$$-\frac{\epsilon_y}{\epsilon_x} = \nu. \tag{6-37}$$

Applying Eq. (6-29) to this state of stress gives

$$\epsilon_y = b_{12}\sigma_x. \tag{6-38}$$

Dividing this equation by Eq. (6-35), we obtain

$$\frac{\epsilon_y}{\epsilon_x} = \frac{b_{12}}{b_{11}},$$

and by comparing this equation to Eq. (6-37), we determine the constant b_{12} in terms of E and ν:

$$b_{12} = -\nu b_{11} = -\frac{\nu}{E}. \tag{6-39}$$

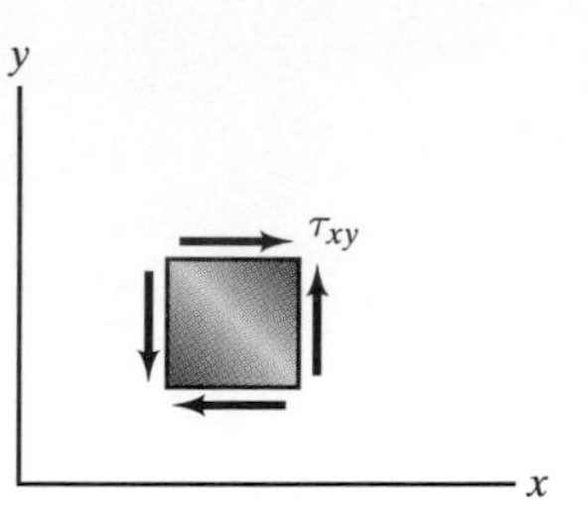

FIGURE 6-26 Applying a shear stress τ_{xy}.

Now we subject the isotropic material to a shear stress τ_{xy} with the other components of stress equal to zero (Fig. 6-26). The ratio of τ_{xy} to the resulting shear strain γ_{xy} is the shear modulus G of the material:

$$\frac{\tau_{xy}}{\gamma_{xy}} = G. \tag{6-40}$$

From Eq. (6 31),

$$\gamma_{xy} = b_{44}\tau_{xy}, \tag{6-41}$$

and by comparing Eqs. (6-40) and (6-41), we determine the constant b_{44} in terms of G:

$$b_{44} = \frac{1}{G}. \tag{6-42}$$

Substituting the expressions (6-36), (6-39), and (6-42) into Eqs. (6-28)–(6-33), we obtain the stress-strain relations for an isotropic linearly elastic material in forms in which they are commonly presented:

$$\epsilon_x = \frac{1}{E}\sigma_x - \frac{\nu}{E}(\sigma_y + \sigma_z), \tag{6-43}$$

$$\epsilon_y = \frac{1}{E}\sigma_y - \frac{\nu}{E}(\sigma_x + \sigma_z), \tag{6-44}$$

$$\epsilon_z = \frac{1}{E}\sigma_z - \frac{\nu}{E}(\sigma_x + \sigma_y), \tag{6-45}$$

$$\gamma_{xy} = \frac{1}{G}\tau_{xy}, \tag{6-46}$$

$$\gamma_{yz} = \frac{1}{G}\tau_{yz}, \tag{6-47}$$

$$\gamma_{xz} = \frac{1}{G}\tau_{xz}. \tag{6-48}$$

RELATING E, ν, AND G

Equations (6-43)–(6-48) express the stress-strain relations for an isotropic linearly elastic material in terms of three coefficients, the modulus of elasticity E, the Poisson's ratio ν, and the shear modulus G. Such a material is actually characterized by only two independent constants, because G can be expressed in terms of E and ν.

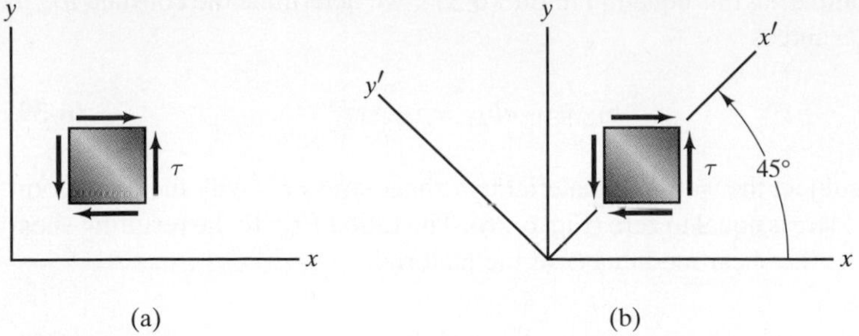

y y x' y' τ τ 45° x x (a) (b)

FIGURE 6-27 (a) Applying a shear stress $\tau_{xy} = \tau$. (b) Introducing a rotated coordinate system $x'y'z'$.

To derive this result, we subject an isotropic material to a shear stress $\tau_{xy} = \tau$ and let other components of stress be zero (Fig. 6-27a). From Eqs. (6-43)–(6-48), the only nonzero component of strain is $\gamma_{xy} = \tau/G$. Now let us consider an $x'y'z'$ coordinate system obtained by rotating the xyz coordinate system 45° about the z axis (Fig. 6-27b). Since we know the state of stress, we can use Eqs. (5-7) and (5-9) to determine the stress components σ'_x and σ'_y:

$$\begin{aligned} \sigma'_x &= \tau \sin 2(45°) = \tau, \\ \sigma'_y &= -\tau \sin 2(45°) = -\tau. \end{aligned} \qquad \textbf{(6-49)}$$

Notice that the stress components $\sigma'_z = \sigma_z = 0$. We also know the state of strain, so we can use Eq. (6-7) to determine the strain component ϵ'_x:

$$\epsilon'_x = \frac{\gamma_{xy}}{2} \sin 2(45°) = \frac{1}{2G}\tau.$$

Because the material is isotropic, the strain component ϵ'_x and the stress components σ'_x, σ'_y, and σ'_z must satisfy Eq. (6-43).

$$\epsilon'_x = \frac{1}{E}\sigma'_x - \frac{\nu}{E}(\sigma'_y + \sigma'_z):$$

$$\frac{1}{2G}\tau = \frac{1}{E}\tau + \frac{\nu}{E}\tau.$$

Solving this equation for G, we determine the shear modulus in terms of the modulus of elasticity and Poisson's ratio:

$$G = \frac{E}{2(1+\nu)}. \qquad \textbf{(6-50)}$$

LAMÉ CONSTANTS AND BULK MODULUS

Equations (6-43)–(6-48) give the components of strain in terms of the components of stress for an isotropic linearly elastic material. When the components of stress are expressed in terms of the components of strain, they can conveniently be written in the forms

$$\sigma_x = (\lambda + 2\mu)\epsilon_x + \lambda(\epsilon_y + \epsilon_z), \tag{6-51}$$

$$\sigma_y = (\lambda + 2\mu)\epsilon_y + \lambda(\epsilon_x + \epsilon_z), \tag{6-52}$$

$$\sigma_z = (\lambda + 2\mu)\epsilon_z + \lambda(\epsilon_x + \epsilon_y), \tag{6-53}$$

$$\tau_{xy} = \mu\gamma_{xy}, \tag{6-54}$$

$$\tau_{yz} = \mu\gamma_{yz}, \tag{6-55}$$

$$\tau_{xz} = \mu\gamma_{xz}, \tag{6-56}$$

where λ and μ are called the *Lamé constants*. The constant $\mu = G$, and the constant λ is given in terms of the modulus of elasticity and Poisson's ratio by

$$\lambda = \frac{\nu E}{(1+\nu)(1-2\nu)}. \tag{6-57}$$

Consider an element of material with volume dV_0 in the reference state (Fig. 6-28a). We subject the element to a pressure p, so that $\sigma_x = -p$, $\sigma_y = -p$, and $\sigma_z = -p$ (Fig. 6-28b). Let the volume of the deformed element be dV. The *bulk modulus* K of the material is defined to be the ratio of $-p$ to the dilatation $e = (dV - dV_0)/dV_0$:

$$K = \frac{-p}{e}. \tag{6-58}$$

From Eq. (6-3), the dilatation of a material subjected to small strains is

$$e = \epsilon_x + \epsilon_y + \epsilon_z,$$

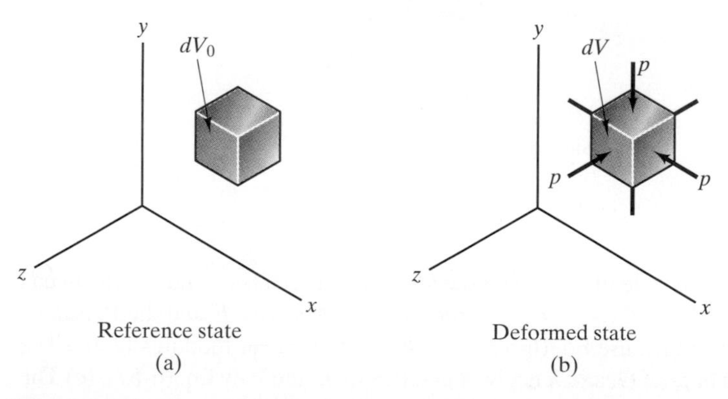

FIGURE 6-28 Element of material in the reference state and when subjected to a pressure p.

and from Eqs. (6-43)–(6-45), the extensional strains of the element are

$$\epsilon_x = \epsilon_y = \epsilon_z = \frac{1 - 2\nu}{E}(-p).$$

Using these relationships, we can express the bulk modulus in the form

$$K = \frac{E}{3(1 - 2\nu)}. \tag{6-59}$$

The bulk modulus is sometimes used instead of the modulus of elasticity E or the Lamé constant λ in expressing the stress-strain relations of an isotropic elastic material.

EXAMPLE 6-5

The cylindrical bar in Fig. 6-29 consists of an isotropic linearly elastic material and is subjected to axial loads. As a result, the bar is subjected to a normal stress $\sigma_x = 420$ MPa and the other components of stress are zero. By measuring the changes in the bar's length and diameter, it is determined that the axial strain is $\epsilon_x = 0.006$ and the lateral strain is $\epsilon_y = \epsilon_z = -0.002$. Determine **(a)** the modulus of elasticity, Poisson's ratio, and shear modulus of the material; **(b)** the Lamé constants of the material; **(c)** the bulk modulus of the material.

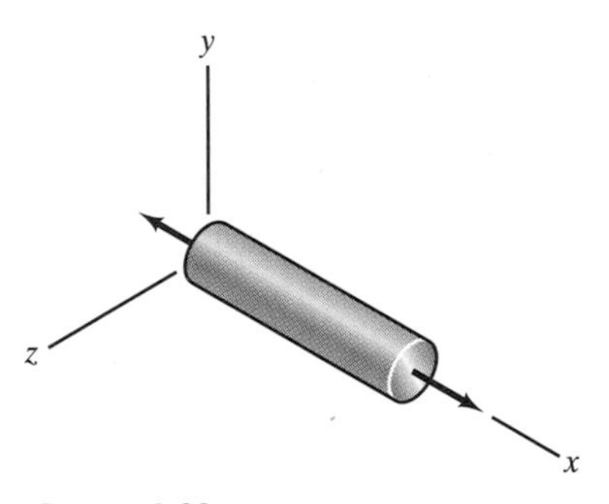

FIGURE 6-29

Strategy

(a) We know the state of stress and the strain components ϵ_x and ϵ_y, so we can solve Eqs. (6-43) and (6-44) for the modulus of elasticity E and the Poisson's ratio ν. We can then use Eq. (6-50) to determine the shear modulus G. **(b)** The Lamé constant $\mu = G$, and λ is given in terms of E and ν by Eq. (6-57). **(c)** The bulk modulus is given in terms of E and ν by Eq. (6-59).

Solution

(a) From Eq. (6-43),

$$\epsilon_x = \frac{1}{E}\sigma_x - \frac{\nu}{E}(\sigma_y + \sigma_z):$$

$$0.006 = \frac{1}{E}(420 \times 10^6),$$

we obtain $E = 70.0$ GPa. Then, from Eq. (6-44),

$$\epsilon_y = \frac{1}{E}\sigma_y - \frac{\nu}{E}(\sigma_x + \sigma_z):$$

$$-0.002 = \frac{-\nu}{70.0 \times 10^9}(420 \times 10^6),$$

we obtain $\nu = 0.333$. From Eq. (6-50), the shear modulus is

$$G = \frac{E}{2(1+\nu)} = \frac{70.0 \times 10^9}{(2)(1+0.333)} = 26.3 \text{ GPa}.$$

(b) The Lamé constant $\mu = G = 26.3$ GPa. From Eq. (6-57),

$$\lambda = \frac{\nu E}{(1+\nu)(1-2\nu)} = \frac{(0.333)(70.0 \times 10^9)}{(1+0.333)[1-(2)(0.333)]} = 52.5 \text{ GPa}.$$

(c) From Eq. (6-59), the bulk modulus is

$$K = \frac{E}{3(1-2\nu)} = \frac{70.0 \times 10^9}{3[1-(2)(0.333)]} = 70.0 \text{ GPa}.$$

Chapter Summary

Components of Strain

In terms of a given coordinate system, the *state of strain* at a point p of a material is defined by the *components of strain*

$$\begin{bmatrix} \epsilon_x & \gamma_{xy} & \gamma_{xz} \\ \gamma_{yx} & \epsilon_y & \gamma_{yz} \\ \gamma_{zx} & \gamma_{zy} & \epsilon_z \end{bmatrix}. \qquad \text{Eq. (6-1)}$$

The components ϵ_x, ϵ_y, and ϵ_z are the extensional strains in the x, y, and z directions. The component $\gamma_{xy} = \gamma_{yx}$ is the shear strain referred to the directions of the x and y axes, and the components $\gamma_{yz} = \gamma_{zy}$ and $\gamma_{xz} = \gamma_{zx}$ are defined similarly.

Consider a volume dV_0 of material in a reference state. Its volume in the deformed state is

$$dV = (1 + \epsilon_x + \epsilon_y + \epsilon_z)\, dV_0. \qquad \text{Eq. (6-2)}$$

The *dilatation* is the change in volume of the material per unit volume:

$$e = \frac{dV - dV_0}{dV_0} = \epsilon_x + \epsilon_y + \epsilon_z. \qquad \text{Eq. (6-3)}$$

Transformations of Plane Strain

The strain at a point p is said to be a state of *plane strain* if it is of the form

$$\begin{bmatrix} \epsilon_x & \gamma_{xy} & 0 \\ \gamma_{yx} & \epsilon_y & 0 \\ 0 & 0 & 0 \end{bmatrix}. \qquad \text{Eq. (6-4)}$$

In terms of a coordinate system $x'y'z'$ oriented as shown in Fig. (a), the components of strain are

$$\epsilon'_x = \frac{\epsilon_x + \epsilon_y}{2} + \frac{\epsilon_x - \epsilon_y}{2} \cos 2\theta + \frac{\gamma_{xy}}{2} \sin 2\theta, \qquad \text{Eq. (6-7)}$$

$$\epsilon'_y = \frac{\epsilon_x + \epsilon_y}{2} - \frac{\epsilon_x - \epsilon_y}{2} \cos 2\theta - \frac{\gamma_{xy}}{2} \sin 2\theta, \qquad \text{Eq. (6-8)}$$

$$\frac{\gamma'_{xy}}{2} = -\frac{\epsilon_x - \epsilon_y}{2} \sin 2\theta + \frac{\gamma_{xy}}{2} \cos 2\theta. \qquad \text{Eq. (6-9)}$$

For an isotropic linearly elastic material, Eqs. (6-7)–(6-9) also apply to the components of strain resulting from a state of plane stress.

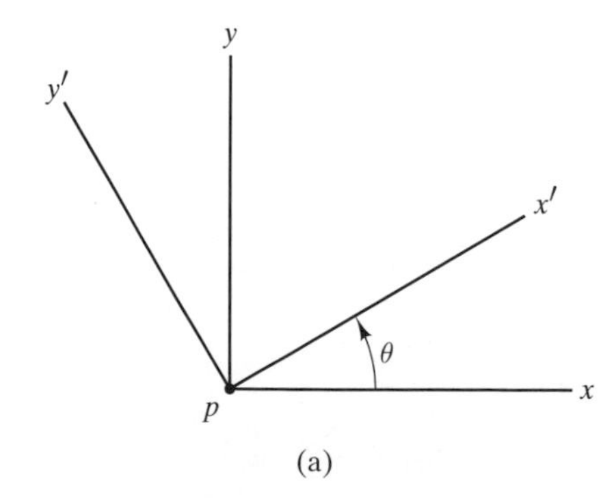

(a)

Strain Gauge Rosette

Suppose that a material is subject to an unknown state of plane strain relative to the x–y coordinate system in Fig. (a). A *strain gauge rosette* measures the extensional strain ϵ'_x in three different directions: θ_a, θ_b, and θ_c. Let the measured strains be ϵ_a, ϵ_b, and ϵ_c. Using Eq. (6-6) to express these strains in terms of the strain components ϵ_x, ϵ_y, and γ_{xy} gives

$$\begin{aligned}\epsilon_a &= \epsilon_x \cos^2\theta_a + \epsilon_y \sin^2\theta_a + \gamma_{xy}\sin\theta_a\cos\theta_a,\\ \epsilon_b &= \epsilon_x \cos^2\theta_b + \epsilon_y \sin^2\theta_b + \gamma_{xy}\sin\theta_b\cos\theta_b,\\ \epsilon_c &= \epsilon_x \cos^2\theta_c + \epsilon_y \sin^2\theta_c + \gamma_{xy}\sin\theta_c\cos\theta_c.\end{aligned} \qquad \text{Eq. (6-10)}$$

This system of equations can be solved for the strain components ϵ_x, ϵ_y, and γ_{xy}.

Maximum and Minimum Strains in Plane Strain

A value of θ for which the extensional strain is a maximum or minimum is determined from the equation

$$\tan 2\theta_\mathrm{p} = \frac{\gamma_{xy}}{\epsilon_x - \epsilon_y}. \qquad \text{Eq. (6-11)}$$

The values of the principal strains can be obtained by substituting θ_p into Eqs. (6-7) and (6-8). Their values can also be determined from the equation

$$\epsilon_1, \epsilon_2 = \frac{\epsilon_x + \epsilon_y}{2} \pm \sqrt{\left(\frac{\epsilon_x - \epsilon_y}{2}\right)^2 + \left(\frac{\gamma_{xy}}{2}\right)^2}, \qquad \text{Eq. (6-14)}$$

although this equation does not indicate their directions. An infinitesimal square element oriented as shown in Fig. (b) is subjected to the principal strains in the x' and y' directions and undergoes no shear strain.

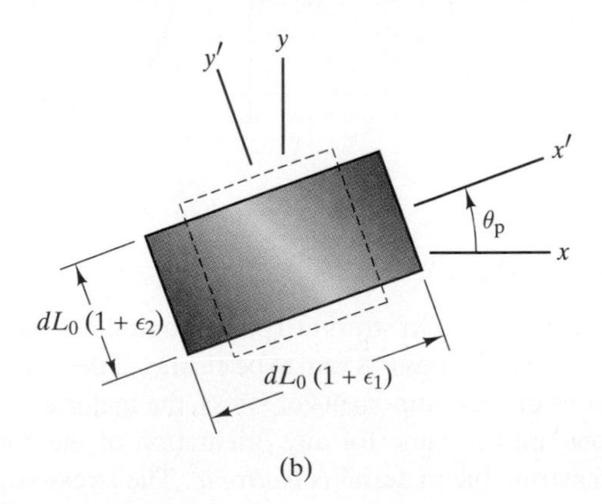

(b)

A value of θ for which the in-plane shear strain is a maximum or minimum is determined from the equation

$$\tan 2\theta_s = -\frac{\epsilon_x - \epsilon_y}{\gamma_{xy}}. \qquad \text{Eq. (6-15)}$$

The corresponding shear strain can be obtained by substituting θ_s into Eq. (6-9). The magnitude of the maximum in-plane shear strain can be determined from the equation

$$\gamma_{max} = \sqrt{(\epsilon_x - \epsilon_y)^2 + \gamma_{xy}^2}. \qquad \text{Eq. (6-18)}$$

The *absolute maximum shear strain* is the largest of the three values

$$|\epsilon_1 - \epsilon_2|, \quad |\epsilon_1|, \quad |\epsilon_2|. \qquad \text{Eq. (6-22)}$$

Mohr's Circle for Plane Strain

Given a state of plane strain ϵ_x, ϵ_y, and γ_{xy}, establish a set of horizontal and vertical axes with normal strain measured to the right along the horizontal axis and shear strain measured downward along the vertical axis. Plot two points, point P with coordinates $(\epsilon_x, \gamma_{xy}/2)$ and point Q with coordinates $(\epsilon_y, -\gamma_{xy}/2)$. Draw a straight line connecting points P and Q. Using the intersection of the straight line with the horizontal axis as the center, draw a circle that passes through the two points [Fig. (c)]. Draw a straight line through the center of the circle at an angle 2θ measured counterclockwise from point P. The point P' at which this line intersects the circle has coordinates $(\epsilon'_x, \gamma'_{xy}/2)$, and the point Q' has coordinates $(\epsilon'_y, -\gamma'_{xy}/2)$, as shown in Fig. (d).

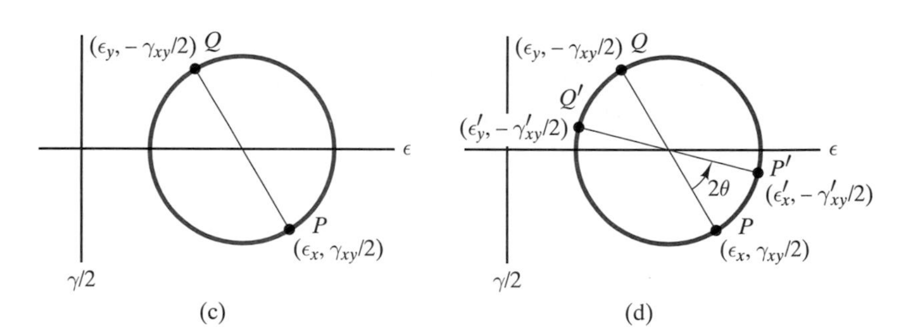

Stress-Strain Relations

A material for which the state of stress at a point is a single-valued function of the current state of strain at that point is said to be *elastic*. If the components of strain are linear functions of the components of stress, the material is *linearly elastic*. If those functions are the same for any orientation of the coordinate system relative to the material, the material is *isotropic*. The stress-strain relations for an isotropic linearly elastic material are

$$\epsilon_x = \frac{1}{E}\sigma_x - \frac{\nu}{E}(\sigma_y + \sigma_z), \qquad \text{Eq. (6-43)}$$

$$\epsilon_y = \frac{1}{E}\sigma_y - \frac{\nu}{E}(\sigma_x + \sigma_z), \qquad \text{Eq. (6-44)}$$

$$\epsilon_z = \frac{1}{E}\sigma_z - \frac{\nu}{E}(\sigma_x + \sigma_y), \qquad \text{Eq. (6-45)}$$

$$\gamma_{xy} = \frac{1}{G}\tau_{xy}, \qquad \text{Eq. (6-46)}$$

$$\gamma_{yz} = \frac{1}{G}\tau_{yz}, \qquad \text{Eq. (6-47)}$$

$$\gamma_{xz} = \frac{1}{G}\tau_{xz}. \qquad \text{Eq. (6-48)}$$

The shear modulus is related to the modulus of elasticity and Poisson's ratio by

$$G = \frac{E}{2(1+\nu)}. \qquad \text{Eq. (6-50)}$$

The stress-strain relations can also be expressed as

$$\sigma_x = (\lambda + 2\mu)\epsilon_x + \lambda(\epsilon_y + \epsilon_z), \qquad \text{Eq. (6-51)}$$

$$\sigma_y = (\lambda + 2\mu)\epsilon_y + \lambda(\epsilon_x + \epsilon_z), \qquad \text{Eq. (6-52)}$$

$$\sigma_z = (\lambda + 2\mu)\epsilon_z + \lambda(\epsilon_x + \epsilon_y), \qquad \text{Eq. (6-53)}$$

$$\tau_{xy} = \mu\gamma_{xy}, \qquad \text{Eq. (6-54)}$$

$$\tau_{yz} = \mu\gamma_{yz}, \qquad \text{Eq. (6-55)}$$

$$\tau_{xz} = \mu\gamma_{xz}, \qquad \text{Eq. (6-56)}$$

where λ and μ are the *Lamé constants*. The constant $\mu = G$, and the constant λ is given in terms of the modulus of elasticity and Poisson's ratio by

$$\lambda = \frac{\nu E}{(1+\nu)(1-2\nu)}. \qquad \text{Eq. (6-57)}$$

Let an isotropic linearly elastic material be subjected to a pressure p so that $\sigma_x = -p$, $\sigma_y = -p$, and $\sigma_z = -p$. The *bulk modulus* K of the material is the ratio of $-p$ to the dilatation:

$$K = \frac{-p}{e}. \qquad \text{Eq. (6-58)}$$

In terms of the modulus of elasticity and Poisson's ratio, the bulk modulus is

$$K = \frac{E}{3(1-2\nu)}. \qquad \text{Eq. (6-59)}$$

PROBLEMS

6-2.1. The components of plane strain at point p are $\epsilon_x = 0.003$, $\epsilon_y = 0$, and $\gamma_{xy} = 0$. If $\theta = 45°$, what are the strains ϵ'_x, ϵ'_y, and γ'_{xy} at point p?

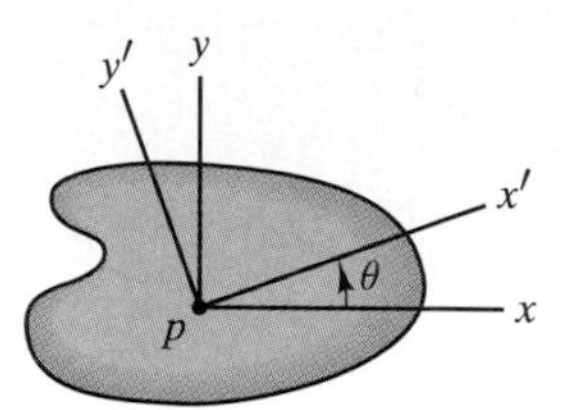

| PROBLEMS 6-2.1–6-2.9

6-2.2. The components of plane strain at point p are $\epsilon_x = 0$, $\epsilon_y = 0$, and $\gamma_{xy} = 0.004$. If $\theta = 45°$, what are the strains ϵ'_x, ϵ'_y, and γ'_{xy} at point p?

6-2.3. The components of plane strain at point p are $\epsilon_x = -0.0024$, $\epsilon_y = 0.0012$, and $\gamma_{xy} = -0.0012$. If $\theta = 25°$, what are the strains ϵ'_x, ϵ'_y, and γ'_{xy} at point p?

6-2.4. The components of plane strain at a point p of a bit during a drilling operation are $\epsilon_x = 0.00400$, $\epsilon_y = -0.00300$, and $\gamma_{xy} = 0.00600$, and the components referred to the $x'y'z'$ coordinate system are $\epsilon'_x = 0.00125$, $\epsilon'_y = -0.00025$, and $\gamma'_{xy} = 0.00910$. What is the angle θ?

6-2.5. The components of plane strain at point p are $\epsilon_x = 0.0024$, $\epsilon_y = -0.0012$, and $\gamma_{xy} = 0.0048$. The extensional strains $\epsilon'_x = 0.00347$ and $\epsilon'_y = -0.00227$. Determine γ'_{xy} and the angle θ.

6-2.6. During liftoff, strain gauges attached to one of the Space Shuttle main engine nozzles determine that the components of plane strain at point p are $\epsilon'_x = 0.00665$, $\epsilon'_y = 0.00825$, and $\gamma'_{xy} = 0.00135$ for a coordinate system oriented at $\theta = 20°$. What are the strains ϵ_x, ϵ_y, and γ_{xy} at that point?

6-2.7. The components of plane strain at point p referred to the $x'y'$ coordinate system are $\epsilon'_x = 0.0066$, $\epsilon'_y = -0.0086$, and $\gamma'_{xy} = 0.0028$. If $\theta = 20°$, what are the strains ϵ_x, ϵ_y, and γ_{xy} at point p?

6-2.8. The strains $\epsilon_x = 0.008$, $\epsilon_y = -0.006$, $\gamma_{xy} = 0.024$, $\epsilon'_x = 0.014$, and $\epsilon'_y = -0.012$. Determine the strain γ'_{xy} and the angle θ.

6-2.9. The strains $\epsilon_x = 0.008$, $\epsilon_y = -0.006$, and $\gamma'_{xy} = 0.024$, and the angle $\theta = 35°$. Determine ϵ'_x, ϵ'_y, and γ_{xy}.

6-2.10. A point p of the bearing's housing is subjected to a state of plane stress, and the strain components $\epsilon_x = -0.0024$, $\epsilon_y = 0.0044$, and $\gamma_{xy} = -0.0030$. If $\theta = 20°$, what are the strains ϵ'_x, ϵ'_y, and γ'_{xy} at p?

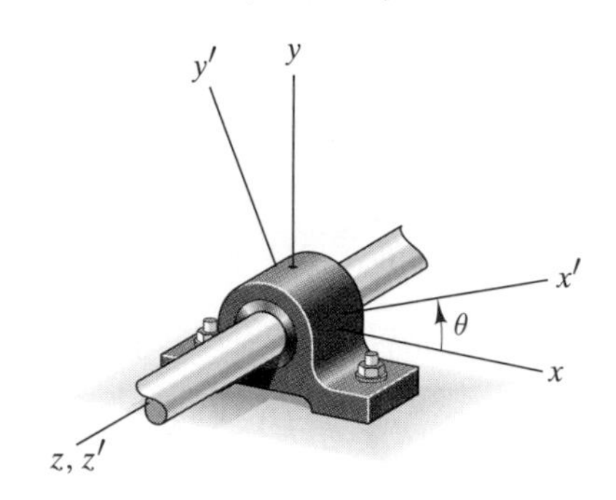

| PROBLEM 6-2.10

6-2.11. A point p of the housing of the bearing shown in Problem 6-2.10 is subjected to the state of plane strain $\epsilon_x = 0.0032$, $\epsilon_y = -0.0026$, and $\gamma_{xy} = 0.0044$. If $\epsilon'_x = 0.0037$ and $\epsilon'_y = -0.0031$, determine the angle θ and the strain γ'_{xy} at p.

6-2.12. Points P and Q are 1 mm apart in the reference state of a material. If the material is subjected to the homogeneous state of plane strain $\epsilon_x = 0.003$, $\epsilon_y = -0.002$, and $\gamma_{xy} = -0.006$, what is the distance between points P and Q in the deformed material?

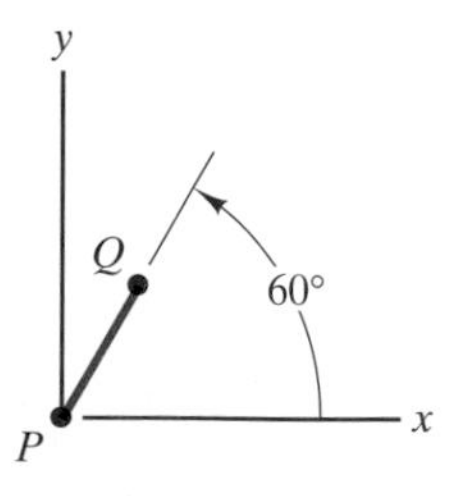

| PROBLEM 6-2.12

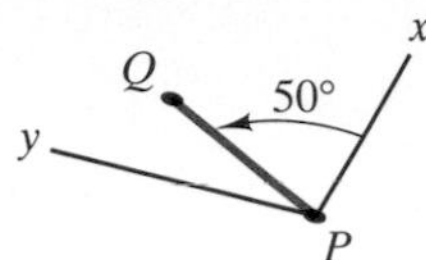

PROBLEM 6-2.13

6-2.13. Two points P and Q of the Concorde's wing are 2 mm apart when the wing is unstressed. In a particular flight condition, the material containing these points is subjected to a homogeneous state of plane stress and the strain components $\epsilon_x = 0.008$, $\epsilon_y = 0.002$, and $\gamma_{xy} = -0.003$. What is the distance between points P and Q?

6-2.14. The points P and Q of the Concorde's wing shown in Problem 6-2.13 are 2 mm apart when the wing is unstressed. In a particular flight condition, they are 1.992 mm apart. The material is in plane stress. If $\epsilon_x = -0.0088$ and $\epsilon_y = 0.0024$, what is γ_{xy}?

6-2.15. Points O and Q are 1 mm apart and points O and P are 2 mm apart in the reference state of a material. If the material is subjected to a homogeneous state of plane strain $\epsilon_x = 0.006$, $\epsilon_y = 0.002$, and $\gamma_{xy} = 0.004$, what is the distance between points P and Q in the deformed material?

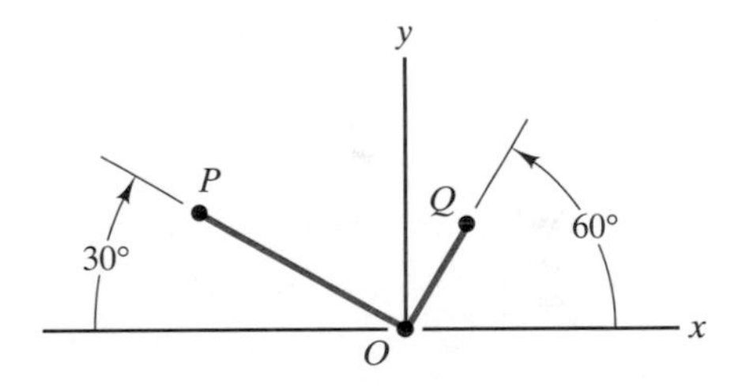

PROBLEM 6-2.15

6-2.16. In Problem 6-2.15, what is the angle between the lines OQ and OP in the deformed material?

6-2.17. Points O and Q shown in Problem 6-2.15 are 1 mm apart and points O and P are 2 mm apart in the reference state of a material. After the material is subjected to a homogeneous state of strain, points O and Q are 1.002 mm apart, points O and P are 1.998 mm apart, and points P and Q are 2.242 mm apart. What are the strain components ϵ_x, ϵ_y, and γ_{xy}?

6-2.18. A bar is subjected to axial forces. The strains measured by a strain gauge rosette oriented as shown are $\epsilon_a = 0.003$, $\epsilon_b = 0.001$, and $\epsilon_c = -0.001$. Determine the shear strain γ_{xy}.

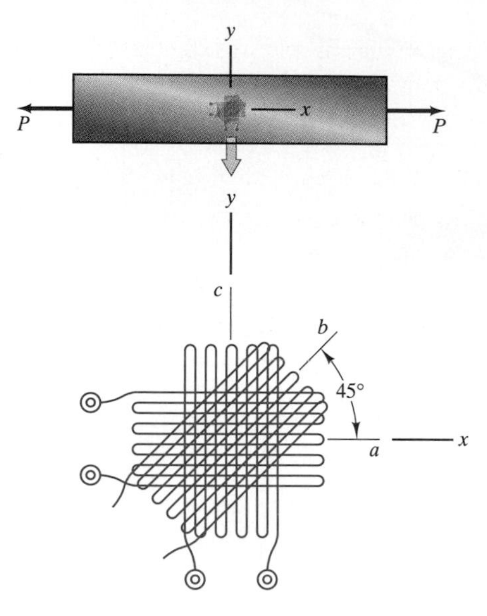

PROBLEM 6-2.18

6-2.19. The strains measured by a strain gauge rosette mounted on the bicycle brake are $\epsilon_a = 0.00220$, $\epsilon_b = -0.00100$, and $\epsilon_c = -0.00360$. Determine the strains ϵ_x, ϵ_y, and γ_{xy}.

6-2.20. The strains measured by a strain gauge rosette oriented as shown are $\epsilon_a = -0.00116$, $\epsilon_b = -0.00065$, and $\epsilon_c = 0.00130$. Determine the strains ϵ_x, ϵ_y, and γ_{xy}.

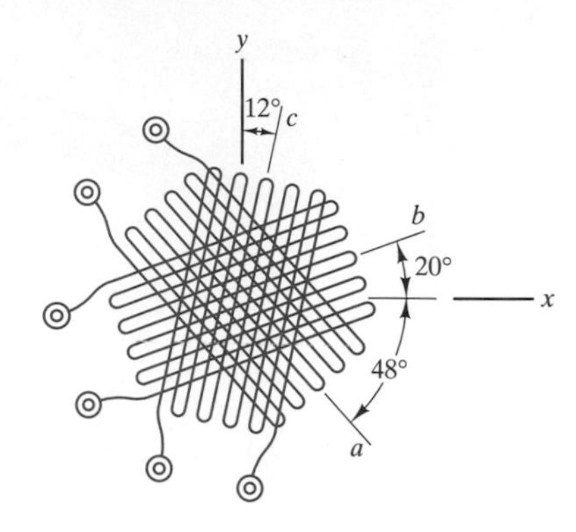

PROBLEM 6-2.20

6-2.21. In Example 6-3, use Eq. (6-18) to calculate the magnitude of the maximum in-plane shear strain.

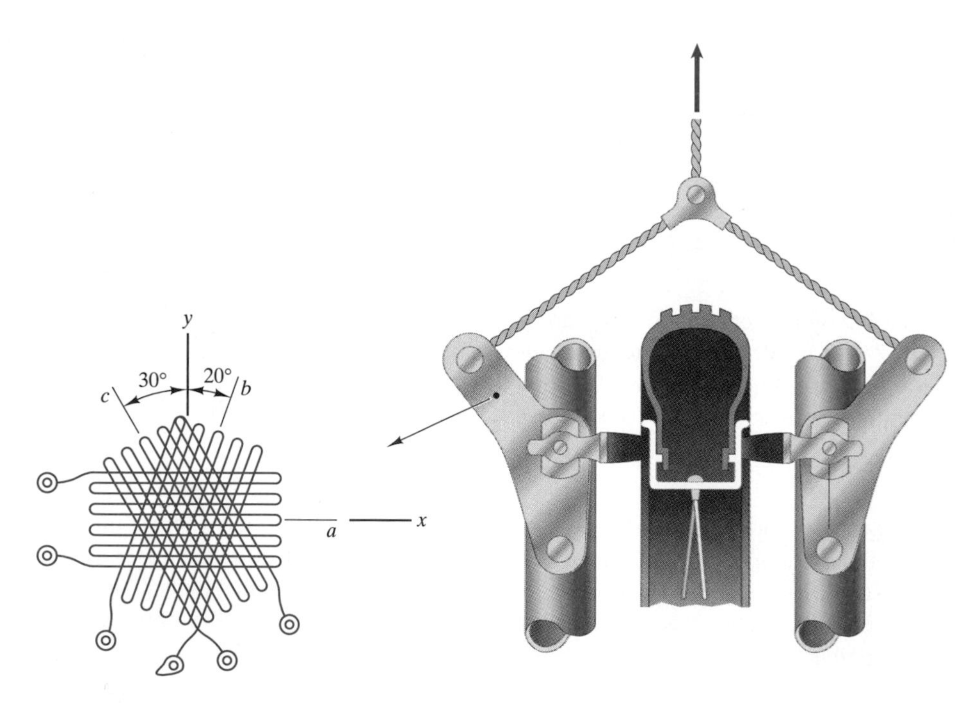

PROBLEM 6-2.19

For the states of plane strain given in Problems 6-2.22–6-2.25, determine the principal strains and the maximum in-plane shear strain and show the orientations of the elements subjected to these strains.

6-2.22. $\epsilon_x = 0.002$, $\epsilon_y = 0.001$, and $\gamma_{xy} = 0$.

6-2.23. $\epsilon_x = 0.0025$, $\epsilon_y = 0$, and $\gamma_{xy} = -0.0050$.

6-2.24. $\epsilon_x = -0.008$, $\epsilon_y = 0.006$, and $\gamma_{xy} = -0.012$.

6-2.25. $\epsilon_x = 0.0024$, $\epsilon_y = -0.0012$, and $\gamma_{xy} = 0.0024$.

6-2.26. At point p of the bearing's housing in Problem 6-2.10, determine the principal strains and the maximum in-plane shear strain.

6-2.27. For the state of plane strain $\epsilon_x = 0.002$, $\epsilon_y = 0.001$, and $\gamma_{xy} = 0$, what is the absolute maximum shear strain?

6-2.28. For the state of plane strain $\epsilon_x = 0.0025$, $\epsilon_y = 0$, and $\gamma_{xy} = -0.005$, what is the absolute maximum shear strain?

6-2.29. For the state of plane strain $\epsilon_x = -0.008$, $\epsilon_y = -0.006$, and $\gamma_{xy} = -0.012$, what is the absolute maximum shear strain?

6-2.30. For the state of plane strain $\epsilon_x = 0.0024$, $\epsilon_y = 0.0012$, and $\gamma_{xy} = 0.0024$, what is the absolute maximum shear strain?

6-2.31. A point p of the MacPherson strut suspension is subjected to the state of plane strain $\epsilon_x = -0.0088$, $\epsilon_y = 0.0024$, $\gamma_{xy} = -0.0036$. Determine the principal strains and the absolute maximum shear strain.

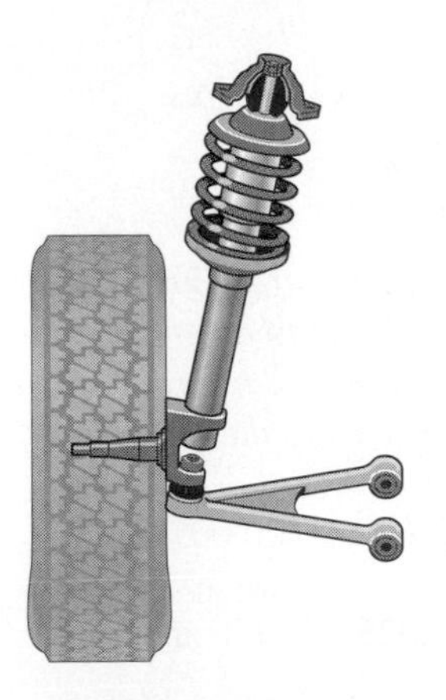

PROBLEM 6-2.31

6-2.32. By setting θ equal to $\theta + 90^\circ$ in Eq. (6-7), derive Eq. (6-8).

6-2.33. Consider an infinitesimal element of material that in the reference state has the triangular shape shown in Fig. (a). In the deformed state [Fig. (b)], the lengths of the sides are known in terms of the extensional strains ϵ_x, ϵ'_x, and ϵ'_y, and the angle between the sides that were perpendicular in the reference state can be expressed in terms of γ'_{xy}. By analyzing the deformed triangle, derive Eq. (6-9).

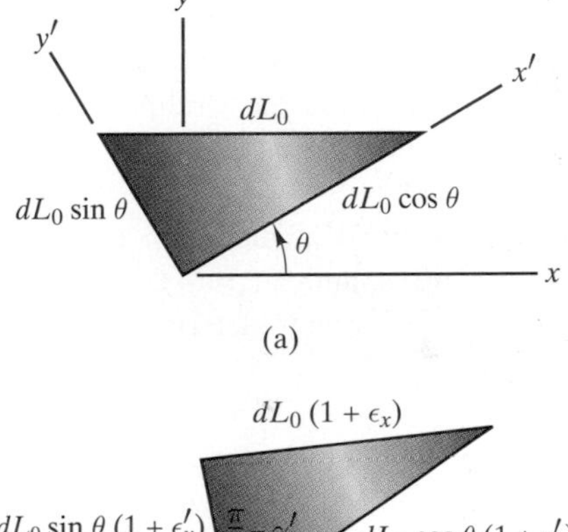

(a)

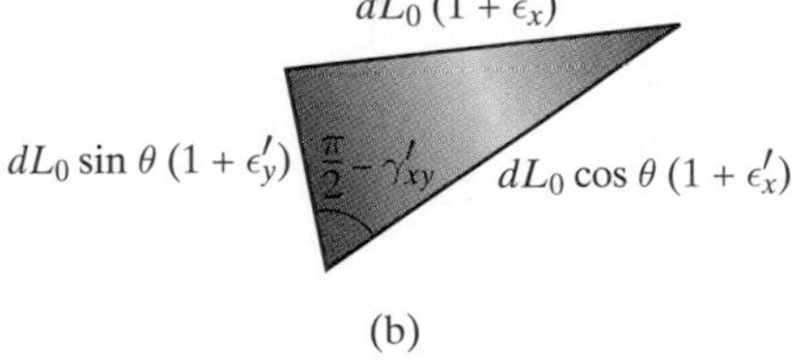

(b)

PROBLEM 6-2.33

6-2.34. A circular line in the x–y plane of circumference $C = 2\pi R$ is drawn in the reference state of a material. If the material is then subjected to a homogeneous strain ϵ_x and other components of strain are zero, show that the length of the deformed line is $C(1 + 0.5\epsilon_x)$.

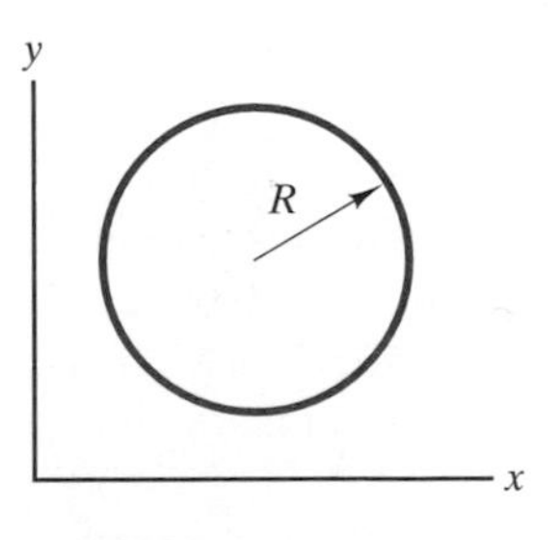

PROBLEM 6-2.34

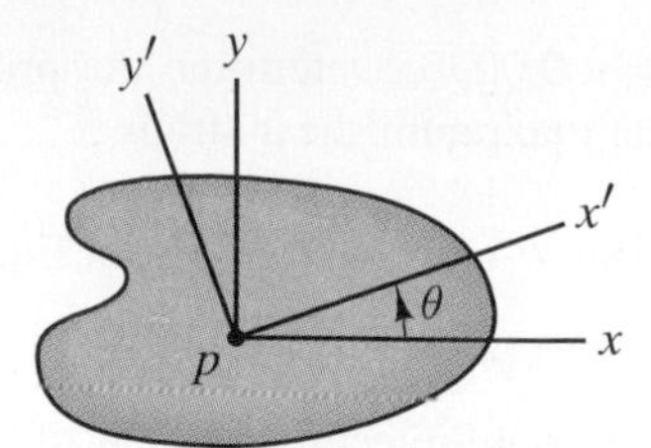

Problems 6-3.1–6-3.8

6-3.1. The components of plane strain at point p are $\epsilon_x = 0.003$, $\epsilon_y = 0$, and $\gamma_{xy} = 0$, and the angle $\theta = 45°$. Use Mohr's circle to determine the strains ϵ'_x, ϵ'_y, and γ'_{xy} at point p.

6-3.2. The components of plane strain at point p are $\epsilon_x = 0$, $\epsilon_y = 0$, and $\gamma_{xy} = 0.004$, and the angle $\theta = 45°$. Use Mohr's circle to determine the strains ϵ'_x, ϵ'_y, and γ'_{xy} at point p.

6-3.3. The components of plane strain at point p are $\epsilon_x = -0.0024$, $\epsilon_y = 0.0012$, and $\gamma_{xy} = -0.0012$, and the angle $\theta = 25°$. Use Mohr's circle to determine the strains ϵ'_x, ϵ'_y, and γ'_{xy} at point p.

6-3.4. The components of plane strain at point p are $\epsilon_x = \epsilon_y = \epsilon_0$, $\gamma_{xy} = 0$. Use Mohr's circle to show that $\epsilon'_x = \epsilon_0$ and $\gamma'_{xy} = 0$ for any value of θ.

6-3.5. The components of plane strain at a point p of a bit during a drilling operation are $\epsilon_x = 0.00400$, $\epsilon_y = -0.00300$, and $\gamma_{xy} = 0.00600$, and the components referred to the $x'y'z'$ coordinate system are $\epsilon'_x = 0.00125$, $\epsilon'_y = -0.00025$, and $\gamma'_{xy} = 0.00910$. What is the angle θ?

6-3.6. The components of plane strain at point p are $\epsilon_x = 0.0024$, $\epsilon_y = -0.0012$, and $\gamma_{xy} = 0.0048$. The extensional strains $\epsilon'_x = 0.00347$ and $\epsilon'_y = -0.00227$. Use Mohr's circle to determine γ'_{xy} and the angle θ.

6-3.7. During liftoff, strain gauges attached to one of the Space Shuttle main engine nozzles determine that the components of plane strain $\epsilon'_x = 0.00727$, $\epsilon'_y = 0.00763$, and $\gamma'_{xy} = 0.00207$ at $\theta = 40°$. Use Mohr's circle to determine the strains ϵ_x, ϵ_y, and γ_{xy} at that point.

Problem 6-3.7

6-3.8. The components of plane strain at point p referred to the $x'y'$ coordinate system are $\epsilon'_x = 0.0066$, $\epsilon'_y = -0.0086$, and $\gamma'_{xy} = 0.0028$, and the angle $\theta = 20°$. Use Mohr's circle to determine the strains ϵ_x, ϵ_y, and γ_{xy} at point p.

6-3.9. At a point p the pliers are subjected to plane stress and the strain components $\epsilon_x = 0.008$, $\epsilon_y = -0.004$, and $\gamma_{xy} = -0.006$. Use Mohr's circle to determine the principal strains and the maximum in-plane shear strain.

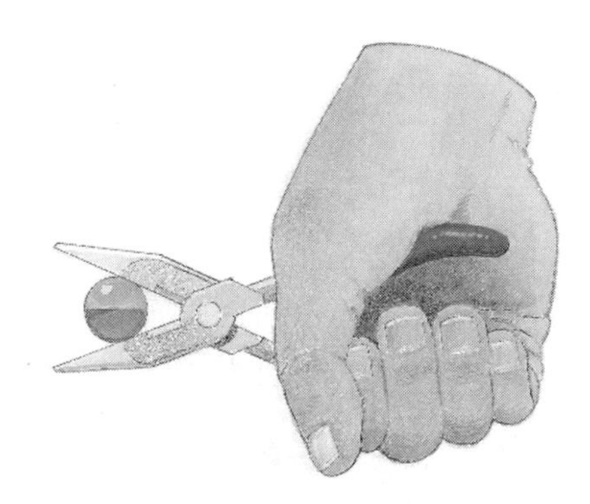

Problem 6-3.9

For the states of plane strain given in Problems 6-3.10–6-3.13, use Mohr's circle to determine the principal strains, the maximum in-plane shear strain, and the orientations of the elements subjected to these strains.

6-3.10. $\epsilon_x = 0.002$, $\epsilon_y = 0.001$, and $\gamma_{xy} = 0$.

6-3.11. $\epsilon_x = 0.0025$, $\epsilon_y = 0$, and $\gamma_{xy} = -0.0050$.

6-3.12. $\epsilon_x = -0.008$, $\epsilon_y = 0.006$, and $\gamma_{xy} = -0.012$.

6-3.13. $\epsilon_x = 0.0024$, $\epsilon_y = -0.0012$, and $\gamma_{xy} = 0.0024$.

6-3.14. Use Mohr's circle to determine the principal strains and the maximum in-plane shear strain to which the Space Shuttle's nozzle is subjected in Problem 6-3.7.

6-3.15. Use Mohr's circle to determine the principal strains and the maximum in-plane shear strain at point p of the MacPherson strut suspension in Problem 6-2.31.

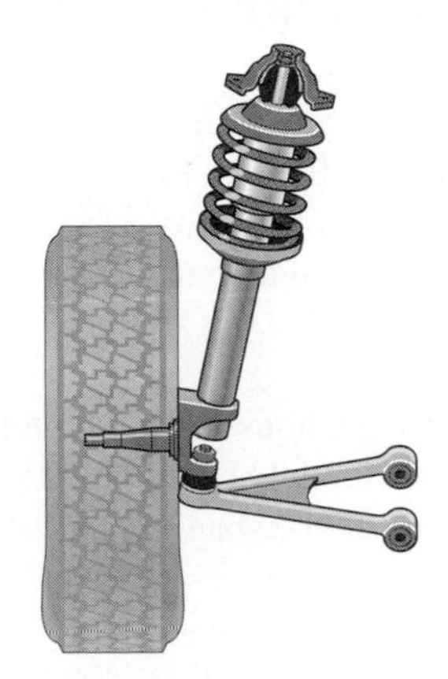

PROBLEM 6-3.15

6-4.1. The state of stress at a point p in a material with modulus of elasticity $E = 28$ GPa and Poisson's ratio $\nu = 0.3$ is

$$\begin{bmatrix} 250 & -20 & 0 \\ -20 & 250 & 40 \\ 0 & 40 & 200 \end{bmatrix} \text{MPa.}$$

What is the state of strain at p?

6-4.2. The state of strain at a point p of the material described in Problem 6-4.1 is

$$\begin{bmatrix} -250 & 125 & 0 \\ 125 & 500 & -125 \\ 0 & -125 & 250 \end{bmatrix} \times 10^{-5}.$$

What is the state of stress at p?

6-4.3. The state of stress at a point p of a nickel pipe in a gaseous diffusion plant is

$$\begin{bmatrix} 35 & -20 & 25 \\ -20 & 45 & 32 \\ 25 & 32 & 40 \end{bmatrix} \text{ksi.}$$

What is the state of strain at p?

6-4.4. Show that a state of plane stress at a point p of an isotropic, linearly elastic material does not necessarily result in a state of plane strain at p. What condition must the state of plane stress satisfy to result in a state of plane strain?

6-4.5. The state of plane stress at a point p in a machine part made of 2014-T6 aluminum alloy is $\sigma_x = 40$ MPa, $\sigma_y = -30$ MPa, and $\tau_{xy} = 30$ MPa. What is the state of strain at p?

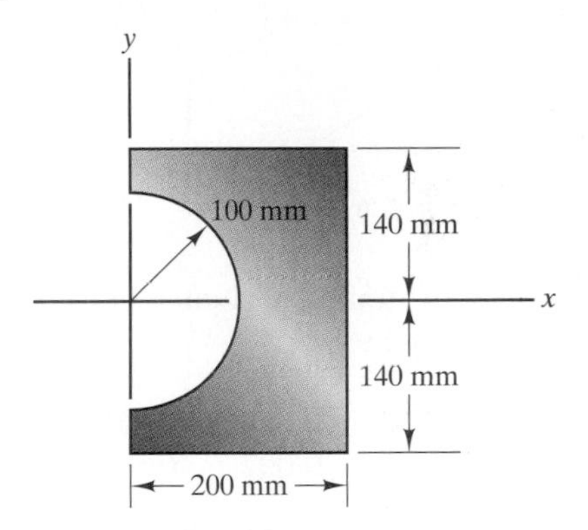

PROBLEM 6-4.5

6-4.6. An arm of a robotic actuator made of 7075-T6 aluminum alloy is subjected to a state of plane stress $\sigma_x, \sigma_y, \tau_{xy}$. Using a strain gauge rosette, it is determined experimentally that $\epsilon_x = 0.00350$, $\epsilon_y = 0.00600$, and $\gamma_{xy} = -0.02400$. What are the components of stress (in ksi)?

6-4.7. A material is subjected to a state of plane stress $\sigma_x = 400$ MPa, $\sigma_y = -200$ MPa, $\tau_{xy} = 300$ MPa. Using a strain gauge rosette, it is determined experimentally that $\epsilon_x = 0.00239$, $\epsilon_y = -0.00162$, and $\gamma_{xy} = 0.00401$. What are the modulus of elasticity and Poisson's ratio of the material?

6-4.8. For the material described in Problem 6-4.1, determine **(a)** the Lamé constants λ and μ; **(b)** the bulk modulus K.

6-4.9. The state of stress at a point p in a material with modulus of elasticity $E = 15 \times 10^6$ psi and Poisson's ratio $\nu = 0.33$ is

$$\begin{bmatrix} 50 & -60 & -40 \\ -60 & 40 & 40 \\ -40 & 40 & -40 \end{bmatrix} \text{ ksi.}$$

What is the state of strain at p?

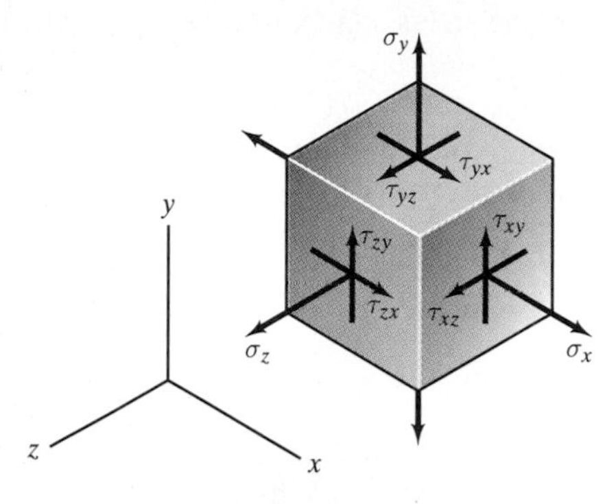

PROBLEM 6-4.9

6-4.10. The state of strain at a point p of the material described in Problem 6-4.9 is

$$\begin{bmatrix} 0.004 & 0.010 & -0.010 \\ 0.010 & -0.002 & 0 \\ -0.010 & 0 & -0.003 \end{bmatrix}.$$

What is the state of stress at p?

6-4.11. For the material described in Problem 6-4.9, determine **(a)** the Lamé constants λ and μ; **(b)** the bulk modulus K.

6-4.12. The cylindrical bar consists of a material with modulus of elasticity E and Poisson's ratio ν and is subjected to axial loads. The resulting state of stress is

$$\begin{bmatrix} \sigma_x & 0 & 0 \\ 0 & 0 & 0 \\ 0 & 0 & 0 \end{bmatrix}.$$

(a) Determine the state of strain in terms of σ_x, E, and ν. **(b)** The length of the unloaded bar is L and its cross-sectional area is A. Determine the volume of the loaded bar in terms of L, A, σ_x, E, and ν.

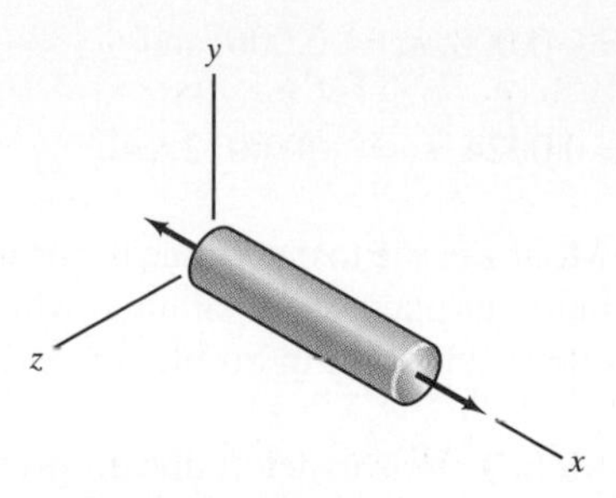

PROBLEM 6-4.12

6-4.13. The bar described in Problem 6-4.12 is subjected to axial loads that cause a normal stress $\sigma_x = 380$ MPa. The axial and lateral extensional strains are measured and determined to be $\epsilon_x = 0.0020$, $\epsilon_y = \epsilon_z = -0.0007$. Determine **(a)** the modulus of elasticity E, Poisson's ratio ν, and shear modulus G of the material; **(b)** the Lamé constants λ and μ of the material.

6-4.14. If the bearing's housing in Problem 6-2.10 is made of steel with elastic modulus $E = 200$ GPa and Poisson's ratio $\nu = 0.28$, what is the state of plane stress at p?

6-4.15. If the bearing's housing in Problem 6-2.10 is made of steel with elastic modulus $E = 200$ GPa and Poisson's ratio $\nu = 0.28$, what is the strain component ϵ_z at p?

6-4.16. At a point of a material subjected to a state of plane stress, the strains measured by the strain gauge rosette are $\epsilon_a = 0.006$, $\epsilon_b = -0.003$, and $\epsilon_c = -0.002$. The modulus of elasticity and Poisson's ratio of the material are $E = 30$ GPa and $\nu = 0.33$. What is the state of stress at the point?

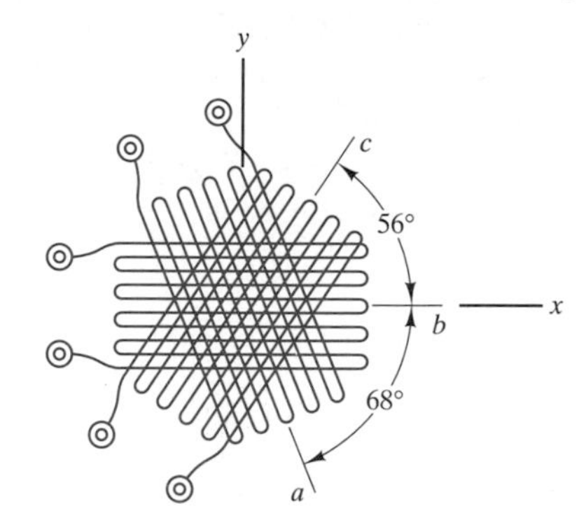

PROBLEM 6-4.16

6-4.17. At a point of a steel hydraulic piston that is subjected to a state of plane stress, the strains measured

by the strain gauge rosette shown in Problem 6-4.16 are $\epsilon_a = -0.00150$, $\epsilon_b = 0.00140$, and $\epsilon_c = 0.00086$. The modulus of elasticity and Poisson's ratio of the material are $E = 200$ GPa and $\nu = 0.33$. What are the principal stresses and the absolute maximum shear stress at the point?

6-4.18. The strain gauge rosette shown in Problem 6-4.16 is mounted on the outer wall of one of the Atlas launch vehicle's nozzles, where the material is in a homogeneous state of plane stress. The strains measured by the rosette are $\epsilon_a = 0.0053$, $\epsilon_b = 0.0038$, and $\epsilon_c = 0.0029$. The modulus of elasticity and Poisson's ratio of the material are $E = 70$ GPa and $\nu = 0.33$. What are the components of plane stress?

PROBLEM 6-4.18

6-4.19. An isotropic material is subjected to a normal stress $\sigma_x = \sigma$ and other components of stress are zero. By writing Eqs. (6-43)–(6-48) and Eqs. (6-51)–(6-56) for this state of stress and using Eq. (6-50), derive the relation

$$\lambda = \frac{\nu E}{(1+\nu)(1-2\nu)}.$$

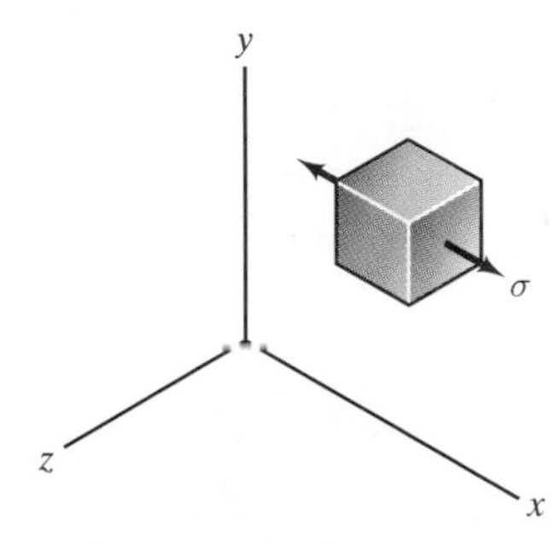

PROBLEM 6-4.19

6-4.20. Show that the bulk modulus is given in terms of the Lamé constants by

$$K = \lambda + \tfrac{2}{3}\mu.$$

6-4.21. An isotropic material is subjected to the same state of stress shown in Fig. 6-22. The $x'y'z'$ coordinate system shown is obtained by rotating the xyz coordinate system in Fig. 6-22 180° about the x axis. In terms of the $x'y'z'$ coordinate system, the only nonzero stress components are $\sigma'_x = \sigma$ and $\tau'_{xy} = -\tau$, so from Eqs. (6-25) the strain components are

$$\begin{aligned} \epsilon'_x &= b_{11}\sigma - b_{14}\tau, & \gamma'_{xy} &= b_{41}\sigma - b_{44}\tau, \\ \epsilon'_y &= b_{21}\sigma - b_{24}\tau, & \gamma'_{yz} &= b_{51}\sigma - b_{54}\tau, \\ \epsilon'_z &= b_{31}\sigma - b_{34}\tau, & \gamma'_{xz} &= b_{61}\sigma - b_{64}\tau. \end{aligned}$$

By comparing these equations to Eqs. (6-26), show that isotropy requires that $b_{14} = 0$, $b_{24} = 0$, $b_{34} = 0$, $b_{41} = 0$, $b_{54} = 0$, and $b_{61} = 0$.

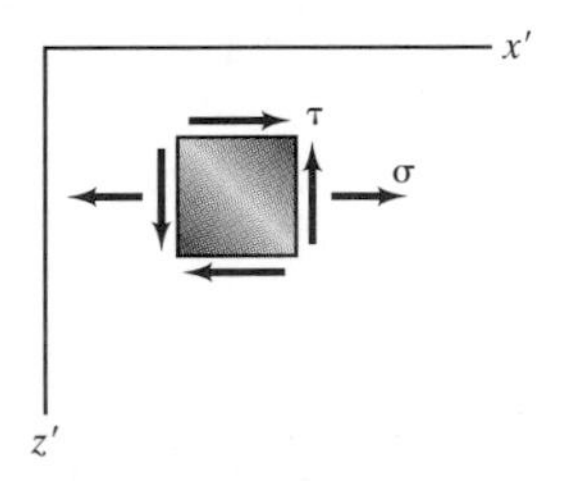

PROBLEM 6-4.21

The frame of Case Study House #22 (Pierre Koenig, 1960) consists of steel beams supported by slender steel columns.

CHAPTER 7

Internal Forces and Moments in Beams

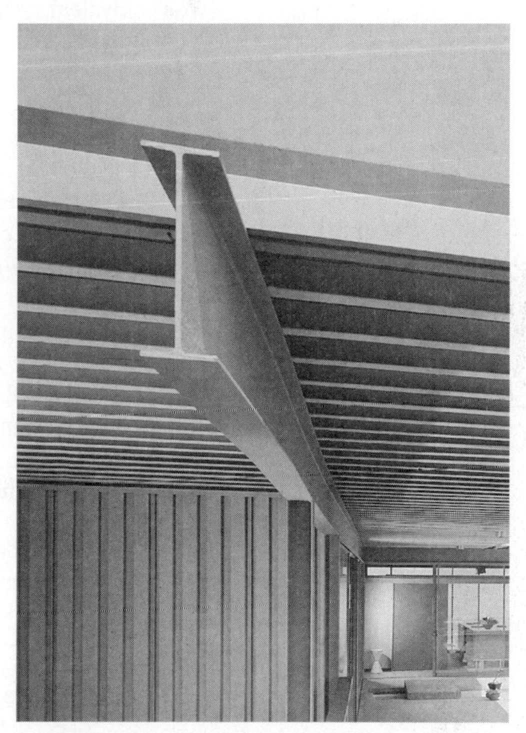

A *beam* is a slender structural member. (The word originally meant either a structural member or a *tree* in the Germanic language that became modern English, because the beams used in constructing buildings and ships were hewn from trees. The word for a tree in modern German is still *baum*.) Beams are the most common structural elements and make up the supporting structures of cars, aircraft, and buildings. In this chapter we begin the task of determining the states of stress and strain in beams by analyzing their internal forces and moments.

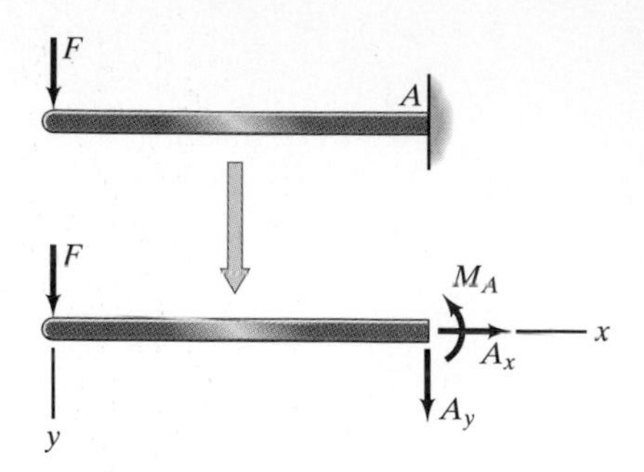

FIGURE 7-1 Beam subjected to a load and reactions.

7-1 | Axial Force, Shear Force, and Bending Moment

Consider the beam subjected to an external load and reactions in Fig. 7-1. Notice that we orient the coordinate system with the y axis downward, which is the traditional orientation for analyzing stresses in beams. To determine the internal forces and moments within the beam, in Fig. 7-2a we cut the beam by a plane perpendicular to the beam's axis and isolate part of it. You can see that the isolated part cannot be in equilibrium unless it is subjected to some system of forces and moments at the plane where it joins the other part of the beam. We know from statics that any system of forces and moments can be represented by an equivalent system consisting of a force acting at a given point and a couple. If the system of external loads and reactions on a beam is two-dimensional, we can represent the internal forces and moments by an equivalent system consisting of two components of force and a couple as shown in Fig. 7-2b. The *axial force* P is parallel to the beam's axis. The force component V normal to the beam's axis is called the *shear force,* and the couple M is called the *bending moment.*

The directions of the axial force, shear force, and bending moment in Fig. 7-2b are the established definitions of the positive directions of these quantities. A positive axial force P subjects the beam to tension. A positive shear force V tends to rotate the axis of the beam clockwise (Fig. 7-3a). Bending moments are defined to be positive when they tend to bend the axis of the beam in the negative y-axis direction (Fig. 7-3b).

In Chapters 8 and 9 we show that knowledge of the internal forces and moment in a beam is essential for evaluating the states of stress and deformations resulting from a given system of loads. Determining the internal forces and moment at a particular cross section of a beam typically involves three steps.

1. Draw the free-body diagram of the entire beam and determine the reactions at its supports.

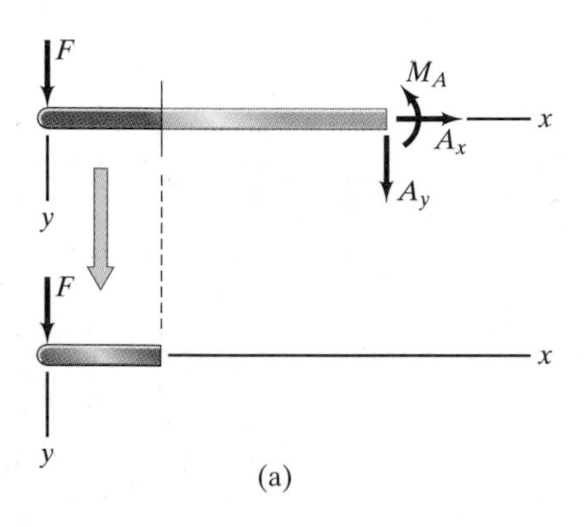

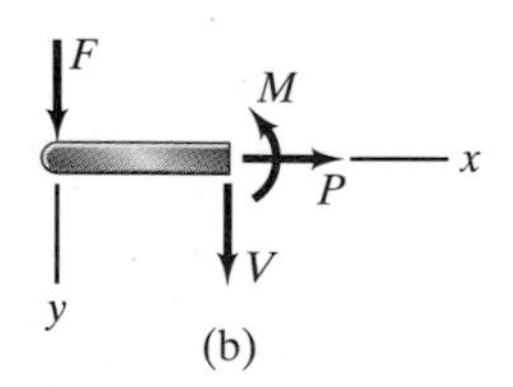

FIGURE 7-2 (a) Isolating part of the beam. (b) Axial force, shear force, and bending moment.

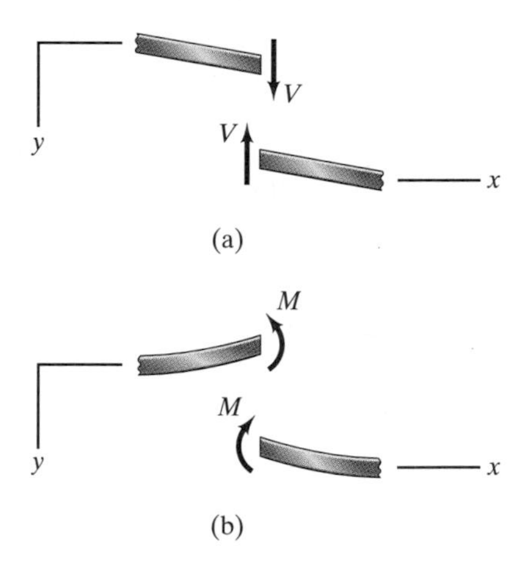

FIGURE 7-3 (a) Positive shear forces tend to rotate the axis of the beam clockwise. (b) Positive bending moments tend to bend the axis of the beam in the negative y-axis direction.

2. Cut the beam where you wish to determine the internal forces and moment and draw the free-body diagram of one of the resulting parts. You can choose the part with the simplest free-body diagram. If your cut divides a distributed load, don't represent the distributed load by an equivalent force until after you have obtained your free-body diagram.
3. Use the equilibrium equations to determine P, V, and M.

EXAMPLE 7-1

For the beam in Fig. 7-4, determine the internal forces and moment at C.

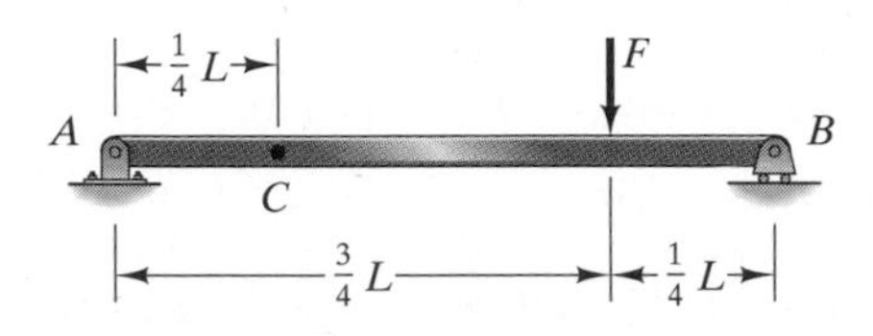

FIGURE 7-4

Solution

Determine the external forces and moments The first step is to draw the free-body diagram of the entire beam and determine the reactions at its supports. We simply show the results of this step in Fig. (a).

Draw the free-body diagram of part of the beam We cut the beam at C [Fig. (a)] and draw the free-body diagram of the left part, including the internal forces and moment P_C, V_C, and M_C in their defined positive directions [Fig. (b)].

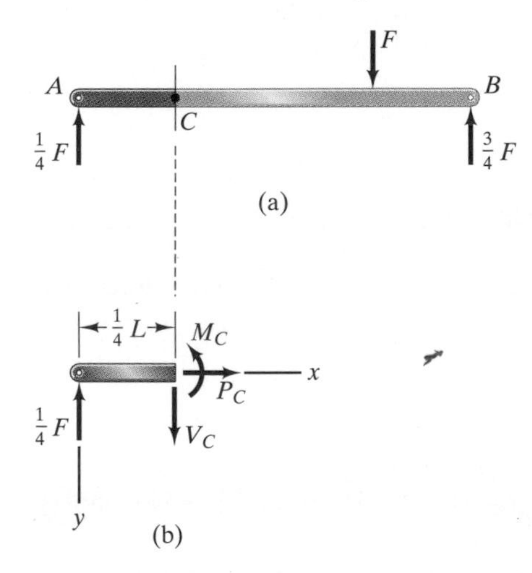

(a) Free-body diagram of the beam and a plane through point C. (b) Free-body diagram of the part of the beam to the left of the plane through point C.

Apply the equilibrium equations From the equilibrium equations

$$\Sigma F_x = P_C = 0,$$
$$\Sigma F_y = V_C - \tfrac{1}{4}F = 0,$$
$$\Sigma M_{\text{point } C} = M_C - \left(\tfrac{1}{4}L\right)\left(\tfrac{1}{4}F\right) = 0,$$

we obtain $P_C = 0$, $V_C = F/4$, and $M_C = LF/16$.

Discussion

We should check our results with the free-body diagram of the other part of the beam [Fig. (c)]. The equilibrium equations are

$$\Sigma F_x = -P_C = 0,$$
$$\Sigma F_y = -V_C + F - \tfrac{3}{4}F = 0,$$
$$\Sigma M_{\text{point } C} = -M_C - \left(\tfrac{1}{2}L\right)F + \left(\tfrac{3}{4}L\right)\left(\tfrac{3}{4}F\right) = 0,$$

which confirm that $P_C = 0$, $V_C = F/4$, and $M_C = LF/16$.

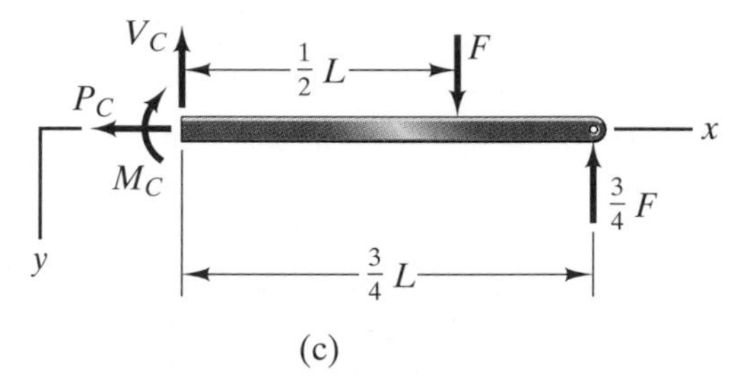

(c)

(c) Free-body diagram of the part of the beam to the right of the plane through point C.

EXAMPLE 7-2

For the beam in Fig. 7-5, determine the internal forces and moment at B and at C.

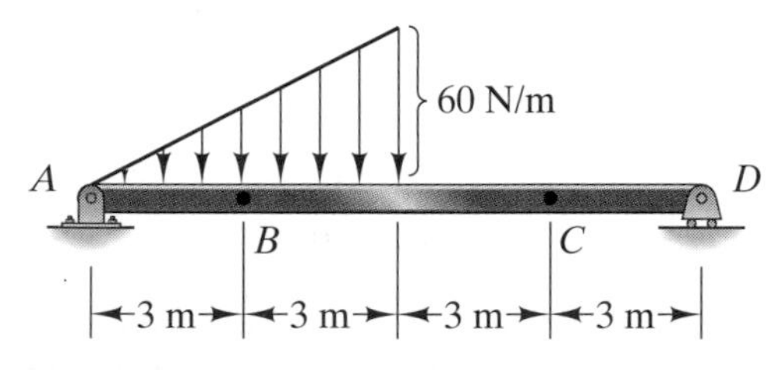

FIGURE 7-5

Solution

Determine the external forces and moments We draw the free-body diagram of the beam and represent the distributed load by an equivalent force in Fig. (a).

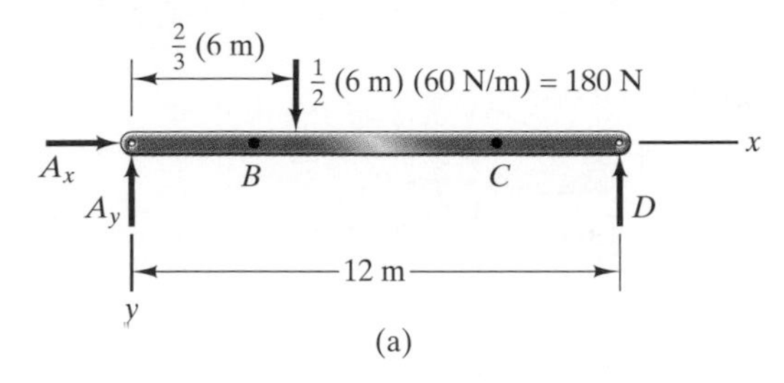

(a) Free-body diagram of the entire beam with the distributed load represented by an equivalent force.

The equilibrium equations are

$$\Sigma F_x = A_x = 0,$$
$$\Sigma F_y = 180 - A_y - D = 0,$$
$$\Sigma M_{\text{point } A} = 12D - (4)(180) = 0.$$

Solving them, we obtain $A_x = 0$, $A_y = 120$ N, and $D = 60$ N.

Draw the free-body diagram of part of the beam We cut the beam at B, obtaining the free-body diagram in Fig. (b). Because point B is at the midpoint of the triangular distributed load, the value of the distributed load at B is 30 N/m. By representing the distributed load in Fig. (b) by an equivalent force, we obtain the free-body diagram in Fig. (c). From the equilibrium equations

$$\Sigma F_x = P_B = 0,$$
$$\Sigma F_y = V_B + 45 - 120 = 0,$$
$$\Sigma M_{\text{point } B} = M_B + (1)(45) - (3)(120) = 0,$$

we obtain $P_B = 0$, $V_B = 75$ N, and $M_B = 315$ N-m.

To determine the internal forces and moment at C, we obtain the simplest free-body diagram by isolating the part of the beam to the right of C [Fig. (d)]. From the equilibrium equations

$$\Sigma F_x = -P_C = 0,$$
$$\Sigma F_y = -V_C - 60 = 0,$$
$$\Sigma M_{\text{point } C} = -M_C + (3)(60) = 0,$$

we obtain $P_C = 0$, $V_C = -60$ N, and $M_C = 180$ N-m.

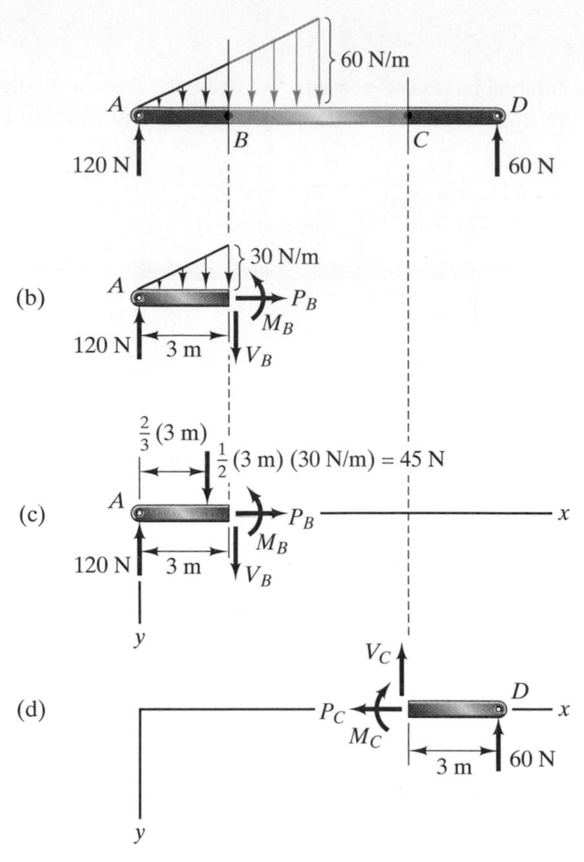

(b), (c) Free-body diagrams of the part of the beam to the left of point B.
(d) Free-body diagram of the part of the beam to the right of point C.

Discussion

If you attempt to determine the internal forces and moment at B by cutting the free-body diagram in Fig. (a) at B, you do *not* obtain correct results. (You can confirm that the resulting free-body diagram of the part of the beam to the left of B gives $P_B = 0$, $V_B = 120$ N, and $M_B = 360$ N-m.) The reason is that you do not account properly for the effect of the distributed load on your free-body diagram. You must wait until *after* you have obtained the free-body diagram of part of the beam before representing distributed loads by equivalent forces.

7-2 | Shear Force and Bending Moment Diagrams

To determine whether a beam will support a given set of loads, the structural designer must know the state of stress throughout the beam. To evaluate the state of stress, the internal forces and moment must be determined throughout

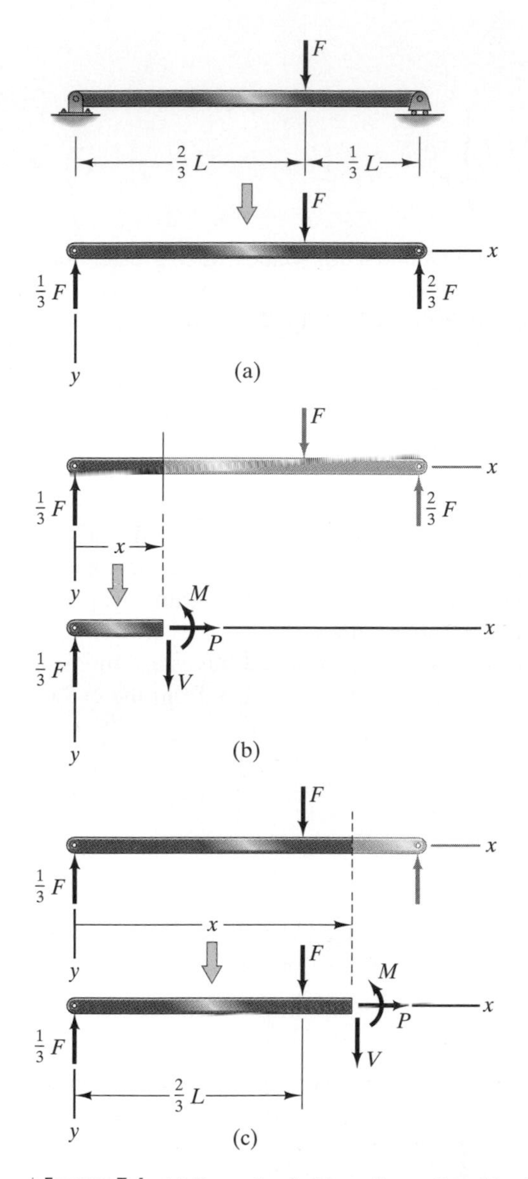

FIGURE 7-6 (a) Beam loaded by a force F and its free-body diagram. (b) Cutting the beam at an arbitrary position x to the left of F. (c) Cutting the beam at an arbitrary position x to the right of F.

the beam's length. In this section we show how the values of P, V, and M can be determined as functions of x and introduce shear force and bending moment diagrams.

Let's consider a simply supported beam loaded by a force (Fig. 7-6a). Instead of cutting the beam at a specific cross section to determine the internal forces and moment, we cut it at an arbitrary position x between the left end of the beam and the load F (Fig. 7-6b). Applying the equilibrium equations to this free-body diagram, we obtain

$$\left.\begin{aligned} P &= 0 \\ V &= \tfrac{1}{3}F \\ M &= \tfrac{1}{3}Fx \end{aligned}\right\} \quad 0 < x < \tfrac{2}{3}L.$$

To determine the internal forces and moment for values of x greater than $\frac{2}{3}L$, we obtain a free-body diagram by cutting the beam at an arbitrary position x between the load F and the right end of the beam (Fig. 7-6c). The results are

$$\left.\begin{aligned} P &= 0 \\ V &= -\tfrac{2}{3}F \\ M &= \tfrac{2}{3}F(L-x) \end{aligned}\right\} \quad \tfrac{2}{3}L < x < L.$$

Shear force and bending moment diagrams are simply graphs of V and M, respectively, as functions of x (Fig. 7-7). They permit you to see the changes in the shear force and bending moment that occur along the beam's length as well as their maximum positive and negative values.

Thus you can determine the distributions of the internal forces and moment in a beam by considering a plane at an arbitrary distance x from the end of

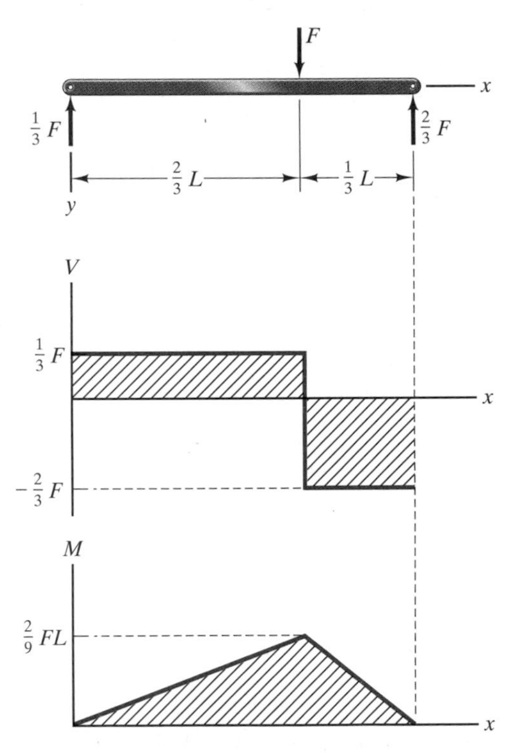

FIGURE 7-7 Shear force and bending moment diagrams indicating the maximum positive and negative values of V and M.

the beam and solving for P, V, and M as functions of x. Depending on the complexity of the loading of the beam, you may have to draw several free-body diagrams to determine the distributions over the entire length of the beam. The resulting equations allow you to determine the maximum positive and negative values of the shear force and bending moment and also draw the shear force and bending moment diagrams.

EXAMPLE 7-3

Worksheet 8

Draw the shear force and bending moment diagrams for the beam in Fig. 7-8.

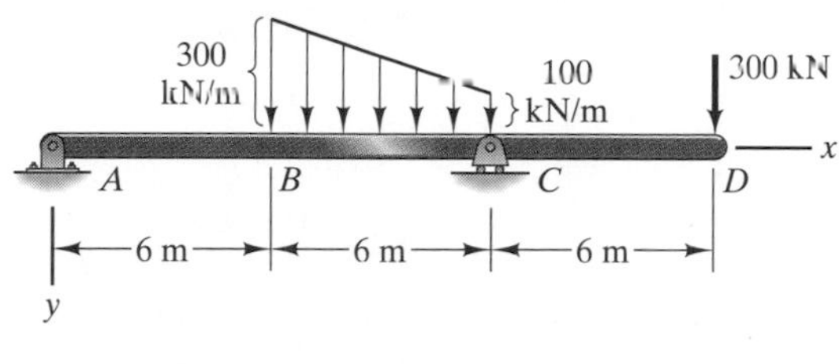

FIGURE 7-8

Strategy

To determine the internal forces and moment as functions of x for the entire beam, we must use three free-body diagrams: one for the range $0 < x < 6$ m, one for $6 < x < 12$ m, and one for $12 < x < 18$ m.

Solution

We begin by drawing the free-body diagram of the entire beam [Fig. (a)]. We treat the distributed load as the sum of uniform and triangular distributed loads and represent these distributed loads by equivalent forces.

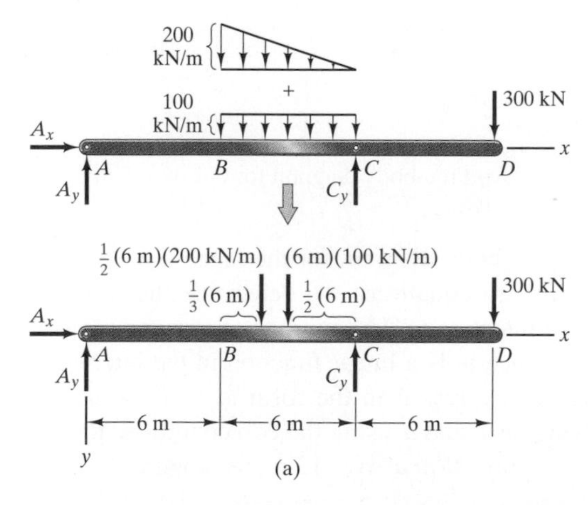

(a) Free-body diagram of the beam representing the distributed load by two equivalent forces.

From the equilibrium equations

$$\Sigma F_x = A_x = 0,$$
$$\Sigma F_y = -A_y - C_y + 600 + 600 + 300 = 0,$$
$$\Sigma M_{\text{point } A} = 12C_y - (8)(600) - (9)(600) - (18)(300) = 0,$$

we obtain the reactions $A_x = 0$, $A_y = 200$ kN, and $C_y = 1300$ kN.

We draw the free-body diagram for the range $0 < x < 6$ m in Fig. (b). From the equilibrium equations

$$\Sigma F_x = P = 0,$$
$$\Sigma F_y = -200 + V = 0,$$
$$\Sigma M_{\text{right end}} = M - 200x = 0,$$

we obtain

$$\left.\begin{aligned} P &= 0 \\ V &= 200 \text{ kN} \\ M &= 200x \text{ kN-m} \end{aligned}\right\} \quad 0 < x < 6 \text{ m}.$$

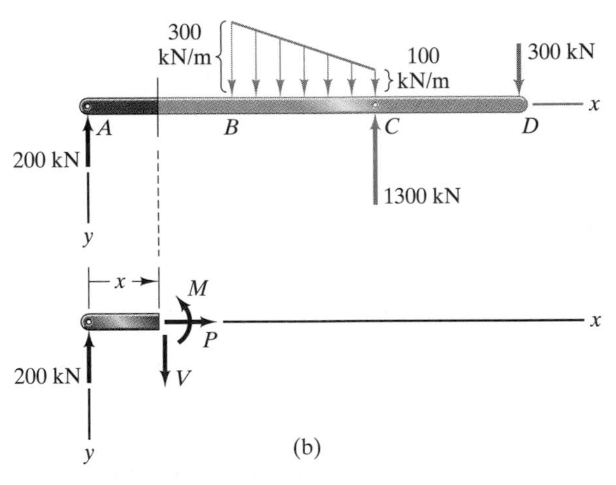

(b) Free-body diagram for $0 < x < 6$ m.

We draw the free-body diagram for the range $6 < x < 12$ m in Fig. (c). To obtain the equilibrium equations, we determine the distributed load w as a function of x and integrate to determine the force and moment exerted by the distributed load. Since w is a linear function in the interval from $x = 6$ m to $x = 12$ m, we can express it in the form $w = cx + d$, where c and d are constants. Solving for c and d using the two conditions that $w = 300$ kN/m at $x = 6$ m and $w = 100$ kN/m at $x = 12$ m, we obtain

$$w = -\tfrac{100}{3}x + 500 \text{ kN/m}.$$

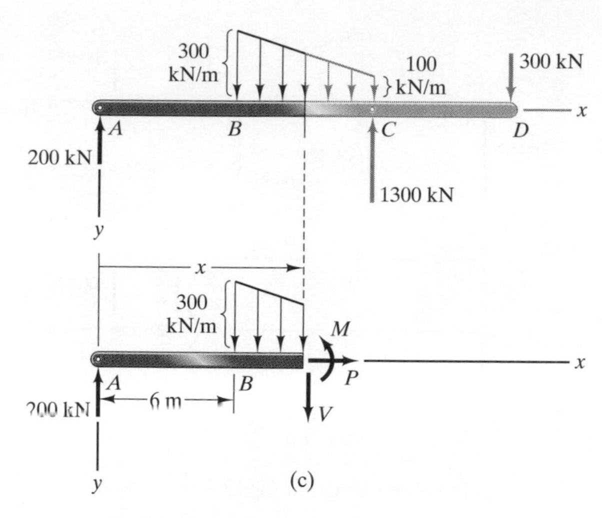

(c) Free-body diagram for $6 < x < 12$ m.

The downward force on the free body in Fig. (c) due to the distributed load is

$$F = \int_L w\,dx = \int_6^x \left(-\tfrac{100}{3}x + 500\right) dx = -\tfrac{50}{3}x^2 + 500x - 2400 \text{ kN}.$$

The clockwise moment about the origin (point A) due to the distributed load is

$$\int_L xw\,dx = \int_6^x \left(-\tfrac{100}{3}x^2 + 500x\right) dx = -\tfrac{100}{9}x^3 + 250x^2 - 6600 \text{ kN-m}.$$

The equilibrium equations are

$$\Sigma F_x = P = 0,$$
$$\Sigma F_y = -200 + V - \tfrac{50}{3}x^2 + 500x - 2400 = 0,$$
$$\Sigma M_{\text{point } A} = M - Vx + \tfrac{100}{9}x^3 - 250x^2 + 6600 = 0.$$

Solving them, we obtain

$$\left.\begin{aligned} P &= 0 \\ V &= \tfrac{50}{3}x^2 - 500x + 2600 \text{ kN} \\ M &= \tfrac{50}{9}x^3 - 250x^2 + 2600x - 6600 \text{ kN-m} \end{aligned}\right\} \quad 6 < x < 12 \text{ m}.$$

For the range $12 < x < 18$ m, we obtain a very simple free-body diagram by using the part of the beam on the right of the cut [Fig. (d)]. From the equilibrium equations,

$$\Sigma F_x = P = 0,$$
$$\Sigma F_y = -V + 300 = 0,$$
$$\Sigma M_{\text{left end}} = -M - 300(18 - x) = 0,$$

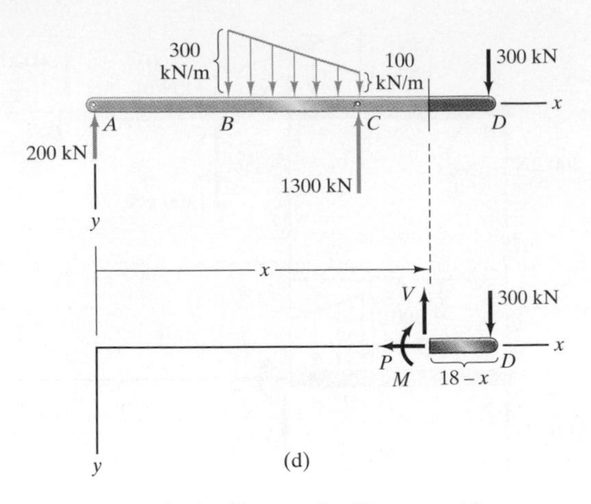

(d) Free-body diagram for $12 < x < 18$ m.

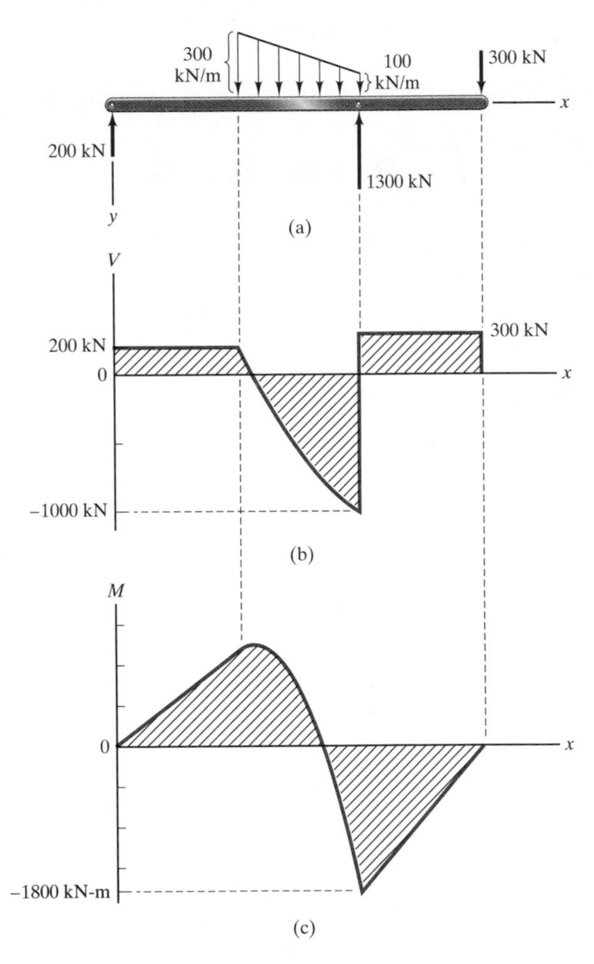

(e) Shear force and bending moment diagrams.

we obtain

$$\left.\begin{aligned} P &= 0 \\ V &= 300 \text{ kN} \\ M &= 300x - 5400 \text{ kN-m} \end{aligned}\right\} \quad 12 < x < 18 \text{ m.}$$

The shear force and bending moment diagrams, obtained by plotting the equations for V and M for the three ranges of x, are shown in Fig. (e).

7-3 | Equations Relating Distributed Load, Shear Force, and Bending Moment

The shear force and bending moment in a beam subjected to a distributed load are governed by simple differential equations. In this section we derive these equations and show that they provide an interesting and enlightening way to obtain the shear force and bending moment diagrams. In Chapters 8 and 9 we show that these equations are also needed for determining the states of stress and the deflections of beams.

Suppose that a portion of a beam is subjected to a distributed load w (Fig. 7-9a). In Fig. 7-9b we obtain a free-body diagram by cutting the beam at x and at $x + \Delta x$. The terms ΔP, ΔV, and ΔM are the changes in the axial force, shear force, and bending moment, respectively, from x to $x + \Delta x$. The sum of the forces in the x direction is

$$\Sigma F_x = P + \Delta P - P = 0.$$

Dividing this equation by Δx and taking the limit as $\Delta x \to 0$, we obtain

$$\frac{dP}{dx} = 0.$$

To sum the forces on the free-body diagram in the y direction, we must determine the downward force exerted by the distributed load. In Fig. 7-9b we introduce a coordinate $\hat{x}$ that measures distance from the left edge of the free-body diagram. In terms of this coordinate, the downward force exerted on the free-body diagram by the distributed load is

$$\int_0^{\Delta x} w(x + \hat{x})\, d\hat{x}, \tag{7-1}$$

where $w(x + \hat{x})$ denotes the value of w at $x + \hat{x}$. To evaluate this integral, we express $w(x + \hat{x})$ as a Taylor series in terms of $\hat{x}$:

$$w(x + \hat{x}) = w(x) + \frac{dw(x)}{dx}\hat{x} + \frac{1}{2}\frac{d^2 w(x)}{dx^2}\hat{x}^2 + \cdots. \tag{7-2}$$

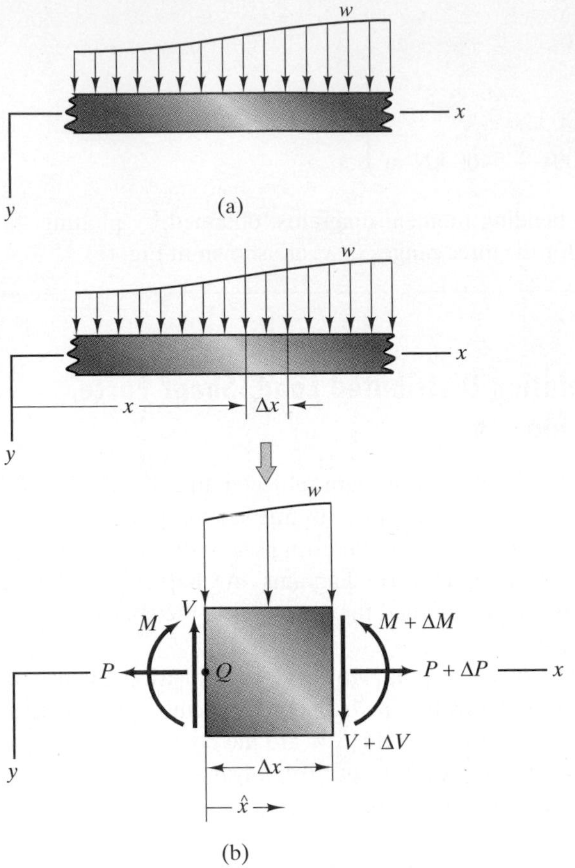

FIGURE 7-9 (a) Portion of a beam subjected to a distributed force w. (b) Obtaining the free-body diagram of an element of the beam.

Substituting this expression into Eq. (7-1) and integrating term by term, the downward force exerted by the distributed load is

$$w(x)\,\Delta x + \frac{1}{2}\frac{dw(x)}{dx}(\Delta x)^2 + \cdots.$$

The sum of the forces on the free-body diagram in the y direction is therefore

$$\Sigma F_y = V + \Delta V - V + w(x)\,\Delta x + \frac{1}{2}\frac{dw(x)}{dx}(\Delta x)^2 + \cdots = 0.$$

Dividing by Δx and taking the limit as $\Delta x \to 0$, we obtain

$$\frac{dV}{dx} = -w,$$

where $w = w(x)$.

Our next step is to sum the moments on the free-body diagram in Fig. 7-9b about point Q. The clockwise moment about Q due to the distributed load is

$$\int_0^{\Delta x} \hat{x} w(x + \hat{x})\, d\hat{x}.$$

Substituting Eq. (7-2) into this expression and integrating term by term, the moment is

$$\frac{1}{2} w(x)(\Delta x)^2 + \frac{1}{3} \frac{dw(x)}{dx} (\Delta x)^3 + \cdots.$$

The sum of the moments on the free-body diagram about Q is therefore

$$\Sigma M_{\text{point } Q} = M + \Delta M - M - (V + \Delta V)\,\Delta x$$
$$- \frac{1}{2} w(x)(\Delta x)^2 - \frac{1}{3} \frac{dw(x)}{dx} (\Delta x)^3 + \cdots = 0.$$

Dividing this equation by Δx and taking the limit as $\Delta x \to 0$ gives

$$\frac{dM}{dx} = V.$$

In summary, we have obtained three differential equations:

$$\frac{dP}{dx} = 0, \quad \textbf{(7-3)}$$

$$\frac{dV}{dx} = -w, \quad \textbf{(7-4)}$$

$$\frac{dM}{dx} = V. \quad \textbf{(7-5)}$$

Equation (7-3) states that the axial force does not depend on x in a portion of a beam subjected only to a lateral distributed load. Equation (7-4) relates the rate of change of the shear force to the distributed load, and Eq. (7-5) relates the rate of change of the bending moment to the shear force. In principle, these equations can be used to determine the distributions of the shear force and bending moment in a beam: We can integrate Eq. (7-4) to determine V as a function of x, then integrate Eq. (7-5) to determine M as a function of x.

However, we derived Eqs. (7-4) and (7-5) for a segment of beam subjected only to a distributed load. To apply them for a more general loading, we must also account for the effects of any point forces and couples acting on the beam. Let us determine what happens to the shear force and bending moment where a beam is subjected to a force F in the positive y direction (Fig. 7-10a). By cutting the beam just to the left and just to the right of the force, we obtain the free-body diagram in Fig. 7-10b, where the subscripts $-$ and $+$ denote values to the left and right of the force. Equilibrium requires that

$$V_+ - V_- = -F, \quad \textbf{(7-6)}$$

$$M_+ - M_- = 0. \quad \textbf{(7-7)}$$

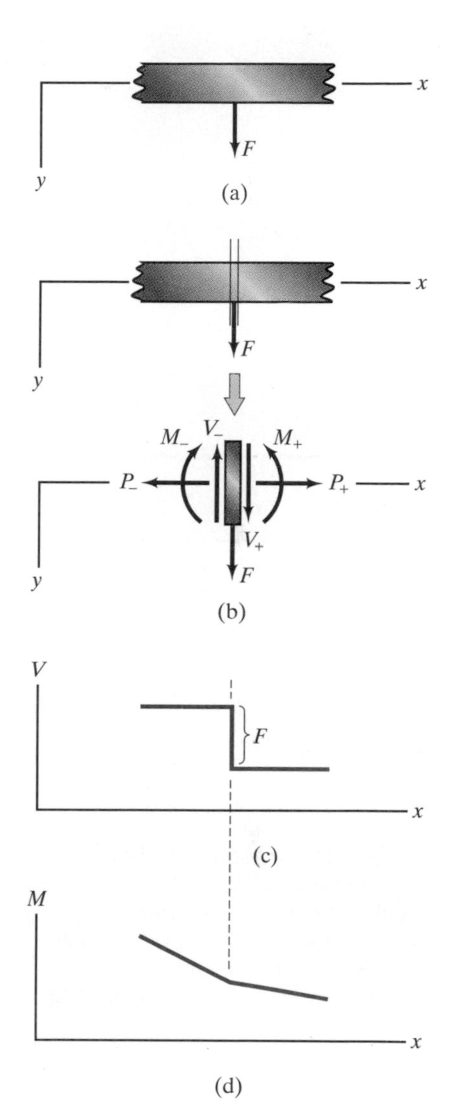

FIGURE 7-10 (a) Portion of a beam subjected to a force F in the positive y direction. (b) Obtaining a free-body diagram by cutting the beam to the left and right of F. (c) The shear force diagram undergoes a decrease of magnitude F. (d) The bending moment diagram is continuous.

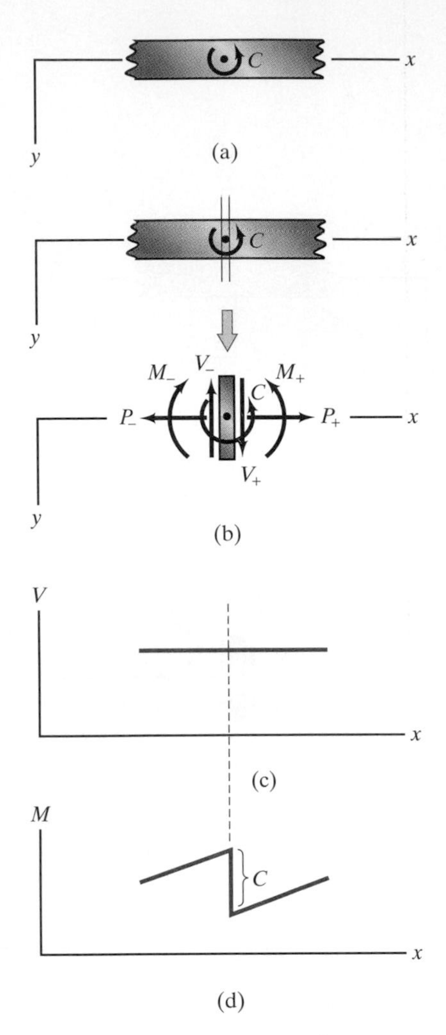

FIGURE 7-11 (a) Portion of a beam subjected to a counterclockwise couple C. (b) Obtaining a free-body diagram by cutting the beam to the left and right of C. (c) The shear force diagram is continuous. (d) The bending moment diagram undergoes a decrease of magnitude C.

We see that the shear force diagram undergoes a decrease of magnitude F (Fig. 7-10c), but the bending moment diagram is continuous (Fig. 7-10d). The change in the shear force is *negative* if the force F is in the positive y direction.

Now we consider what happens to the shear force and bending moment diagrams where a beam is subjected to a counterclockwise couple C (Fig. 7-11a). Cutting the beam just to the left and just to the right of the couple (Fig. 7-11b), we determine that

$$V_+ - V_- = 0, \quad \textbf{(7-8)}$$

$$M_+ - M_- = -C. \quad \textbf{(7-9)}$$

The shear force diagram is continuous (Fig. 7-11c), but the bending moment diagram undergoes a decrease of magnitude C (Fig. 7-11d) where a beam is subjected to a couple. The change in the bending moment is *negative* if the couple is in the counterclockwise direction.

Summarizing these results:

1. A point force results in a jump discontinuity in the shear force but no discontinuity in the bending moment. A force F in the positive y direction causes a decrease in the shear force of magnitude F. Observe in Fig. 7-10 that the shear force distribution changes in the same direction as the force.
2. A couple results in a jump discontinuity in the bending moment but no discontinuity in the shear force. A counterclockwise couple C causes a decrease in the bending moment of magnitude C.

We can demonstrate these results with the cantilever beam in Fig. 7-12a. To determine the shear force diagram, we first observe that the force F at $x = 0$ results in a decrease in V of magnitude F (Fig. 7-12b). Since there is no distributed load on the beam, Eq. (7-4) states that $dV/dx = 0$. The couple at $x = L/2$ does not affect the shear force, so the shear force remains constant, $V = -F$, in the interval $0 < x < L$ (Fig. 7-12c).

To determine the bending moment diagram, we begin again at $x = 0$. There is no couple at $x = 0$, so the value of the bending moment there is zero. In the interval $0 < x < L/2$, the shear force $V = -F$. Integrating Eq. (7-5) from $x = 0$ to an arbitrary value of x within the interval $0 < x < L/2$,

$$\int_0^M dM = \int_0^x V\,dx = \int_0^x -F\,dx,$$

we determine M as a function of x in this interval:

$$M = -Fx, \qquad 0 < x < L/2.$$

The bending moment diagram in this interval is shown in Fig. 7-12d. The value of the bending moment just to the left of the couple at $x = L/2$ is $M = -FL/2$. The counterclockwise couple C causes a decrease in the value of the bending moment of magnitude C (Fig. 7-12e), so the bending moment just to the right of the couple C is $M = -(FL/2) - C$.

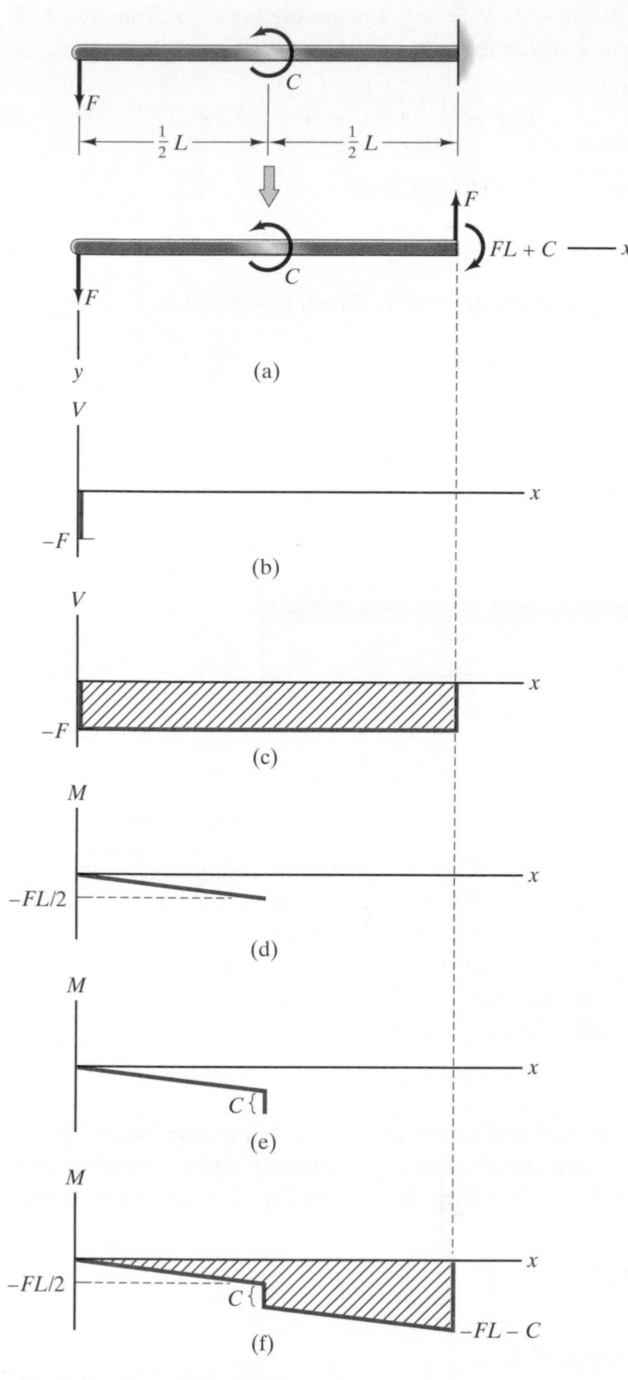

FIGURE 7-12 (a) Beam loaded by a force and a couple. (b) The shear force undergoes a decrease of magnitude F at $x = 0$. (c) The shear force is constant in the interval $0 < x < L$. (d) Bending moment diagram from $x = 0$ to $x = L/2$. (e) The bending moment undergoes a decrease of magnitude C at $x = L/2$. (f) Complete bending moment diagram.

In the interval $L/2 < x < L$, $V = -F$. Integrating Eq. (7-5) from $x = L/2$ to an arbitrary value of x within the interval $L/2 < x < L$,

$$\int_{-(FL/2)-C}^{M} dM = \int_{L/2}^{x} V\,dx = \int_{L/2}^{x} -F\,dx,$$

we obtain M as a function of x in this interval:

$$M = -Fx - C, \qquad L/2 < x < L.$$

The completed bending moment diagram is shown in Fig. 7-12f.

EXAMPLE 7-4

Use Eqs. (7-4) and (7-5) to determine the shear force and bending moment diagrams for the beam in Fig. 7-13.

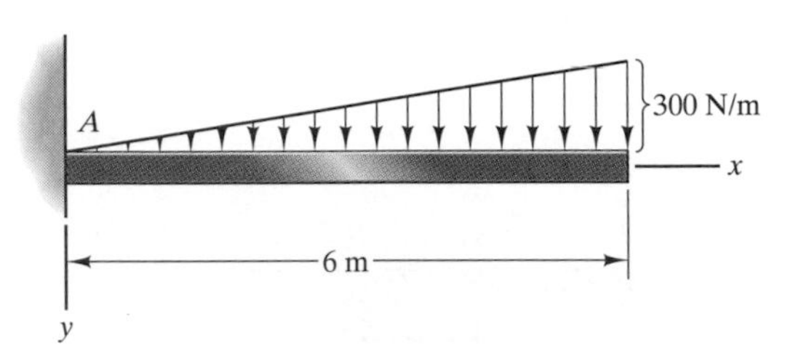

FIGURE 7-13

Strategy

We must first draw the free-body diagram of the entire beam and determine the reactions at the built-in support A. The result of this step is shown in Fig. (a). We can then use Eq. (7-4) to determine the shear force as a function of x, accounting for the effect of the 900-N force at A. Once the shear force is known, we can use Eq. (7-5) to determine the bending moment as a function of x, accounting for the effect of the 3600 N-m couple at A.

Solution

Shear force diagram The force at A causes an increase in the shear force of 900-N magnitude at $x = 0$. Expressing the linearly distributed load as a function of x, we obtain $w = (x/6)300 = 50x$ N/m. Integrating Eq. (7-4) from $x = 0$ to an arbitrary value of x,

$$\int_{900}^{V} dV = \int_{0}^{x} -w\,dx = \int_{0}^{x} -50x\,dx,$$

we obtain V as a function of x:

$$V = 900 - 25x^2\text{N}.$$

The shear force diagram is shown in Fig. (b).

Bending moment diagram The counterclockwise couple at A causes a decrease in the bending moment of 3600 N-m magnitude at $x = 0$. Integrating Eq. (7-5) from $x = 0$ to an arbitrary value of x,

$$\int_{-3600}^{M} dM = \int_0^x V\,dx = \int_0^x (900 - 25x^2)\,dx,$$

we obtain M as a function of x:

$$M = -3600 + 900x - \tfrac{25}{3}x^3 N - m.$$

The bending moment diagram is shown in Fig. (c).

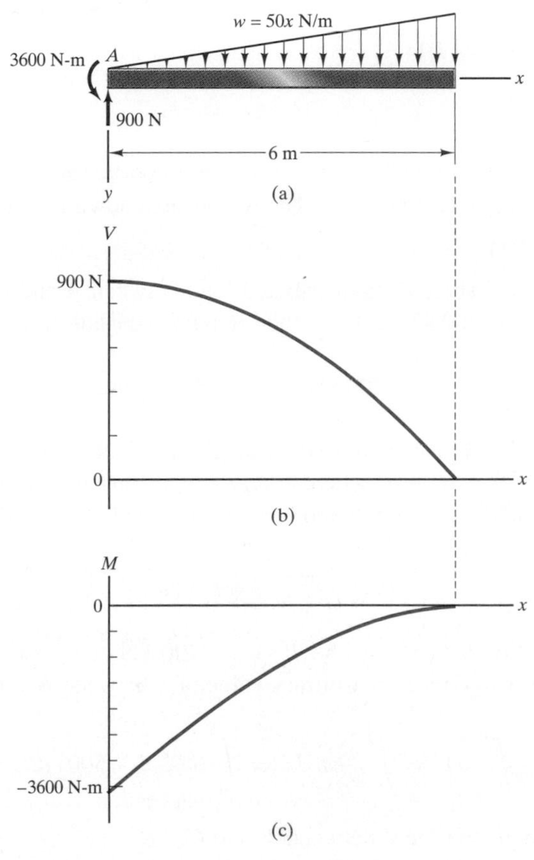

(a) Free-body diagram of the beam. (b) Shear force diagram. (c) Bending moment diagram.

Discussion

Once you have determined V and M as functions of x, a useful check is to confirm that your results satisfy the equation $dM/dx = V$. In this example,

$$\frac{dM}{dx} = \frac{d}{dx}\left(-3600 + 900x - \tfrac{25}{3}x^3\right) = 900 - 25x^2 = V.$$

EXAMPLE 7-5

Worksheet 9

Use Eqs. (7-4) and (7-5) to determine the shear force and bending moment diagrams for the beam in Fig. 7-14.

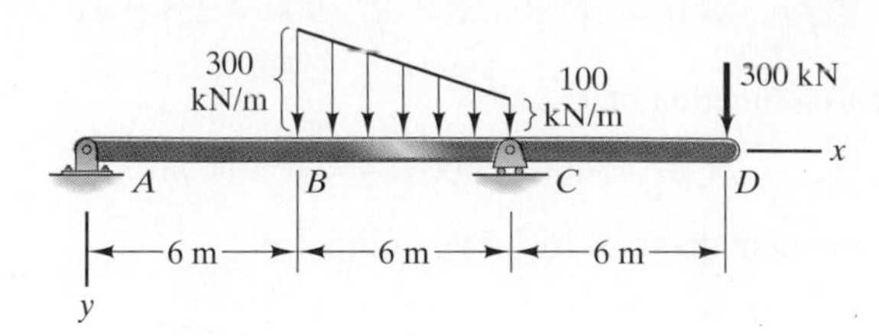

FIGURE 7-14

Solution

The first step, determining the reactions at the supports, was carried out for this beam and loading in Example 7-3. The results are shown in Fig. (a).

Shear force diagram

From A to B There is no distributed load between A and B, so the shear force increases by 200 kN at A and then remains constant from A to B:

$$V = 200 \text{ kN}, \qquad 0 < x < 6 \text{ m}.$$

From B to C We can express the linearly distributed load between B and C in the form $w = cx + d$, where c and d are constants. Using the conditions $w = 300$ kN/m at $x = 6$ m and $w = 100$ kN/m at $x = 12$ m, we obtain the equation

$$w = -\tfrac{100}{3}x + 500 \text{ kN/m}.$$

From our solution between A and B, $V = 200$ kN at $x = 6$ m. Integrating Eq. (7-4) from $x = 6$ m to an arbitrary value of x between B and C,

$$\int_{200}^{V} dV = \int_{6}^{x} -w\, dx = \int_{6}^{x} \left(\tfrac{100}{3}x - 500\right) dx,$$

we obtain an equation for V between B and C:

$$V = \tfrac{100}{6}x^2 - 500x + 2600 \text{ kN}, \qquad 6 < x < 12 \text{ m}.$$

From C to D At C, V undergoes an increase of 1300-N magnitude due to the force exerted by the pin support. Adding this change to the value of V at C obtained from our solution from B to C, the value of V just to the right of C is

$$1300 + \tfrac{100}{6}(12)^2 - 500(12) + 2600 = 300 \text{ kN}.$$

There is no loading between C and D, so V remains constant from C to D:

$$V = 300 \text{ kN}, \qquad 12 < x < 18 \text{ m}.$$

The shear force diagram is shown in Fig. (b).

Bending moment diagram

From A to B Integrating Eq. (7-3) from $x = 0$ to an arbitrary value of x between A and B,

$$\int_0^M dM = \int_0^x V\,dx = \int_0^x 200\,dx,$$

we obtain

$$M = 200x \text{ kN-m}, \qquad 0 < x < 6 \text{ m}.$$

At $x = 6$ m, $M = 1200$ kN-m.

From B to C Integrating Eq. (7-5) from $x = 6$ m to an arbitrary value of x between B and C,

$$\int_{1200}^M dM = \int_6^x V\,dx = \int_6^x \left(\tfrac{100}{6}x^2 - 500x + 2600\right) dx,$$

we obtain

$$M = \tfrac{50}{9}x^3 - 250x^2 + 2600x - 6600 \text{ kN-m}, \qquad 6 < x < 12 \text{ m}.$$

At $x = 12$ m, $M = -1800$ kN-m.

From C to D Integrating Eq. (7-5) from $x = 12$ m to an arbitrary value of x between C and D,

$$\int_{-1800}^M dM = \int_{12}^x V\,dx = \int_{12}^x 300\,dx,$$

we obtain

$$M = 300x - 5400 \text{ kN-m}, \qquad 12 < x < 18 \text{ m}.$$

The bending moment diagram is shown in Fig. (c).

Discussion

Compare this example with Example 7-3, in which we use free-body diagrams to determine the shear force and bending moment as functions of x for this beam and loading.

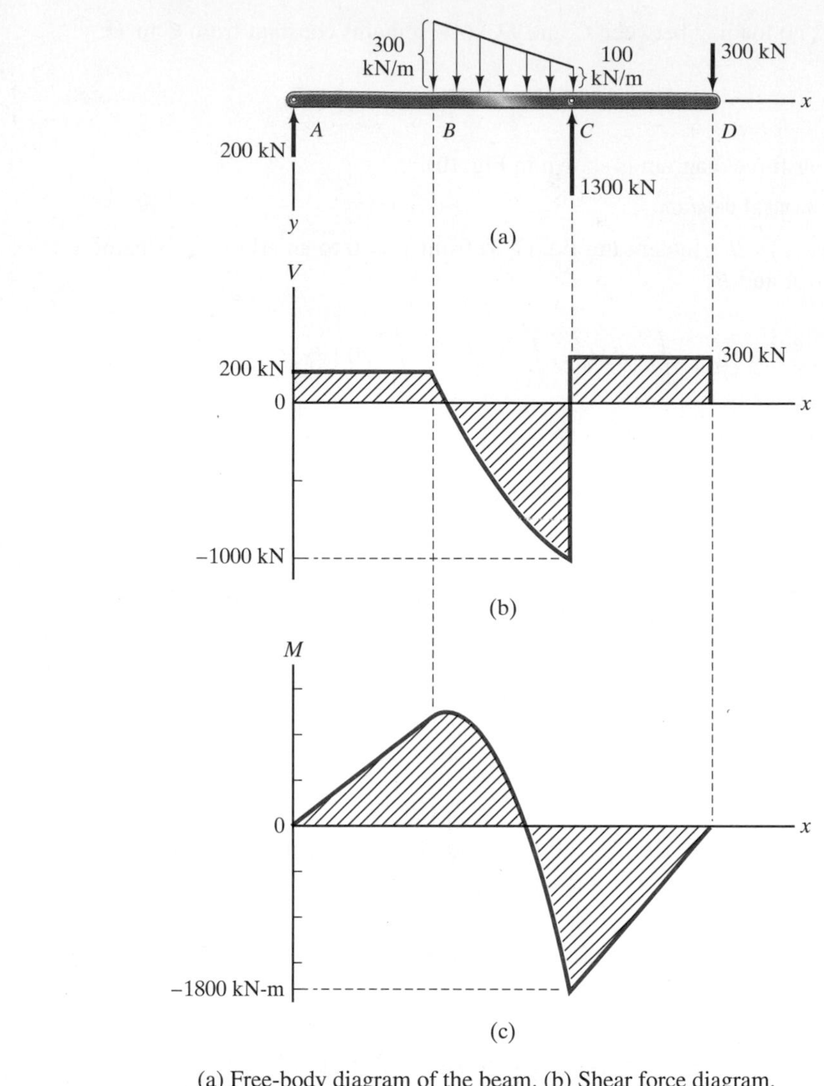

(a) Free-body diagram of the beam. (b) Shear force diagram. (c) Bending moment diagram.

Chapter Summary

Axial Force, Shear Force, and Bending Moment

If the system of external loads and reactions on a beam is two-dimensional, the internal forces and moments can be represented by an equivalent system consisting of the *axial force* P, the *shear force* V, and the *bending moment* M [Fig. (a)]. The directions of P, V, and M in Fig. (a) are the established positive directions of these quantities. The *shear force and bending moment diagrams* are simply the graphs of V and M, respectively, as functions of x.

Equations Relating Distributed Load, Shear Force, and Bending Moment

In a portion of a beam subjected to a distributed load w [Fig. (b)], the shear force and bending moment satisfy the differential equations

$$\frac{dV}{dx} = -w, \qquad \text{Eq. (7-4)}$$

$$\frac{dM}{dx} = V. \qquad \text{Eq. (7-5)}$$

A point force results in a jump discontinuity in the shear force but no discontinuity in the bending moment. A force F in the positive y direction causes a decrease in the shear force of magnitude F [Fig. (c)]. A couple results in a jump discontinuity in the bending moment but no discontinuity in the shear force. A counterclockwise couple C causes a decrease in the bending moment of magnitude C [Fig. (d)].

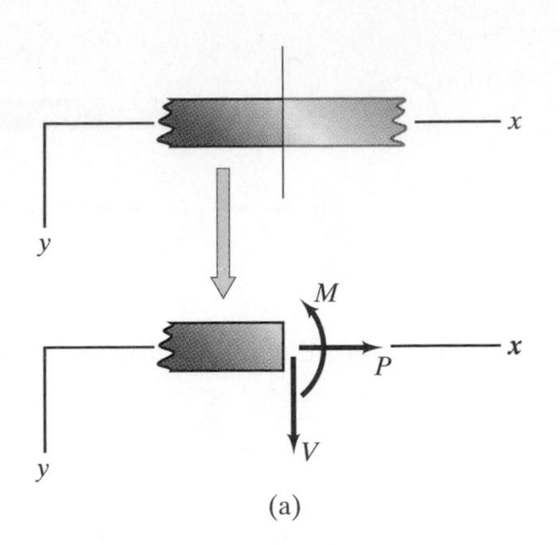

(a)

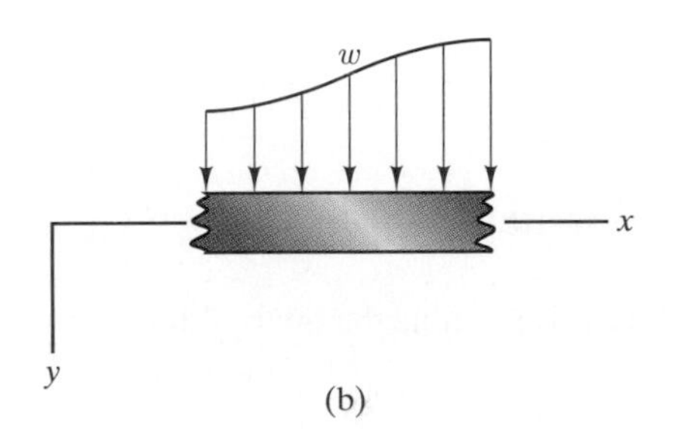

(b)

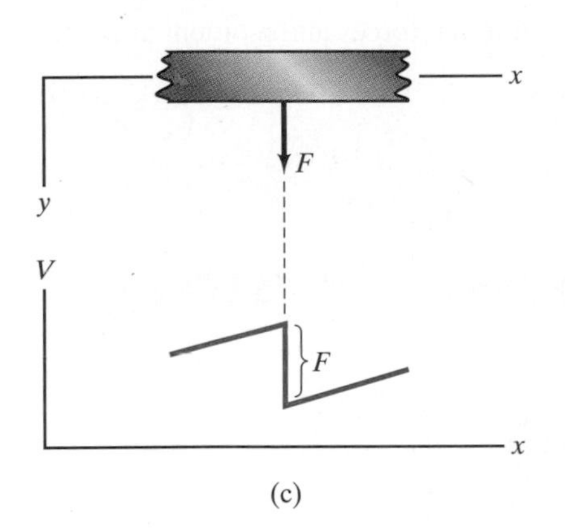

(c)

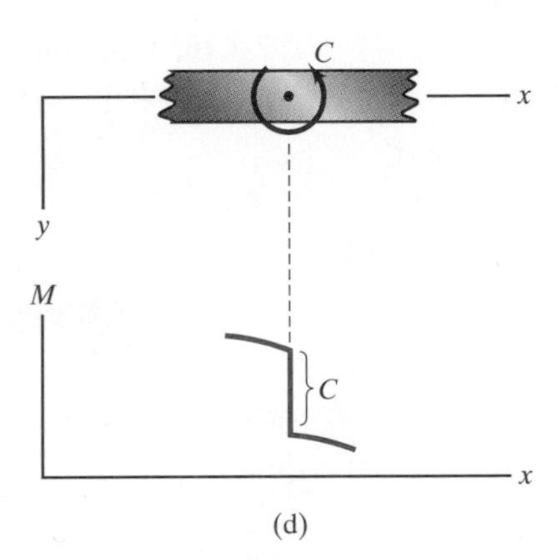

(d)

PROBLEMS

7-1.1. In Example 7-1, determine the internal forces and moment at C if the distance from A to C is $L/2$.

7-1.2. Determine the internal forces and moment at A, B, and C. (*Strategy:* In this case you don't need to determine the reactions at the built-in support. Cut the beam at the point where you want to determine the internal forces and moment and draw the free-body diagram of the part of the beam to the left of your cut. Remember that P, V, and M must be in their defined positive directions in your free-body diagrams.)

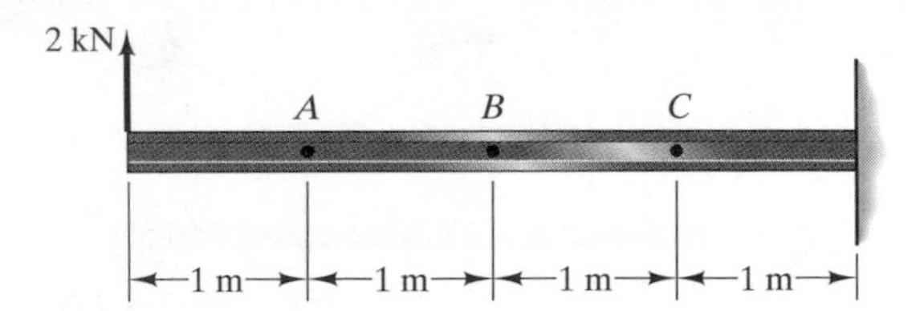

PROBLEM 7-1.2

7-1.3. Determine the internal forces and moment at A, B, and C.

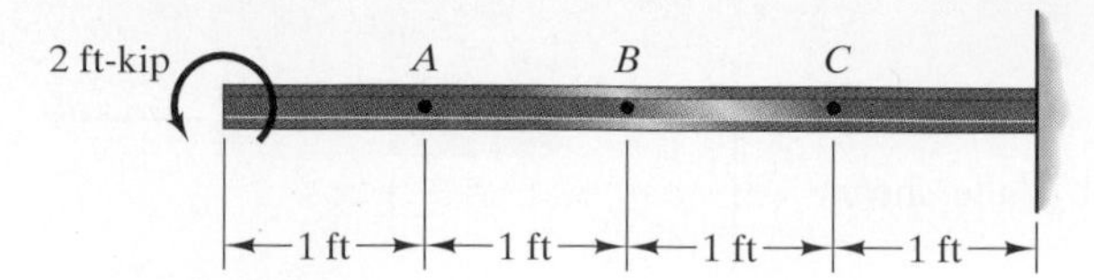

PROBLEM 7-1.3

7-1.4. Determine the internal forces and moment at A.

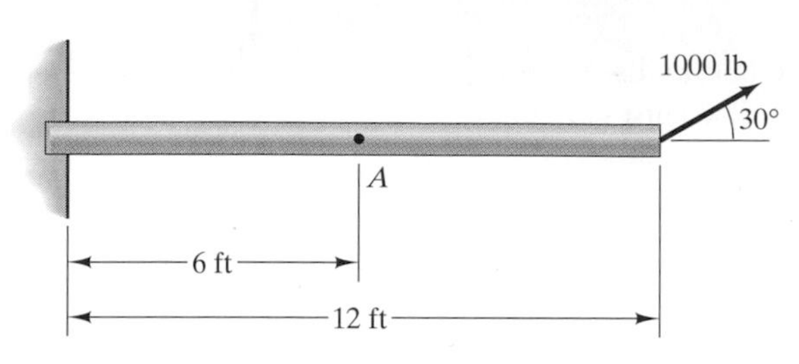

PROBLEM 7-1.4

7-1.5. Determine the internal forces and moment at A.

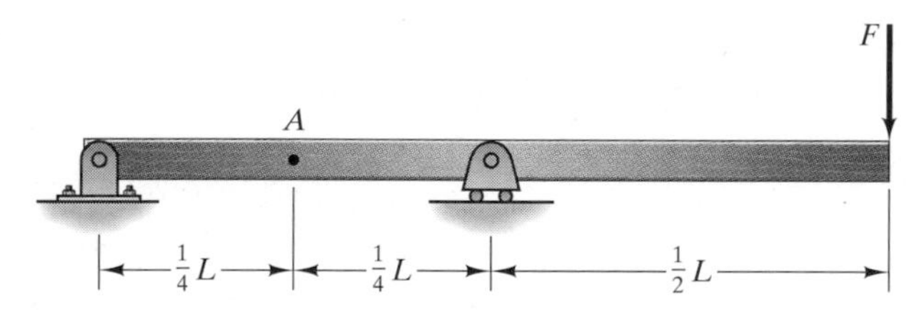

PROBLEM 7-1.5

7-1.6. In Problem 7-1.5, determine the internal forces and moment at A if point A is **(a)** just to the left of the roller support; **(b)** just to the right of the roller support.

7-1.7. Determine the internal forces and moment **(a)** at B; **(b)** at C.

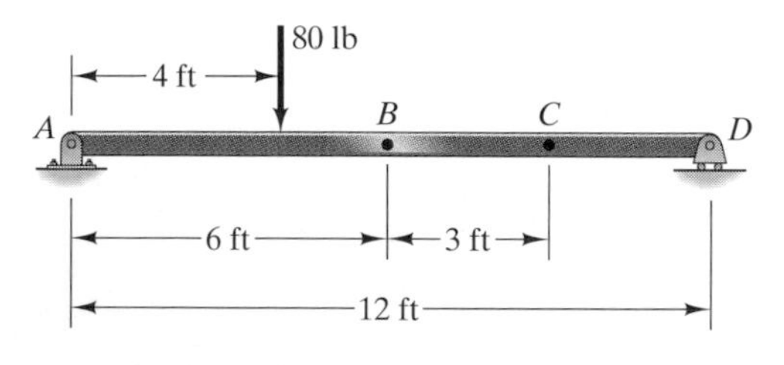

PROBLEM 7-1.7

7-1.8. Determine the internal forces and moment at B **(a)** if $x = 250$ mm; **(b)** if $x = 750$ mm.

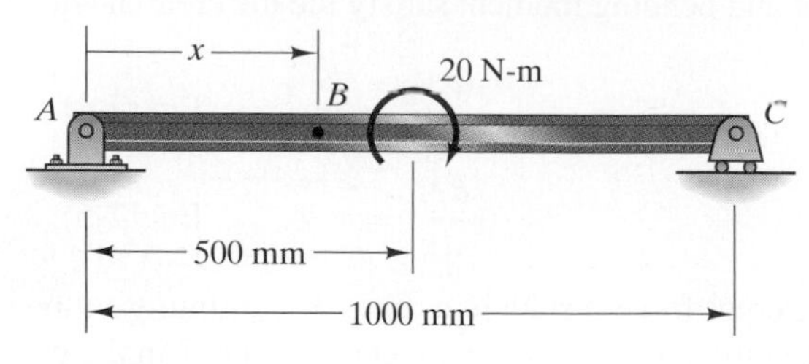

PROBLEM 7-1.8

7-1.9. In Example 7-2, determine the internal forces and moment at B if the distance from A to B is 4 m.

7-1.10. Determine the internal forces and moment at A for each loading.

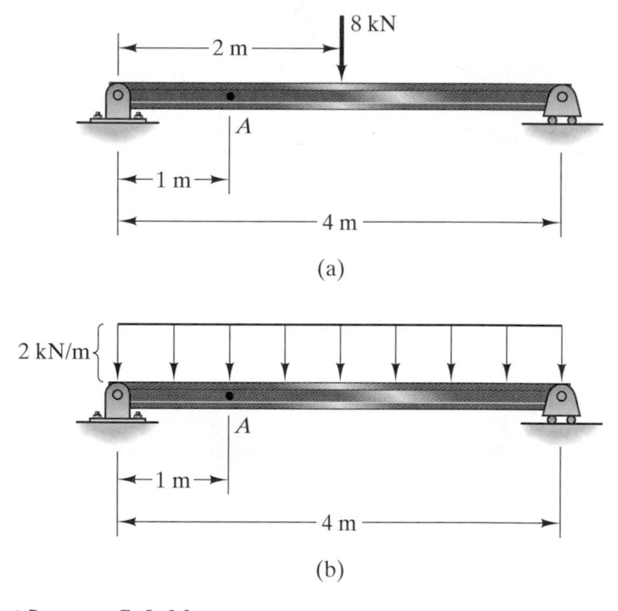

PROBLEM 7-1.10

7-1.11. Model the ladder rung as a simply supported (pin-supported) beam and assume that the 200-lb load exerted by the person's shoe is uniformly distributed. Determine the internal forces and moment at A.

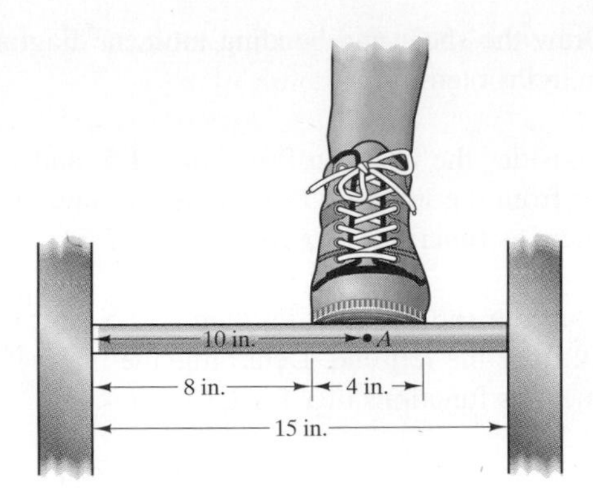

PROBLEM 7-1.11

7-1.12. In Problem 7-1.11, determine the internal forces and moment at A if the distance from A to the left edge of the rung is 9 in.

7-1.13. If $x = 3$ m, what are the internal forces and moment at A?

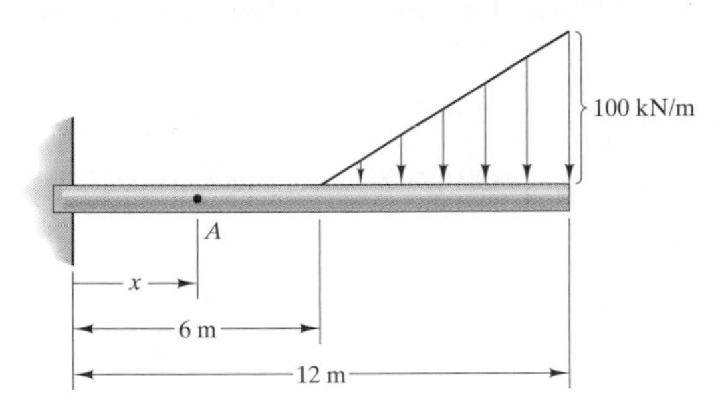

PROBLEM 7-1.13

7-1.14. If $x = 9$ m in Problem 7-1.13, what are the internal forces and moment at A?

7-1.15. Determine the internal forces and moment at A.

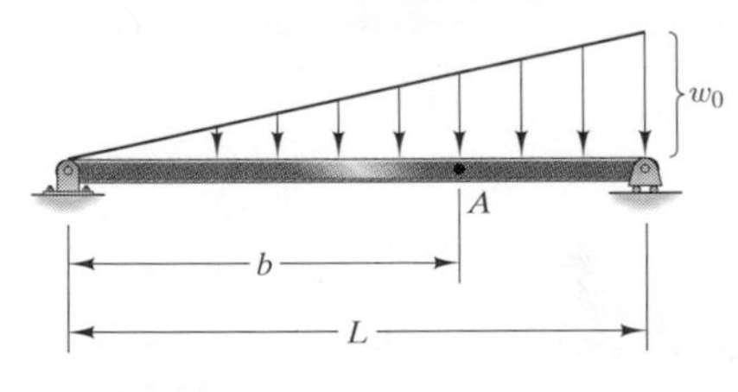

PROBLEM 7-1.15

7-1.16. If $x = 3$ ft, what are the internal forces and moment at A?

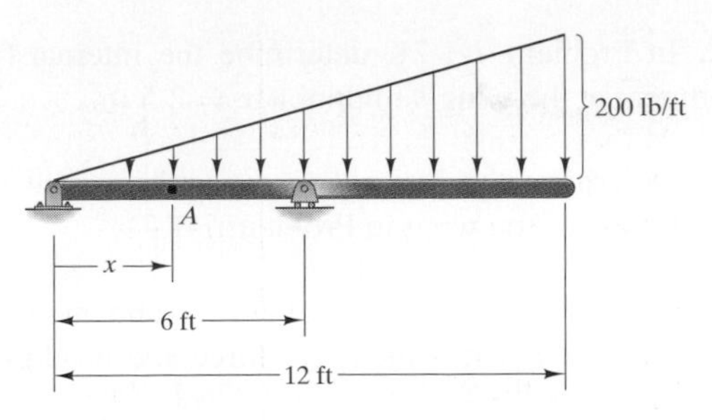

PROBLEM 7-1.16

7-1.17. If $x = 9$ ft in Problem 7-1.16, what are the internal forces and moment at A?

7-1.18. Determine the internal forces and moment at A.

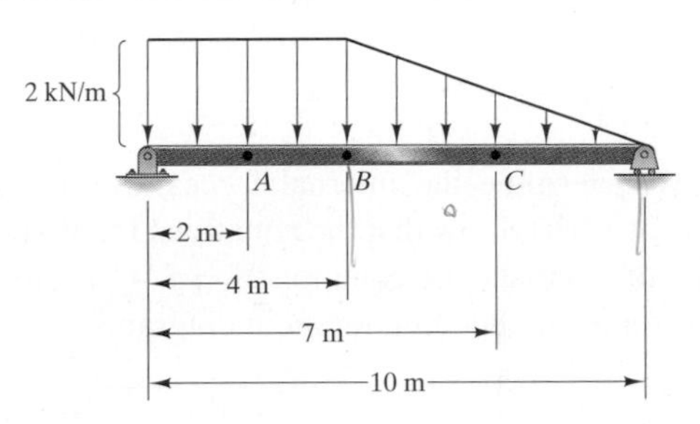

PROBLEM 7-1.18

7-1.19. Determine the internal forces and moment at point B of the beam in Problem 7-1.18.

7-1.20. Determine the internal forces and moment at point C of the beam in Problem 7-1.18.

7-1.21. The lift force on the airplane's wing is given by the distributed load $w = -15(1 - 0.04x^2)$ kN/m and the weight of the wing is given by the distributed load $w = 5 - 0.5x$ kN/m. Determine the internal forces and moment at the wing root R.

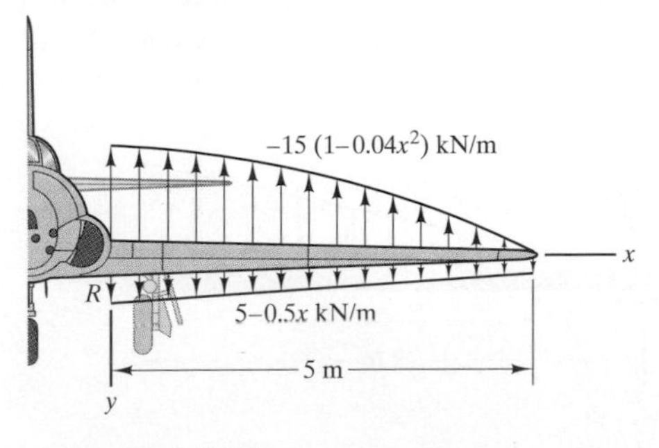

PROBLEM 7-1.21

7-1.22. In Problem 7-1.21, determine the internal forces and moment at the wing's midpoint $x = 2.5$ m.

7-1.23. Determine the internal forces and moment at the center of mass of the wing in Problem 7-1.21.

7-2.1. **(a)** Determine the internal forces and moment as functions of x. **(b)** Draw the shear force and bending moment diagrams.

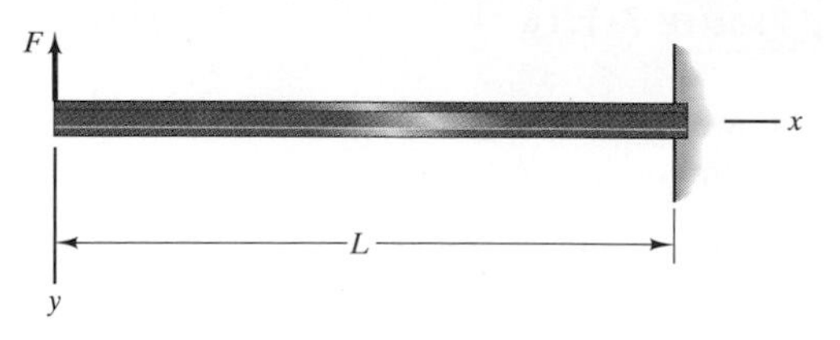

PROBLEM 7-2.1

7-2.2. **(a)** Determine the internal forces and moment as functions of x. **(b)** Show that the equations for V and M as functions of x satisfy the equation $V = dM/dx$. **(c)** Draw the shear force and bending moment diagrams.

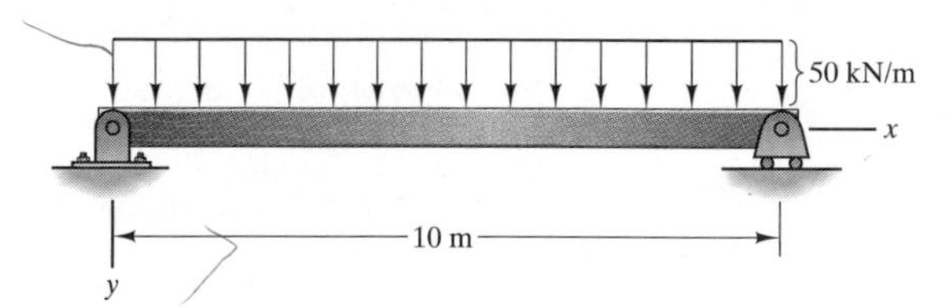

PROBLEM 7-2.2

7-2.3. The beam in Problem 7-2.2 will safely support a bending moment of 1 MN-m (meganewton-meter) at any cross section. Based on this criterion, what is the maximum safe value of the uniformly distributed load?

7-2.4. **(a)** Determine the internal forces and moment as functions of x. **(b)** Show that the equations for V and M as functions of x satisfy the equation $V = dM/dx$. **(c)** Determine the maximum bending moment in the beam and the value of x where it occurs.

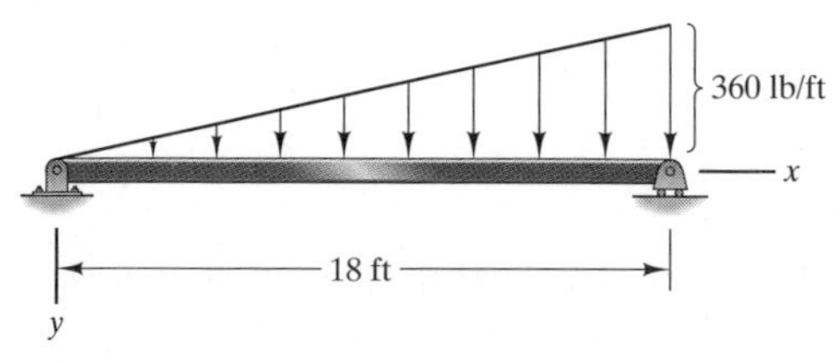

PROBLEM 7-2.4

7-2.5. Draw the shear and bending moment diagrams for the beam in Problem 7-2.4.

7-2.6. Consider the beam in Problem 7-1.5, and let x be measured from the left end. Determine the internal forces and moment as functions of x for $0 < x < L/2$.

7-2.7. Consider the beam in Problem 7-1.5, and let x be measured from the left end. Determine the internal forces and moment as functions of x for $L/2 < x < L$.

7-2.8. Consider the beam in Problem 7-1.8. Determine the internal forces and moment as functions of x for $0 < x < 0.5$ m.

7-2.9. Consider the beam in Problem 7-1.8. Determine the internal forces and moment as functions of x for $0.5 < x < 1$ m.

7-2.10. Consider the beam in Problem 7-1.16. Determine the internal forces and moment as functions of x for $0 < x < 6$ ft.

7-2.11. Consider the beam in Problem 7-1.16. Determine the internal forces and moment as functions of x for $6 < x < 12$ ft.

7-2.12. **(a)** Determine the internal forces and moment as functions of x. **(b)** Draw the shear force and bending moment diagrams.

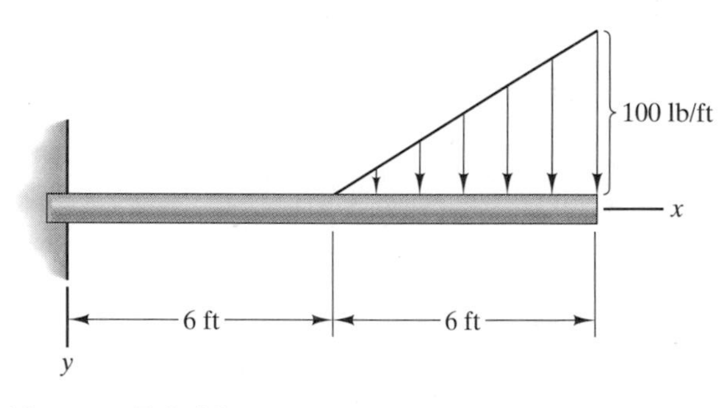

PROBLEM 7-2.12

7-2.13. The loads $F = 200$ N and $C = 800$ N-m. **(a)** Determine the internal forces and moment as functions of x. **(b)** Draw the shear force and bending moment diagrams.

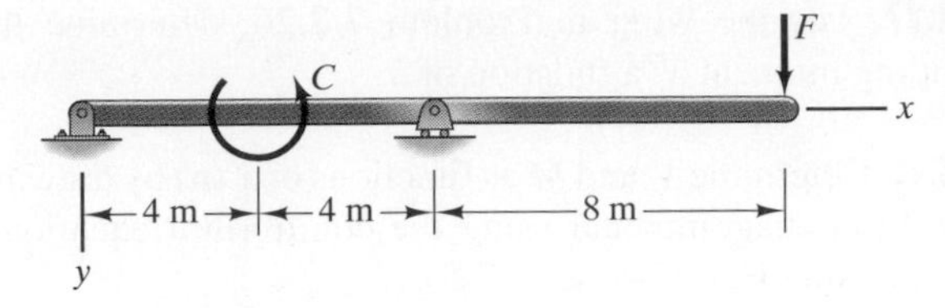

PROBLEM 7-2.13

7-2.14. The beam in Problem 7-2.13 will safely support shear forces and bending moments of magnitudes 2 kN and 6.5 kN-m, respectively. Based on this criterion, can it safely be subjected to the loads $F = 1$ kN, $C = 1.6$ kN-m?

7-2.15. The bar *BD* is rigidly fixed to the beam *ABC* at *B*. Draw the shear force and bending moment diagrams for the beam *ABC*.

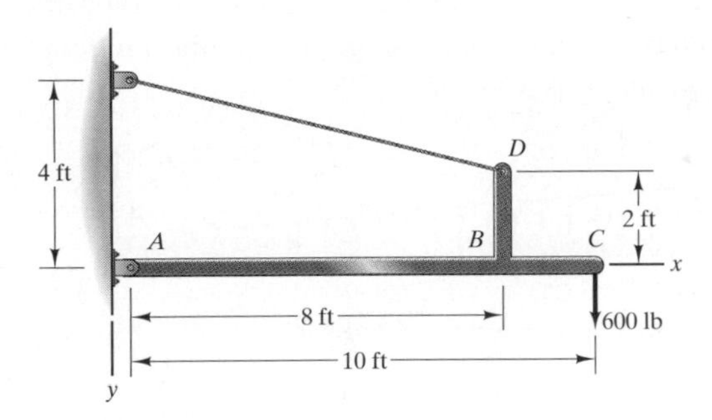

PROBLEM 7-2.15

7-2.16. Draw the shear force and bending moment diagrams for beam *ABC*.

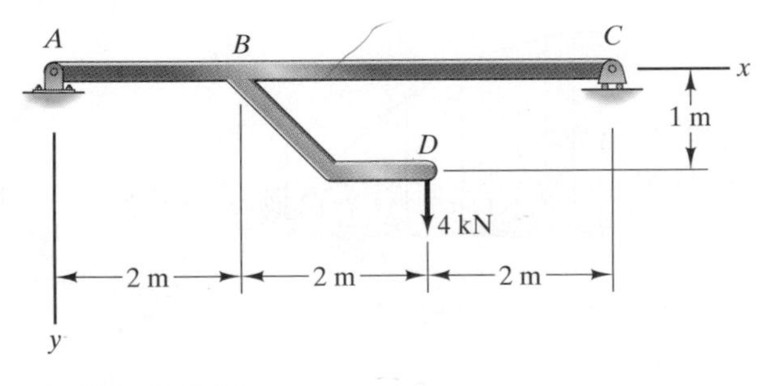

PROBLEM 7-2.16

7-2.17. Model the ladder rung as a simply supported (pin-supported) beam and assume that the 200-lb load exerted by the person's shoe is uniformly distributed. Draw the shear force and bending moment diagrams for the rung.

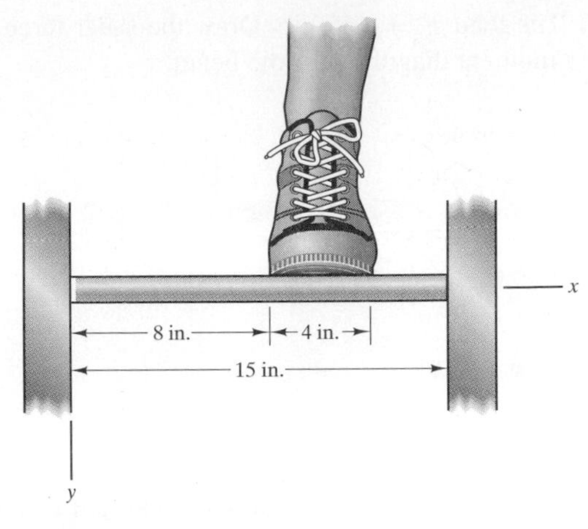

PROBLEM 7-2.17

7-2.18. What is the maximum bending moment in the ladder rung in Problem 7-2.17, and where does it occur?

7-2.19. Assume that the surface on which the beam rests exerts a uniformly distributed load on the beam. Draw the shear force and bending moment diagrams.

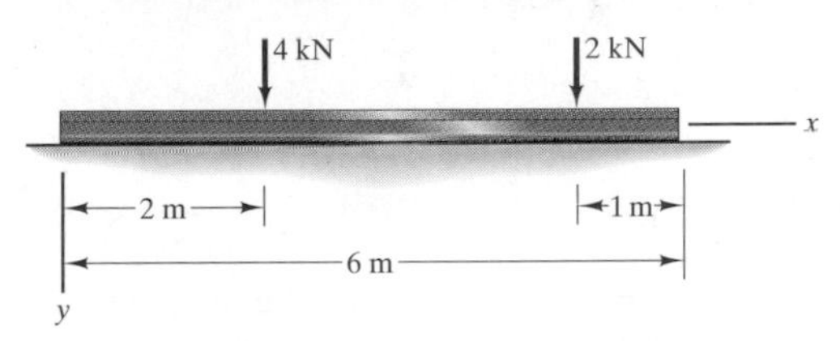

PROBLEM 7-2.19

7-2.20. The homogeneous beams *AB* and *CD* weigh 600 lb and 500 lb, respectively. Draw the shear force and bending moment diagrams for beam *CD*. (Remember that the beam's weight is a distributed load.)

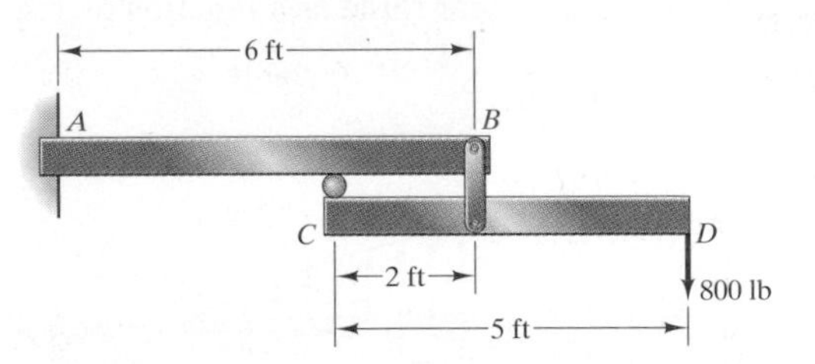

PROBLEM 7-2.20

7-2.21. Draw the shear force and bending moment diagrams for beam *AB* in Problem 7-2.20, including the beam's weight.

7-2.22. The load $F = 4650$ lb. Draw the shear force and bending moment diagrams for the beam.

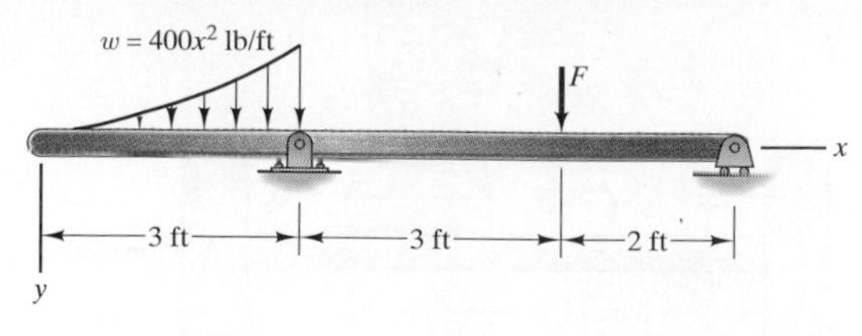

PROBLEM 7-2.22

7-2.23. If the load $F = 2150$ lb in Problem 7-2.22, what are the maximum positive and negative values of the shear force and bending moment, and at what values of x do they occur?

7-2.24. Draw the shear force and bending moment diagrams.

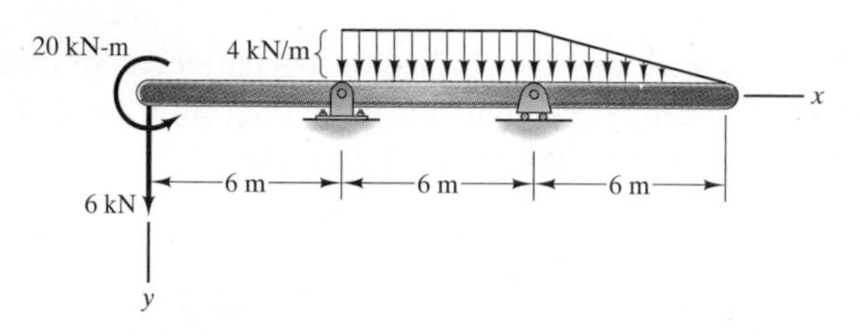

PROBLEM 7-2.24

7-2.25. In Problem 7-2.24, what are the maximum positive and negative values of the shear force and bending moment, and at what values of x do they occur?

7-2.26. The lift force on the airplane's wing is given by the distributed load $w = -15(1-0.04x^2)$ kN/m and the weight of the wing is given by the distributed load $w = 5 - 0.5x$ kN/m. Determine the shear force as a function of x.

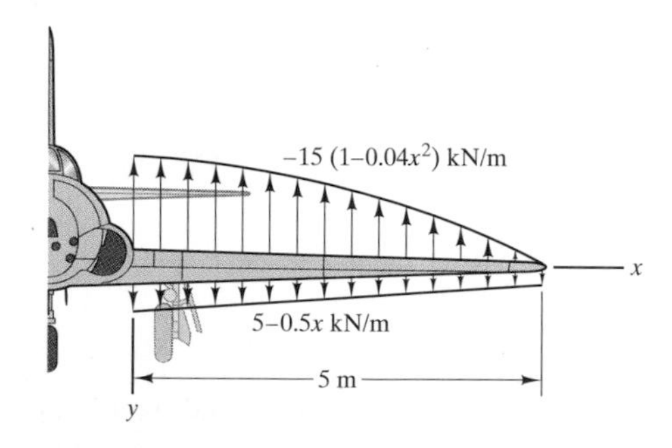

PROBLEM 7-2.26

7-2.27. For the wing in Problem 7-2.26, determine the bending moment as a function of x.

7-3.1. Determine V and M as functions of x **(a)** by drawing free-body diagrams and using the equilibrium equations; **(b)** by using Eqs. (7-4) and (7-5).

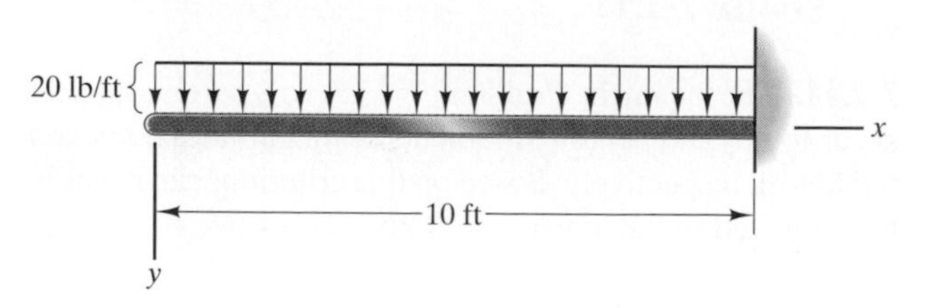

PROBLEM 7-3.1

7-3.2. Determine V and M as functions of x **(a)** by drawing free-body diagrams and using the equilibrium equations; **(b)** by using Eqs. (7-4) and (7-5).

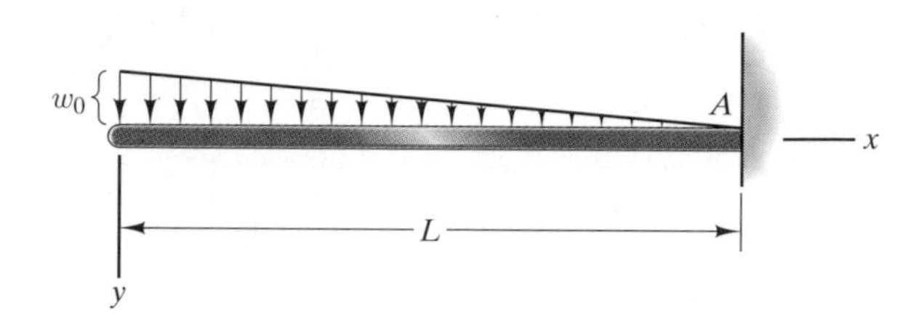

PROBLEM 7-3.2

7-3.3. Determine V and M as functions of x by using Eqs. (7-4) and (7-5).

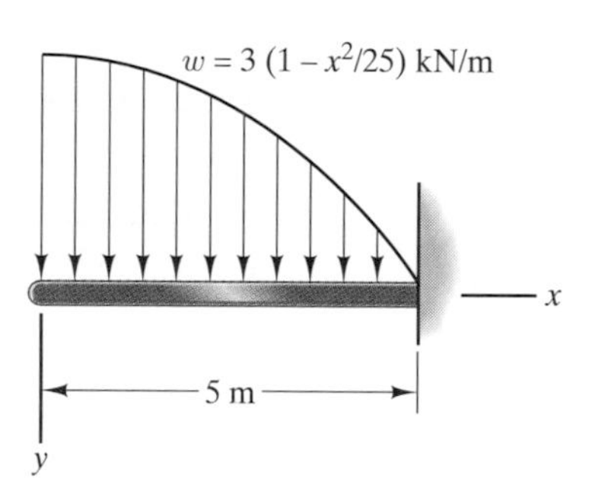

PROBLEM 7-3.3

7-3.4. Use Eqs. (7-4) and (7-5) to determine the internal forces and moment as functions of x for the beam in Problem 7-2.1.

7-3.5. Use Eqs. (7-4) and (7-5) to determine the internal forces and moment as functions of x for the beam in Problem 7-2.2.

7-3.6. Determine V and M as functions of x by using Eqs. (7-4) and (7-5).

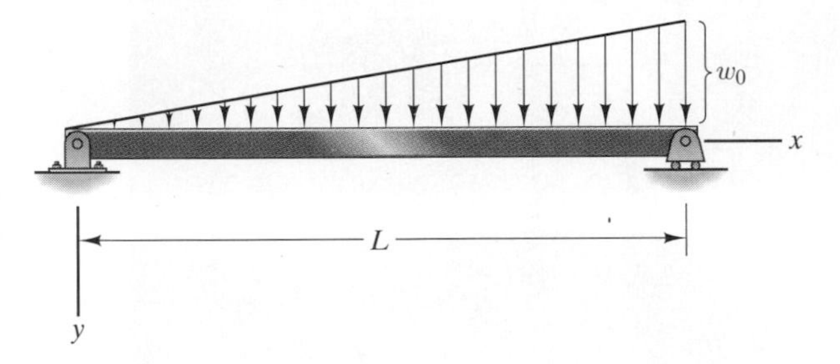

PROBLEM 7-3.6

7-3.7. Use Eqs. (7-4) and (7-5) to determine the internal forces and moment as functions of x for the beam in Problem 7-1.16.

7-3.8. Use Eqs. (7-4) and (7-5) to determine the internal forces and moment as functions of x for the beam in Problem 7-2.15.

7-3.9. **(a)** Determine V and M as functions of x by using Eqs. (7-4) and (7-5). **(b)** Draw the shear force and bending moment diagrams.

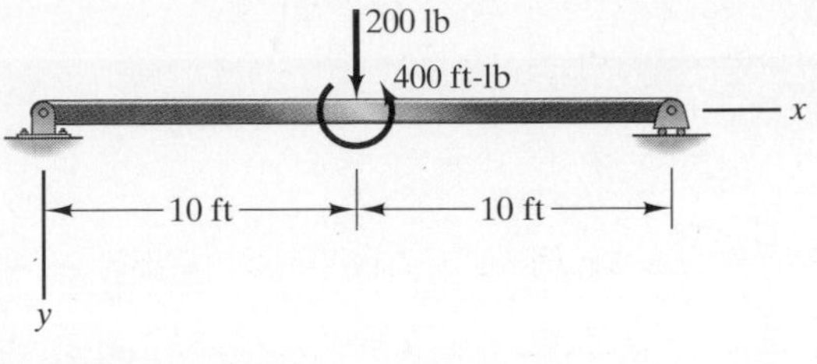

PROBLEM 7-3.9

7-3.10. Use Eqs. (7-4) and (7-5) to solve Problem 7-2.12.

7-3.11. Use Eqs. (7-4) and (7-5) to solve Problem 7-2.13.

7-3.12. Use Eqs. (7-4) and (7-5) to solve Problem 7-2.16.

7-3.13. Use Eqs. (7-4) and (7-5) to solve Problem 7-2.17.

7-3.14. Use Eqs. (7-4) and (7-5) to determine the internal forces and moment as functions of x for the beam in Problem 7-2.19.

7-3.15. Use Eqs. (7-4) and (7-5) to solve Problem 7-2.22.

7-3.16. Use Eqs. (7-4) and (7-5) to solve Problem 7-2.24.

7-3.17. Use Eqs. (7-4) and (7-5) to solve Problem 7-2.26.

Gymnast performing on a balance beam made of laminated wood.

CHAPTER 8

Stresses in Beams

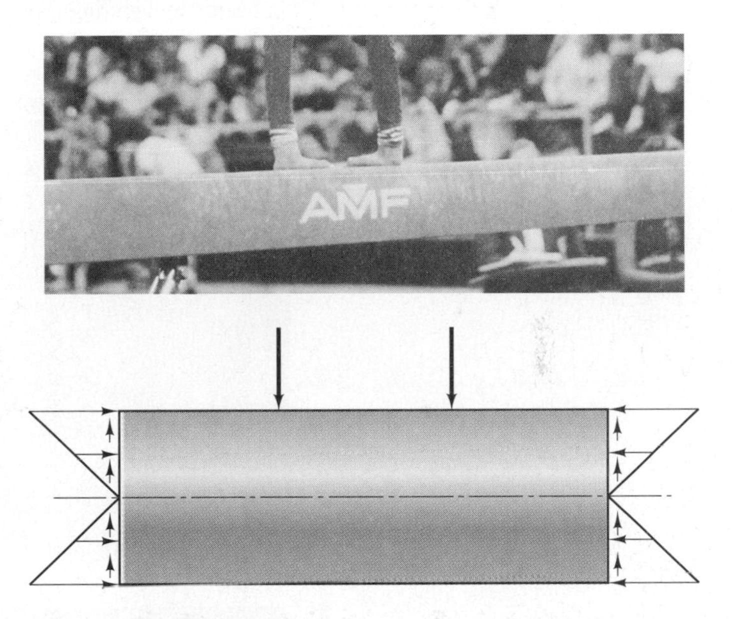

Because beams are utilized in so many ways and in so many types of structures, determining their states of stress is an important part of structural analysis and design. In this chapter we show that the stresses at a given cross section of a beam can be expressed in terms of the values of the axial force, shear force, and bending moment at that cross section.

Normal Stress

8-1 | Distribution of the Stress

You know that the easiest way to break a small piece of firewood is to bend it (Fig. 8-1a), causing the wood to fracture (Fig. 8-1b). In doing so, you subject the stick to couples at the ends, inducing stresses within the wood that cause it to fail. In the same way, stresses induced by bending moments in the members of a structure can cause the members to become permanently deformed or even lead to collapse of the structure. In this section we analyze the stresses induced in beams by bending moments.

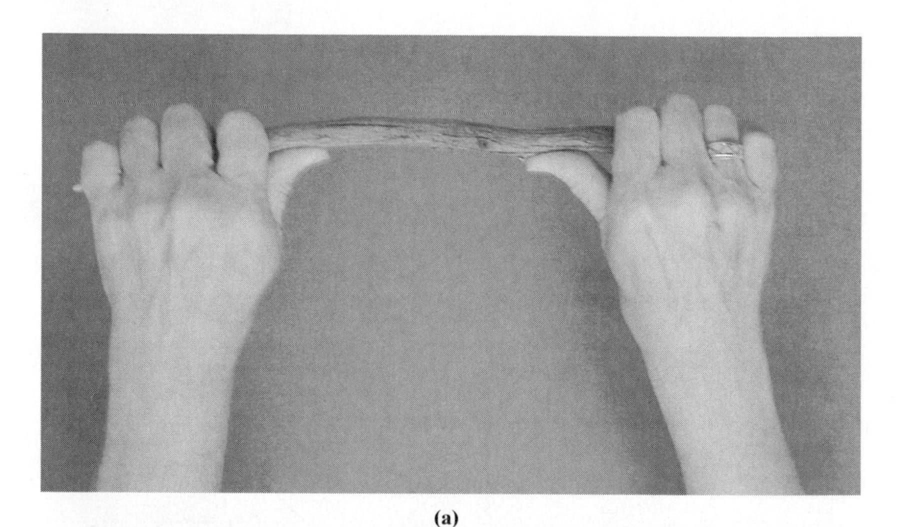

(a)

(b)

FIGURE 8-1 (a) Bending a stick. (b) The resulting stresses can cause the stick to break.

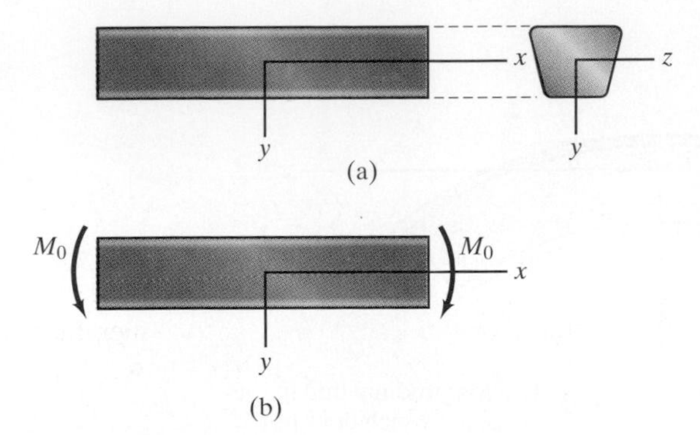

FIGURE 8-2 (a) Prismatic beam with a symmetrical cross section. (b) Subjecting the beam to couples at its ends.

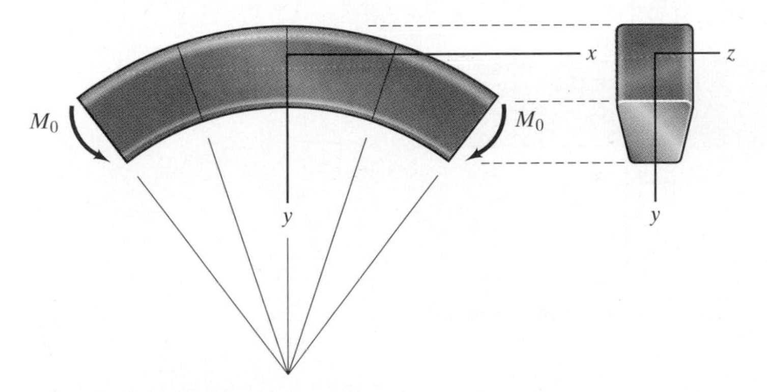

FIGURE 8-3 Deformation resulting from the applied couples. Each cross section of the beam remains plane.

Geometry of Deformation

Let us consider a prismatic beam of isotropic elastic material (Fig. 8-2a). The x axis of the coordinate system is parallel to the beam's longitudinal axis, and we assume that the beam's cross section is symmetric about the y axis. Suppose that we were to carry out an experiment in which we subject this beam to couples M_0 at its ends as shown in Fig. 8-2b. The resulting deformation of the beam is shown in Fig. 8-3. Each line of the beam that is parallel to the longitudinal axis before the couples are applied deforms into a circular arc parallel to the x–y plane. Each cross section of the beam remains plane and perpendicular to the beam's curved longitudinal axis. (The latter result is traditionally expressed by the Henry Higgins–sounding statement "Plane sections remain plane.")

When the beam bends, longitudinal lines toward the top of its cross section become longer and those near the bottom become shorter (Fig. 8-4). The longitudinal line in the x–y plane that does not change length when the beam bends is called the *neutral axis*. (We will demonstrate presently that the neutral axis is coincident with the centroid of the beam's cross section.) Let us assume that the

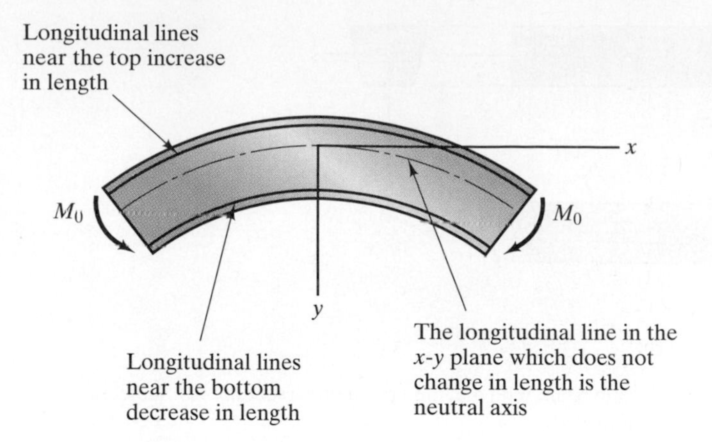

FIGURE 8-4 Changes in the lengths of longitudinal lines.

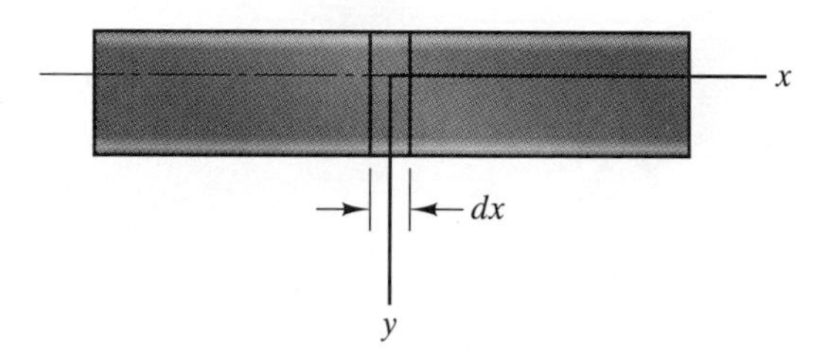

FIGURE 8-5 Element of the beam of width dx.

origin of the coordinate system lies on the neutral axis, and consider an element of the beam that is of width dx before the couples are applied (Fig. 8-5). We show this element after the couples are applied in Fig. 8-6, isolating it so that we can analyze its geometry. The radius of the beam's neutral axis is denoted by ρ. Since we assume that the origin coincides with the neutral axis, the width of the element at $y = 0$ equals dx after the couples are applied. The width of the element decreases below the neutral axis and increases above it. Let dx' be the width of the element at a distance y from the neutral axis (Fig. 8-6). In terms of the width before and after the deformation, the extensional strain in the x direction is

$$\epsilon_x = \frac{dx' - dx}{dx} = \frac{dx'}{dx} - 1. \tag{8-1}$$

We can express dx and dx' in terms of the radius ρ and the angle $d\theta$ shown in Fig. 8-6:

$$dx = \rho\, d\theta,$$

$$dx' = (\rho - y)\, d\theta.$$

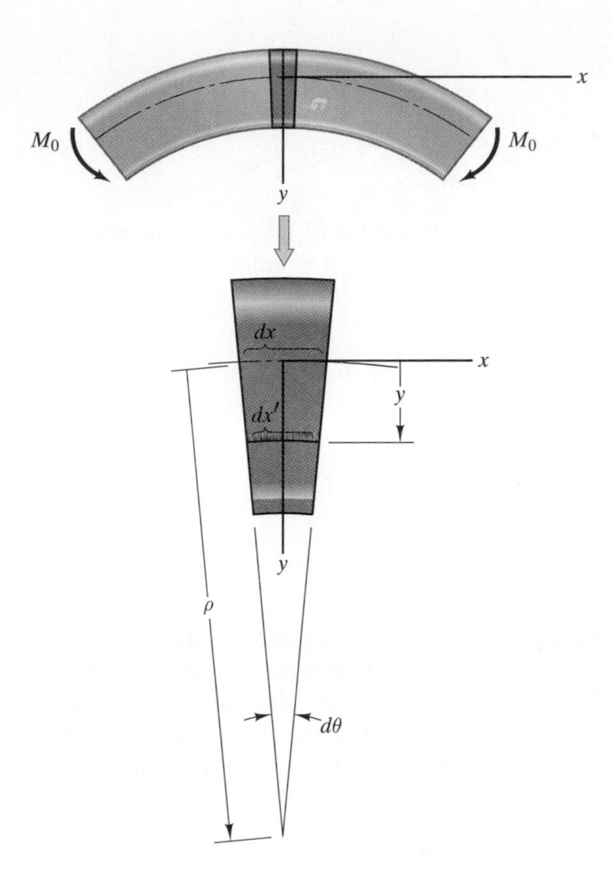

| **FIGURE 8-6** Element after the couples are applied.

Dividing the second equation by the first,

$$\frac{dx'}{dx} = 1 - \frac{y}{\rho},$$

and substituting this result into Eq. (8-1), we obtain

$$\epsilon_x = -\frac{y}{\rho}. \qquad \textbf{(8-2)}$$

The extensional strain in the direction parallel to the beam's axis is a linear function of y, which can also be seen from the shape of the element in Fig. 8-6. The negative sign confirms that the width of the beam decreases in the positive y direction (below the neutral axis) and increases in the negative y direction (above the neutral axis).

The deformation of the element in Fig. 8-6 implies the presence of normal stresses on the vertical faces of the element, causing the material to be stretched above the neutral axis and compressed below it. From Eq. (6-43), the extensional strain ϵ_x in an isotropic elastic material is given in terms of the components of stress by

$$\epsilon_x = \frac{1}{E}\sigma_x - \frac{\nu}{E}(\sigma_y + \sigma_z). \tag{8-3}$$

Since the beam we are considering is subjected to no loads perpendicular to its axis, let us assume that the normal stresses σ_y and σ_z are zero. Then Eq. (8-3) states that

$$\sigma_x = E\epsilon_x. \tag{8-4}$$

Substituting Eq. (8-2) into this expression, we obtain

$$\sigma_x = -\frac{Ey}{\rho}. \tag{8-5}$$

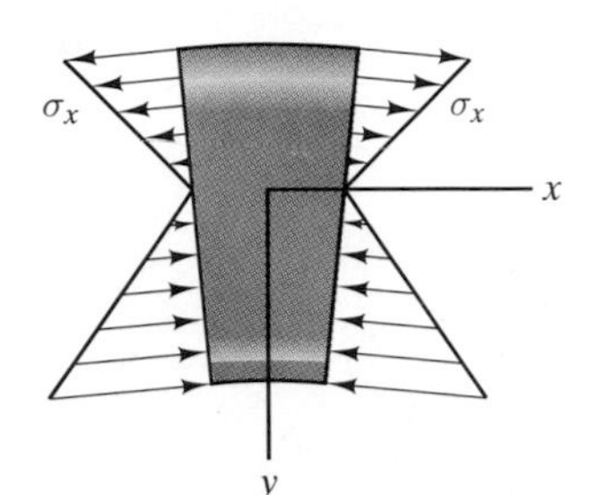

FIGURE 8-7 Normal stresses on the vertical faces of the element.

We see that the material is subjected to a normal stress σ_x which is a linear function of y. The result of our "effect–cause" analysis is shown in Fig. 8-7. The normal stress is negative (compressive) for positive values of y, causing the width of the element to decrease, and positive (tensile) for negative values of y, causing the width of the element to increase. Since the same analysis can be applied to any element of the beam, the normal stress at every cross section is described by Eq. (8-5).

Relation between Normal Stress and Bending Moment

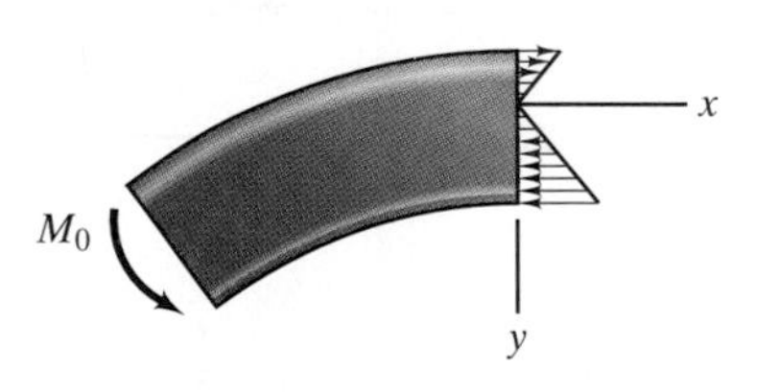

FIGURE 8-8 Free-body diagram obtained by passing a plane perpendicular to the beam's axis.

Equations (8-2) and (8-5) indicate that the extensional strain ϵ_x and normal stress σ_x are linear functions of the distance y from the neutral axis, and that they are inversely proportional to the radius of curvature ρ. But we have not determined the location of the neutral axis, and we don't know the relationship between ρ and the applied couple M_0.

In Fig. 8-8 we obtain a free-body diagram by passing a plane perpendicular to the beam's axis. Since the couple M_0 exerts no net force, the horizontal force exerted on the free-body diagram by the distribution of stress σ_x must be zero if the beam is in equilibrium. Letting dA be an element of the beam's cross-sectional area, the horizontal force is

$$\int_A \sigma_x \, dA = 0. \tag{8-6}$$

Substituting Eq. (8-5) into this equation, we find that the horizontal force exerted on the free-body diagram is zero only if

$$\int_A y \, dA = 0. \tag{8-7}$$

From the equation for the y coordinate of the centroid of the beam's cross section,

$$\bar{y} = \frac{\int_A y\,dA}{\int_A dA},$$

we see that Eq. (8-7) implies that the origin of the coordinate system coincides with the centroid. Since we had assumed the origin of the coordinate system to be at the neutral axis, *the neutral axis coincides with the centroid of the beam's cross section.*

We have determined the location of the neutral axis from the condition that the sum of the forces on the free-body diagram in Fig. 8-8 must equal zero. We can relate the radius of curvature ρ of the neutral axis to the couple M_0, and thus determine the distribution of the normal stress in terms of M_0, from the condition that the sum of the moments on the free-body diagram must equal zero. From Fig. 8-9, the moment about the z axis due to the normal stress acting on an element dA of the beam's cross section is $-y\sigma_x\,dA$. The total moment about the z axis due to the stress distribution and the couple M_0 in Fig. 8-8 is therefore

$$\int_A -y\sigma_x\,dA - M_0 = 0.$$

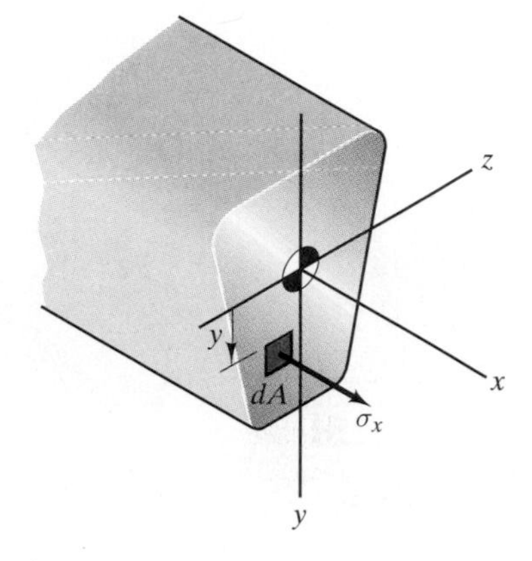

FIGURE 8-9 Determining the moment due to the normal stress acting on an element dA.

By substituting Eq. (8-5) into this expression, we can obtain a relation between M_0 and ρ. We write the resulting equation as

$$\frac{1}{\rho} = \frac{M_0}{EI}, \qquad \textbf{(8-8)}$$

where

$$I = \int_A y^2\,dA$$

is the moment of inertia of the beam's cross-sectional area about the z axis. We now substitute Eq. (8-8) into Eq. (8-5) to obtain the normal stress distribution in terms of M_0:

$$\sigma_x = -\frac{M_0 y}{I}. \qquad \textbf{(8-9)}$$

Although it has been convenient to derive these results for a beam subjected to couples M_0 in the directions shown in Fig. 8-2, those couples do not conform to our convention for the positive direction of the bending moment. We can apply our results to a beam subjected to positive couples M (Fig. 8-10) simply by making the substitution $M_0 = -M$.

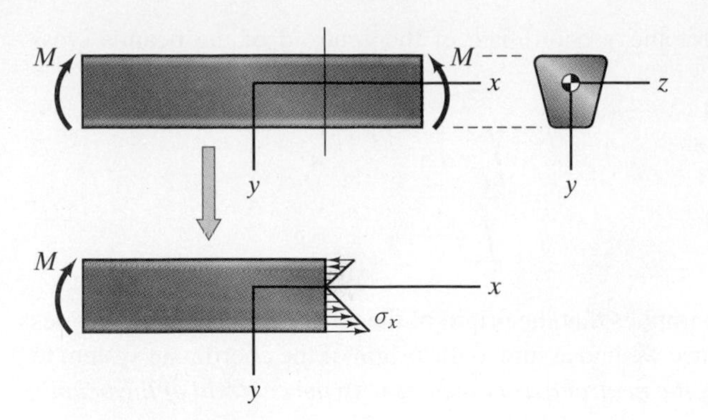

FIGURE 8-10 Beam subjected to positive couples M.

We summarize the results for a beam subjected to the couples shown in Fig. 8-10:

1. Radius of curvature:

$$\frac{1}{\rho} = -\frac{M}{EI}. \tag{8-10}$$

The sign of ρ indicates the direction of the curvature of the neutral axis. If ρ is positive, the positive y axis is on the concave side of the neutral axis. The product EI is called the flexural rigidity *of the beam.*

2. Distribution of extensional strain:

$$\epsilon_x = -\frac{y}{\rho} = \frac{My}{EI}. \tag{8-11}$$

3. Distribution of normal stress:

$$\sigma_x = \frac{My}{I}. \tag{8-12}$$

Beams Subjected to Arbitrary Loads

The internal bending moment in a beam loaded as shown in Fig. 8-10 has the same value M at every cross section. We have seen that the internal bending moment in a beam subjected to arbitrary loading is a function of position along the beam's axis. If such a beam is slender (its cross-sectional dimensions are small in comparison with its length), the radius of curvature and distributions of extensional strain and normal stress *at a given cross section* can be approximated using Eqs. (8-10)–(8-12). When this is done, M is the value of the bending moment at the given cross section and ρ is the radius of curvature of the beam's neutral axis in the neighborhood of the cross section.

EXAMPLE 8-1

Worksheet 10

The beam in Fig. 8-11 is subjected to couples $M = 4$ kN-m. It consists of aluminum alloy with modulus of elasticity $E = 70$ GPa. The shape and dimensions of the cross section are shown. Determine the resulting radius of curvature of the beam's neutral axis. What are the maximum tensile and compressive normal stresses, and where do they occur?

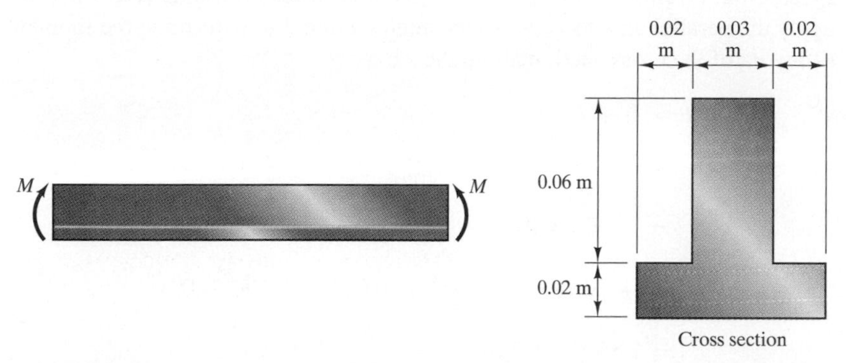

FIGURE 8-11

Strategy

The radius of curvature is given by Eq. (8-10). To apply that equation, we must determine the position of the neutral axis (the centroid of the beam's cross section), then use the parallel axis theorem to determine the moment of inertia I. The distribution of normal stress is given by Eq. (8-12), from which we can determine the maximum tensile and compressive stresses.

Solution

We can use any convenient coordinate system to locate the centroid of the cross section [Fig. (a)].

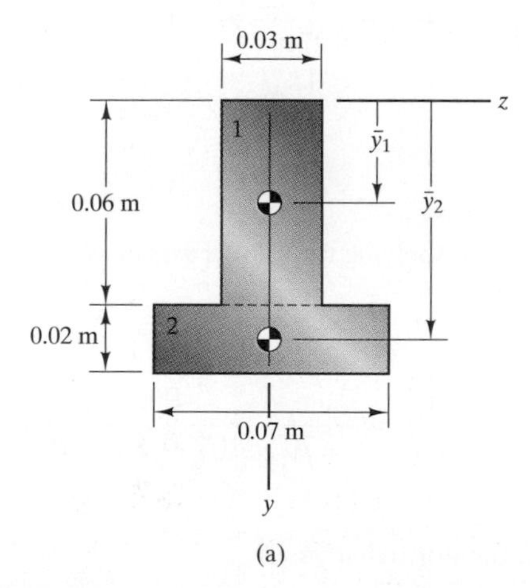

(a) Coordinate system for determining the position of the neutral axis.

Dividing the cross section into the rectangles 1 and 2 shown in Fig. (a), the y coordinate of its centroid is

$$\bar{y} = \frac{\bar{y}_1 A_1 + \bar{y}_2 A_2}{A_1 + A_2} = \frac{(0.03)(0.03)(0.06) + (0.07)(0.07)(0.02)}{(0.03)(0.06) + (0.07)(0.02)} = 0.0475 \text{ m}.$$

Placing the origin of the coordinate system at the neutral axis [Fig. (b)], we apply the parallel axis theorem to rectangles 1 and 2 to determine the moment of inertia of the cross section about the z axis:

$$\begin{aligned} I &= I_1 + I_2 \\ &= \left(\tfrac{1}{12} b_1 h_1^3 + d_1^2 A_1\right) + \left(\tfrac{1}{12} b_2 h_2^3 + d_2^2 A_2\right) \\ &= \left[\tfrac{1}{12}(0.03)(0.06)^3 + (0.0175)^2(0.03)(0.06)\right] \\ &\quad + \left[\tfrac{1}{12}(0.07)(0.02)^3 + (0.0225)^2(0.07)(0.02)\right] \\ &= 1.85 \times 10^{-6} \text{ m}^4. \end{aligned}$$

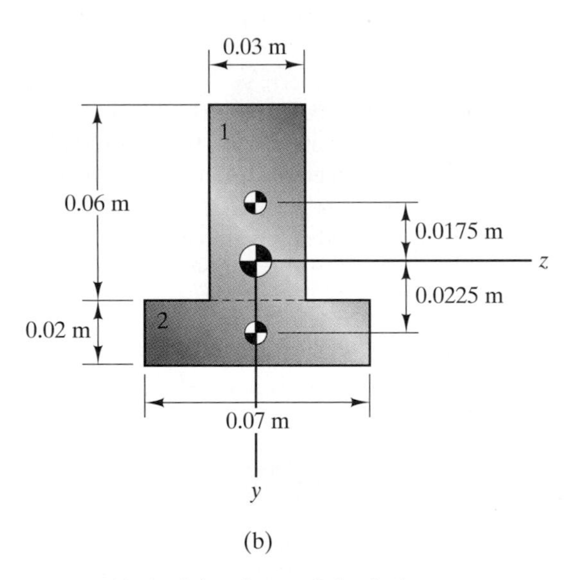

(b) Applying the parallel axis theorem.

From Eq. (8-10),

$$\frac{1}{\rho} = -\frac{M}{EI} = -\frac{4000}{(70 \times 10^9)(1.85 \times 10^{-6})},$$

we obtain $\rho = -32.3$ m.

From Eq. (8-12), the normal stress is

$$\sigma_x = \frac{My}{I} = \frac{4000y}{1.85 \times 10^{-6}} = 2.17 \times 10^9 y.$$

The distribution of normal stress is shown in Fig. (c). The maximum tensile

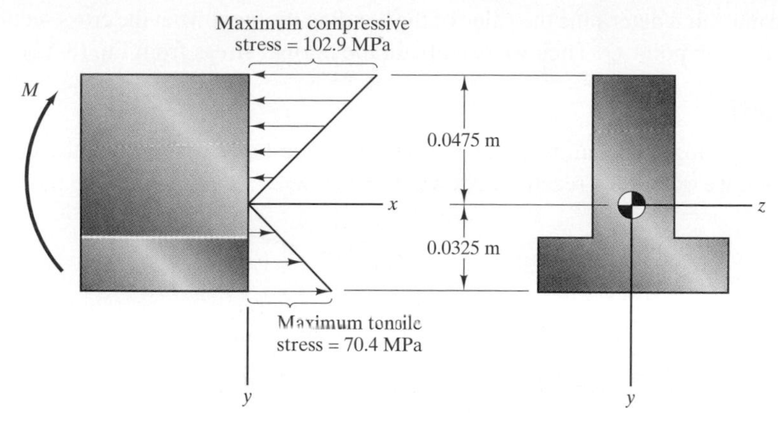

(c) Normal stress.

stress occurs at the bottom of the beam:

$$\sigma_x = (2.17 \times 10^9)(0.0325) = 70.4 \text{ MPa}.$$

The maximum compressive stress occurs at the top:

$$\sigma_x = (2.17 \times 10^9)(-0.0475) = -102.9 \text{ MPa}.$$

Discussion

Notice that the maximum tensile stress is smaller in magnitude than the maximum compressive stress. This occurs in this example because the shape of the cross section causes the neutral axis to be closer to the bottom of the beam, where the maximum tensile stress occurs [Fig. (c)]. For this reason, this type of cross section is often used in designing beams made of brittle materials such as concrete, which are relatively weak in tension and strong in compression.

EXAMPLE 8-2

For the beam in Fig. 8-12, determine the normal stress due to bending at point Q.

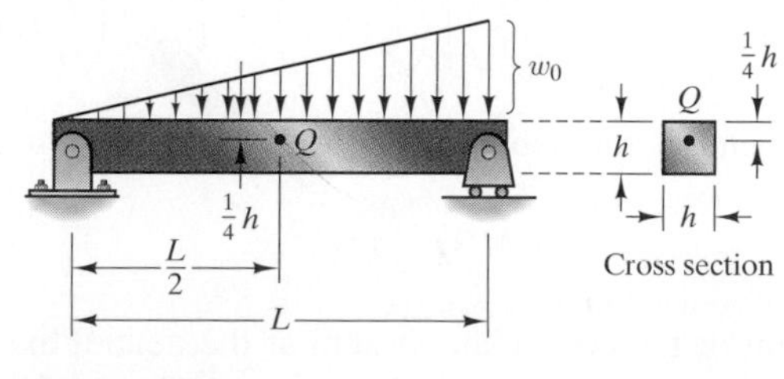

FIGURE 8-12

Strategy

We must first determine the value of the bending moment M at the cross section containing point Q. Then we can obtain the normal stress from Eq. (8-12).

Solution

By applying the equilibrium equations to a free-body diagram of the entire beam, we obtain the reactions shown in Fig. (a).

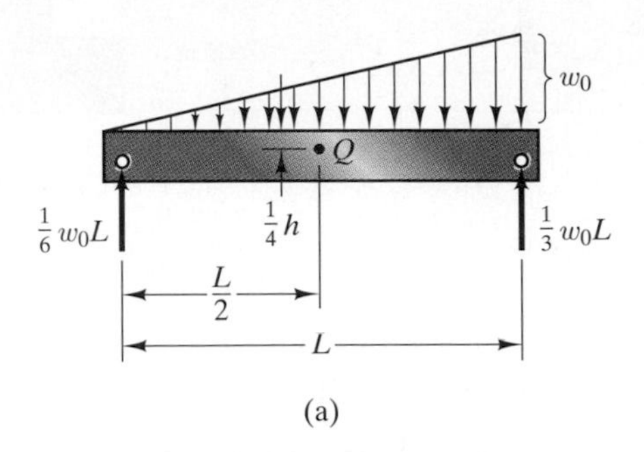

(a) Reactions at the supports.

We obtain a free-body diagram by passing a plane through Q [Fig. (b)] and represent the distributed load by an equivalent force [Fig. (c)].

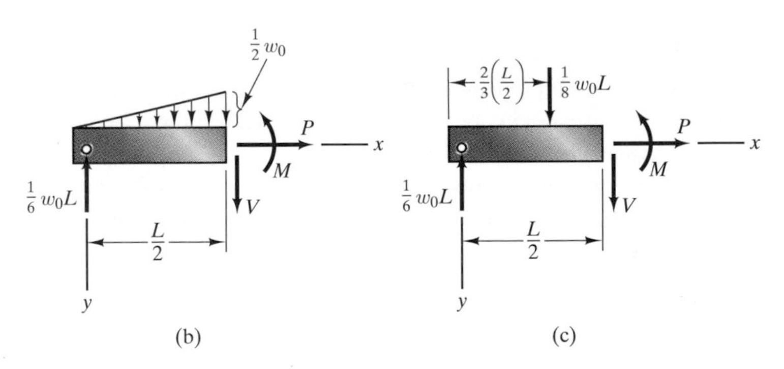

(b), (c) Free-body diagram of the part of the beam to the left of Q.

From the equilibrium equation

$$\Sigma M_{\text{right end}} = M + \left[\frac{1}{3}\left(\frac{L}{2}\right)\right]\left(\frac{1}{8}w_0 L\right) - \left(\frac{L}{2}\right)\left(\frac{1}{6}w_0 L\right) = 0,$$

we find that the bending moment at the cross section containing Q is

$$M = \tfrac{1}{16} w_0 L^2.$$

Placing the origin of the coordinate system at the neutral axis [Fig. (d)], the moment of inertia of the cross section about the z axis is $I = (1/12)h^4$, and the y coordinate of point Q is $y = -h/4$.

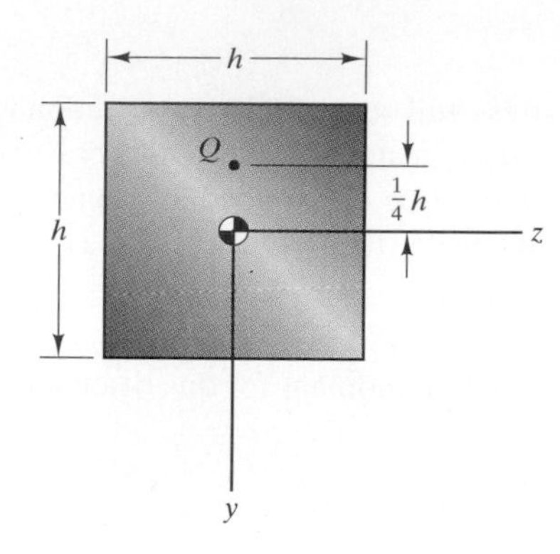

(d) Position of Q relative to the neutral axis.

The normal stress at Q is

$$\sigma_x = \frac{My}{I} = \frac{(\frac{1}{16}w_0L^2)(-h/4)}{\frac{1}{12}h^4} = -\frac{3w_0L^2}{16h^3}.$$

EXAMPLE 8-3

For the beam in Fig. 8-13, what is the maximum tensile stress due to bending, and where does it occur?

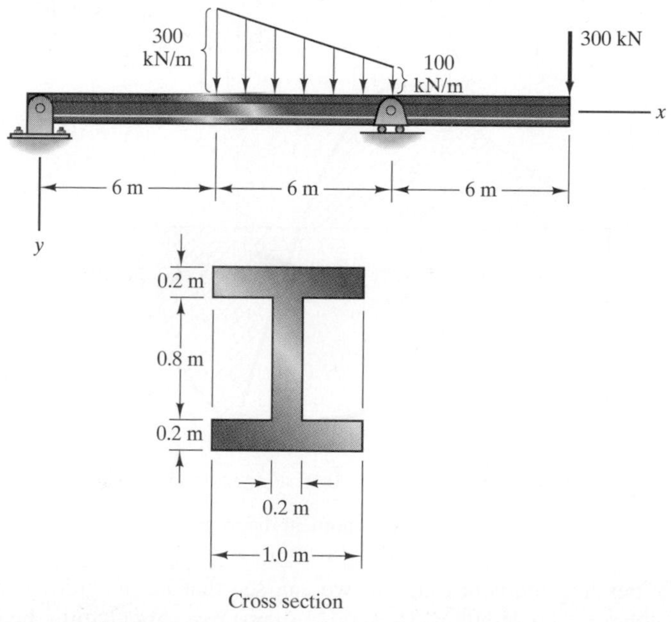

FIGURE 8-13

Strategy

The maximum tensile stress will occur at the cross section where the magnitude of the bending moment M is greatest. To locate this cross section and calculate the bending moment, M must be determined as a function of x. We can then determine the maximum tensile stress from Eq. (8-12).

Solution

The distribution of the bending moment for this beam and loading were determined in Example 7-3:

$$0 < x < 6 \text{ m}, \qquad M = 200x \text{ kN-m},$$

$$6 < x < 12 \text{ m}, \qquad M = \tfrac{50}{9}x^3 - 250x^2 + 2600x - 6600 \text{ kN-m},$$

$$12 < x < 18 \text{ m}, \qquad M = 300x - 5400 \text{ kN-m}.$$

The bending moment diagram is shown in Fig. (a).

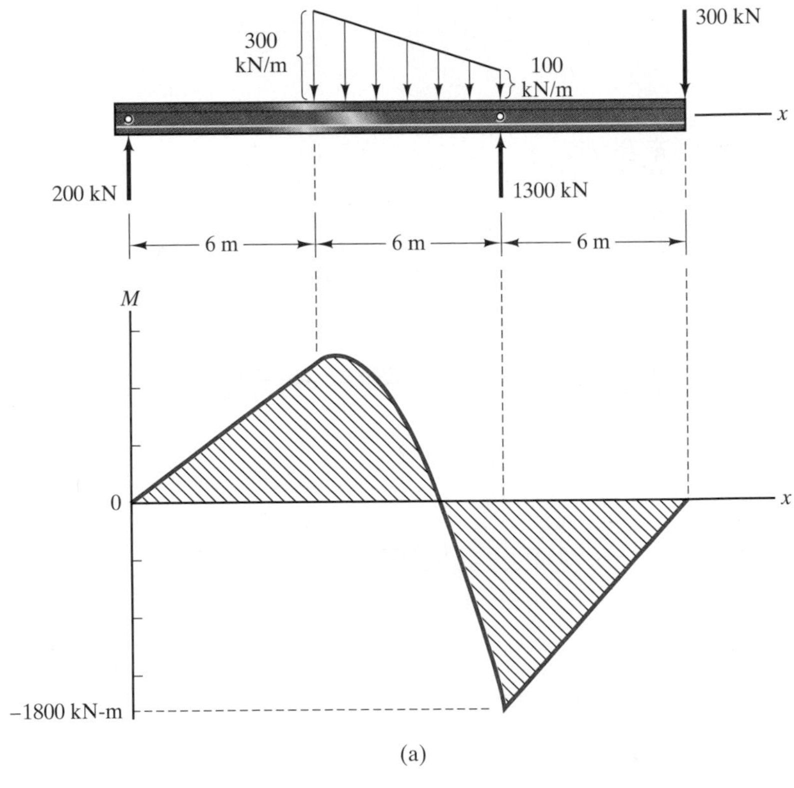

(a) Bending-moment diagram.

From the bending moment diagram we can see that the magnitude of M is greatest either where M attains its maximum positive value within the interval

$6 < x < 12$ m or at $x = 12$ m. To determine the maximum value of M within the interval $6 < x < 12$ m, we equate the derivative of M with respect to x in that interval to zero:

$$\frac{dM}{dx} = \tfrac{50}{3}x^2 - 500x + 2600 = 0.$$

Solving this equation, we find that the maximum occurs at $x = 6.69$ m. Substituting this value of x into the equation for M, we obtain $M = 1270$ kN-m. Therefore, the greatest magnitude of the bending moment occurs at $x = 12$ m, where $M = -1800$ kN-m $= -1.8 \times 10^6$ N-m.

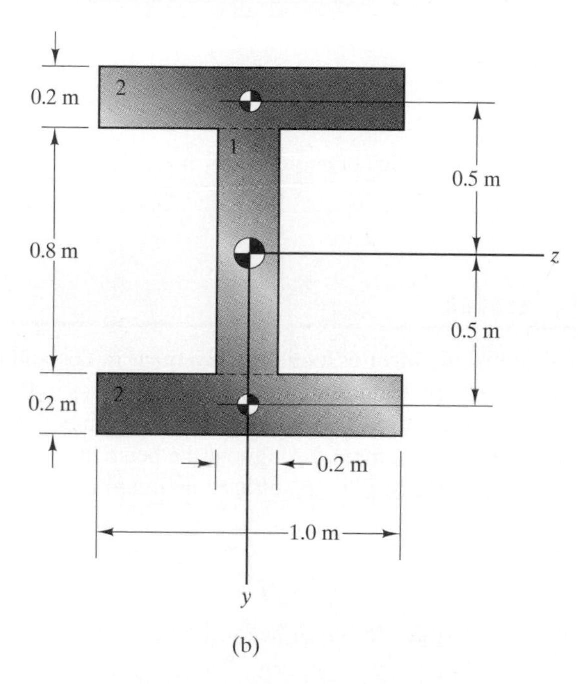

(b) Applying the parallel axis theorem.

Applying the parallel axis theorem to the cross section [Fig. (b)], the moment of inertia about the z axis is

$$\begin{aligned} I &= I_1 + 2I_2 \\ &= \tfrac{1}{12}(0.2)(0.8)^3 + 2\left[\tfrac{1}{12}(1.0)(0.2)^3 + (0.5)^2(1.0)(0.2)\right] \\ &= 0.110 \text{ m}^4. \end{aligned}$$

From Eq. (8-12), the distribution of normal stress at $x = 12$ m is

$$\begin{aligned} \sigma_x &= \frac{My}{I} = \frac{-1.8 \times 10^6 y}{0.110} \\ &= -16.4 \times 10^6 y \text{ Pa}. \end{aligned}$$

The maximum tensile stress occurs at $y = -0.6$ m [Fig. (c)]:

$$\sigma_x = (-16.4 \times 10^6)(-0.6) = 9.83 \times 10^6 \text{ Pa}.$$

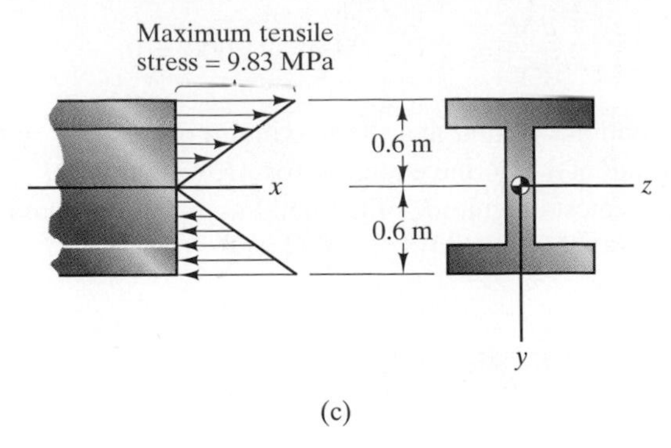

(c)

(c) Distribution of normal stress at $x = 12$ m.

8-2 Design Issues

For many applications the most essential requirement in designing a beam is to ensure that the maximum tensile and compressive stresses induced by bending moments do not exceed allowable values. This requires determining the distribution of the bending moment throughout the beam due to the maximum anticipated loads and evaluating the resulting maximum normal stresses with Eq. (8-12):

$$\sigma_x = \frac{My}{I}. \tag{8-13}$$

In this section we discuss the design of beams based on this criterion.

Cross Sections

The fact that the normal stress on a beam's cross section is inversely proportional to I explains in large part the designs of many of the beams you see in use, for example in highway overpasses and in the frames of buildings: Their cross sections are configured to increase their moments of inertia. (See the I-beam used to support the roof of Case Study House #22 at the beginning of Chapter 7.) The cross sections in Fig. 8-14 all have the same area. The numbers are the ratios of the value of the moment of inertia I about the z axis to the value for the solid square cross section. In some cases the cross section is also tailored to alter the position of the neutral axis, for example when the beam's material is stronger in compression than in tension (Fig. 8-15).

Using a cross section that increases I generally permits given loads to be supported with a smaller, lighter beam, or enhances the load-carrying capacity of a beam of a given weight. But configuring the cross section to increase its moment of inertia can be carried too far. Figure 8-16a shows a cantilever beam

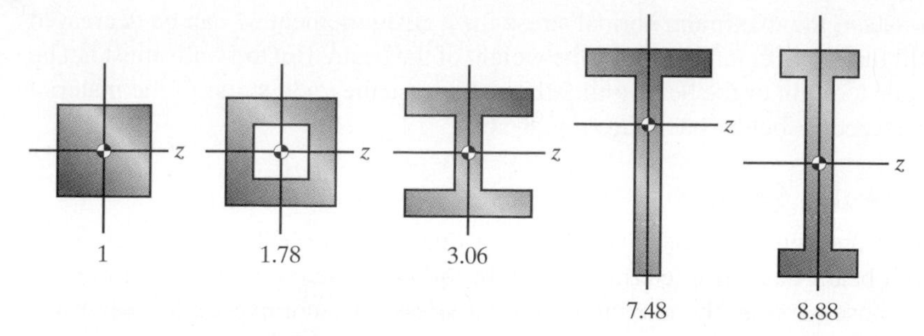

FIGURE 8-14 Typical beam cross sections and the ratio of I to the value for a solid square beam of equal cross-sectional area.

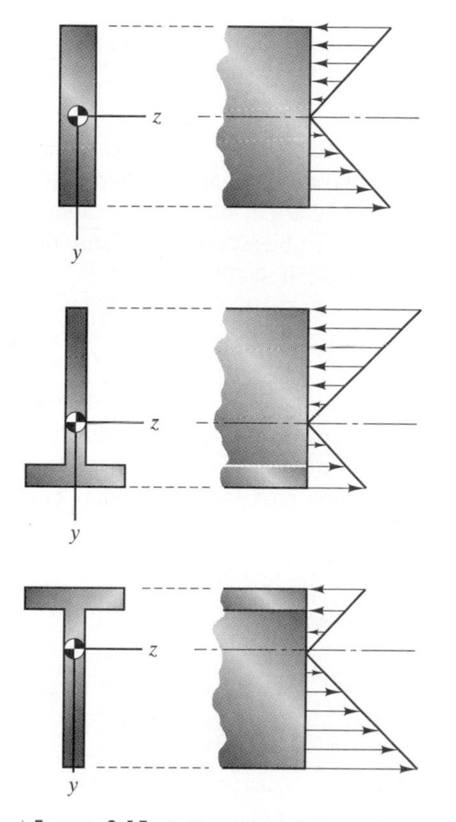

FIGURE 8-15 A T cross section can be used to decrease either the maximum tensile stress or the maximum compressive stress to which a beam is subjected.

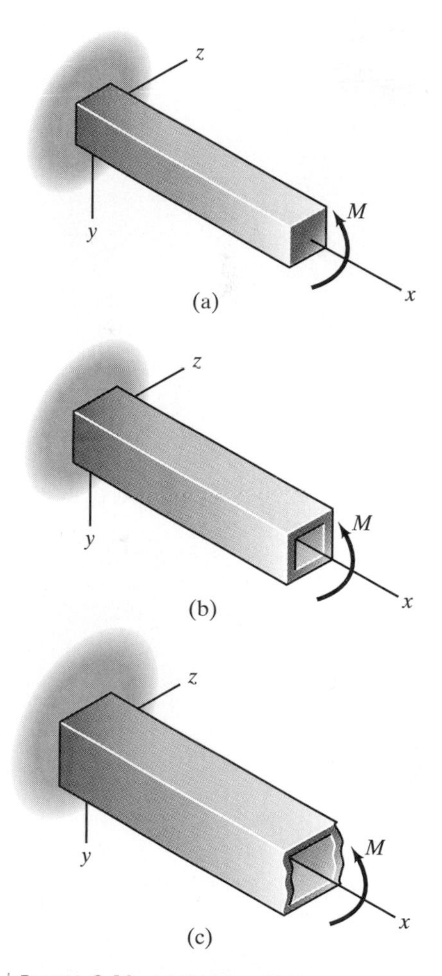

FIGURE 8-16 (a) Beam with a square cross section. (b) Box beam with the same cross-sectional area as the beam in (a). (c) A box beam with thin walls can buckle.

with a square cross section subjected to a moment M. In Fig. 8-16b, a beam with a "box" cross section having the same area as the beam in Fig. 8-16a is subjected to the moment M. If the two beams are made of the same material, they weigh the same. But the magnitude of the maximum normal stress in the box beam is substantially smaller, due to its larger moment of inertia. By making the walls of the box beam still thinner while holding the cross-sectional area constant, the maximum normal stress for a given moment M can be decreased still further without increasing the weight of the beam. But the walls must not be made too thin or the beam will fail, not because the yield stress of the material is exceeded, but by buckling (Fig. 8-16c).

Allowable Stress

Once the bending moment distribution due to the expected maximum loads on a beam has been determined, the beam's material and cross section must be chosen so that the maximum normal stress does not exceed the allowable stress σ_{allow}. (Depending on the properties of the material, it may be necessary to specify allowable stresses in both tension and compression.) The ratio of the material's yield stress to the allowable stress determines the factor of safety for the beam:

$$S = \frac{\sigma_{\text{Y}}}{\sigma_{\text{allow}}}.$$

It is essential to keep in mind that Eq. (8-13), and therefore the design procedure we describe, does not apply near a beam's supports or near locations where loads are applied. A more detailed stress analysis is necessary in those regions. Other considerations of which the structural designer must be aware in choosing the beam's material and factor of safety were discussed in Section 3-7.

EXAMPLE 8-4

The maximum anticipated magnitude of the load on the beam in Fig. 8-17 is $w_0 = 150$ kN/m. The beam's length is $L = 3$ m. The two candidate cross sections (a) and (b) have approximately the same cross-sectional area. The

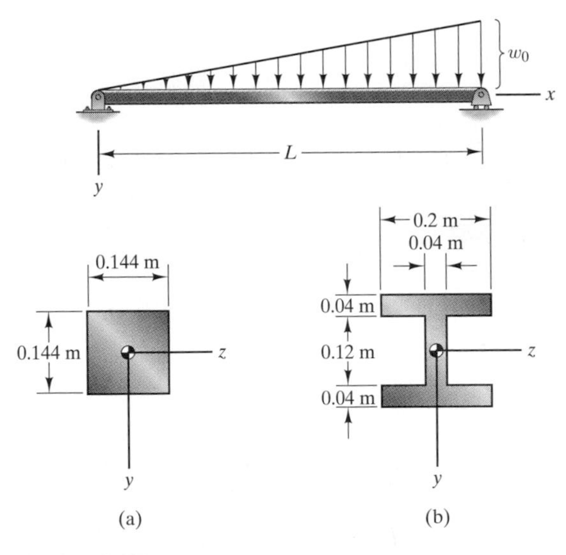

FIGURE 8-17

beam is to be made of material with yield stress $\sigma_Y = 700$ MPa. Compare the beam's factor of safety if it has cross section (a) to its factor of safety with cross section (b).

Strategy

We must determine the maximum normal stress resulting from the given load for each cross section. This requires determining the maximum bending moment in the beam and applying Eq. (8-13) for each cross section.

Solution

The distribution of the bending moment in the beam is

$$M = \frac{1}{6}w_0\left(Lx - \frac{x^3}{L}\right).$$

To determine where the maximum value of M occurs, we equate the derivative of M with respect to x to zero,

$$\frac{dM}{dx} = \frac{1}{6}w_0\left(L - \frac{3x^2}{L}\right) = 0,$$

and solve for x, obtaining $x = L/\sqrt{3}$. Substituting this value into the equation for M, the maximum value of the bending moment is

$$\begin{aligned} M_{\max} &= \frac{w_0 L^2}{9\sqrt{3}} \\ &= \frac{(150{,}000)(3)^2}{9\sqrt{3}} \\ &= 86{,}600 \text{ N-m.} \end{aligned}$$

(a) Square cross section The moment of inertia of the square cross section about the z axis is

$$I_x = \tfrac{1}{12}(0.144)(0.144)^3 = 3.58 \times 10^{-5} \text{ m}^4.$$

The maximum normal stress is

$$\begin{aligned} \sigma_{\max} &= \frac{My}{I} \\ &= \frac{(86{,}600)(0.072)}{3.58 \times 10^{-5}} \\ &= 174 \text{ MPa,} \end{aligned}$$

so the beam's factor of safety is

$$S = \frac{\sigma_Y}{\sigma_{allow}} = \frac{700}{174} = 4.02.$$

(b) I-beam cross section The moment of inertia of the vertical rectangle about the z axis is

$$I_v = \tfrac{1}{12}(0.04)(0.12)^3 = 0.576 \times 10^{-5}\ \text{m}^4.$$

Applying the parallel axis theorem, the moment of inertia of each horizontal rectangle about the z axis is

$$I_h = \tfrac{1}{12}(0.2)(0.04)^3 + (0.06 + 0.02)^2(0.2)(0.04) = 5.227 \times 10^{-5}\ \text{m}^4.$$

Therefore, the moment of inertia of the cross section is

$$I = I_v + 2I_h = 11.03 \times 10^{-5}\ \text{m}^4.$$

The maximum normal stress is

$$\sigma_{max} = \frac{My}{I} = \frac{(86{,}600)(0.06 + 0.04)}{11.03 \times 10^{-5}} = 78.5\ \text{MPa}.$$

The beam's factor of safety is

$$S = \frac{\sigma_Y}{\sigma_{allow}} = \frac{700}{78.5} = 8.91.$$

Discussion

The advantage of a cross section that increases the beam's moment of inertia is apparent in this example. But in particular applications, other factors, such as the availability of stock with a desired cross section, or the cost of manufacturing it, may be overriding considerations.

8-3 Composite Beams

If the walls of a box beam are made too thin in an effort to obtain a large moment of inertia, the beam can fail by buckling as shown in Fig. 8-18a. To help prevent buckling of beams with thin walls, a light "filler" material can be used (Fig. 8-18b). The filler material does not add significant weight but helps stabilize the walls, resulting in a light, strong beam. This is an example of a *composite beam,* a beam consisting of two or more materials. By taking advantage of the different properties of materials, structural engineers can design composite beams with characteristics that are not possible with a single material. Our objective is to explain how to determine the distribution of normal stress at a given cross section of a prismatic composite beam consisting of two materials.

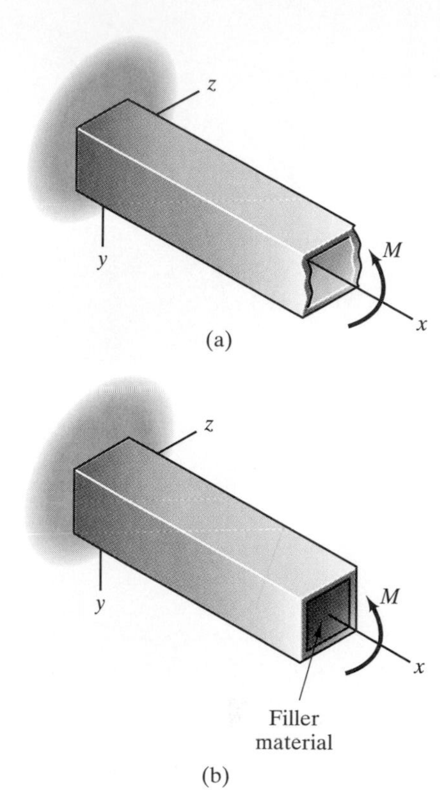

FIGURE 8-18 (a) A box beam with thin walls can buckle. (b) Using a filler material to stabilize thin walls.

Consider a given cross section of a prismatic composite beam, and let the materials be denoted A and B (Fig. 8-19). We make no assumption about the geometry of the cross section except that the parts consisting of materials A and B are each symmetric about the y axis. Notice that as in the case of a beam consisting of a single material, we do not know the location of the neutral axis and must determine it as part of our analysis.

If the two materials are bonded together and we retain the fundamental assumption that plane sections remain plane, the geometric arguments leading to Eq. (8-2) hold and the extensional strain in both materials is given by

$$\epsilon_x = -\frac{y}{\rho},$$

where ρ is the radius of curvature of the neutral axis. Since the materials may have different elastic moduli, we must write Eq. (8-4) for each material, obtaining

$$\begin{aligned} \sigma_A &= E_A \epsilon_x = -\frac{E_A y}{\rho}, \\ \sigma_B &= E_B \epsilon_x = -\frac{E_B y}{\rho}. \end{aligned} \tag{8-14}$$

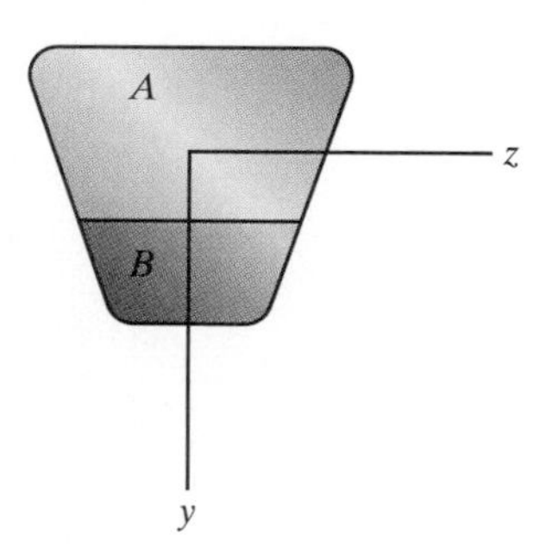

FIGURE 8-19 Cross section of a composite beam consisting of materials A and B.

Equation (8-6), which expresses the equilibrium requirement that the normal stress distribution exerts no net force, now becomes

$$\int_A \sigma_x \, dA = \int_{A_A} \sigma_A \, dA + \int_{A_B} \sigma_B \, dA = 0,$$

where A_A and A_B denote the cross sections of the materials. Substituting the expressions (8-14) leads to the equation

$$E_A \int_{A_A} y \, dA + E_B \int_{A_B} y \, dA = 0.$$

We can express this equation as

$$E_A A_A \bar{y}_A + E_B A_B \bar{y}_B = 0, \tag{8-15}$$

where $\bar{y}_A$ and $\bar{y}_B$ are the coordinates of the centroids of A_A and A_B relative to the neutral axis. As we demonstrate in Example 8-5, this equation can be used to determine the location of the neutral axis.

The moment about the z axis due to the normal stresses must equal the bending moment M at the given cross section:

$$M = \int_{A_A} y\sigma_A \, dA + \int_{A_B} y\sigma_B \, dA.$$

Substituting the expressions (8-14) into this equation gives

$$\begin{aligned} M &= -\frac{1}{\rho}\left(E_A \int_{A_A} y^2 \, dA + E_B \int_{A_B} y^2 \, dA\right) \\ &= -\frac{1}{\rho}(E_A I_A + E_B I_B), \end{aligned}$$

where I_A and I_B are the moments of inertia of A_A and A_B about the z axis. By solving this equation for the radius of curvature ρ and substituting the result into Eqs. (8-14), we obtain the distributions of normal stress in the individual materials:

$$\begin{aligned} \sigma_A &= \frac{My}{I_A + (E_B/E_A)I_B}, \\ \sigma_B &= \frac{My}{(E_A/E_B)I_A + I_B}. \end{aligned} \tag{8-16}$$

Now that we know the stress distributions in the individual materials, we can use them to determine the part of the bending moment M supported by each material:

$$\begin{aligned} M_A &= \int_{A_A} y\sigma_A \, dA = \frac{I_A}{I_A + (E_B/E_A)I_B} M, \\ M_B &= \int_{A_B} y\sigma_B \, dA = \frac{I_B}{(E_A/E_B)I_A + I_B} M. \end{aligned} \tag{8-17}$$

Observe that $M_A + M_B = M$.

EXAMPLE 8-5

Figure 8-20 is the cross section of a prismatic beam made of steel (material A) and aluminum alloy (material B) with elastic moduli $E_A = 200$ GPa and $E_B = 72$ GPa. If $M = 12$ kN-m at a given cross section, determine the distributions of normal stress in the steel and in the aluminum.

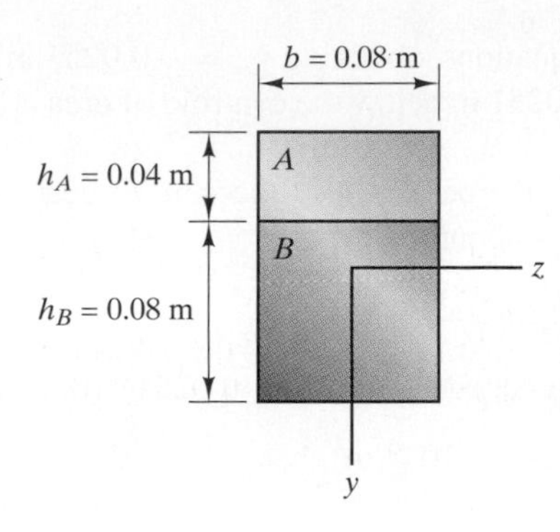

FIGURE 8 20

Strategy

We must first determine the position of the neutral axis by using Eq. (8-15). We can then determine the moments of intertia I_A and I_B of the cross sections of the two materials about the z axis, and the distributions of the normal stress are given by Eqs. (8-16).

Solution

We begin by placing the coordinate system at an arbitrary position [Fig. (a)].

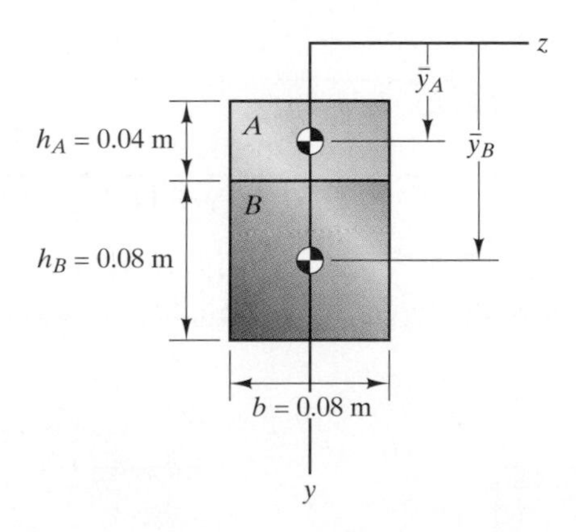

(a) Coordinate system for determining the position of the neutral axis.

Equation (8-15) is

$$E_A A_A \bar{y}_A + E_B A_B \bar{y}_B = 0:$$

$$(200 \times 10^9)(0.04)(0.08)\bar{y}_A + (72 \times 10^9)(0.08)(0.08)\bar{y}_B = 0.$$

From Fig. (a) we see that

$$\bar{y}_B - \bar{y}_A = \tfrac{1}{2}(h_A + h_B) = 0.06 \text{ m}.$$

We solve these two equations, obtaining $\bar{y}_A = -0.0251$ m and $\bar{y}_B = 0.0349$ m. The neutral axis is 0.0251 m below the centroid of area A (0.0349 m above the centroid of area B).

We can now apply the parallel axis theorem to determine the moments of inertia of the two areas about the z axis:

$$\begin{aligned} I_A &= \tfrac{1}{12} b h_A^3 + \bar{y}_A^2 A_A \\ &= \tfrac{1}{12}(0.08)(0.04)^3 + (-0.0251)^2(0.04)(0.08) \\ &= 2.45 \times 10^{-6}\ \text{m}^4, \end{aligned}$$

$$\begin{aligned} I_B &= \tfrac{1}{12} b h_B^3 + \bar{y}_B^2 A_B \\ &= \tfrac{1}{12}(0.08)(0.08)^3 + (0.0349)^2(0.08)(0.08) \\ &= 11.20 \times 10^{-6}\ \text{m}^4. \end{aligned}$$

From Eqs. (8-16), the distributions of normal stress in the two materials are

$$\begin{aligned} \sigma_A &= \frac{My}{I_A + (E_B/E_A) I_B} \\ &= \frac{12{,}000y}{2.45 \times 10^{-6} + (72/200)(11.20 \times 10^{-6})} \\ &= 1.852y\ \text{GPa}, \end{aligned}$$

$$\begin{aligned} \sigma_B &= \frac{My}{(E_A/E_B) I_A + I_B} \\ &= \frac{12{,}000y}{(200/72)(2.45 \times 10^{-6}) + 11.20 \times 10^{-6}} \\ &= 0.667y\ \text{GPa}. \end{aligned}$$

The position of the neutral axis and the distributions of normal stress are shown in Fig. (b).

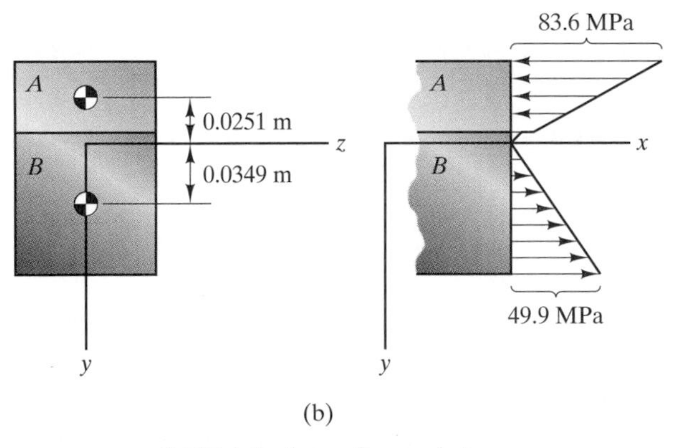

(b) Distributions of normal stress.

8-4 | Elastic–Perfectly Plastic Beams

If the bending moment at a given cross section of a beam becomes sufficiently large, the magnitude of the maximum normal stress will equal the yield stress of the material. For larger values of the bending moment, the material will undergo plastic deformation. Engineers normally design structural elements so that stresses remain well below the values at which yield occurs. But in some cases beams are intentionally designed to support their loads in a partially yielded state, and in safety and failure analyses of structures it is often necessary to understand the plastic behavior of beams.

Although more realistic models must usually be used in actual design, we can provide insight into the plastic behavior of beams by assuming that the material is elastic–perfectly plastic (Fig. 8-21). Let M be the bending moment at a given location on the axis of a beam with a rectangular cross section (Fig. 8-22a). When M is sufficiently large, the magnitudes of the normal stresses at the top and bottom of the cross section are equal to the yield stress (Fig. 8-22b). What

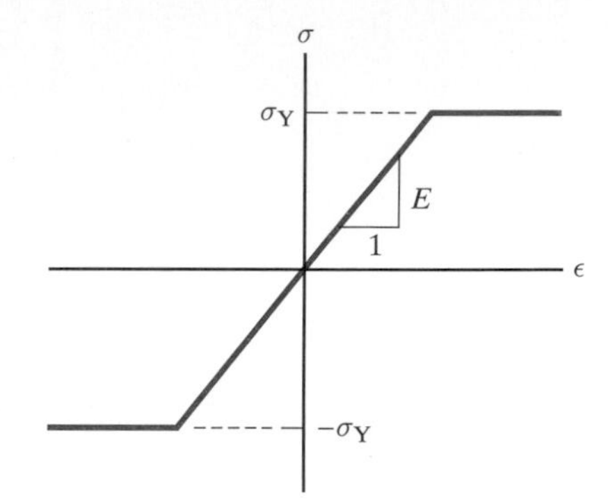

FIGURE 8-21 Model of an elastic–perfectly plastic material.

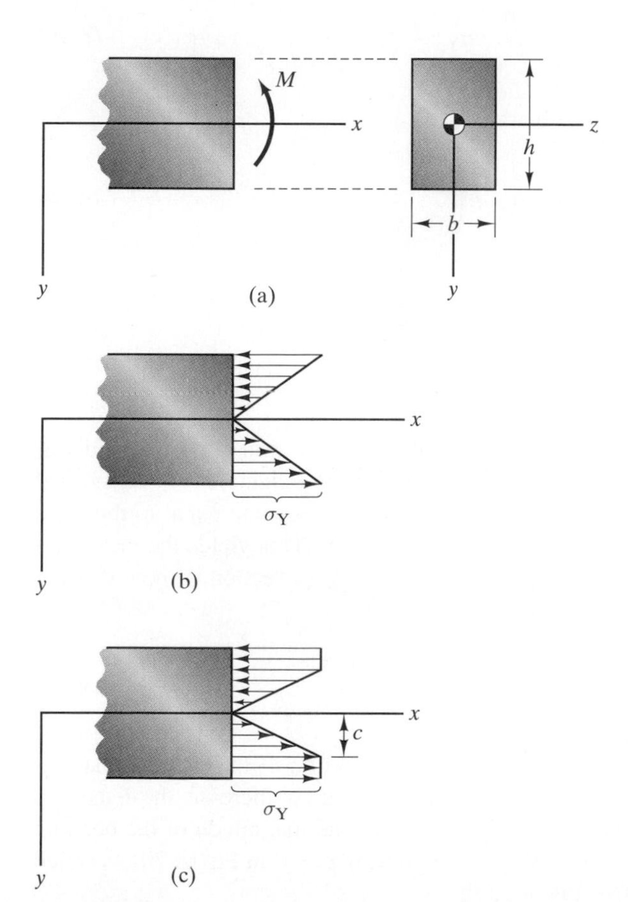

FIGURE 8-22 (a) Bending moment at a cross section of a rectangular beam. (b) For a sufficiently large bending moment, the magnitude of the maximum stress equals the yield stress. (c) Stress distribution as the bending moment continues to increase.

will the distribution of stress be for a still larger value of M? If we retain the assumption that plane sections remain plane, the distribution of the extensional strain ϵ_x continues to be a linear function of y. But the magnitude of the normal stress cannot exceed the yield stress σ_Y, resulting in the distribution shown in Fig. 8-22c. The magnitude of the normal stress increases linearly until it reaches the yield stress at some distance c from the neutral axis, then remains constant.

We can calculate the distance c for a given bending moment, and thereby determine the distribution of the normal stress, by equating M to the moment exerted about the z axis by the normal stress:

$$M = \int_A y\sigma_x \, dA. \tag{8-18}$$

To evaluate the integral, let the element of area dA be a horizontal strip of infinitesimal height: $dA = b\,dy$. For values of y in the range $-c < y < c$, the normal stress is a linear function of y: $\sigma_x = \sigma_Y(y/c)$. Therefore,

$$M = \int_{-h/2}^{-c} y(-\sigma_Y)b\,dy + \int_{-c}^{c} y\sigma_Y\left(\frac{y}{c}\right)b\,dy + \int_{c}^{h/2} y\sigma_Y b\,dy. \tag{8-19}$$

We evaluate these integrals, obtaining

$$M = \sigma_Y b\left(\frac{h^2}{4} - \frac{c^2}{3}\right), \tag{8-20}$$

and solve for c:

$$c = \sqrt{3\left(\frac{h^2}{4} - \frac{M}{\sigma_Y b}\right)}. \tag{8-21}$$

With these relationships we can describe the evolution of the stress distribution as M increases. The value of M at which the magnitudes of the normal stresses at the top and bottom of the cross section become equal to the yield stress is obtained by setting $c = h/2$ in Eq. (8-20). This yields the maximum bending moment that can be applied at a given cross section without causing yielding of the material (Fig. 8-23a):

$$M = \frac{\sigma_Y bh^2}{6}. \tag{8-22}$$

When M exceeds this value (Fig. 8-23b), the portion of the material that has yielded is determined by Eq. (8-21). As M continues to increase, the distance c decreases until all of the material has yielded. The magnitude of the bending moment at which this occurs, obtained by setting $c = 0$ in Eq. (8-20), is called the *ultimate moment* M_U (Fig. 8-23c):

$$M_U = \frac{\sigma_Y bh^2}{4}. \tag{8-23}$$

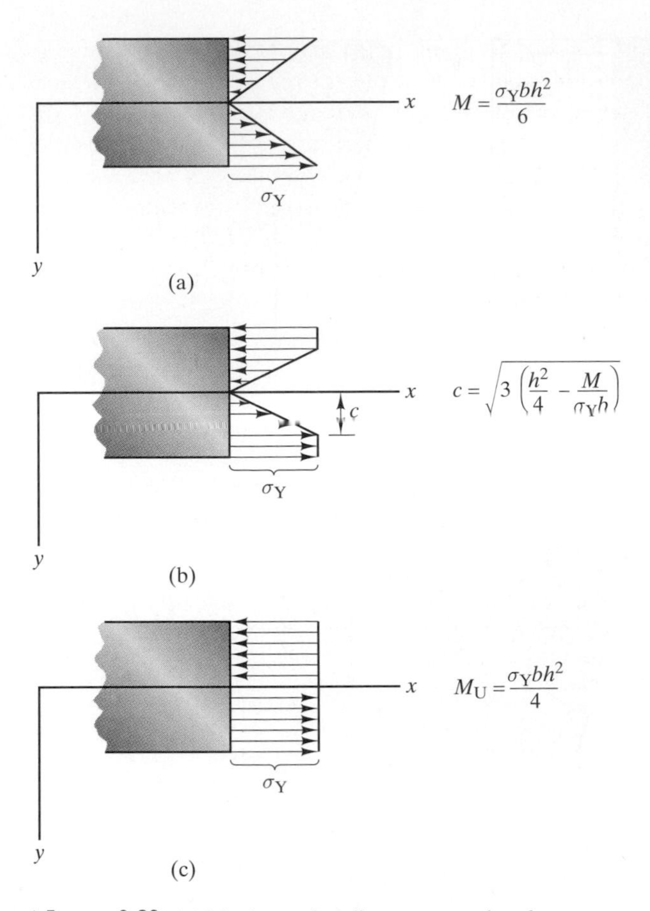

FIGURE 8-23 (a) Maximum bending moment that does not cause yielding of the material. (b) Stress distribution when the material is partially yielded. (c) Bending moment when the material is completely yielded.

When the moment at a given cross section reaches this value, there is no resistance to further bending and the beam is said to form a *plastic hinge*.

Our analysis thus far has been limited to a beam with a rectangular cross section. For other cross sections, determining the distribution of the normal stress when the material is partially yielded usually requires a numerical solution, but calculating the ultimate moment M_U is straightforward. Let us assume that the material at a given cross section is fully yielded, and let A_T and A_C be the areas of the cross section that are subjected to tensile and compressive stress (Fig. 8-24a). From the equilibrium requirement that the distribution of normal stress must exert no net axial force,

$$\int_A \sigma_x \, dA = \int_{A_T} \sigma_Y \, dA + \int_{A_C} (-\sigma_Y) \, dA$$
$$= \sigma_Y (A_T - A_C) = 0,$$

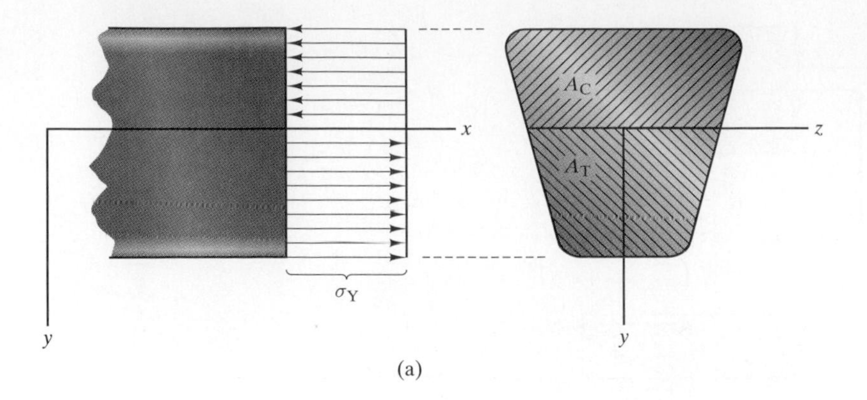

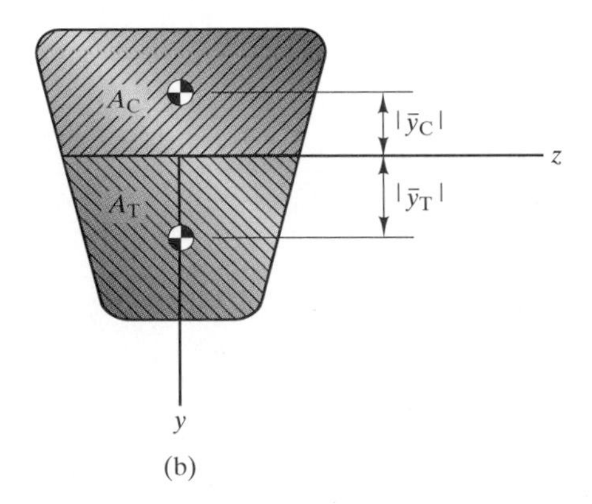

FIGURE 8-24 (a) Parts of the cross section subject to tensile and compressive stress. (b) Distances from the neutral axis to the centroids of A_C and A_T.

we see that

$$A_T = A_C. \tag{8-24}$$

The areas of the cross section subjected to tensile and compressive stress must be equal. This condition can be used to determine the distribution of the normal stress. The ultimate moment is

$$M_U = \int_A y\sigma_x \, dA = \int_{A_T} y\sigma_Y \, dA + \int_{A_C} y(-\sigma_Y) \, dA.$$

We can express this result as

$$M_U = \sigma_Y(\bar{y}_T A_T - \bar{y}_C A_C), \tag{8-25}$$

where $\bar{y}_T$ and $\bar{y}_C$ are the coordinates of the centroids of A_T and A_C relative to the neutral axis (Fig. 8-24b).

EXAMPLE 8-6

The beam in Fig. 8-25 consists of elastic–perfectly plastic material with yield stress $\sigma_Y = 340$ MPa. **(a)** Sketch the distribution of normal stress at $x = 3$ m if $w_0 = 70{,}000$ N/m. **(b)** If w_0 is increased progressively, at what value will the beam fail by formation of a plastic hinge? Where does the plastic hinge occur?

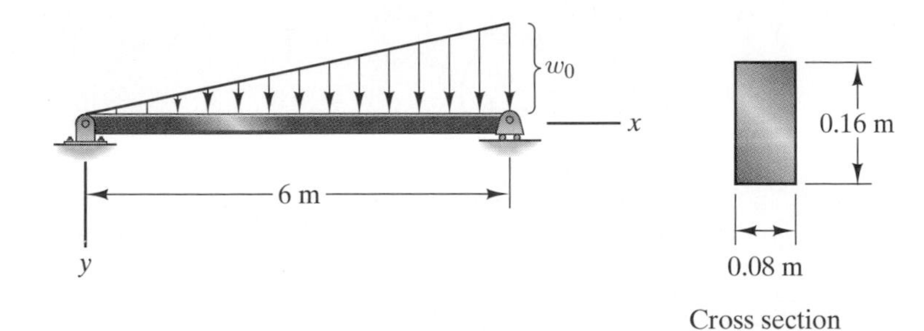

FIGURE 8-25

Strategy

(a) With w_0 known, we can determine the bending moment M at $x = 3$ m. Then we can use Eqs. (8-22) and (8-23) to determine whether the material is partially yielded. If there is partial yielding, calculating the distance c from Eq. (8-21) establishes the distribution of normal stress. **(b)** As w_0 increases, the material will first become completely yielded and a plastic hinge will form at the cross section where the magnitude of the bending moment is a maximum. The value of the ultimate moment is given by Eq. (8-23).

Solution

Determining the bending moment in the beam as a function of x, we find that

$$M = w_0\left(x - \frac{x^3}{36}\right). \qquad \textbf{(8-26)}$$

(a) If $w_0 = 70{,}000$ N/m, the bending moment at $x = 3$ m is

$$M = 70{,}000\left[3 - \frac{(3)^3}{36}\right] = 157{,}500 \text{ N-m}.$$

From Eq. (8-22), the maximum moment that will not cause yielding of the material is

$$\frac{\sigma_Y bh^2}{6} = \frac{(340 \times 10^6)(0.08)(0.16)^2}{6} = 116{,}100 \text{ N-m},$$

and from Eq. (8-23), the ultimate moment when the material is completely yielded is

$$M_U = \frac{\sigma_Y b h^2}{4} = \frac{(340 \times 10^6)(0.08)(0.16)^2}{4} = 174{,}100 \text{ N-m}. \qquad \textbf{(8-27)}$$

Therefore, the material is partially yielded. From Eq. (8-21), the distance c is

$$\begin{aligned} c &= \sqrt{3\left(\frac{h^2}{4} - \frac{M}{\sigma_Y b}\right)} \\ &= \sqrt{3\left[\frac{(0.16)^2}{4} - \frac{157{,}500}{(340 \times 10^6)(0.08)}\right]} \\ &= 0.0428 \text{ m}. \end{aligned}$$

Figure (a) shows the distribution of the normal stress.

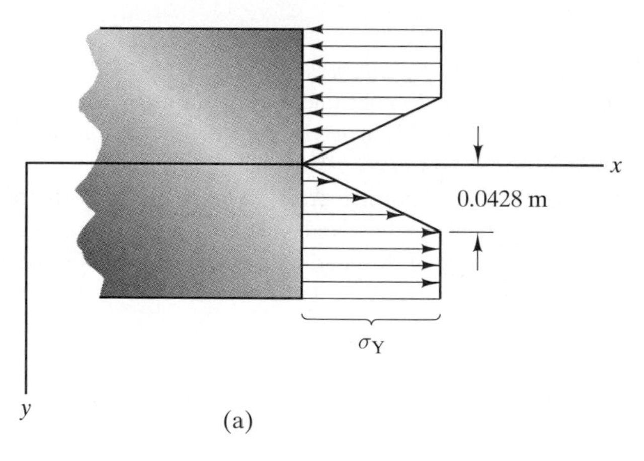

(a) Normal stress at $x = 3$ m.

(b) To determine the cross section at which the bending stress is a maximum, we equate the derivative of Eq. (8-26) to zero.

$$\frac{dM}{dx} = w_0\left(1 - \frac{x^2}{12}\right) = 0.$$

Solving this equation, we find that the bending moment is a maximum, and the plastic hinge will occur, at $x = \sqrt{12} = 3.46$ m. Substituting this value of x into Eq. (8-26), the value of the maximum bending moment in the beam in terms of w_0 is $M_{max} = 4w_0/\sqrt{3}$. The plastic hinge forms when the maximum bending moment is equal to the ultimate moment determined in Eq. (8-27):

$$M_{max} = M_U,$$

$$\frac{4w_0}{\sqrt{3}} = 174{,}100 \text{ N-m}.$$

Solving, we obtain $w_0 = 75{,}400$ N/m.

EXAMPLE 8-7

The beam in Fig. 8-26 is subjected to a moment M. If the material is elastic–perfectly plastic with yield stress $\sigma_Y = 500$ MPa, what is the ultimate moment M_U?

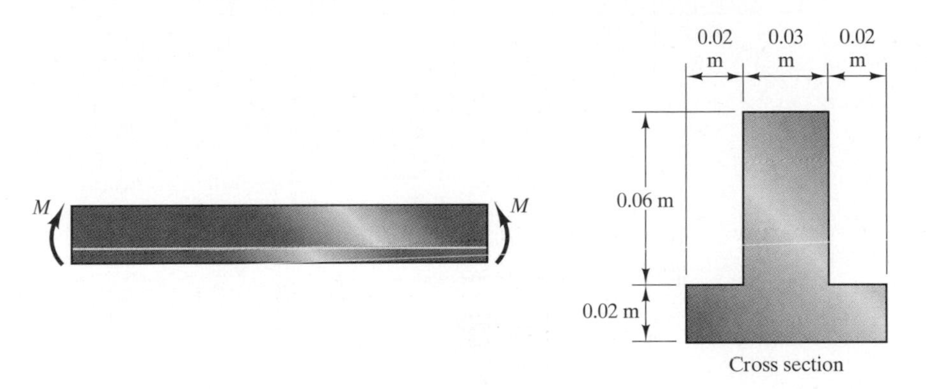

FIGURE 8-26

Strategy

The ultimate moment is given by Eq. (8-25) in terms of the areas A_T and A_C of the cross section that are subjected to tensile and compressive stress and the coordinates of their centroids relative to the neutral axis. We must begin by determining the location of the neutral axis from the condition that $A_T = A_C$.

Solution

Let H be the unknown distance from the top of the cross section to the neutral axis [Fig. (a)]. In terms of H, the areas A_C and A_T are

$$A_C = 0.03H,$$

$$A_T = (0.03)(0.06 - H) + (0.07)(0.02).$$

From the condition

$$A_T = A_C:$$
$$(0.03)(0.06 - H) + (0.07)(0.02) = 0.03H,$$

we find that $H = 0.0533$ m and $A_T = A_C = 0.00160\ \text{m}^2$.

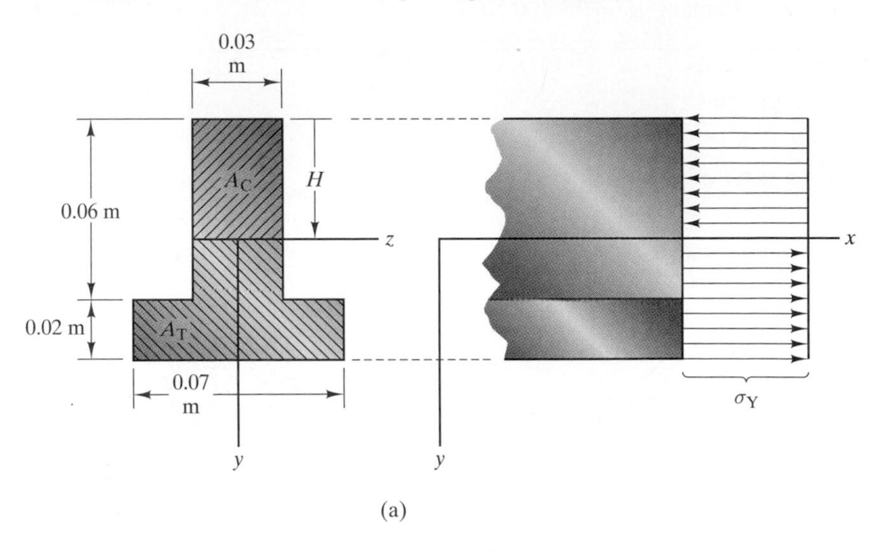

(a)

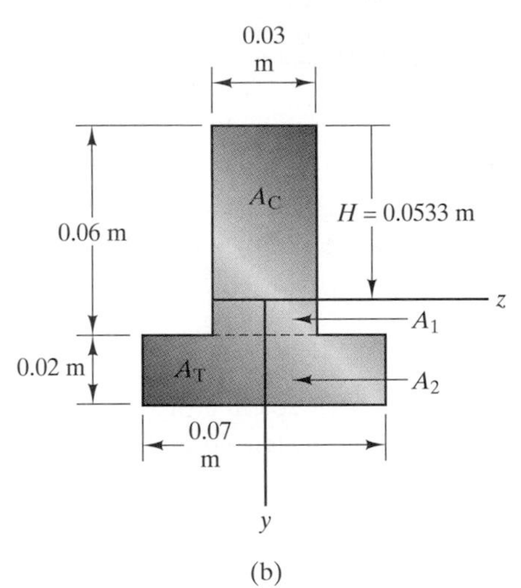

(b)

(a) Distance H to the neutral axis.
(b) Determining the centroid of A_T.

The coordinate of the centroid of A_C is

$$\bar{y}_C = -\frac{H}{2} = -\frac{0.0533}{2} = -0.0267\ \text{m}.$$

Treating A_T as a composite of two rectangles [Fig. (b)], the coordinate of its centroid is

$$\bar{y}_T = \frac{\bar{y}_1 A_1 + \bar{y}_2 A_2}{A_1 + A_2}$$

$$= \frac{\left(\frac{0.06 - H}{2}\right)(0.03)(0.06 - H) + \left(0.06 - H + \frac{0.02}{2}\right)(0.07)(0.02)}{(0.03)(0.06 - H) + (0.07)(0.02)}$$

$$= 0.0150 \text{ m}.$$

From Eq. (8-25), the ultimate moment is

$$M_U = \sigma_Y(\bar{y}_T A_T - \bar{y}_C A_C)$$
$$= (500 \times 10^6)[(0.0150)(0.00160) - (-0.0267)(0.00160)]$$
$$= 33{,}300 \text{ N-m}.$$

8-5 | Unsymmetric Cross Sections

In Section 8-1 we derived the distribution of the normal stress in a beam due to bending under the assumption that the beam's cross section was symmetric about the y axis. Let us now consider a beam with an arbitrary cross section that is subjected to couples at the ends (Fig. 8-27). We label the coordinate axes $x'y'z'$ and assume that the moments are exerted about the z' axis.

Moment Exerted about a Principal Axis

If we make the same geometric assumptions regarding the beam's deformation that we made in Section 8-1, the steps leading to Eq. (8-5) are unchanged and we conclude that the distribution of normal stress at a given cross section is

$$\sigma_x = -\frac{Ey'}{\rho}, \qquad \textbf{(8-28)}$$

where E is the modulus of elasticity and ρ is the radius of curvature of the neutral axis. In Fig. 8-28 we obtain a free-body diagram by passing a plane

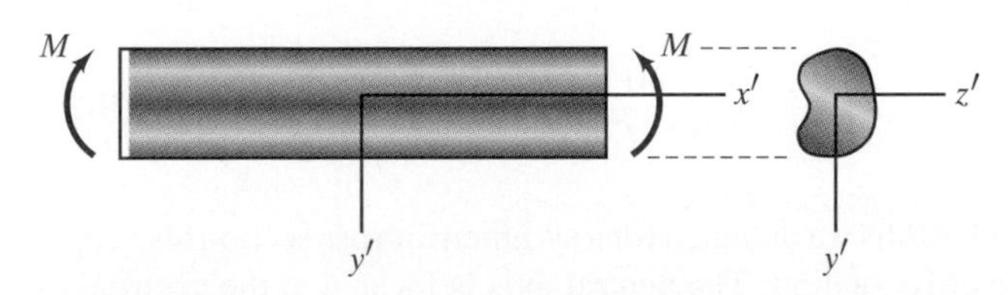

FIGURE 8-27 Subjecting a beam with an arbitrary cross section to a bending moment M.

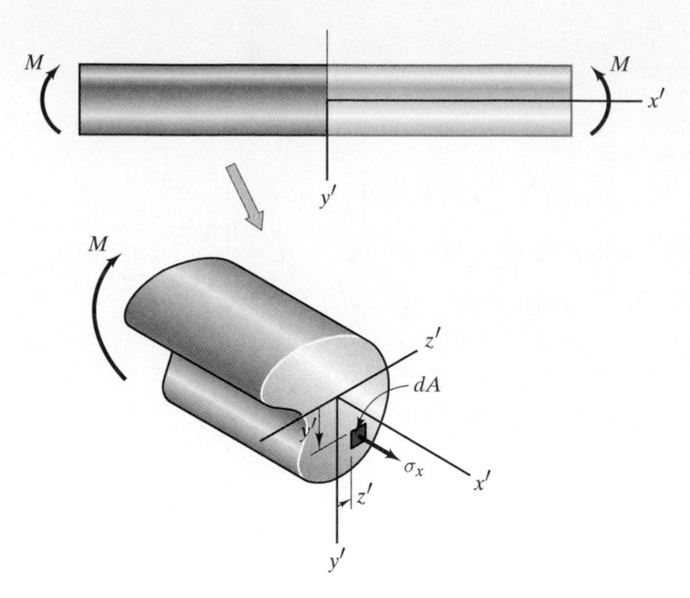

FIGURE 8-28 Obtaining a free-body diagram by passing a plane perpendicular to the beam's axis.

perpendicular to the beam's axis and show the normal stress acting on an element dA of the cross section. What conditions are necessary for this free-body diagram to be in equilibrium? The force exerted in the x' direction is

$$\int_A \sigma_x \, dA = 0.$$

Substituting Eq. (8-28) into this expression confirms that the neutral axis must coincide with the centroid of the cross section. The sum of the moments about the z' axis is

$$M - \int_A y' \sigma_x \, dA = 0.$$

Just as in Section 8-1, substituting Eq. (8-28) into this expression yields the equation for the distribution of the normal stress in terms of the bending moment and the moment of inertia of the cross section about the z' axis:

$$\sigma_x = \frac{My'}{I_{z'}}. \qquad \textbf{(8-29)}$$

The two essential results we obtained for a symmetric cross section also apply to an unsymmetric cross section: The neutral axis is located at the centroid of the cross section, and the distribution of the normal stress is given by Eq. (8-29). But a third condition is necessary for the free-body diagram in Fig. 8-28 to be in equilibrium. The sum of the moments about the y' axis is

$$\int_A z'\sigma_x \, dA = 0. \quad \textbf{(8-30)}$$

Substituting Eq. (8-28) into this expression, we conclude that

$$\int_A y'z' \, dA = 0. \quad \textbf{(8-31)}$$

This requirement is satisfied if the beam's cross section is symmetric about the y' axis. Since our discussion in Section 8-1 was limited to such cross sections, it was not necessary to consider this additional equilibrium condition. Equation (8-31) is also satisfied if the cross section is symmetric about the z' axis. More generally, Eq. (8-31) is satisfied, and the distribution of stress is given by Eq. (8-29), only if the z' axis about which the bending moment M is exerted *is a principal axis of the cross section.* (We discuss principal axes in Appendix C.)

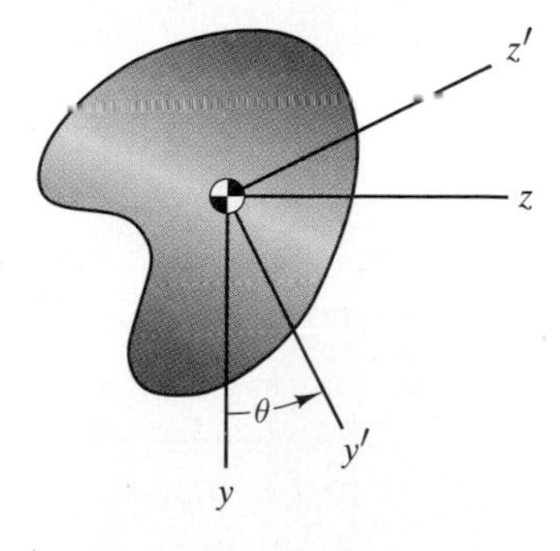

FIGURE 8-29 Orientation of the $y'z'$ coordinate system relative to the yz coordinate system.

Suppose that we know the moments and product of inertia of a given cross section in terms of a coordinate system yz with its origin at the centroid. In Fig. 8-29 the $y'z'$ coordinate system is rotated through an angle θ relative to the yz coordinate system. The coordinates of a point of the cross section in the $y'z'$ system are given in terms of the coordinates of the point in the yz system by

$$y' = y\cos\theta + z\sin\theta, \quad \textbf{(8-32)}$$

$$z' = -y\sin\theta + z\cos\theta. \quad \textbf{(8-33)}$$

The moments and product of inertia of the cross section in terms of the $y'z'$ system are given in terms of the moments and product of inertia in terms of the yz system by the expressions (see Section C-5)

$$I_{y'} = \frac{I_y + I_z}{2} + \frac{I_y - I_z}{2}\cos 2\theta - I_{yz}\sin 2\theta, \quad \textbf{(8-34)}$$

$$I_{z'} = \frac{I_y + I_z}{2} - \frac{I_y - I_z}{2}\cos 2\theta + I_{yz}\sin 2\theta, \quad \textbf{(8-35)}$$

$$I_{y'z'} = \frac{I_y - I_z}{2}\sin 2\theta + I_{yz}\cos 2\theta. \quad \textbf{(8-36)}$$

A value of θ for which y' and z' are principal axes, which we denote by θ_p, satisfies the equation

$$\tan 2\theta_p = \frac{2I_{yz}}{I_z - I_y}. \quad \textbf{(8-37)}$$

We can determine the orientation of the principal axes by solving this equation for θ_p, and evaluate the moments of inertia about the principal axes from Eqs. (8-34) and (8-35). Then the distribution of the normal stress due to a bending moment M exerted about the z' axis is given by Eq. (8-29).

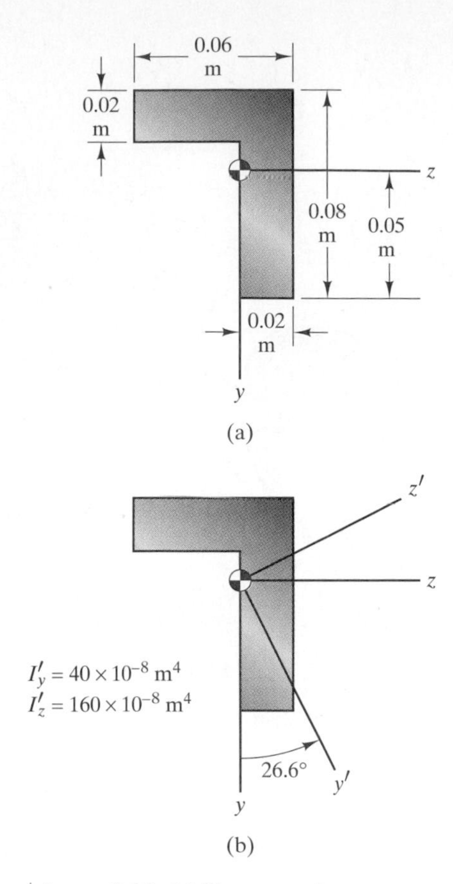

FIGURE 8-30 (a) Unsymmetric cross section. (b) Orientation of the principal axes.

For example, consider the unsymmetric cross section in Fig. 8-30a. Its moments of inertia about the y and z axes are $I_y = 64 \times 10^{-8}$ m^4, $I_z = 136 \times 10^{-8}$ m^4, and its product of inertia is $I_{yz} = 48 \times 10^{-8}$ m^4. From Eq. (8-37),

$$\tan 2\theta_p = \frac{(2)(48 \times 10^{-8})}{(136 \times 10^{-8}) - (64 \times 10^{-8})} = 1.33,$$

we obtain $\theta_p = 26.6°$. Substituting this angle and the values of the moments and product of intertia into Eqs. (8-34) and (8-35), we obtain $I_{y'} = 40 \times 10^{-8}$ m^4 and $I_{z'} = 160 \times 10^{-8}$ m^4. Figure 8-30b shows the orientation of the principal axes and the associated moments of inertia. Based on this information, we can use Eq. (8-29) to determine the normal stress due to a bending moment M exerted about the z' axis for each of the orientations of the beam's cross section shown in Fig. 8-31.

Thus we can use Eq. (8-29) to determine the distribution of normal stress for an unsymmetric cross section when the z' axis about which the bending moment M is exerted is a principal axis. We next consider the distribution of normal stress when M is exerted about an arbitrary axis.

Moment Exerted about an Arbitrary Axis

Suppose that the axis about which M acts is not a principal axis of the cross section (Fig. 8-32a). By representing the couple M by a vector **M** (Fig. 8-32b), we can resolve it into components in terms of a coordinate system $y'z'$ that is aligned with the principal axes (Fig. 8-32c). Then we can obtain the distribution of the normal stress by superimposing the normal stresses due to the moments about the principal axes:

$$\sigma_x = \frac{M_{y'}z'}{I_{y'}} - \frac{M_{z'}y'}{I_{z'}}. \quad \textbf{(8-38)}$$

To understand the signs of the terms in this equation, notice from Fig. 8-32c that if $M_{y'}$ is positive, it results in positive normal stresses for positive values of z', whereas if $M_{z'}$ is positive, it results in negative normal stresses for positive values of y'.

By setting $\sigma_x = 0$ in Eq. (8-38), we obtain an equation for a straight line in the y'–z' plane along which the normal stress equals zero. This is the beam's neutral axis:

$$z' = \frac{M_{z'}I_{y'}}{M_{y'}I_{z'}}y'. \quad \textbf{(8-39)}$$

Notice that the neutral axis does not coincide with the axis about which the moment M is applied unless $I_{y'} = I_{z'}$.

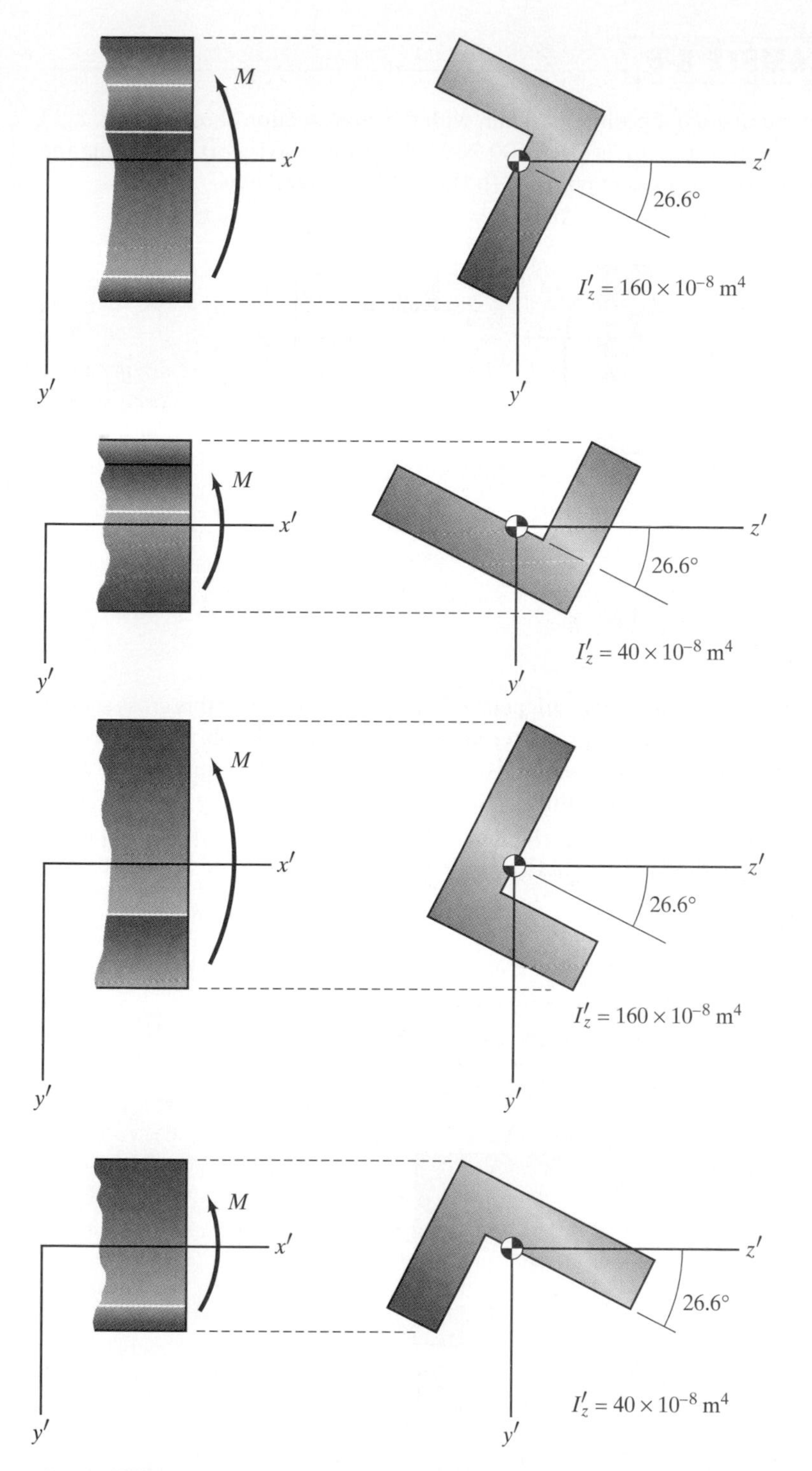

FIGURE 8-31 Bending moment M applied about a principal axis of the cross section.

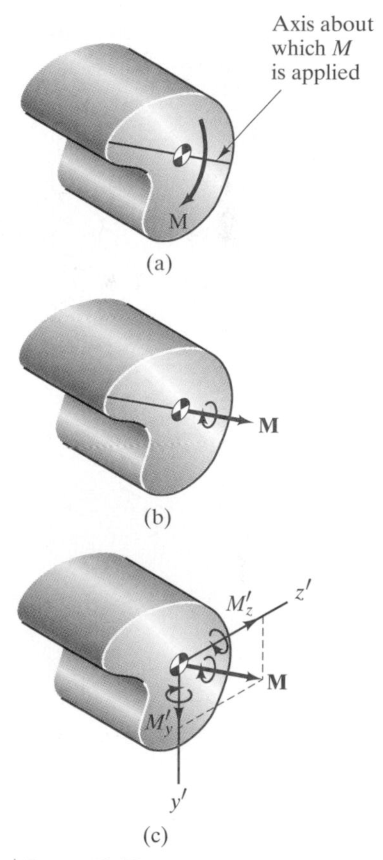

FIGURE 8-32 (a) Moment M about an arbitrary axis. (b) Representing the moment by a vector. (c) Resolving the vector into components parallel to the principal axes.

EXAMPLE 8-8

At a particular axial position, a beam with the cross section shown in Fig. 8-33 is subjected to a moment $M = 400$ N-m about the z axis. **(a)** Determine the resulting normal stress at point P. **(b)** Locate the neutral axis.

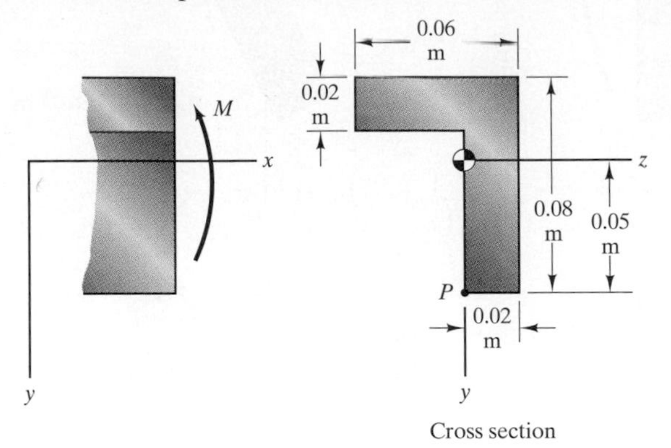

FIGURE 8-33

Strategy

(a) A $y'z'$ coordinate system aligned with the principal axes of this cross section and the associated moments of inertia are shown in Fig. 8-30b. If we represent the moment M as a vector and resolve it into components in terms of the $y'z'$ coordinate system, the distribution of the normal stress is given by Eq. (8-38). We can use Eqs. (8-32) and (8-33) to determine the coordinates of point P in terms of the $y'z'$ coordinate system. **(b)** The neutral axis is described by Eq. (8-39).

Solution

(a) We represent the moment M as a vector $\mathbf{M}$ in Fig. (a). Its components in terms of the $y'z'$ coordinate system are

$$M_{y'} = -400 \sin 26.6^\circ = -179 \text{ N-m},$$

$$M_{z'} = -400 \cos 26.6^\circ = -358 \text{ N-m}.$$

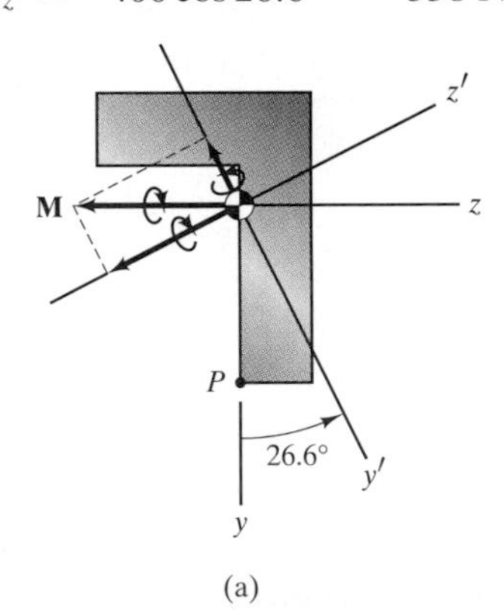

(a)

(a) Resolving the moment into components parallel to the principal axes.

The moments of inertia about the principal axes are $I_{y'} = 40 \times 10^{-8}$ m^4 and $I_{z'} = 160 \times 10^{-8}$ m^4. From Eq. (8-38), the distribution of the normal stress is

$$\begin{aligned}\sigma_x &= \frac{M_{y'}z'}{I_{y'}} - \frac{M_{z'}y'}{I_{z'}}\\ &= \frac{(-179)z'}{40 \times 10^{-8}} - \frac{(-358)y'}{160 \times 10^{-8}}\\ &= (-447z' + 224y') \times 10^6 \text{ Pa.}\end{aligned} \tag{8-40}$$

The coordinates of point P in terms of the yz coordinate system are $y = 0.05$ m, $z = 0$. Substituting these values into Eqs. (8-32) and (8-33), we obtain

$$\begin{aligned}y' &= y\cos\theta + z\sin\theta\\ &= (0.05)\cos 26.6^\circ\\ &= 0.0447 \text{ m,}\end{aligned}$$

$$\begin{aligned}z' &= -y\sin\theta + z\cos\theta\\ &= -(0.05)\sin 26.6^\circ\\ &= -0.0224 \text{ m.}\end{aligned}$$

Substituting these values into Eq. (8-40), the normal stress at point P is

$$\begin{aligned}\sigma_x &= [(-447)(-0.0224) + (224)(0.0447)] \times 10^6\\ &= 20 \times 10^6 \text{ Pa.}\end{aligned}$$

(b) From Eq. (8-39), the equation describing the neutral axis is

$$\begin{aligned}z' &= \frac{M_{z'}I_{y'}}{M_{y'}I_{z'}}y'\\ &= \frac{(-358)(40 \times 10^{-8})}{(-179)(160 \times 10^{-8})}y'\\ &= 0.5y'.\end{aligned}$$

The neutral axis is shown in Fig. (b). The angle $\beta = \arctan(0.5) = 26.6^\circ$.

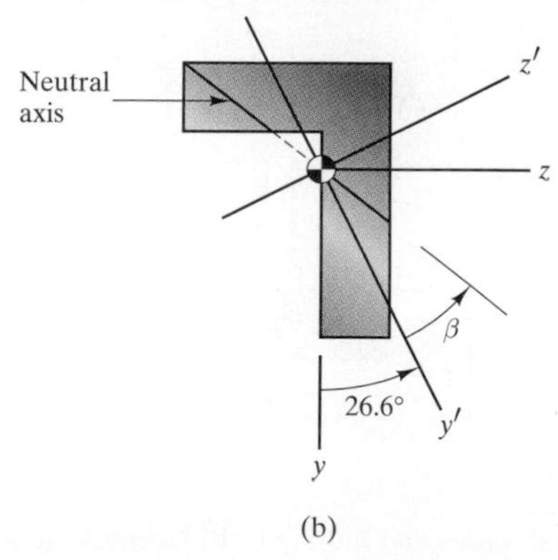

(b) Orientation of the neutral axis.

Shear Stress

The internal forces and moments in a beam include the axial force P, shear force V, and bending moment M (Fig. 8-34a). In Chapter 3 we discussed the uniform normal stress distribution associated with the axial force (Fig. 8-34b). In the previous section we discussed the normal stress distribution associated with the bending moment (Fig. 8-34c). If the shear force is not zero at a given cross section, there must be a distribution of shear stress on the cross section that exerts a force in the y direction equal to V (Fig. 8-35). In the following sections we analyze the shear stresses in beams.

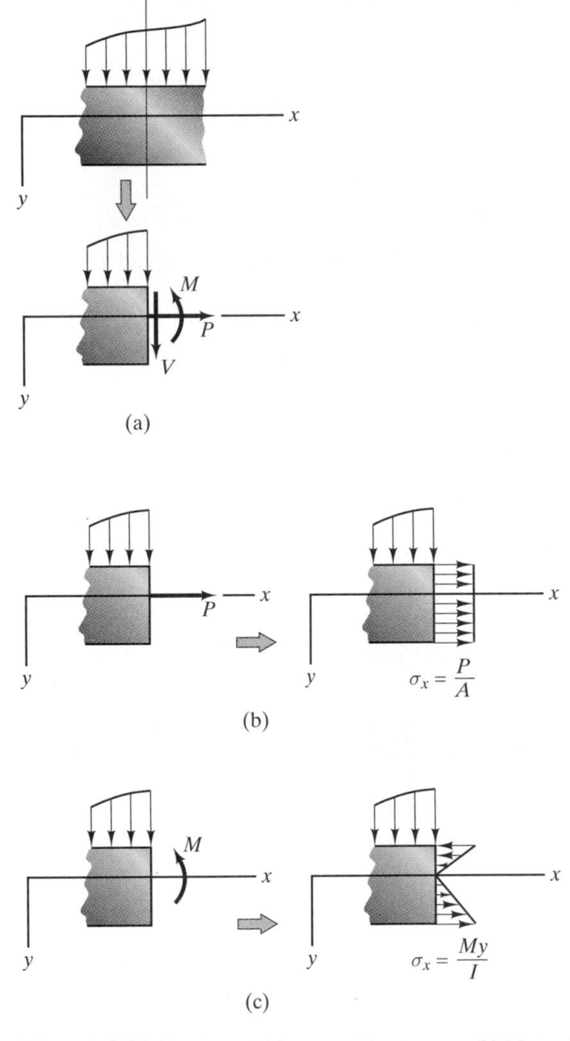

FIGURE 8-34 (a) Internal forces and moment. (b) Normal stress distribution associated with the axial load. (c) Normal stress distribution associated with the bending moment.

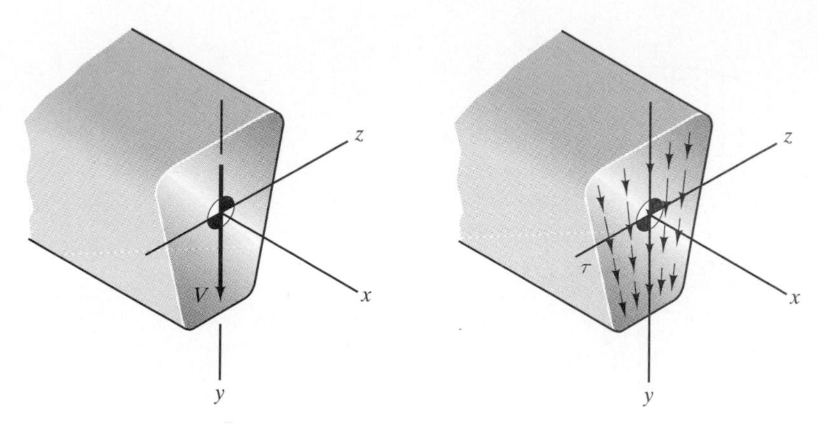

FIGURE 8-35 The shear load results from a distribution of shear stress.

8-6 | Distribution of the Average Stress

Determining the distribution of the shear stress over a beam's cross section generally requires advanced methods of analysis or the use of a numerical solution. But we can obtain some information about the shear stress by an interesting indirect deduction that leads to a result called the shear formula.

Shear Formula

From Eq. (7-5), the shear force is related to the bending moment by

$$\frac{dM}{dx} = V, \tag{8-41}$$

which states that the shear force at a given cross section is related to the rate of change of the bending moment with respect to x. Let us consider a beam whose cross section is symmetric about the vertical (y) axis. If we isolate an element of the beam of width dx, the normal stress distributions on its faces are different if the bending moment varies with respect to x (Fig. 8-36). Let us pass a horizontal plane through this element at a position y' relative to the neutral axis and draw the free-body diagram of the part of the element below the plane (Fig. 8-37). Because of the different normal stresses on the opposite faces of the element isolated in Fig. 8-37, the element can be in equilibrium only if shear stress acts on its top surface. We denote the average value of this shear stress by τ_{av}. By passing a second horizontal plane through this element at $y = y' + dy$ and considering the shear stresses on the resulting element (Fig. 8-38), we can see that equilibrium requires that the shear stress τ_{av} also act on the vertical faces of the element. This is the shear stress on the beam's cross section whose distribution we are seeking. Notice that if the shear force V is positive, the direction of τ_{av} on the beam cross section is such that it points *into* the area A'.

From this analysis we cannot determine the distribution of the shear stress across the width b of the element. However, we can determine the dependence of τ_{av} on y' from the free-body diagram of the element in Fig. 8-37. The area

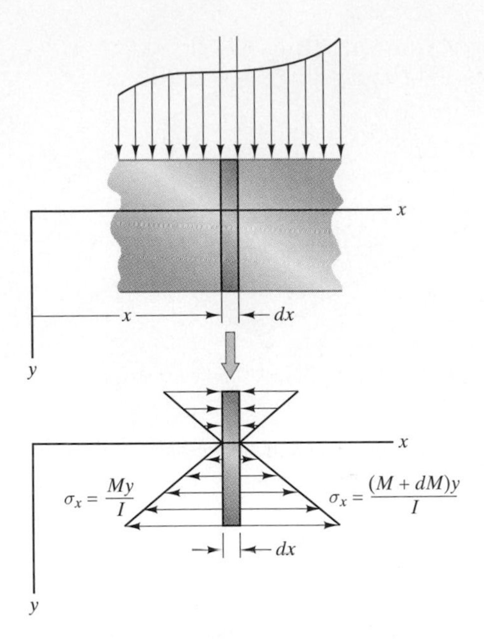

FIGURE 8-36 Element of a beam of width dx showing the normal stresses on the faces.

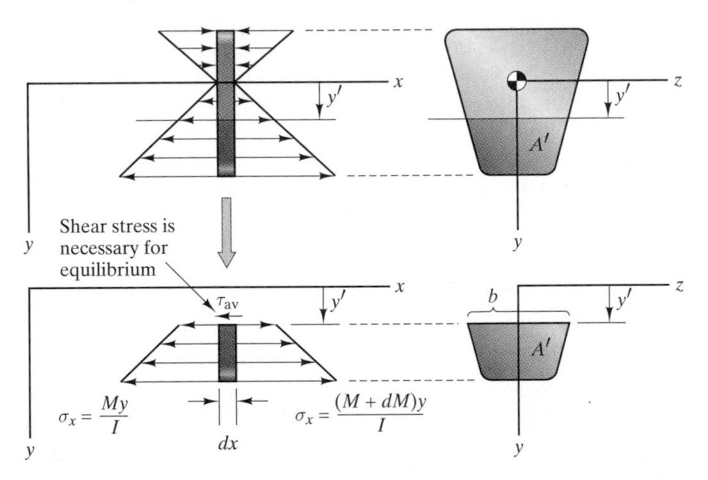

FIGURE 8-37 Isolating the part of the element below a horizontal plane at position y'.

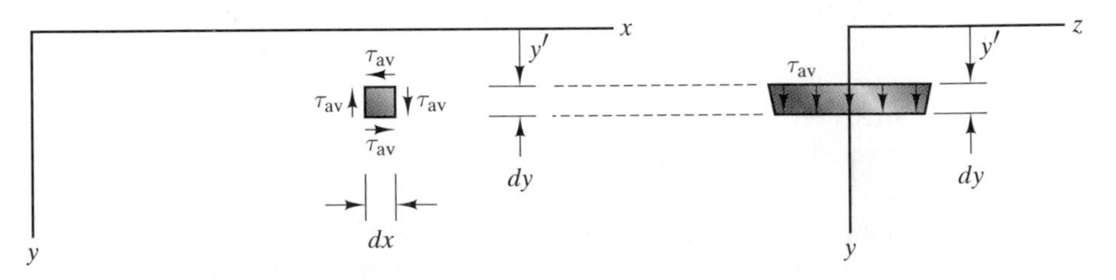

FIGURE 8-38 State of shear stress on an element of infinitesimal height dy obtained from the top of the element in Fig. 8-37.

acted upon by τ_{av} is $b\,dx$. Denoting the area of the part of the beam's cross section below $y = y'$ by A', the sum of the forces on the element is

$$-\tau_{av} b\,dx - \int_{A'} \frac{My}{I} dA + \int_{A'} \frac{(M + dM)y}{I} dA = 0.$$

We solve this equation for τ_{av}, obtaining

$$\tau_{av} = \frac{1}{bI}\frac{dM}{dx}\int_{A'} y\,dA.$$

From Eq. (8-41), $dM/dx = V$, so we obtain an equation for the shear stress in terms of the shear force:

$$\tau_{av} = \frac{VQ}{bI}, \qquad \textbf{(8-42)}$$

where

$$Q = \int_{A'} y\,dA.$$

Equation (8-42) is called the *shear formula*. It determines τ_{av} for a given cross section of a beam at a given position y' relative to the neutral axis. To apply it, we must determine the moment of inertia I of the beam's cross section and the shear force V at the cross section under consideration. We must also determine b and evaluate Q.

We can express Q in terms of the area A' and the position $\bar{y}'$ of the centroid of A' relative to the neutral axis (Fig. 8-39a). The definition of the position of the centroid of A' is

$$\bar{y}' = \frac{\int_{A'} y\,dA}{A'},$$

so Q is given by

$$Q = \bar{y}'A'. \qquad \textbf{(8-43)}$$

It is sometimes convenient to express Q in terms of the area complementary to A'. We denote the complementary area and its centroid by A'' and $\bar{y}''$ in Fig. 8-39b. Because the origin of the coordinate system coincides with the centroid of the entire cross section, which we can express as

$$\bar{y} = \frac{\bar{y}'A' + \bar{y}''A''}{A} = 0,$$

we see that $\bar{y}''A'' = -\bar{y}'A'$, so

$$Q = |\bar{y}''|A''.$$

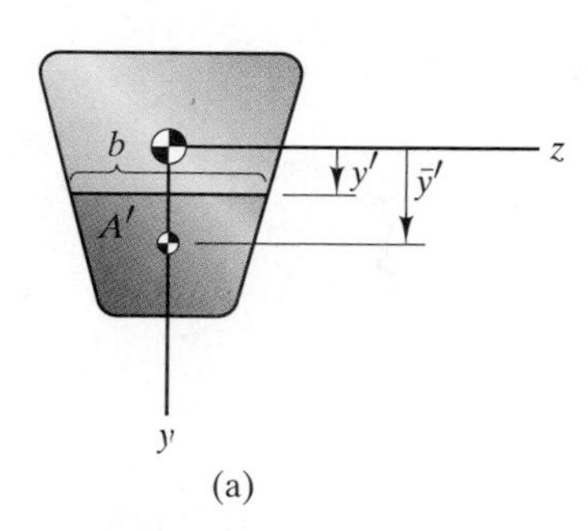

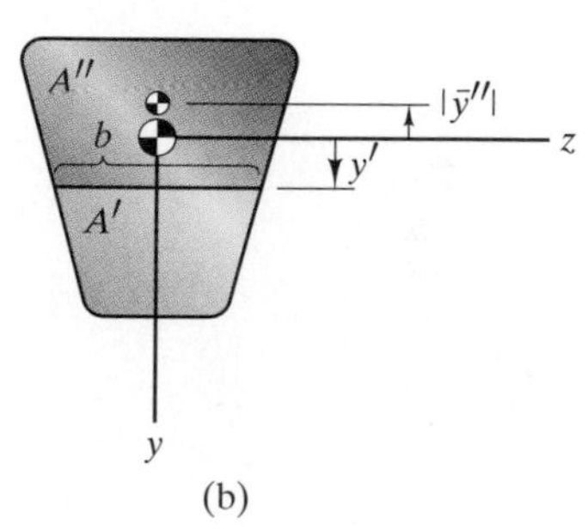

FIGURE 8-39 Determining Q using (a) the area A'; (b) the complementary area A''.

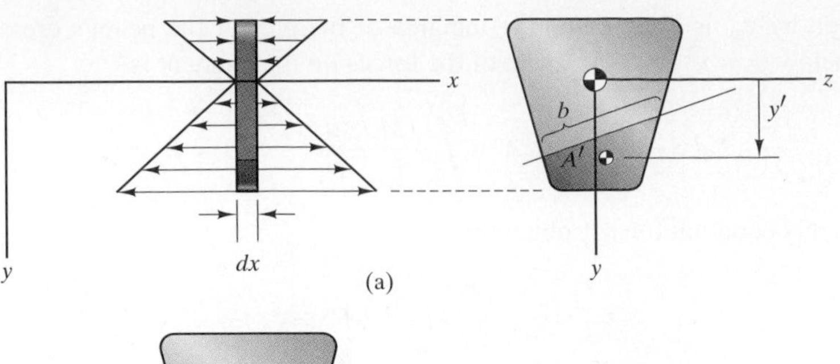

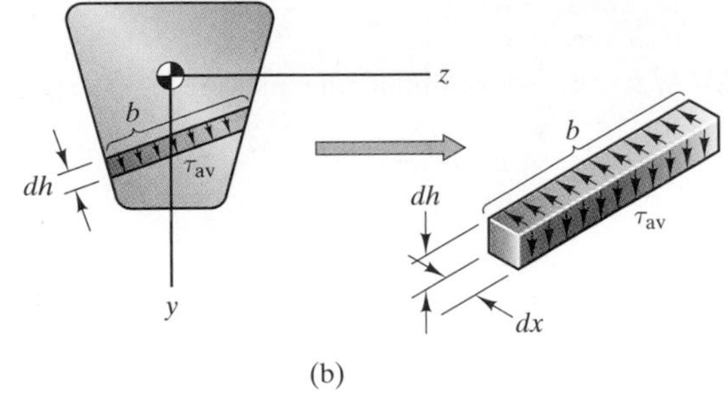

FIGURE 8-40 (a) Determining the average shear stress on an arbitrary plane. (b) Average shear stress.

In Fig. 8-37 we passed a plane parallel to the z axis through the element shown to derive an equation for the average shear stress τ_{av}, obtaining Eq. (8-42). If we pass a plane at an arbitrary angle relative to the z axis through the element as shown in Fig. 8-40a, the derivation of the shear formula is unaltered and the average shear stress, shown on an infinitesimal element of the cross section in Fig. 8-40b, is still given by Eqs. (8-42) and (8-43). The result is the average of the component of the shear stress perpendicular to the line of length b (see Example 8-10). We again observe that if the shear force V is positive, the direction of τ_{av} on the beam cross section is such that it points into the area A'.

Rectangular Cross Section

For a beam with a rectangular cross section (Fig. 8-41a), we can obtain a simple expression for the dependence of the average shear stress on y'. From Fig. 8-41b, the area $A' = b(h/2 - y')$ and the position of the centroid of A' is $\bar{y}' = y' + \frac{1}{2}(h/2 - y')$, so

$$Q = \bar{y}'A' = \frac{b}{2}\left[\left(\frac{h}{2}\right)^2 - (y')^2\right].$$

The moment of inertia of the rectangular cross section about the z axis is $I = \frac{1}{12}bh^3$. From Eq. (8-42), the shear stress is

$$\tau_{\text{av}} = \frac{VQ}{bI} = \frac{6V}{bh^3}\left[\left(\frac{h}{2}\right)^2 - (y')^2\right]. \qquad \textbf{(8-44)}$$

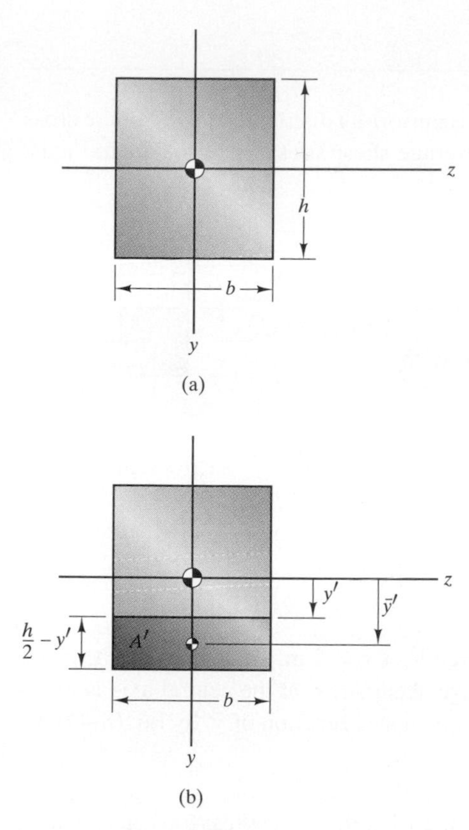

| **FIGURE 8-41** (a) Rectangular cross section. (b) Determining Q.

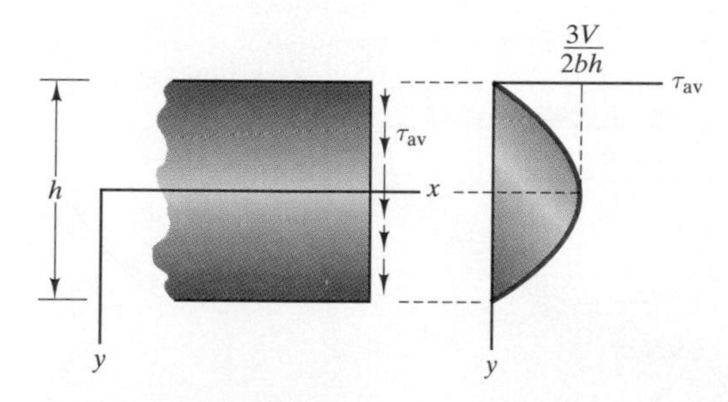

| **FIGURE 8-42** Distribution of τ_{av} on a rectangular cross section.

From this equation we see that the average shear stress on a rectangular cross section is a parabolic function of y' (Fig. 8-42). Its value is zero at the top of the cross section ($y' = -h/2$). At the neutral axis it reaches its maximum magnitude,

$$(\tau_{av})_{y'=0} = \frac{3V}{2bh} = \frac{3V}{2A}, \tag{8-45}$$

and its value decreases to zero at the bottom of the cross section ($y' = h/2$).

EXAMPLE 8-9

The beam in Fig. 8-43 is subjected to a uniformly distributed load. For the cross section at $x = 2$ m, determine the average shear stress **(a)** at the neutral axis; **(b)** at $y' = 0.1$ m.

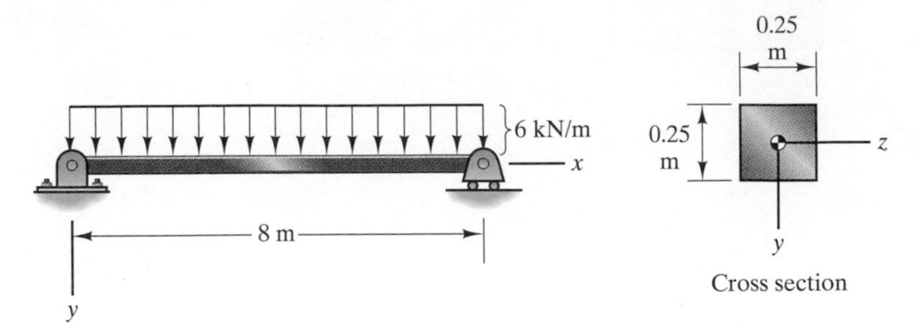

FIGURE 8-43

Strategy

We must first determine the shear force V at $x = 2$ m. Then, because the beam has a square cross section, the average shear stress at the neutral axis is given by Eq. (8-45) and the shear stress is given as a function of y' by Eq. (8-44).

Solution

In Fig. (a) we draw a free-body diagram to determine the shear force at $x = 2$ m, obtaining $V = 12$ kN.

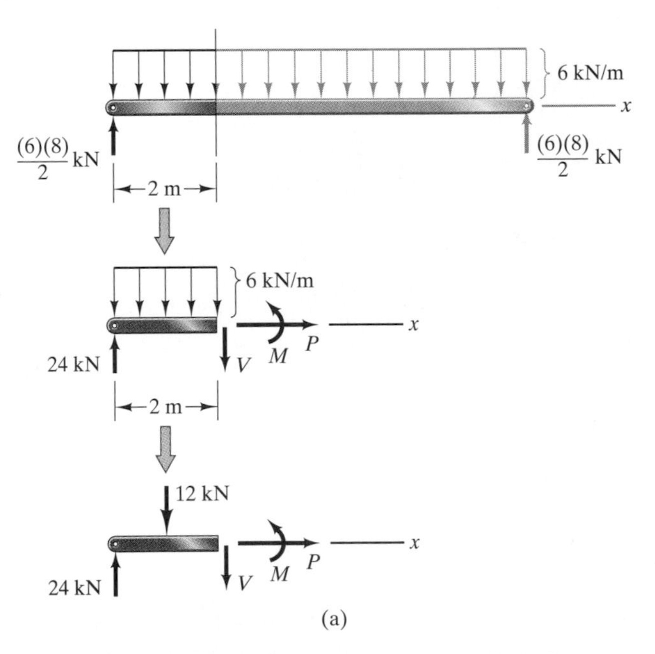

(a) Free-body diagram obtained by passing a plane through the beam at $x = 2$ m.

(a) From Eq. (8-45), the average shear stress at the neutral axis is

$$(\tau_{av})_{y'=0} = \frac{3V}{2A} = \frac{(3)(12{,}000)}{(2)(0.25)(0.25)} = 288 \text{ kPa}.$$

(b) From Eq. (8-44), the average shear stress at $y' = 0.1$ m is

$$\tau_{av} = \frac{6V}{bh^3}\left[\left(\frac{h}{2}\right)^2 - (y')^2\right] = \frac{6(12{,}000)}{(0.25)^4}\left[\left(\frac{0.25}{2}\right)^2 - (0.1)^2\right] = 104 \text{ kPa}.$$

EXAMPLE 8-10

Worksheet 11

The beam whose cross section is shown in Fig. 8-44 consists of five planks of wood glued together. At a given axial position the beam is subjected to a shear force $V = 6000$ lb. **(a)** What is the average shear stress at the neutral axis $y' = 0$? **(b)** What are the magnitudes of the average shear stresses acting on each glued joint?

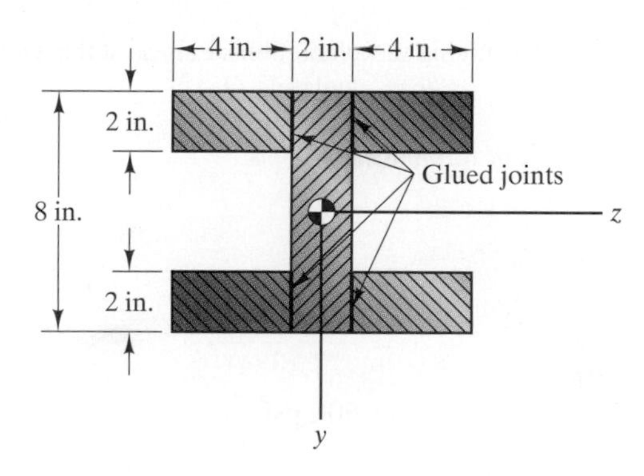

FIGURE 8-44

Strategy

The average shear stress is given by the shear formula, Eq. (8-42), where I is the moment of intertia of the entire cross section about the z axis. We must also determine the appropriate values of b and Q for parts (a) and (b).

Solution

We can obtain the moment of inertia of the entire cross section about the z axis by summing the moments of inertia of the planks about the z axis:

$$\begin{aligned} I &= \tfrac{1}{12}(2)(8)^3 + 4\left[\tfrac{1}{12}(4)(2)^3 + (3)^2(2)(4)\right] \\ &= 384 \text{ in}^4. \end{aligned}$$

(a) We determine the average shear stress at the neutral axis ($y' = 0$) by using the area A' and dimension b shown in Fig. (a). We can calculate Q by summing the contributions of the individual planks [Fig. (b)]:

$$\begin{aligned} Q = \bar{y}'A' &= \bar{y}_1'A_1' + \bar{y}_2'A_2' + \bar{y}_3'A_3' \\ &= (2)(2)(4) + (3)(4)(2) + (3)(4)(2) \\ &= 64 \text{ in}^3. \end{aligned}$$

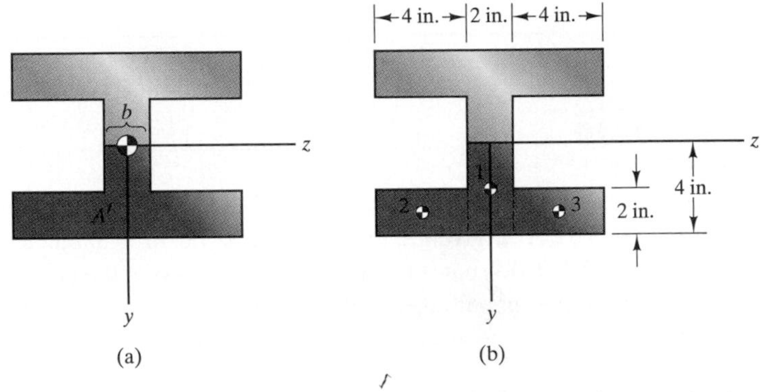

(a) Area A' for determining the average shear stress at the neutral axis.
(b) Calculating Q.

The average shear stress at the neutral axis is

$$\begin{aligned} \tau_{av} &= \frac{VQ}{bI} \\ &= \frac{(6000)(64)}{(2)(384)} \\ &= 500 \text{ psi}. \end{aligned}$$

(b) We can determine the average shear stress acting on the lower-right glued joint by using the area A' and dimension b shown in Fig. (c). The value of Q is

$$\begin{aligned} Q &= \bar{y}'A' \\ &= (3)(4)(2) \\ &= 24 \text{ in}^3. \end{aligned}$$

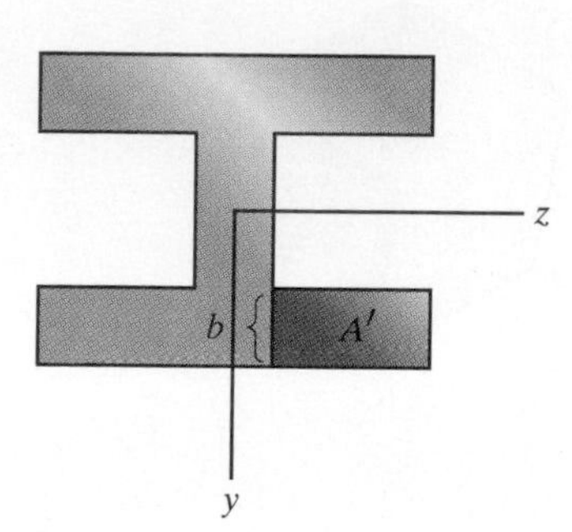

(c)

(c) Area A' for determining the average shear stress on a glued joint.

The average shear stress is

$$\tau_{av} = \frac{VQ}{bI} = \frac{(6000)(24)}{(2)(384)} = 188 \text{ psi.}$$

We leave it as an exercise to show that each glued joint is subjected to the same average shear stress.

8-7 | Thin-Walled Cross Sections

In Section 8-6 we used the shear formula to determine average values of shear stresses on beam cross sections. Although a numerical solution is generally required to determine the detailed distribution of the shear stress, we can obtain analytical solutions for beams with thin-walled cross sections. For such cross sections the shear stress can be approximated as being parallel to the wall and uniformly distributed across its width. Although proof of this result is beyond our scope, from consideration of an element of the beam wall (Fig. 8-45) it is clear that the component of the shear stress perpendicular to the wall must approach zero as the thickness of the wall decreases since the wall surfaces are free of stress. We can use the shear formula to determine the magnitude of the shear stress,

$$\tau = \frac{VQ}{bI}, \qquad \textbf{(8-46)}$$

and its direction is indicated by the rule of thumb that the stress points into the area A' when V is positive (Fig. 8-46). In this way we can determine the distribution of the shear stress, called the *shear flow,* throughout the cross section (Fig. 8-47). This process is demonstrated in Examples 8-11 and 8-12.

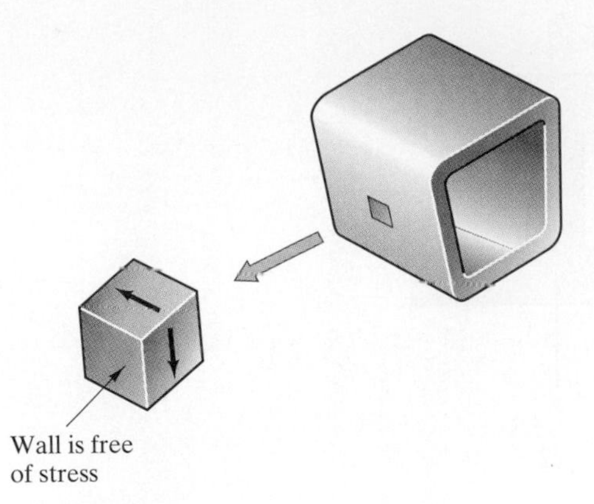

| **FIGURE 8-45** Shear stress on an element of a thin-walled beam.

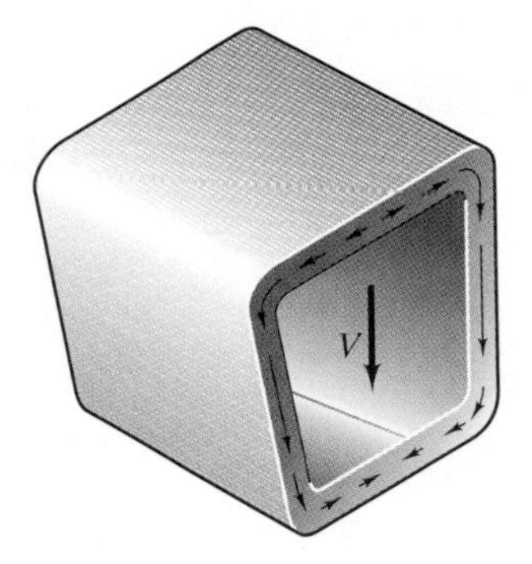

| **FIGURE 8-47** Shear flow.

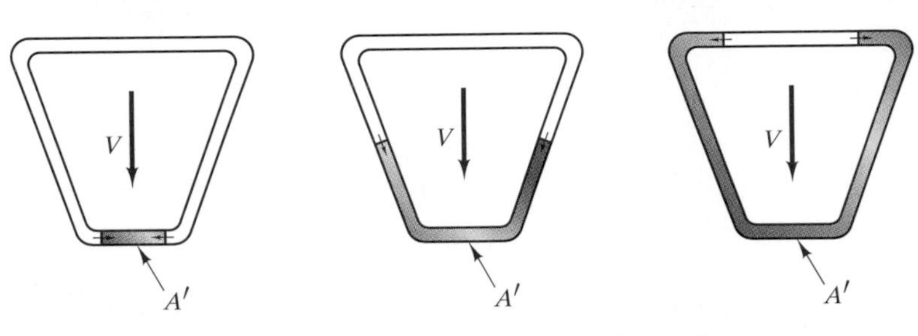

| **FIGURE 8-46** The shear stress points into A' when V is positive.

EXAMPLE 8-11

A beam with the thin-walled cross section in Fig. 8-48 is subjected to a shear force $V = 5$ kN. (The wall thickness is not shown to scale.) Determine the distribution of shear stress.

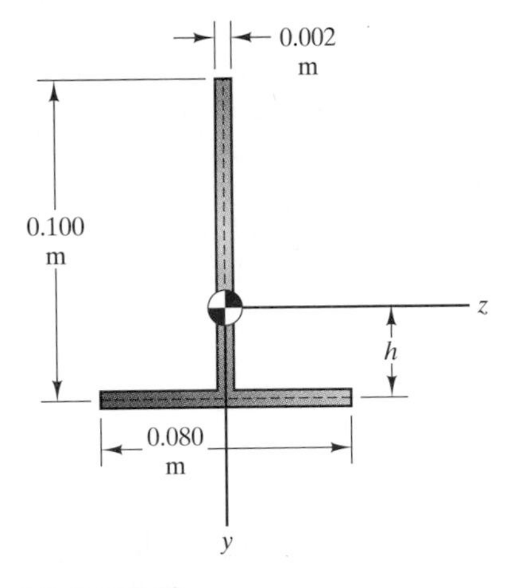

| **FIGURE 8-48**

Strategy

We must determine the position of the neutral axis and the moment of inertia I about the z axis. Then we can use Eq. (8-46) to determine the distribution of shear stress.

Solution

The vertical distance h from the midline of the beam's horizontal web to the neutral axis is

$$h = \frac{(0)(0.08)(0.002) + (0.0505)(0.002)(0.099)}{(0.08)(0.002) + (0.002)(0.099)} = 0.0279 \text{ m}.$$

Denoting the horizontal and vertical parts of the cross section as webs 1 and 2, respectively, the moment of inertia of the cross section about the z axis is

$$\begin{aligned} I &= I_1 + I_2 \\ &= \tfrac{1}{12}(0.08)(0.002)^3 + h^2(0.08)(0.002) \\ &\quad + \tfrac{1}{12}(0.002)(0.099)^3 + (0.0505 - h)^2(0.002)(0.099) \\ &= 3.87 \times 10^{-7} \text{ m}^4. \end{aligned}$$

We will first determine the distribution of shear stress in the horizontal web. Introducing the variable η in Fig. (a) to specify position in the web, the area $A' = 0.002\eta$ and

$$\begin{aligned} Q &= \bar{y}'A' \\ &= h(0.002\eta) \\ &= 5.59 \times 10^{-5}\eta. \end{aligned}$$

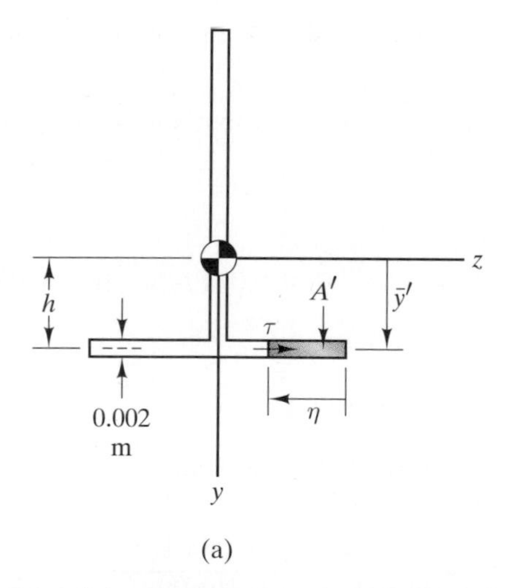

(a) Determining the shear stress in the horizontal web.

Applying the shear formula, we obtain the shear stress in the right half of the horizontal web as a function of η:

$$\begin{aligned}\tau &= \frac{VQ}{bI}\\ &= \frac{(5000)(5.59 \times 10^{-5}\eta)}{(0.002)(3.87 \times 10^{-7})}\\ &= (3.60 \times 10^{8})\eta \text{ Pa} \quad (\eta \text{ in meters}). \qquad \textbf{(8-47)}\end{aligned}$$

We will now determine the distribution of shear stress in the vertical web in terms of the distance y' from the neutral axis [Fig. (b)].

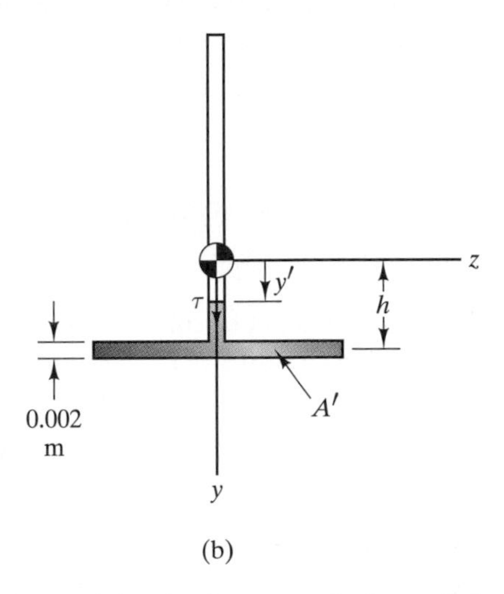

(b) Determining the shear stress in the vertical web.

We can evaluate Q by summing the contributions of the vertical and horizontal parts of A':

$$\begin{aligned}Q &= Q_{\text{horiz}} + Q_{\text{vert}}\\ &= h(0.08)(0.002) + \left[y' + \tfrac{1}{2}(h - 0.001 - y')\right](h - 0.001 - y')(0.002)\\ &= 5.19 \times 10^{-6} - 0.001(y')^2.\end{aligned}$$

The shear stress in the vertical web is

$$\begin{aligned}\tau &= \frac{VQ}{bI}\\ &= \frac{(5000)[5.19 \times 10^{-6} - 0.001(y')^2]}{(0.002)(3.87 \times 10^{-7})}\\ &= [0.335 - 64.5(y')^2] \times 10^{8} \text{ Pa} \quad (y' \text{ in meters}). \qquad \textbf{(8-48)}\end{aligned}$$

Equations (8-47) and (8-48) determine the shear stress throughout the cross section. Figure (c) indicates the direction and magnitude of the distribution.

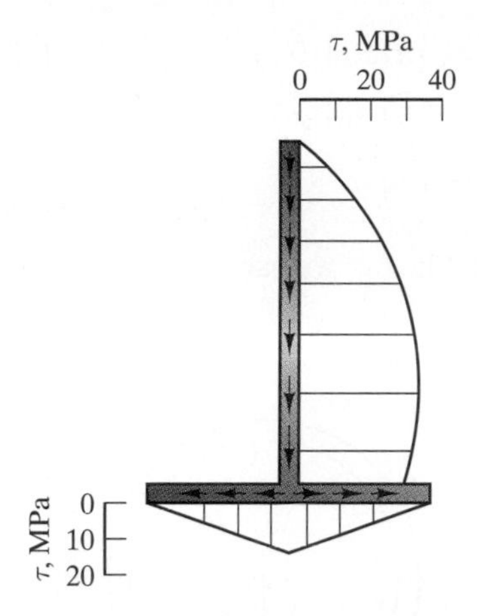

(c) Graph of the stress distribution.

EXAMPLE 8-12

A beam with the circular thin-walled cross section in Fig. 8-49 is subjected to a shear force $V = 20{,}000$ lb. The radius $R = 6$ in. and the wall thickness $t = \frac{1}{4}$ in. (The wall thickness is not shown to scale.) Determine the distribution of the shear stress.

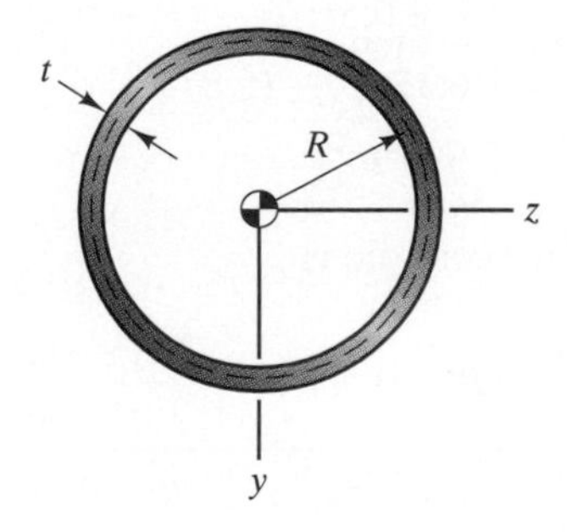

FIGURE 8-49

Strategy

We can use the shear formula, Eq. (8-46), to determine the distribution of shear stress.

Solution

The moment of inertia of the cross section about the z axis is

$$
\begin{aligned}
I &= \tfrac{1}{4}\pi(R+0.5t)^4 - \tfrac{1}{4}\pi(R-0.5t)^4 \\
&= \tfrac{1}{4}\pi[(6.125)^4 - (5.875)^4] \\
&= 170 \text{ in}^4.
\end{aligned}
$$

By using the area A' in Fig. (a), we can determine the shear stress as a function of the angle α.

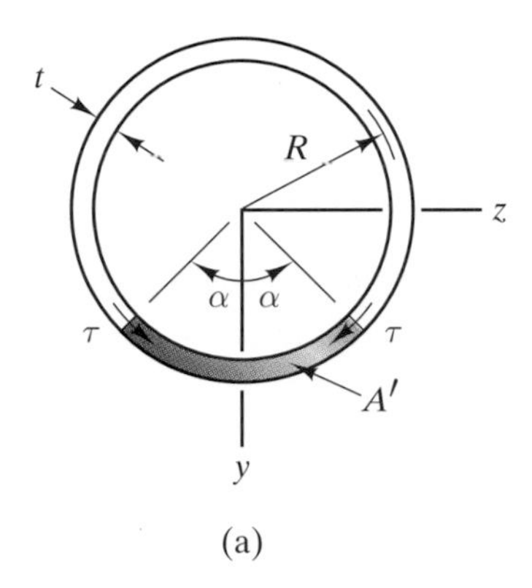

(a) Area A' for determining the shear stress.

We can determine the area and centroid of A' by using the results in Appendix D for a circular sector. The area is

$$
\begin{aligned}
A' &= (R+0.5t)^2\alpha - (R-0.5t)^2\alpha \\
&= [(6.125)^2 - (5.875)^2]\alpha \\
&= 3.00\alpha \text{ in}^2,
\end{aligned}
$$

and the y coordinate of the centroid is

$$
\begin{aligned}
\bar{y}' &= \frac{\dfrac{2(R+0.5t)\sin\alpha}{3\alpha}(R+0.5t)^2\alpha - \dfrac{2(R-0.5t)\sin\alpha}{3\alpha}(R-0.5t)^2\alpha}{(R+0.5t)^2\alpha - (R-0.5t)^2\alpha} \\
&= \frac{(R+0.5t)^3 - (R-0.5t)^3}{(R+0.5t)^2 - (R-0.5t)^2}\,\frac{2\sin\alpha}{3\alpha} \\
&= \frac{(6.125)^3 - (5.875)^3}{(6.125)^2 - (5.875)^2}\,\frac{2\sin\alpha}{3\alpha} \\
&= \frac{6.00\sin\alpha}{\alpha} \text{ in.}
\end{aligned}
$$

Therefore,

$$Q = \bar{y}'A' = 18.0 \sin\alpha \text{ in}^3.$$

The distribution of the shear stress is

$$\begin{aligned}\tau &= \frac{VQ}{bI} \\ &= \frac{(20{,}000)(18.0\sin\alpha)}{\left[(2)\left(\frac{1}{4}\right)\right](170)} \\ &= 4240 \sin\alpha \text{ psi.}\end{aligned} \qquad \textbf{(8-49)}$$

Figure (b) indicates the direction and magnitude of the distribution.

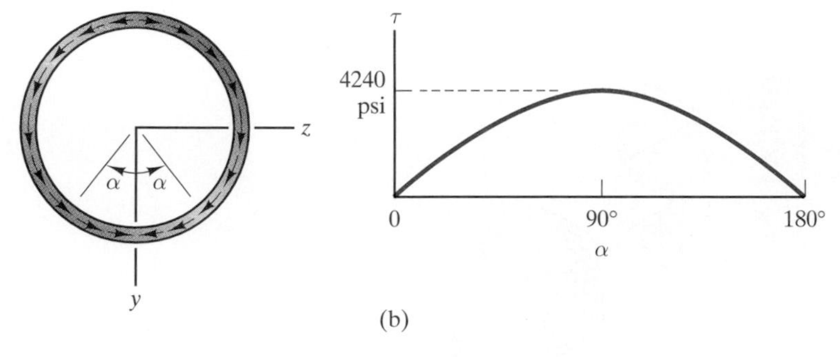

(b) Graph of the stress distribution.

8-8 | Shear Center

In Fig. 8-50a a lateral force F acts at the end of a thin-walled cantilever beam. At a given cross section, the beam is subjected to a shear force $V = F$ and a bending moment $M = -dF$ (Fig. 8-50b). Although the beam's cross section is not symmetric about the vertical (y) axis, the moment M acts about a principal axis of the cross section (see Section 8-5). If we make the same geometric assumptions regarding the beam's deformation that we made in Section 8-1, the normal stress due to the bending moment is given by the familiar equation $\sigma_x = My/I$. As a consequence, our derivation of the shear formula applies, and we can use it to determine the distribution of shear stress throughout the beam's cross section as we did in Section 8-7.

But there is a shortcoming in this analysis. With the force F applied at the centroid of the cross section as shown in Fig. 8-50a, the free-body diagram in Fig. 8-50b is not in equilibrium. This can be seen in the oblique view in Fig. 8-51, which shows the shear flow on the free-body diagram. The shear stress exerts an unbalanced moment about the neutral axis. If the beam were to be loaded in this way, its end would rotate and the neutral axis would bend out

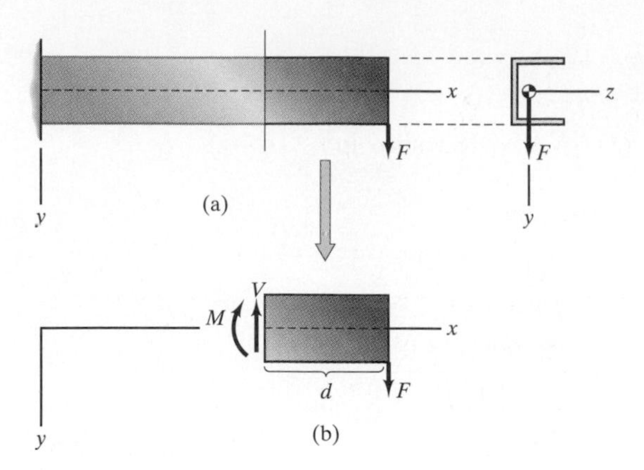

FIGURE 8-50 (a) Beam subjected to a lateral load. (b) Resulting shear force and bending moment.

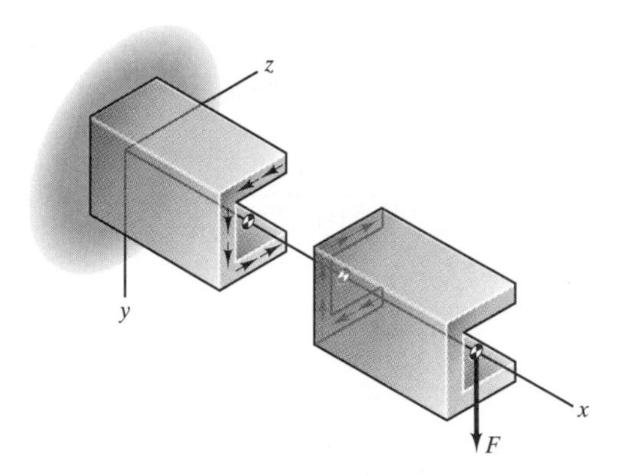

FIGURE 8-51 The shear flow exerts a moment about the x axis.

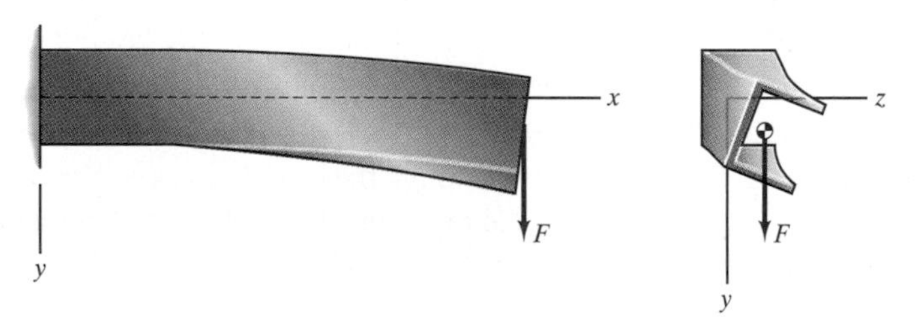

FIGURE 8-52 Rotation of the beam's end and distortion of the neutral axis out of the x–y plane.

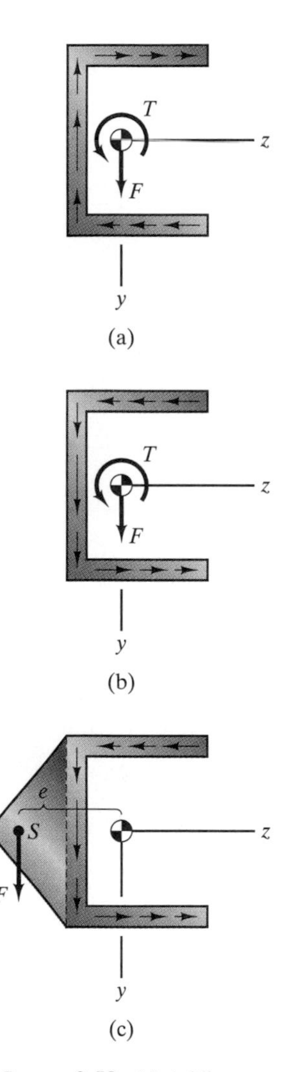

FIGURE 8-53 (a) Adding a torque T to achieve equilibrium. (b) The torque must equal the moment of the shear flow about the centroid. (c) Applying F at the shear center S.

of the x–y plane, violating our geometric assumptions (Fig. 8-52).

We can achieve equilibrium by applying a torque T to the end of the beam that balances the moment exerted about the neutral axis by the shear stress in addition to the force F (Fig. 8-53a). Stated in terms of the shear stress on the left part of Fig. 8-51, the torque T must equal the moment exerted by the shear flow about the neutral axis (Fig. 8-53b). Alternatively, we can load the beam

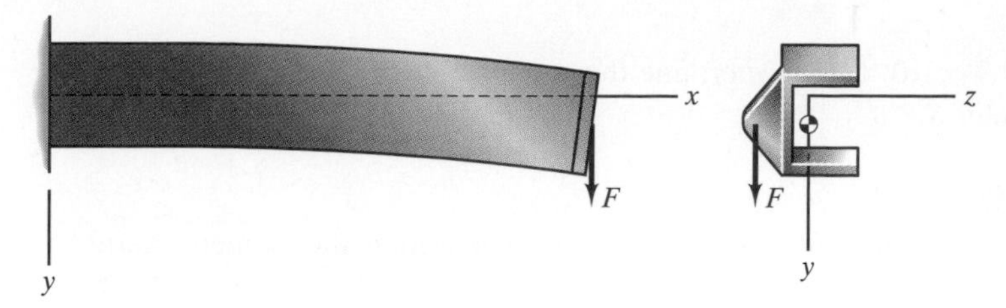

FIGURE 8-54 Deformation of the beam when F is applied at the shear center.

by the force F alone by adding a suitable flange to the end and moving F to the left a distance e from the neutral axis such that $eF = T$ (Fig. 8-53c). The point S at which F is applied is called the *shear center*. Loaded in this way, the beam bends in the x–y plane (Fig. 8-54), our geometrical assumptions hold, and the distributions of normal and shear stress can be determined as we have described.

The position of the shear center can be determined by calculating the moment exerted about any given point by the distribution of shear stress. The force F must be placed so that it exerts the same moment about that point. We determine shear centers of beams subjected to lateral forces in Examples 8-13 and 8-14. In general, whenever a distribution of shear stress is represented by an equivalent force, a point at which the force can be assumed to act is called a shear center for that distribution.

EXAMPLE 8-13

The cantilever beam in Fig. 8-55 is subjected to a lateral force F applied at the shear center. The dimensions of the cross section are $d = 0.06$ m, $h = 0.08$ m, and $t = 0.002$ m. (The wall thickness is not shown to scale.) The horizontal distance c from the midline of the beam's vertical web to the neutral axis is $c = 0.018$ m, and the moment of inertia of the cross section about the z axis

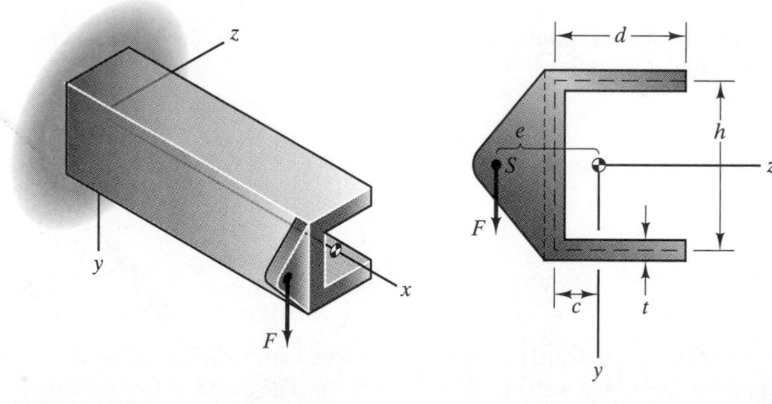

FIGURE 8-55

is $I = 4.7 \times 10^{-7}$ m^4. Determine the distance e from the neutral axis to the shear center S.

Strategy

We can use the shear formula with $V = F$ to determine the moment exerted by the distribution of shear stress about the neutral axis. The moment eF exerted by the lateral force about the neutral axis must equal the moment due to the shear flow about the neutral axis. From this condition we can determine e.

Solution

We first determine the shear stress in the bottom horizontal web. In terms of the variable η in Fig. (a), the area $A' = t\eta$ and the term Q is

$$
\begin{aligned}
Q &= \bar{y}'A' \\
&= \frac{h}{2}t\eta.
\end{aligned}
$$

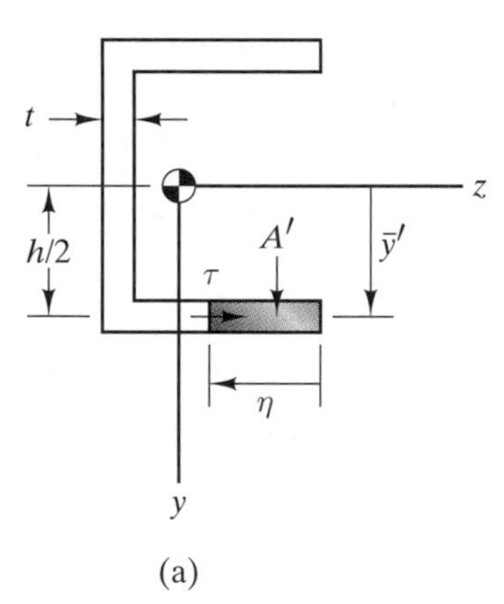

(a) Determining the shear stress in the bottom horizontal web.

Applying the shear formula, we obtain the shear stress in the bottom horizontal web as a function of η:

$$
\begin{aligned}
\tau &= \frac{VQ}{bI} \\
&= \frac{F(h/2)t\eta}{tI} \\
&= \frac{hF}{2I}\eta. \qquad \textbf{(8-50)}
\end{aligned}
$$

The force exerted (toward the right) by the shear stress acting on an element of the bottom horizontal web of width $d\eta$ is $\tau t\, d\eta$. The counterclockwise moment about the neutral axis due to the shear stress is therefore

$$\begin{aligned} M_{\text{bottom web}} &= \int_0^d \frac{h}{2}\tau t \, d\eta \\ &= \int_0^d \frac{h^2 t F}{4I}\eta \, d\eta \\ &= \frac{h^2 d^2 t F}{8I} \\ &= 0.0123F. \end{aligned}$$

We leave it as an exercise to show that the same counterclockwise moment is exerted about the neutral axis by the shear stress in the top horizontal web:

$$M_{\text{top web}} = 0.0123F.$$

We now determine the shear stress in the vertical web in terms of the distance y' from the neutral axis [Fig. (b)].

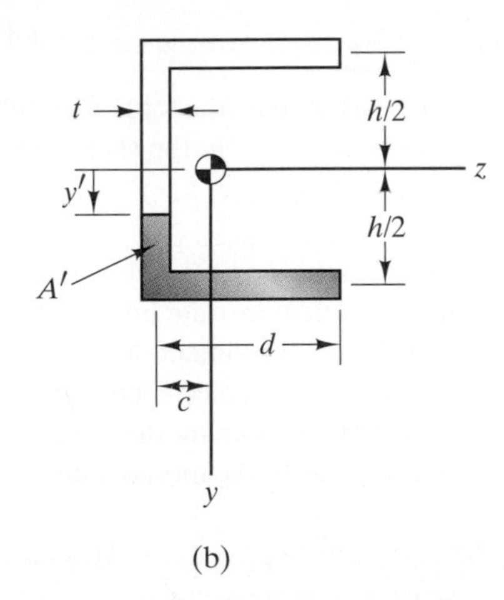

(b) Determining the shear stress in the vertical web.

We can evaluate Q by summing the contributions of the vertical and horizontal parts of A':

$$\begin{aligned} Q &= Q_{\text{horiz}} + Q_{\text{vert}} \\ &= \frac{h}{2}t\left(d + \frac{t}{2}\right) + \left[y' + \frac{1}{2}\left(\frac{h}{2} - \frac{t}{2} - y'\right)\right]t\left(\frac{h}{2} - \frac{t}{2} - y'\right) \\ &= [0.00320 - 0.5(y')^2]t. \end{aligned}$$

The shear stress in the vertical web is

$$\tau = \frac{VQ}{bI} = \frac{F[0.00320 - 0.5(y')^2]t}{tI} = \frac{F[0.00320 - 0.5(y')^2]}{I}. \quad \textbf{(8-51)}$$

The (downward) force exerted by the shear stress acting on an element of the vertical web of height dy' is $\tau t\, dy'$. The counterclockwise moment about the neutral axis due to the shear stress is therefore

$$M_{\text{vertical web}} = \int_{-h/2}^{h/2} c\tau t\, dy' = \int_{-h/2}^{h/2} \frac{ctF}{I}[0.00320 - 0.5(y')^2]\, dy' = 0.0180F.$$

The total moment exerted about the neutral axis by the shear flow is

$$M_{\text{bottom web}} + M_{\text{top web}} + M_{\text{vertical web}} = 0.0425F.$$

Equating this to the moment eF exerted by the force F about the neutral axis, we see that the distance from the neutral axis to the shear center is $e = 0.0425$ m.

Discussion

We determined the shear stress in the vertical web and calculated the resulting moment about the neutral axis to demonstrate how to do it, but we didn't actually need to. If we calculate the total moment due to the distribution of shear stress about the point where the z axis intersects the midline of the vertical web, the shear stress in the vertical web exerts no moment. Equating the moment due to the lateral force F about this point to the moment due to the shear stress, we obtain

$$(e - c)F = M_{\text{bottom web}} + M_{\text{top web}}:$$
$$(e - 0.018)F = 2(0.0123F).$$

Solving, we again obtain $e = 0.0425$ m.

EXAMPLE 8-14

The cantilever beam in Fig. 8-56 has a semicircular thin-walled cross section and is subjected to a lateral force F applied at the shear center. The radius is $R = 0.06$ m and the wall thickness is $t = 0.002$ m. (The wall thickness is not shown to scale.) The horizontal distance c from the center of the semicircle to the neutral axis is $c = 0.0382$ m, and the moment of inertia of the cross section about the z axis is $I = 6.79 \times 10^{-7}$ m^4. Determine the distance e from the neutral axis to the shear center S.

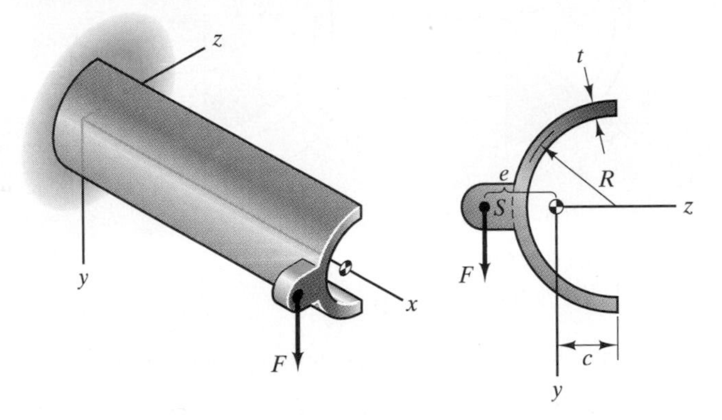

FIGURE 8-56

Strategy

We can use the shear formula with $V = F$ to determine the moment exerted by the distribution of shear stress about the center of the semicircular cross section. The moment $(c + e)F$ exerted by the lateral force about the center of the cross section must equal the moment due to the shear flow.

Solution

We can use the area A' in Fig. (a) to determine the shear stress as a function of the angle θ.

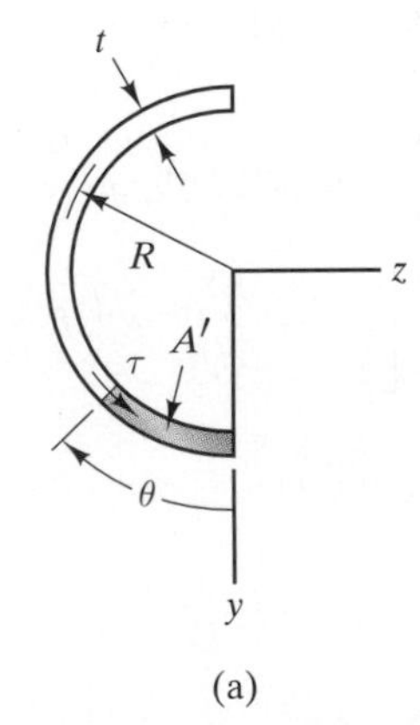

(a)

(a) Area A' for determining the shear stress.

To determine the term Q for A', we will use the areas A'_1 and A'_2 in Fig. (b) and apply the results in Appendix D for a circular sector.

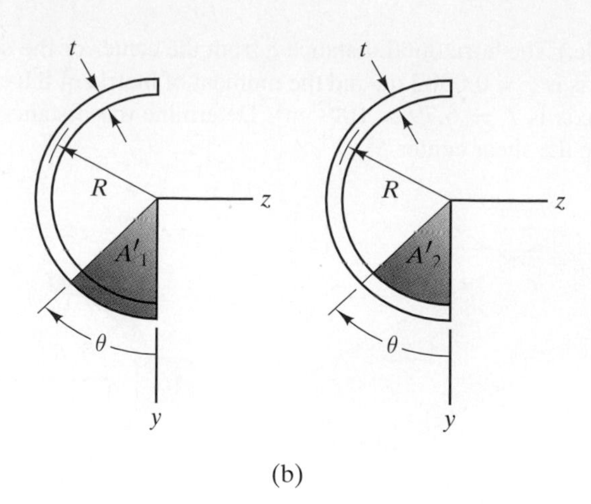

(b) Areas A_1' and A_2' used to determine the value of Q for A'.

The areas A_1' and A_2' are

$$A_1' = \frac{1}{2}\theta\left(R + \frac{t}{2}\right)^2, \qquad A_2' = \frac{1}{2}\theta\left(R - \frac{t}{2}\right)^2,$$

and the radial distances to their centroids are

$$\bar{r}_1' = \frac{2(R + t/2)\sin\frac{1}{2}\theta}{\frac{3}{2}\theta}, \qquad \bar{r}_2' = \frac{2(R - t/2)\sin\frac{1}{2}\theta}{\frac{3}{2}\theta}.$$

Therefore, the value of Q for the area A' is

$$\begin{aligned} Q = \bar{y}_1'A_1' - \bar{y}_2'A_2' &= \left(\bar{r}_1'\cos\frac{1}{2}\theta\right)A_1' - \left(\bar{r}_2'\cos\frac{1}{2}\theta\right)A_2' \\ &= \frac{2}{3}\left[\left(R + \frac{t}{2}\right)^3 - \left(R - \frac{t}{2}\right)^3\right]\sin\frac{1}{2}\theta\cos\frac{1}{2}\theta. \end{aligned}$$

The shear stress as a function of θ is

$$\begin{aligned} \tau &= \frac{VQ}{bI} \\ &= \frac{2F}{3tI}\left[\left(R + \frac{t}{2}\right)^3 - \left(R - \frac{t}{2}\right)^3\right]\sin\frac{1}{2}\theta\cos\frac{1}{2}\theta. \end{aligned} \qquad \textbf{(8-52)}$$

The moment about the center of the semicircular cross section due to the shear stress acting on an element of angular dimension $d\theta$ is $R\tau\,dA = R\tau tR\,d\theta$ [Fig. (c)].

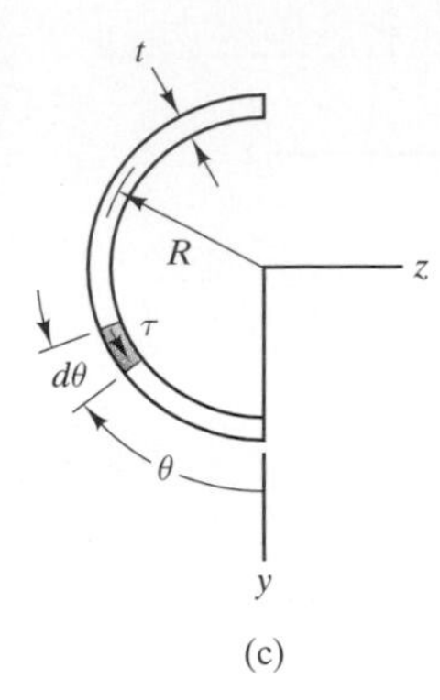

(c)

(c) Element of area for calculating the moment due to the shear stress.

The total moment is therefore

$$M = \int_0^{\pi} R^2 \tau t \, d\theta.$$

Substituting Eq. (8-52) and integrating, we obtain

$$M = \frac{2FR^2}{3I}\left[\left(R + \frac{t}{2}\right)^3 - \left(R - \frac{t}{2}\right)^3\right].$$

We equate this expression to the moment about the center of the semicircular cross section due to the lateral force F acting at the shear center (Fig. 8-56):

$$(c + e)F = \frac{2FR^2}{3I}\left[\left(R + \frac{t}{2}\right)^3 - \left(R - \frac{t}{2}\right)^3\right].$$

Solving for e, we obtain $e = 0.0382$ m.

Chapter Summary

Normal Stress

Distribution of the Stress

Consider a slender prismatic beam of isotropic linearly elastic material subjected to arbitrary loads. At the cross section with axial coordinate x [Fig. (a)], the radius of curvature of the beam's neutral axis is given by the equation

$$\frac{1}{\rho} = -\frac{M}{EI}, \qquad \text{Eq. (8-10)}$$

where I is the moment of inertia of the beam's cross section about the z axis and M is the value of the bending moment at x. If ρ is positive, the positive

(a)

y axis is on the concave side of the neutral axis. The product EI is the beam's *flexural rigidity*. The distribution of the extensional strain is

$$\epsilon_x = -\frac{y}{\rho} = \frac{My}{EI}, \qquad \text{Eq. (8-11)}$$

and the distribution of the normal stress is

$$\sigma_x = \frac{My}{I}. \qquad \text{Eq. (8-12)}$$

Composite Beams

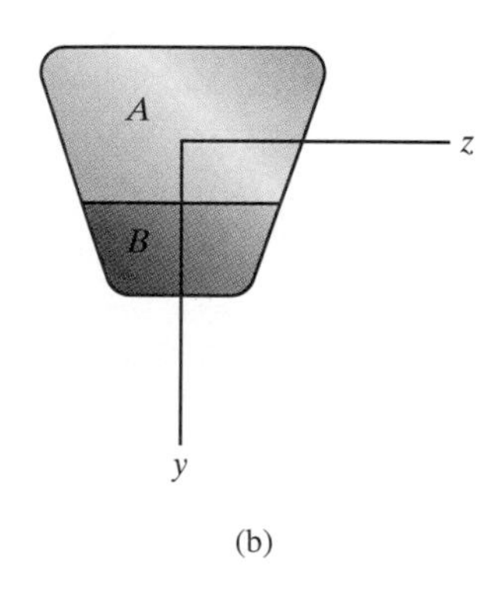

(b)

Consider a given cross section of a prismatic composite beam consisting of materials A and B that are each symmetric about the y axis [Fig. (b)]. The location of the neutral axis can be determined from the relation

$$E_A A_A \bar{y}_A + E_B A_B \bar{y}_B = 0, \qquad \text{Eq. (8-15)}$$

where $\bar{y}_A$ and $\bar{y}_B$ are the coordinates of the centroids of the areas A_A and A_B relative to the neutral axis. The distributions of normal stress in the individual materials are

$$\sigma_A = \frac{My}{I_A + (E_B/E_A)I_B}, \qquad \sigma_B = \frac{My}{(E_A/E_B)I_A + I_B}. \qquad \text{Eq. (8-16)}$$

Elastic–Perfectly Plastic Beams

Let M be the bending moment at a given location of a beam of elastic-perfectly plastic material with a rectangular cross section. When M exceeds the value that causes the maximum normal stress to equal the yield stress, the normal stress increases linearly until it reaches the yield stress at some distance c from the neutral axis, then remains constant [Fig. (c)]. The distance c is given by

$$c = \sqrt{3\left(\frac{h^2}{4} - \frac{M}{\sigma_Y b}\right)}. \qquad \text{Eq. (8-21)}$$

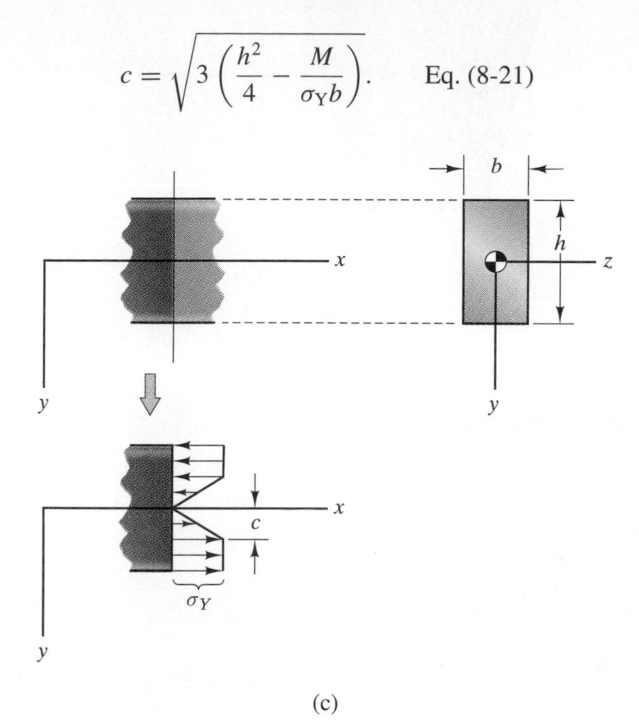

(c)

The maximum bending moment that can be applied without causing yielding of the material is

$$M = \frac{\sigma_Y b h^2}{6}. \qquad \text{Eq. (8-22)}$$

The magnitude of the bending moment at which all the material is yielded ($c = 0$) is the *ultimate moment:*

$$M_U = \frac{\sigma_Y b h^2}{4}. \qquad \text{Eq. (8-23)}$$

When the moment reaches this value there is no resistance to further bending and the beam forms a *plastic hinge*.

For other cross sections, the distribution of the normal stress when the material is completely yielded can be determined from the condition that the areas A_T and A_C that are subjected to tensile and compressive stress are equal. The ultimate moment is

$$M_U = \sigma_Y(\bar{y}_T A_T - \bar{y}_C A_C), \qquad \text{Eq. (8-25)}$$

where $\bar{y}_T$ and $\bar{y}_C$ are the coordinates of the centroids of A_T and A_C relative to the neutral axis [Fig. (d)].

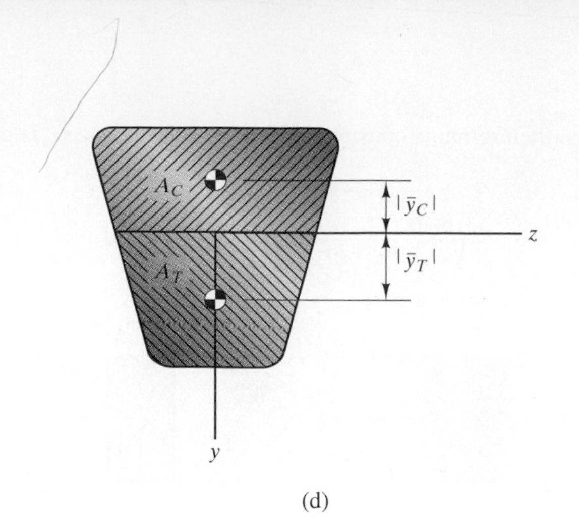

(d)

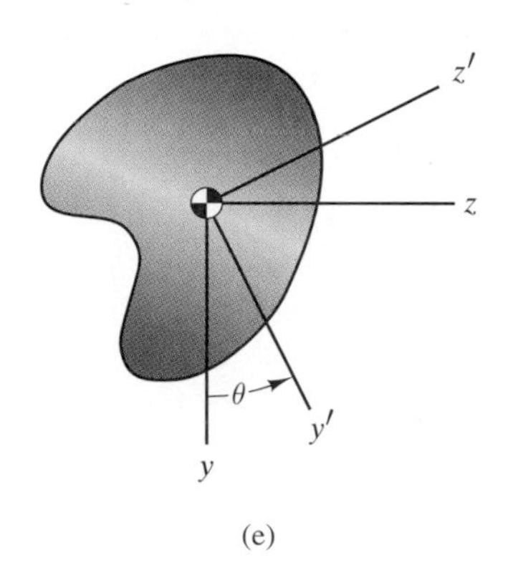

(e)

Unsymmetric Cross Sections

The moments and product of inertia of the cross section in Fig. (e) in terms of the $y'z'$ system are given in terms of the moments and product of inertia in terms of the yz system by

$$I_{y'} = \frac{I_y + I_z}{2} + \frac{I_y - I_z}{2}\cos 2\theta - I_{yz}\sin 2\theta, \qquad \text{Eq. (8-34)}$$

$$I_{z'} = \frac{I_y + I_z}{2} - \frac{I_y - I_z}{2}\cos 2\theta + I_{yz}\sin 2\theta, \qquad \text{Eq. (8-35)}$$

$$I_{y'z'} = \frac{I_y - I_z}{2}\sin 2\theta + I_{yz}\cos 2\theta. \qquad \text{Eq. (8-36)}$$

A value of θ for which y' and z' are principal axes satisfies

$$\tan 2\theta_p = \frac{2I_{yz}}{I_z - I_y}. \qquad \text{Eq. (8-37)}$$

If z' is a principal axis, the distribution of the normal stress due to a bending moment M exerted about z' [Fig. (f)] is

$$\sigma_x = \frac{My'}{I_{z'}}. \qquad \text{Eq. (8-29)}$$

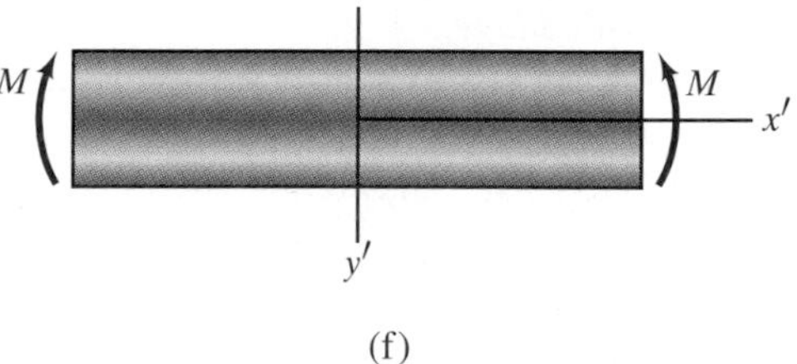

(f)

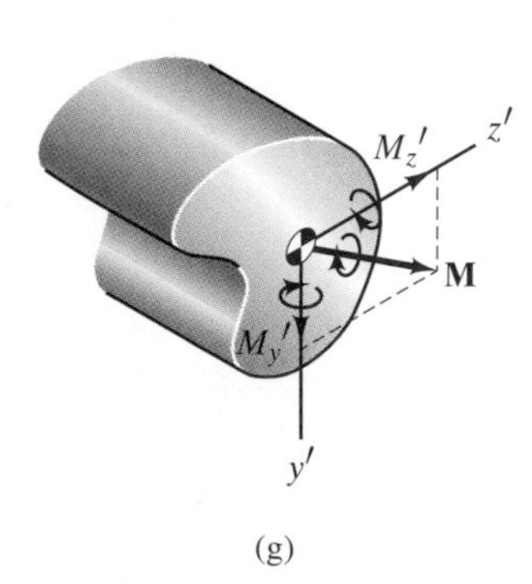

(g)

If the axis about which M acts is not a principal axis, the vector representing M can be resolved into components in terms of a coordinate system $y'z'$ aligned with the principal axes [Fig. (g)]. The resulting distribution of normal stress is

$$\sigma_x = \frac{M_{y'}z'}{I_{y'}} - \frac{M_{z'}y'}{I_{z'}}. \qquad \text{Eq. (8-38)}$$

The beam's neutral axis is given by

$$z' = \frac{M_{z'}I_{y'}}{M_{y'}I_{z'}}y'. \qquad \text{Eq. (8-39)}$$

Shear Stress

Distribution of the Average Stress

Consider a slender prismatic beam of isotropic linearly elastic material subjected to arbitrary loads. At a given cross section, the average of the component of the shear stress perpendicular to the line of length b in Fig. (h) is given by the *shear formula*

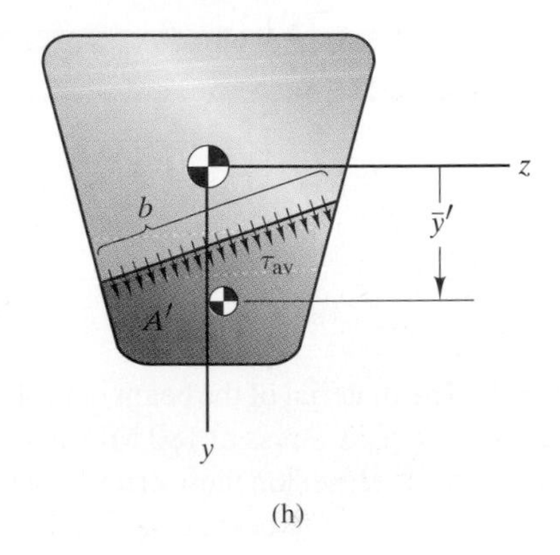

(h)

$$\tau_{\text{av}} = \frac{VQ}{bI}, \qquad \text{Eq. (8-42)}$$

where I is the moment of inertia of the beam's cross section about the z axis, V is the value of the shear force at the given cross section, and

$$Q = \bar{y}'A'. \qquad \text{Eq. (8-43)}$$

For a beam with a rectangular cross section, the average of the shear stress over the horizontal line in Fig. (i) as a function of y' is

$$\tau_{\text{av}} = \frac{VQ}{bI} = \frac{6V}{bh^3}\left[\left(\frac{h}{2}\right)^2 - (y')^2\right]. \qquad \text{Eq. (8-44)}$$

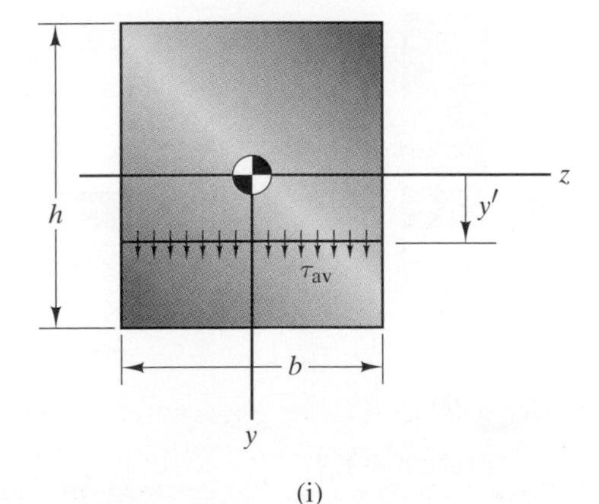

(i)

PROBLEMS

8-1.1. The beam consists of material with modulus of elasticity $E = 70$ GPa and is subjected to couples $M = 250$ kN-m at its ends. **(a)** What is the resulting radius of curvature of the neutral axis? **(b)** Determine the maximum tensile stress due to bending.

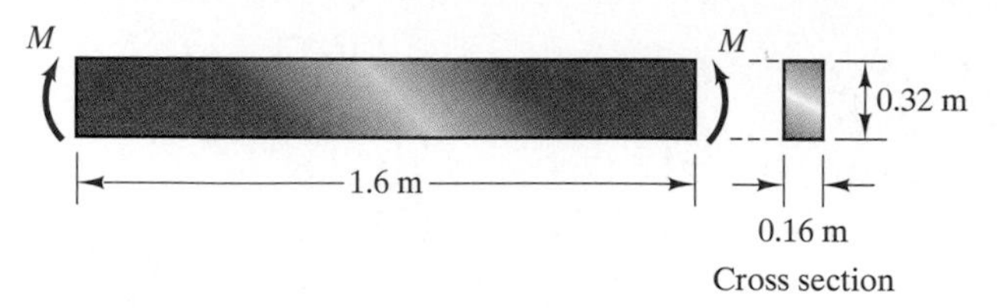

PROBLEM 8-1.1

8-1.2. The material of the beam in Problem 8-1.1 will safely support a tensile stress of 180 MPa and a compressive stress of 200 MPa. Based on these criteria, what is the largest couple M to which the beam can be subjected?

8-1.3. The material of the beam in Problem 8-1.1 will safely support a tensile stress of 180 MPa and a compressive stress of 200 MPa. Suppose that the beam is rotated 90° about its axis, so that the width of its cross section is 0.32 m and its height is 0.16 m. What is the largest couple M to which the beam can be subjected? Compare your answer to the answer to Problem 8-1.2.

8-1.4. The beam consists of material with modulus of elasticity $E = 14 \times 10^6$ psi and is subjected to couples $M = 150,000$ in-lb at its ends. **(a)** What is the resulting radius of curvature of the neutral axis? **(b)** Determine the maximum tensile stress due to bending.

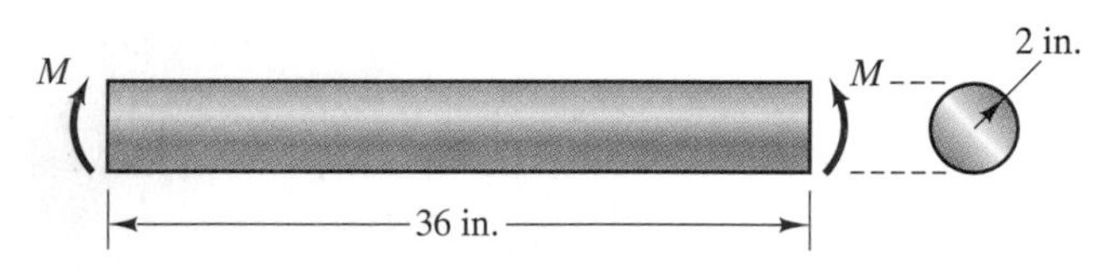

PROBLEM 8-1.4

8-1.5. The material of the beam in Problem 8-1.4 will safely support a tensile or compressive stress of 30,000 psi. Based on this criterion, what is the largest couple M to which the beam can be subjected?

8-1.6. The material of the beam in Problem 8-1.4 will safely support a tensile or compressive stress of 30,000 psi. If the beam has a hollow circular cross section, with 2-in. outer radius and 1-in. inner radius, what is the largest couple M to which the beam can be subjected?

8-1.7. Suppose that the beam in Example 8-1 is made of a brittle material that will safely support a tensile stress of 20 MPa or a compressive stress of 50 MPa. What is the largest couple M to which the beam can be subjected?

8-1.8. What is the maximum tensile stress due to bending in the beam in Example 8-2, and where does it occur?

8-1.9. The beam consists of material that will safely support a tensile or compressive stress of 350 MPa. Based on this criterion, determine the largest force F the beam will safely support if it has the cross section (a); if it has the cross section (b). (The two cross sections have approximately the same area.)

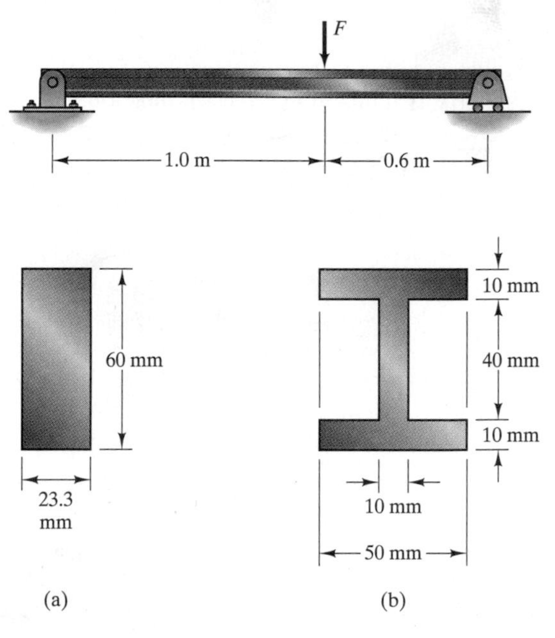

PROBLEM 8-1.9

8-1.10. If the beam in Problem 8-1.9 is subjected to a force $F = 6$ kN, what is the maximum tensile stress due to bending at the cross section midway between the beam's supports in cases (a) and (b)?

8-1.11. The beam in Problem 8-1.9 consists of material that will safely support a tensile or compressive stress of 350 MPa. If it has the cross section (a) and is subjected to a force $F = 17$ kN, what is the maximum distance from the ends of the beam at which F can be applied?

8-1.12. The beam is subjected to a uniformly distributed load $w_0 = 300$ lb/in. Determine the maximum tensile stress due to bending at $x = 20$ in. if the beam has the cross section (a); if it has the cross section (b). (The two cross sections have approximately the same area.)

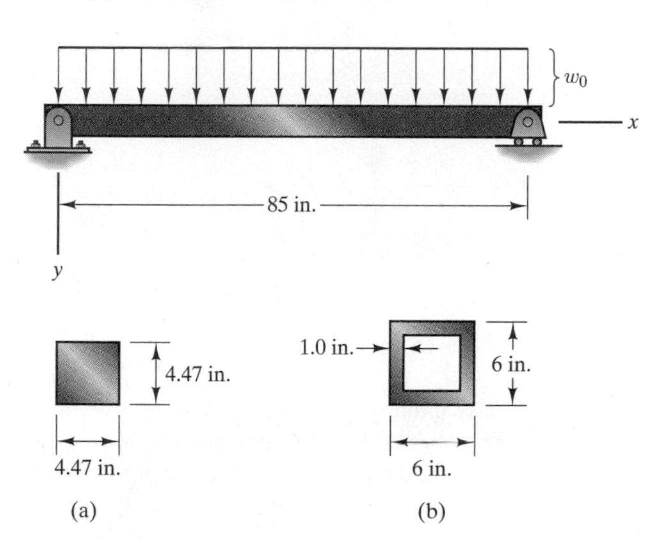

PROBLEM 8-1.12

8-1.13. The beam in Problem 8-1.12 consists of material that will safely support a tensile or compressive stress of 30 ksi. Based on this criterion, determine the largest distributed load w_0 (in lb/in.) the beam will safely support if it has the cross section (a); if it has the cross section (b).

8-1.14. A bandsaw blade with 2-mm thickness and 20-mm width is wrapped around a pulley with 160-mm radius. The blade is made of steel with modulus of elasticity $E = 200$ GPa. What maximum tensile stress is induced in the blade as a result of being wrapped around the pulley?

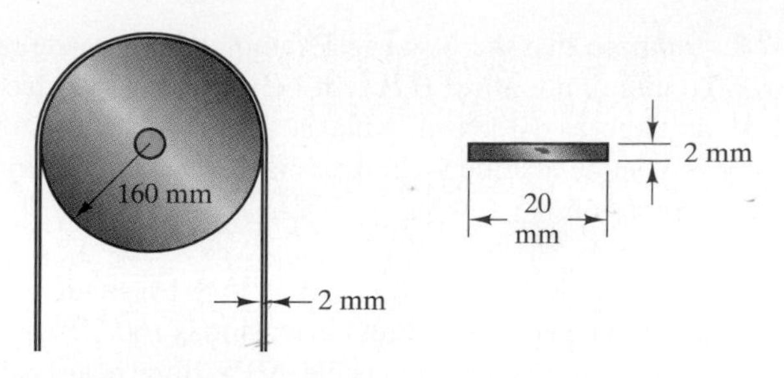

PROBLEM 8-1.14

8-1.15. In Problem 8-1.14, what is the magnitude of the bending moment induced in the blade as a result of being wrapped around the pulley?

8-1.16. Assume that the surface on which the beam rests exerts a uniformly distributed load on the beam. Determine the maximum tensile and compressive stresses due to bending at $x = 3$ m.

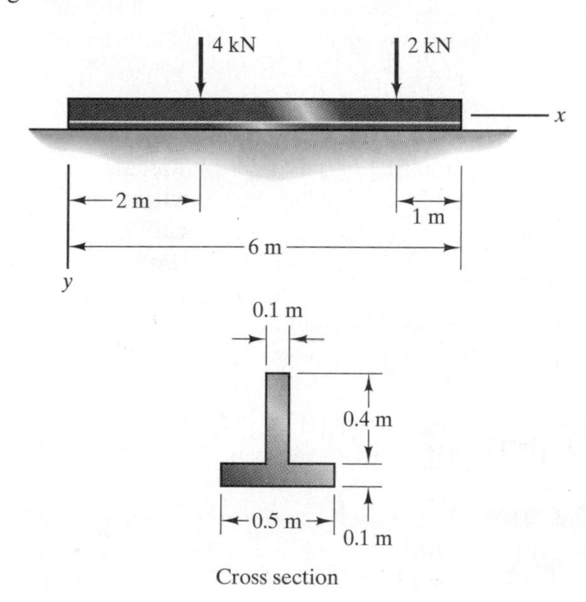

PROBLEM 8-1.16

8-1.17. If you are selecting a material for the beam in Problem 8-1.16, what maximum tensile and compressive stresses must the material be able to support?

8-2.1. If the beam in Example 8-1 is made of a material for which the allowable stress in tension and compression is $\sigma_{\text{allow}} = 120$ MPa, what is the largest allowable magnitude of the couple M?

8-2.2. Suppose that the beam in Example 8-1 is made of 7075-T6 aluminum alloy. If it will be subjected to values of M as large as 10 kN-m, what is the beam's factor of safety? (Assume that the yield stress is the same in tension and compression.)

8-2.3. Suppose that the beam in Example 8-1 is made of a material for which the yield stress in tension is 160 MPa and the yield stress in compression is 200 MPa. If the beam will be subjected to (positive) values of M as large as 4 kN-m, what is the beam's factor of safety?

8-2.4. Suppose that the length of the beam in Example 8-2 is $L = 8$ ft and it is made of ASTM-A36 structural steel. The maximum anticipated magnitude of the distributed load is $w_0 = 2400$ lb/ft. Determine the dimension h so that the beam has a factor of safety $S = 3$.

8-2.5. Suppose that the loads on the beam in Example 8-3 are the maximum anticipated loads and the beam is made of wood with yield stress $\sigma_Y = 40$ MPa. What is the beam's factor of safety?

8-2.6. A beam made of 7075-T6 aluminum alloy will be subjected to anticipated bending moments as large as 1500 N-m. Determine the beam's factor of safety for two cases: **(a)** It has a solid circular cross section with 20-mm radius. **(b)** It has a hollow circular cross section with 30-mm outer radius and the inner radius chosen so that the beam has the same weight as the beam in case (a).

8-2.7. Design a cross section for the beam in Example 8-4 so that the beam's factor of safety is $S = 2$.

8-2.8. The device shown is a playground seesaw. Make a conservative estimate of the maximum weight to which it will be subjected at each end when in use. (Consider contingent situations such as an adult sitting with a child.) Choose a material from Appendix B and design a cross section for the 4.8-m beam so that it has a factor of safety $S = 4$.

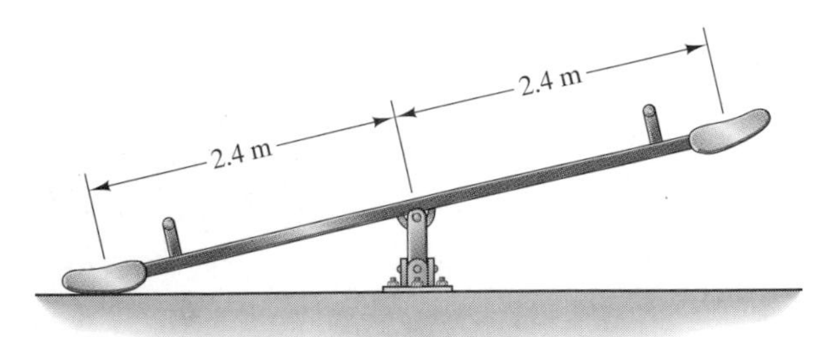

PROBLEM 8-2.8

8-2.9. The maximum anticipated value of the uniformly distributed load on the 8-ft-long segment of a building's frame is 2000 lb/ft. The beam is to be made of ASTM-A572 steel. Design the I-beam's cross section so that its height is 1 ft and its factor of safety is $S = 4$.

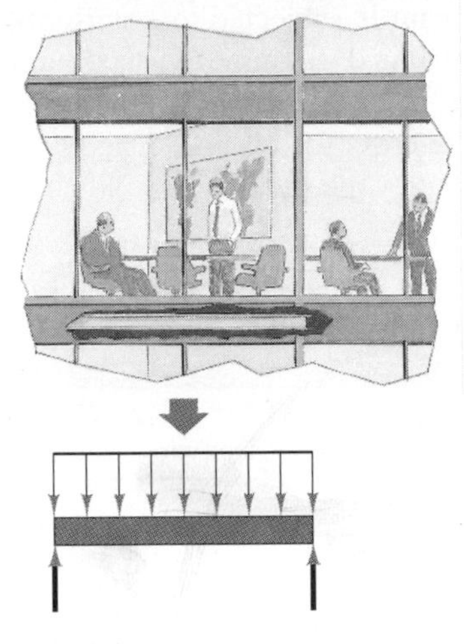

PROBLEM 8-2.9

8-2.10. The maximum anticipated load on the beam is shown. Choose a material from Appendix B and design a cross section for the beam so that it has a factor of safety $S = 2$.

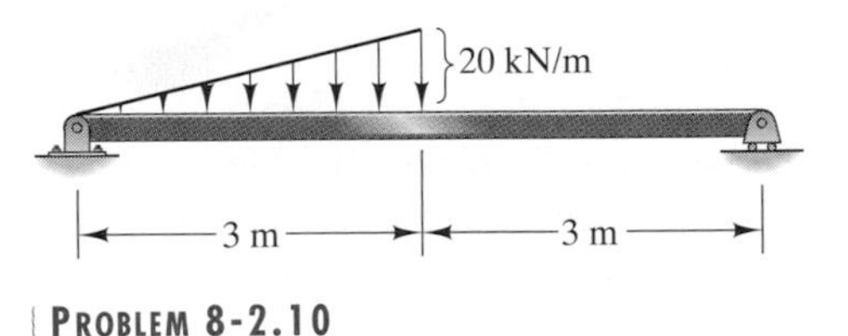

PROBLEM 8-2.10

8-2.11. For a preliminary design of the ladder rung, assume that it has pin supports at the ends. Make a conservative estimate of the maximum weight to which it will be subjected when in use. (Assume that the maximum weight acts as a point force at the center of the rung.) The rung is to be made of 6061-T6 aluminum alloy. Design a cross section so that it has a factor of safety $S = 4$. Consider the appropriate width the rung should have.

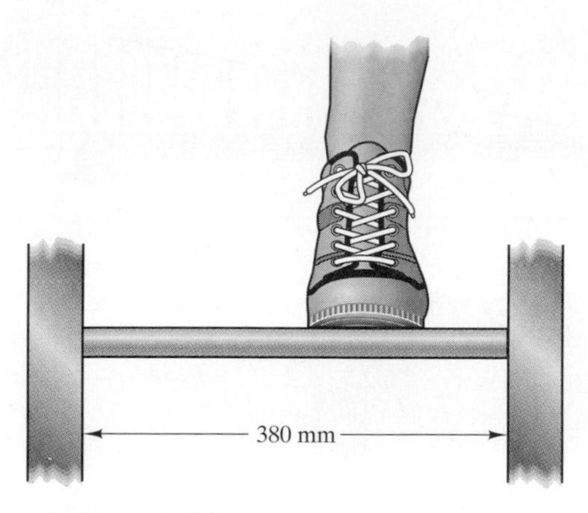

PROBLEM 8-2.11

8-3.1. A composite beam consists of two concentric cylinders. The inner (solid) cylinder A has a 1-in. radius and is made of wood with modulus of elasticity 1.6×10^6 psi. The outer cylinder B has a 1-in. inner radius and a 1.2-in. outer radius and is made of aluminum alloy with modulus of elasticity 10.4×10^6 psi. If the beam is subjected to a bending moment $M = 2000$ in-lb at a particular cross section, what is the maximum tensile stress in each material?

8-3.2. In Example 8-5, calculate the maximum tensile and compressive stresses in the beam and confirm the values shown in Fig. (b) of Example 8-5.

8-3.3. In Example 8-5, calculate the maximum tensile and compressive stresses in the beam if the vertical dimensions of parts A and B are $h_A = 0.08$ m and $h_B = 0.04$ m.

8-3.4. The figure shows the cross section of a prismatic beam made of material A with elastic modulus $E_A = 80$ GPa and material B with elastic modulus $E_B = 16$ GPa. The dimensions $b = 0.10$ m and $h = 0.03$ m. Determine the distance H from the top of the cross section to the neutral axis.

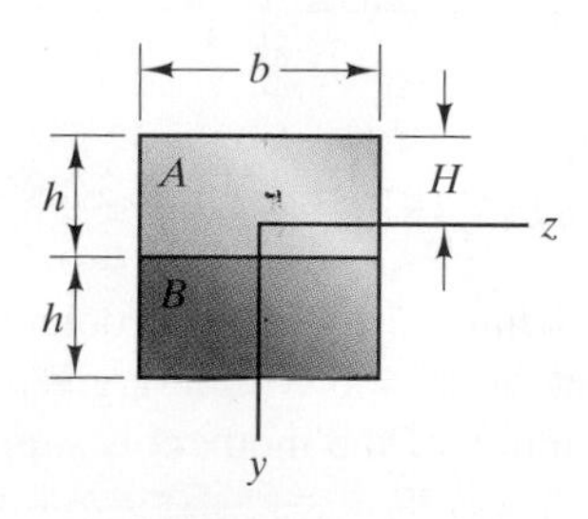

PROBLEMS 8-3.4 – 8-3.8

8-3.5. For the beam described in Problem 8-3.4, if the bending moment $M = 720$ N-m at a given cross section, determine the maximum tensile and compressive stresses.

8-3.6. The figure shows the cross section of a prismatic beam made of material A with elastic modulus $E_A = 12 \times 10^6$ psi and material B with elastic modulus $E_B = 2.4 \times 10^6$ psi. The dimensions $b = 12$ in. and $h = 4$ in. Determine the distance H from the top of the cross section to the neutral axis.

8-3.7. For the beam described in Problem 8-3.6, if the bending moment $M = 8000$ in-lb at a given cross section, determine how much of the moment is supported by each material.

8-3.8. The figure shows the cross section of a prismatic beam made of material A with elastic modulus $E_A = 72$ GPa and a material B. The dimensions $b = 0.05$ m and $h = 0.02$ m. If the distance H from the top of the cross section to the neutral axis is 0.025 m, what is the elastic modulus of material B?

8-3.9. For the beam described in Problem 8-3.8, if the bending moment $M = 100$ N-m at a given cross section, determine the maximum tensile and compressive stresses.

8-3.10. Figure (a) is the cross section of a steel beam with elastic modulus $E = 220$ GPa. Figure (b) is the cross section of a steel box beam (material A) with elastic modulus $E = 220$ GPa whose walls are stabilized by a light filler material (material B) with elastic modulus $E = 6$ GPa. (The cross-sectional area of steel is approximately the same in each cross section.) If the bending moment $M = 400$ N-m, determine the magnitude of the maximum normal stress for each cross section.

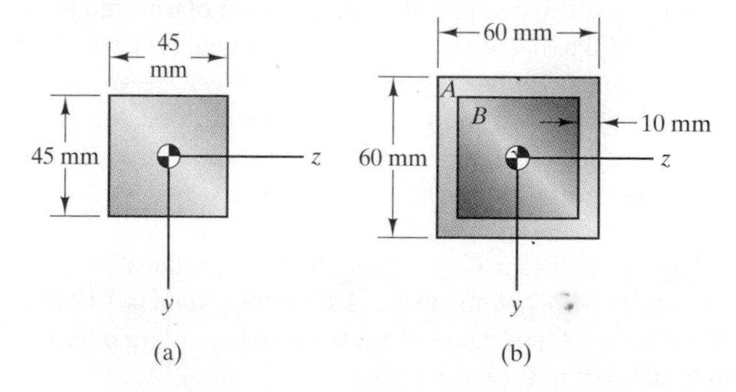

PROBLEM 8-3.10

8-3.11. The beam is subjected to a uniformly distributed load $w_0 = 600$ lb/in. In case (a), the cross section is square and consists of steel with elastic modulus $E = 30 \times 10^6$ psi. In case (b), the cross section consists of a box beam of steel (material A) with elastic modulus $E_A = 30 \times 10^6$ psi and a light filler material (material B) with elastic modulus $E_B = 1.2 \times 10^6$ psi. (The cross-sectional area of steel is approximately the same in each cross section.) Determine the magnitude of the maximum normal stress at $x = 20$ in. for each case.

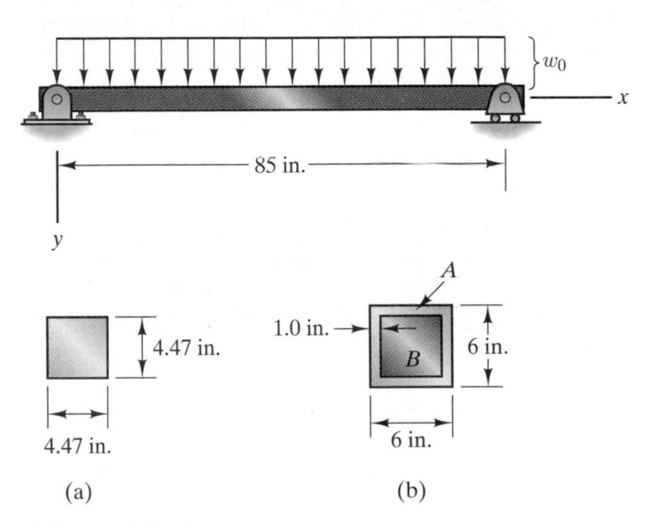

| **PROBLEM 8-3.11**

8-3.12. In Problem 8-3.11, the yield stress of the steel is $\sigma_Y = 80$ ksi. For cases (a) and (b), what is the largest value of w_0 for which yielding of the steel will not occur?

8-3.13. The value of the triangular distributed load at $x = 4$ m is $w_0 = 2$ kN/m. The composite beam consists of two aluminum alloy beams (material A) with elastic modulus $E_A = 70$ GPa that are bonded to a wood beam (material B) with elastic modulus $E_B = 12$ GPa. What are the magnitudes of the maximum normal stresses in the aluminum alloy and in the wood at $x = 2$ m? Draw a graph of the distribution of normal stress at $x = 2$ m.

8-3.14. In Problem 8-3.13, the yield stress of the aluminum alloy is 35 MPa and the yield stress of the wood is 14 MPa. What is the largest value of w_0 for which yielding of either material will not occur?

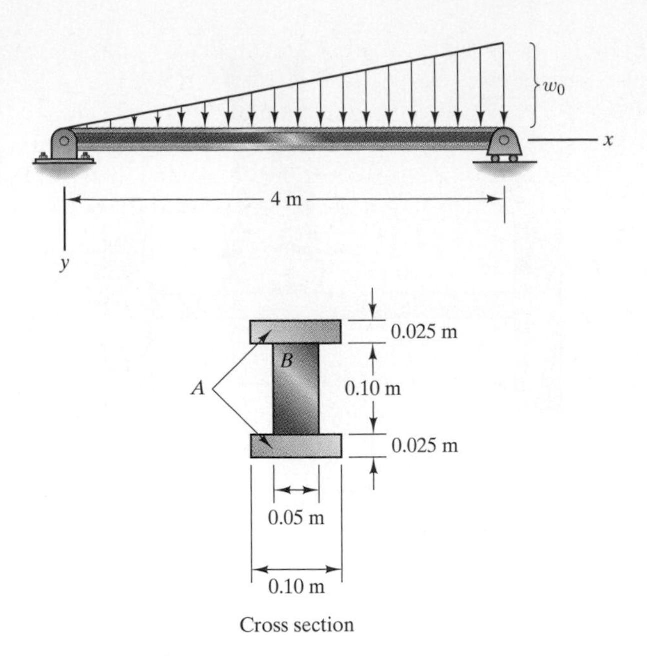

| **PROBLEM 8-3.13**

8-3.15. The figure shows the cross section of a prismatic beam made of material A with elastic modulus $E_A = 200$ GPa and material B with elastic modulus $E_B = 120$ GPa. Determine the distance H from the top of the cross section to the neutral axis.

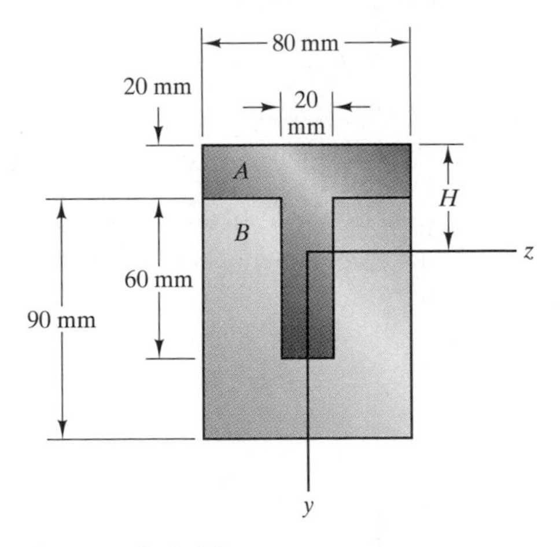

| **PROBLEM 8-3.15**

8-3.16. For the beam described in Problem 8-3.15, if the bending moment $M = 400$ N-m at a given cross section, determine how much of the moment is supported by each material.

8-4.1. A beam with the cross section shown consists of elastic–perfectly plastic material with yield stress $\sigma_Y =$ 400 MPa. **(a)** What is the maximum bending moment that can be applied at a given cross section without causing yielding of the material? **(b)** What is the ultimate moment that will cause formation of a plastic hinge at a given cross section?

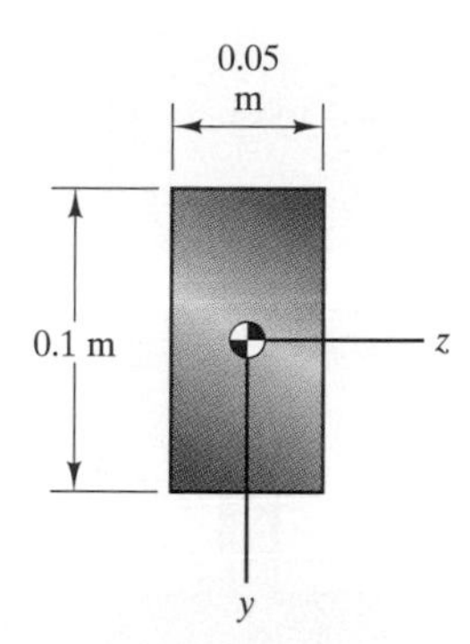

PROBLEM 8-4.1

8-4.2. For the beam described in Problem 8-4.1, determine the distance c and sketch the distribution of the normal stress at a given cross section if the moment $M = 45{,}000$ N-m.

8-4.3. For the beam described in Problem 8-4.1, determine the bending moment M that will cause 50% of the beam's cross section to be yielded.

8-4.4. A beam with the cross section shown consists of elastic–perfectly plastic material with yield stress $\sigma_Y =$ 60,000 psi. **(a)** What is the maximum bending moment that can be applied at a given cross section without causing yielding of the material? **(b)** What is the ultimate moment that will cause formation of a plastic hinge at a given cross section?

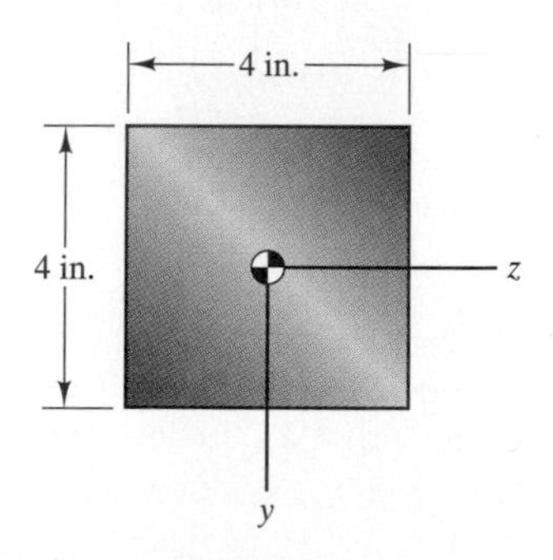

PROBLEM 8-4.4

8-4.5. For the beam described in Problem 8-4.4, determine the distance c and sketch the distribution of the normal stress at a given cross section if the moment $M = 880{,}000$ in-lb.

8-4.6. For the beam in Example 8-6, what is the largest value of w_0 that will not cause yielding of the material?

8-4.7. In Example 8-6, sketch the distribution of normal stress at $x = 2.2$ m if $w_0 = 70{,}000$ N/m.

8-4.8. For the beam in Example 8-6, let $w_0 = 72{,}000$ N/m. At what axial position x is the distance c a minimum? What is the minimum value of c?

8-4.9. The beam consists of elastic–perfectly plastic material with yield stress $\sigma_Y = 700$ MPa. Its dimensions are $L = 1.2$ m, $b = 18$ mm, $h = 36$ mm. Determine the distance c and sketch the distribution of normal stress at $x = 0.4$ m if $w_0 = 22{,}500$ N/m.

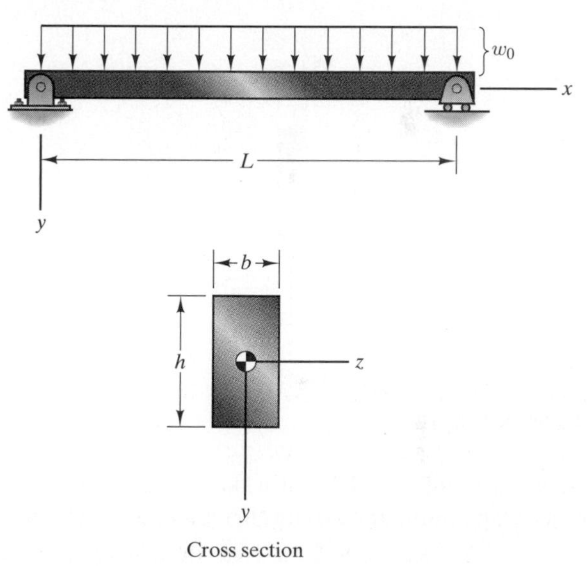

PROBLEM 8-4.9

8-4.10. For the beam described in Problem 8-4.9, if w_0 is progressively increased, at what value will the beam fail by formation of a plastic hinge? Where does the plastic hinge occur?

8-4.11. Suppose that the beam shown in Problem 8-4.9 consists of elastic–perfectly plastic material with yield stress $\sigma_Y = 150{,}000$ psi, and its dimensions are $L = 36$ in., $b = 1/2$ in, $h = 1$ in. Determine the distance c and sketch the distribution of normal stress at $x = 12$ in. if $w_0 =$ 115 lb/in.

8-4.12. For the beam described in Problem 8-4.11, if w_0 is progressively increased, at what value will the beam fail by formation of a plastic hinge? Where does the plastic hinge occur?

8-4.13. The beam consists of elastic-perfectly plastic material with yield stress $\sigma_Y = 250$ MPa. Determine the distance c and sketch the distribution of normal stress at $x = 2$ m if the force $F = 80$ kN.

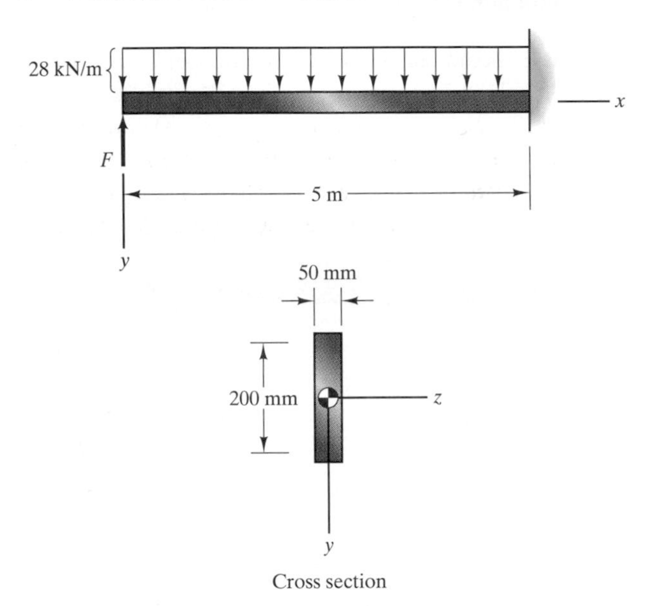

PROBLEM 8-4.13

8-4.14. In Problem 8-4.13, at what axial position x is the distance c a minimum? What is the minimum value of c?

8-4.15. In Problem 8-4.13, if the force F is progressively increased from its initial value of 80 kN, at what value will the beam fail by formation of a plastic hinge? Where does the plastic hinge occur?

8-4.16. By evaluating the integrals in Eq. (8-19), derive Eq. (8-21) for the distance from the neutral axis at which the magnitude of the normal stress equals the yield stress.

8-4.17. By applying Eq. (8-25) for the ultimate moment to a rectangular cross section with width b and height h, confirm Eq. (8-23).

8-4.18. The beam is subjected to a moment M. If the material is elastic–perfectly plastic with yield stress $\sigma_Y =$ 350 MPa, what is the ultimate moment M_U?

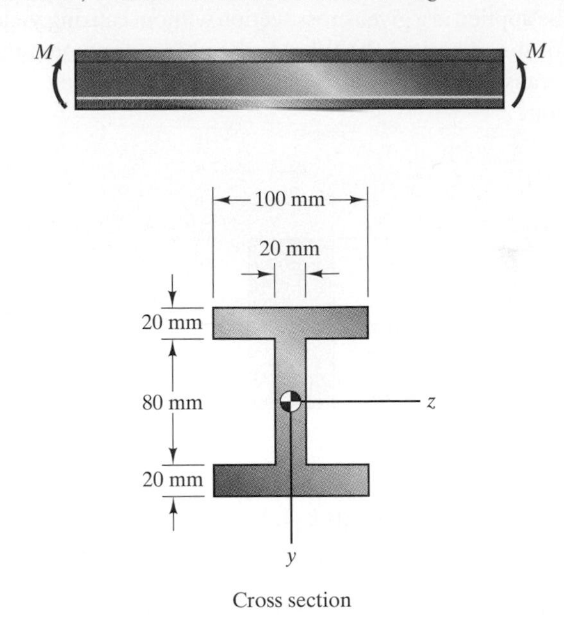

PROBLEM 8-4.18

8-4.19. The beam is subjected to a moment M. If the material is elastic–perfectly plastic with yield stress $\sigma_Y =$ 450 MPa, what is the ultimate moment M_U?

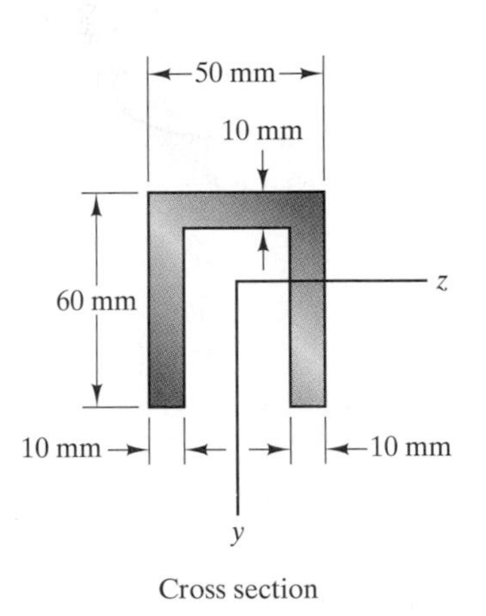

PROBLEM 8-4.19

8-5.1. The beam is subjected to a bending moment $M = 400$ N-m about the z' axis, which is a principal axis. The moment of inertia of the cross section about z' is $I_{z'} = 40 \times 10^{-8}\ \text{m}^4$. Determine the normal stress at point P, which has coordinates $y' = 0.0313$ m, $z' = -0.0269$ m.

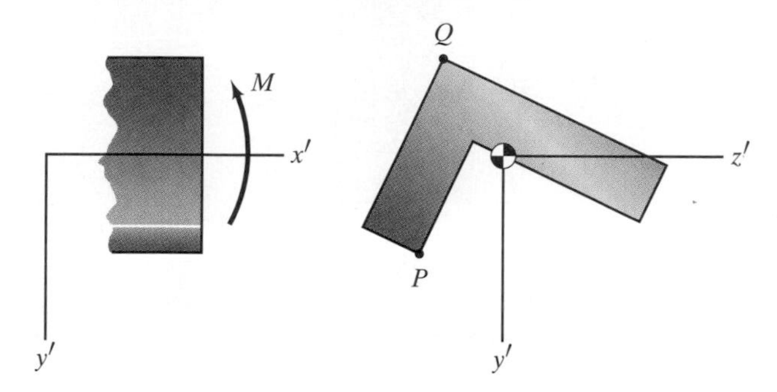

PROBLEM 8-5.1

8-5.2. In Problem 8-5.1, determine the normal stress at point Q, which has coordinates $y' = -0.0313$ m, $z' = -0.0179$ m.

8-5.3. Suppose that the 400-N-m bending moment in Problem 8-5.1 is applied about the y' axis instead of the z' axis, in a direction such that points of the cross section with positive z' coordinates are subjected to tensile stress. The moment of inertia of the cross section about y' is $I_{y'} = 160 \times 10^{-8}\ \text{m}^4$. Determine the normal stress at point P, which has coordinates $y' = 0.0313$ m, $z' = -0.0269$ m.

8-5.4. In Example 8-8, what is the normal stress at the point of the cross section with coordinates $y = 0.05$ m, $z = 0.02$ m?

8-5.5. In Example 8-8, what is the normal stress at the point of the cross section with coordinates $y = -0.03$ m, $z = -0.04$ m?

8-5.6. A beam with the cross section shown is subjected to a bending moment $M = 80{,}000$ in-lb about the principal axis with the larger moment of inertia. What is the maximum magnitude of the resulting distribution of normal stress?

8-5.7. In Problem 8-5.6, determine the magnitude of the normal stress at the point of the cross section with coordinates $y = -1$ in., $z = 2$ in.

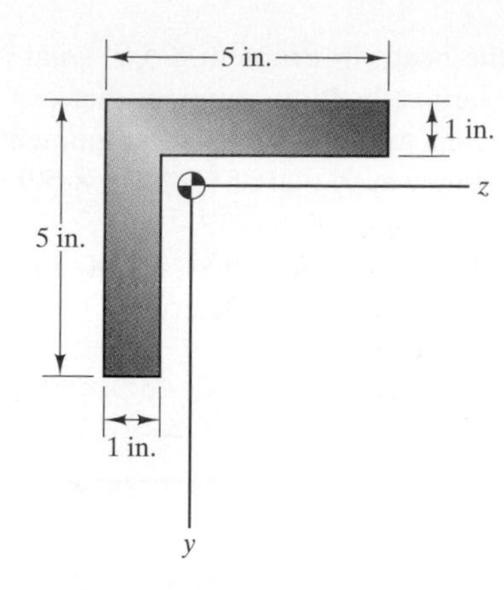

PROBLEM 8-5.6

8-5.8. If the beam in Problem 8-5.6 is subjected to a moment $M = 80{,}000$ in-lb about the principal axis with the smaller moment of inertia, what is the maximum magnitude of the resulting distribution of normal stress? Compare your answer to that of Problem 8-5.6.

8-5.9. A beam with the cross section shown consists of material that will safely support a normal stress (tensile or compressive) of 5 MPa. Based on this criterion, what is the magnitude of the largest bending moment that can be applied about the principal axis with the larger moment of inertia?

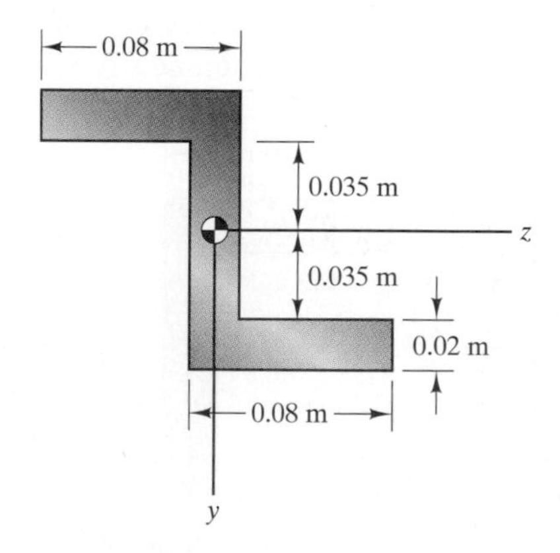

PROBLEM 8-5.9

8-5.10. If the bending moment determined in Problem 8-5.9 is applied about the principal axis with the larger moment of inertia, what is the magnitude of the normal stress at the point of the cross section with coordinates $y = 0.045$ m, $z = 0.030$ m?

8-5.11. For the beam in Problem 8-5.9, what is the magnitude of the largest bending moment that can be applied about the principal axis with the smaller moment of inertia? Compare your answer to that of Problem 8-5.9.

8-5.12. The beam is subjected to a moment $M = 80{,}000$ in-lb about the z axis. Determine the normal stress at point P.

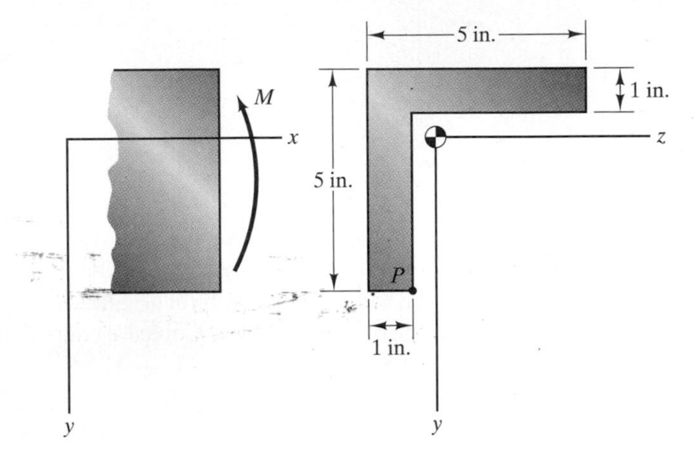

PROBLEM 8-5.12

8-5.13. For the beam in Problem 8-5.12, draw a sketch indicating the location of the neutral axis.

8-5.14. The beam is subjected to a moment $M = 200$ N-m about the z axis. Determine the normal stress at point P.

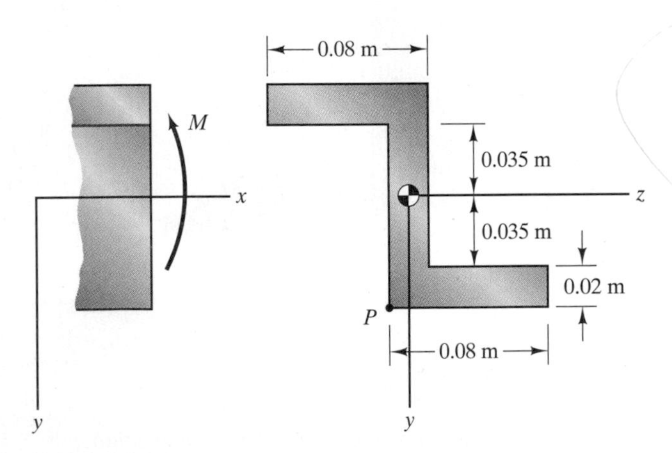

PROBLEM 8-5.14

8-5.15. For the beam in Problem 8-5.14, draw a sketch indicating the location of the neutral axis.

8-6.1. A beam with the cross section shown is subjected to a shear force $V = 8$ kN. What is the average shear stress at the neutral axis ($y' = 0$)?

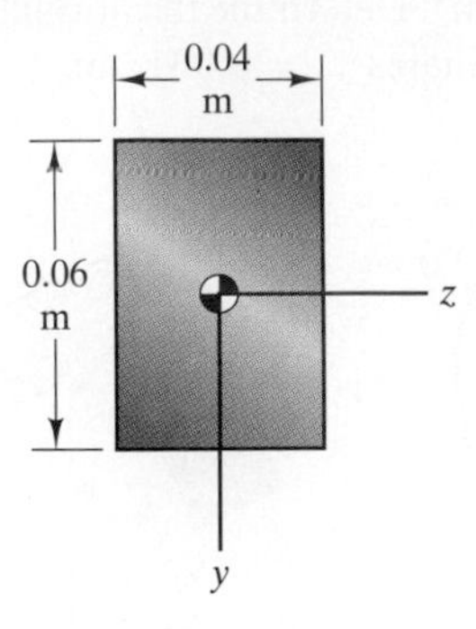

PROBLEM 8-6.1

8-6.2. In Problem 8-6.1, determine the average shear stress **(a)** at $y' = 0.01$ m; **(b)** at $y' = -0.02$ m.

8-6.3. In Example 8-9, consider the cross section at $x = 3$ m. What is the average shear stress at $y' = 0.05$ m?

8-6.4. What is the maximum magnitude of the average shear stress in the beam in Example 8-9, and where does it occur?

8-6.5. The beam is subjected to a distributed load. For the cross section at $x = 40$ in., determine the average shear stress **(a)** at the neutral axis; **(b)** at $y' = 1.5$ in.

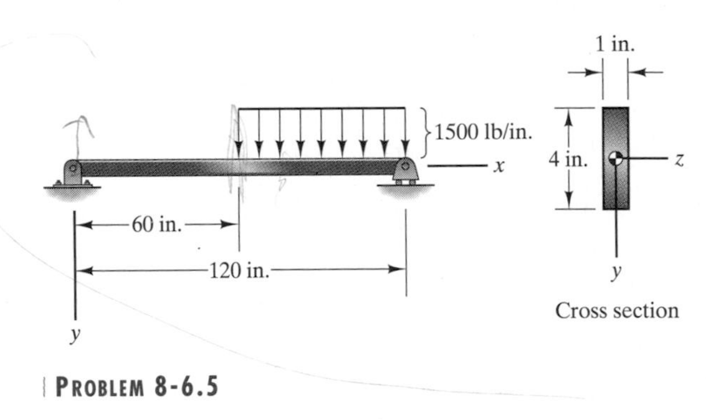

PROBLEM 8-6.5

8-6.6. Solve Problem 8-6.5 for the cross section at $x = 80$ in.

8-6.7. What is the maximum magnitude of the average shear stress in the beam in Problem 8-6.5, and where does it occur?

8-6.8. The beam is subjected to a distributed load. For the cross section at $x = 0.6$ m, determine the average shear stress **(a)** at the neutral axis; **(b)** at $y' = 0.02$ m.

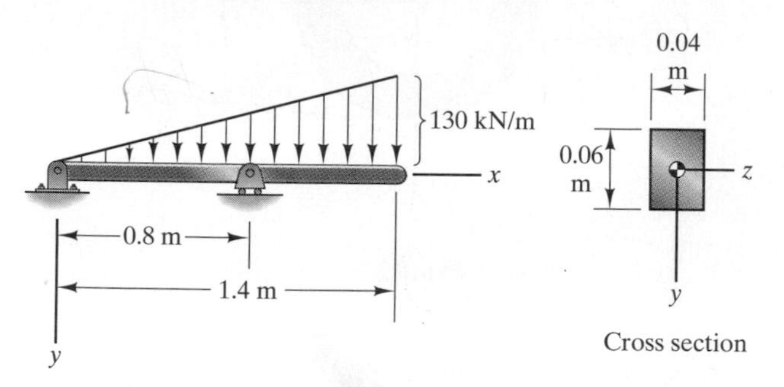

PROBLEM 8-6.8

8-6.9. Solve Problem 8-6.8 for the cross section at $x = 1.0$ m.

8-6.10. By integrating the stress distribution given by Eq. (8-44), confirm that the total force exerted on the rectangular cross section by the shear stress is equal to V.

8-6.11. Prove that the quantity Q defined by Eq. (8-43) is a maximum at the neutral axis ($y' = 0$).

8-6.12. At a particular axial position, the beam whose cross section is shown is subjected to a shear force $V = 20$ kN. Determine the average shear stress acting on the slanted infinitesimal element.

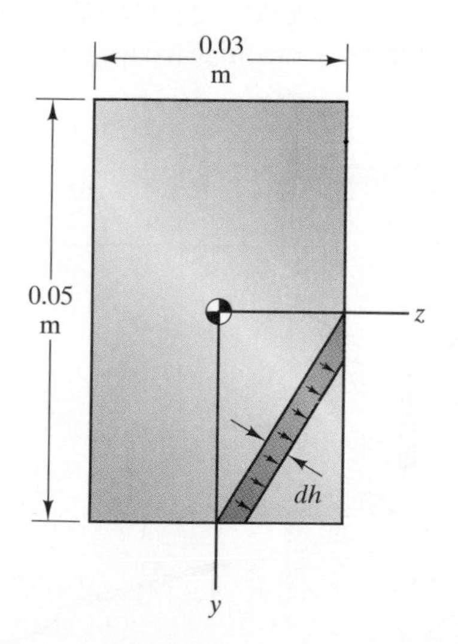

PROBLEM 8-6.12

8-6.13. In Example 8-10, determine the average shear stress at $y' = 1$ in.

8-6.14. In Example 8-10, determine the average shear stress in the upper-right glued joint.

8-6.15. The beam whose cross section is shown consists of three planks of wood glued together. At a given axial position it is subjected to a shear force $V = 2400$ lb. What is the average shear stress at the neutral axis $y' = 0$?

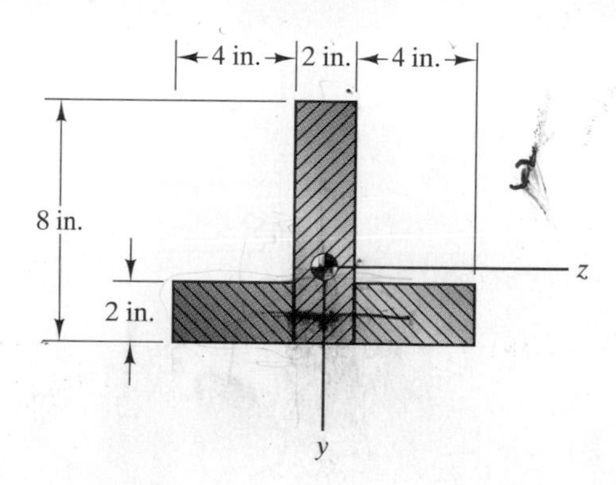

PROBLEM 8-6.15

8-6.16. In Problem 8-6.15, what are the magnitudes of the average shear stresses acting on each glued joint?

8-6.17. For the cross section at $x = 8$ ft, determine the average shear stress **(a)** at the neutral axis; **(b)** at $y' = 2$ in.

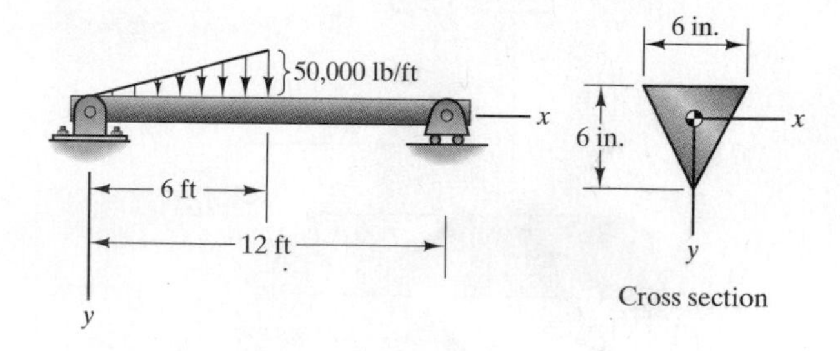

PROBLEM 8-6.17

8-6.18. In Problem 8-6.17, determine the value of y' at which the magnitude of the average shear stress is a maximum. (Notice that the maximum magnitude does *not* occur at the neutral axis.) What is the maximum magnitude?

8-6.19. Solve Problem 8-6.17 for the cross section at $x = 4$ ft.

8-6.20. At a particular axial position, the beam whose cross section is shown is subjected to a shear force $V = 15$ kN. Determine the average shear stress **(a)** at the neutral axis $y' = 0$; **(b)** at $y' = 0.025$ m.

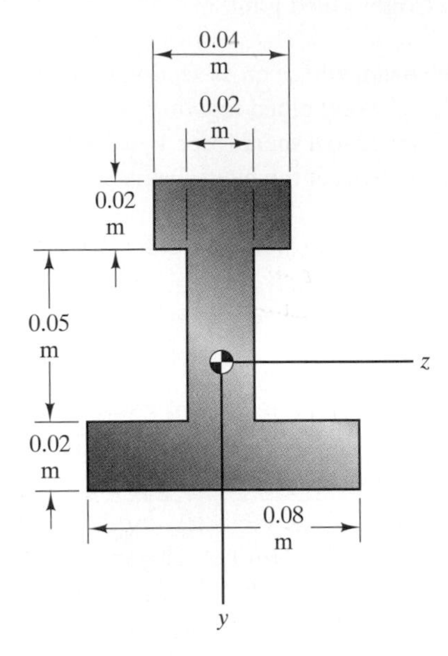

PROBLEM 8-6.20

8-6.21. For the beam in Problem 8-6.20, determine the average shear stress on the infinitesimal element shown.

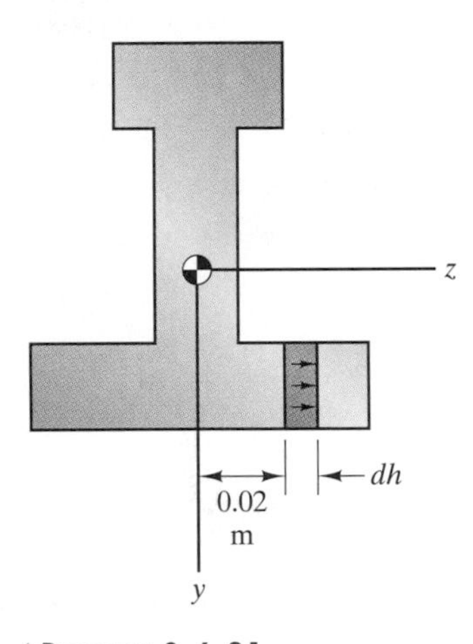

PROBLEM 8-6.21

8-6.22. At a particular axial position, the beam whose cross section is shown is subjected to a shear force $V = 40$ kN. What is the average shear stress at the neutral axis ($y' = 0$)?

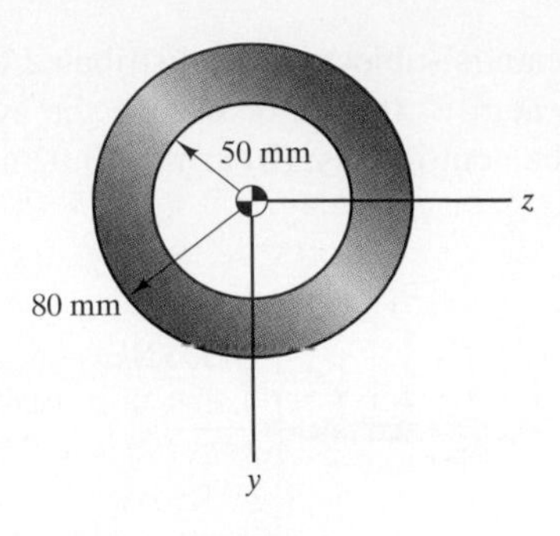

PROBLEM 8-6.22

8-6.23. For the beam in Problem 8-6.22, what is the average shear stress at $y' = 50$ mm?

8-6.24. For the beam in Problem 8-6.22, what is the average shear stress at $y' = 25$ mm?

8-7.1. In Example 8-11, determine the magnitude of the shear stress at the neutral axis.

8-7.2. In Example 8-11, determine the magnitude of the shear stress at $y' = -0.02$ m.

8-7.3. In Example 8-12, what is the shear stress at $\alpha = 45°$?

8-7.4. In Example 8-12, determine the shear stress at $\alpha = 45°$ if the radius $R = 8$ in.

8-7.5. A beam with the thin-walled cross section shown is subjected to a shear force $V = 2.4$ kN. (The wall thickness is not shown to scale.) Determine the magnitude of the shear

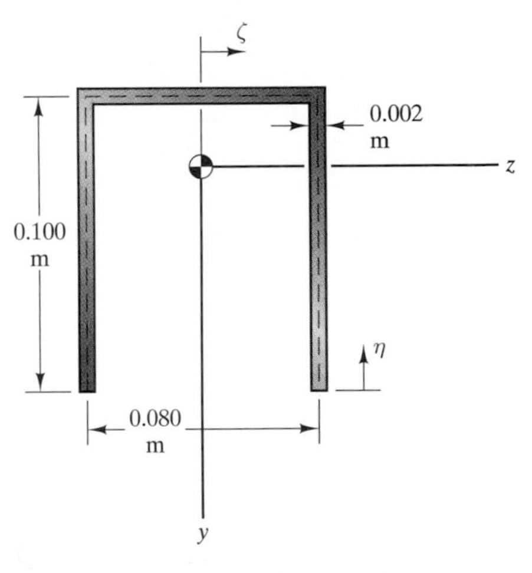

PROBLEM 8-7.5

stress in the vertical web of the beam as a function of the variable η shown.

8-7.6. For the beam in Problem 8-7.5, determine the magnitude of the shear stress in the horizontal web of the beam as a function of the variable ζ shown.

8-7.7. A beam with the semicircular thin-walled cross section shown is subjected to a shear force $V = 20{,}000$ lb. The radius $R = 6$ in. and the wall thickness $t = \frac{1}{4}$ in. (The wall thickness is not shown to scale.) What is the magnitude of the shear stress at the neutral axis, and at what value of the angle α does it occur?

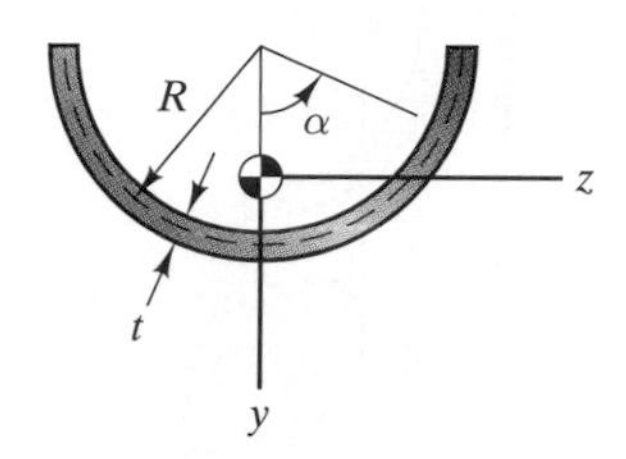

PROBLEM 8-7.7

8-7.8. For the beam in Problem 8-7.7, draw a graph of the shear stress as a function of the angle α.

8-7.9. A beam with the thin-walled cross section shown is subjected to a shear force $V = 4.5$ kN. (The wall thickness is not shown to scale.) Determine the magnitude of the shear stress in the vertical web of the beam as a function of the variable η shown.

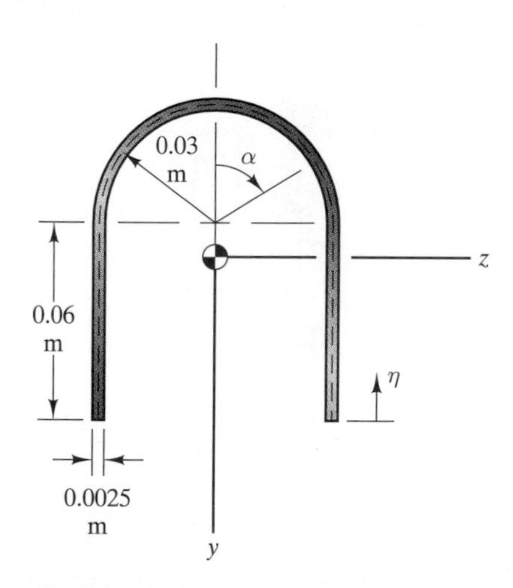

PROBLEM 8-7.9

8-7.10. For the beam in Problem 8-7.9, draw a graph of the shear stress in the circular web of the beam as a function of the angle α shown.

8-8.1. In Example 8-13, show that the moment exerted about the neutral axis by the shear stress in the beam's top horizontal web is

$$M_{\text{top web}} = 0.0123F.$$

Cantilever beams with the cross sections shown in Problems 8-8.2–8-8.13 are subjected to a lateral force F at the end. Determine the distance e from the neutral axis to the shear center S.

8-8.2. The dimensions are $d = 0.50$ m, $h = 1.00$ m, and $t = 0.03$ m. The horizontal distance c from the midline of the beam's vertical web to the neutral axis is $c = 0.125$ m, and the moment of inertia of the cross section about the z axis is $I = 0.01$ m^4. (*Strategy:* Equate the moment due to F about the point where the z axis intersects the midline of the beam's vertical web to the total moment due to the distribution of shear stress about that point.)

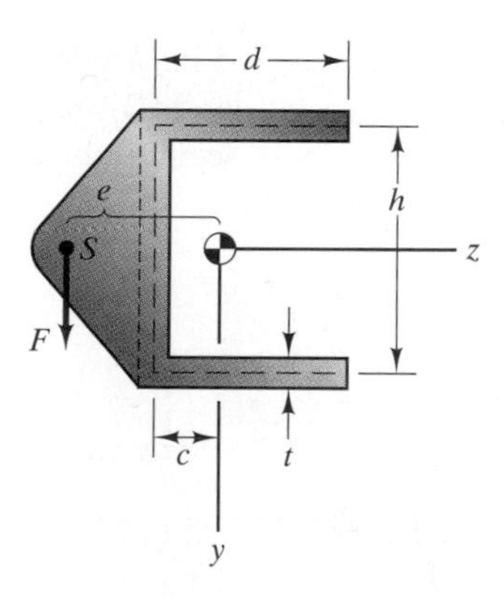

PROBLEM 8-8.2

8-8.3. The dimensions of the cross section shown in Problem 8-8.2 are $d = 5$ in., $h = 9$ in., and $t = \frac{1}{8}$ in. The horizontal distance c from the midline of the beam's vertical web to the neutral axis is $c = 1.315$ in., and the moment of inertia of the cross section about the z axis is $I = 32.9$ in^4.

8-8.4. The dimensions are $d_L = 0.04$ m, $d_R = 0.08$ m, $h = 0.08$ m, and $t = 0.002$ m. The horizontal distance c from the midline of the beam's vertical web to the neutral axis is $c = 0.0151$ m, and the moment of inertia of the cross section about the z axis is $I = 8.47 \times 10^{-7}$ m^4.

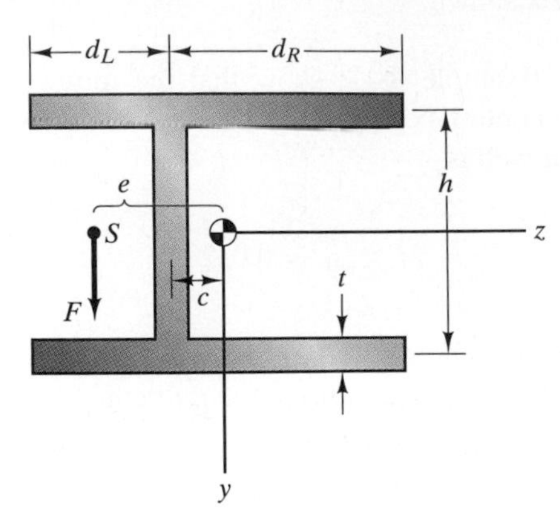

PROBLEM 8-8.4

8-8.5. The dimensions of the cross section shown in Problem 8-8.4 are $d_L = 3$ in., $d_R = 8$ in., $h = 8$ in., and $t = \frac{1}{8}$ in. The horizontal distance c from the midline of the beam's vertical web to the neutral axis is $c = 1.84$ in., and the moment of inertia of the cross section about the z axis is $I = 49.1$ in^4.

8-8.6. The dimensions are $d = 5$ in., $h = 9$ in., and $t = \frac{1}{8}$ in. The horizontal distance c is 4.62 in., and the moment of inertia of the cross section about the z axis is $I = 116$ in^4.

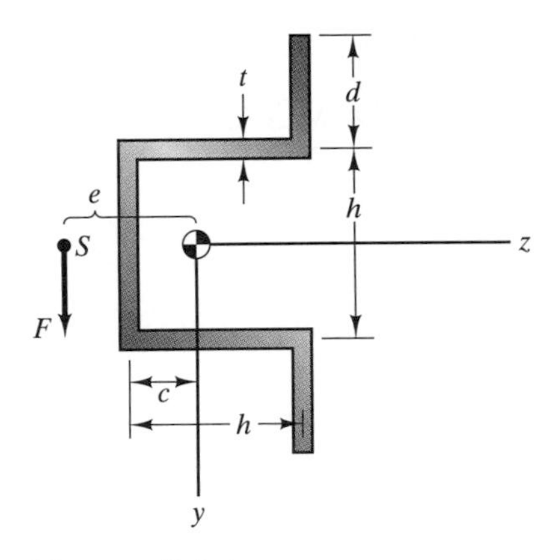

PROBLEM 8-8.6

8-8.7. The dimensions of the cross section shown in Problem 8-8.6 are $d = 0.08$ m, $h = 0.12$ m, and $t = 0.004$ m. The horizontal distance c is 0.065 m, and the moment of inertia of the cross section about the z axis is $I = 1.06 \times 10^{-5}$ m^4.

8-8.8. The radius is $R = 0.1$ m and the wall thickness is $t = 0.003$ m. The horizontal distance c from the center of the semicircle to the neutral axis is $c = 0.0637$ m, and the moment of inertia of the cross section about the z axis is $I = 4.71 \times 10^{-6}$ m^4.

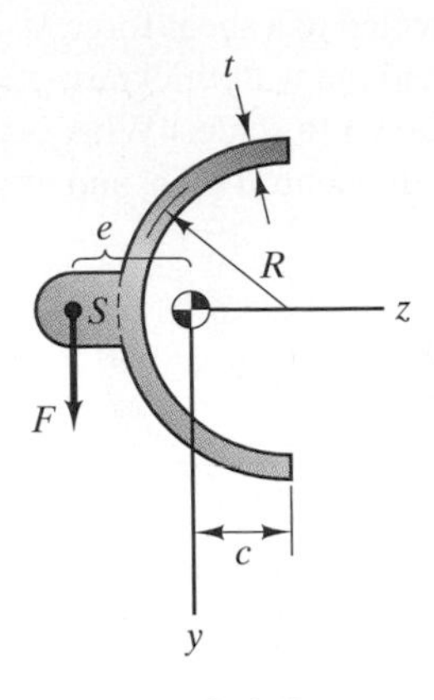

PROBLEM 8-8.8

8-8.9. The dimensions of the cross section shown in Problem 8-8.8 are $R = 4$ in. and $t = \frac{1}{16}$ in. The horizontal distance c from the center of the semicircle to the neutral axis is $c = 2.55$ in., and the moment of inertia of the cross section about the z axis is $I = 6.28$ in^4.

8-8.10. The radius is $R = 4$ in., $d = 2$ in., and the wall thickness is $t = \frac{1}{8}$ in. The horizontal distance c from the center of the semicircle to the neutral axis is $c = 1.95$ in., and the moment of inertia of the cross section about the z axis is $I = 25.0$ in^4.

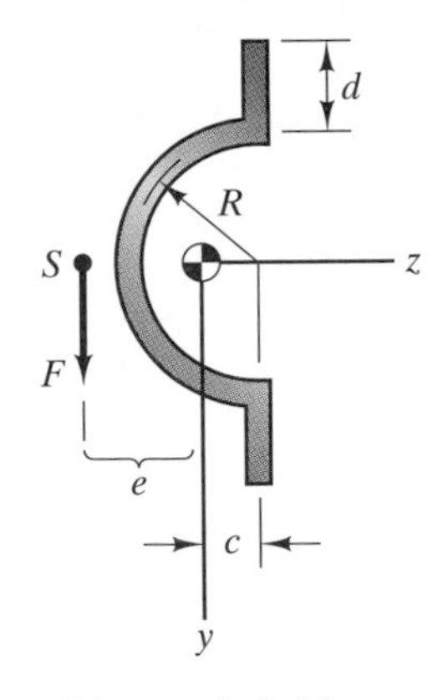

PROBLEM 8-8.10

8-8.11. The dimensions of the cross section shown in Problem 8-8.10 are $R = 0.035$ m, $d = 0.020$ m, and $t = 0.003$ m. The horizontal distance c from the center of the semicircle to the neutral axis is $c = 0.0163$ m, and the moment of inertia of the cross section about the z axis is $I = 4.66 \times 10^{-7}$ m^4.

8-8.12. The radius is $R = 0.035$ m, $d = 0.020$ m, and the wall thickness is $t = 0.003$ m. The horizontal distance c from the center of the semicircle to the neutral axis is $c = 0.0137$ m, and the moment of inertia of the cross section about the z axis is $I = 3.5 \times 10^{-7}$ m^4.

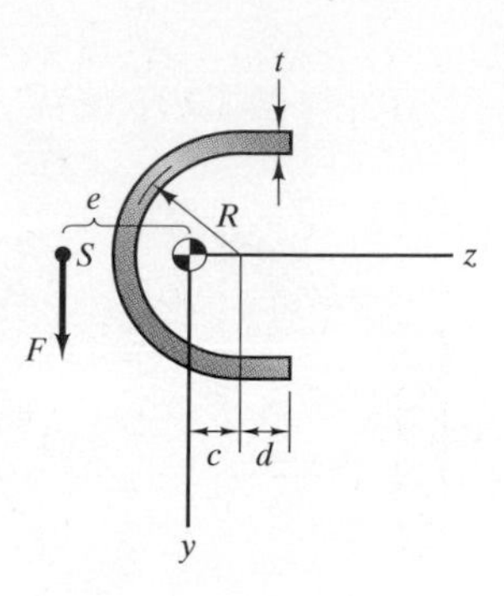

PROBLEM 8-8.12

8-8.13. The dimensions of the cross section shown in Problem 8-8.12 are $R = 4$ in., $d = 2$ in., and $t = \frac{1}{8}$ in. The horizontal distance c from the center of the semicircle to the neutral axis is $c = 1.69$ in., and the moment of inertia of the cross section about the z axis is $I = 20.6$ in^4.

Deflection of a layered beam called a leaf spring isolates the vehicle's frame from the effects of road irregularities.

CHAPTER 9

Deflections of Beams

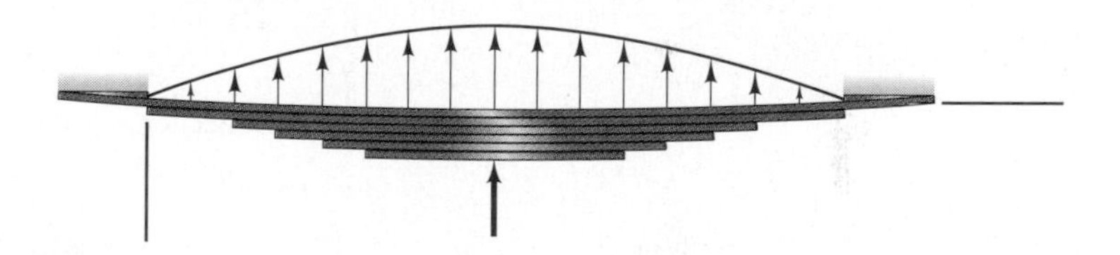

Deflections of some beams, including leaf springs, diving boards, and vaulting poles, are central to their function. In contrast, many structural beams are designed simply to support loads; deflections are not important considerations in their design. But even in such cases, we show in this chapter that calculation of deflections can be of crucial importance for another reason: It is through calculating deflections that we can determine the reactions on statically indeterminate beams.

9-1 | Determination of the Deflection

Let v be the deflection of a beam's neutral axis relative to the x axis, and let θ be the angle between the neutral axis and the x axis (Fig. 9-1). Our objective is to determine v and θ as functions of x for a beam with given loads.

Differential Equation

By considering the deflection at x and at $x + dx$ (Fig. 9-2), we see that v and θ are related by

$$\frac{dv}{dx} = \tan\theta = \theta + \frac{1}{3}\theta^3 + \cdots,$$

where we express $\tan\theta$ in terms of its Taylor series. We restrict our analysis to beams and loadings for which θ is small enough to neglect terms of second and higher orders, so that

$$\frac{dv}{dx} = \theta. \tag{9-1}$$

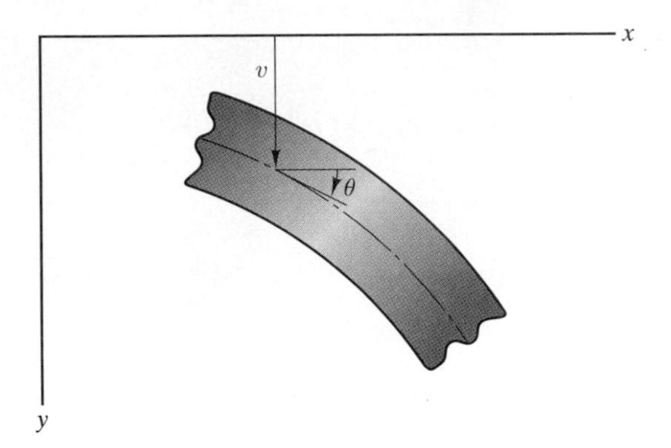

FIGURE 9-1 Deflection v and the angle θ between the neutral and x axes.

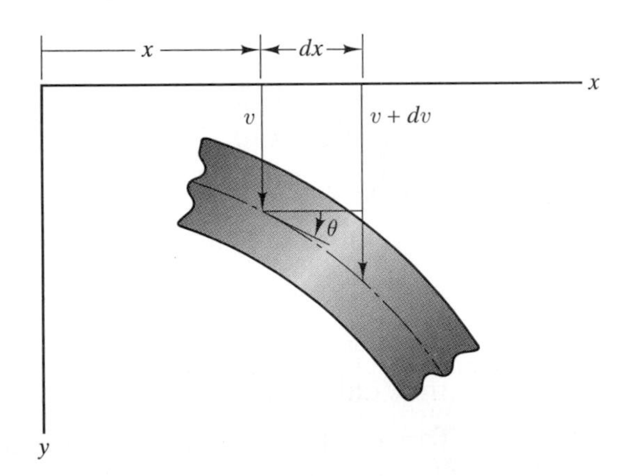

FIGURE 9-2 Determining the relation between v and θ.

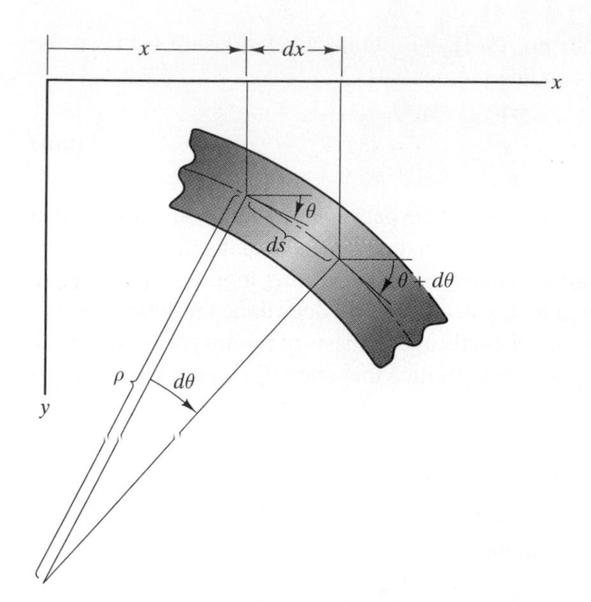

FIGURE 9-3 Relating θ to the radius of curvature of the neutral axis.

(The magnitude of θ in the figures is greatly exaggerated.) In Fig. 9-3 we draw lines perpendicular to the neutral axis at x and at $x + dx$. The angle $d\theta$ between these lines equals the change in θ from x to $x + dx$. In terms of the radius of curvature of the neutral axis ρ and the distance ds, the angle $d\theta$ is

$$d\theta = \frac{1}{\rho}\, ds. \qquad \textbf{(9-2)}$$

Notice from Fig. 9-3 that

$$dx = ds \cos\theta = ds \left(1 - \frac{1}{2}\theta^2 + \cdots\right).$$

Therefore, $ds = dx$ when we neglect terms of second order and higher in θ, so we can express Eq. (9-2) as

$$\frac{d\theta}{dx} = \frac{1}{\rho}.$$

Substituting Eq. (9-1) into this expression, we obtain

$$\frac{d^2 v}{dx^2} = \frac{1}{\rho}. \qquad \textbf{(9-3)}$$

In Chapter 8 we obtained an equation relating the beam's radius of curvature, the bending moment, and the flexural rigidity. From Eq. (8-10),

$$\frac{1}{\rho} = -\frac{M}{EI}.$$

Substituting this result into Eq. (9-3), we obtain a relationship between the beam's deflection and the bending moment:

$$v'' = -\frac{M}{EI}, \tag{9-4}$$

where the primes denote derivatives with respect to x. With this equation we can determine deflections of beams. Although there are several details that we must discuss, the basic procedure is to determine the bending moment M in a beam as a function of x, then integrate Eq. (9-4) twice to determine the deflection v as a function of x. (If the beam is prismatic and consists of homogeneous material, the flexural rigidity EI is a constant.) Notice that once v is known as a function of x, we can use Eq. (9-1) to determine θ as a function of x.

Boundary Conditions

The beam in Fig. 9-4 has pin and roller supports and is subjected to a uniformly distributed load. The bending moment M as a function of x is

$$M = \frac{w_0}{2}(Lx - x^2).$$

We substitute this expression into Eq. (9-4):

$$v'' = \frac{w_0}{2EI}(-Lx + x^2).$$

Integrating, we obtain

$$v' = \frac{w_0}{2EI}\left(-\frac{Lx^2}{2} + \frac{x^3}{3}\right) + A,$$

where A is an integration constant. Integrating again yields

$$v = \frac{w_0}{2EI}\left(-\frac{Lx^3}{6} + \frac{x^4}{12}\right) + Ax + B, \tag{9-5}$$

where B is a second integration constant.

To complete our determination of the deflection, we must evaluate the integration constants A and B by using the boundary conditions imposed by the beam's supports (Fig. 9-5). The deflection is zero at the pin support ($x = 0$).

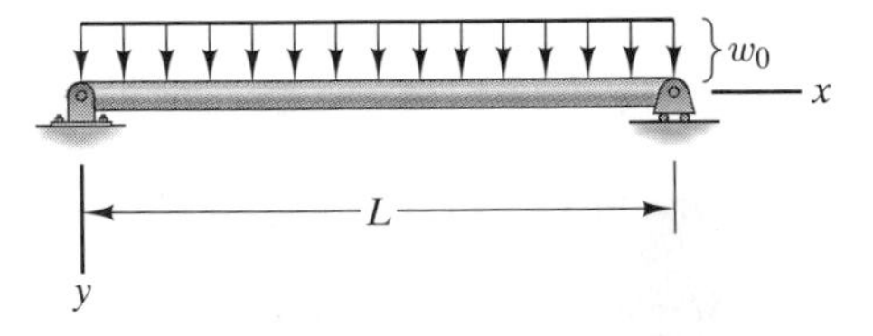

FIGURE 9-4 Beam subjected to a uniformly distributed load.

We substitute this condition into Eq. (9-5), obtaining the value of B:

$$v|_{x=0} = B = 0.$$

The deflection is also zero at the roller support ($x = L$):

$$v|_{x=L} = \frac{w_0}{2EI}\left(-\frac{L^4}{6} + \frac{L^4}{12}\right) + AL = 0.$$

From this equation we obtain $A = w_0L^3/24EI$. Substituting the values of A and B into Eq. (9-5), we obtain the solution for the beam's deflection:

$$v = \frac{w_0 x}{24EI}(L^3 - 2Lx^2 + x^3).$$

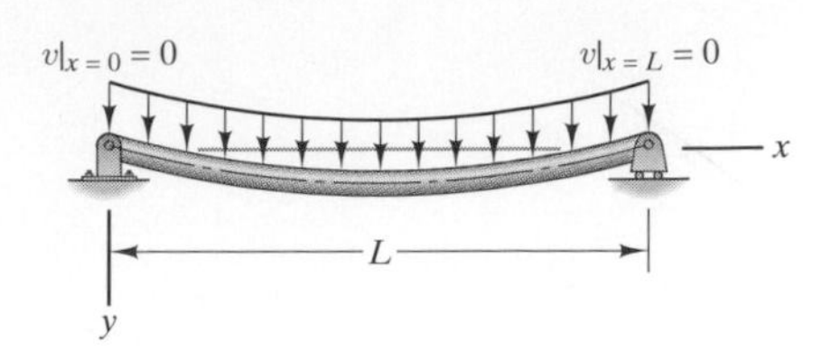

FIGURE 9-5 The deflection equals zero at each support.

In the previous example we determined the beam's deflection by substituting the expression for the bending moment as a function of x into Eq. (9-4), integrating twice, and using the boundary conditions at the supports to evaluate the integration constants. A new consideration arises when we apply this procedure to the beam in Fig. 9-6a. When we determine the bending moment as a function of x, we obtain one expression M_1 that applies for $0 \le x \le L/2$ and a different expression M_2 that applies for $L/2 \le x \le L$ (Fig. 9-6b). We need to determine the deflection independently for each of these regions. Let v_1 denote the beam's deflection from $x = 0$ to $x = L/2$. We substitute the expression for M_1 into Eq. (9-4):

$$v_1'' = -\frac{M_1}{EI} = \frac{w_0L}{2EI}\left(\frac{3L}{4} - x\right).$$

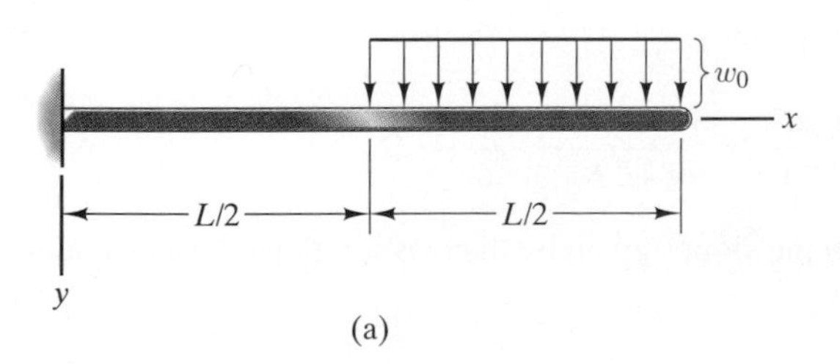

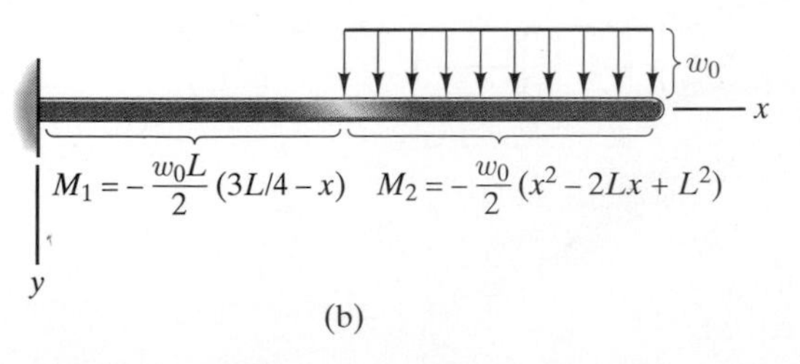

FIGURE 9-6 (a) Beam subjected to a uniformly distributed load over part of its length. (b) Functions describing the distribution of the bending moment.

Integrating twice gives

$$v_1' = \frac{w_0 L}{2EI}\left(\frac{3Lx}{4} - \frac{x^2}{2}\right) + A,$$

$$v_1 = \frac{w_0 L}{2EI}\left(\frac{3Lx^2}{8} - \frac{x^3}{6}\right) + Ax + B,$$

where A and B are integration constants. Letting v_2 denote the beam's deflection from $x = L/2$ to $x = L$, we substitute the expression for M_2 into Eq. (9-4) and integrate twice:

$$v_2'' = -\frac{M_2}{EI} = \frac{w_0}{2EI}(x^2 - 2Lx + L^2),$$

$$v_2' = \frac{w_0}{2EI}\left(\frac{x^3}{3} - Lx^2 + L^2x\right) + C,$$

$$v_2 = \frac{w_0}{2EI}\left(\frac{x^4}{12} - \frac{Lx^3}{3} + \frac{L^2x^2}{2}\right) + Cx + D,$$

where C and D are integration constants.

We must now identify four boundary conditions with which to evaluate the integration constants A, B, C, and D. Two conditions are imposed by the beam's built-in support. At $x = 0$, the deflection and slope equal zero (Fig. 9-7):

$$v_1|_{x=0} = B = 0,$$

$$v_1'|_{x=0} = A = 0.$$

We also know that the deflections given by the two solutions must be equal at $x = L/2$ (Fig. 9-7):

$$v_1|_{x=L/2} = v_2|_{x=L/2}:$$

$$\frac{14w_0L^4}{384EI} = \frac{17w_0L^4}{384EI} + \frac{CL}{2} + D.$$

The fourth condition is that the slopes given by the two solutions must be equal

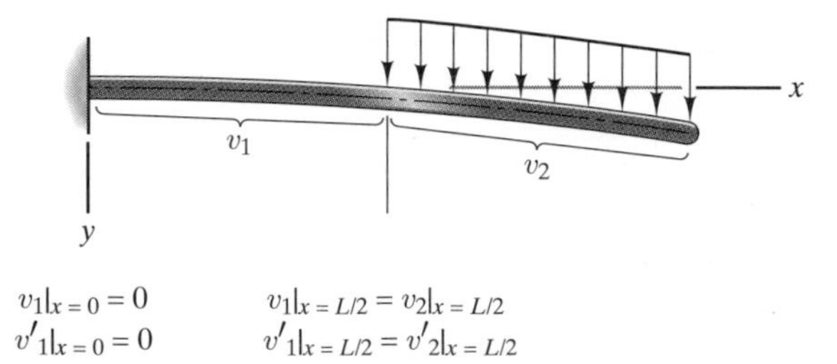

FIGURE 9-7 Boundary conditions at the built-in support and where the two solutions meet.

at $x = L/2$.

$$v_1'|_{x=L/2} = v_2'|_{x=L/2}:$$

$$\frac{6w_0L^3}{48EI} = \frac{7w_0L^3}{48EI} + C.$$

From the four boundary conditions we find that $A = 0$, $B = 0$, $C = -w_0L^3/48EI$, and $D = w_0L^4/384EI$. Substituting these results into our expressions for v_1 and v_2, the beam's deflection is

$$v = \begin{cases} \dfrac{w_0L}{48EI}(9Lx^2 - 4x^3), & 0 \le x \le L/2 \quad \textbf{(9-6)} \\ \dfrac{w_0}{384EI}(16x^4 - 64Lx^3 + 96L^2x^2 - 8L^3x + L^4), & L/2 \le x \le L. \quad \textbf{(9-7)} \end{cases}$$

The examples we have discussed indicate the steps required to determine a beam's deflection:

1. Determine the bending moment as a function of x. As in our second example, this may result in two or more functions $M_1, M_2 \ldots$, each of which applies to a different segment of the beam's length.
2. For each segment, integrate Eq. (9-4) twice to determine $v_1, v_2 \ldots$. If there are N segments, this step will result in $2N$ unknown integration constants.
3. Use the boundary conditions to determine the integration constants. Our two examples illustrate the most common boundary conditions encountered in determining deflections of beams. These boundary conditions are summarized in Fig. 9-8.

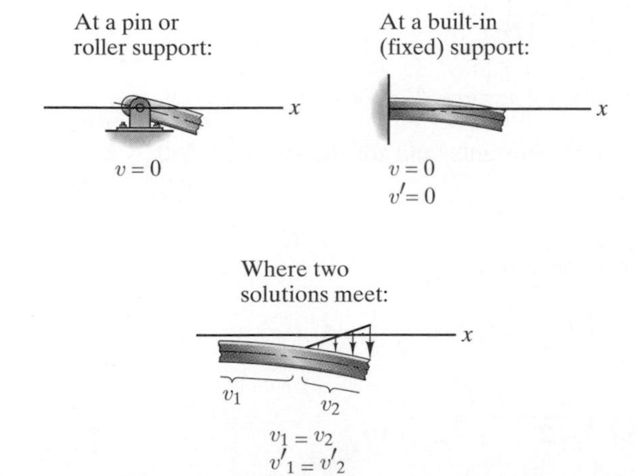

FIGURE 9-8 Common boundary conditions.

EXAMPLE 9-1

Worksheet 12

The beam in Fig. 9-9 consists of material with modulus of elasticity $E = 72$ GPa, the moment of inertia of its cross section is $I = 1.6 \times 10^{-7}$ m^4, and its length is $L = 4$ m. If $w_0 = 100$ N/m, what is the deflection at the right end of the beam?

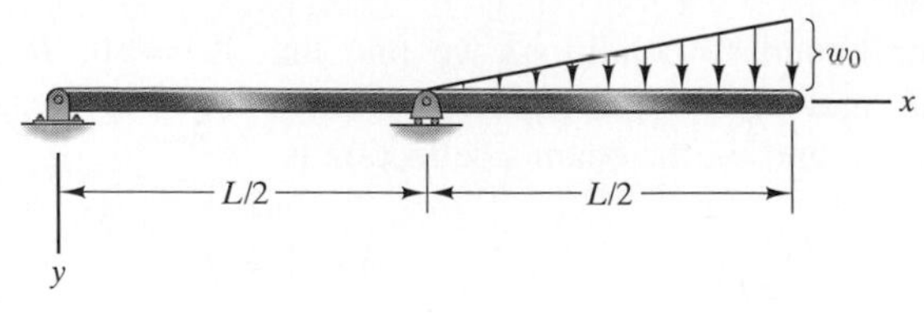

FIGURE 9-9

Solution

Determine the bending moment as a function of *x* We leave it as an exercise to show that the distributions of the bending moment in the left and right halves of the beam are given by the expressions in Fig. (a).

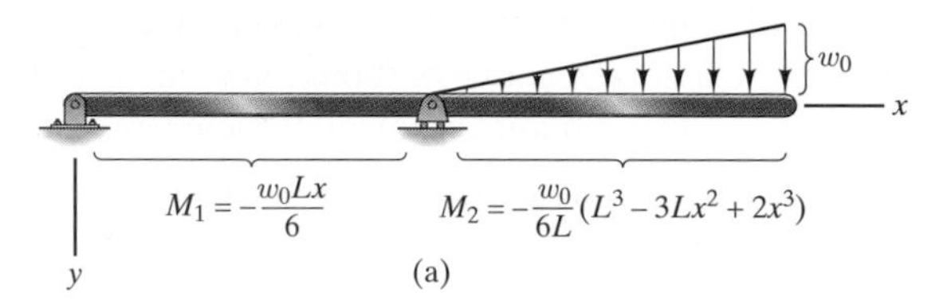

(a) Equations for the bending moment.

For each segment, integrate Eq. (9-4) For the segment $0 \leq x \leq L/2$, we obtain

$$v_1'' = -\frac{M_1}{EI} = \frac{w_0 L x}{6EI},$$

$$v_1' = \frac{w_0 L x^2}{12EI} + A,$$

$$v_1 = \frac{w_0 L x^3}{36EI} + Ax + B,$$

where A and B are integration constants, and for the segment $L/2 \leq x \leq L$ we obtain

$$v_2'' = -\frac{M_2}{EI} = \frac{w_0}{6LEI}(L^3 - 3Lx^2 + 2x^3),$$

$$v_2' = \frac{w_0}{6LEI}\left(L^3 x - Lx^3 + \frac{x^4}{2}\right) + C,$$

$$v_2 = \frac{w_0}{6LEI}\left(\frac{L^3 x^2}{2} - \frac{Lx^4}{4} + \frac{x^5}{10}\right) + Cx + D,$$

where C and D are integration constants.

Use the boundary conditions to determine the integration constants We can apply the four boundary conditions shown in Fig. (b).

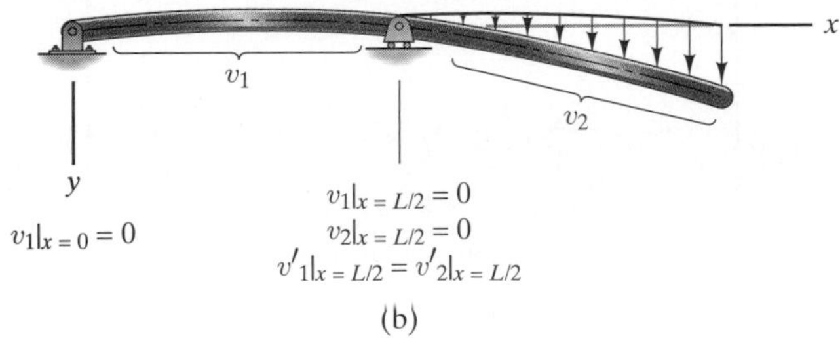

(b) Boundary conditions.

The deflection is zero at $x = 0$:

$$v_1|_{x=0} = B = 0.$$

The deflection is zero at $x = L/2$, a condition that applies to both v_1 and v_2:

$$v_1|_{x=L/2} = \frac{w_0 L^4}{288EI} + \frac{AL}{2} = 0,$$

$$v_2|_{x=L/2} = \frac{3w_0 L^4}{160EI} + \frac{CL}{2} + D = 0.$$

The slopes of the two solutions must be equal at $x = L/2$.

$$v'_1|_{x=L/2} = v'_2|_{x=L/2}:$$

$$\frac{w_0 L^3}{48EI} + A = \frac{13w_0 L^3}{192EI} + C.$$

Substituting the values of w_0, L, E, and I into these four equations and solving, we obtain $A = -0.00386$, $B = 0$, $C = -0.0299$, and $D = 0.0181$. Now we can use the equation for v_2 to obtain the deflection at the right end of the beam:

$$\begin{aligned} v_2 &= \frac{w_0}{6LEI}\left(\frac{L^3x^2}{2} - \frac{Lx^4}{4} + \frac{x^5}{10}\right) + Cx + D \\ &= \frac{100}{(6)(4)(72 \times 10^9)(1.6 \times 10^{-7})}\left[\frac{(4)^3(4)^2}{2} - \frac{(4)(4)^4}{4} + \frac{(4)^5}{10}\right] \\ &\quad + (-0.0299)(4) + 0.0181 \\ &= 0.0282 \text{ m}. \end{aligned}$$

Figure (c) is a graph of the beam's deflection.

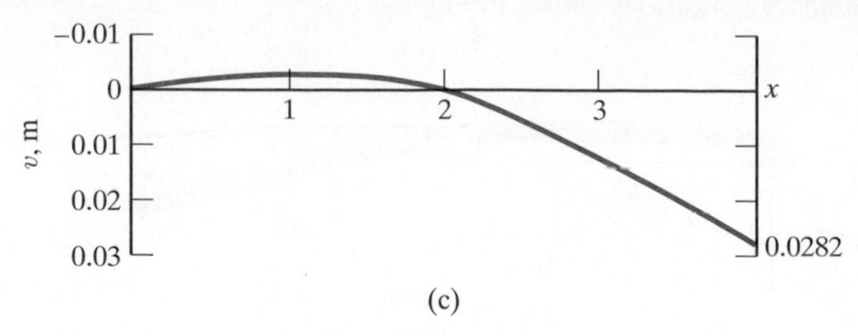

(c) Deflection as a function of x.

Discussion

From Fig. 9-8 you would conclude that the beam in this example has three boundary conditions at the roller support: $v_1|_{x=L/2} = 0$, $v_2|_{x=L/2} = 0$, and $v_1|_{x=L/2} = v_2|_{x=L/2}$. But notice that only two of these conditions are independent. Once any two of them are chosen, the third one is implied.

9-2 Statically Indeterminate Beams

Remarkably, the procedure we used in Section 9-1 to determine the deflections of statically determinate beams also works for statically indeterminate beams, yielding both the deflection and the unknown reactions.

Consider the beam in Fig. 9-10a. From the free-body diagram (Fig. 9-10b) we obtain the equilibrium equations

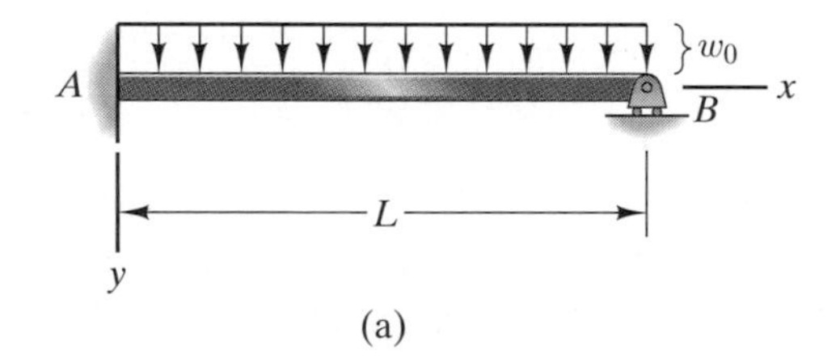

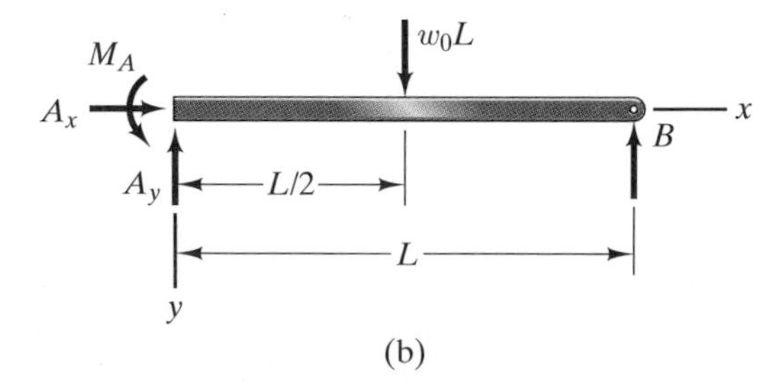

FIGURE 9-10 (a) Statically indeterminate beam. (b) Free-body diagram of the entire beam.

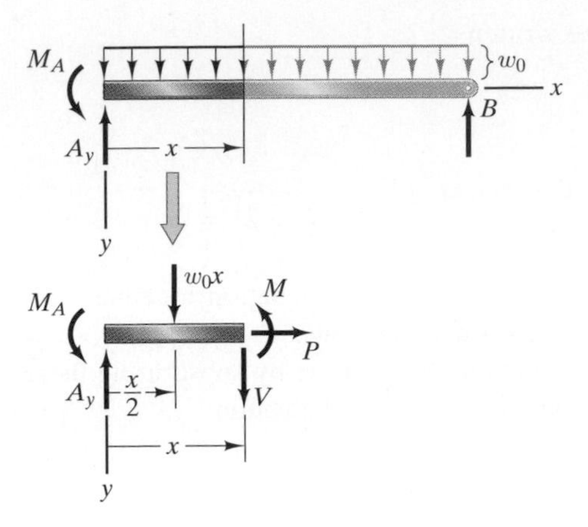

FIGURE 9-11 Determining the distribution of the bending moment.

$$\Sigma F_x = A_x = 0, \tag{9-8}$$

$$\Sigma F_y = -A_y - B + w_0 L = 0, \tag{9-9}$$

$$\Sigma M_{\text{point } A} = M_A + LB - \frac{L}{2} w_0 L = 0. \tag{9-10}$$

We can't solve Eqs. (9-9) and (9-10) for the reactions A_y, M_A, and B. The beam is statically indeterminate.

We ignore this setback and proceed to determine the deflection in terms of the unknown reactions. From Fig. 9-11, the distribution of the bending moment is

$$M = -\frac{w_0 x^2}{2} + A_y x - M_A.$$

Substituting this expression into Eq. (9-4) and integrating gives

$$EIv'' = \frac{w_0 x^2}{2} - A_y x + M_A,$$

$$EIv' = \frac{w_0 x^3}{6} - \frac{A_y x^2}{2} + M_A x + C,$$

$$EIv = \frac{w_0 x^4}{24} - \frac{A_y x^3}{6} + \frac{M_A x^2}{2} + Cx + D,$$

where C and D are integration constants.

Our next step is to use the boundary conditions to evaluate the integration constants. But while there are two integration constants, we see from Fig. 9-12 that there are three boundary conditions. This is the key to the solution: *There are more boundary conditions than there are unknown integration constants.* (The boundary conditions are compatibility conditions imposed on the beam's deflection.) We can use the three equations obtained from the boundary conditions together with the two equilibrium equations (9-9) and (9-10) to determine the integration constants C and D and the unknown reactions A_y, M_A, and B.

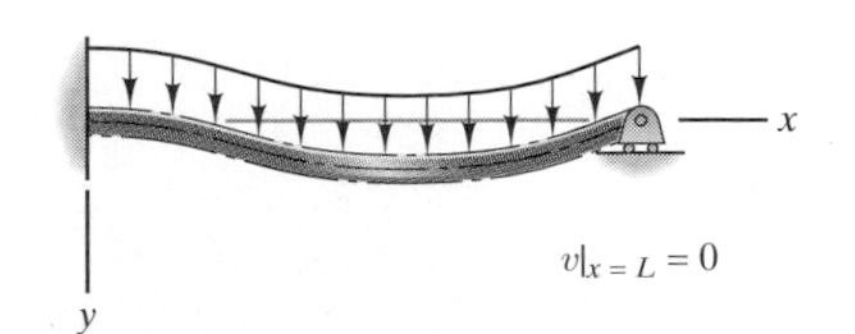

FIGURE 9-12 Boundary conditions.

The boundary conditions can be written

$$EIv|_{x=0} = D = 0,$$
$$EIv'|_{x=0} = C = 0,$$
$$EIv|_{x=L} = \frac{w_0 L^4}{24} - \frac{A_y L^3}{6} + \frac{M_A L^2}{2} + CL + D = 0.$$

We see that $C = 0$ and $D = 0$. Solving the third equation together with Eqs. (9-9) and (9-10) yields the unknown reactions: $A_y = 5w_0L/8$, $B = 3w_0L/8$, and $M_A = w_0L^2/8$. We complete the solution by substituting these results into the expression for v to obtain the beam's deflection:

$$v = \frac{w_0}{48EI}(2x^4 - 5Lx^3 + 3L^2x^2).$$

This example sheds light on the analysis of statically indeterminant structures in general. An object is statically indeterminate if it has more supports than are necessary for equilibrium. If the roller support is removed from the beam in Fig. 9-10a, it remains in equilibrium and is statically determinate. The roller support introduces an additional reaction, which makes the beam statically indeterminate *but also introduces an additional boundary (compatibility) condition.* Each redundant support added to a structure introduces a new compatibility condition. As a consequence, the number of combined equilibrium equations and compatibility conditions remains sufficient to determine the reactions.

In the following example we demonstrate the steps required to analyze statically indeterminate beams:

1. *Draw a free-body diagram of the entire beam and write the equilibrium equations.*
2. *Determine the bending moment as a function of x in terms of the unknown reactions. This may result in two or more functions* $M_1, M_2 \ldots,$ *each of which applies to a different segment of the beam's length.*
3. *For each segment, integrate Eq. (9-4) twice to determine* $v_1, v_2 \ldots.$
4. *Use the boundary conditions together with the equilibrium equations to determine the integration constants and unknown reactions.*

EXAMPLE 9-2

The beam in Fig. 9-13 has length $L = 4$ m and $w_0 = 5$ kN/m. What are the reactions at A, B, and C?

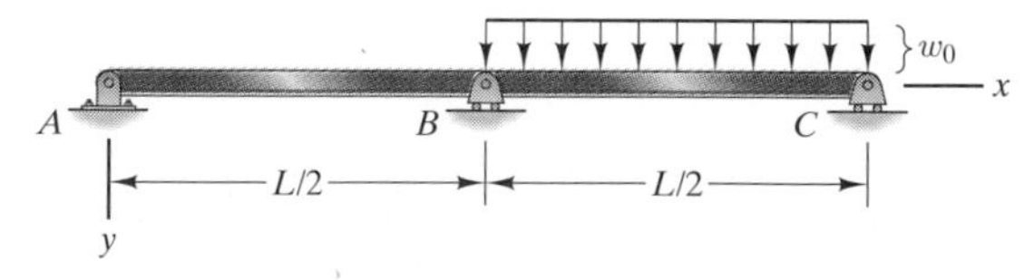

FIGURE 9-13

Solution

Write the equilibrium equations From the free-body diagram in Fig. (a), we obtain the equations

$$\Sigma F_x = A_x = 0,$$

$$\Sigma F_y = -A_y - B - C + \frac{w_0 L}{2} = 0, \quad \textbf{(9-11)}$$

$$\Sigma M_{\text{point } A} = \frac{L}{2}B + LC - \left(\frac{3}{4}L\right)\frac{w_0 L}{2} = 0. \quad \textbf{(9-12)}$$

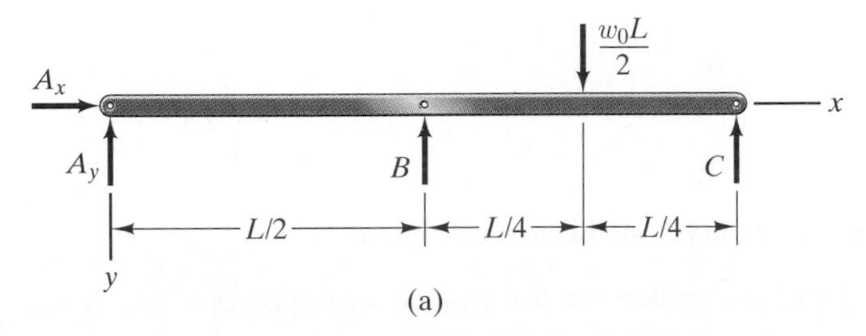

(a)

(a) Free-body diagram of the entire beam.

Determine the bending moment as a function of x From Fig. (b), the distributions of the bending moment in the left and right halves of the beam are

$$M_1 = A_y x,$$

$$M_2 = -\frac{w_0 x^2}{2} + (w_0 L - C)x - \left(\frac{w_0 L}{2} - C\right)L.$$

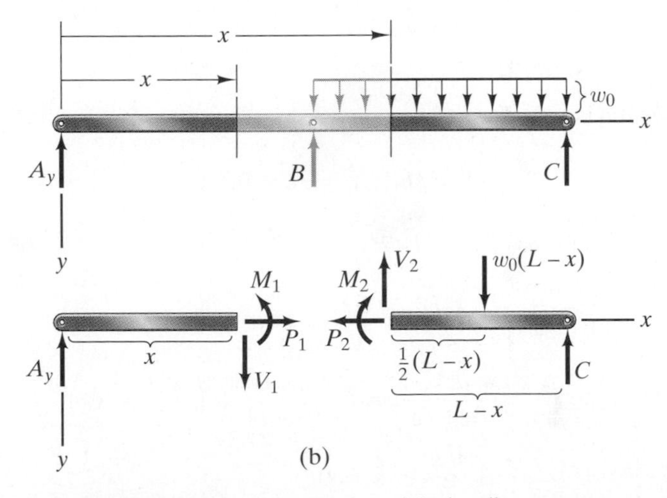

(b)

(b) Determining the distribution of the bending moment.

For each segment, integrate Eq. (9-4) Substituting the expressions for M_1 and M_2 and integrating, we obtain

$$EIv_1'' = -M_1 = -A_y x,$$

$$EIv_1' = -\frac{A_y x^2}{2} + G,$$

$$EIv_1 = -\frac{A_y x^3}{6} + Gx + H,$$

$$EIv_2'' = -M_2 = \frac{w_0 x^2}{2} - (w_0 L - C)x + \left(\frac{w_0 L}{2} - C\right) L,$$

$$EIv_2' = \frac{w_0 x^3}{6} - (w_0 L - C)\frac{x^2}{2} + \left(\frac{w_0 L}{2} - C\right) Lx + J,$$

$$EIv_2 = \frac{w_0 x^4}{24} - (w_0 L - C)\frac{x^3}{6} + \left(\frac{w_0 L}{2} - C\right) \frac{Lx^2}{2} + Jx + K,$$

where G, H, J, and K are integration constants.

Use the boundary conditions together with the equilibrium equations We can apply the five boundary conditions shown in Fig. (c).

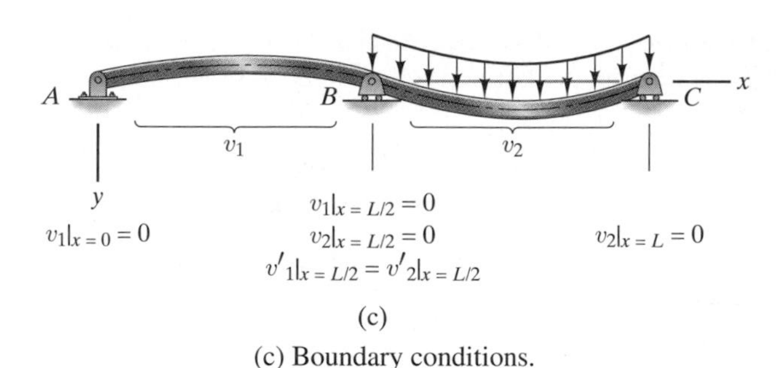

(c) Boundary conditions.

The deflection is zero at $x = 0$:

$$EIv_1|_{x=0} = H = 0. \tag{9-13}$$

Both v_1 and v_2 are zero at $x = L/2$:

$$EIv_1|_{x=L/2} = -\frac{A_y L^3}{48} + \frac{GL}{2} + H = 0, \tag{9-14}$$

$$EIv_2|_{x=L/2} = \frac{w_0 L^4}{384} - (w_0 L - C)\frac{L^3}{48} + \left(\frac{w_0 L}{2} - C\right)\frac{L^3}{8} + \frac{JL}{2} + K = 0. \tag{9-15}$$

The slopes of the two solutions must be equal at $x = L/2$.

$$EIv_1'|_{x=L/2} = EIv_2'|_{x=L/2}:$$

$$-\frac{A_y L^2}{8} + G = \frac{w_0 L^3}{48} - (w_0 L - C)\frac{L^2}{8} + \left(\frac{w_0 L}{2} - C\right)\frac{L^2}{2} + J. \quad \textbf{(9-16)}$$

The deflection is zero at $x = L$:

$$EIv_2|_{x=L} = \frac{w_0 L^4}{24} - (w_0 L - C)\frac{L^3}{6} + \left(\frac{w_0 L}{2} - C\right)\frac{L^3}{2} + JL + K = 0. \quad \textbf{(9-17)}$$

We can solve the equilibrium equations (9-11) and (9-12) together with Eqs. (9-13)–(9-17) for the integration constants G, H, J, and K and the unknown reactions A_y, B, and C. Substituting the values of L and w_0 and solving, the solutions for the reactions are $A_y = -625$ N, $B = 6250$ N, and $C = 4375$ N.

9-3 | Deflections Using the Fourth-Order Equation

In this section we describe an alternative method for determining deflections of beams.

Differential Equation

In Section 9-1 we determined deflections by using Eq. (9-4):

$$\frac{d^2 v}{dx^2} = -\frac{M}{EI}. \quad \textbf{(9-18)}$$

The distributed load w, shear load V, and bending moment M are related by Eqs. (7-4) and (7-5):

$$\frac{dV}{dx} = -w, \quad \textbf{(9-19)}$$

$$\frac{dM}{dx} = V. \quad \textbf{(9-20)}$$

Differentiating Eq. (9-18) and using Eq. (9-20), we obtain

$$\frac{d^3 v}{dx^3} = -\frac{V}{EI}. \quad \textbf{(9-21)}$$

Differentiating this equation and using Eq. (9-19) yields

$$v'''' = \frac{w}{EI}, \quad \textbf{(9-22)}$$

where the primes indicate derivatives with respect to x. We will demonstrate that deflections of beams can be determined by integration of this equation.

Boundary Conditions

In determining beam deflections using the second-order equation (9-18), the boundary conditions are expressed in terms of the deflection v and its first derivative v', the slope. When beam deflections are determined using the fourth-order equation (9-22), the boundary conditions are expressed in terms of the dependent variable v and its first, second, and third derivatives. From Eq. (9-18), the second derivative of v is related to the bending moment by

$$M = -EIv'', \tag{9-23}$$

and from Eq. (9-21), the third derivative of v is related to the shear force by

$$V = -EIv'''. \tag{9-24}$$

Therefore, the boundary conditions for Eq. (9-22) can be expressed in terms of the deflection v, the slope v', the bending moment M, and the shear force V. Some common boundary conditions are summarized in Fig. 9-14.

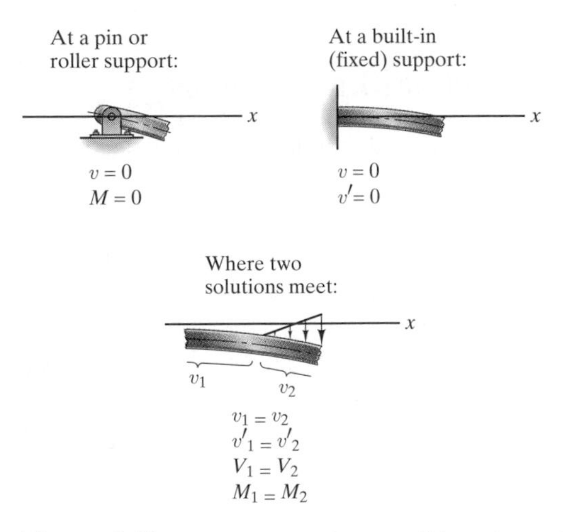

FIGURE 9-14 Common boundary conditions for applying the fourth-order equation.

EXAMPLE 9-3

The beam in Fig. 9-15 has a built-in support and is subjected to a uniformly distributed load. Use Eq. (9-22) to determine its deflection.

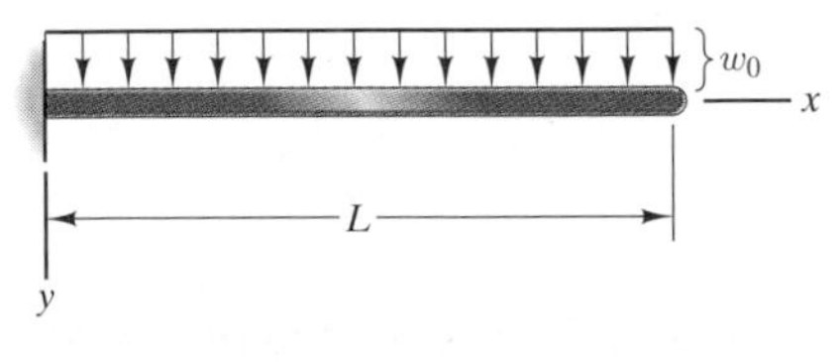

FIGURE 9-15

Solution

Integrate Eq. (9-22) We obtain

$$EIv'''' = w_0,$$
$$EIv''' = w_0 x + A,$$
$$EIv'' = \frac{w_0 x^2}{2} + Ax + B,$$
$$EIv' = \frac{w_0 x^3}{6} + \frac{Ax^2}{2} + Bx + C,$$
$$EIv = \frac{w_0 x^4}{24} + \frac{Ax^3}{6} + \frac{Bx^2}{2} + Cx + D,$$

where A, B, C, and D are integration constants.

Use the boundary conditions to determine the integration constants The deflection and slope are zero at the built-in support, but we need two more boundary conditions to determine the four integration constants. Since no force or couple act at the right end of the beam, V and M equal zero at $x = L$ [Fig. (a)]. From Eqs. (9-23) and (9-24) this implies that v'' and v''' are zero at $x = L$, which provides the two additional boundary conditions.

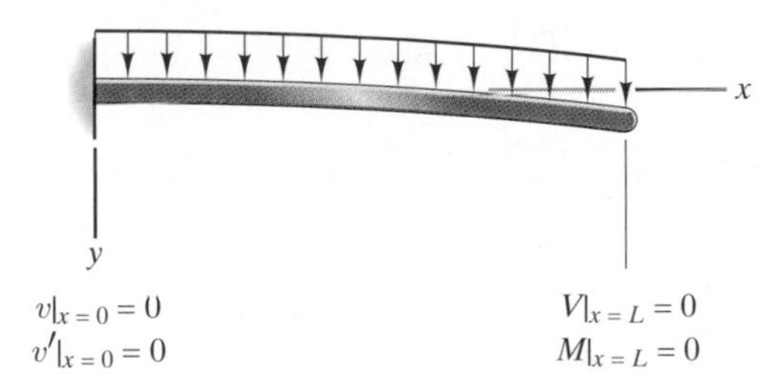

(a) Boundary conditions.

Using the equations

$$EIv|_{x=0} = D = 0,$$
$$EIv'|_{x=0} = C = 0,$$
$$EIv''|_{x=L} = \frac{w_0 L^2}{2} + AL + B = 0,$$
$$EIv'''|_{x=L} = w_0 L + A = 0$$

to evaluate A, B, C, and D, the beam's deflection is

$$v = \frac{w_0 x^2}{24EI}(6L^2 - 4Lx + x^2).$$

EXAMPLE 9-4

Use Eq. (9-22) to determine to determine the deflection of the beam in Fig. 9-16.

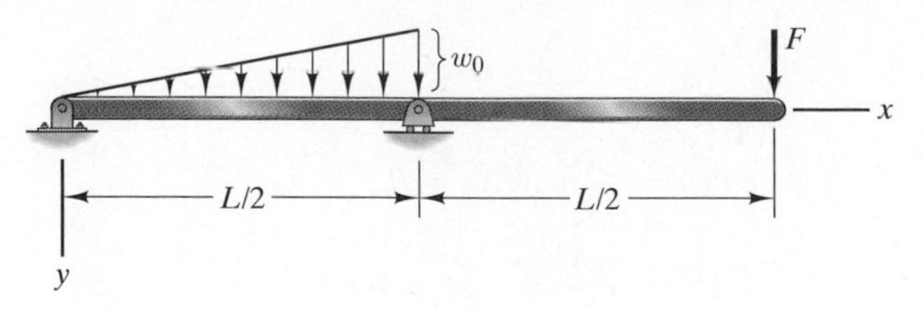

FIGURE 9-16

Solution

For each segment, integrate Eq. (9-22) The distributed load on the left half of the beam is $w = 2w_0x/L$. For the segment $0 \leq x \leq L/2$, we obtain

$$
\begin{aligned}
EIv_1'''' &= \frac{2w_0x}{L}, \\
EIv_1''' &= \frac{w_0x^2}{L} + A, \\
EIv_1'' &= \frac{w_0x^3}{3L} + Ax + B, \\
EIv_1' &= \frac{w_0x^4}{12L} + \frac{Ax^2}{2} + Bx + C, \\
EIv_1 &= \frac{w_0x^5}{60L} + \frac{Ax^3}{6} + \frac{Bx^2}{2} + Cx + D,
\end{aligned}
$$

where A, B, C and D are integration constants. For the segment $L/2 \leq x \leq L$, we obtain

$$
\begin{aligned}
EIv_2'''' &= 0, \\
EIv_2''' &= G, \\
EIv_2'' &= Gx + H, \\
EIv_2' &= \frac{Gx^2}{2} + Hx + J, \\
EIv_2 &= \frac{Gx^3}{6} + \frac{Hx^2}{2} + Jx + K,
\end{aligned}
$$

where G, H, J and K are integration constants.

Use the boundary conditions to determine the integration constants Obtaining a free-body diagram by passing a plane just to the left of the force F [Fig. (a)], we see that $V = -EIv''' = F$ and $M = -EIv'' = 0$ at $x = L$. We can determine the integration constants by applying the eight boundary conditions shown in Fig. (b).

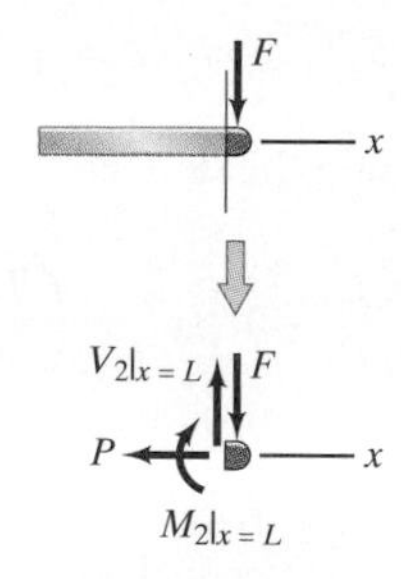

(a) Determining the boundary conditions at $x = L$.

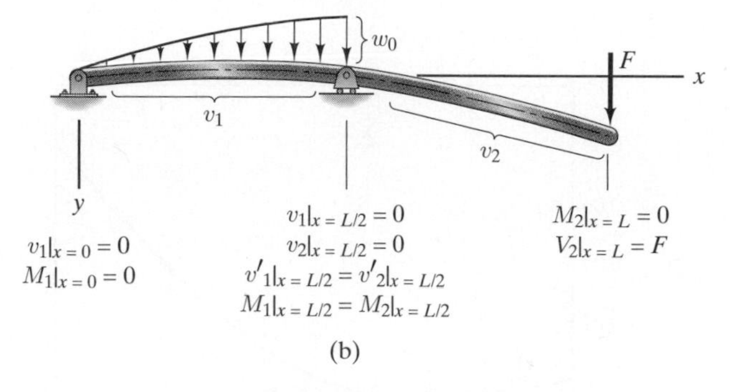

(b) Boundary conditions.

The boundary conditions at $x = 0$ are

$$(1) \quad EIv_1|_{x=0} = D = 0,$$

$$(2) \quad EIv_1''|_{x=0} = B = 0.$$

The boundary conditions at $x = L/2$ are

$$(3) \quad EIv_1|_{x=L/2} = \frac{w_0L^4}{1920} + \frac{AL^3}{48} + \frac{BL^2}{8} + \frac{CL}{2} + D = 0,$$

$$(4) \quad EIv_2|_{x=L/2} = \frac{GL^3}{48} + \frac{HL^2}{8} + \frac{JL}{2} + K = 0,$$

$$(5) \quad EIv_1'|_{x=L/2} = EIv_2'|_{x=L/2}:$$

$$\frac{w_0L^3}{192} + \frac{AL^2}{8} + \frac{BL}{2} + C = \frac{GL^2}{8} + \frac{HL}{2} + J,$$

$$(6) \quad EIv_1''|_{x=L/2} = EIv_2''|_{x=L/2}:$$

$$\frac{w_0L^2}{24} + \frac{AL}{2} + B = \frac{GL}{2} + H.$$

The boundary conditions at $x = L$ are

$$(7) \quad EIv_2''|_{x=L} = GL + H = 0,$$

$$(8) \quad EIv_2'''|_{x=L} = G = -F.$$

Solving these equations, we obtain the deflections

$$v_1 = \frac{Fx}{6EI}\left[\frac{w_0 x^4}{10FL} - \left(\frac{w_0 L}{12F} - 1\right)x^2 + \left(\frac{7w_0 L^3}{480F} - \frac{L^2}{4}\right)\right],$$

$$v_2 = \frac{F}{2EI}\left[-\frac{x^3}{3} + Lx^2 - \left(\frac{w_0 L^3}{180F} + \frac{7L^2}{12}\right)x + \left(\frac{L^3}{12} + \frac{w_0 L^4}{360F}\right)\right].$$

In Fig. (c) a diminsionless measure of the deflection is plotted as a function of x/L for $w_0 L/F = 1$.

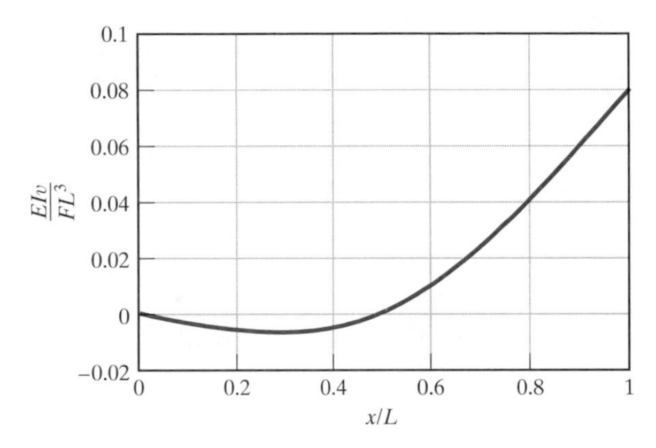

(c) Graph of the deflection.

9-4 | Method of Superposition

Determining deflections of beams is a time-consuming process even when the loads are relatively simple. For the convenience of structural engineers, deflections of prismatic beams with typical supports and simple loads are available in tables such as the one in Appendix E. Furthermore, we will show that the solutions in such tables can be superimposed to obtain deflections of beams with more complicated loads.

Consider the beam and loading in Fig. 9-17a. Let the bending moment in the beam be denoted by M_a, and let v_a be its deflection. The bending moment and deflection satisfy Eq. (9-4):

$$v_a'' = -\frac{M_a}{EI}. \tag{9-25}$$

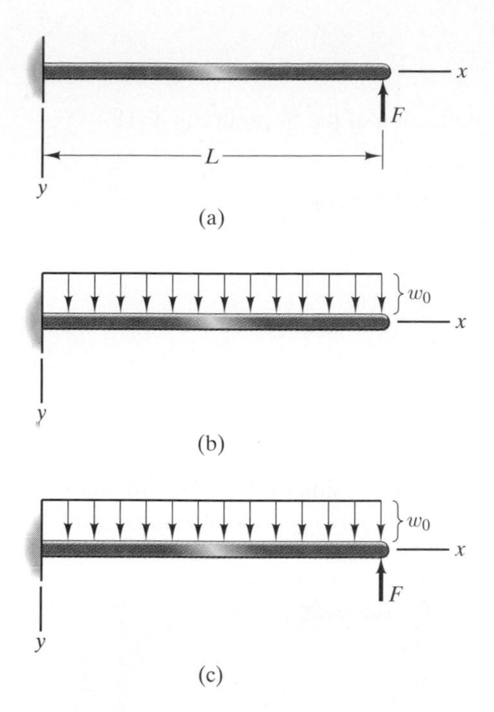

FIGURE 9-17 Superimposing the loads (a) and (b) results in (c).

In Fig. 9-17b we subject the same beam to a different load. Let M_b and v_b denote the resulting bending moment and deflection. They also satisfy Eq. (9-4):

$$v_b'' = -\frac{M_b}{EI}. \tag{9-26}$$

In Fig. 9-17c we superimpose the loads in Figs. 9-17a and b. Summing Eqs. (9-25) and (9-26), we conclude that the superimposed deflections and bending moments also satisfy Eq. (9-4):

$$(v_a + v_b)'' = -\frac{M_a + M_b}{EI}.$$

This result confirms that we can obtain the deflection resulting from the loads in Fig. 9-17c by summing the deflections resulting from the loads in Fig. 9-17a and b. Using the deflections given in Appendix E, we obtain

$$v_a + v_b = \frac{-Fx^2}{6EI}(3L - x) + \frac{w_0 x^2}{24EI}(6L^2 - 4Lx + x^2).$$

The reduction in effort achieved by this approach is evident. Notice that to determine the deflection of a given beam by superposition, the loading must be matched by the superimposed loads and the boundary conditions must be satisfied by the superimposed displacements.

EXAMPLE 9-5

Use superposition to determine the deflection of the beam in Fig. 9-18.

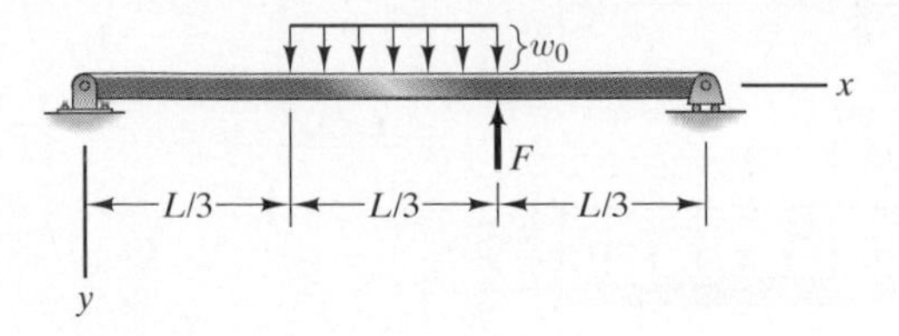

FIGURE 9-18

Strategy

We obtain this loading by superimposing the loads in Figs. (a), (b), and (c), so we can determine the deflection by superimposing the deflections for these three loads.

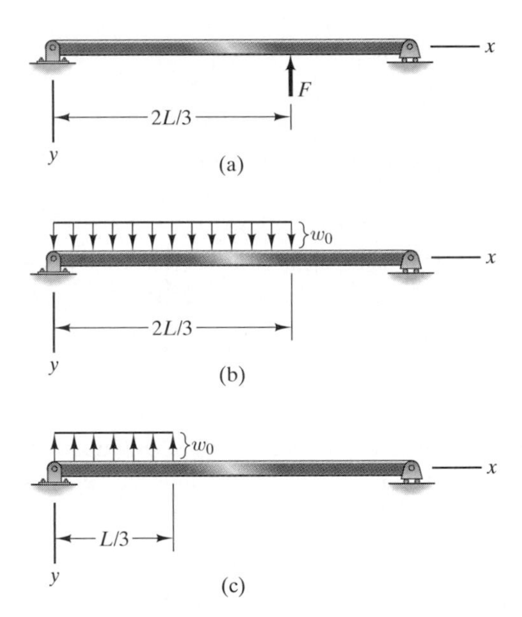

(a),(b),(c) Three loads that can be superimposed to obtain the desired load.

Solution

Using the results in Appendix E, the deflections due to the loads in Figs. (a), (b), and (c) are

$$v_a = \begin{cases} \dfrac{-Fx}{162EI}(8L^2 - 9x^2), & 0 \le x \le 2L/3 \\[2ex] \dfrac{-F}{162EI}(18x^3 - 54Lx^2 + 44L^2x - 8L^3), & 2L/3 \le x \le L \end{cases}$$

$$v_b = \begin{cases} \dfrac{w_0 x}{1944EI}(64L^3 - 144Lx^2 + 81x^3), & 0 \le x \le 2L/3 \\ \dfrac{w_0 L}{1944EI}(72x^3 - 216Lx^2 + 160L^2x - 16L^3), & 2L/3 \le x \le L \end{cases}$$

$$v_c = \begin{cases} \dfrac{-w_0 x}{1944EI}(25L^3 - 90Lx^2 + 81x^3), & 0 \le x \le L/3 \\ \dfrac{-w_0 L}{1944EI}(18x^3 - 54Lx^2 + 37L^2x - L^3). & L/3 \le x \le L \end{cases}$$

Superimposing these results, we obtain the beam's deflection.

$0 \le x \le L/3$:

$$v = \frac{-Fx}{162EI}(8L^2 - 9x^2) + \frac{w_0 x}{1944EI}(39L^3 - 54Lx^2).$$

$L/3 \le x \le 2L/3$:

$$v = \frac{-Fx}{162EI}(8L^2 - 9x^2) + \frac{w_0}{1944EI}(81x^4 - 162Lx^3 + 54L^2x^2 + 27L^3x + L^4).$$

$2L/3 \le x \le L$:

$$v = \frac{-F}{162EI}(18x^3 - 54Lx^2 + 44L^2x - 8L^3) + \frac{w_0 L}{1944EI}(54x^3 - 162Lx^2 + 123L^2x - 15L^3).$$

A nondimensional measure of the beam's deflection when $F = w_0L/3$ is shown in Fig. (d).

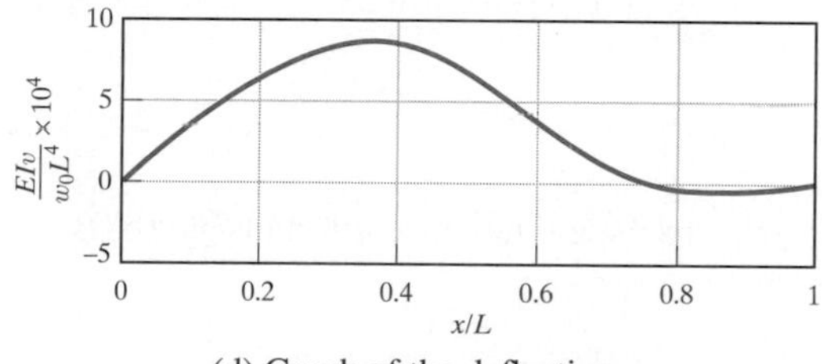

(d) Graph of the deflection.

EXAMPLE 9-6

Use superposition to determine the deflection of the beam in Fig. 9-19.

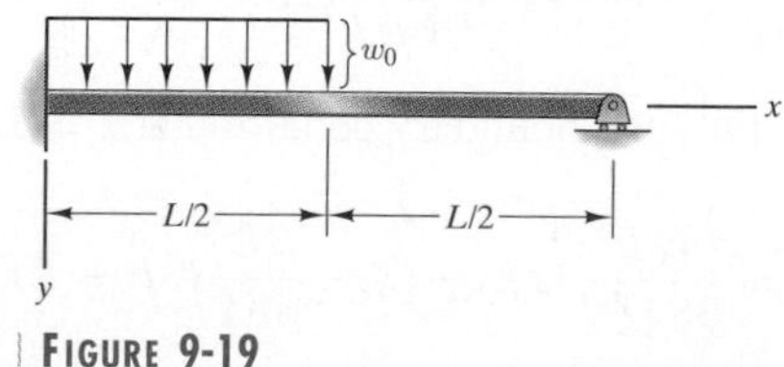

FIGURE 9-19

Strategy

The beam is statically indeterminate. If we superimpose the loads in Figs. (a) and (b), we can express the beam's deflection in terms of the unknown reaction B exerted by the roller support. Then we can determine B from the condition that the deflection equals zero at $x = L$.

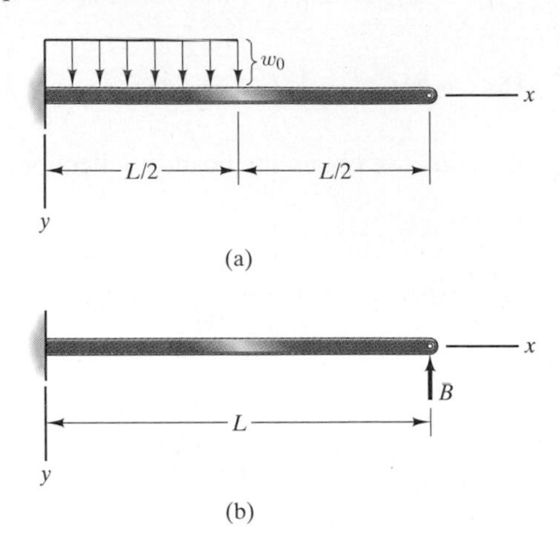

(a),(b) Superimposing two loads to determine the deflection in terms of B.

Solution

Using the results in Appendix E, the deflection due to the load in Fig. (a) is

$$v_a = \begin{cases} \dfrac{w_0 x^2}{48EI}(3L^2 - 4Lx + 2x^2), & 0 \le x \le L/2 \\ \dfrac{w_0 L^3}{384EI}(8x - L), & L/2 \le x \le L \end{cases}$$

and the deflection due to the unknown reaction in Fig. (b) is

$$v_b = \frac{-Bx^2}{6EI}(3L - x).$$

The beam's deflection in terms of B is

$$v = v_a + v_b = \begin{cases} \dfrac{w_0 x^2}{48EI}(3L^2 - 4Lx + 2x^2) - \dfrac{Bx^2}{6EI}(3L - x), & 0 \le x \le L/2 \\ \dfrac{w_0 L^3}{384EI}(8x - L) - \dfrac{Bx^2}{6EI}(3L - x). & L/2 \le x \le L \end{cases}$$

To determine B, we apply the boundary condition at $x = L$:

$$v|_{x=L} = \frac{w_0 L^3}{384EI}(8L - L) - \frac{BL^2}{6EI}(3L - L) = 0.$$

Solving, we obtain $B = 7w_0L/128$. Substituting this result, the beam's deflection is

$$v = \begin{cases} \dfrac{w_0x^2}{768EI}(27L^2 - 57Lx + 32x^2), & 0 \le x \le L/2 \\ \dfrac{w_0L}{768EI}(7x^3 - 21Lx^2 + 16L^2x - 2L^3). & L/2 \le x \le L \end{cases}$$

Chapter Summary

Determination of the Deflection

Let v be the deflection of a beam's neutral axis relative to the x axis, and let θ be the angle between the neutral axis and the x axis [Fig. (a)]. For small deflections, v and θ are related by

$$\frac{dv}{dx} = \theta. \qquad \text{Eq. (9-1)}$$

The deflection is related to the bending moment by

$$v'' = -\frac{M}{EI}, \qquad \text{Eq. (9-4)}$$

where the primes denote derivatives with respect to x. Determining a beam's deflection using Eq. (9-4) requires three steps:

1. *Determine the bending moment as a function of x in terms of the loads and reactions acting on the beam. This may result in two or more functions $M_1, M_2 \ldots$, each of which applies to a different segment of the beam's length.*
2. *For each segment, integrate Eq. (9-4) twice to determine $v_1, v_2, \ldots$. If there are N segments, this step will result in 2N unknown integration constants.*

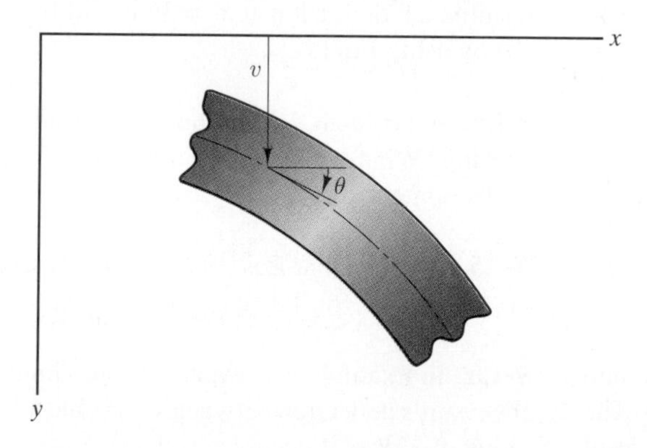

(a)

3. *Use the boundary conditions to determine the integration constants and, if the beam is statically indeterminate, the unknown reactions. Common boundary conditions are summarized in Fig. 9-8.*

Fourth-Order Equation

An alternative method of determining beam deflections is by integration of the equation

$$v'''' = \frac{w}{EI}. \qquad \text{Eq. (9-22)}$$

When this equation is used, the boundary conditions are expressed in terms of the dependent variable v and its first, second, and third derivatives. Because of the relations

$$M = -EIv'', \qquad \text{Eq. (9-23)}$$

$$V = -EIv''', \qquad \text{Eq. (9-24)}$$

the boundary conditions can be expressed in terms of the deflection v, the slope v', the bending moment M, and the shear force V. Some common boundary conditions for applying the fourth-order equation are summarized in Fig. 9-14.

Method of Superposition

Deflections of prismatic beams with typical supports and simple loads are available in tables such as the one in Appendix E. The solutions in such tables can be superimposed to obtain deflections of beams with more complicated loads. To determine the deflection of a given beam by superposition, the loading must be matched by the superimposed loads and the boundary conditions must be satisfied by the superimposed displacements.

PROBLEMS

9-1.1. The beam in Fig. 9-4 has length $L = 4$ m and the magnitude of the distributed load is $w_0 = 12$ kN/m. The beam has a solid circular cross section with 80-mm radius and is made of material with modulus of elasticity $E = 200$ GPa. Determine the beam's deflection and slope at $x = 1$ m.

9-1.2. In Problem 9-1.1, what is the beam's maximum deflection? What is the slope of the beam where the maximum deflection occurs?

9-1.3. The beam in Fig. 9-6a has length $L = 72$ in. and the magnitude of the distributed load is $w_0 = 14$ lb/in. The beam's moment of inertia $I = 2.4$ in^4 and it is made of material with modulus of elasticity $E = 30 \times 10^6$ psi. Determine the deflection at $x = 36$ in. **(a)** by using Eq. (9-6); **(b)** by using Eq. (9-7).

9-1.4. In Problem 9-1.3, what is the beam's maximum deflection? What is the slope of the beam where the maximum deflection occurs?

9-1.5. Confirm that Eqs. (9-6) and (9-7) satisfy the boundary conditions for the beam in Fig. 9-6a.

9-1.6. In Example 9-1, what is the maximum magnitude of the beam's deflection between $x = 0$ and $x = 2$ m? At what value of x does it occur?

For the beams shown in Problems 9-1.7–9-1.19, determine the deflection v as a function of x and confirm the results in Appendix E.

9-1.7.

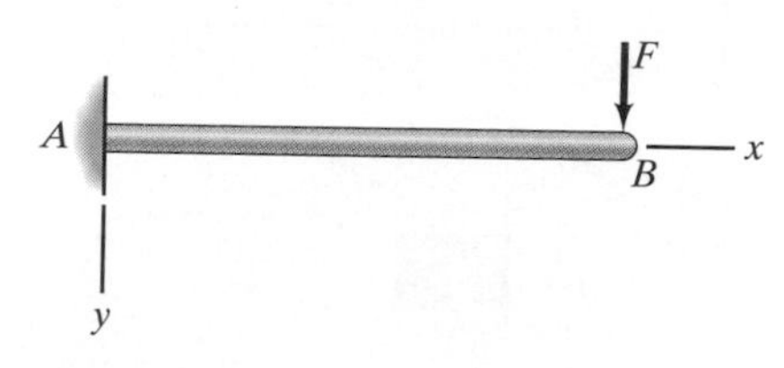

PROBLEM 9-1.7

9-1.8.

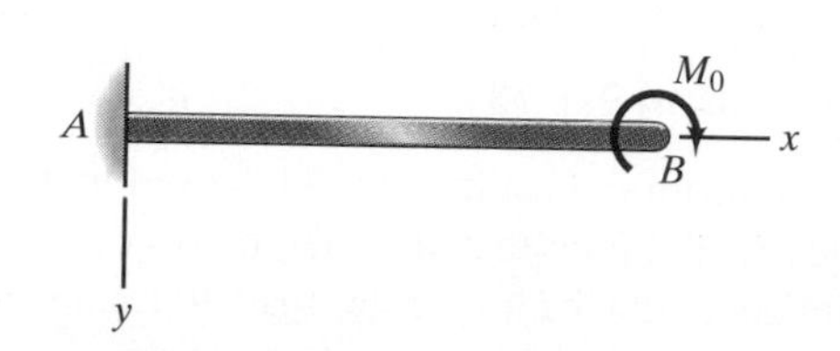

PROBLEM 9-1.8

9-1.9.

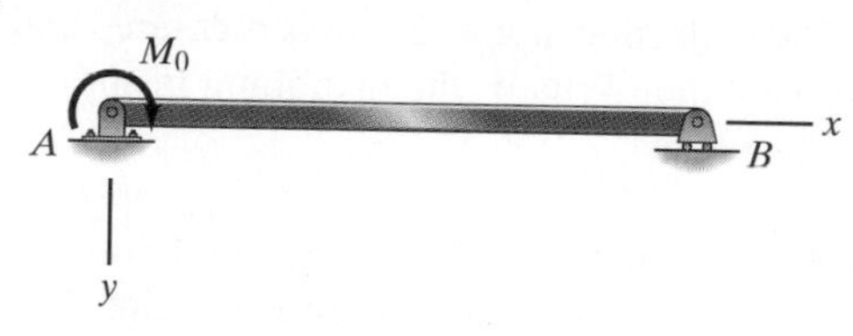

PROBLEM 9-1.9

9-1.10.

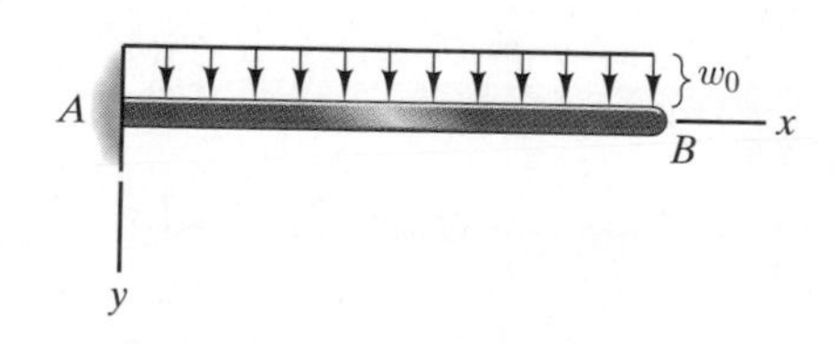

PROBLEM 9-1.10

9-1.11.

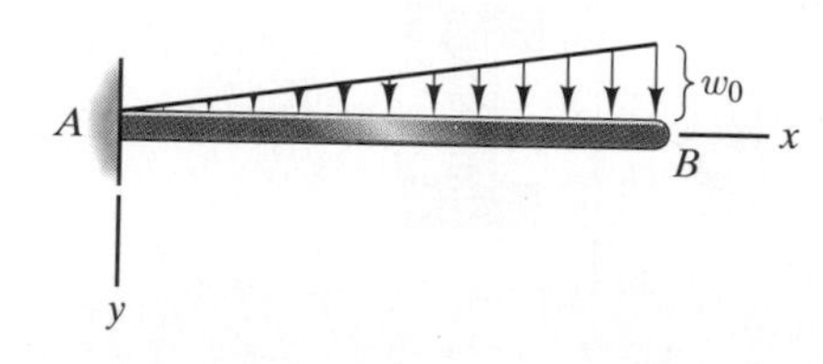

PROBLEM 9-1.11

9-1.12.

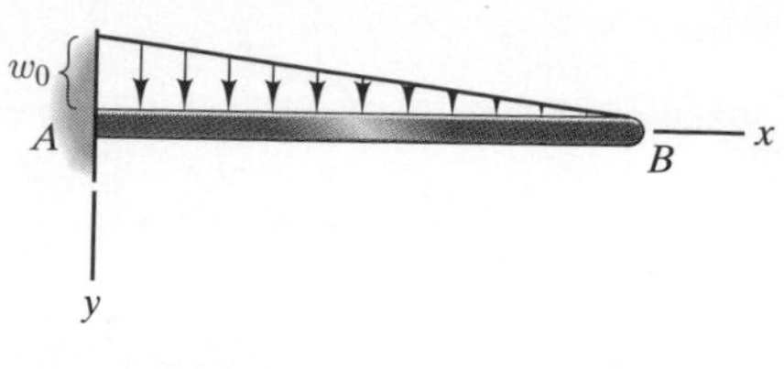

PROBLEM 9-1.12

9-1.13.

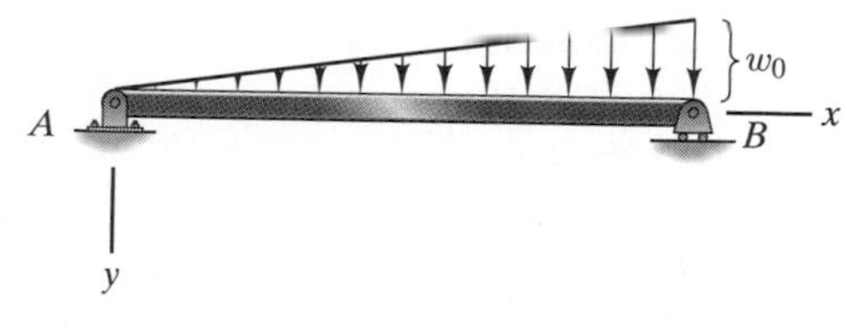

PROBLEM 9-1.13

9-1.14.

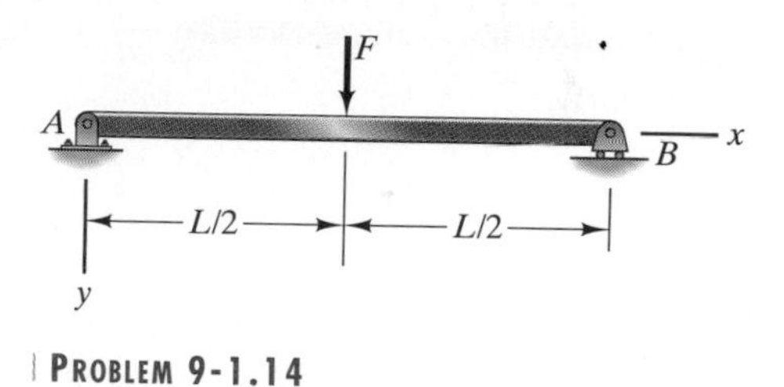

PROBLEM 9-1.14

9-1.15.

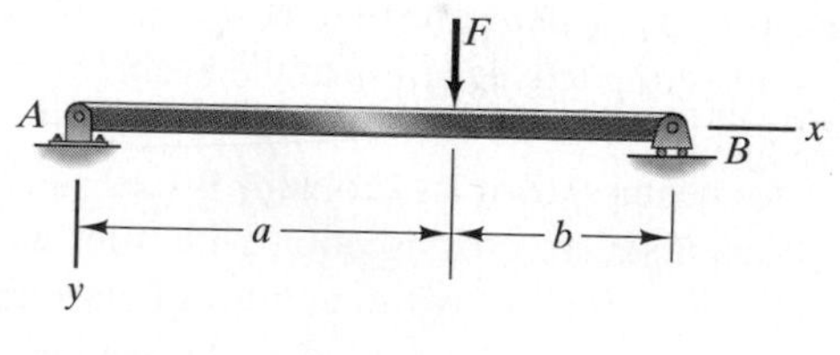

PROBLEM 9-1.15

9-1.16.

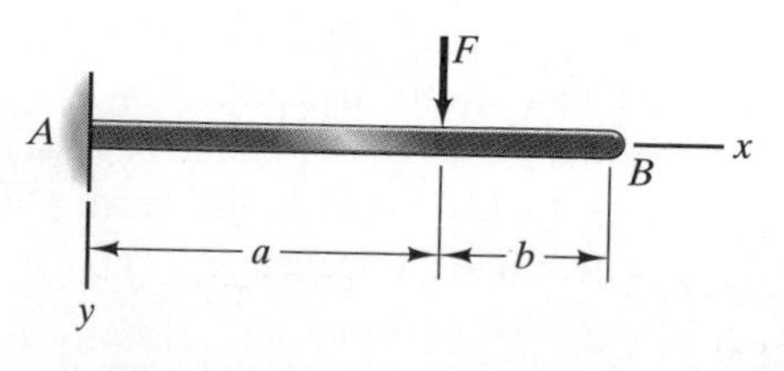

PROBLEM 9-1.16

9-1.17.

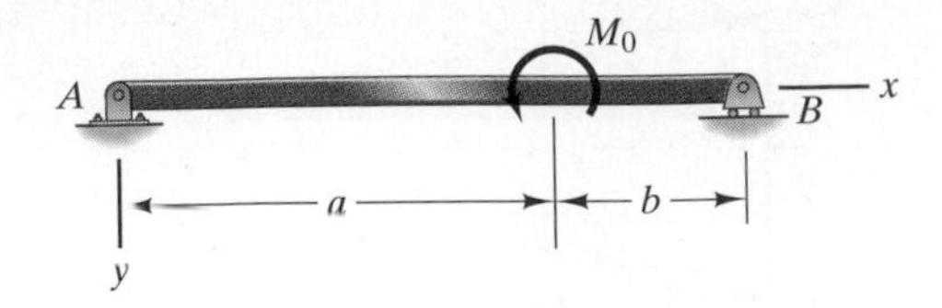

| PROBLEM 9-1.17

9-1.18.

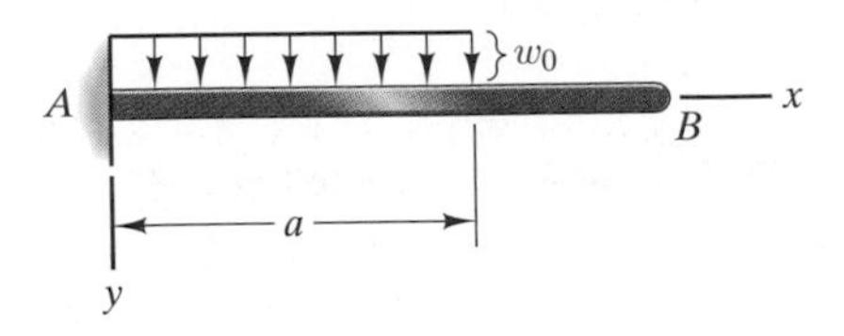

| PROBLEM 9-1.18

9-1.19.

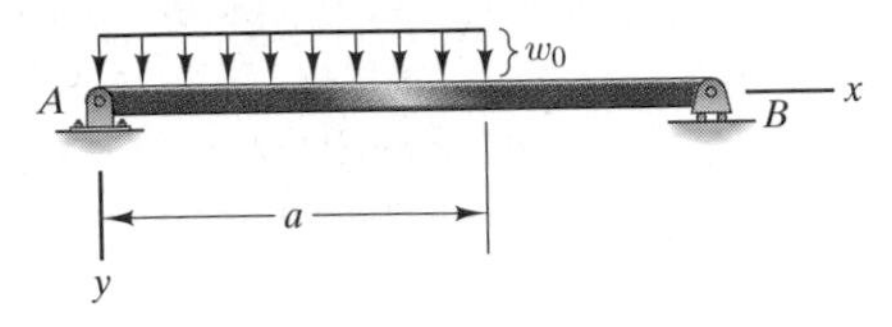

| PROBLEM 9-1.19

9-1.20. The beam shown in Problem 9-1.7 is 2 m in length. It has a solid circular cross section with 20-mm radius and is made of aluminum alloy with modulus of elasticity $E = 70$ GPa. When the force F is applied, the deflection at B is measured and determined to be 4.8 mm. What is the maximum resulting tensile stress in the beam?

9-1.21. The beam shown in Problem 9-1.10 is 96 in. in length. It has a square cross section with 4-in. width and is made of titanium alloy with modulus of elasticity $E = 15 \times 10^6$ psi. When the distributed load is applied, the deflection at B is measured and determined to be 2 in. What is the maximum resulting tensile stress in the beam?

9-1.22. The beam shown in Problem 9-1.15 is 3 m in length and $a = 2$ m. Its moment of inertia is $I = 2 \times 10^{-5}$ m^4 and it is made of material with modulus of elasticity $E = 120$ GPa. If $F = 15$ kN, what is the beam's maximum deflection, and where does it occur?

9-1.23. The beam consists of material with elastic modulus $E = 190$ GPa. Determine the axial position x at which the magnitude of the deflection due to the couple M_0 is a maximum.

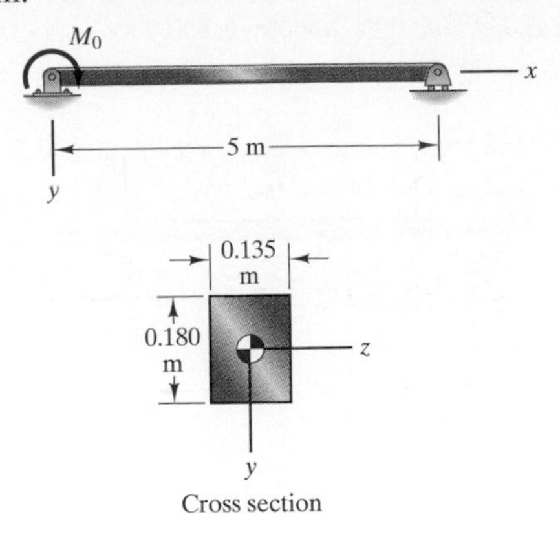

| PROBLEM 9-1.23

9-1.24. The beam in Problem 9-1.23 is used in a structure whose design requires that the magnitude of the beam's maximum deflection be no greater than 20 mm. Determine the maximum couple M_0 that can be applied.

9-1.25. A couple M_0 is applied to the beam in Problem 9-1.23. The deflection at $x = 2.5$ m is measured and determined to be 5 mm. What is the maximum resulting tensile stress in the beam?

9-1.26. The titanium beam has elastic modulus $E = 16 \times 10^6$ psi. Determine the axial position x at which the magnitude of the deflection is a maximum.

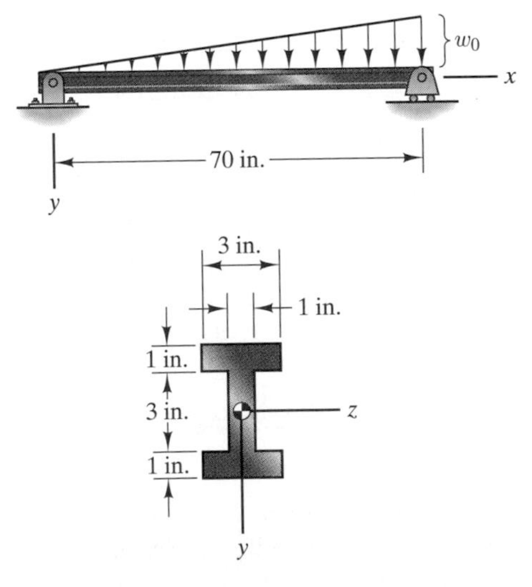

| PROBLEM 9-1.26

9-1.27. The beam in Problem 9-1.26 is used in a structure whose design requires that the magnitude of the beam's maximum deflection be no greater than 1.0 in. Determine the maximum value of w_0 that can be applied.

9-2.1. The prismatic beam has elastic modulus E and moment of inertia I. Determine the reactions at A and B.

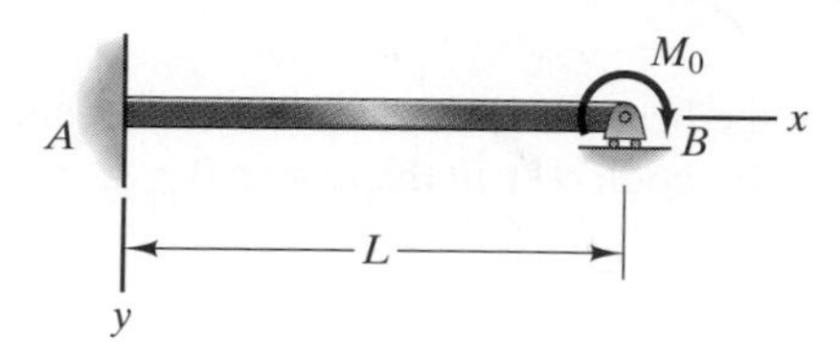

PROBLEM 9-2.1

9-2.2. Determine the deflection as a function of x for the beam in Problem 9-2.1.

9-2.3. The beam in Problem 9-2.1 has length $L = 4$ m, a solid circular cross section with 80-mm radius, and is made of material with modulus of elasticity $E = 200$ GPa. If the couple $M_0 = 2$ kN-m, what is the maximum tensile stress in the beam?

9-2.4. The prismatic beam has elastic modulus E and moment of inertia I. Determine the reactions at A and B.

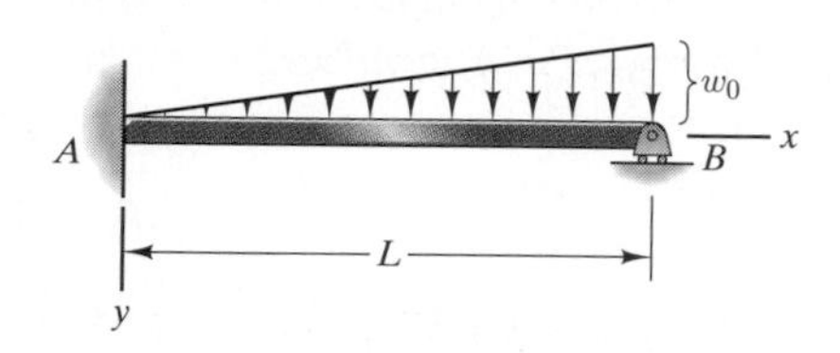

PROBLEM 9-2.4

9-2.5. Determine the deflection as a function of x for the beam in Problem 9-2.4.

9-2.6. The beam in Problem 9-2.4 has length $L = 60$ in., elastic modulus $E = 20 \times 10^6$ psi, moment of inertia $I = 12$ in^4, and $w_0 = 6000$ lb/in. **(a)** Draw a graph of the beam's deflection as a function of x. **(b)** Estimate the value of the maximum deflection and the axial position at which it occurs.

9-2.7. The prismatic beam has elastic modulus E and moment of inertia I. Determine the reactions at the left wall. (Assume that the supports exert no axial forces on the beam.)

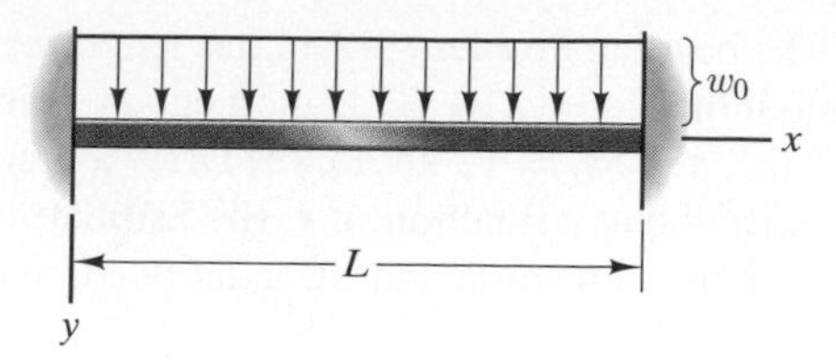

PROBLEM 9-2.7

9-2.8. Determine the deflection as a function of x for the beam in Problem 9-2.7.

9-2.9. The beam in Problem 9-2.7 has a square cross section with 4-in. width and modulus of elasticity $E = 15 \times 10^6$ psi. The beam's length is $L = 96$ in. and it is made of material that will safely support a normal stress of 110 ksi. Based on this criterion, what is the maximum safe value of w_0?

9-2.10. The beam in Problem 9-2.7 has the cross section shown. The beam's length is $L = 4$ m and it is made of material that will safely support a normal stress of 220 MPa. What is the maximum safe value of w_0?

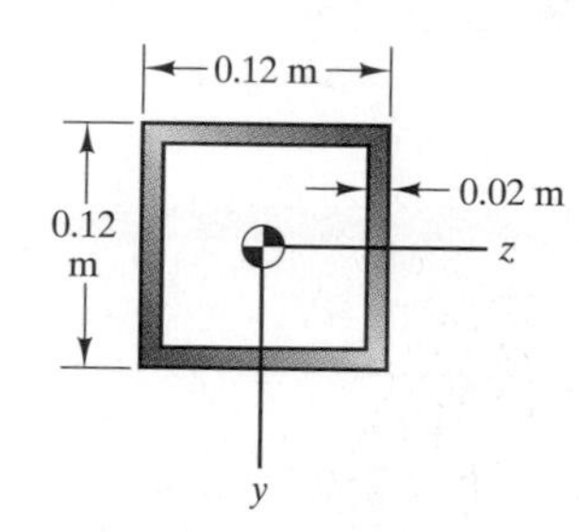

PROBLEM 9-2.10

9-2.11. The prismatic beam has elastic modulus E and moment of inertia I. Determine the reactions at the walls. (Assume that the supports exert no axial forces on the beam.)

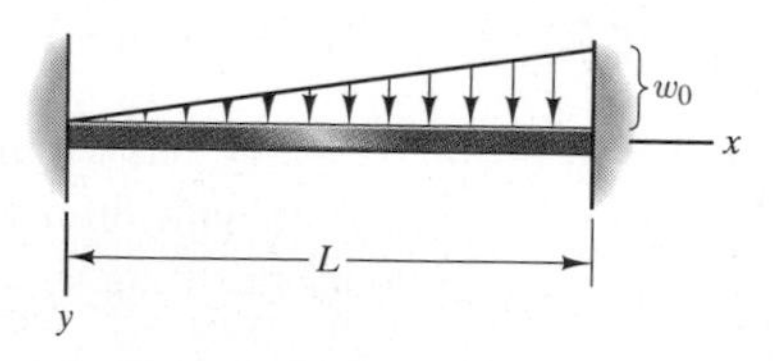

PROBLEM 9-2.11

9-2.12. Determine the deflection as a function of x for the beam in Problem 9-2.11.

9-2.13. The beam in Problem 9-2.11 has length $L = 10$ m, elastic modulus $E = 210$ GPa, moment of inertia $I = 8 \times 10^{-6}$ m^4, and $w_0 = 12$ kN/m. **(a)** Draw a graph of the beam's deflection as a function of x. **(b)** Estimate the value of the maximum deflection and the axial position at which it occurs.

9-2.14. The length of the beam in Example 9-2 is $L = 4$ m and $w_0 = 5$ kN/m. The beam consists of material with modulus of elasticity $E = 72$ GPa and the moment of inertia of its cross section is $I = 1.6 \times 10^{-7}$ m^4. Determine the beam's deflection at $x = 1$ m.

9-2.15. The length of the beam in Example 9-2 is $L = 4$ m and $w_0 = 5$ kN/m. The beam consists of material with modulus of elasticity $E = 72$ GPa and the moment of inertia of its cross section is $I = 1.6 \times 10^{-7}$ m^4. What is the maximum magnitude of the beam's deflection between $x = 0$ and $x = 2$ m? At what value of x does it occur?

9-2.16. The prismatic beam has elastic modulus E and moment of inertia I. Determine the reactions at A and B.

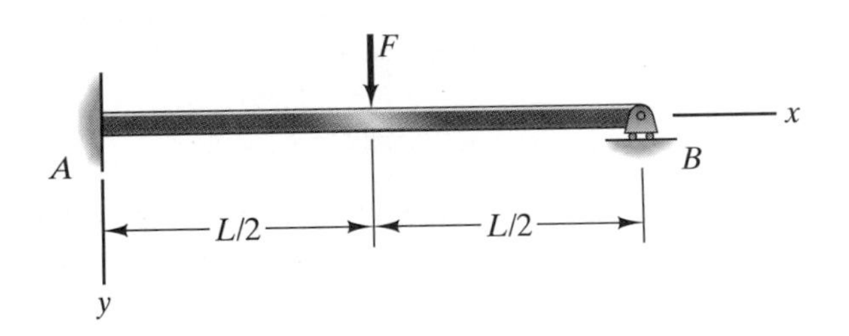

| **PROBLEM 9-2.16**

9-2.17. For the beam in Problem 9-2.16, determine the deflection as a function of x in the region $0 \leq x \leq L/2$.

9-2.18. For the beam in Problem 9-2.16, determine the deflection as a function of x in the region $L/2 \leq x \leq L$.

9-2.19. The beam in Problem 9-2.16 has length $L = 4$ m and a solid circular cross section with 80-mm radius. If $F = 5$ kN, what is the maximum resulting tensile stress in the beam?

9-2.20. The prismatic beam has elastic modulus E and moment of inertia I. Determine the reactions at A and B. (Assume that the supports exert no axial forces on the beam.)

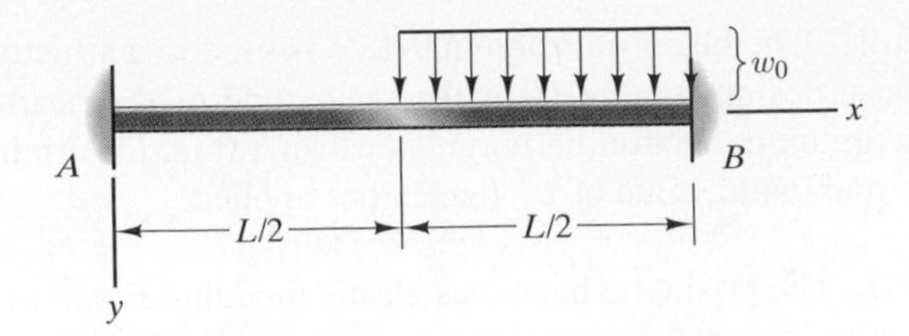

| **PROBLEM 9-2.20**

9-2.21. For the beam in Problem 9-2.20, determine the deflection as a function of x in the region $0 \leq x \leq L/2$.

9-2.22. For the beam in Problem 9-2.20, determine the deflection as a function of x in the region $L/2 \leq x \leq L$.

9-2.23. The beam in Problem 9-2.20 has length $L = 6$ m. It has a square cross section with 60-mm width and is made of material that will safely support a normal stress of 280 MPa. Based on this criterion, what is the maximum safe value of w_0?

9-3.1. If you use Eq. (9-22) to determine the deflection of the beam in Fig. 9-4, what are the boundary conditions? What is the deflection?

For the beams shown in Problems 9-3.2–9-3.9, use Eq. (9-22) to determine the deflection v as a function of x and confirm the results in Appendix E.

9-3.2.

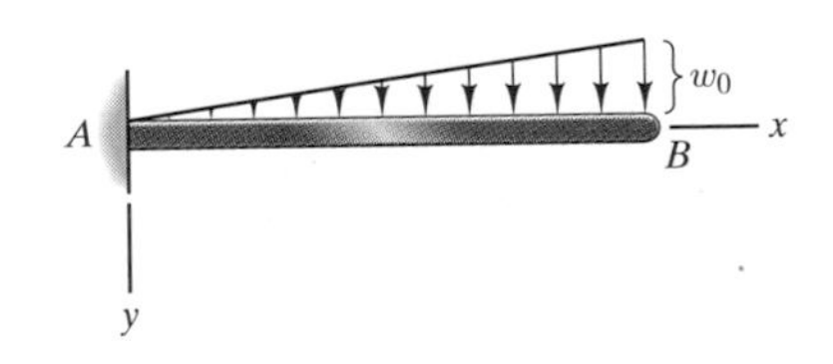

| **PROBLEM 9-3.2**

9-3.3.

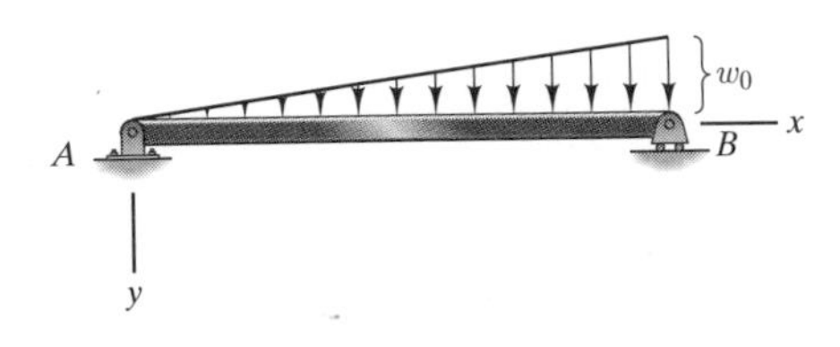

| **PROBLEM 9-3.3**

9-3.4.

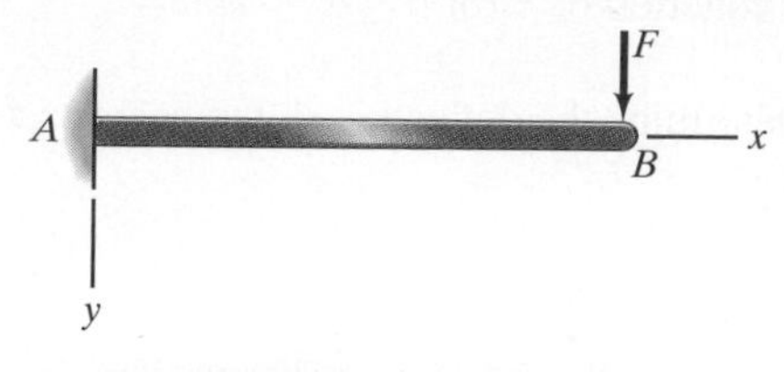

PROBLEM 9-3.4

9-3.5.

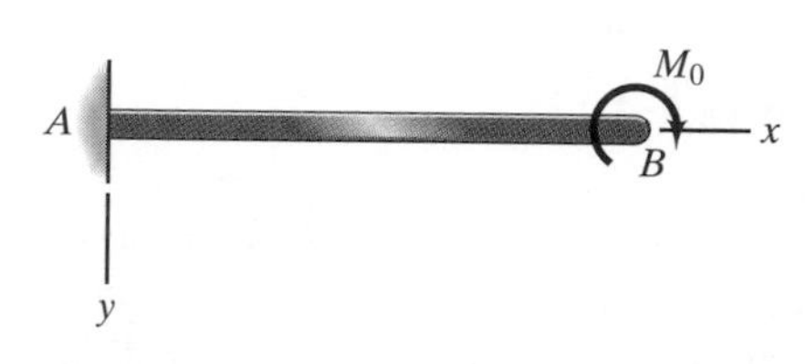

PROBLEM 9-3.5

9-3.6.

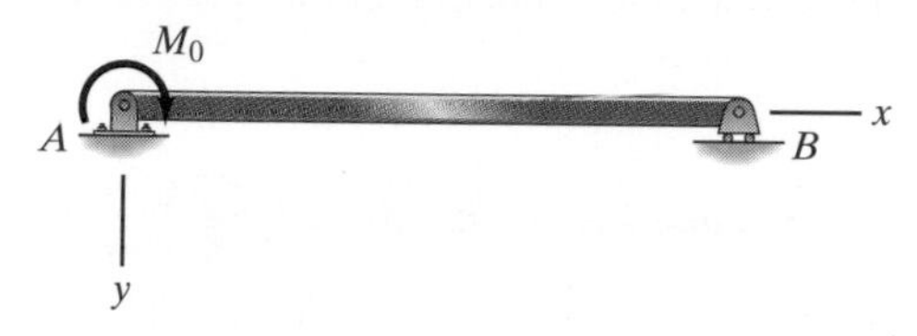

PROBLEM 9-3.6

9-3.7.

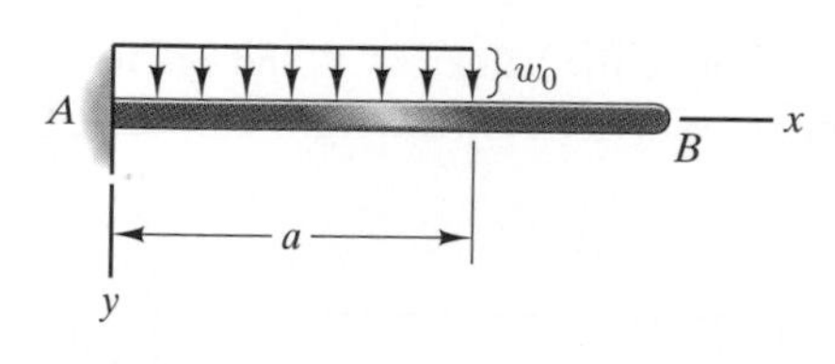

PROBLEM 9-3.7

9-3.8.

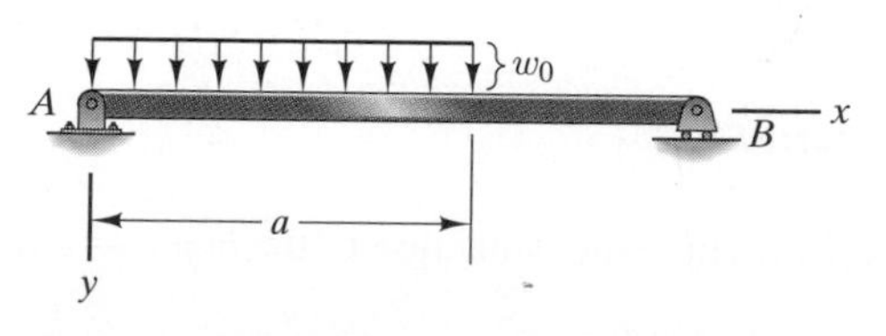

PROBLEM 9-3.8

9-3.9.

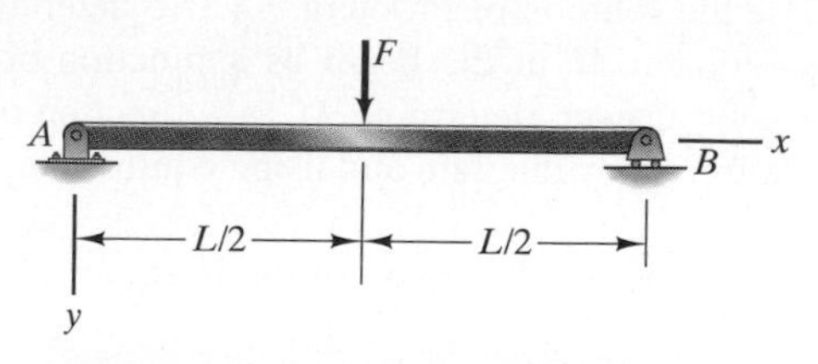

PROBLEM 9-3.9

9-3.10. Use Eq. (9-22) to determine the deflection of the right end of the beam in Example 9-1. Compare the boundary conditions you use to those used in Example 9-1.

9-3.11. Use Eq. (9-22) to solve Problem 9-1.23.

9-3.12. Use Eq. (9-22) to solve Problem 9-2.1.

9-3.13. Use Eq. (9-22) to solve Problem 9-2.4.

9-3.14. Use Eq. (9-22) to determine the deflection as a function of x for the beam in Problem 9-2.7.

9-3.15. Use Eq. (9-22) to determine the deflection as a function of x for the beam in Problem 9-2.11.

9-3.16. Use Eq. (9-22) to solve Problem 9-2.16.

Solve Problems 9-4.1–9-4.24 by superposition using the results in Appendix E. Assume that beams are prismatic with elastic modulus E and moment of inertia I.

9-4.1. Determine the deflection of the beam as a function of x.

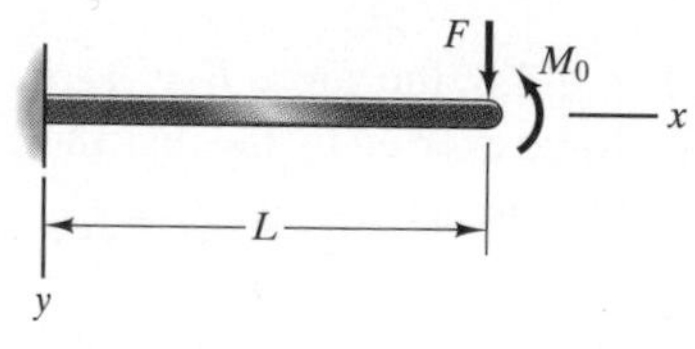

PROBLEM 9-4.1

9-4.2. For the beam in Problem 9-4.1, determine the value of the couple M_0 for which the right end of the beam has no deflection.

9-4.3. For the beam in Problem 9-4.1, determine the value of the couple M_0 for which the slope of the right end of the beam is zero.

9-4.4. Use the solution of Problem 9-4.1 to determine the bending moment M in the beam as a function of x. To confirm your answer, determine M as a function of x by drawing a free-body diagram and using equilibrium.

9-4.5. Determine the deflection of the beam as a function of x.

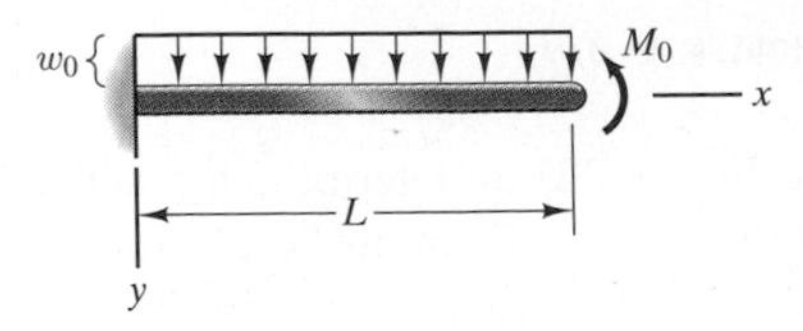

PROBLEM 9-4.5

9-4.6. For the beam in Problem 9-4.5, determine the value of the couple M_0 for which the right end of the beam has no deflection.

9-4.7. For the beam in Problem 9-4.5, determine the value of the couple M_0 for which the slope of the right end of the beam is zero.

9-4.8. Determine the deflection of the beam as a function of x for $0 \le x \le L/2$.

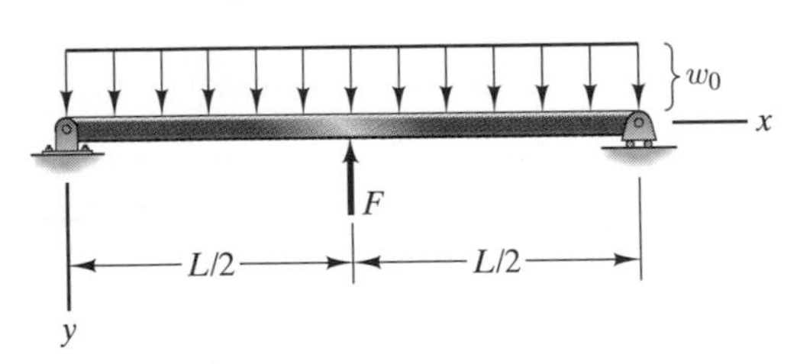

PROBLEM 9-4.8

9-4.9. In Problem 9-4.8, the force $F = w_0 L$. That is, F is equal to the total force exerted by the distributed load. What is the beam's deflection at $x = L/2$?

9-4.10. Determine the reactions at A, B, and C.

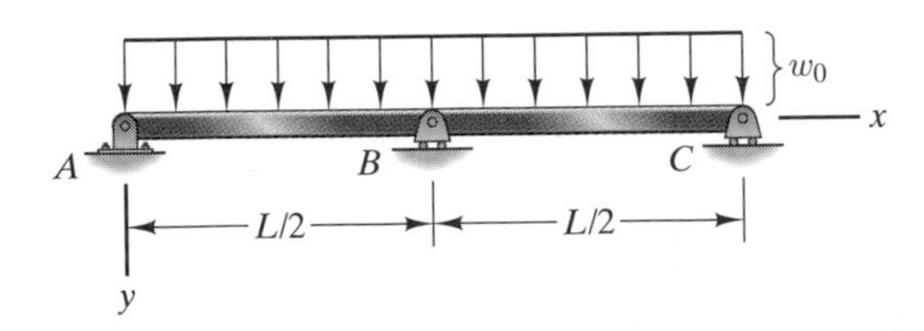

PROBLEM 9-4.10

9-4.11. In Problem 9-4.10, determine the deflection in the beam as a function of x for $0 \le x \le L/2$.

9-4.12. Determine the deflection of the beam as a function of x for $0 \le x \le L/3$.

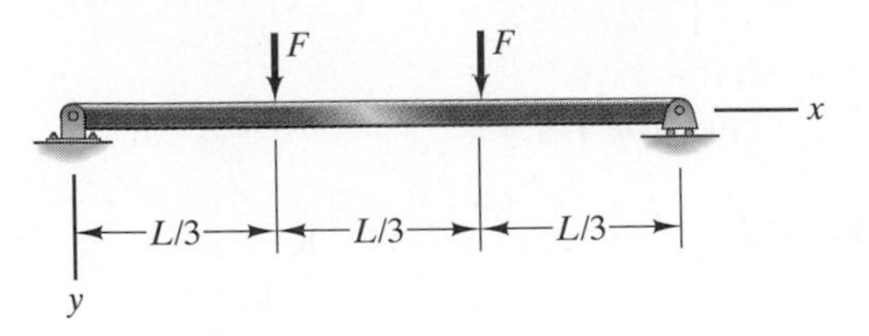

PROBLEM 9-4.12

9-4.13. The beam in Problem 9-4.12 consists of material with modulus of elasticity $E = 72$ GPa, the moment of inertia of its cross section is $I = 1.6 \times 10^{-7}$ m^4, and its length is $L = 6$ m. If $F = 100$ N, what is the deflection at $x = 3$ m?

9-4.14. Determine the deflection of the beam as a function of x.

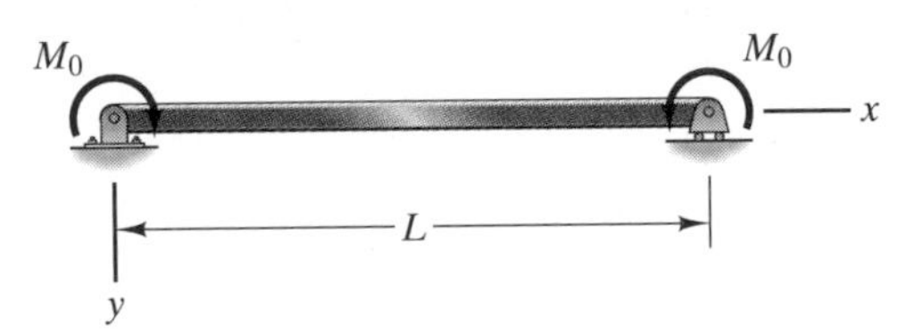

PROBLEM 9-4.14

9-4.15. Determine the values of F and M_0 for which both the deflection and slope of the right end of the beam are zero.

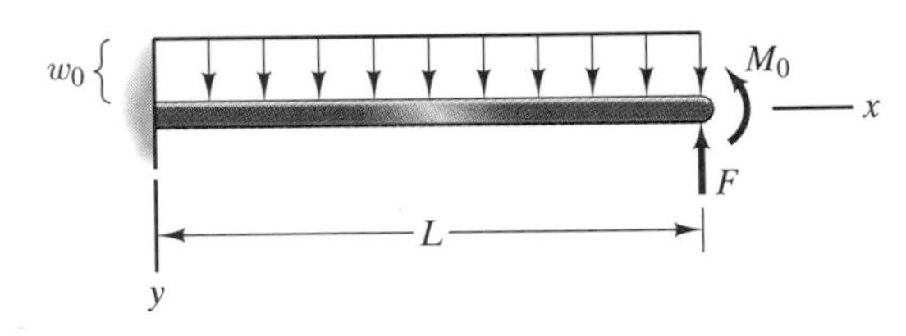

PROBLEM 9-4.15

9-4.16. Determine the deflection of the beam as a function of x.

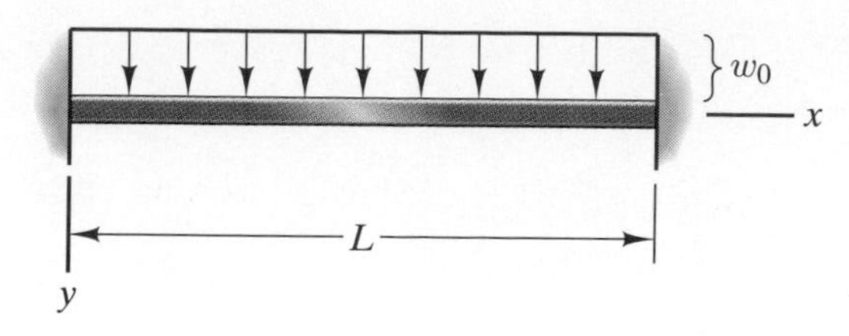

PROBLEM 9-4.16

9-4.17. Use the solution of Problem 9-4.16 to determine the bending moment M in the beam as a function of x.

9-4.18. Use superposition to solve Problem 9-2.1.

9-4.19. Use superposition to solve Problem 9-2.4.

9-4.20. Use superposition to solve Problem 9-2.16.

9-4.21. What force is exerted on the beam by the spring?

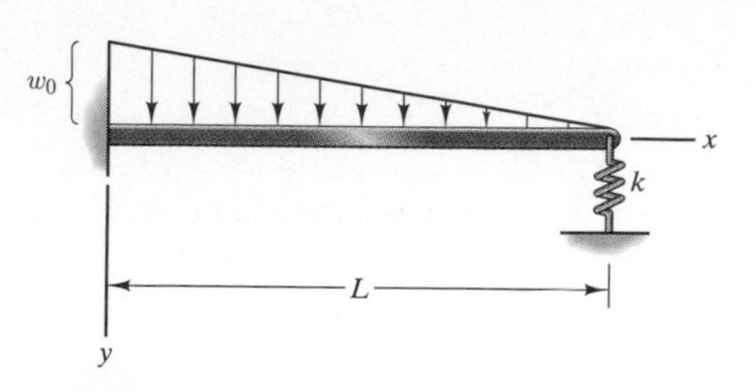

PROBLEM 9-4.21

9-4.22. Determine the reactions at B and C.

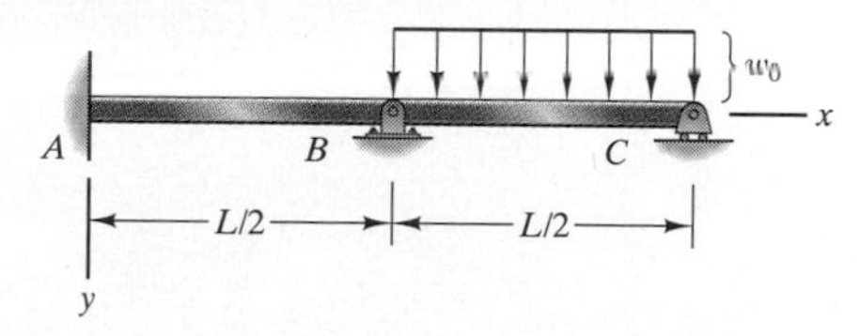

PROBLEM 9-4.22

9-4.23. In Problem 9-4.22, determine the deflection in the beam as a function of x for $0 \leq x \leq L/2$.

9-4.24. In Problem 9-4.22, determine the deflection in the beam as a function of x for $L/2 \leq x \leq x$.

The facade of the John F. Kennedy Center for the Performing Arts in Washington, DC (Edward Durrel Stone, 1971) is dominated by columns that are both functional and aesthetic.

CHAPTER 10

Buckling of Columns

Beams subjected to compressive loads must be designed so that they do not fail by buckling laterally. We analyze the buckling of axially loaded beams in this chapter. Because load-bearing columns of buildings must support large compressive axial loads, this subject is traditionally called *buckling of columns*.

10-1 | Euler Buckling Load

Suppose that a beam with the dimensions shown in Fig. 10-1 is subjected to a compressive axial load P. If the material is steel with yield stress $\sigma_Y = 520$ MPa, the value of P that can be applied without exceeding the yield stress is

$$\begin{aligned} P &= \sigma_Y A \\ &= (520 \times 10^6)(0.012)(0.0005) \\ &= 3120 \text{ N} \quad (701 \text{ lb}). \end{aligned}$$

The beam we have described is a common hacksaw blade. If you hold one between your palms (Fig. 10-2a) and exert an increasing compressive force, it will quickly collapse into a bowed shape (Fig. 10-2b). Although the axial force you exert is tiny compared to the force necessary to exceed the yield stress of the material, the blade fails as a structural element; it will not support the compressive load you exert.

Clearly, the criterion for preventing failure that we have used until now—making sure that loads do not cause the yield stress of the material (or some other

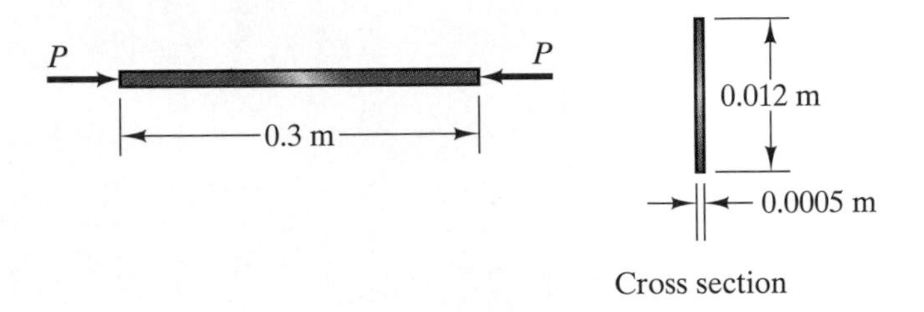

FIGURE 10-1 Applying compressive axial load to a beam.

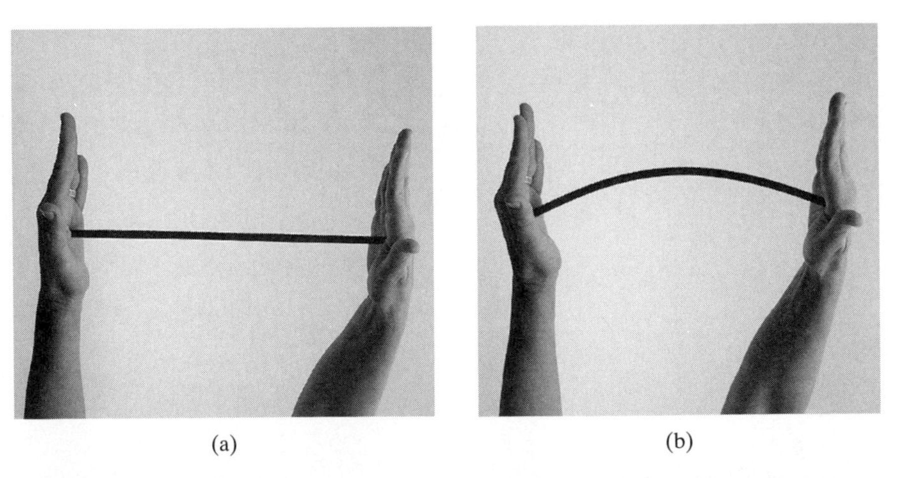

FIGURE 10-2 (a) Applying compressive axial load to a hacksaw blade. (b) A small compressive load causes the blade to collapse into a bowed shape.

defined allowable stress) to be exceeded—does not apply in this situation. The hacksaw blade fails by geometric instability, or *buckling*. Buckling can occur whenever a slender structural member—a thin beam or a thin-walled plate—is subjected to compression.

In this section we derive the buckling load for a prismatic beam subjected to axial forces at the ends. We begin by assuming that the beam has already buckled (Fig. 10-3a) and seek to determine the value of P necessary to hold it in equilibrium. We can accomplish this by proceeding to determine the distribution of the beam's deflection in terms of P. In Fig. 10-3b we introduce a coordinate system and obtain a free-body diagram by passing a plane through the beam at an arbitrary position x. Solving for the bending moment yields $M = Pv$, where v is the beam's deflection at x. We substitute this expression into Eq. (9-4), $v'' = -M/EI$, and write the resulting equation as

$$v'' + \lambda^2 v = 0, \qquad \textbf{(10-1)}$$

where

$$\lambda^2 = \frac{P}{EI}. \qquad \textbf{(10-2)}$$

The general solution of the second-order differential equation (10-1) is

$$v = B \sin \lambda x + C \cos \lambda x, \qquad \textbf{(10-3)}$$

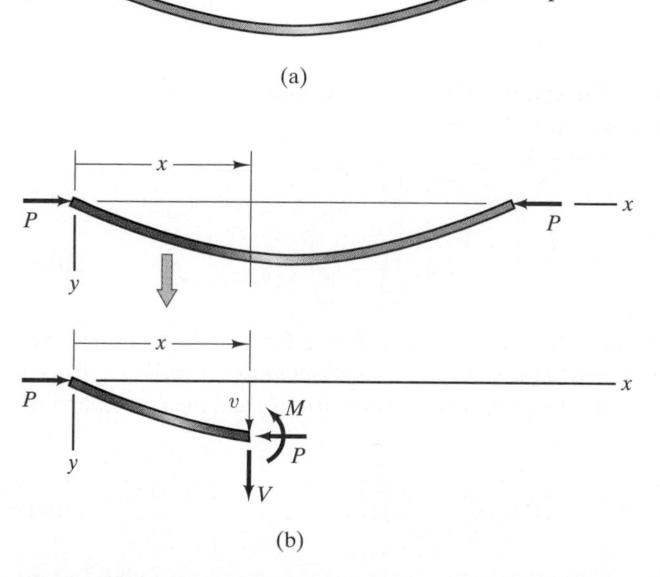

FIGURE 10-3 (a) Buckled beam in equilibrium. (b) Determining the bending moment as a function of x.

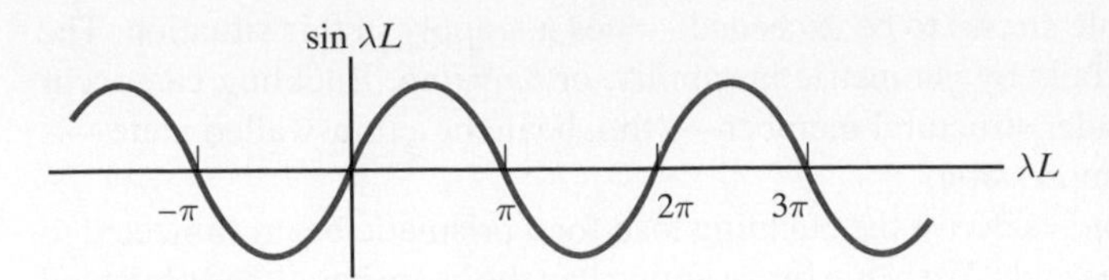

FIGURE 10-4 Roots of $\sin \lambda L = 0$.

where B and C are constants we must determine from the boundary conditions. [You should confirm that this expression satisfies Eq. (10-1).] From the boundary condition that the deflection equals zero at $x = 0$, $v|_{x=0} = 0$, we see that $C = 0$. The deflection is also zero at $x = L$.

$$v|_{x=L} = 0:$$
$$B \sin \lambda L = 0.$$

This boundary condition is satisfied if $B = 0$, but in that case the solution for the deflection reduces to $v = 0$. [Notice that this solution does indeed satisfy Eq. (10-1) and the boundary conditions, but we are seeking the buckled solution.] If $B \neq 0$, the second boundary condition requires that

$$\sin \lambda L = 0. \tag{10-4}$$

The parameter λ depends on P, so it is from this condition that we can determine the axial load. Here something interesting happens. As Fig. 10-4 indicates, Eq. (10-4) has not one but an infinite number of roots for λL. It is satisfied if

$$\lambda = \frac{n\pi}{L},$$

where n is any integer. Substituting this expression into Eqs. (10-2) and (10-3), we obtain the axial load and deflection:

$$P = \frac{n^2\pi^2 EI}{L^2}, \tag{10-5}$$

$$v = B \sin \frac{n\pi x}{L}. \tag{10-6}$$

What are all these solutions? Consider the solution for $n = 1$. As x increases from O to L, the argument of the sine in Eq. (10-6) increases from 0 to π. This is the buckled solution we have been seeking (Fig. 10-5a), and the corresponding value of P is

$$P = \frac{\pi^2 EI}{L^2}. \tag{10-7}$$

Notice that we have not determined the value of B, which is the beam's maximum deflection. The load P given by Eq. (10-7) is the force necessary to hold

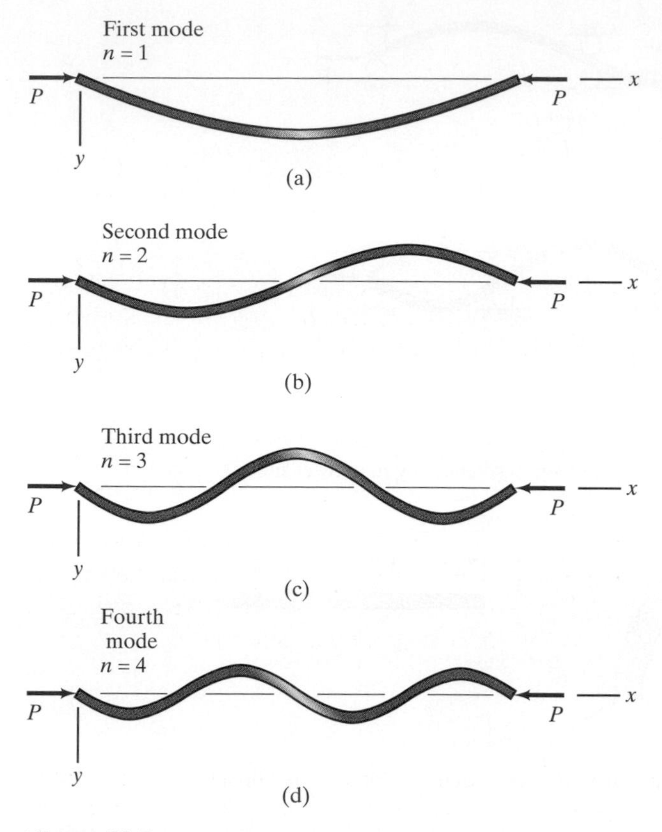

FIGURE 10-5 Deflection distributions for increasing values of n.

the beam in the buckled state for an arbitrarily small value of B, so it is interpreted as the buckling load. This result was obtained by Leonhard Euler in 1744 and is called the *Euler buckling load.*

When $n = 2$, the argument of the sine in Eq. (10-6) increases from 0 to 2π as x increases from 0 to L, resulting in the deflection in Fig. 10-5b. The deflections for $n = 3$ and $n = 4$ are shown in Figs. 10-5c and d. (Our analysis requires that the beam's slope remain small, so the deflections in Fig. 10-5 are exaggerated.) The solutions for the various values of n are all valid in the sense that they satisfy Eq. (10-1) and the boundary conditions. They are called the *buckling modes,* and are referred to as the first mode ($n = 1$), second mode ($n = 2$), and so on. Although the buckled shape of the beam could theoretically correspond to any of these modes, we know from experience that it will buckle in the first mode. The higher modes are unstable. But they are not merely of academic interest. The beam can be provided with supports so that the lowest mode in which it can buckle is the second or a higher mode (Fig. 10-6). As the figure emphasizes, this greatly increases the beam's buckling load.

We can now return to the example with which we began this discussion, the hacksaw blade in Fig. 10-1. The value of the axial load P necessary to exceed the yield stress of the material was 3120 N, or 701 lb. Let us calculate

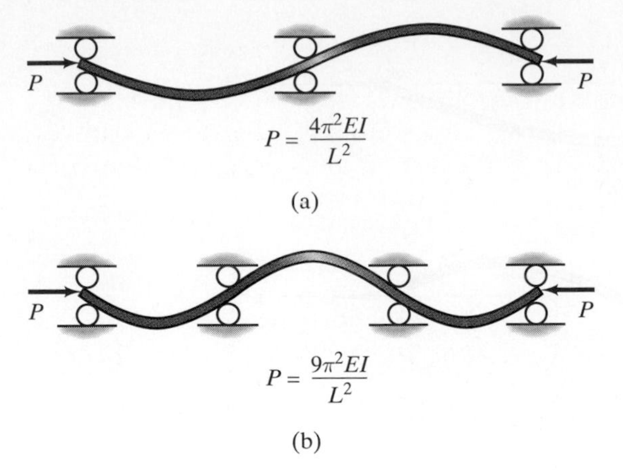

FIGURE 10-6 Preventing a beam from buckling in a lower mode.

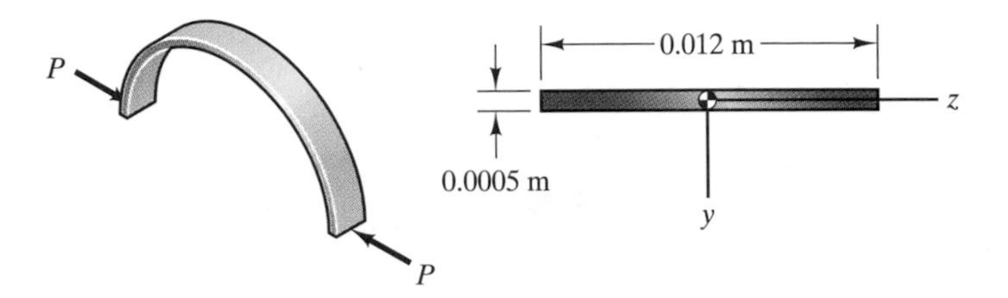

FIGURE 10-7 Orientation of the cross section of the buckled blade.

the buckling load. When the blade buckles, the axis of its cross section about which it bends is obvious (Fig. 10-7), so the moment of inertia about the z axis is

$$\begin{aligned} I &= \tfrac{1}{12}bh^3 \\ &= \tfrac{1}{12}(0.012)(0.0005)^3 \\ &= 1.25 \times 10^{-13}\ \text{m}^4. \end{aligned}$$

(Generally, if it is free to do so, *a buckling beam will bend about the principal axis of its cross section which has the smaller moment of inertia*.) If the elastic modulus of the steel blade is $E = 200$ GPa, its Euler buckling load is

$$\begin{aligned} P &= \frac{\pi^2 EI}{L^2} \\ &= \frac{\pi^2(200 \times 10^9)(1.25 \times 10^{-13})}{(0.3)^2} \\ &= 2.74\ \text{N}\ \ (0.616\ \text{lb}). \end{aligned}$$

The buckling load of the hacksaw blade is one-tenth of 1% of the compressive load necessary to exceed the yield stress of the material. This illustrates the relative vulnerability to buckling of beams subjected to compression.

EXAMPLE 10-1

Worksheet 13

The truss in Fig. 10-8 consists of bars with the cross section shown. The material has modulus of elasticity $E = 72$ GPa and will safely support a normal stress of 270 MPa in both tension and compression. What is the largest force F that can be applied to the truss?

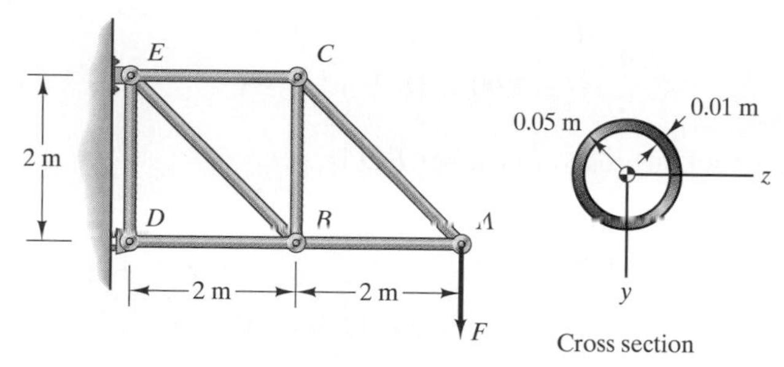

FIGURE 10-8

Strategy

The truss must first be analyzed to determine the axial loads in the members in terms of F. We can then identify the member in which the magnitude of the axial load is greatest and determine the value of F that will subject that member to the allowable normal stress. We must then identify the members that are subject to compression and determine the smallest value of F that will cause a member to buckle.

Solution

We leave it as an exercise to show that the axial loads in the members are

AB: F (C)
AC: $\sqrt{2}F$ (T)
BC: F (C)
BD: $2F$ (C)
BE: $\sqrt{2}F$ (T)
CE: F (T)
DE: zero

where (T) and (C) denote tension and compression. The magnitude of the axial load is greatest in member BD. Setting the magnitude of the normal stress in member BD equal to 270 MPa,

$$\frac{2F}{A} = \frac{2F}{\pi(0.05)^2 - \pi(0.04)^2} = 270 \times 10^6,$$

we obtain $F = 382$ kN. This is the largest value of F that will not exceed the normal stress the material will safely support.

We must now consider buckling. Three members, AB, BC, and BD, are subjected to compression. These three members are of equal length and the compressive load is greatest in member BD, so it is member BD with which we must be concerned with regard to buckling. The moment of inertia of the cross section is

$$\begin{aligned} I &= \tfrac{1}{4}\pi(0.05)^4 - \tfrac{1}{4}\pi(0.04)^4 \\ &= 2.90 \times 10^{-6}\ \text{m}^4, \end{aligned}$$

so the Euler buckling load of member BD is

$$\begin{aligned} P &= \frac{\pi^2 EI}{L^2} \\ &= \frac{\pi^2(72 \times 10^9)(2.90 \times 10^{-6})}{(2)^2} \\ &= 515\ \text{kN}. \end{aligned}$$

Equating this value to $2F$ (the compressive load in member BD), we determine that member BD buckles when $F = 257$ kN.

We have found that a force $F = 382$ kN causes the maximum normal stress in the truss to equal the allowable value, while a force $F = 257$ kN will cause member BD to buckle. If an increasing force F is applied, the truss fails, not by failure of the material but by geometric instability (Fig. 10-9). Thus it is the value $F = 257$ kN, with a suitable factor of safety imposed, that determines the largest force that can be applied to the truss.

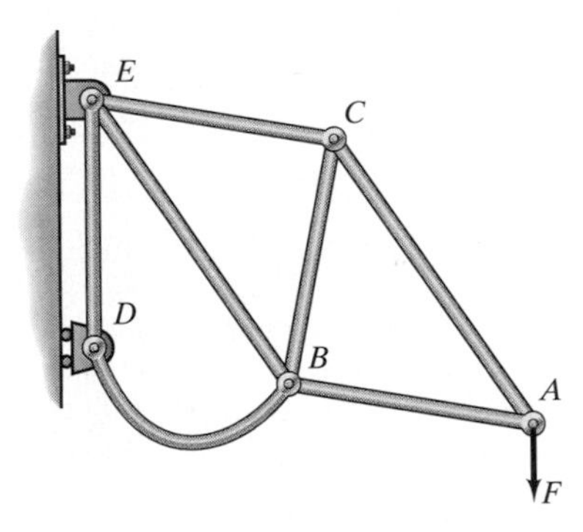

FIGURE 10-9 Collapse of the truss by buckling of member BD.

Discussion

The largest force the truss in this example could support was determined by the buckling load of the members subjected to compression. To achieve light structures while preventing buckling, trusses are sometimes designed with compression members that are relatively thick in comparison to the tension members.

This approach is strikingly evident in R. Buckminster Fuller's *tensegrity* structures (Fig. 10-10). The compression members, supported by tension members consisting of nearly invisible cords, appear to be suspended in the air.

FIGURE 10-10 Tensegrity structure with thick compression members and slender tension members. (Model by Design Science Toys, Ltd.)

10-2 | Other End Conditions

The Euler buckling load is derived under the assumption that the beam is free to rotate at the ends where the compressive axial loads are applied. That is, there are no reactions at the ends other than the applied axial loads. Various types of end supports are used in applications of beams subjected to compressive axial loads, and the choice of supports can greatly affect the resulting buckling load. In some of these cases the buckling load can be determined by analyzing the beam's deflection as we did in Section 10-1.

Analysis of the Deflection

The beam in Fig. 10-11a is free at the left end and has a built-in support at the right end. We begin by assuming that it is buckled (Fig. 10-11b) and seek to determine the value of axial force P necessary to hold it in equilibrium. In Fig. 10-11c we introduce a coordinate system with its origin at the beam's left end and obtain a free-body diagram by passing a plane through the beam

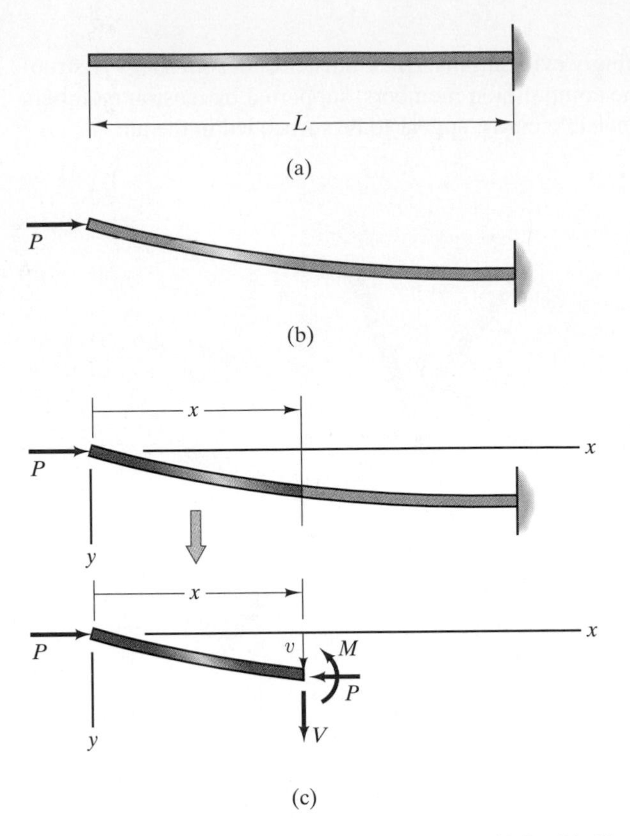

FIGURE 10-11 (a) Beam with built-in support. (b) Buckled beam in equilibrium. (c) Determining the bending moment as a function of x.

at an arbitrary position x. This free-body diagram is identical to the one we obtained in deriving the Euler buckling load (Fig. 10-3b), so the steps leading to Eq. (10-3) are unchanged. The deflection is governed by

$$v = B \sin \lambda x + C \cos \lambda x, \tag{10-8}$$

where B and C are constants we must determine from the boundary conditions and

$$\lambda^2 = \frac{P}{EI}. \tag{10-9}$$

Substituting the boundary condition $v|_{x=0} = 0$ into Eq. (10-8), we find that the constant $C = 0$. Because of the built-in support at the right end, the slope is zero at $x = L$.

$$v'|_{x=L} = 0:$$
$$\lambda B \cos \lambda L = 0.$$

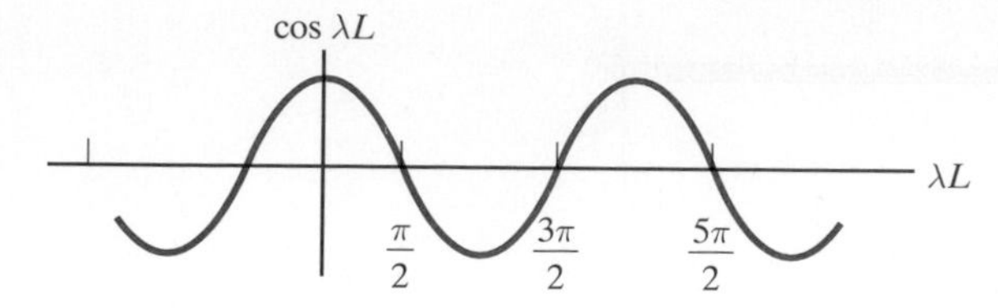

FIGURE 10-12 Roots of $\cos \lambda L = 0$.

This boundary condition requires that

$$\cos \lambda L = 0. \tag{10-10}$$

From Fig. 10-12, we see that this equation is satisfied if

$$\lambda = \frac{\pi}{2L}, \frac{3\pi}{2L}, \frac{5\pi}{2L}, \ldots.$$

Substituting $\lambda = \pi/2L$ into Eqs. (10-9) and (10-8), we obtain

$$P = \frac{\pi^2 EI}{4L^2}, \tag{10-11}$$

$$v = B \sin \frac{\pi x}{2L}. \tag{10-12}$$

As x increases from O to L, the argument of the sine in Eq. (10-12) increases from 0 to $\pi/2$. This is the buckled solution shown in Fig. 10-11b, and Eq. (10-11) is the buckling load. (Notice that the buckling load is one-fourth of the Euler buckling load for a beam of equal length.) This is the first buckling mode for a prismatic beam supported in this way. We leave it as an exercise to determine the buckling loads and distributions of the deflection for higher modes (see Problem 10-2.6).

As another example, the beam in Fig. 10-13a has a roller support that prevents lateral deflection at the left end and a built-in support at the right end. We assume that it is buckled (Fig. 10-13b) and seek to determine the value of axial force P necessary to hold it in equilibrium. In Fig. 10-13c we introduce a coordinate system and obtain a free-body diagram by passing a plane through the beam at an arbitrary position x. The force V_0 is the unknown reaction at the roller support, which equals the shear force at x. Solving for the bending moment and substituting the resulting expression into the equation $v'' = -M/EI$, we write the resulting equation as

$$v'' + \lambda^2 v = -\frac{V_0}{EI} x, \tag{10-13}$$

where

$$\lambda^2 = \frac{P}{EI}. \tag{10-14}$$

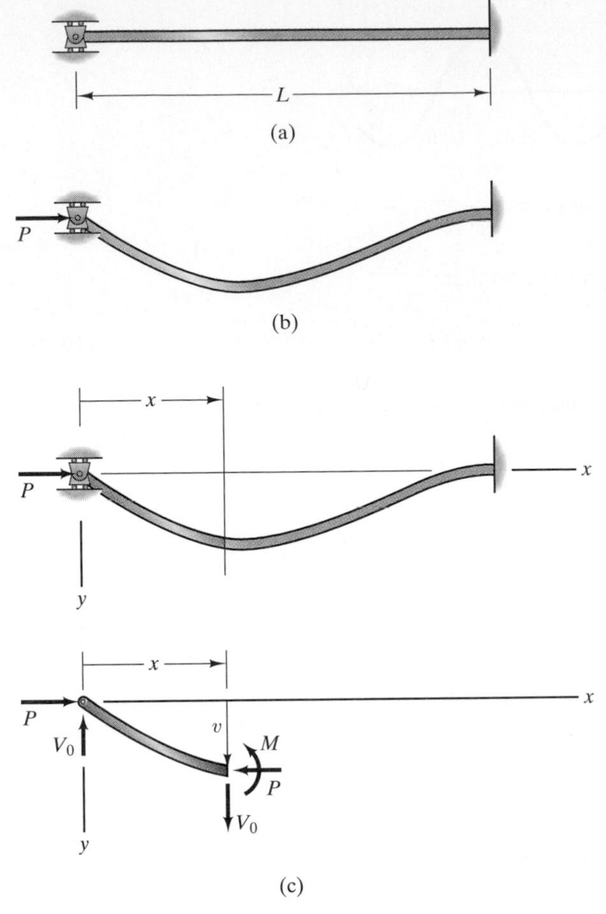

FIGURE 10-13 (a) Beam with roller and built-in supports. (b) Buckled beam in equilibrium. (c) Determining the bending moment as a function of x.

Equation (10-13) is nonhomogeneous, because the term on the right side does not contain v or one of its derivatives. Its general solution consists of the sum of the homogeneous and particular solutions:

$$v = v_{\mathrm{h}} + v_{\mathrm{p}}.$$

The homogeneous solution is the general solution of Eq. (10-13) with the right side set equal to zero, which we introduced in Section 10-1:

$$v_{\mathrm{h}} = B \sin \lambda x + C \cos \lambda x,$$

where B and C are constants. The particular solution is one that satisfies Eq. (10-13). The nonhomogeneous term is a polynomial in x, so we seek a particular solution in the form of a polynomial of the same order: $v_{\mathrm{p}} = a_0 + a_1 x$, where a_0 and a_1 are constants we must determine. Substituting this expression into Eq. (10-13), we write the resulting equation as

$$\lambda^2 a_0 + \left(\lambda^2 a_1 + \frac{V_0}{EI}\right) x = 0.$$

This equation is satisfied over an interval of x only if $a_0 = 0$ and $a_1 = -V_0/\lambda^2 EI$, yielding the particular solution

$$v_{\mathrm{p}} = -\frac{V_0}{\lambda^2 EI} x.$$

[You should confirm that this is a particular solution by substituting it into Eq. (10-13).] The general solution of Eq. (10-13) is

$$v = v_{\mathrm{h}} + v_{\mathrm{p}} = B \sin \lambda x + C \cos \lambda x - \frac{V_0}{\lambda^2 EI} x. \tag{10-15}$$

From the boundary condition $v|_{x=0} = 0$, we obtain $C = 0$. The boundary conditions at $x = L$ are

$$v|_{x=L} = B \sin \lambda L - \frac{V_0 L}{\lambda^2 EI} = 0, \tag{10-16}$$

$$v'|_{x=L} = \lambda B \cos \lambda L - \frac{V_0}{\lambda^2 EI} = 0. \tag{10-17}$$

We solve the first of these equations for the reaction V_0:

$$V_0 = \frac{\lambda^2 EI \sin \lambda L}{L} B. \tag{10-18}$$

Substituting this result into Eq. (10-17), we find that the boundary conditions at $x = L$ are satisfied only if

$$\sin \lambda L - \lambda L \cos \lambda L = 0. \tag{10-19}$$

We also substitute Eq. (10-18) into Eq. (10-15), obtaining the distribution of the deflection in the form

$$v = B\left(\sin \lambda x - \frac{x}{L} \sin \lambda L\right). \tag{10-20}$$

Equation (10-19) is called the *characteristic equation* for this problem. For each value of λ that satisfies the characteristic equation, Eq. (10-14) determines the buckling load and Eq. (10-20) determines the shape of the buckled beam. [Equation (10-10) is the characteristic equation for a beam with free and fixed ends.] In this case we must determine the roots of the characteristic equation numerically. Figure 10-14 is a graph of the characteristic function $f(\lambda L) = \sin \lambda L - \lambda L \cos \lambda L$. The first four roots, determined numerically, are $\lambda L = 4.493$, 7.725, 10.904, and 14.066. The shapes of the resulting buckling modes and the associated buckling loads are shown in Fig. 10-15.

The common end conditions that can be analyzed in this way are shown in Fig. 10-16 together with their buckling loads.

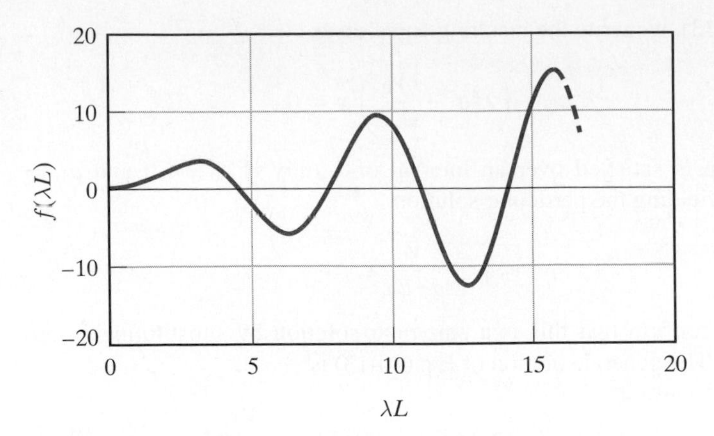

| **FIGURE 10-14** Graph of the characteristic function.

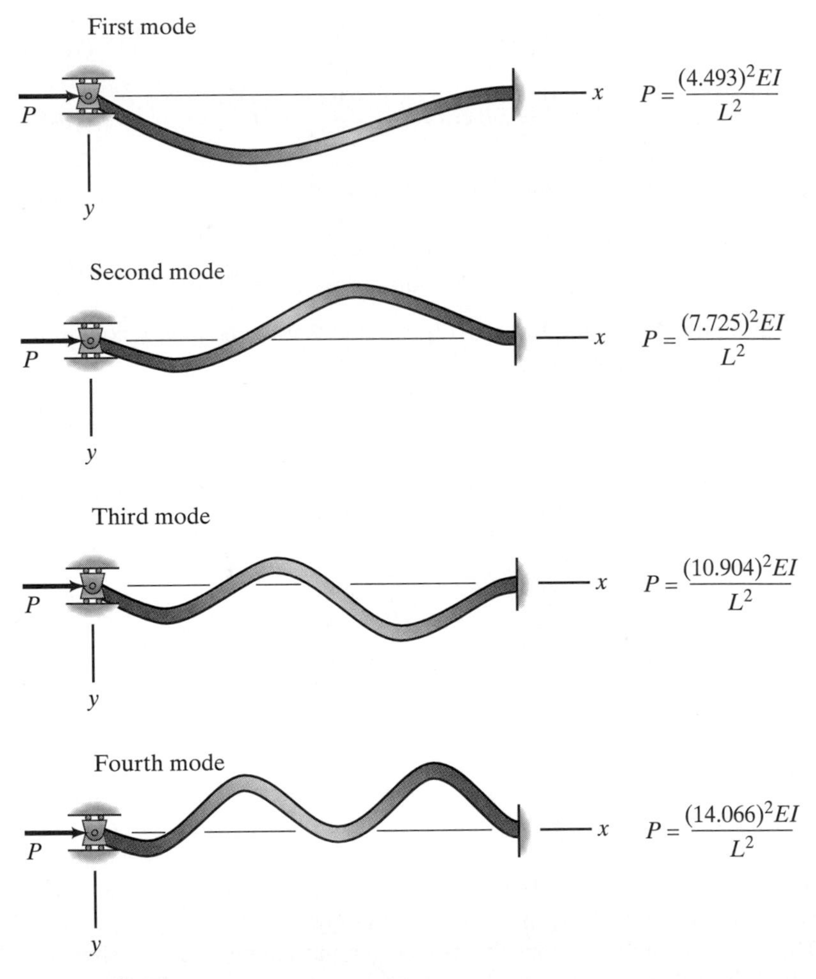

| **FIGURE 10-15** First four buckling modes.

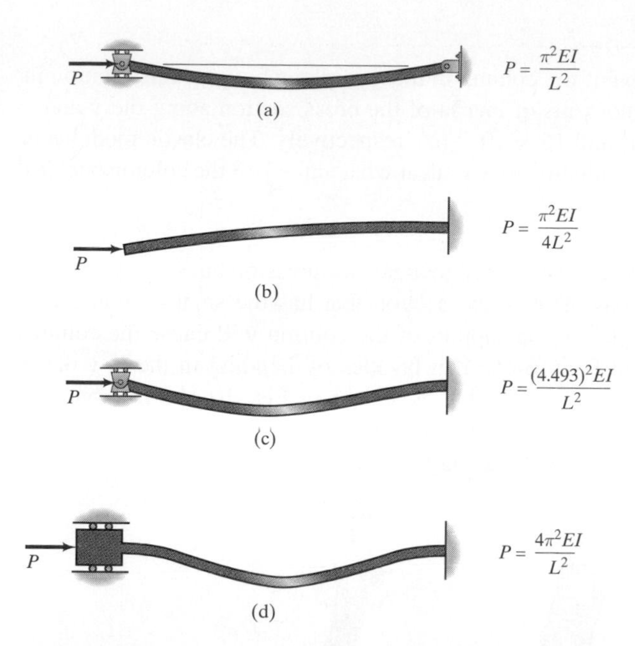

FIGURE 10-16 First-mode buckling loads of beams with common end conditions.

EXAMPLE 10-2

The column in Fig. 10-17 supports an axial load P. The base of the column is built in. The support at the top prevents lateral deflection. The support at

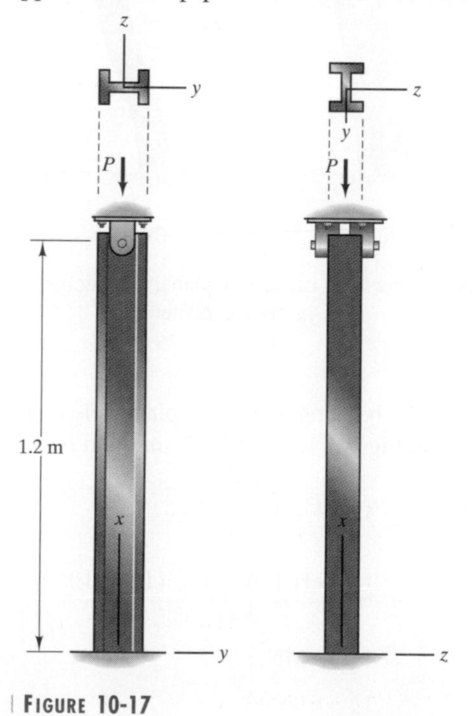

FIGURE 10-17

the top allows rotation of the column in the x–y plane but prevents rotation in the x–z plane. The moments of inertia of the cross section about the y and z axes are 5×10^{-6} m^4 and 15×10^{-6} m^4, respectively. The elastic modulus is $E = 70$ GPa. If P is gradually increased, at what value will the column buckle?

Strategy

If it is free to do so, a beam subjected to axial compression buckles by bending about the principal axis of the cross section that has the smaller moment of inertia. In this example the top support of the column will cause the column to buckle as shown in Fig. 10-16c if it buckles by bending in the x–y plane [Fig. (a)], but will cause it to buckle as shown in Fig. 10-16d if it buckles by bending in the x–z plane [Fig. (b)]. We must determine which of these possibilities yields the lowest buckling load.

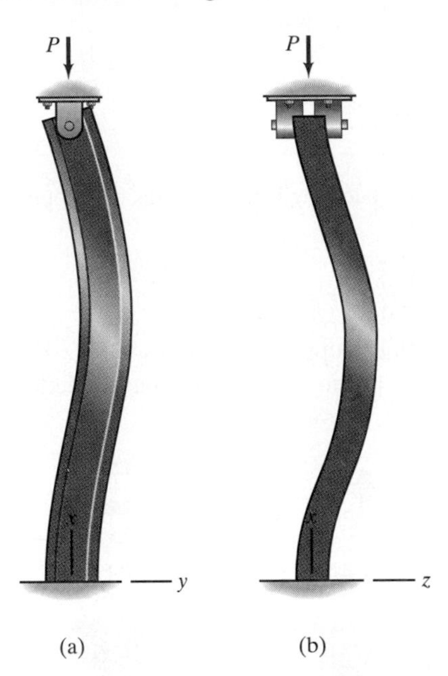

(a) Geometry of buckling in the x–y plane. (b) Geometry of buckling in the x–z plane.

Solution

If the column buckles by bending in the x–y plane [Fig. (a)], it fails by bending about the z axis. From Fig. 10-16c, the buckling load is

$$\begin{aligned} P &= \frac{(4.493)^2 EI}{L^2} \\ &= \frac{(4.493)^2(70 \times 10^9)(15 \times 10^{-6})}{(1.2)^2} \\ &= 14.72 \text{ MN}. \end{aligned}$$

If the column buckles by bending in the x–z plane [Fig. (b)], it fails by bending about the y axis. From Fig. 10-16d, the buckling load is

$$\begin{aligned} P &= \frac{4\pi^2 EI}{L^2} \\ &= \frac{4\pi^2(70 \times 10^9)(5 \times 10^{-6})}{(1.2)^2} \\ &= 9.60 \text{ MN}. \end{aligned}$$

We see that the column will buckle as shown in Fig. (b) and the buckling load is 9.60 MN. That is, the column does buckle by bending about the principal axis of the cross section that has the smaller moment of inertia (see Problem 10-2.6).

EXAMPLE 10-3

The beam in Fig. 10-18 has a support that prevents lateral deflection and rotation at the left end and a built-in support at the right end. Determine its buckling load.

| FIGURE 10-18

Solution

In Fig. (a) we introduce a coordinate system and obtain a free-body diagram by passing a plane through the buckled beam at an arbitrary position x.

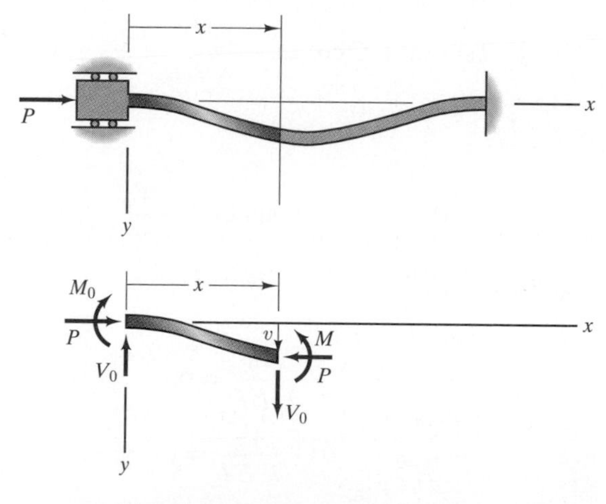

(a) Determining the bending moment.

Solving for the bending moment and substituting it into the equation $v'' = -M/EI$, we write the resulting equation as

$$v'' + \lambda^2 v = -\frac{V_0}{EI}x - \frac{M_0}{EI}, \tag{10-21}$$

where

$$\lambda^2 = \frac{P}{EI}. \tag{10-22}$$

The homogeneous solution of Eq. (10-21) is

$$v_{\mathrm{h}} = B \sin \lambda x + C \cos \lambda x,$$

where B and C are constants. The nonhomogeneous term in Eq. (10-21) is a polynomial in x, so we seek a particular solution in the form of a polynomial of the same order: $v_{\mathrm{p}} = a_0 + a_1 x$. Substituting this expression into Eq. (10-21) and writing the resulting equation as

$$\left(\lambda^2 a_0 + \frac{M_0}{EI}\right) + \left(\lambda^2 a_1 + \frac{V_0}{EI}\right) x = 0,$$

we find that $a_0 = -M_0/\lambda^2 EI$ and $a_1 = -V_0/\lambda^2 EI$, yielding the particular solution

$$v_{\mathrm{p}} = -\frac{M_0}{\lambda^2 EI} - \frac{V_0}{\lambda^2 EI}x.$$

The general solution of Eq. (10-21) is therefore

$$v = v_{\mathrm{h}} + v_{\mathrm{p}} = B \sin \lambda x + C \cos \lambda x - \frac{M_0}{\lambda^2 EI} - \frac{V_0}{\lambda^2 EI}x, \tag{10-23}$$

and the slope is

$$v' = \lambda B \cos \lambda x - \lambda C \sin \lambda x - \frac{V_0}{\lambda^2 EI}. \tag{10-24}$$

The boundary conditions at $x = 0$ are

$$v|_{x=0} = C - \frac{M_0}{\lambda^2 EI} = 0, \tag{10-25}$$

$$v'|_{x=0} = \lambda B - \frac{V_0}{\lambda^2 EI} = 0. \tag{10-26}$$

We solve these equations for M_0 and V_0 and substitute the results into Eqs. (10-23) and (10-24), obtaining

$$v = (\sin \lambda x - \lambda x)B + (\cos \lambda x - 1)C, \qquad \textbf{(10-27)}$$

$$v' = (\lambda \cos \lambda x - \lambda)B - \lambda C \sin \lambda x. \qquad \textbf{(10-28)}$$

The boundary conditions at $x = L$ are

$$v|_{x=L} = (\sin \lambda L - \lambda L)B + (\cos \lambda L - 1)C = 0, \qquad \textbf{(10-29)}$$

$$v'|_{x=L} = (\lambda \cos \lambda L - \lambda)B - \lambda C \sin \lambda L = 0. \qquad \textbf{(10-30)}$$

Solving the first of these equations for B gives

$$B = \frac{1 - \cos \lambda L}{\sin \lambda L - \lambda L} C. \qquad \textbf{(10-31)}$$

Substituting this result into Eq. (10-30) yields the characteristic equation:

$$\lambda L \sin \lambda L + 2 \cos \lambda L - 2 = 0. \qquad \textbf{(10-32)}$$

We also substitute Eq. (10-31) into Eq. (10-27), obtaining the distribution of the deflection in the form

$$v = C\left[\frac{1 - \cos \lambda L}{\sin \lambda L - \lambda L}(\sin \lambda x - \lambda x) + \cos \lambda x - 1\right]. \qquad \textbf{(10-33)}$$

Figure 10-19 is a graph of the characteristic function $f(\lambda L) = \lambda L \sin \lambda L + 2 \cos \lambda L - 2$. The first four roots, determined numerically, are $\lambda L = 2\pi$, 8.987, 12.566, and 15.451. The shapes of the first three buckling modes [obtained from

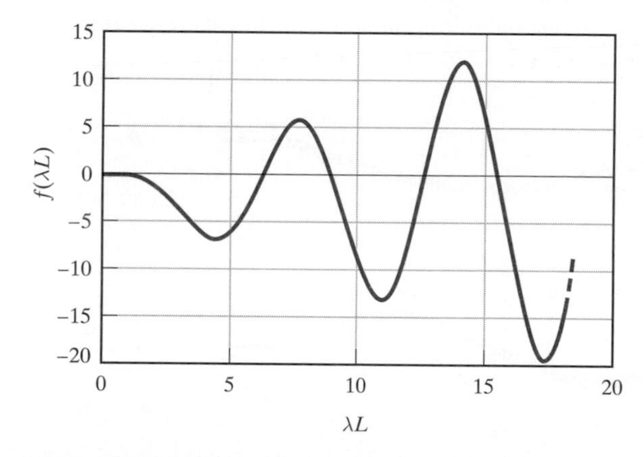

FIGURE 10-19 Graph of the characteristic function.

Eq. (10-33)] and the associated buckling loads [obtained from Eq. (10-22)] are shown in Fig. 10-20.

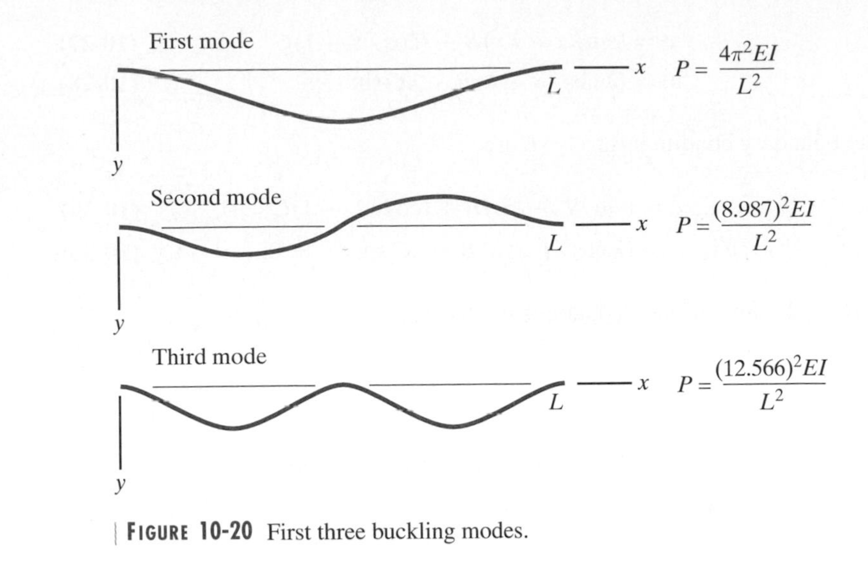

FIGURE 10-20 First three buckling modes.

Effective Length

In Section 10-1 we analyzed a prismatic beam that is free at the ends and subjected to compressive axial load, obtaining the Euler buckling load (Fig. 10-21a). We also obtained the buckling load of the second mode from the same analysis (Fig. 10-21b), but there is a simpler way to determine the buckling load of the second mode.

Let us assume that the beam is buckled in the second mode. In Fig. 10-22 we cut it by a plane at the midpoint and draw the resulting free-body diagrams.

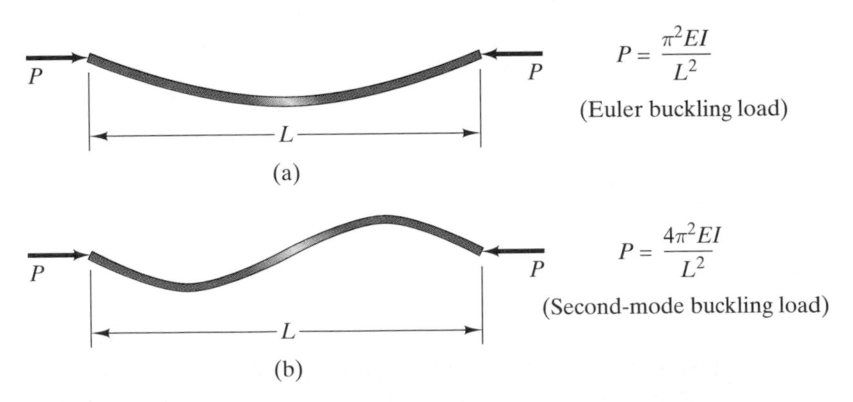

FIGURE 10-21 First two buckling modes of a beam that is free at the ends. The buckling load of the first mode is the Euler buckling load.

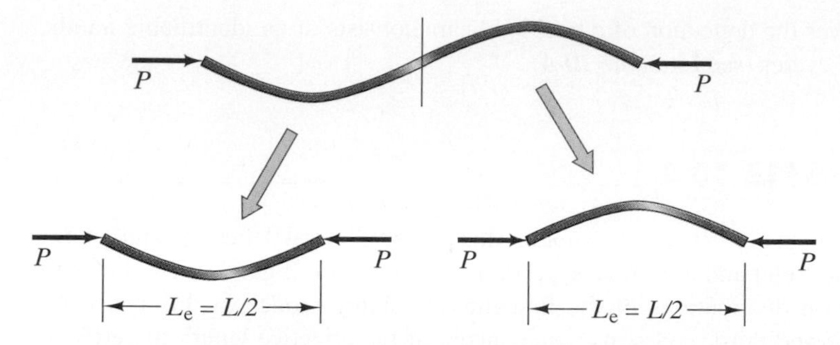

FIGURE 10-22 Obtaining free-body diagrams by dividing the beam at its midpoint.

Each of these free-body diagrams is buckled in the first mode. We can therefore determine the second-mode buckling load by calculating the Euler buckling load of a beam of length $L_e = L/2$:

$$\begin{aligned} P &= \frac{\pi^2 EI}{L_e^2} \\ &= \frac{\pi^2 EI}{(L/2)^2} \\ &= \frac{4\pi^2 EI}{L^2}. \end{aligned}$$

We see that the second-mode buckling load of the beam of length L is equal to the Euler buckling load of a beam of length $L_e = L/2$.

This example motivates a concept called the effective length. Suppose that a given beam buckles in a particular mode, and the buckling load is P. The *effective length* L_e is defined to be the length of a beam of the same flexural rigidity whose Euler buckling load equals P:

$$P = \frac{\pi^2 EI}{L_e^2}. \tag{10-34}$$

If P is known for a given beam and buckling mode, this equation can be used to determine the effective length. Alternatively, in some cases the effective length can be determined or approximated by observation and Eq. (10-34) can be used to determine P.

How can we determine the effective length by observation? The distribution of the deflection of a beam that is free at the ends and buckled in the first mode, from which the Euler buckling load is derived (Fig. 10-21a), is a half-cycle of a sine function. When such a beam buckles in the second mode (Fig. 10-21b), its deflection forms two half-cycles, each of length $L/2$. The buckling load of the beam in the second mode equals the Euler buckling load of these half-cycles (Fig. 10-22), so the effective length is $L/2$. Thus we obtain the effective length by dividing L by the number of half-cycles. This approach can be applied whenever the deflection of a buckled beam consists of an identifiable number of half-cycles (see Example 10-4).

EXAMPLE 10-4

The beam in Fig. 10-23 has a support that prevents lateral deflection and rotation at the left end and a built-in support at the right end. Figure 10-20 shows the deflection distributions of the first three buckling modes of this beam. For the first and third modes, use the concept of the effective length to verify the buckling loads in Fig. 10-20.

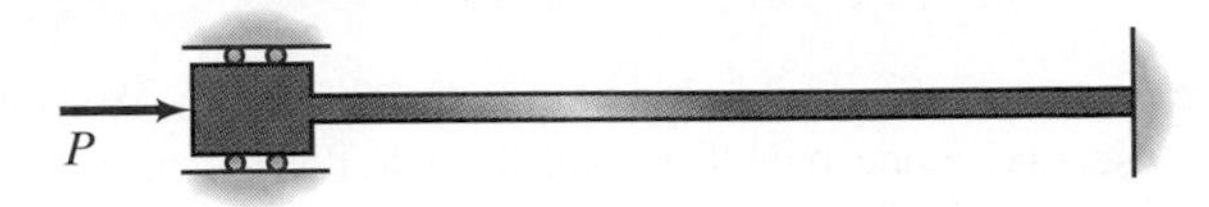

FIGURE 10-23

Strategy

By counting the number of half-cycles of sine functions in the deflection distribution for each mode, we can determine its effective length. The buckling load is the Euler buckling load for a beam of that length.

Solution

First mode The distribution of the deflection consists of four quarter-cycles of a sine function [Fig. (a)] or two half-cycles. The effective length is therefore $L_e = L/2$, and the buckling load is

$$\begin{aligned} P &= \frac{\pi^2 EI}{L_e^2} \\ &= \frac{\pi^2 EI}{(L/2)^2} \\ &= \frac{4\pi^2 EI}{L^2}. \end{aligned}$$

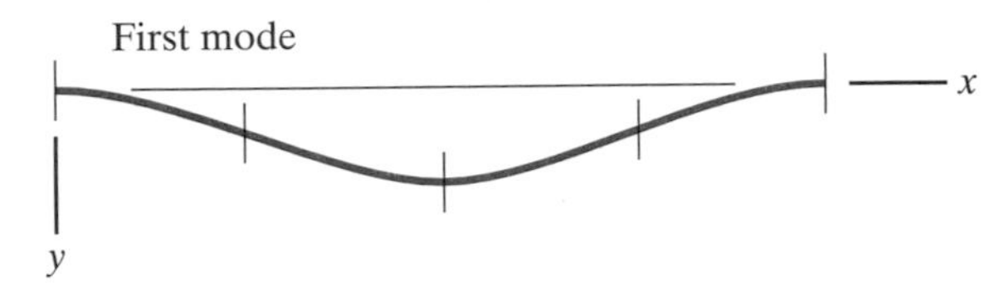

(a) First mode divided into quarter-cycles.

Third mode The distribution of the deflection consists of eight quarter-cycles of a sine function [Fig. (b)], or four half-cycles. The effective length is $L_e = L/4$, and the buckling load is

$$P = \frac{\pi^2 EI}{L_e^2} = \frac{\pi^2 EI}{(L/4)^2} = \frac{(12.566)^2 EI}{L^2}.$$

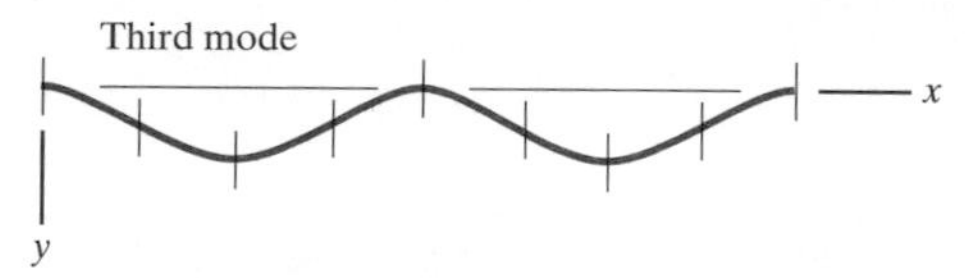

(b) Third mode divided into quarter-cycles.

10-3 | Eccentric Loads

In determining buckling loads of beams in Section 10-2, we assumed that compressive loads were applied precisely at the centroids of the cross sections of perfectly prismatic beams. In practice, the lines of action of the loads will often be offset from the centroid to some extent, either unintentionally or to satisfy some design requirement, and the beam's neutral axis will not be perfectly straight. These variations in the geometry of the classical buckling problem have important quantitative and qualitative effects on a beam's response to compressive loads. In this section we illustrate these effects by analyzing a prismatic beam that is free to rotate at the ends and is subjected to *eccentric loads*, loads that are offset relative to the centroid of the cross section.

Analysis of the Deflection

The prismatic beam in Fig. 10-24 is subjected to axial loads P that are displaced a distance e from the beam's neutral axis. The couples resulting from the displaced loads will cause the beam to bend, and our objective is to determine the resulting deflection. In Fig. 10-25 we introduce a coordinate system and obtain a free-body diagram by passing a plane through the bent beam at an arbitrary position x. Solving for the bending moment, we obtain

$$M = P(v + e). \qquad \textbf{(10-35)}$$

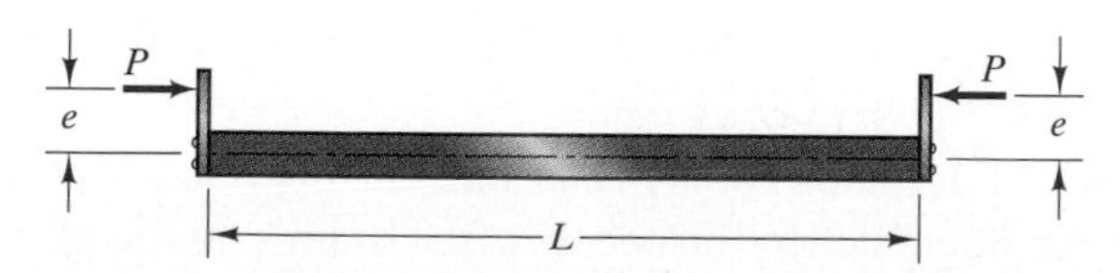

FIGURE 10-24 Beam subjected to eccentric axial loads.

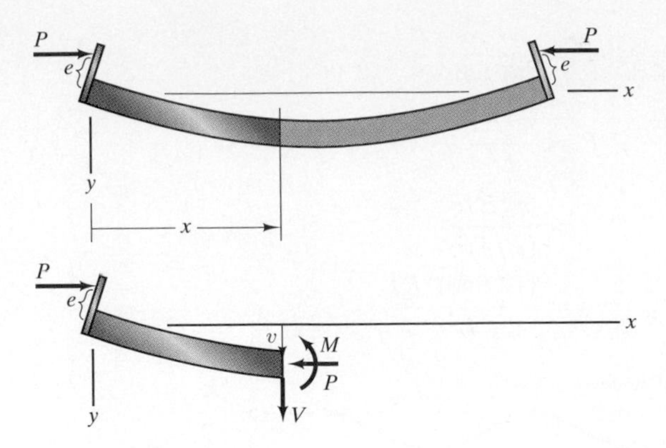

FIGURE 10-25 Determining the bending moment.

(The exact bending moment is $M = P[v + e\cos(\theta|_{x=0})]$, but the beam's slope is assumed small, so we can make the approximation $\cos(\theta|_{x=0}) = 1$.) We substitute Eq. (10-35) into the equation $v'' = -M/EI$ and write the resulting equation as

$$v'' + \lambda^2 v = -\lambda^2 e, \tag{10-36}$$

where

$$\lambda^2 = \frac{P}{EI}. \tag{10-37}$$

The general solution of Eq. (10-36) is

$$v = B\sin\lambda x + C\cos\lambda x - e. \tag{10-38}$$

Using the boundary conditions $v|_{x=0} = 0$ and $v|_{x=L} = 0$ to evaluate the constants B and C, we obtain the deflection:

$$v = e\left(\frac{1-\cos\lambda L}{\sin\lambda L}\sin\lambda x + \cos\lambda x - 1\right). \tag{10-39}$$

By substituting the trigonometric identities

$$\begin{aligned} \sin\lambda L &= 2\sin\frac{\lambda L}{2}\cos\frac{\lambda L}{2}, \\ \cos\lambda L &= \cos^2\frac{\lambda L}{2} - \sin^2\frac{\lambda L}{2} \end{aligned} \tag{10-40}$$

into Eq. (10-39), we can write it as

$$v = e\left(\tan\frac{\lambda L}{2}\sin\lambda x + \cos\lambda x - 1\right). \tag{10-41}$$

For a given value of P, Eqs. (10-37) and (10-41) determine the distribution of the beam's deflection. The maximum deflection occurs at the beam's midpoint:

$$v_{\max} = v|_{x=L/2} = e\left(\sec\frac{\lambda L}{2} - 1\right). \tag{10-42}$$

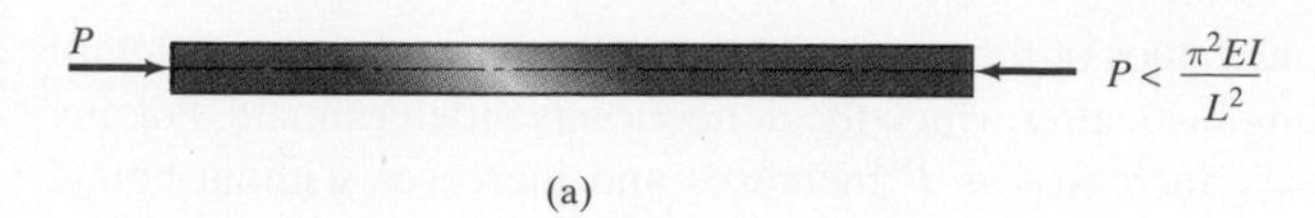

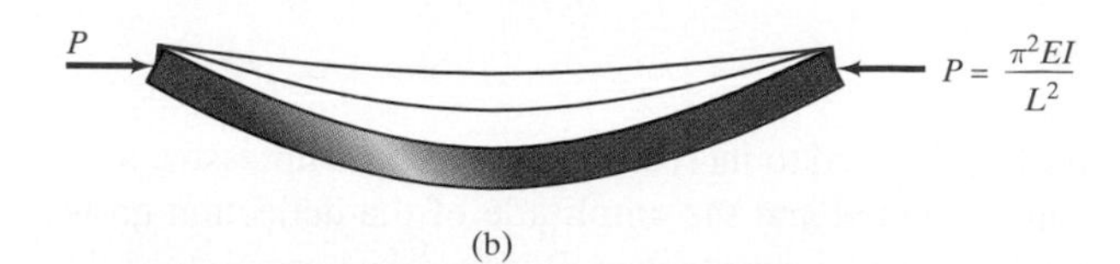

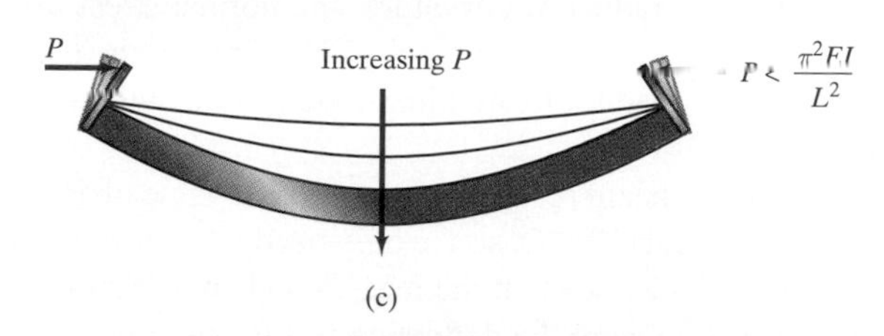

FIGURE 10-26 (a) Beam subjected to axial loads applied at the centroid of the cross section. (b) At the buckling load, the deflection is indeterminate. (c) With eccentric loads, the amplitude of the deflection increases as P increases.

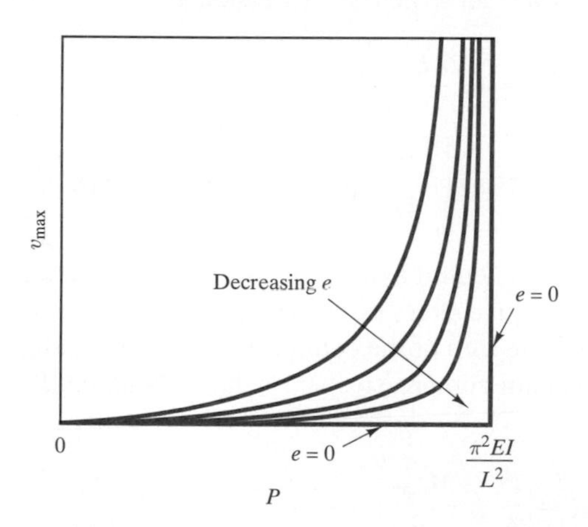

FIGURE 10-27 Graph of the maximum deflection as a function of the axial load.

But what is the relationship between these results and the beam's buckling load? If increasing compressive loads are applied to the beam at the centroid of its cross section (Fig. 10-26a), there is no deflection until the value of P reaches the buckling load. When P equals the buckling load, the deflection is indeterminate because the beam is in equilibrium for any value of its amplitude (Fig. 10-26b). (Of course, the magnitude is limited by the underlying assumption that the beam's slope is small.) In contrast, when increasing eccentric loads are applied, the beam begins bending immediately and the amplitude of the deflection increases as P increases (Fig. 10-26c). These phenomena are illustrated by Fig. 10-27, in which the maximum deflection is plotted as a function of P

for different constant values of the eccentricity e. When $e = 0$, $v_{max} = 0$ until P equals the buckling load, after which the deflection is indeterminate. For any finite value of e, v_{max} increases as P increases and increases without bound as P approaches the buckling load. As $e \to 0$, the beam's behavior predictably approaches the behavior observed when the loads are applied at the centroid.

Secant Formula

We have seen that a beam subjected to increasing eccentric compressive loads deflects as soon as load is applied and the amplitude of the deflection grows without bound as the buckling load is approached. Because the normal stress due to bending depends on the beam's radius of curvature, the normal stress also grows without bound as the buckling load is approached. As a result, the focus in design shifts from the buckling load to the maximum deflection and normal stress that can be allowed.

The maximum deflection of the beam resulting from a given value of P can be determined from Eqs. (10-37) and (10-42), so we only need to determine the maximum stress. From Eq. (10-35) we see that the maximum bending moment in the beam occurs at the midpoint where the deflection is a maximum:

$$M_{max} = P(v_{max} + e).$$

By using Eqs. (10-37) and (10-42) we can write this expression as

$$\begin{aligned} M_{max} &= Pe \sec \frac{\lambda L}{2} \\ &= Pe \sec \sqrt{\frac{PL^2}{4EI}}. \end{aligned} \qquad \textbf{(10-43)}$$

Let c be the distance from the beam's neutral axis to the location on the cross section where the maximum compressive stress occurs (Fig. 10-28). The maximum normal stress in the beam is the sum of the compressive stress resulting from the axial load and the maximum compressive stress due to bending. Its magnitude is

$$\sigma_{max} = \frac{P}{A} + \frac{M_{max}c}{I}.$$

Substituting Eq. (10-43) and introducing the *radius of gyration* $k = \sqrt{I/A}$, we can write the maximum stress as

$$\sigma_{max} = \frac{P}{A}\left[1 + \frac{ec}{k^2} \sec\left(\frac{L}{2k}\sqrt{\frac{P}{EA}}\right)\right]. \qquad \textbf{(10-44)}$$

This equation is called the *secant formula*. It expresses the magnitude of the maximum stress as a function of P/A, E, and two dimensionless parameters, the *eccentricity ratio* ec/k^2 and the *slenderness ratio* L/k.

When the beam's dimensions, the elastic modulus, and the eccentricity are known and the maximum allowable compressive stress is specified, Eq. (10-44)

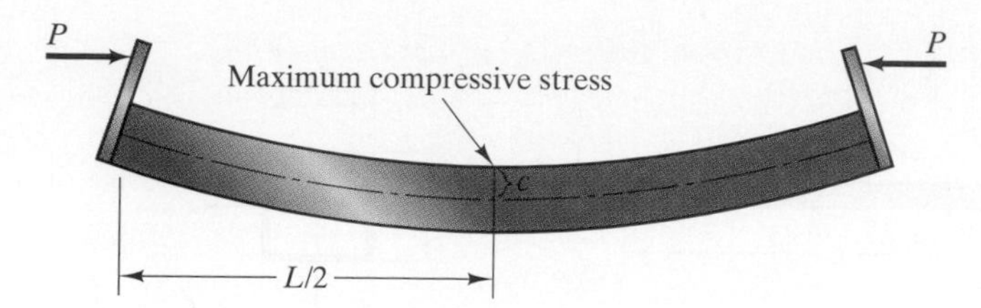

FIGURE 10-28 Location of the maximum compressive stress.

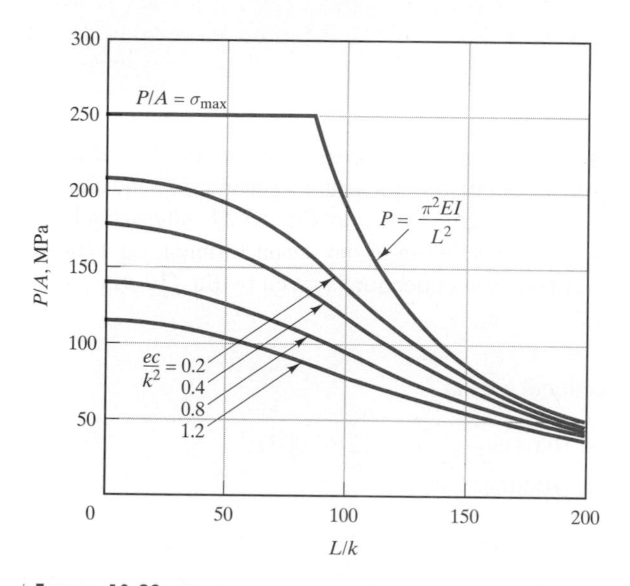

FIGURE 10-29 Graph of P/A as a function of the slenderness ratio for $E = 200$ GPa and $\sigma_{max} = 250$ MPa.

can be solved numerically for the maximum allowable axial load. As an example, we consider steel beams with elastic modulus $E = 200$ GPa and assume that the maximum allowable compressive stress is $\sigma_{max} = 250$ MPa. In Fig. 10-29 the solution for P/A obtained from Eq. (10-44) is plotted as a function of the slenderness ratio for different constant values of the eccentricity ratio. When the eccentricity is zero, the allowable values of P/A are bounded by the buckling load and σ_{max}.

EXAMPLE 10-5

The beam in Fig. 10-30 has the cross section shown, length $L = 0.76$ m, and is subjected to compressive loads with eccentricity $e = 0.0046$ m. The material is steel with modulus of elasticity $E = 200$ GPa. **(a)** If the allowable compressive stress is 250 MPa, what is the largest allowable value of P? **(b)** If the compressive load determined in part (a) is applied, what is the beam's maximum deflection?

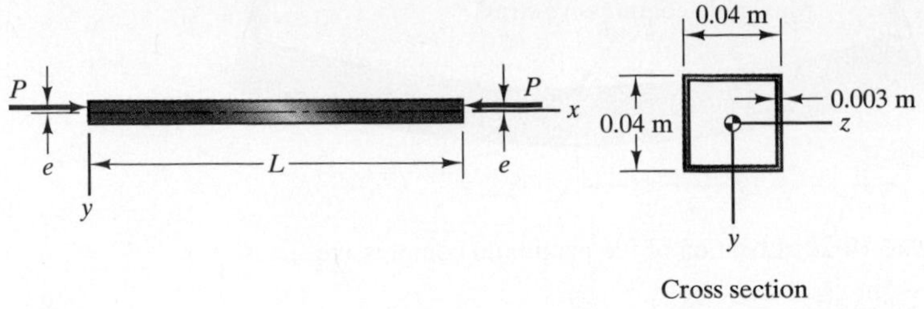

| FIGURE 10-30

Strategy

(a) The elastic modulus and allowable compressive stress are the values on which Fig. 10-29 is based. By calculating the beam's eccentricity and slenderness ratios, we can estimate the value of P/A from Fig. 10-29. Alternatively, we can determine P/A by numerical solution of the secant formula, Eq. (10-44). **(b)** Once P is known, the maximum deflection is given by Eq. (10-42).

Solution

(a) The beam's cross-sectional area is

$$A = (0.04)^2 - [0.04 - (2)(0.003)]^2 = 0.000444 \text{ m}^2,$$

and the moment of inertia about the z axis is

$$I = \tfrac{1}{12}(0.04)^4 - \tfrac{1}{12}[0.04 - (2)(0.003)]^4 = 1.020 \times 10^{-7} \text{ m}^4.$$

The radius of gyration is

$$k = \sqrt{\frac{I}{A}} = 0.0152 \text{ m}.$$

Using this result, we calculate the eccentricity ratio

$$\frac{ec}{k^2} = \frac{(0.0046)(0.02)}{(0.0152)^2} = 0.401$$

and the slenderness ratio

$$\frac{L}{k} = \frac{0.76}{0.0152} = 50.1.$$

With these values, we estimate from Fig. 10-29 that $P/A = 165$ MPa. By solving Eq. (10-44) numerically, we obtain $P/A = 163.3$ MPa. Using the latter value, the largest allowable value of P is

$$P = (163 \times 10^6)(0.000444)$$
$$= 72.5 \text{ kN}.$$

(b) The value of the parameter λ is

$$\lambda = \sqrt{\frac{P}{EI}}$$
$$= \sqrt{\frac{72.5 \times 10^3}{(200 \times 10^9)(1.020 \times 10^{-7})}}$$
$$= 1.89 \text{ m}^{-1}.$$

From Eq. (10-42), the maximum deflection is

$$v_{\max} = e\left(\sec\frac{\lambda L}{2} - 1\right)$$
$$= 0.0046\left[\sec\frac{(1.89)(0.76)}{2} - 1\right]$$
$$= 0.00150 \text{ m}.$$

Chapter Summary

Buckling Loads

The buckling load for a prismatic beam of length L that is free at the ends and buckles as shown in Fig. (a) is called the *Euler buckling load*:

$$P = \frac{\pi^2 EI}{L^2}. \qquad \text{Eq. (10-7)}$$

P P x y

(a)

Figure (a) is the first buckling mode for a beam that is free at the ends. The distributions of the deflection for the second, third, and fourth modes are shown in Fig. 10-5. The buckling load of the nth mode is

$$P = \frac{n^2\pi^2 EI}{L^2}. \qquad \text{Eq. (10-5)}$$

First-mode buckling loads of beams with other common end conditions are shown in Fig. 10-16.

Effective Length

Suppose that a given beam buckles in a particular mode, and the buckling load is P. The *effective length* L_e is the length of a beam of the same flexural rigidity whose Euler buckling load equals P:

$$P = \frac{\pi^2 EI}{L_e^2}. \qquad \text{Eq. (10-34)}$$

For example, if a beam of length L that is free at the ends buckles in the second mode, the effective length is $L_e = L/2$ [Fig. (b)] and the buckling load is

$$P = \frac{\pi^2 EI}{L_e^2} = \frac{4\pi^2 EI}{L^2}.$$

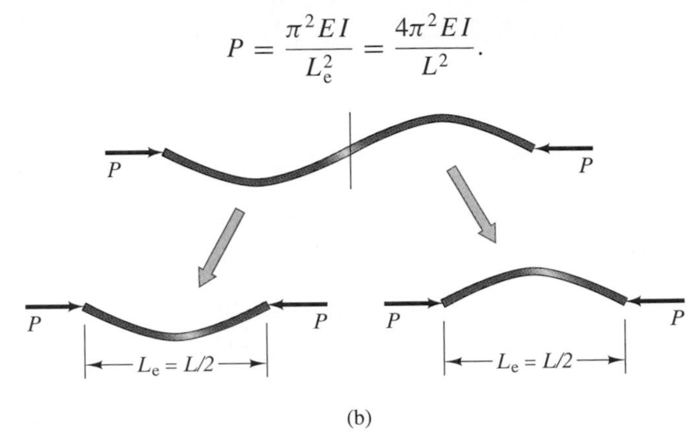

(b)

In some cases the effective length of a buckled beam can be determined or approximated by observation and Eq. (10-34) can be used to determine the buckling load.

Eccentric Loads and the Secant Formula

Consider a prismatic beam that is free to rotate at the ends and is subjected to eccentric loads [Fig. (c)]. The distribution of the deflection is

$$v = e\left(\tan\frac{\lambda L}{2}\sin\lambda x + \cos\lambda x - 1\right) \qquad \text{Eq. (10-41)}$$

and the maximum deflection is

$$v_{\max} = v|_{x=L/2} = e\left(\sec\frac{\lambda L}{2} - 1\right), \qquad \text{Eq. (10-42)}$$

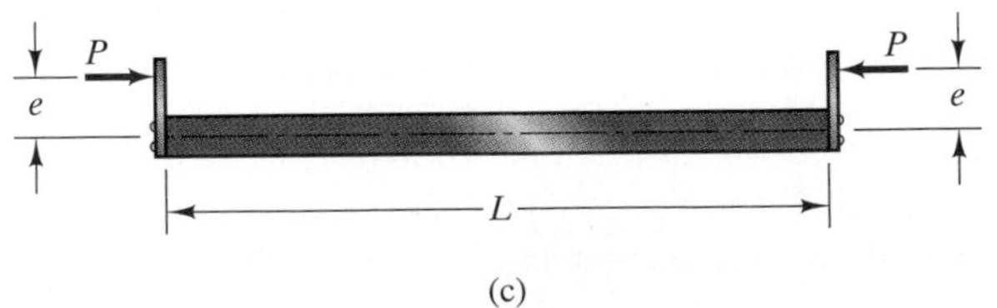

(c)

where $\lambda^2 = P/EI$.

The maximum compressive stress in the beam in Fig. (c) is given by the *secant formula*

$$\sigma_{\max} = \frac{P}{A}\left[1 + \frac{ec}{k^2}\sec\left(\frac{L}{2k}\sqrt{\frac{P}{EA}}\right)\right], \qquad \text{Eq. (10-44)}$$

where $k = \sqrt{I/A}$ is the radius of gyration of the cross section. The parameter ec/k^2 is the *eccentricity ratio* and L/k is the *slenderness ratio*. When the beam's dimensions, the elastic modulus, and the eccentricity are known and the maximum allowable compressive stress is specified, Eq. (10-44) can be solved numerically for the maximum allowable axial load.

PROBLEMS

10-1.1. The beam is steel with elastic modulus $E = 28 \times 10^6$ psi. Determine its Euler buckling load.

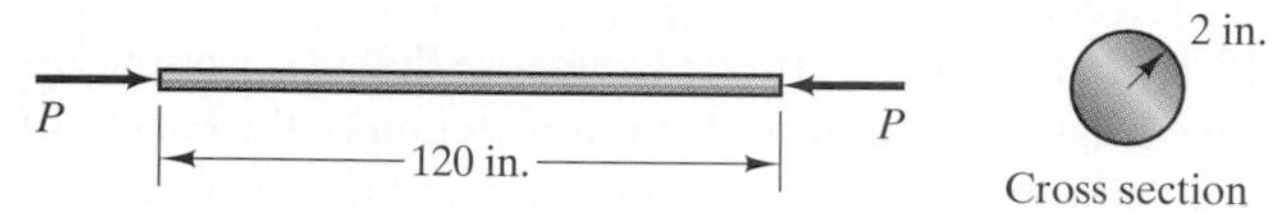

PROBLEM 10-1.1

10-1.2. What is the second-mode buckling load of the beam in Problem 10-1.1?

10-1.3. If you want to design the beam in Problem 10-1.1 so that its length is 120 in. and its Euler buckling load is 380 kip, what should the radius of its cross section be?

10-1.4. The beam is aluminum alloy with elastic modulus $E = 70$ GPa. Determine its Euler buckling load.

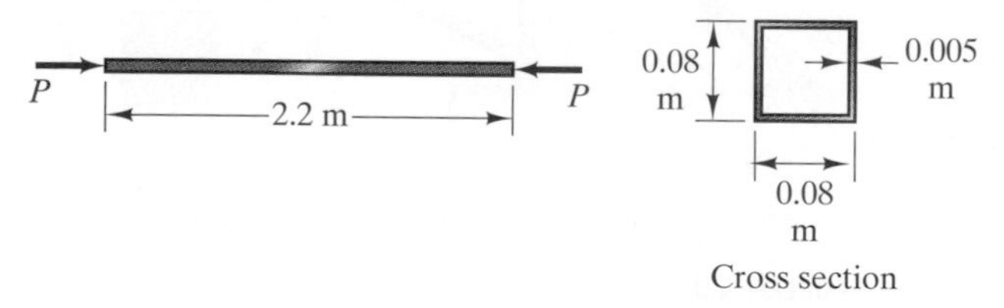

PROBLEM 10-1.4

10-1.5. If you want to increase the wall thickness of the beam in Problem 10-1.4 so that its Euler buckling load is 300 kN, what wall thickness is required?

10-1.6. The beam in Problem 10-1.4 is provided with lateral supports as shown. What is the beam's buckling load?

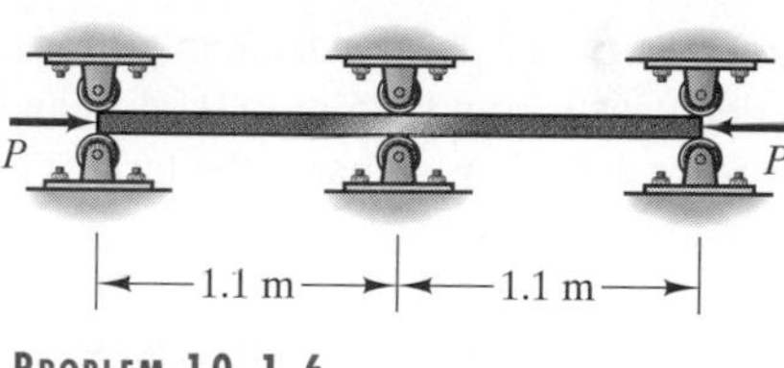

PROBLEM 10-1.6

10-1.7. Confirm that if λ is given by Eq. (10-2), then Eq. (10-3) satisfies Eq. (10-1) for any values of the constants B and C.

10-1.8. The beam has a solid circular cross section with radius R. The material will safely support an allowable normal stress σ_{allow}. Suppose that you want to achieve an optimal design in the sense that the compressive axial load P that subjects the material to a normal stress of magnitude σ_{allow} is equal to the beam's Euler buckling load. Show that this is achieved by choosing the dimensions R and L so that

$$\frac{R}{L} = \frac{2}{\pi}\sqrt{\frac{\sigma_{\text{allow}}}{E}}$$

P P L

PROBLEM 10-1.8

10-1.9. The bar AB has a solid circular cross section with 20-mm radius. It consists of material with elastic modulus $E = 14$ GPa. If the force F is gradually increased, at what value will bar AB buckle?

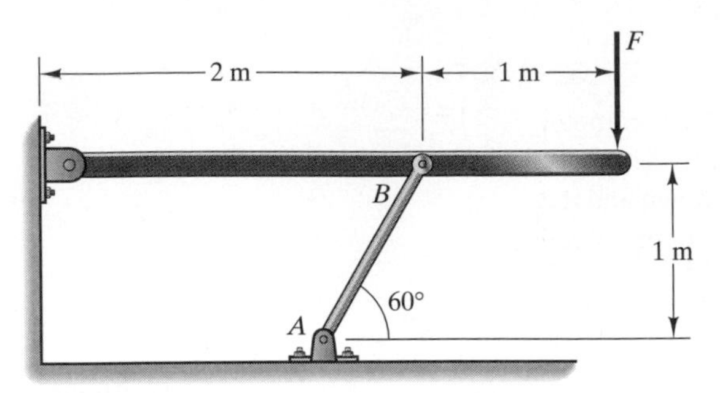

| PROBLEM 10-1.9

10-1.10. If you want to design the bar AB in Problem 10-1.9 so that it doesn't buckle until the force $F = 20$ kN, what should its radius be?

10-1.11. In Example 10-1, suppose that the outer radius of the bars is decreased from 0.05 m to 0.04 m and their wall thickness is kept at 0.01 m. What is the largest force F that can be applied to the truss?

10-1.12. The bars of the truss have the cross section shown and consist of material with elastic modulus $E = 2 \times 10^6$ psi. If the force F is gradually increased, at what value will the structure fail due to buckling?

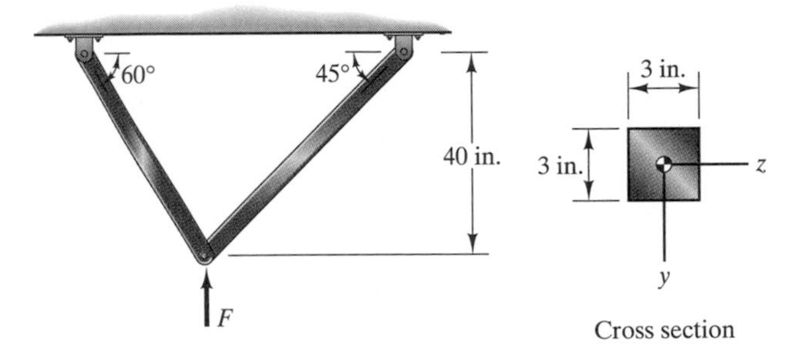

| PROBLEM 10-1.12

10-1.13. If the bars of the truss in Problem 10-1.12 consist of 2014-T6 Aluminum alloy, at what value of the force F will the structure fail due to buckling?

10-1.14. The bars of the truss have the cross section shown and consist of material with elastic modulus $E = 70$ GPa. If the force F is gradually increased, at what value will the structure fail due to buckling?

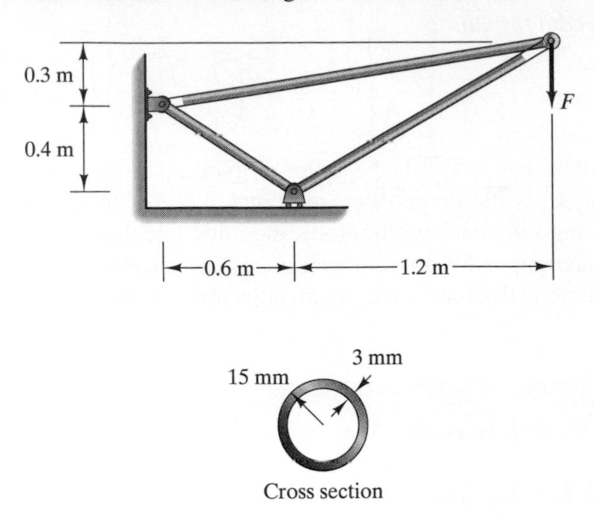

| PROBLEM 10-1.14

10-1.15. In Problem 10-1.14, suppose that you want to increase the outer radius of the cross section of the bars with the wall thickness kept at 3 mm so that the structure does not fail due to buckling until $F = 7.5$ kN. What should the outer radius be?

10-1.16. The bars of the truss have a solid circular cross section with 30-mm radius and consist of material with elastic modulus $E = 70$ GPa. What is the smallest value of the mass m that will cause the structure to fail due to buckling?

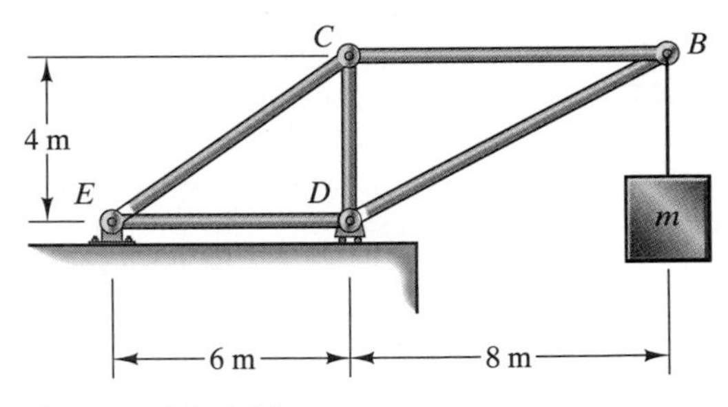

| PROBLEM 10-1.16

10-1.17. The bars of the truss have the cross section shown and consist of material with elastic modulus $E = 2 \times 10^6$ psi. If $b = 4$ in. and the force F is gradually increased, at what value of F will the structure fail due to buckling?

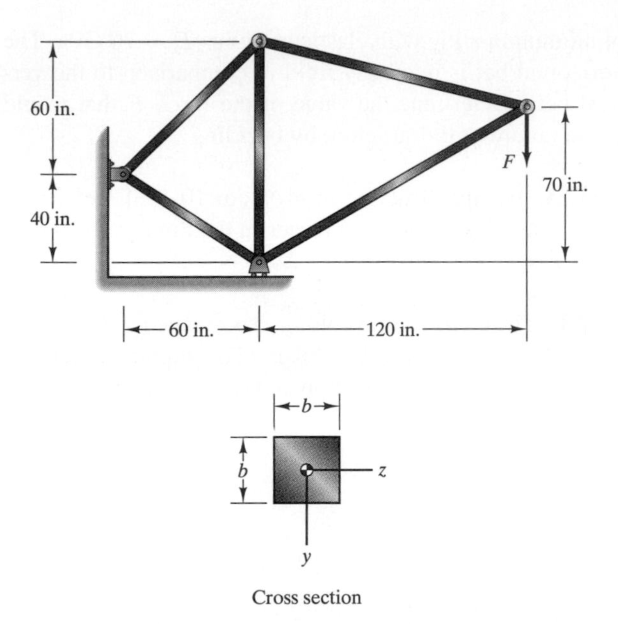

PROBLEM 10-1.17

10-1.18. The truss in Problem 10-1.17 consists of material that will safely support an allowable normal stress $\sigma_{\text{allow}} = 1000$ psi in tension or compression. Suppose that you want to achieve an optimal design in the sense that the force F which causes the maximum normal stress in the truss to equal σ_{allow} is equal to the smallest force F that causes buckling of a member. Determine the necessary value of the dimension b and the value of F at which failure occurs.

10-1.19. Bars AB and CD have a solid circular cross section with 20-mm radius. They consist of material with elastic modulus $E = 14$ GPa. If the force F is gradually increased, at what value does the structure fail due to buckling?

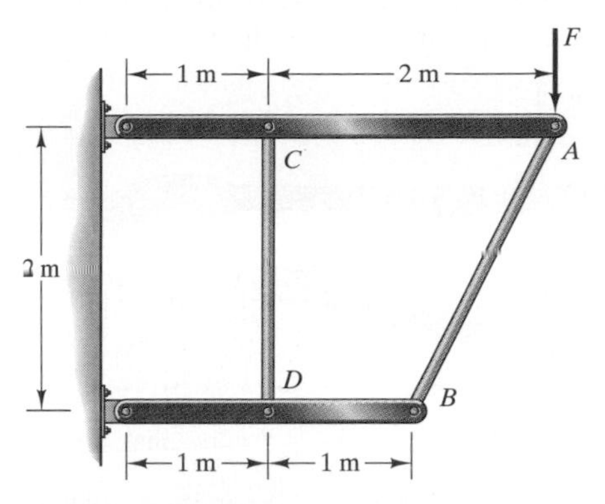

PROBLEM 10-1.19

10-1.20. The bars in Problem 10-1.19 will safely support a normal stress (in tension or compression) of 12 GPa. Suppose that you don't want the structure to fail until the force F reaches 5 kN, but you want the radius of each of the bars AB and CD to be as small as possible. What are the radii of the two bars?

10-1.21. The system supports half of the weight of the 680-kg excavator. Member AC has the cross section shown and consists of material with elastic modulus $E = 73$ GPa.

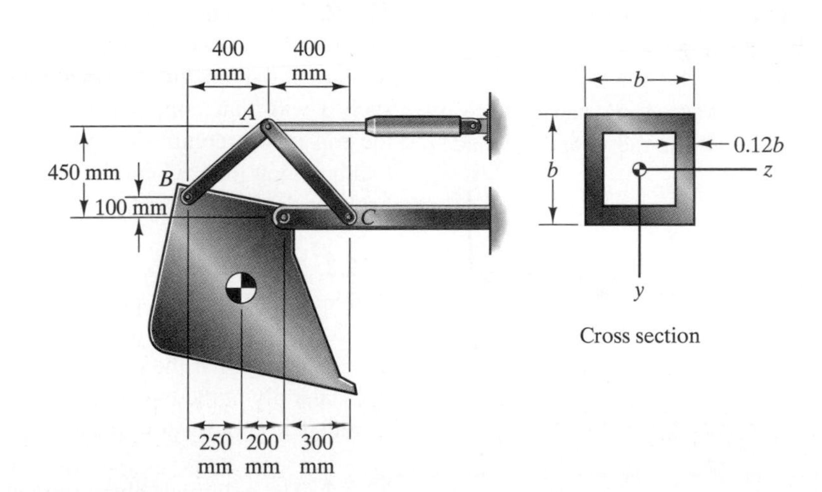

PROBLEM 10-1.21

What value of the dimension b would cause member AC to buckle with the stationary system in the position shown?

10-1.22. The link AB of the pliers has the cross section shown and is made of steel with elastic modulus $E = 190$ GPa. Determine the value of the force F that would cause failure of the link by buckling.

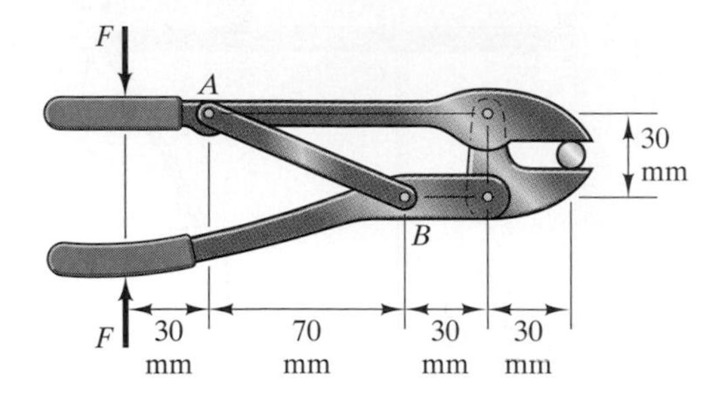

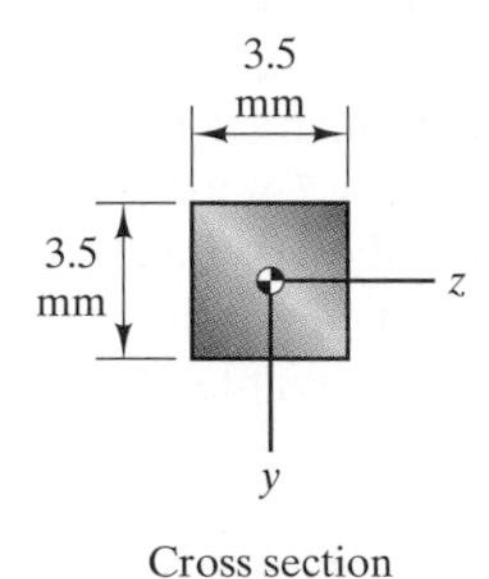

PROBLEM 10-1.22

10-1.23. If the link AB of the pliers in Problem 10-1.22 is made of 7075-T6 aluminum alloy, what force F would cause failure of the link by buckling?

10-1.24. The identical vertical bars A, B, and C have a solid circular cross section with 5-mm radius and are made of aluminum alloy with elastic modulus $E = 70$ GPa. The horizontal bar is relatively rigid in comparison to the vertical bars. Determine the value of the force F that would cause failure of the structure by buckling.

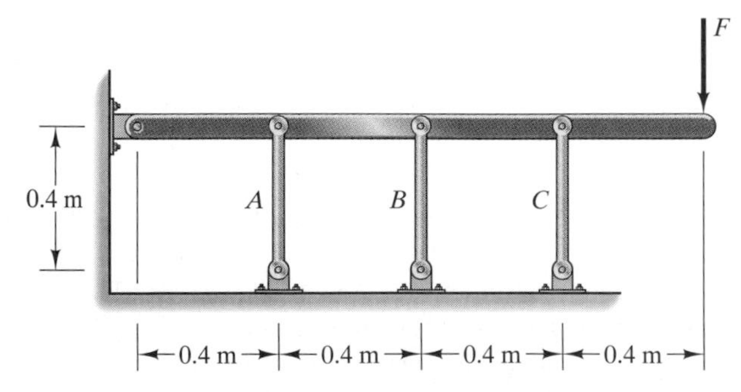

PROBLEM 10-1.24

10-1.25. For the structure in Problem 10-1.24, determine the magnitudes of the axial forces in the three vertical bars **(a)** if $F = 2200$ N; **(b)** if $F = 2800$ N.

10-2.1. The architectural column has elastic modulus $E = 1.8 \times 10^6$ psi. Its base is built in. The support at the top prevents both lateral deflection and rotation. Determine the buckling load P.

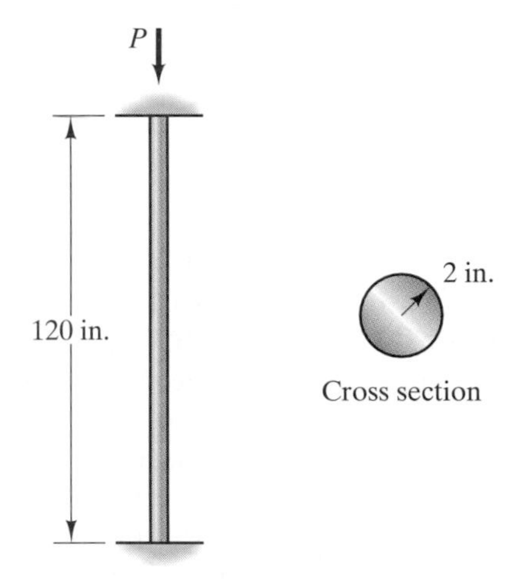

PROBLEM 10-2.1

10-2.2. The architect's specifications for the column in Problem 10-2.1 would prevent both lateral deflection and rotation at the top, but the contractor installs the column in such a way that only lateral deflection is prevented and the top of the column is free to rotate. What is the actual buckling load? Compare your answer to the answer to Problem 10-2.1.

10-2.3. The architect's specifications for the column in Problem 10-2.1 would prevent both lateral deflection and rotation at the top, but the contractor installs the column in such a way that the column is free to rotate and to deflect laterally at the top. What is the actual buckling load? Compare your answer to the answer to Problem 10-2.1.

10-2.4. The column is aluminum alloy with elastic modulus $E = 70$ GPa. Its base is built in. The support at the top prevents lateral deflection but allows rotation in the x–z plane. Determine the buckling load P.

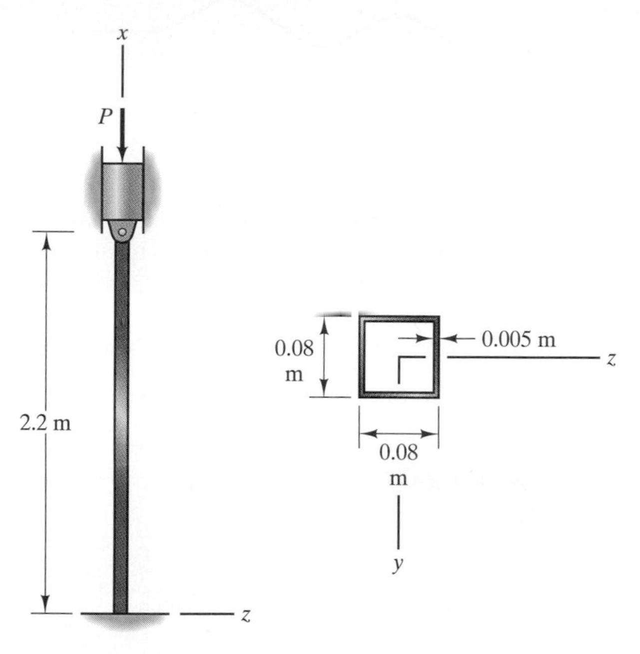

PROBLEM 10-2.4

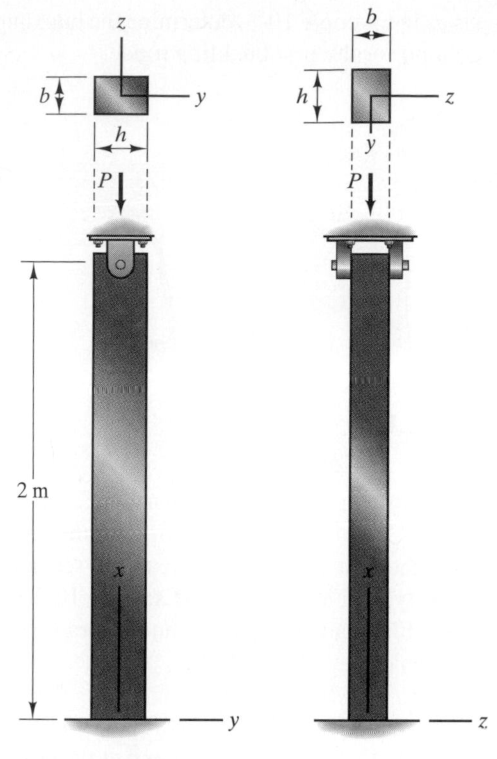

PROBLEM 10-2.7

10-2.5. Suppose that you want to increase the wall thickness of the cross section of the column in Problem 10-2.4, while keeping the 0.08-m dimension fixed, so that the column's buckling load is 700 kN. What is the necessary wall thickness?

10-2.6. In Example 10-2, suppose that the base of the column is supported in the same way as the top. Then the supports at the top and bottom prevent lateral deflection and prevent rotation in the x–z plane but allow rotation in the x–y plane. What is the column's buckling load? Does it buckle by bending in the x–y plane or the x–z plane?

10-2.7. The column supports an axial load P. The base of the column is built in. The support at the top prevents lateral deflection. The support at the top allows rotation of the column in the x–y plane but prevents rotation in the x–z plane. The dimensions $b = 0.08$ m and $h = 0.14$ m. The elastic modulus is $E = 12$ GPa. What is the column's buckling load? Does it buckle by bending in the x–y plane or the x–z plane?

10-2.8. Solve Problem 10-2.7 if $b = 0.12$ m and $h = 0.14$ m.

10-2.9. For the column shown in Problem 10-2.7, determine the ratio b/h for which the buckling load if the column buckles in the x–y plane is equal to the buckling load if it buckles in the x–z plane.

10-2.10. For the beam in Fig. 10-11, determine the buckling loads for buckling modes two, three, and four, and draw graphs of the distributions of the deflection.

10-2.11. For the beam in Example 10-3, determine the buckling load for the fourth buckling mode and draw a graph of the distribution of the deflection.

10-2.12. The rectangular platform in Fig. (a) is supported by four identical columns of length L. The platform is loaded with weights in such a way that the axial force on each column is the same until the columns buckle [Fig. (b)]. The connections of the columns to the floor and the platform behave like built-in supports. By using the same type of analysis as in Example 10-3, determine the buckling load of each column for the first buckling mode.

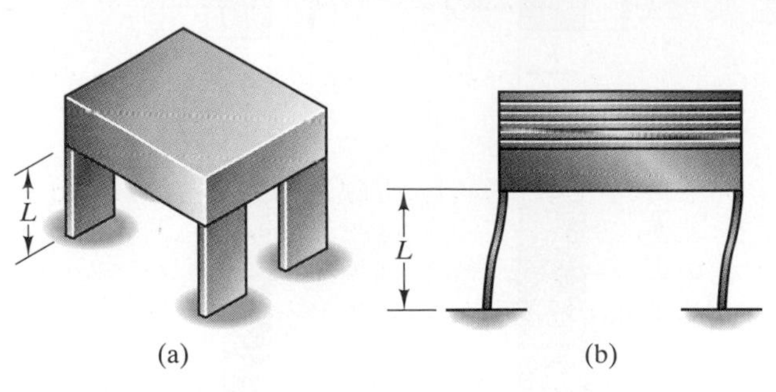

PROBLEM 10-2.12

10-2.13. Suppose that the rectangular platform in Problem 10-2.12 is provided with supports in the horizontal plane so that the four columns buckle in the second mode. Use the same type of analysis as in Example 10-3 to determine the buckling load of each column and draw a graph of the distribution of the deflection.

10-2.14. Figure 10-5 shows the first four buckling modes of a beam of length L that is free to bend at the ends. What are the effective lengths of the four modes?

10-2.15. In Fig. 10-20, the distributions of the deflection and the buckling loads are given for the first three modes of the beam in Fig. 10-18. Use the expressions for the buckling loads to calculate the effective lengths of the first three modes.

10-2.16. The prismatic beam of length L has a built-in support at the right end and is free at the left end. It is shown buckled in the first mode. What is its effective length? Use the effective length to determine the buckling load.

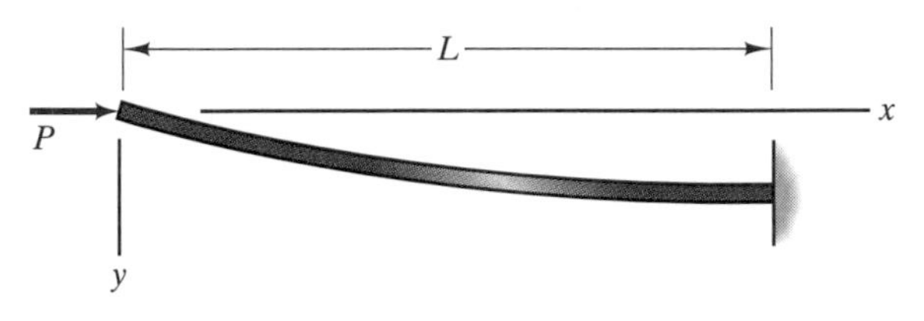

PROBLEM 10-2.16

10-2.17. The figure shows the fourth buckling mode of the beam in Fig. 10-18. **(a)** Use the figure to determine the approximate effective length. **(b)** Use the approximate effective length to determine the buckling load.

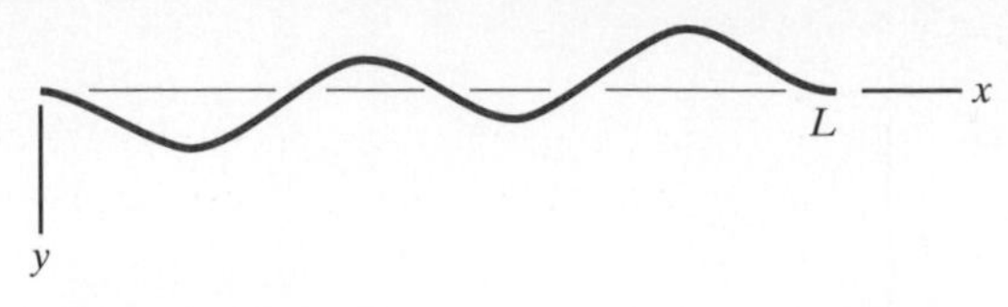

PROBLEM 10-2.17

10-3.1. The beam is subjected to compressive loads $P = 400$ kN with eccentricity $e = 0.5$ m. The beam's length is $L = 6$ m, its cross-sectional area is $A = 0.122$ m^2, and its moment of inertia is $I = 0.00125$ m^4. The elastic modulus of the material is 72 GPa. What is the beam's maximum deflection?

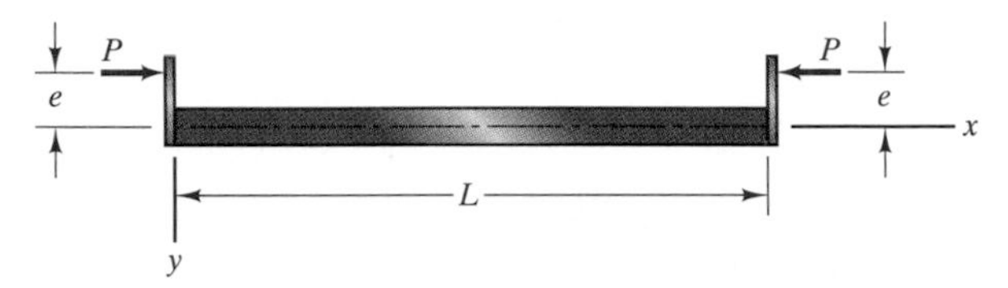

PROBLEM 10-3.1

10-3.2. For the beam in Problem 10-3.1, draw a graph of the deflection as a function of x.

10-3.3. Consider the eccentrically loaded beam in Problem 10-3.1. The distance from the beam's neutral axis to the location on the cross section where the maximum compressive stress occurs is $c = 0.175$ m. What is the maximum normal stress in the beam?

10-3.4. The beam shown in Problem 10-3.1 is subjected to compressive loads $P = 100$ kip with eccentricity $e = 20$ in. The beam's length is $L = 240$ in, its cross-sectional area is $A = 140$ in^2, and its moment of inertia is $I = 1730$ in^4. The elastic modulus of the material is 11×10^6 psi. What is the beam's maximum deflection?

10-3.5. Consider the eccentrically loaded beam described in Problem 10-3.4. The distance from the beam's neutral axis to the location on the cross section where the maximum compressive stress occurs is $c = 6$ in. What is the maximum normal stress in the beam?

10-3.6. If the compressive load on the beam in Example 10-5 is $P = 40$ kN, what is the beam's maximum deflection?

10-3.7. Suppose that the eccentricity in Example 10-5 is increased to $e = 0.0092$ m. Use Fig. 10-29 to estimate the largest allowable value of P.

10-3.8. The beam has the cross section shown, length $L = 2.5$ m, and is subjected to compressive loads with eccentricity $e = 0.00625$ m. The material is steel with modulus of elasticity $E = 200$ GPa. If the compressive load is $P = 200$ kN, what is the beam's maximum deflection?

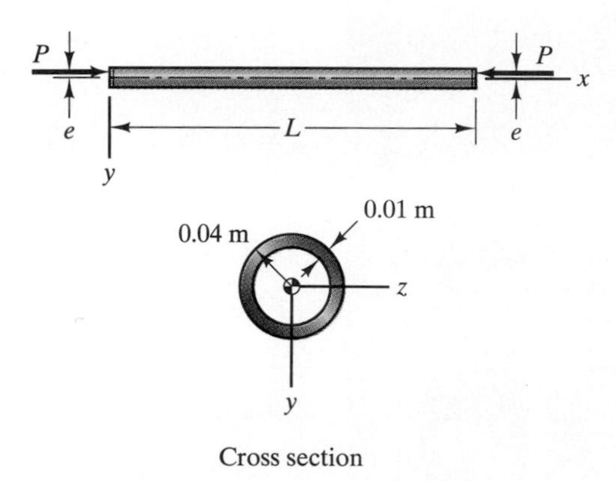

PROBLEM 10-3.8

10-3.9. What is the maximum normal stress in the beam in Problem 10-3.8?

10-3.10. Consider the eccentrically loaded beam described in Problem 10-3.8. If the allowable compressive stress is 250 MPa, what is the largest allowable value of P? (*Strategy:* Use Fig. 10-29 to estimate the value of P/A.)

10-3.11. By using the boundary conditions $v|_{x=0} = 0$ and $v|_{x=L} = 0$ to evaluate the constants B and C in Eq. (10-38), derive Eq. (10-39) for the deflection.

10-3.12. By substituting the trigonometric identities (10-40) into Eq. (10-39), derive Eq. (10-41) for the beam's deflection. Show that the deflection at $x = L/2$ is given by Eq. (10-42).

Energy is stored in the bent bow. When the archer releases the bowstring, the bow's energy is transferred to the arrow, producing its velocity.

CHAPTER 11

Energy Methods

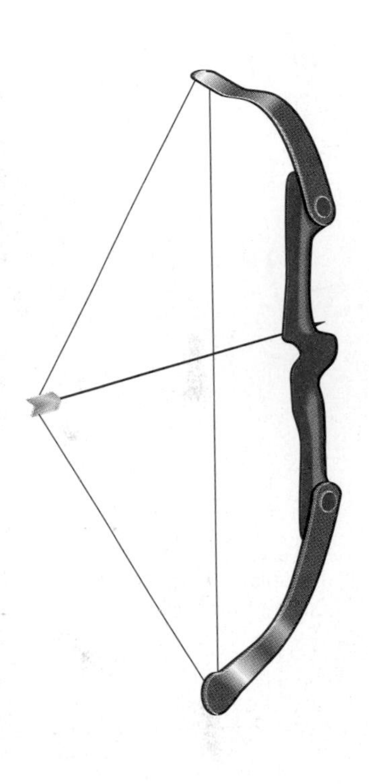

As in the case of a bent bow, energy can be stored in an object when work is done to deform it relative to a reference state. Many advanced techniques used in the mechanics of materials, including finite elements, are based on energy methods. In this chapter we introduce some of the fundamental concepts underlying these methods and present simple applications.

11-1 | Work and Energy

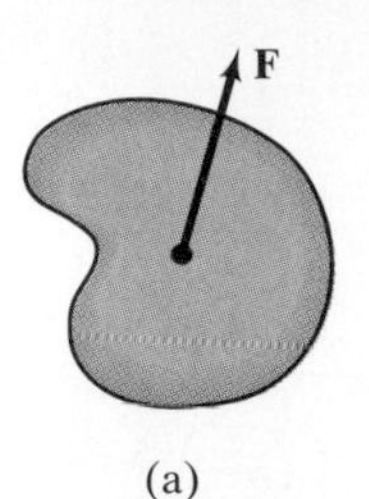

(a)

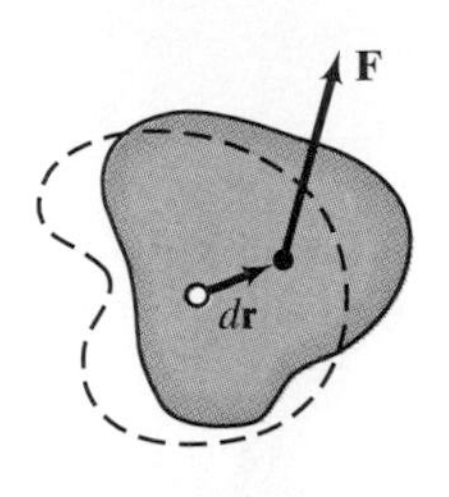

(b)

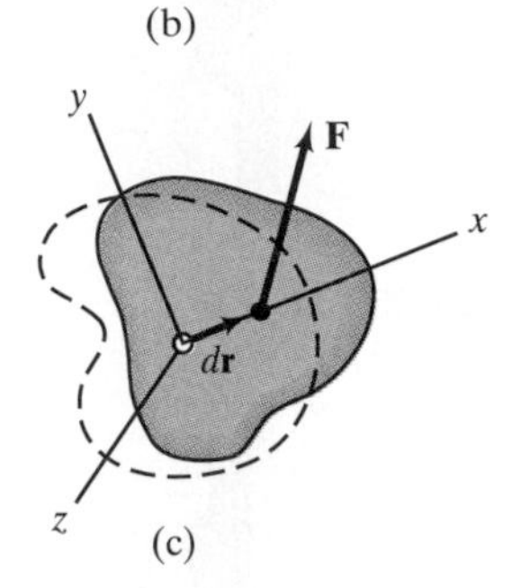

(c)

FIGURE 11-1 (a) Object subjected to a force. (b) Displacement of the point of application. (c) Introducing a coordinate system.

In this section we describe a class of problems in solid mechanics that can be analyzed by considering the energy stored in deformed elastic materials. To do so, we must first define the work done by a force or couple.

Work

Suppose that an object is subjected to a force **F** (Fig. 11-1a). If the point of the object to which **F** is applied undergoes an infinitesimal displacement represented by a vector $d\mathbf{r}$ (Fig. 11-1b), the *work* done on the object by the force is defined to be

$$dW = \mathbf{F} \cdot d\mathbf{r}.$$

The dimensions of work are (force)×(length). In U.S. Customary units, work is usually expressed in in-lb or ft-lb. In SI units, work is expressed in N-m, or joules (J). Let us introduce a coordinate system oriented as shown in Fig. 11-1c, with the x axis parallel to $d\mathbf{r}$. Expressing the force and displacement vectors in terms of their components as $\mathbf{F} = F_x\,\mathbf{i} + F_y\,\mathbf{j} + F_z\,\mathbf{k}$ and $d\mathbf{r} = dx\,\mathbf{i}$, the work is

$$\begin{aligned} dW &= (F_x\,\mathbf{i} + F_y\,\mathbf{j} + F_z\,\mathbf{k}) \cdot (dx\,\mathbf{i}) \\ &= F_x\,dx. \end{aligned}$$

We see that *only the component of the force parallel to the displacement does work.*

Since we will only be concerned with the component of **F** parallel to the displacement, let $F_x = F$. If the point to which F is applied moves a finite distance along the x axis from $x = 0$ to $x = x_0$ (Fig. 11-2), the work done is

$$W = \int_0^{x_0} F\,dx.$$

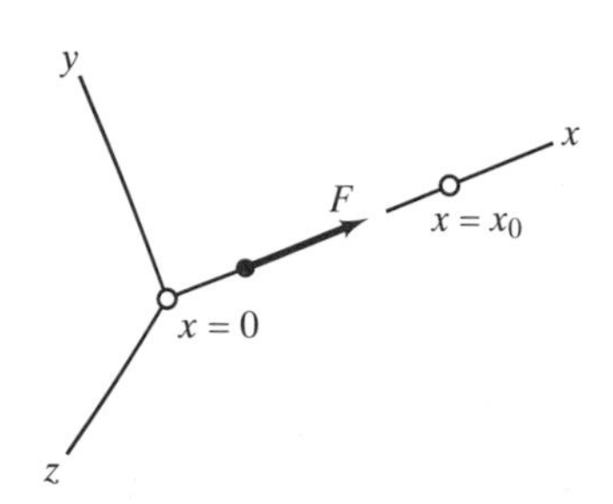

FIGURE 11-2 Path of the point of application along the x axis.

In a graph of F as a function of x, the work done equals the "area" between the force and the x axis (Fig. 11-3a). Two particular cases are of interest:

Constant Force If the force is a constant $F = F_0$, the work is simply the product of F_0 and the magnitude of the displacement (Fig. 11-3b):

$$W = F_0 x_0.$$

Linear Force Suppose that F is proportional to x and its value is F_0 when $x = x_0$. Then $F = (F_0/x_0)x$, and the work is

$$\begin{aligned} W &= \int_0^{x_0} \frac{F_0}{x_0} x\,dx \\ &= \frac{1}{2} F_0 x_0. \end{aligned}$$

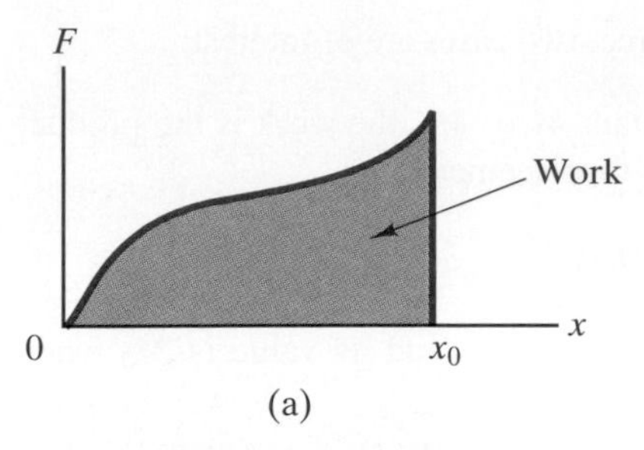

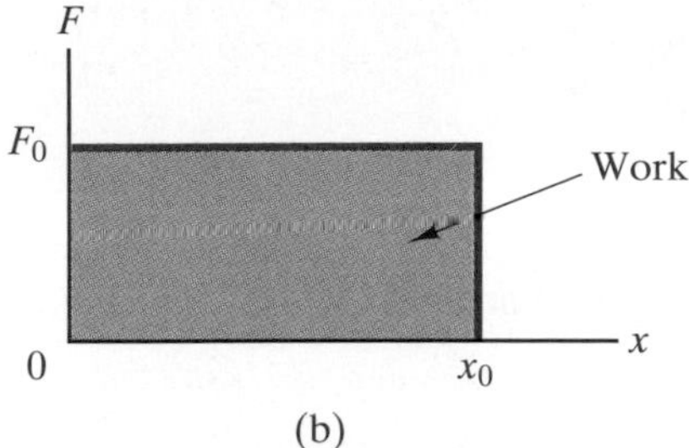

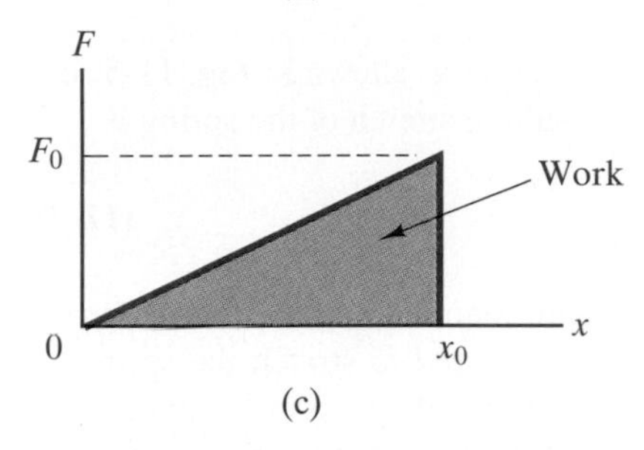

FIGURE 11-3 (a) The work equals the area defined by the graph of F as a function of x. (b) Work done by a constant force. (c) Work done by a linear force.

In this case the work is one-half the product of the value of F at $x = x_0$ and the magnitude of the displacement (Fig. 11-3c).

Now suppose that an object is subjected to a couple (Fig. 11-4a). Both forces are contained in the plane of the page, so the couple exerts a counterclockwise moment $M = Fh$. If the object undergoes a motion that causes the couple to rotate through a counterclockwise angle $d\theta$ (Fig. 11-4b), the work done is $dW = F(\frac{1}{2}h\,d\theta) + F(\frac{1}{2}h\,d\theta) = M\,d\theta$. Thus when a couple of moment M rotates through an angle $d\theta$ in the same direction as the couple (Fig. 11-4c), the work done is

$$dW = M\,d\theta. \tag{11-1}$$

If the angle varies from $\theta = 0$ to $\theta = \theta_0$, the work done is

$$W = \int_0^{\theta_0} M\,d\theta.$$

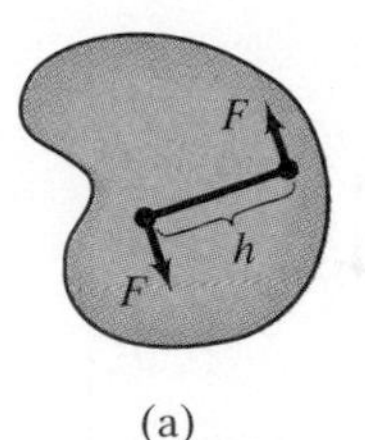

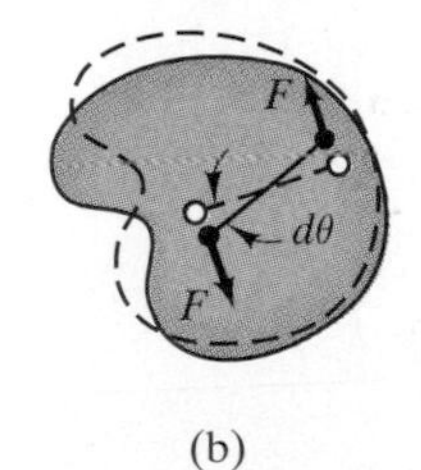

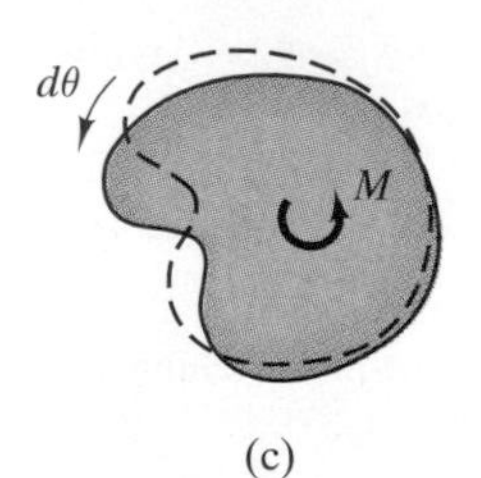

FIGURE 11-4 (a) Couple acting on an object. (b) Infinitesimal rotation of the object. (c) Object acted upon by a couple of moment M that rotates through an angle $d\theta$.

As in the case of the work done by a force, two cases are of interest:

Constant Couple If the moment is a constant $M = M_0$, the work is the product of M_0 and the magnitude of the angular displacement:

$$W = M_0\theta_0.$$

Linear Couple Suppose that M is proportional to θ and its value is M_0 when $\theta = \theta_0$. Then $M = (M_0/\theta_0)\theta$, and the work is

$$\begin{aligned} W &= \int_0^{\theta_0} \frac{M_0}{\theta_0}\theta\, d\theta \\ &= \tfrac{1}{2}M_0\theta_0. \end{aligned}$$

The work is one-half the product of the value of M at $\theta = \theta_0$ and the magnitude of the angular displacement.

Strain Energy

Suppose that we apply a force F to a linear spring as shown in Fig. 11-5. By definition, the relation between F and the resulting stretch of the spring is

$$F = kx, \tag{11-2}$$

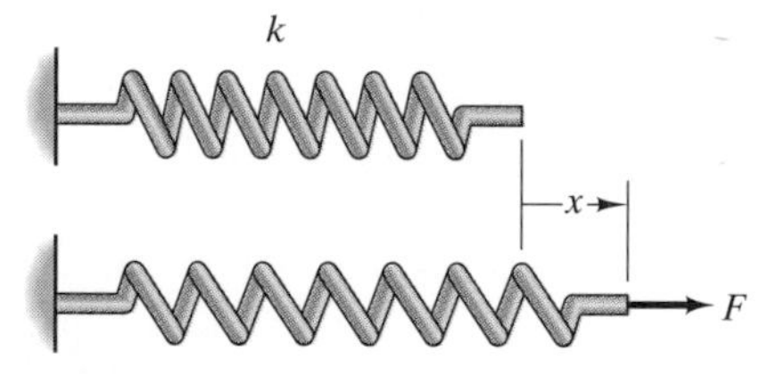

FIGURE 11-5 Stretching a linear spring.

where k is the spring constant. Let the force required to stretch the spring an amount x_0 be $F_0 = kx_0$. Since the force required to stretch the spring is a linear function of x, we have seen that the work W done in stretching the spring an amount x_0 is one-half the product of the force necessary to cause the displacement and the magnitude of the displacement (Fig. 11-6):

$$W = \tfrac{1}{2}F_0x_0.$$

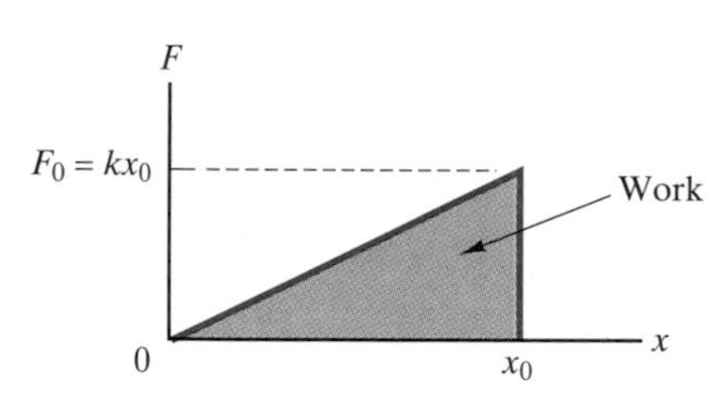

FIGURE 11-6 Work done in stretching a linear spring an amount x_0.

The work done is stored in the spring in the sense that the spring is capable of doing an equal amount of work as it contracts to its original length. We say that the work can be stored in the form of potential energy. We will show that potential energy can be stored within any object consisting of elastic material when the object is deformed. This potential energy is called *strain energy*.

We begin with a cube of elastic material with dimensions $b \times b \times b$ (Fig. 11-7a) and subject it to a normal stress σ_x (Fig. 11-7b). The tensile force exerted on the cube is $\sigma_x(b)^2$, and the change in its length in the x direction is $\epsilon_x b$. The work done, which equals the strain energy stored in the cube, is one-half the product of the force necessary to cause the displacement and the displacement:

$$\begin{aligned} \text{strain energy} = U &= \tfrac{1}{2}[\sigma_x(b)^2]\epsilon_x b \\ &= \tfrac{1}{2}\sigma_x\epsilon_x b^3. \end{aligned}$$

Let the strain energy per unit volume of material be denoted by u. The volume

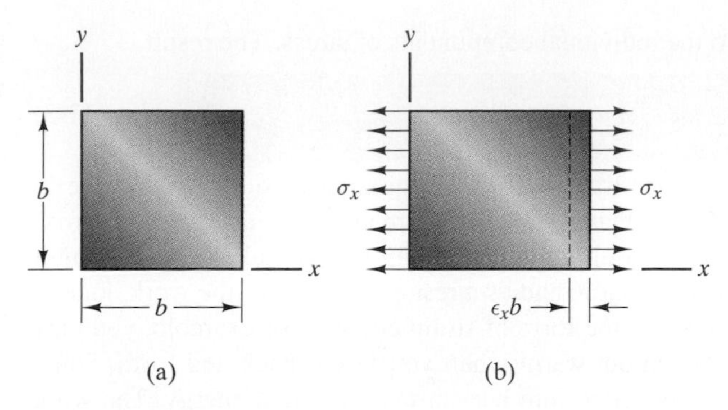

| **FIGURE 11-7** Subjecting a cube of material to normal stress.

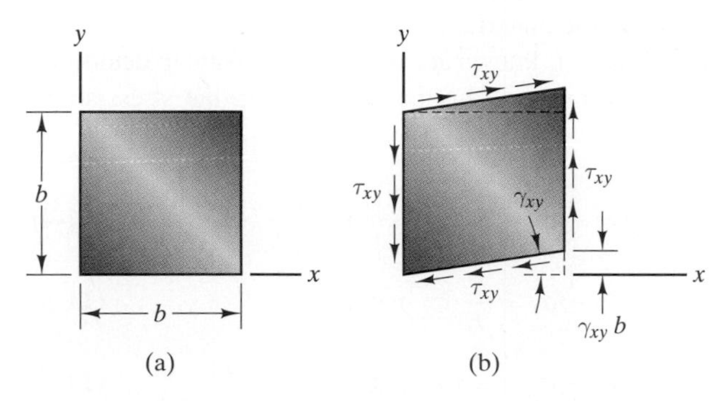

| **FIGURE 11-8** Subjecting the cube to shear stress.

of the cube is b^3, so

$$u = \tfrac{1}{2}\sigma_x \epsilon_x.$$

Now let us subject the cube to a shear stress τ_{xy} (Fig. 11-8) and determine the strain energy per unit volume. How much work is done on the cube? Comparing the deformed shape of the cube to its original shape as shown in Fig. 11-8b, we see that work is done only by the shear stress acting on the cube's right face. (The force exerted on each infinitesimal element of the top and bottom faces is perpendicular to the motion of the element and so does no work.) The force exerted on the right face is $\tau_{xy}(b)^2$, and its displacement in the y direction is $\gamma_{xy} b$. The resulting strain energy is

$$\begin{aligned} U &= \tfrac{1}{2}[\tau_{xy}(b)^2]\gamma_{xy} b \\ &= \tfrac{1}{2}\tau_{xy}\gamma_{xy} b^3, \end{aligned}$$

so the strain energy per unit volume of material is

$$u = \tfrac{1}{2}\tau_{xy}\gamma_{xy}.$$

We can obtain the strain energy per unit volume of an isotropic linearly elastic material subjected to a general state of stress by superimposing the strain energies due to the individual components of stress. The result is

$$u = \tfrac{1}{2}(\sigma_x \epsilon_x + \sigma_y \epsilon_y + \sigma_z \epsilon_z + \tau_{xy}\gamma_{xy} + \tau_{yz}\gamma_{yz} + \tau_{xz}\gamma_{xz}). \quad \textbf{(11-3)}$$

We have calculated the strain energy by equating it to the work done in deforming a linearly elastic material. Although many materials closely approximate elastic behavior under appropriate conditions, all real materials exhibit some internal energy dissipation, and as a result only part of the work done in deforming them is stored in the form of strain energy. For example, you may have noticed a wire becoming warm when you flex it back and forth. Some of the work you do is converted into heat instead of strain energy. The work done in permanently deforming an object is obviously not stored as recoverable strain energy. In fact, the existence of a recoverable strain energy is sometimes used as the definition of an elastic material.

Suppose that the state of stress is known at a point and we want to determine the strain energy per unit volume u at that point. By using the stress-strain relations [Eqs. (6-43)–(6-48)]

$$\epsilon_x = \frac{1}{E}\sigma_x - \frac{\nu}{E}(\sigma_y + \sigma_z), \quad \textbf{(11-4)}$$

$$\epsilon_y = \frac{1}{E}\sigma_y - \frac{\nu}{E}(\sigma_x + \sigma_z), \quad \textbf{(11-5)}$$

$$\epsilon_z = \frac{1}{E}\sigma_z - \frac{\nu}{E}(\sigma_x + \sigma_y), \quad \textbf{(11-6)}$$

$$\gamma_{xy} = \frac{1}{G}\tau_{xy}, \quad \textbf{(11-7)}$$

$$\gamma_{yz} = \frac{1}{G}\tau_{yz}, \quad \textbf{(11-8)}$$

$$\gamma_{xz} = \frac{1}{G}\tau_{xz}, \quad \textbf{(11-9)}$$

we can determine the state of strain and then use Eq. (11-3) to determine u. Or, if we know the state of strain at a point, we can solve Eqs. (11-4)–(11-9) for the state of stress and then use Eq. (11-3) to determine u.

Applications

Deflections of trusses and beams can be determined by equating the work done by external forces and couples during their application to the resulting strain energy. Here we consider applications in which only a single external force or couple does work. Although this restriction allows us to introduce energy methods in a simple context, it greatly limits the situations we can consider. We remove this restriction in Section 11-2.

Axially Loaded Bars

In Fig. 11-9 an axial load P acts on a prismatic bar of length L, causing its length to increase an amount δ. Suppose that we know P and want to determine δ.

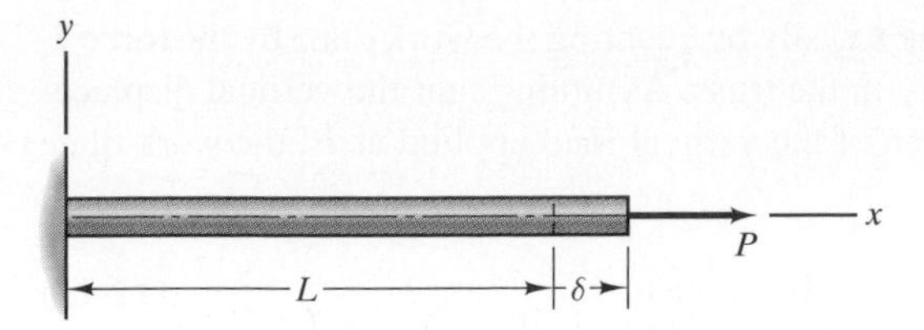

FIGURE 11-9 Applying an axial load to a bar.

Since the change in length of the bar is a linear function of the axial load, the work done by the external load P is one-half the product of the force and the change in the bar's length:

$$W = \tfrac{1}{2} P\delta. \tag{11-10}$$

If we can determine the strain energy in the bar, we can equate it to the external work and solve for δ. In terms of the coordinate system shown, we know that the only nonzero stress component in the bar is $\sigma_x = P/A$, where A is the bar's cross-sectional area. From Eq. (11-4), the strain component $\epsilon_x = (1/E)\sigma_x$. Applying Eq. (11-3), the strain energy per unit volume of the bar is

$$\begin{aligned} u &= \frac{1}{2}\sigma_x \epsilon_x \\ &= \frac{1}{2E}\sigma_x^2 \\ &= \frac{P^2}{2EA^2}. \end{aligned}$$

Multiplying this expression by the bar's volume AL, the strain energy of the axially loaded bar is

$$U = \frac{P^2 L}{2EA}. \tag{11-11}$$

Equating this result to Eq. (11-10) and solving for δ, we obtain

$$\delta = \frac{PL}{EA}.$$

You are probably yawning at this point, because we already obtained this result in a simpler way in Chapter 3. But we are now in a position to consider an example that dramatically demonstrates the power of energy methods.

In Fig. 11-10 a vertical load F acts at point B of the truss. Let the vertical component of the resulting displacement of point B be denoted by v. Suppose that we know F and want to determine v. We can analyze the truss in the usual way to determine the axial loads in the individual members (assuming that the change in the geometry of the truss due to the application of F is small) and thereby determine the change in length of each member. But using that information to determine the displacement v would be very difficult.

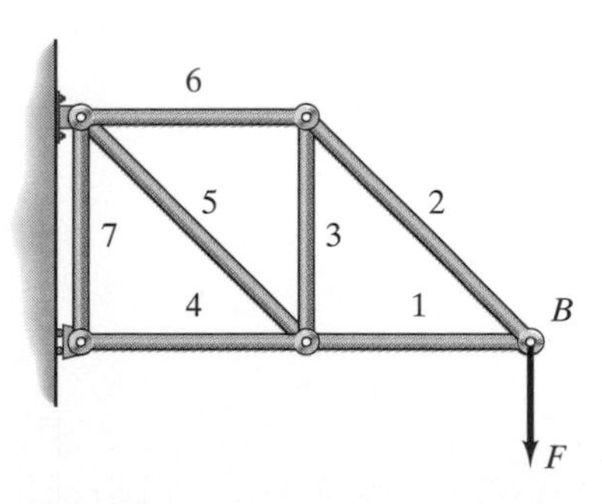

FIGURE 11-10 Subjecting a truss to a force F.

Instead, we can determine v easily by equating the work done by the force F to the resulting strain energy in the truss. Assuming that the vertical displacement at B is a linear function of the vertical load applied at B, the work done by F is

$$W = \tfrac{1}{2} F v. \qquad \textbf{(11-12)}$$

Let the axial loads in the members be $P_1, P_2, \ldots, P_7$. Once we have determined the axial loads, we can calculate the strain energy in each member. From Eq. (11-11), the strain energy in the ith member is

$$U_i = \frac{P_i^2 L_i}{2E_i A_i}.$$

We equate the work done by the external force to the total strain energy,

$$\frac{1}{2} F v = \sum_{i=1}^{7} \frac{P_i^2 L_i}{2E_i A_i},$$

and solve for v:

$$v = \frac{1}{F} \sum_{i=1}^{7} \frac{P_i^2 L_i}{E_i A_i}.$$

Although this result was easy to obtain, and the example is an important type of problem, notice that we have determined the displacement of the point where F is applied, and F is the only external force or couple that does work. The procedure we use in this section cannot be used to determine the displacements of other points of the truss or be applied to more complicated loadings. We consider such problems in Section 11-2.

EXAMPLE 11-1

The truss in Fig. 11-11 is subjected to a vertical force F at B. The members each have elastic modulus E and cross-sectional area A. What is the vertical component of the resulting displacement of point B?

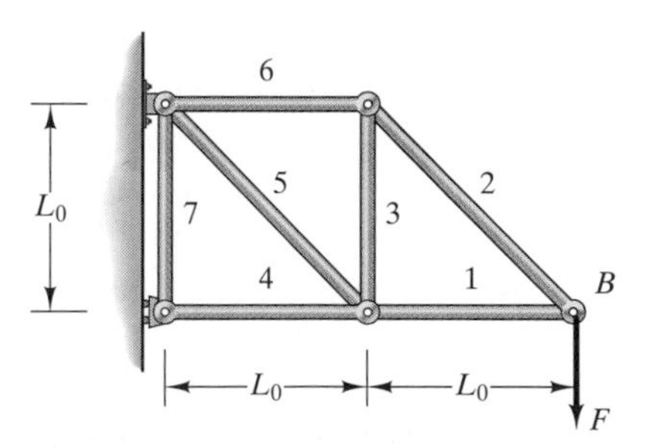

FIGURE 11-11

Strategy

Let the vertical component of the displacement of B be denoted by v. We can determine v by equating the work done by the external load F to the sum of the strain energies in the members. The strain energy in an axially loaded bar is given by Eq. (11-11).

Solution

The work done by F is

$$W = \frac{1}{2} F v.$$

Let P_i be the axial load in the ith member. From Eq. (11-11), the strain energy in the ith member is

$$U_i = \frac{P_i^2 L_i}{2EA}.$$

We equate the work to the sum of the strain energies,

$$\frac{1}{2} F v = \sum_{i=1}^{7} \frac{P_i^2 L_i}{2EA},$$

and solve for v:

$$v = \frac{1}{FEA} \sum_{i=1}^{7} P_i^2 L_i. \qquad \textbf{(11-13)}$$

Determining the axial loads in the members and their lengths, we obtain

$$\begin{aligned}
P_1 &= -F, & L_1 &= L_0 \\
P_2 &= \sqrt{2}F, & L_2 &= \sqrt{2}L_0 \\
P_3 &= -F, & L_3 &= L_0 \\
P_4 &= -2F, & L_4 &= L_0 \\
P_5 &= \sqrt{2}F, & L_5 &= \sqrt{2}L_0 \\
P_6 &= F, & L_6 &= L_0 \\
P_7 &= 0, & L_7 &= L_0.
\end{aligned}$$

Substituting these values into Eq. (11-13), the vertical displacement is

$$v = \frac{(7 + 4\sqrt{2}) F L_0}{EA}.$$

BEAMS

Consider a prismatic beam whose cross section is symmetric about the vertical (y) axis. Our objective is to determine the strain energy in the beam resulting from the normal stress due to bending. Figure 11-12 shows an element of a

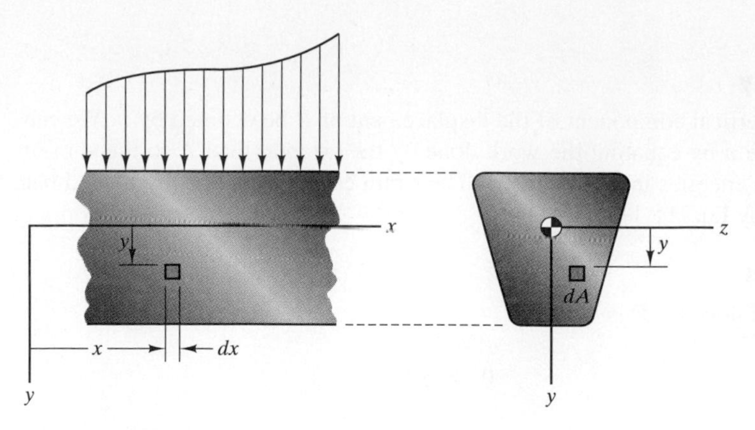

FIGURE 11-12 Element of a beam.

beam with length dx and cross-sectional area dA. This element is subjected to normal stress

$$\sigma_x = \frac{My}{I}, \tag{11-14}$$

where y is the element's position relative to the neutral axis and M is the bending moment at the axial position x. From Eqs. (11-3) and (11-4), the strain energy per unit volume associated with the normal stress is

$$\begin{aligned} u &= \frac{1}{2}\sigma_x \epsilon_x \\ &= \frac{1}{2E}\sigma_x^2. \end{aligned}$$

Multiplying this expression by the volume of the element and substituting Eq. (11-14) give the strain energy of the element:

$$u\,dx\,dA = \frac{M^2 y^2}{2EI^2}\,dx\,dA.$$

Integrating this result with respect to A yields the strain energy of a "slice" of the beam of width dx, which we denote by dU:

$$dU = \int_A u\,dx\,dA = \frac{M^2}{2EI^2}\,dx \int_A y^2\,dA.$$

Recognizing that $\int_A y^2\,dA = I$, we obtain the strain energy of an element of the beam of width dx (Fig. 11-13) in the form

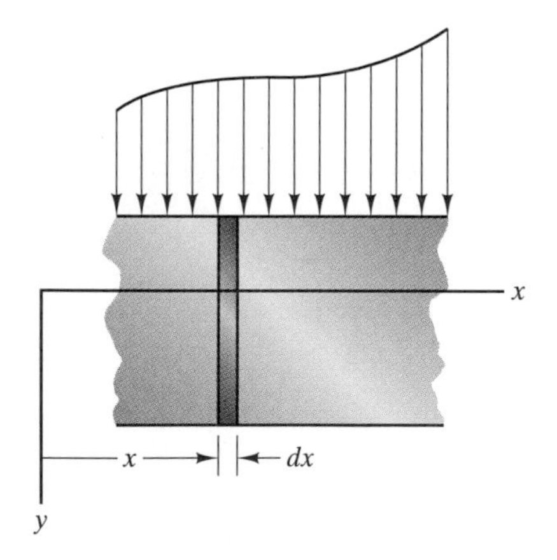

FIGURE 11-13 Element of a beam of width dx.

$$dU = \frac{M^2}{2EI}\,dx. \tag{11-15}$$

In general, the strain energy in a beam also includes contributions due to the stresses caused by the axial load P and the shear force V. But we will consider only beams that are not subjected to axial loads, and for slender beams the strain energy due to the shear force is normally negligible in comparison to that resulting from the bending moment.

To determine the strain energy of an entire beam, we must integrate Eq. (11-15) with respect to x over the beam's length. For example, the bending moment in the prismatic cantilever beam in Fig. 11-14 is

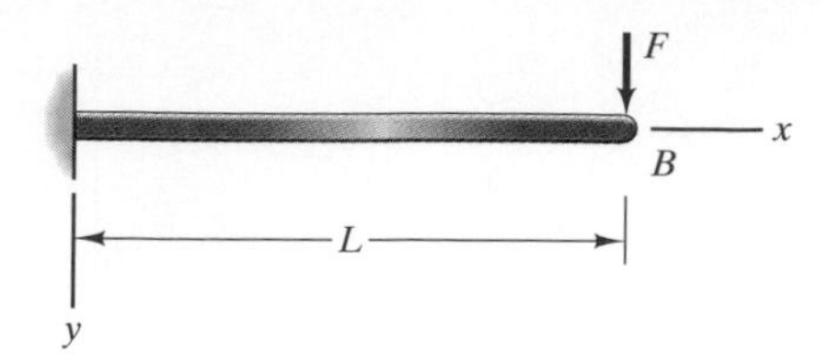

FIGURE 11-14 Cantilever beam subjected to a force.

$$M = F(x - L).$$

By substituting this expression into Eq. (11-15) and integrating from $x = 0$ to $x = L$, we obtain the strain energy:

$$U = \int_0^L \frac{[F(x - L)]^2}{2EI}\,dx$$

$$= \frac{F^2L^3}{6EI}.$$

Let v be the beam's deflection at B. The work done by the external force F is one-half the product of the force and the deflection at B:

$$W = \tfrac{1}{2}Fv.$$

Equating the work and the strain energy and solving for v, we obtain

$$v = \frac{FL^3}{3EI},$$

which agrees with the value given in Appendix E.

EXAMPLE 11-2

The simply supported beam in Fig. 11-15 is subjected to a vertical force F. What is the beam's deflection at the point where F is applied?

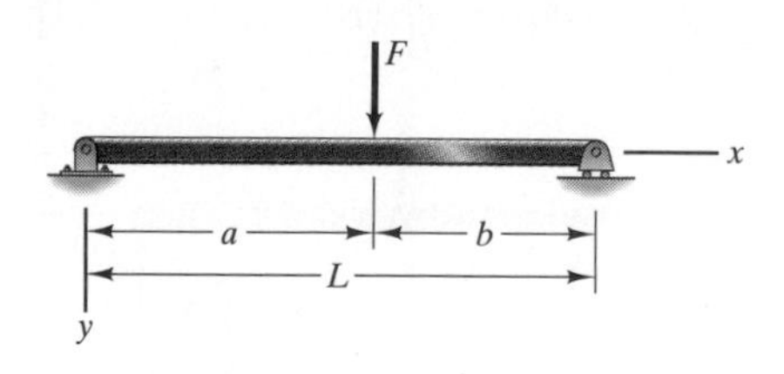

FIGURE 11-15

Strategy

We can determine the deflection by equating the work done by the external load F to the strain energy in the beam. Since the bending moment in the beam is described by different equations in the regions $0 \le x \le a$ and $a \le x \le L$, we must determine the strain energy by integrating Eq. (11-15) over these two regions separately.

Solution

From Fig. (a), the bending moment in the region $0 \leq x \leq a$ is

$$M = \frac{bF}{L}x.$$

(a) Determining the bending moment in the region $0 \leq x \leq a$.

Integrating Eq. (11-15), the strain energy contained in this region of the beam is

$$\begin{aligned} U_1 &= \frac{1}{2EI}\int_0^a M^2\,dx \\ &= \frac{1}{2EI}\int_0^a \frac{b^2F^2}{L^2}x^2\,dx \\ &= \frac{a^3b^2F^2}{6EIL^2}. \end{aligned}$$

From Fig. (b), the bending moment in the region $a \leq x \leq L$ is

$$M = \frac{aF}{L}(L - x).$$

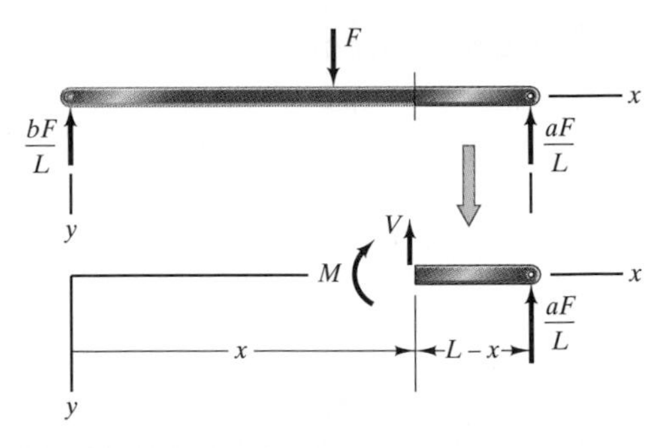

(b) Determining the bending moment in the region $a \leq x \leq L$.

The strain energy contained in this region of the beam is

$$\begin{aligned} U_2 &= \frac{1}{2EI}\int_a^L M^2\,dx \\ &= \frac{1}{2EI}\int_a^L \frac{a^2F^2}{L^2}(L-x)^2\,dx \\ &= \frac{a^2F^2}{2EIL^2}\left(\frac{L^3}{3} - L^2a + La^2 - \frac{a^3}{3}\right). \end{aligned}$$

Let v be the beam's deflection at the point where F is applied. The work done by F equals the total strain energy:

$$\begin{aligned} \frac{1}{2}Fv &= U_1 + U_2 \\ &= \frac{a^2F^2}{2EIL^2}\left(\frac{ab^2}{3} + \frac{L^3}{3} - L^2a + La^2 - \frac{a^3}{3}\right). \end{aligned}$$

Solving for v, the beam's deflection at the point where F is applied is

$$v = \frac{a^2F}{EIL^2}\left(\frac{ab^2}{3} + \frac{L^3}{3} - L^2a + La^2 - \frac{a^3}{3}\right).$$

11-2 | Castigliano's Second Theorem

In applying strain energy in Section 11-1, we were limited to situations in which only a single external force or couple did work, and we were able to determine the resulting deflection or angle only at the point at which the force or couple was applied. We can apply strain energy to a much broader class of problems using a result presented by Italian engineer Alberto Castigliano in 1879. We will sketch a proof of his result and then present some applications.

Derivation

Consider an object consisting of elastic material that is subjected to N external forces $F_1, F_2, \ldots, F_N$ (Fig. 11-16). Let u_i be the component of the deflection of the point of application of the ith force F_i in the direction of F_i, and assume that u_i depends linearly on F_i. Our objective is to determine u_i. We assume that the strain energy U resulting from the application of the N forces can be expressed as a function of the forces:

$$U = U(F_1, F_2, \ldots, F_N). \qquad \textbf{(11-16)}$$

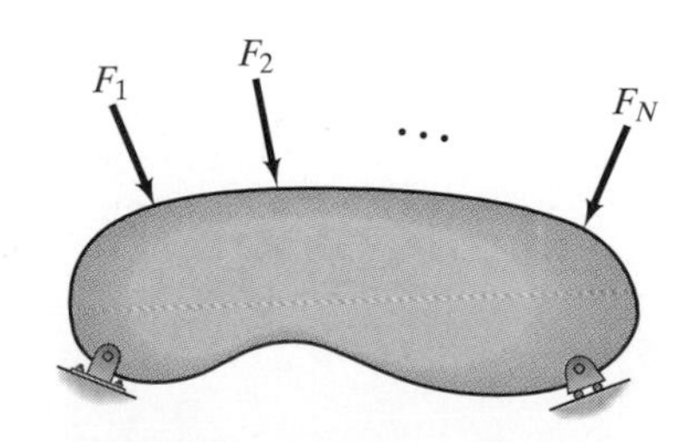

FIGURE 11-16 The N forces $F_1, F_2, \ldots, F_N$.

If we increase the magnitude of the ith force F_i by an amount ΔF_i and assume that Eq. (11-16) can be expressed as a Taylor series in terms of each of its arguments, we can write the increased strain energy as

$$\begin{aligned} &U(F_1, F_2, \ldots, F_i + \Delta F_i, \ldots, F_N) \\ &\quad = U(F_1, F_2, \ldots, F_N) + \frac{\partial U}{\partial F_i}\Delta F_i + O\left(\Delta F_i^2\right), \end{aligned} \qquad \textbf{(11-17)}$$

where the notation $O(\Delta F_i^2)$ means "terms of order two or greater in ΔF_i."

Now suppose that the force ΔF_i is applied before the forces $F_1, F_2, \ldots, F_N$. If Δu_i is the component of the displacement of the point of application of ΔF_i in the direction of ΔF_i, the work done is $\frac{1}{2}\Delta F_i\, \Delta u_i$. (Notice that $\Delta u_i \rightarrow 0$ as $\Delta F_i \rightarrow 0$.) If we then apply the forces $F_1, F_2, \ldots, F_N$, the force ΔF_i undergoes the additional displacement u_i, doing work $\Delta F_i\, u_i$. The resulting strain energy must equal the work done by the force ΔF_i plus the strain energy associated with the application of the forces $F_1, F_2, \ldots, F_N$:

$$\text{strain energy} = \tfrac{1}{2}\Delta F_i\, \Delta u_i + \Delta F_i\, u_i + U(F_1, F_2, \ldots, F_N). \qquad \textbf{(11-18)}$$

But the strain energy does not depend on the order in which the forces are applied, so Eqs. (11-17) and (11-18) must be equal:

$$\begin{aligned} &U(F_1, F_2, \ldots, F_N) + \frac{\partial U}{\partial F_i}\Delta F_i + O\left(\Delta F_i^2\right) \\ &\quad = \tfrac{1}{2}\Delta F_i \Delta u_i + \Delta F_i u_i + U(F_1, F_2, \ldots, F_N). \end{aligned}$$

Dividing this equation by ΔF_i and taking the limit as $\Delta F_i \rightarrow 0$, we obtain

$$u_i = \frac{\partial U}{\partial F_i}. \qquad \textbf{(11-19)}$$

This result is called *Castigliano's second theorem*. It can be used to determine the component of the deflection of the point of application of a given force F_i in the direction of F_i. In the following section we show how this result can also be used to determine the deflection of a point at which no force acts.

We have simplified the derivation of this theorem by assuming that the object was subjected only to forces. But as we will demonstrate in the following section, it can also be used when both forces and couples act on an object. In that case, the angle of rotation of the point of application of a given couple M_i in the direction of M_i is given by

$$\theta_i = \frac{\partial U}{\partial M_i}. \qquad \textbf{(11-20)}$$

Castigliano's second theorem is based on the assumption that the work done by external forces and couples is equal to the resulting strain energy. In the case of a structure, this requires that no net work be done at connections between members when forces and couples are applied. For example, the theorem does not hold if work is done by friction in pinned joints.

Applications

We can use Castigliano's second theorem to determine deflections of trusses and beams and we are not limited to applications in which only a single external force or couple does work. Furthermore, we can determine the deflection or rotation at a point at which no force or couple acts.

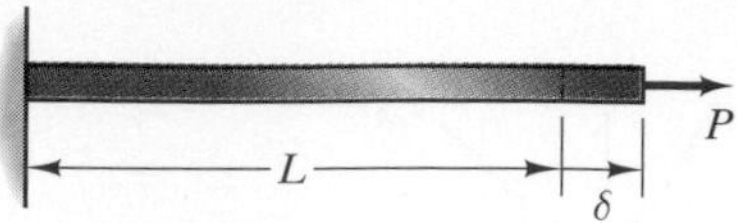

FIGURE 11-17 Subjecting a bar to an axial load.

AXIALLY LOADED BARS

In Fig. 11-17 an elastic bar of length L is subjected to an axial load P. From Eq. (11-11), the strain energy stored in the bar is

$$U = \frac{P^2 L}{2EA}.$$

Applying Castigliano's second theorem, we obtain the familiar expression for the displacement of the right end of the bar:

$$\delta = \frac{\partial U}{\partial P} = \frac{PL}{EA}.$$

Suppose that the truss in Fig. 11-18a consists of bars with the same elastic modulus and cross-sectional area, and that a downward force F is applied at joint A, resulting in horizontal and vertical deflections u and v (Fig. 11-18b). We can use Castigliano's second theorem to determine the vertical deflection v. From the free-body diagram of joint A (Fig. 11-19), we obtain the equilibrium equations

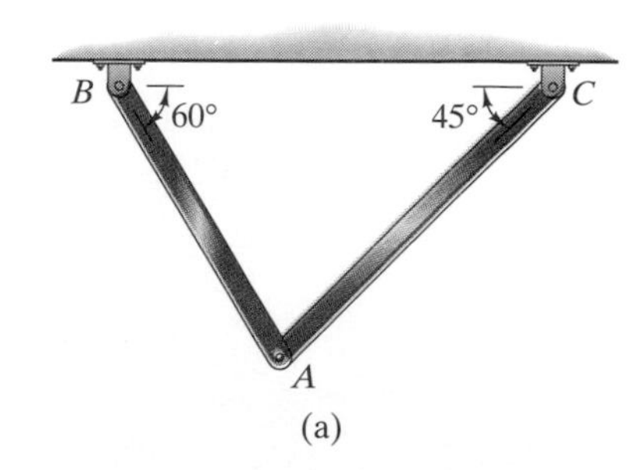

(a)

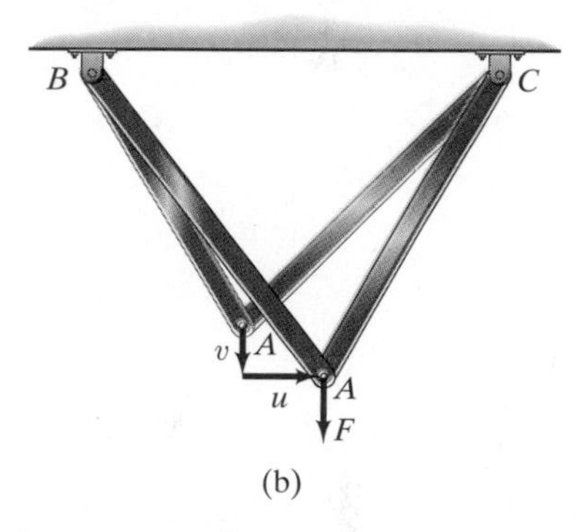

(b)

FIGURE 11-18 Subjecting a truss to a downward force F.

$$\Sigma F_x = -P_{AB}\cos 60^\circ + P_{AC}\cos 45^\circ = 0,$$

$$\Sigma F_y = P_{AB}\sin 60^\circ + P_{AC}\sin 45^\circ - F = 0.$$

The solutions of these equations for the axial loads in the bars are

$$P_{AB} = \frac{F\cos 45^\circ}{D}, \qquad P_{AC} = \frac{F\cos 60^\circ}{D},$$

where $D = \sin 45^\circ \cos 60^\circ + \cos 45^\circ \sin 60^\circ$. Using these expressions for the axial loads, the strain energy in the truss is

$$U = \frac{P_{AB}^2 L_{AB}}{2EA} + \frac{P_{AC}^2 L_{AC}}{2EA}$$

$$= \frac{F^2(L_{AB}\cos^2 45^\circ + L_{AC}\cos^2 60^\circ)}{2EAD^2}.$$

Now we can apply Castigliano's second theorem to determine v:

$$v = \frac{\partial U}{\partial F} = \frac{F(L_{AB}\cos^2 45^\circ + L_{AC}\cos^2 60^\circ)}{EAD^2}.$$

This is the same result as that obtained by a different method in Example 3-6.

By employing a bit of sleight of hand, we can also use Castigliano's second theorem to determine the horizontal deflection u. There is no horizontal external

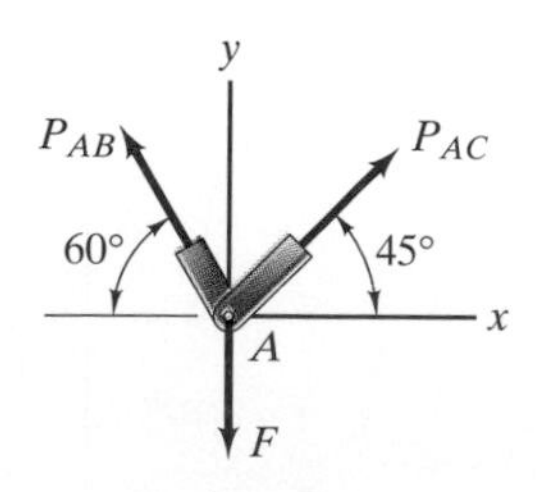

FIGURE 11-19 Joint A.

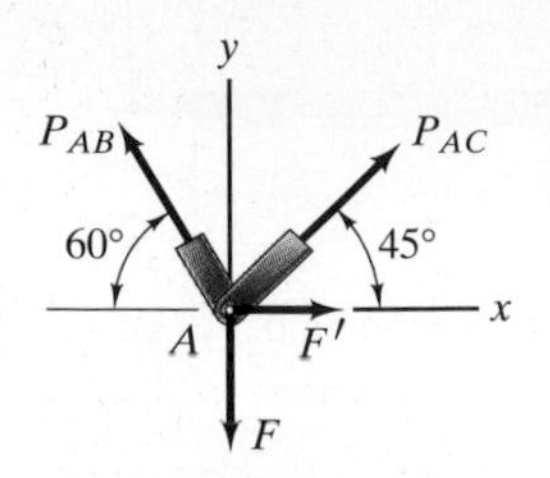

FIGURE 11-20 Joint A with an assumed horizontal force F'.

force at joint A. But we can assume that a horizontal force F' acts at A in addition to the downward force F, use Castigliano's second theorem to determine u, and then set $F' = 0$. From the free-body diagram of joint A (Fig. 11-20) we now obtain the axial loads

$$P_{AB} = \frac{F\cos 45^\circ + F'\sin 45^\circ}{D}, \qquad P_{AC} = \frac{F\cos 60^\circ - F'\sin 60^\circ}{D}.$$

The strain energy in the truss is

$$\begin{aligned} U &= \frac{P_{AB}^2 L_{AB}}{2EA} + \frac{P_{AC}^2 L_{AC}}{2EA} \\ &= \frac{(F\cos 45^\circ + F'\sin 45^\circ)^2 L_{AB} + (F\cos 60^\circ - F'\sin 60^\circ)^2 L_{AC}}{2EAD^2}. \end{aligned}$$

Applying Castigliano's second theorem, the deflection u is

$$\begin{aligned} u = \frac{\partial U}{\partial F'} = &\ \frac{(F\cos 45^\circ + F'\sin 45^\circ)L_{AB}\sin 45^\circ}{EAD^2} \\ &- \frac{(F\cos 60^\circ - F'\sin 60^\circ)L_{AC}\sin 60^\circ}{EAD^2}. \end{aligned}$$

Setting $F' = 0$ in this expression, the horizontal displacement due to the vertical force F alone is

$$u = \frac{F(L_{AB}\sin 45^\circ\cos 45^\circ - L_{AC}\sin 60^\circ\cos 60^\circ)}{EAD^2},$$

which is the same result obtained in Example 3-6.

EXAMPLE 11-3

Worksheet 14

The truss in Fig. 11-21 is subjected to a vertical force F at B. The members each have elastic modulus E and cross-sectional area A. Use Castigliano's second theorem to determine the vertical component of the resulting displacement at point C.

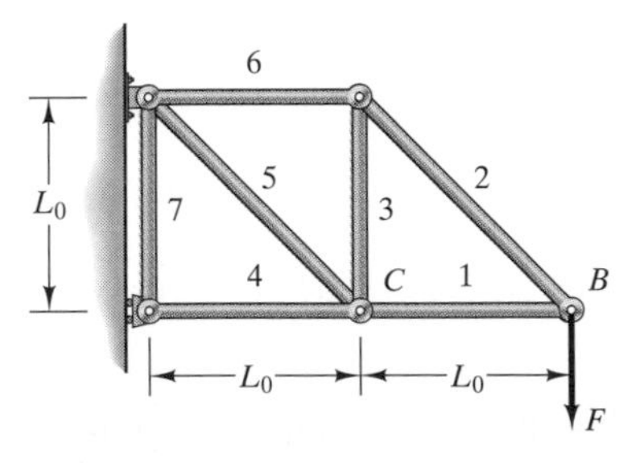

FIGURE 11-21

Strategy

There is no external force at C, but we can determine the vertical displacement by assuming that a vertical force F' acts at C, applying Castigliano's second theorem, and then setting $F' = 0$ in the resulting expression for the displacement.

Solution

Placing a downward force F' at C [Fig. (a)], the axial loads in the members and their lengths are

$$
\begin{aligned}
P_1 &= -F, & L_1 &= L_0 \\
P_2 &= \sqrt{2}F, & L_2 &= \sqrt{2}L_0 \\
P_3 &= -F, & L_3 &= L_0 \\
P_4 &= -(2F + F'), & L_4 &= L_0 \\
P_5 &= \sqrt{2}(F + F'), & L_5 &= \sqrt{2}L_0 \\
P_6 &= F, & L_6 &= L_0 \\
P_7 &= 0, & L_7 &= L_0.
\end{aligned}
$$

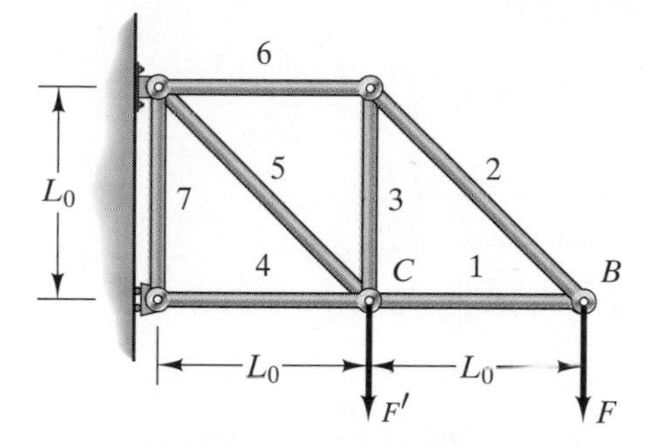

(a) Assuming that a vertical force F' acts at C.

Using these values, the strain energy in the truss is

$$
\begin{aligned}
U &= \sum_{i=1}^{7} \frac{P_i^2 L_i}{2EA} \\
&= \frac{1}{2EA}[F^2 L_0 + 2\sqrt{2}F^2 L_0 + F^2 L_0 + (2F + F')^2 L_0 \\
&\quad + 2\sqrt{2}(F + F')^2 L_0 + F^2 L_0].
\end{aligned}
$$

Let v be the downward displacement at C. Applying Castigliano's second theorem yields

$$
v = \frac{\partial U}{\partial F'} = \frac{1}{2EA}[2(2F + F')L_0 + 4\sqrt{2}(F + F')L_0].
$$

Setting $F' = 0$ in this expression, we obtain

$$v = \frac{2(1+\sqrt{2})FL_0}{EA}.$$

BEAMS

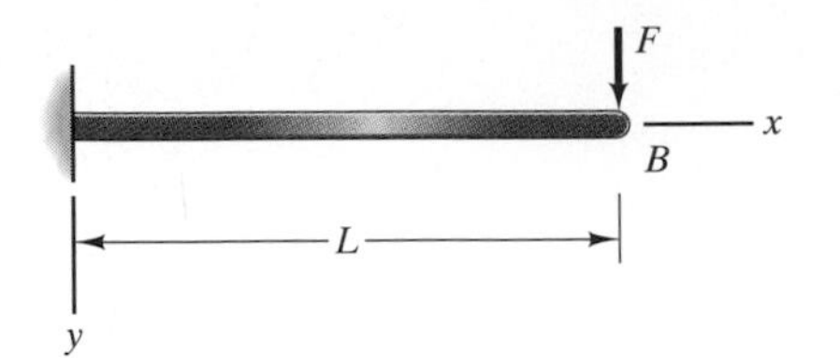

FIGURE 11-22 Cantilever beam subjected to a force.

The cantilever beam in Fig. 11-22 is loaded by a force at the right end. The deflection and the clockwise slope at the right end of this beam given in Appendix E are

$$v_B = \frac{FL^3}{3EI}, \qquad \theta_B = \frac{FL^2}{2EI}.$$

We can use Castigliano's second theorem to obtain these results. The bending moment in the beam is

$$M = F(x - L).$$

Substituting this expression into Eq. (11-15) and integrating from $x = 0$ to $x = L$, the strain energy is

$$U = \int_0^L \frac{[F(x-L)]^2}{2EI}\,dx$$
$$= \frac{F^2L^3}{6EI}.$$

Applying Castigliano's second theorem, the deflection at the right end of the beam is

$$v_B = \frac{\partial U}{\partial x} = \frac{FL^3}{3EI}.$$

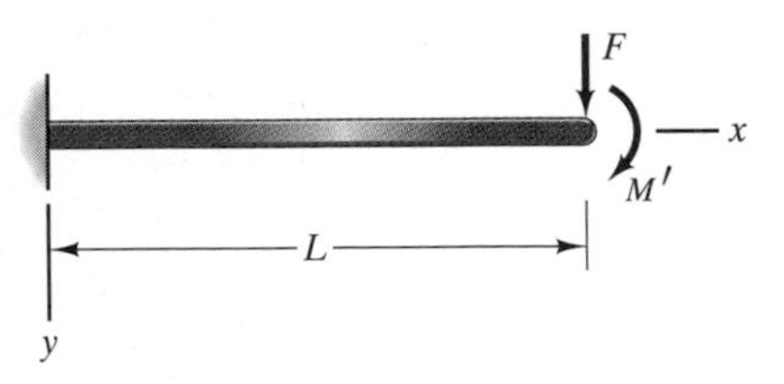

FIGURE 11-23 Applying a couple M' at the right end of the beam.

To determine the slope at the right end, we can assume that a couple M' acts at the right (Fig. 11-23), calculate the slope, and then set $M' = 0$. In this case the bending moment is

$$M = F(x - L) - M'.$$

Substituting this expression into Eq. (11-15), the strain energy is

$$U = \int_0^L \frac{[F(x-L) - M']^2}{2EI}\,dx$$
$$= \frac{1}{2EI}\left[\frac{F^2L^3}{3} + FL^2M' + L(M')^2\right].$$

Applying Eq. (11-20), the clockwise slope at the right end of the beam is

$$\theta_B = \frac{\partial U}{\partial M'} = \frac{1}{2EI}(FL^2 + 2LM').$$

Setting $M' = 0$ in this expression, we obtain

$$\theta_B = \frac{FL^2}{2EI}.$$

The cantilever beam in Fig. 11-24 is loaded by a distributed force. The deflection at the right end of this beam given in Appendix E is

$$v_B = \frac{w_0 L^4}{8EI}.$$

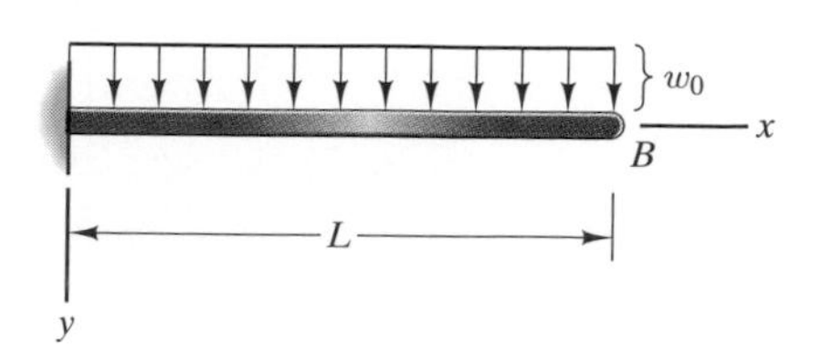

FIGURE 11-24 Cantilever beam subjected to a distributed load.

We can obtain this result by assuming that a force F' acts at the end of the beam (Fig. 11-25), applying Castigliano's second theorem to determine the deflection, and then setting $F' = 0$. The bending moment in the beam is

$$M = -\tfrac{1}{2}w_0(L-x)^2 - F'(L-x).$$

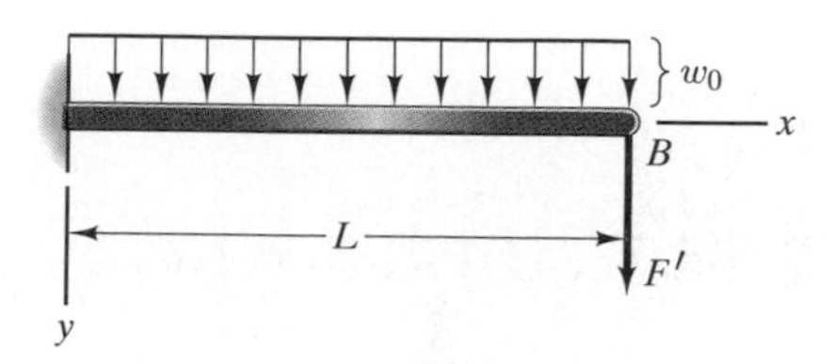

FIGURE 11-25 Assuming that a force F' acts at the right end.

From Eq. (11-15), the strain energy is

$$U = \int_0^L \frac{\left[-\tfrac{1}{2}w_0(L-x)^2 - F'(L-x)\right]^2}{2EI}\,dx$$

$$= \frac{1}{2EI}\left[\frac{w_0^2 L^5}{20} + \frac{w_0 L^4 F'}{4} + \frac{L^3 (F')^2}{3}\right].$$

The deflection at the right end of the beam is

$$v_B = \frac{\partial U}{\partial F'} = \frac{1}{2EI}\left(\frac{w_0 L^4}{4} + \frac{2L^3 F'}{3}\right). \qquad \textbf{(11-21)}$$

Setting $F' = 0$ in this expression, we obtain

$$v_B = \frac{w_0 L^4}{8EI}.$$

EXAMPLE 11-4

The beam in Fig. 11-26 is statically indeterminate. Use Castigliano's second theorem to determine the reaction at B.

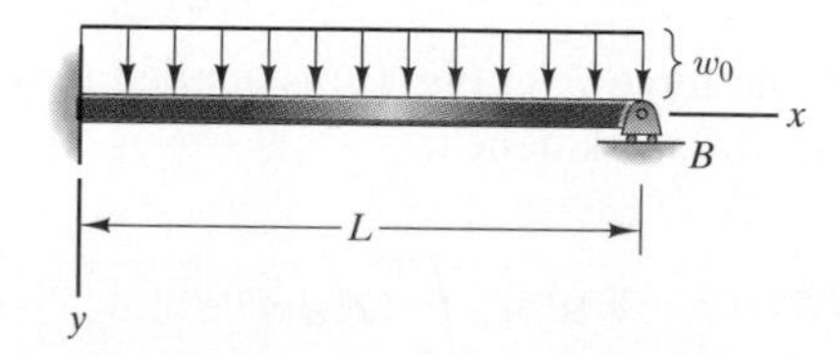

FIGURE 11-26

Strategy

In Fig. (a) we show the unknown reaction B exerted on the beam by the roller support. We can use Castigliano's second theorem to determine the deflection at the right end of the beam due to the force B, then determine B by applying the boundary (compatibility) condition that the deflection of the right end of the beam is zero.

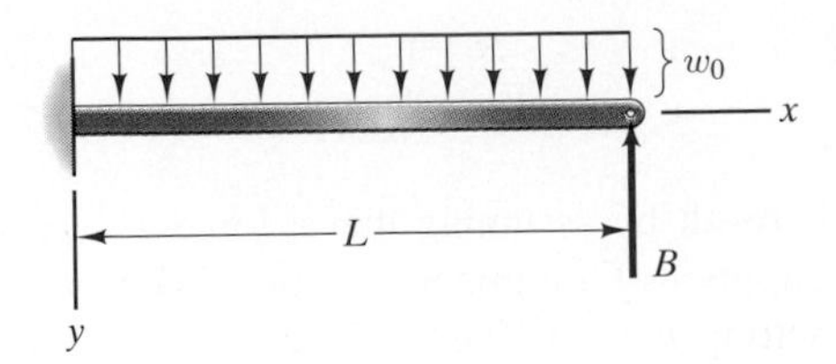

(a) Showing the reaction exerted by the roller support.

Solution

We have already applied Castigliano's second theorem to a cantilever beam subjected to a uniformly distributed load and a point force at the right end (Fig. 11-25). The resulting deflection of the right end of the beam is given by Eq. (11-21). By setting $F' = -B$ in that expression, we obtain the deflection of the right end of the beam in Fig. (a):

$$v_B = \frac{1}{2EI}\left(\frac{w_0 L^4}{4} - \frac{2L^3 B}{3}\right).$$

Then from the boundary condition $v_B = 0$, we obtain the unknown reaction at the roller support:

$$B = \frac{3w_0 L}{8}.$$

Discussion

Compare the solution given in this example with the solution in Section 9-2. Energy methods often provide quicker solutions even for relatively simple problems.

Chapter Summary

Work

If the point to which the force F in Fig. (a) is applied moves along the x axis from $x = 0$ to $x = x_0$, the work done is

$$W = \int_0^{x_0} F\,dx.$$

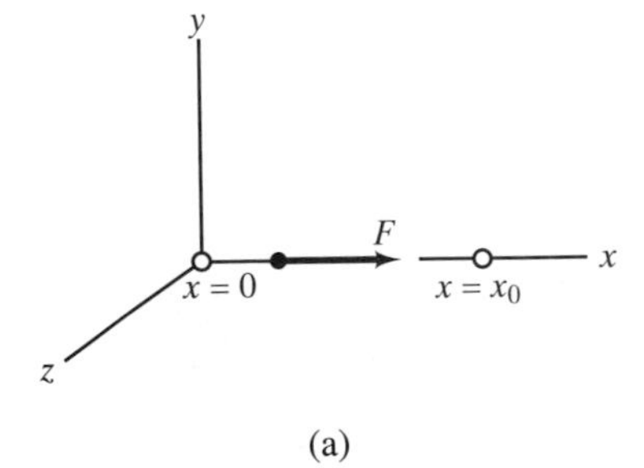

(a)

If the force is a constant $F = F_0$, the work is $W = F_0 x_0$. If F is proportional to x and its value is F_0 when $x = x_0$, the work is $W = \frac{1}{2} F_0 x_0$.

When a couple of moment M rotates through an angle from $\theta = 0$ to $\theta = \theta_0$ in the same direction as the couple, the work done is

$$W = \int_0^{\theta_0} M \, d\theta.$$

If the moment is a constant $M = M_0$, the work is $W = M_0 \theta_0$. If M is proportional to θ and its value is M_0 when $\theta = \theta_0$, the work is $W = \frac{1}{2} M_0 \theta_0$.

Strain Energy

The strain energy per unit volume of an isotropic linearly elastic material subjected to a general state of stress is

$$u = \tfrac{1}{2}(\sigma_x \epsilon_x + \sigma_y \epsilon_y + \sigma_z \epsilon_z + \tau_{xy} \gamma_{xy} + \tau_{yz} \gamma_{yz} + \tau_{xz} \gamma_{xz}). \qquad \text{Eq. (11-3)}$$

Axially Loaded Bars

The strain energy of a prismatic bar of length L and cross-sectional area A that is subjected to an axial load P is

$$U = \frac{P^2 L}{2EA}. \qquad \text{Eq. (11-11)}$$

Beams

The strain energy of an element of a beam of width dx [(Fig. b)] is

$$dU = \frac{M^2}{2EI} dx. \qquad \text{Eq. (11-15)}$$

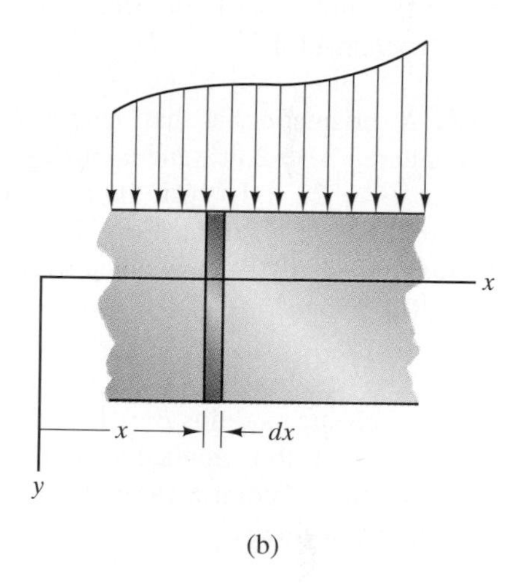

(b)

Castigliano's Second Theorem

Consider an object of elastic material that is subjected to N external forces $F_1, F_2, \ldots, F_N$ [Fig. (c)]. Let u_i be the component of the deflection of the point of application of the ith force F_i in the direction of F_i. Then

$$u_i = \frac{\partial U}{\partial F_i}, \qquad \text{Eq. (11-19)}$$

where U is the strain energy resulting from the application of the N forces.

If couples also act on an object, the angle of rotation of the point of application of the ith couple M_i in the direction of M_i is given by

$$\theta_i = \frac{\partial U}{\partial M_i}. \qquad \text{Eq. (11-20)}$$

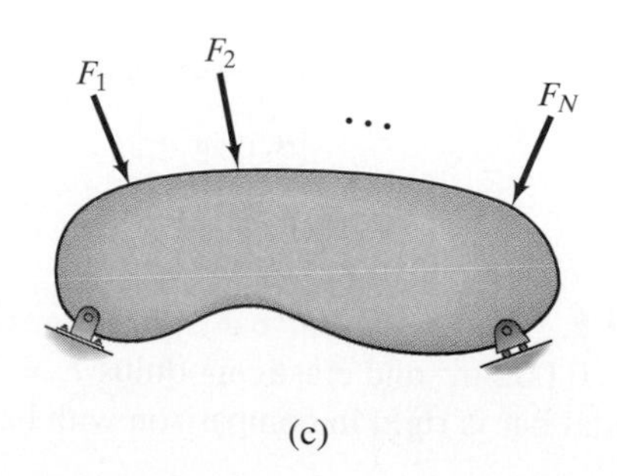

(c)

PROBLEMS

11-1.1. The bar in Fig. 11-9 is elastic and is subjected to an axial force $P = 20$ kN. As a result, it stretches a distance $\delta = 3$ mm. How much strain energy is stored in the bar?

11-1.2. A prismatic bar has length $L = 2$ m and cross-sectional area $A = 0.0004\ \text{m}^2$ and is made of aluminum alloy with elastic modulus $E = 70$ GPa. Suppose that the bar is subjected to a 200-N axial load. **(a)** How much strain energy is stored in the bar? **(b)** How much work is done on the bar by the 200-N load?

11-1.3. If the bar in Problem 11-1.2 consists of rubber with elastic modulus $E = 0.002$ GPa, how much work is done on the bar by the 200-N load? What is the ratio of the work done on the rubber bar to the work done on the aluminum bar in Problem 11-1.2?

11-1.4. A prismatic bar has length $L = 48$ in., cross-sectional area $A = 2\ \text{in}^2$, and is made of steel with elastic modulus $E = 28 \times 10^6$ psi. Suppose that the bar is subjected to a 30-kip axial load. **(a)** How much strain energy is stored in the bar? **(b)** How much work is done on the bar by the 30-kip load?

11-1.5. The wooden bar AB has cross-sectional area $A = 0.8\ \text{in}^2$ and elastic modulus $E = 1.4 \times 10^6$ psi. A downward force $F = 3500$ lb is applied at B. Determine the resulting deflection v of point B **(a)** without using strain energy; **(b)** using strain energy.

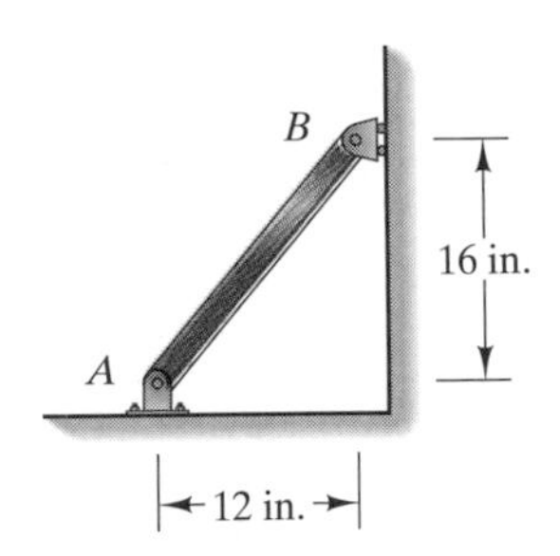

PROBLEM 11-1.5

11-1.6. The aluminum bar AB has cross-sectional area $A = 0.002\ \text{m}^2$ and elastic modulus $E = 72$ GPa. The horizontal bar is rigid in comparison with bar AB. A counterclockwise couple $M = 1.2$ MN-m is applied at C. Use strain energy to determine the angle through which the horizontal bar rotates.

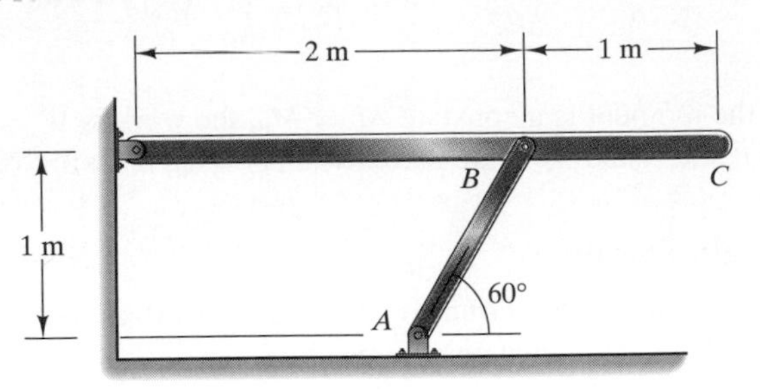

PROBLEM 11-1.6

11-1.7. Each bar has length L, cross-sectional area A, and elastic modulus E. If a downward force F is applied at B, determine the resulting displacement v of point B **(a)** without using strain energy; **(b)** using strain energy.

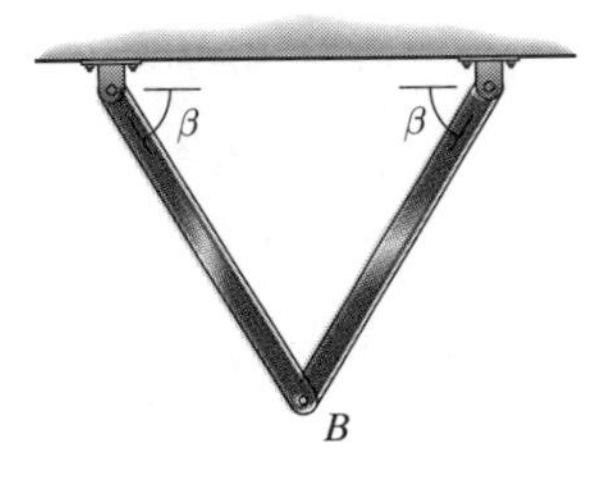

PROBLEM 11-1.7

11-1.8. In Problem 11-1.7, $\beta = 60°$, $L = 2$ m, $A = 0.0004\ \text{m}^2$, and $E = 70$ GPa. What downward force applied at B will cause B to move 4 mm?

11-1.9. Each bar has cross-sectional area $A = 2\ \text{in}^2$ and elastic modulus $E = 1.8 \times 10^6$ psi. If a downward force

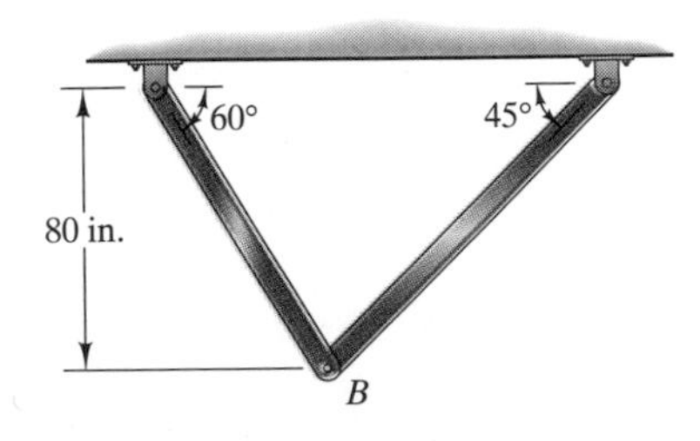

PROBLEM 11-1.9

$F = 10$ kip is applied at B, use strain energy to determine the vertical component v of the displacement of point B.

11-1.10. In Example 11-1, suppose that the vertical force F is applied at the joint where members 1, 3, 4, and 5 are connected instead of at joint B. What is the vertical component of the resulting displacement of the joint where F is applied?

11-1.11. The truss is made of steel bars with cross-sectional area $A = 0.003\ \text{m}^2$ and elastic modulus $E = 220$ GPa. Determine the vertical component v of the displacement of point B if the suspended mass $m = 5000$ kg.

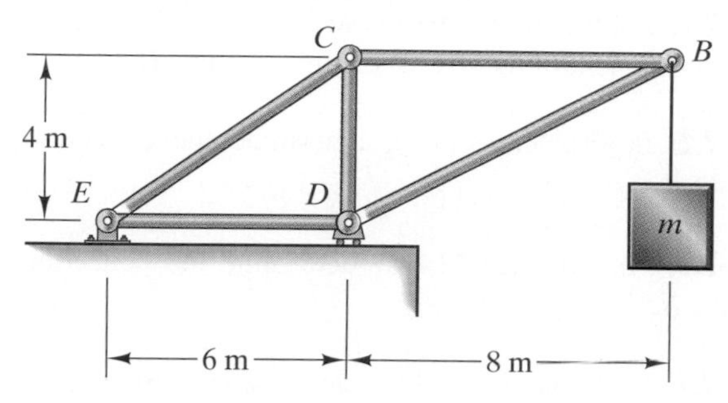

PROBLEM 11-1.11

11-1.12. Each member of the steel truss has cross-sectional area $A = 0.026\ \text{m}^2$ and elastic modulus $E = 200$ GPa. If a downward force $F = 600$ kN is applied at B, what is the vertical component v of the resulting displacement of point B?

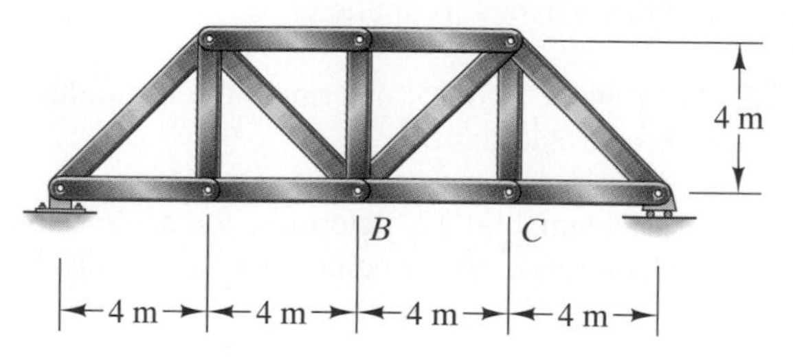

PROBLEM 11-1.12

11-1.13. In Problem 11-1.12, if the downward force $F = 600$ kN is applied at C instead, what is the vertical component v of the resulting displacement of point C?

11-1.14. Use strain energy to determine the beam's deflection at the point where F is applied.

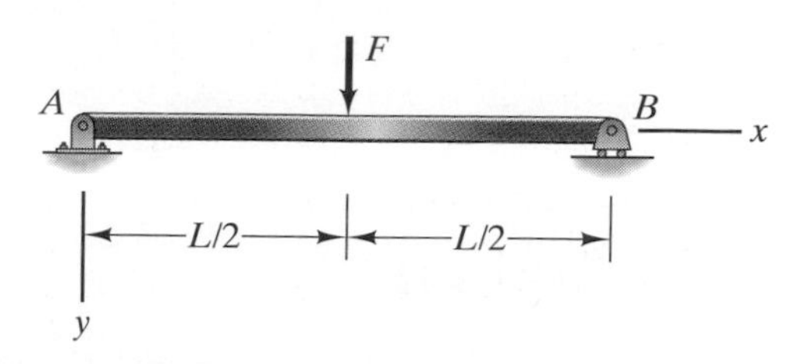

PROBLEM 11-1.14

11-1.15. Use strain energy to determine the angle through which the beam rotates at B and confirm the value given in Appendix E.

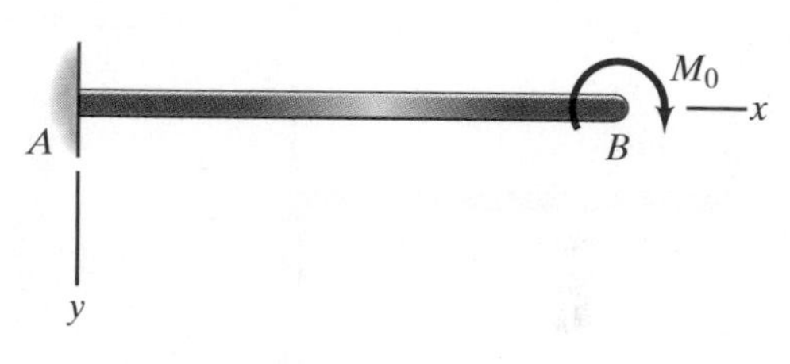

PROBLEM 11-1.15

11-1.16. Use strain energy to determine the angle through which the beam rotates at A and confirm the value given in Appendix E.

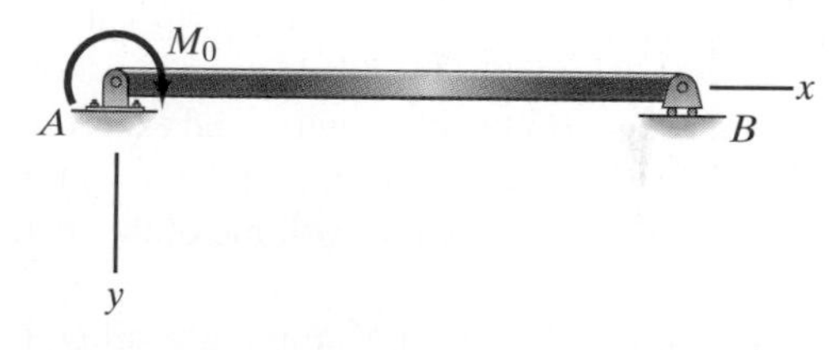

PROBLEM 11-1.16

11-1.17. Use strain energy to determine the angle through which the beam rotates at the point where the couple M_0 is applied.

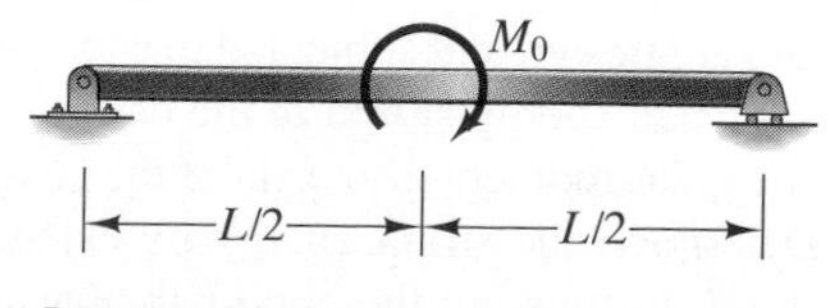

PROBLEM 11-1.17

11-1.18. Use strain energy to determine the beam's deflection at the point where F is applied.

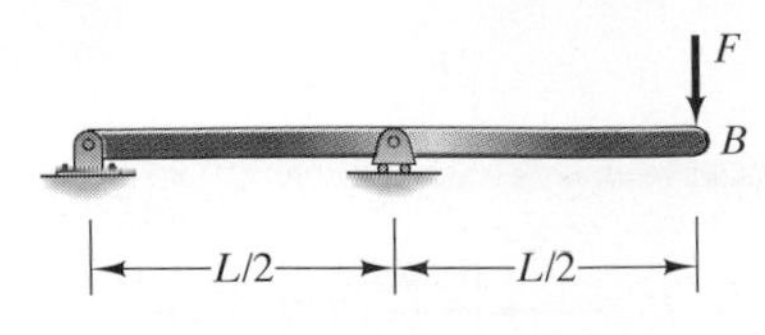

PROBLEM 11-1.18

11-1.19. The horizontal beam has moment of inertia I and elastic modulus E. The vertical bar has cross-sectional area A and elastic modulus E. A downward force F is applied at B. Use strain energy to determine the deflection at B. (*Strategy:* Draw separate free-body diagrams of the beam and bar and use strain energy to determine the deflection of the beam and the change in length of the bar.)

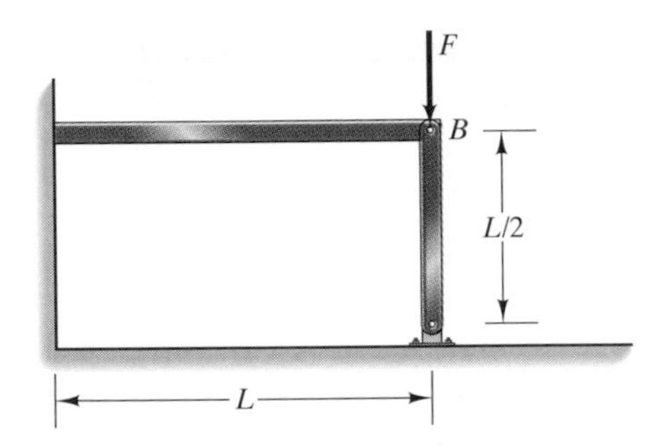

PROBLEM 11-1.19

11-1.20. A material is subjected to a uniform state of plane stress $\sigma_x = 400$ MPa, $\sigma_y = -200$ MPa, $\tau_{xy} = 300$ MPa. Using a strain gauge rosette, it is determined experimentally that $\epsilon_x = 0.00239$, $\epsilon_y = -0.00162$, and $\gamma_{xy} = 0.00401$. What is the strain energy per unit volume of the material?

11-1.21. A sample of 2014-T6 aluminum alloy is subjected to a uniform state of plane stress $\sigma_x = 40$ MPa, $\sigma_y = -30$ MPa, and $\tau_{xy} = 30$ MPa. What is the strain energy per unit volume of the material?

11-1.22. Consider an elastic bar of length L with a solid circular cross section and shear modulus G. Show that if the bar is fixed at one end and subjected to a torque T at the other end, the strain energy stored in the bar is $T^2L/2GJ$, where J is the polar moment of inertia of the cross section. [*Strategy:* Determine the strain energy by calculating the work done by T in twisting the end of the bar and using Eq. (4-7).]

11-1.23. From Eq. (4-9), the shear stress in the bar in Problem 11-1.22 as a function of distance r from the bar's axis is $\tau = Tr/J$. From Eq. (11-3), the strain energy per unit volume in the bar is therefore

$$u = \frac{1}{2}\tau\gamma = \frac{1}{2G}\tau^2 = \frac{T^2r^2}{2GJ^2}.$$

Integrate this expression over the bar's volume to determine the strain energy stored in the bar and confirm the answer to Problem 11-1.22.

Solve Problems 11-2.1–11-2.13 by using Castigliano's second theorem.

11-2.1. For the truss in Example 11-3, determine the vertical component of the displacement of point B.

11-2.2. In Problem 11-1.5, determine the deflection of point B.

11-2.3. In Problem 11-1.7, determine the deflection of point B.

11-2.4. In Problem 11-1.11, determine the vertical component v of the deflection of point B.

11-2.5. In Problem 11-1.11, determine the horizontal component u of the deflection of point B.

11-2.6. For the beam in Fig. 11-24, determine the slope at the right end of the beam.

11-2.7. In Problem 11-1.14, determine the beam's deflection at the point where F is applied.

11-2.8. In Problem 11-1.15, determine the angle through which the beam rotates at B.

11-2.9. In Problem 11-1.17, determine the angle through which the beam rotates at the point where the couple M_0 is applied.

11-2.10. In Problem 11-1.18, determine the beam's deflection at the point where F is applied.

11-2.11. In Problem 11-1.18, use Castigliano's second theorem to determine the angle through which the beam rotates at the point where F is applied.

11-2.12. Determine the reaction at the roller support of the statically indeterminate beam.

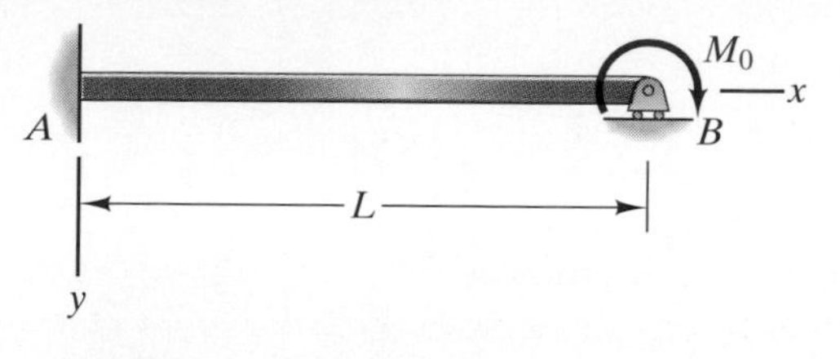

PROBLEM 11-2.12

11-2.13. Determine the reaction at the roller support of the statically indeterminate beam.

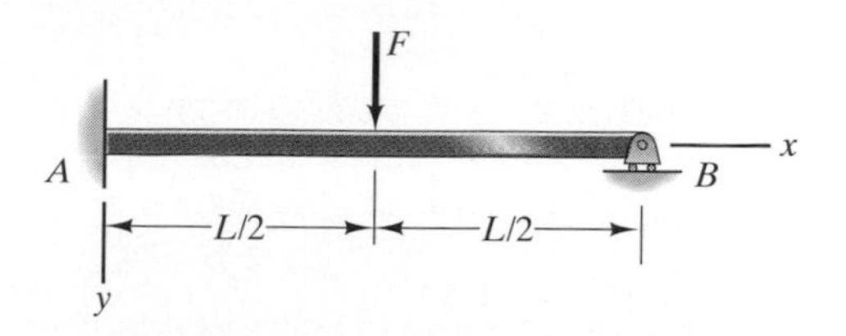

PROBLEM 11-2.13

The hull of this T-2 tanker cracked in two while the ship was moored in calm water.

CHAPTER 12

Criteria for Failure and Fracture

In this chapter we discuss failures of structural components. The failures we consider may be due to a single overload or smaller repeated, or cyclic loads. In both cases we distinguish between situations in which a structural component does or does not contain preexisting flaws or cracks. When it does not, it is sufficient to rely on stress analyses of the kind developed in previous chapters and apply failure criteria based on them. When a structural component does contain a crack, as in the case of the T-2 tanker's hull, we show how to use fracture mechanics to determine the state of stress and predict when failure will occur.

12-1 | Failure

The structural designer may compromise in satisfying many requirements and constraints, including material availability, machinability, cost, weight, and aesthetic concerns, but avoiding failure is essential. In previous chapters we have presented many examples of the design of structural members to prevent failure under specified loads. The members we analyzed—bars subjected to axial loads and torques, pressure vessels, and beams subjected to lateral loads—had simple geometries and simple states of stress, and we were able to apply simple failure criteria. In contrast, many structures have complicated geometries that require advanced analytical or numerical methods of analysis and lead to general states of stress. In this section we discuss criteria that have been introduced to prevent failure in materials subjected to general states of stress. We first consider overloads, or monotonic loading, and then repeated loads, or fatigue.

Overloads

The mechanism of failure resulting from an overload depends on the material. Broadly dividing structural metals into those that are brittle and those that are ductile, failure in brittle materials occurs when a component breaks or fractures. In contrast, failure in ductile materials is usually defined to occur with the onset of yielding.

In previous chapters we have introduced failure criteria based on simple states of stress. As we discussed in Chapter 3, yielding in a bar subjected to axial load takes place when the yield stress σ_Y has been exceeded, and fracture occurs when the ultimate stress σ_U occurs. We will now describe criteria that can be used to predict when failure will occur in objects subjected to more complex states of stress, such as those in bars subjected to combined axial load and torsion or in the walls of pressure vessels.

Maximum Normal Stress Criterion

The simplest way to extend the concept of failure used in bars subjected to axial loads to more complex states of stress is to compare the maximum principal stress at a point to the ultimate stress determined under axial loading. This is called the *maximum normal stress criterion*. (See our discussions of principal stresses in Sections 5-2–5-4.) Expressed mathematically, this criterion states that failure will occur when

$$\text{Max}(|\sigma_1|, |\sigma_2|, |\sigma_3|) = \sigma_U. \quad \textbf{(12-1)}$$

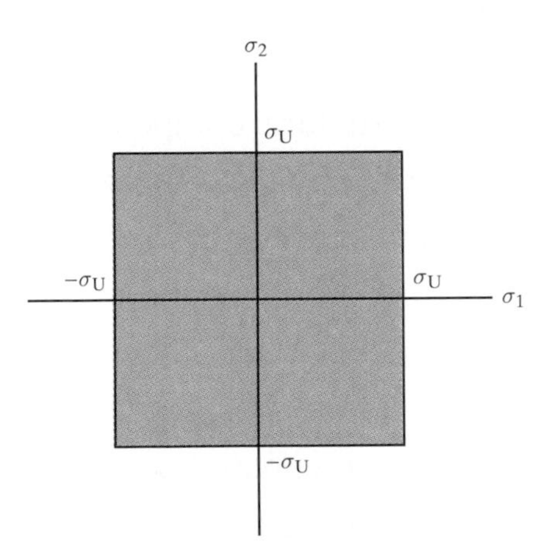

FIGURE 12-1 Graphical representation of the maximum normal stress criterion under plane stress.

The use of absolute values means that materials that follow this criterion have the same strength in tension and compression. Since the implied mechanism of failure in the material is one of separation rather than sliding (shear), this criterion applies to brittle materials, which do not yield or undergo much plastic deformation.

In the case of plane stress ($\sigma_3 = 0$) the maximum normal stress criterion can be described in a simple graphical manner (Fig. 12-1). The values of σ_1 and σ_2 for which failure will not occur are bounded by the square region defined by the lines

$$\sigma_1 = \pm\sigma_U, \qquad \sigma_2 = \pm\sigma_U. \tag{12-2}$$

Notice that there is no interaction between the principal stresses in this criterion—its boundaries are horizontal and vertical. In a three-dimensional stress state, the safe stress regime is enclosed in the cube bounded by the planes

$$\sigma_1 = \pm\sigma_U, \qquad \sigma_2 = \pm\sigma_U, \qquad \sigma_3 = \pm\sigma_U. \tag{12-3}$$

EXAMPLE 12-1

In Fig. 12-2 a bar of ultra high-strength steel is subjected to an axial stress σ and a lateral pressure p. It is a brittle material with an ultimate normal stress of 300 ksi in tension or compression. Determine the pressure level p at which the bar will fail if the pressure and axial stress are applied proportionally with the ratio $\sigma = 2p$.

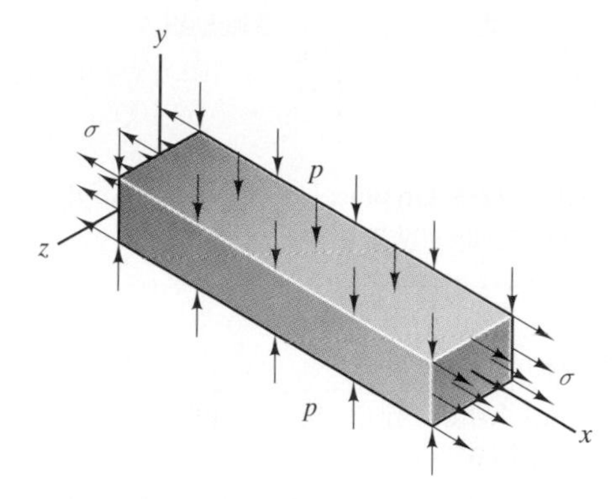

FIGURE 12-2

Strategy

Because the material is brittle and fails at the same stress in tension and compression, the maximum normal stress criterion applies. We can determine the principal stresses and use Eq. (12-1) to determine the level of p at which failure occurs. We will also determine the pressure at which failure occurs graphically.

Solution

In terms of the coordinate system in Fig. 12-2, the only nonzero components of stress in the bar are $\sigma_x = \sigma$, $\sigma_y = -p$. The bar is in a state of plane stress and the principal stresses are $\sigma_1 = \sigma$, $\sigma_2 = -p$, $\sigma_3 = 0$. Since the axial stress and pressure are applied in the ratio $\sigma = 2p$, we can write the failure criterion (12-1) as

$$\text{Max}(|\sigma_1|, |\sigma_2|, |\sigma_3|) = \sigma_U :$$
$$\text{Max}(|2p|, |-p|, 0) = 300 \text{ ksi}.$$

We see that failure occurs when $p = 150$ ksi. This is demonstrated in Fig. 12-3, which shows the loading path and failure in σ_1 versus σ_2 space. Because $\sigma_2 = -p$ and $\sigma_1 = 2p$ as the loading is applied, the loading path follows the line $\sigma_2 = -\frac{1}{2}\sigma_1$. The loading path reaches the boundary, and failure occurs, when $\sigma_1 = 2p = 300$ ksi.

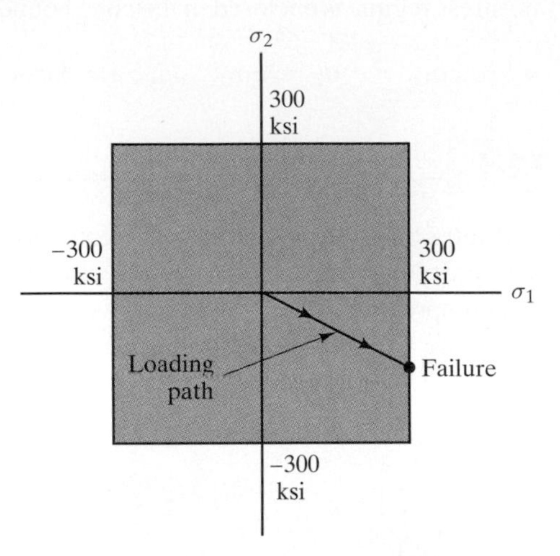

FIGURE 12-3 Graphical representation of the loading path and failure.

MOHR'S FAILURE CRITERION

There are many materials whose fracture strengths in tension and compression differ, which requires a modified version of Fig. 12-1. The failure boundaries in the first and third quadrants are qualitatively the same as before, but the ultimate tensile stress σ_U^t and ultimate compressive stress σ_U^c are different (Fig. 12-4a). The question that then arises is the nature of the failure boundaries in the second and fourth quadrants. Mohr suggested that they should be the straight lines joining the boundaries in the first and third quadrants in Fig. 12-4a. The resulting failure boundary is *Mohr's failure criterion*. The rationale for his choice is shown in Fig. 12-4b. The right circle is the Mohr's circle mapping the states of stress when $\sigma_1 = \sigma_U^t$, $\sigma_2 = 0$. The left circle is the Mohr's circle mapping the states of stress when $\sigma_1 = 0$, $\sigma_2 = -\sigma_U^c$. Mohr postulated that the failure boundary in σ–τ space is defined by these two circles and the straight lines tangent to them in Fig. 12-4b. When σ_1 and σ_2 are of opposite sign, values that lie on the straight lines in the second and fourth quadrants of Fig. 12-4a yield Mohr's circles that are tangent to the straight lines in Fig. 12-4b.

Figure 12-4a indicates that while there is no interaction between the values of σ_1 and σ_2 at which fracture occurs in the first and third quadrants, interaction does exist in the second and fourth quadrants. In the first quadrant failure occurs when

$$\sigma_1 = \sigma_U^t \quad \text{or} \quad \sigma_2 = \sigma_U^t, \tag{12-4}$$

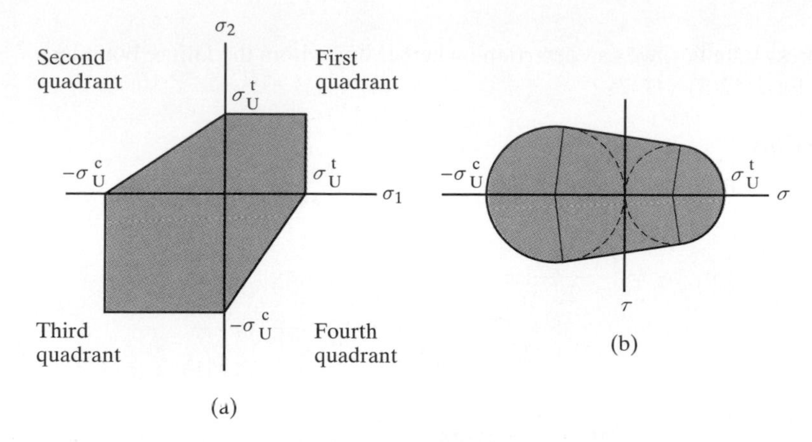

FIGURE 12-4 (a) Graphical representation of Mohr's stress criterion under plane stress conditions. (b) Rationale for the failure boundaries in the second and fourth quadrants.

and in the third quadrant it occurs when

$$\sigma_1 = -\sigma_U^c \quad \text{or} \quad \sigma_2 = -\sigma_U^c. \tag{12-5}$$

In the second quadrant, failure occurs when σ_1 and σ_2 lie on the straight line

$$\sigma_2 = \sigma_U^t + \frac{\sigma_U^t}{\sigma_U^c}\sigma_1, \tag{12-6}$$

and in the fourth quadrant it occurs when σ_1 and σ_2 lie on the straight line

$$\sigma_2 = -\sigma_U^c + \frac{\sigma_U^c}{\sigma_U^t}\sigma_1. \tag{12-7}$$

As in the case of the maximum normal stress criterion, the failure mechanism being modeled is fracture, so Mohr's failure criterion also applies to brittle materials.

EXAMPLE 12-2

Preliminary design of a mechanism indicates that a point on the surface of a component will be in the state of plane stress $\sigma_x = \sigma_y = -75$ ksi, $\tau_{xy} = 175$ ksi. The material fails according to Mohr's stress criterion with ultimate tensile and compressive strengths of $\sigma_U^t = 150$ ksi and $\sigma_U^c = 300$ ksi, respectively. Is the design safe with respect to this state of stress?

Strategy

We can use Eq. (5-15) to determine the principal stresses σ_1 and σ_2 for the given state of plane stress. Then, knowing the quadrant of Fig. 12-4a in which the stress state lies, we can determine whether it is within the failure boundary from Eqs. (12-4)–(12-7).

Solution

The principal stresses are

$$\begin{aligned}\sigma_1, \sigma_2 &= \frac{\sigma_x + \sigma_y}{2} \pm \sqrt{\left(\frac{\sigma_x - \sigma_y}{2}\right)^2 + \tau_{xy}^2} \\ &= \frac{-75 - 75}{2} \pm \sqrt{\left(\frac{-75 + 75}{2}\right)^2 + (175)^2} \\ &= 100 \text{ ksi}, \ -250 \text{ ksi}.\end{aligned} \quad \textbf{(12-8)}$$

The state of stress lies in the fourth quadrant. Substituting $\sigma_1 = 100$ ksi into Eq. (12-7), the value of σ_2 corresponding to the failure boundary is

$$\sigma_2|_{\text{boundary}} = -\sigma_U^c + \frac{\sigma_U^c}{\sigma_U^t}\sigma_1 = -300 + \left(\frac{300}{150}\right)100 = -100 \text{ ksi}.$$

Since the design value $\sigma_2 = -250$ ksi is below this value, the design state of stress lies outside the failure boundary and the material would fail if the design were implemented.

TRESCA CRITERION

In this section we consider a failure criterion that is valid for ductile materials, which yield before reaching their ultimate tensile strengths. Although materials do not fracture upon reaching their yield strengths, permanent deformation can occur and in many structural components proper function can be lost at that point. At the same time there are structures, such as oil pipelines in deep water, that are intentionally allowed to undergo significant plastic deformation during installation or fabrication.

The plastic deformation that is initiated when the yield strength is reached takes place through a process of slip, or shear deformation. It is therefore natural to expect criteria for failure by yielding to be expressed in terms of shear stress. Conceptually, then, the simplest experiment to determine yielding in ductile structural materials would be a shear test, such as subjecting a tube to torsion. If a thin-walled tube of median radius R and wall thickness t is subjected to a torque T, the resulting shear stress is

$$\tau = \frac{T}{2\pi R^2 t}. \quad \textbf{(12-9)}$$

In this way the yield stress in shear τ_Y can be determined. If we define the onset of yield as failure, τ_Y is also the maximum shear strength of the material.

However, not all materials are easily available in the form of thin-walled cylinders, and subjecting bars to axial loads has become the common benchmark for determining yield strength. As we have seen, when a bar is subjected to a normal stress σ by axial load, the maximum resulting shear stress is $\sigma/2$, so the shear stress at the onset of yielding is related to the normal stress by

$$\tau_{\mathrm{Y}} = \frac{\sigma_{\mathrm{Y}}}{2}. \tag{12-10}$$

This criterion can be extended to an arbitrary state of stress by assuming that yielding occurs when the absolute maximum shear stress is equal to τ_{Y}:

$$\mathrm{Max}\left(\left|\frac{\sigma_1 - \sigma_2}{2}\right|, \left|\frac{\sigma_2 - \sigma_3}{2}\right|, \left|\frac{\sigma_1 - \sigma_3}{2}\right|\right) = \tau_{\mathrm{Y}}. \tag{12-11}$$

This is the *Tresca criterion* for failure, which is also called the *maximum shear stress criterion*. By using the relationship (12-10), the Tresca criterion can be expressed in terms of the tensile yield stress:

$$\mathrm{Max}(|\sigma_1 - \sigma_2|, |\sigma_2 - \sigma_3|, |\sigma_1 - \sigma_3|) = \sigma_{\mathrm{Y}}. \tag{12-12}$$

This criterion can be visualized easily in the case of plane stress. Then Eq. (12-12) reduces to

$$\mathrm{Max}(|\sigma_1 - \sigma_2|, |\sigma_2|, |\sigma_1|) = \sigma_{\mathrm{Y}}. \tag{12-13}$$

Consequently, in a plot of σ_2 versus σ_1, the safe region is bounded by the lines

$$\sigma_1 - \sigma_2 = \pm\sigma_{\mathrm{Y}}, \qquad \sigma_1 = \pm\sigma_{\mathrm{Y}}, \qquad \sigma_2 = \pm\sigma_{\mathrm{Y}}. \tag{12-14}$$

These lines form the hexagon shown in Fig. 12-5. (We discuss the ellipse circumscribing the hexagon below.) Stress states that give rise to values of σ_1 and σ_2 lying on or outside the hexagon cause yielding. This boundary therefore defines the maximum shear stress criterion in plane stress.

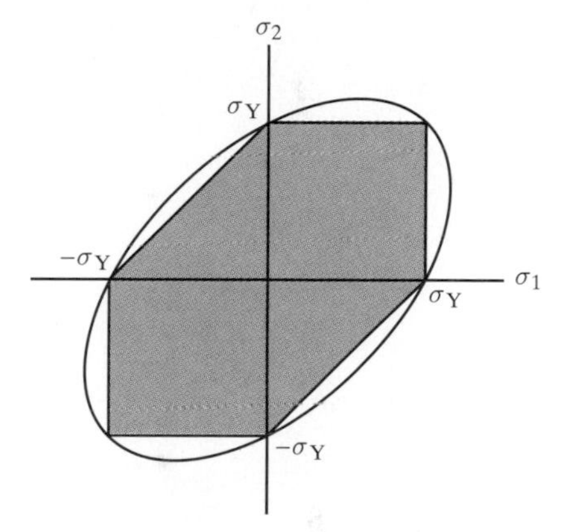

FIGURE 12-5 Failure boundaries for the Tresca (hexagon) and von Mises (ellipse) failure criteria under plane stress.

Notice from Eq. (12-12) that for a given state of stress, the normal stress

$$\sigma_{\mathrm{T}} = \mathrm{Max}(|\sigma_1 - \sigma_2|, |\sigma_2 - \sigma_3|, |\sigma_1 - \sigma_3|) \tag{12-15}$$

can be compared to the yield stress σ_{Y} to obtain a measure of how close the material is to failure. Called the *Tresca equivalent stress*, it motivates the definition of the *Tresca factor of safety:*

$$S_{\mathrm{T}} = \frac{\sigma_{\mathrm{Y}}}{\sigma_{\mathrm{T}}}. \tag{12-16}$$

As we have seen in previous chapters, a given safety factor can be prescribed as a design objective. Calculating the safety factor for a given state of stress is also useful because it provides a quick measure of how near a component is to failure. (Failure occurs when $S = 1$.)

von Mises Criterion

This criterion for failure of ductile materials states that yielding occurs when

$$\tfrac{1}{2}[(\sigma_1-\sigma_2)^2+(\sigma_2-\sigma_3)^2+(\sigma_1-\sigma_3)^2]=\sigma_Y^2. \tag{12-17}$$

This relation can be derived from strain energy considerations and is also known as the *distortional energy criterion*. In plane stress the failure boundary simplifies to

$$\sigma_1^2-\sigma_1\sigma_2+\sigma_2^2=\sigma_Y^2, \tag{12-18}$$

which describes the ellipse in Fig. 12-5. The Tresca and von Mises criteria coincide under conditions of uniaxial tension and compression and in equibiaxial tension and compression. Otherwise, the Tresca criterion is slightly more conservative (that is, the stresses required to cause failure are generally smaller).

We see from Eq. (12-17) that the *von Mises equivalent stress,* defined by

$$\sigma_M=\frac{1}{\sqrt{2}}\sqrt{(\sigma_1-\sigma_2)^2+(\sigma_2-\sigma_3)^2+(\sigma_1-\sigma_3)^2}, \tag{12-19}$$

can be compared to σ_Y as a measure of how close the material is to failure. It can be expressed in terms of the components of stress as

$$\sigma_M=\frac{1}{\sqrt{2}}\sqrt{(\sigma_x-\sigma_y)^2+(\sigma_y-\sigma_z)^2+(\sigma_x-\sigma_z)^2+6\left(\tau_{xy}^2+\tau_{yz}^2+\tau_{xz}^2\right)}. \tag{12-20}$$

Thus, if we have a solution to a particular problem, we can form the von Mises equivalent stress directly from the components of stress and map regions of yielding within a component without first having to determine the principal stresses. The *von Mises factor of safety* is defined by

$$S_M=\frac{\sigma_Y}{\sigma_M}. \tag{12-21}$$

EXAMPLE 12-3

M

Worksheet 15

A cylindrical pressure vessel with hemispherical ends has radius $R=2$ m, wall thickness $t=10$ mm, and is made of steel with yield stress $\sigma_Y=1800$ MPa. It is internally pressurized at $p=2$ MPa. Compare the Tresca and von Mises safety factors.

Strategy

We must calculate the principal stresses in the vessel (see Section 5-5), then determine the factors of safety from Eqs. (12-16) and (12-21).

Solution

Neglecting the stress normal to the vessel wall, the principal stresses are the hoop and axial stresses

$$\sigma_1 = \frac{pR}{t}, \qquad \sigma_2 = \frac{pR}{2t}.$$

From Eq. (12-15), the Tresca equivalent stress is

$$\begin{aligned}
\sigma_T &= \text{Max}(|\sigma_1 - \sigma_2|, |\sigma_2 - \sigma_3|, |\sigma_1 - \sigma_3|) \\
&= \text{Max}\left(\left|\frac{pR}{t} - \frac{pR}{2t}\right|, \left|\frac{pR}{2t}\right|, \left|\frac{pR}{t}\right|\right) \\
&= \frac{pR}{t} \\
&= \frac{(2 \times 10^6)(2)}{0.01} \\
&= 400 \text{ MPa},
\end{aligned}$$

so the Tresca safety factor is

$$S_T = \frac{\sigma_Y}{\sigma_T} = \frac{1800}{400} = 4.50.$$

From Eq. (12-19), the von Mises equivalent stress is

$$\begin{aligned}
\sigma_M &= \frac{1}{\sqrt{2}}\sqrt{(\sigma_1 - \sigma_2)^2 + (\sigma_2 - \sigma_3)^2 + (\sigma_1 - \sigma_3)^2} \\
&= \frac{1}{\sqrt{2}}\sqrt{\left(\frac{pR}{t} - \frac{pR}{2t}\right)^2 + \left(\frac{pR}{2t}\right)^2 + \left(\frac{pR}{t}\right)^2} \\
&= \frac{\sqrt{3}\,pR}{2t} \\
&= \frac{\sqrt{3}(2 \times 10^6)(2)}{2(0.01)} \\
&= 346 \text{ MPa}.
\end{aligned}$$

Therefore, the von Mises safety factor is

$$S_M = \frac{\sigma_Y}{\sigma_M} = \frac{1800}{346} = 5.20.$$

The von Mises safety factor is 15.6% larger than that based on the Tresca equivalent stress. This reflects the fact that the von Mises failure envelope generally lies outside the Tresca envelope (Fig. 12-34).

Repeated Loads

On the basis of our discussion in the preceding section, you might think that safe design of a component requires only that stresses remain below a suitable failure criterion. But it has been found that components subjected to repeated loads may fail even though the associated stress levels are well below the yield strength. Basically, a small amount of damage is produced each time a repeated load is applied. Although the amount of damage done in each repetition, or *cycle,* is insufficient to cause failure, damage can accumulate and eventually result in failure. The term *fatigue* was first applied in the mid-nineteenth century by engineers dealing with this type of failure in stagecoach and railroad car axles.

Fatigue is generally subdivided into three categories: low cycle, high cycle, and fatigue crack growth. In *low-cycle fatigue,* stress levels exceed the yield strength, with the result that the number of cycles to failure is relatively low ($<10^3$). If you have ever broken a piece of wire by bending it back and forth, you have experienced a case of low-cycle fatigue. (If you have not, you should carry out this experiment with a paper clip.) *High-cycle fatigue* can occur when stress levels are lower than the yield strength, and failure may require 10^3 to 10^6 cycles. As we discuss in Section 12-3, *fatigue crack growth* refers to the growth of discrete cracks under repeated loading. In this section we consider high-cycle fatigue.

S–N CURVES

The resistance of materials to high-cycle fatigue is determined by subjecting them to stress levels of constant frequency and amplitude in axial load or bending experiments. Of course, in applications, loads may be random in frequency and amplitude, but the use of constant values in testing is desirable both for simplicity and from the point of view of making comparisons.

Figure 12-6 shows two examples of the stress resulting from cyclical loading with constant frequency and amplitude. In case (a) the mean value of the stress is zero and in case (b) it is not. Denoting the maximum and minimum values

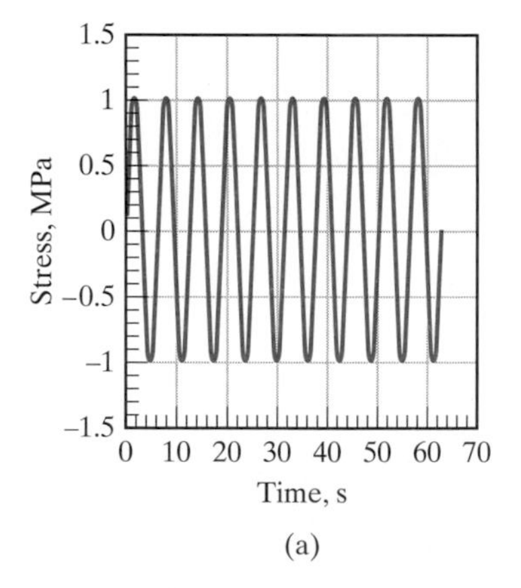

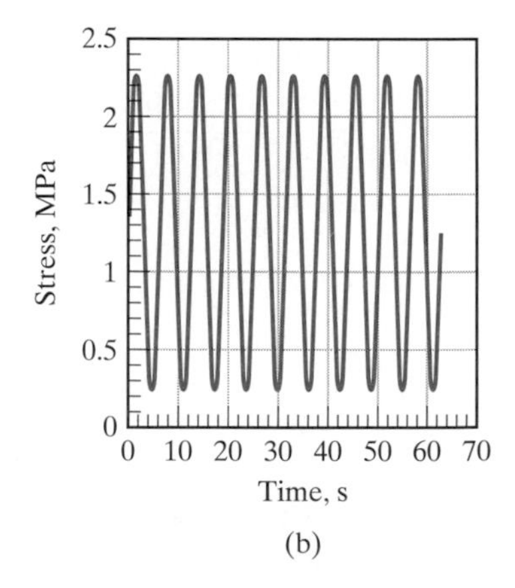

FIGURE 12-6 Constant stress amplitude loading: (a) with zero mean stress; (b) with nonzero mean stress.

of the stress by σ_{max} and σ_{min}, the *stress amplitude* σ_a and *mean stress* σ_m are

$$\sigma_a = \frac{\sigma_{max} - \sigma_{min}}{2}, \qquad \sigma_m = \frac{\sigma_{max} + \sigma_{min}}{2}. \tag{12-22}$$

To determine the fatigue strength of a specimen subjected to cyclic loading with zero mean value, the number of cycles N required to fail the specimen is recorded for a given stress amplitude or *fatigue strength*. The fatigue strength is plotted as a function of N (Fig. 12-7), yielding what is called an S–N or *endurance curve*. Notice that the fatigue strength decreases as the number of cycles is increased. For some materials, notably steels, there is a distinct stress limit σ_{fat} below which the specimen or component will have essentially infinite fatigue life. This is known as the *fatigue limit* or *endurance limit,* and becomes another stress level to be considered in design. Keeping stress levels below the fatigue limit is known as *safe life design*. In some situations safe life design is incompatible with other design constraints (such as weight), and a finite fatigue life must be accepted.

An empirical expression used to fit S–N curves is

$$\sigma_a = \sigma_{fat} + \frac{b}{N^c}, \tag{12-23}$$

where the fatigue limit σ_{fat} and the parameters b and c are material properties that measure a material's resistance to high-cycle fatigue. (For the example shown in Fig. 12-7, $\sigma_{fat} = 60$ ksi, $b = 2000$ ksi, and $c = 0.4$.) Once these values have been determined by fatigue testing, the fatigue strength or allowable stress level for a specified lifetime can be determined from Eq. (12-23). Alternatively, if the stress amplitude is the specified quantity, Eq. (12-23) can be solved for the

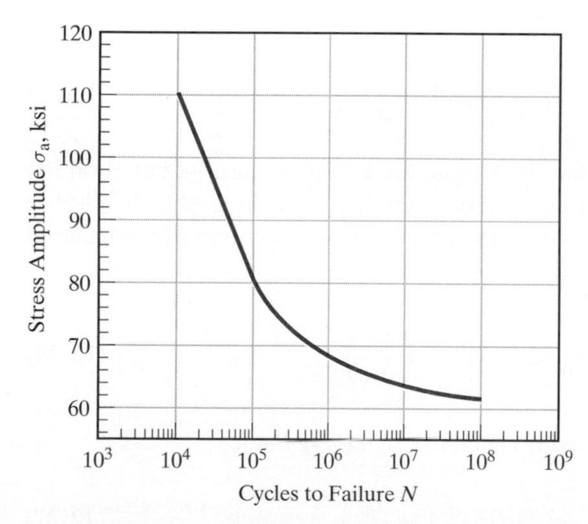

FIGURE 12-7 S–N (endurance) curve.

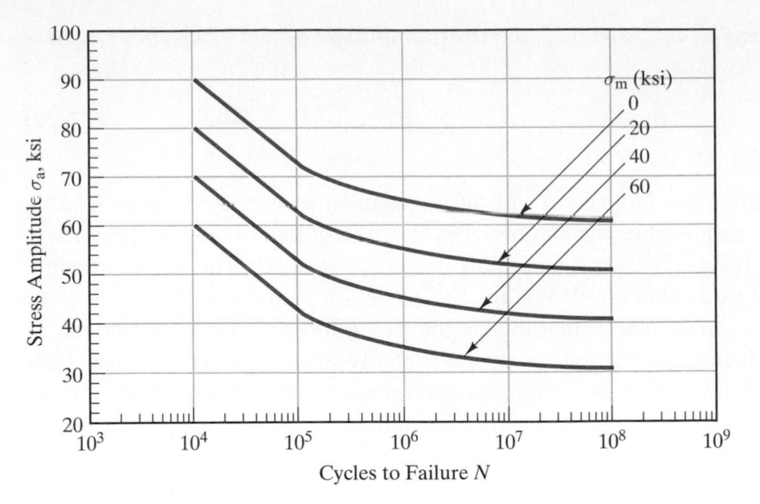

FIGURE 12-8 Effect of mean stress on an *S–N* curve.

corresponding lifetime:

$$N = \left(\frac{b}{\sigma_a - \sigma_{fat}}\right)^{1/c}. \tag{12-24}$$

Fatigue life is also affected by the mean stress level. When there is a mean stress σ_m, the maximum stress in each cycle is brought closer to the yield stress, which leads to more damage per cycle and decreases fatigue life. The main effect of mean stress is to depress the fatigue strength, so the *S–N* curves shift downward with increasing mean stress level (Fig. 12-8).

The manner in which the fatigue strength is affected by mean stress depends on the material, and as a consequence three different empirical relationships have been developed. Denoting the fatigue limit in the presence of mean stress by σ'_{fat}, the *Goodman relation* is

$$\sigma'_{fat} = \sigma_{fat}\left(1 - \frac{\sigma_m}{\sigma_U}\right). \tag{12-25}$$

According to this expression the fatigue limit reduces linearly, based on the ultimate tensile strength of the material. Another equation used is the *Gerber parabola,* which is also based on the ultimate tensile strength of the material but uses a quadratic expression:

$$\sigma'_{fat} = \sigma_{fat}\left[1 - \left(\frac{\sigma_m}{\sigma_U}\right)^2\right]. \tag{12-26}$$

The third relation that is used is called the *Soderberg line:*

$$\sigma'_{fat} = \sigma_{fat}\left(1 - \frac{\sigma_m}{\sigma_Y}\right). \tag{12-27}$$

It uses the yield strength of the material as the basis for adjusting the endurance limit in the presence of mean stress. These relationships have been derived from what are called *constant life diagrams,* in which the normalized stress amplitude is plotted as a function of mean stress with lifetime as the third parameter. Depending on the material, such plots are usually either approximately linear or approximately quadratic, which explains the origin of these empirical expressions. Once σ'_{fat} has been determined for a given material and value of mean stress, it can be substituted for σ_{fat} in Eq. (12-23) to determine the stress amplitude as a function of the number of cycles to failure.

EXAMPLE 12-4

Fatigue testing of a steel alloy at zero mean stress results in the *S–N* data shown in Table 12-1. This material is to be used in a 1-in.-diameter rod that will be subjected to an axial load varying between equal levels in tension and compression. The rod must withstand 500,000 cycles. Determine the maximum load level that can be applied.

Table 12-1 *S–N* data

Cycles to failure (1000 cycles)	Stress amplitude (ksi)
1.0	190.7
10.0	135.1
100.0	100.0
1000.0	77.9

Strategy

By drawing a graph of the measured stress amplitude as a function of the number of cycles to failure, we can estimate the stress amplitude corresponding to 500,000 cycles and thereby determine the maximum load amplitude.

Solution

We plot the *S–N* curve in semilog form in Fig. 12-9. The fatigue strength (stress amplitude) corresponding to 500,000 cycles is approximately $\sigma_a = 84$ ksi. Since this is a case of zero mean stress, the maximum and minimum stresses and therefore the maximum and minimum loads have the same amplitude. Therefore, the maximum load amplitude is

$$\begin{aligned} P_{max} = -P_{min} &= \pi R^2 \sigma_a \\ &= \pi (0.5)^2 (84 \times 10^3) \\ &= 66{,}000 \text{ lb.} \end{aligned}$$

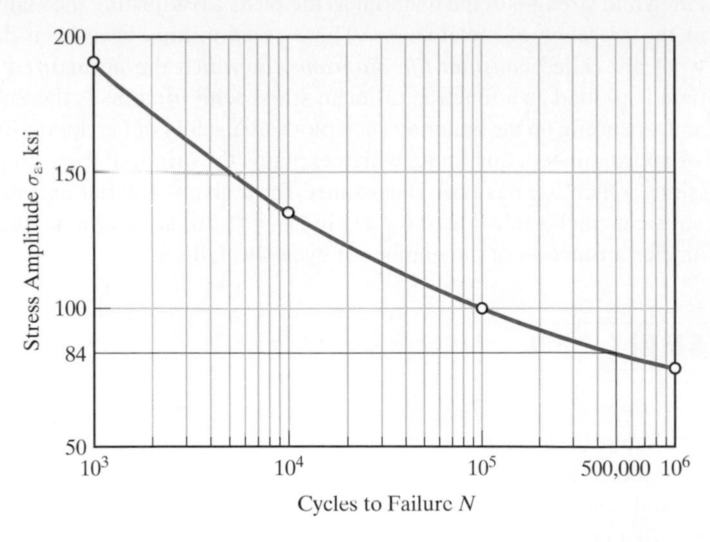

FIGURE 12-9 S–N curve.

EXAMPLE 12-5

The I-beam of a gantry crane (Fig. 12-10) is used for loading and unloading 80,000-lb containers at a factory. The beam has built-in supports. Its dimensions and moment of intertia are $L = 20$ ft, $h = 12.5$ in., and $I = 285$ in^4. The beam's material has ultimate stress $\sigma_U = 80$ ksi and its S–N curve is described by Eq. (12-23) with $\sigma_{fat} = 10$ ksi, $b = 120$ ksi, and $c = 0.15$. Mean stress effects can be accounted for through the Goodman relation, Eq. (12-25). Determine how many years the crane can be used if the beam is the critical structural element and 20 containers per day are lifted in a 200-working-day year.

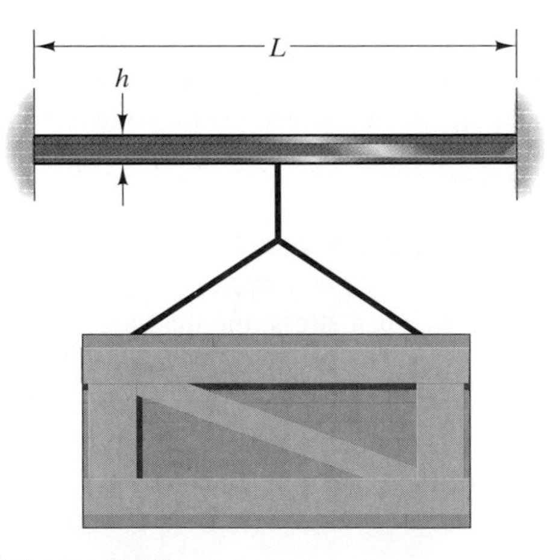

FIGURE 12-10

Strategy

We must first determine the maximum range of stresses to which the beam is subjected, and calculate its amplitude and mean value. Then we can determine σ'_{fat} from the Goodman relation and substitute it into Eq. (12-24) to obtain the number of cycles.

Solution

We leave it as an exercise to show that the maximum bending moment in the beam due to the weight W of a container supported at the beam's midpoint is $M = WL/8$. The resulting maximum tensile stress occurs at $y = h/2$:

$$\begin{aligned}\sigma_{\max} &= \frac{My}{I} \\ &= \frac{WLh}{(8)(2)I} \\ &= \frac{(80{,}000)(240)(12.5)}{(8)(2)(285)} \\ &= 52{,}600 \text{ psi.}\end{aligned}$$

The stress cycles between this value and zero (when the container is released), so the stress amplitude and mean stress are

$$\sigma_{\text{a}} = \sigma_{\text{m}} = \frac{\sigma_{\max}}{2} = 26{,}320 \text{ psi.}$$

Applying the Goodman relation yields

$$\begin{aligned}\sigma'_{\text{fat}} &= \sigma_{\text{fat}}\left(1 - \frac{\sigma_{\text{m}}}{\sigma_{\text{U}}}\right) \\ &= 10{,}000\left(1 - \frac{26{,}320}{80{,}000}\right) \\ &= 6710 \text{ psi.}\end{aligned}$$

We determine the number of cycles from Eq. (12-24) with σ'_{fat} substituted for σ_{fat}:

$$\begin{aligned}N &= \left(\frac{b}{\sigma_{\text{a}} - \sigma'_{\text{fat}}}\right)^{1/c} \\ &= \left(\frac{120{,}000}{26{,}320 - 6710}\right)^{1/0.15} \\ &= 176{,}000 \text{ cycles.}\end{aligned}$$

Dividing N by 4000 cycles/year gives a useful lifetime of 44 years.

Cumulative Damage

It is actually quite rare for a structural component to be subjected to cyclic loading with a single amplitude. Components are generally subjected to a spectrum of loads. But it is often possible to divide the loading into segments, or blocks, each of which is of approximately constant amplitude. A certain amount of damage will occur during each loading block, and the question that arises is how to quantify the cumulative damage.

Suppose that N is the number of cycles to failure of a particular component at a given stress amplitude and mean stress. Let the *damage* D done in subjecting the component to n cycles at that stress amplitude and mean stress be defined by

$$D = \frac{n}{N}.$$

Notice that failure occurs at $D = 1$ if the loading continues until $n = N$. This is called *Miner's law,* which also states that the cumulative damage resulting from blocks of loading of different amplitudes and mean stresses can be obtained by summing the damages of the individual blocks, irrespective of the order in which they are applied, and that failure occurs when $D = 1$. In practice, the sequence in which blocks of loads are applied does have an effect in certain materials. Nevertheless, this simple approach works surprisingly well. Suppose that a component is subjected to a sequence of blocks of loading. Let n_i be the number of cycles applied in the ith block, and let N_i be the number of cycles required to fail the undamaged component at that stress amplitude and mean stress. The resulting damage is $D_i = n_i/N_i$. The accumulated damage after k blocks of loading is

$$D = \sum_{i=1}^{k} D_i = \sum_{i=1}^{k} \frac{n_i}{N_i}. \qquad \text{(12-28)}$$

We demonstrate the application of Miner's law in the following example.

EXAMPLE 12-6

The numbers of cycles applied to a structural component at several levels of stress amplitude are listed in Table 12-2. The loading blocks all consisted of stress cycles with zero mean stress, and the S–N curve is described by

Table 12-2 Cycles applied to a structural component

Loading block i	Stress amplitude (ksi)	Cycles applied (1000 cycles)
1	80	2.79
2	75	12.59
3	70	58.37
4	65	124.93

Eq. (12-23) with $\sigma_{fat} = 60$ ksi, $b = 1200$ ksi, and $c = 0.4$. How many additional cycles of loading could be applied to the component at a stress amplitude of 85 ksi?

Strategy

The first step is to determine how many cycles would be required to fail the component at each of the stress levels in the table. We can then use Eq. (12-28) to determine the accumulated damage. Knowing how much damage remains to be accumulated before failure occurs, we can calculate the necessary number of cycles at a stress amplitude of 85 ksi.

Solution

From Eq. (12-24), the number of cycles to failure for the first block of loading is

$$N_1 = \left(\frac{b}{\sigma_a - \sigma_{fat}}\right)^{1/c} = \left(\frac{1{,}200{,}000}{80{,}000 - 60{,}000}\right)^{1/0.4} = 27{,}900 \text{ cycles.}$$

The damage done in this loading block is therefore

$$D_1 = \frac{n_1}{N_1} = \frac{2790}{27{,}900} = 0.1.$$

Repeating these steps for each loading block, the results are summarized in Table 12-3. After the four load blocks have been applied, the accumulated damage is

$$D = \sum_{i=1}^{4} D_i = 0.83.$$

The damage remaining before failure is $1 - D = 0.17$. The number of cycles to failure at 85 ksi is

$$N = \left(\frac{1{,}200{,}000}{85{,}000 - 60{,}000}\right)^{1/0.4} = 16{,}000 \text{ cycles.}$$

Table 12-3 Loading block damages

Loading block i	Stress amplitude (ksi)	Cycles to failure N_i (1000 cycles)	Cycles applied n_i (1000 cycles)	Damage D_i
1	80	27.9	2.79	0.1
2	75	57.2	12.59	0.22
3	70	157.7	58.37	0.37
4	65	892.3	124.93	0.14

To determine the number of cycles n that can be applied until failure, we set

$$\frac{n}{N} = 0.17,$$

obtaining $n = 2710$ cycles.

12-2 | Stress Concentrations

Either intentionally, to satisfy specific design requirements, or as a result of processing or manufacturing flaws, structural members may have geometric features (including holes, notches, and corners) that result in localized regions of high stress called *stress concentrations*. They are of great concern in structural design, because the stresses in these regions can be of much larger magnitude than the stresses to which the member would be subjected in the absence of the geometric features causing them. For example, in the stepped bar subjected to tension in Fig. 12-11, the magnitude of the maximum stress near the step substantially exceeds the value predicted by the equation $\sigma = P/A$, although the stress approaches this value roughly one diameter away from the step. In this section we give examples of geometric features that give rise to stress concentrations and present empirical information available to the structural designer for analyzing them.

Axially Loaded Bars

Structural members may be designed with holes, for example to decrease weight or to allow a bolt to be inserted, or holes may be present inadvertently due to manufacturing flaws. In Fig. 12-12a, a bar of rectangular cross section containing a hole of radius a is subjected to a tensile stress σ. The hole creates a stress concentration (Fig. 12-12b). The ratio of the maximum stress σ_{max} to the nominal stress σ is called the *stress concentration factor:*

$$C = \frac{\sigma_{max}}{\sigma}. \qquad \textbf{(12-29)}$$

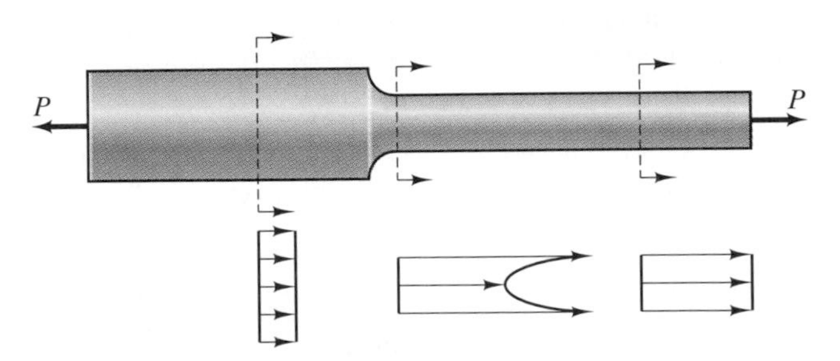

FIGURE 12-11 Stepped rod in tension showing the normal stress distributions before, after, and at the step.

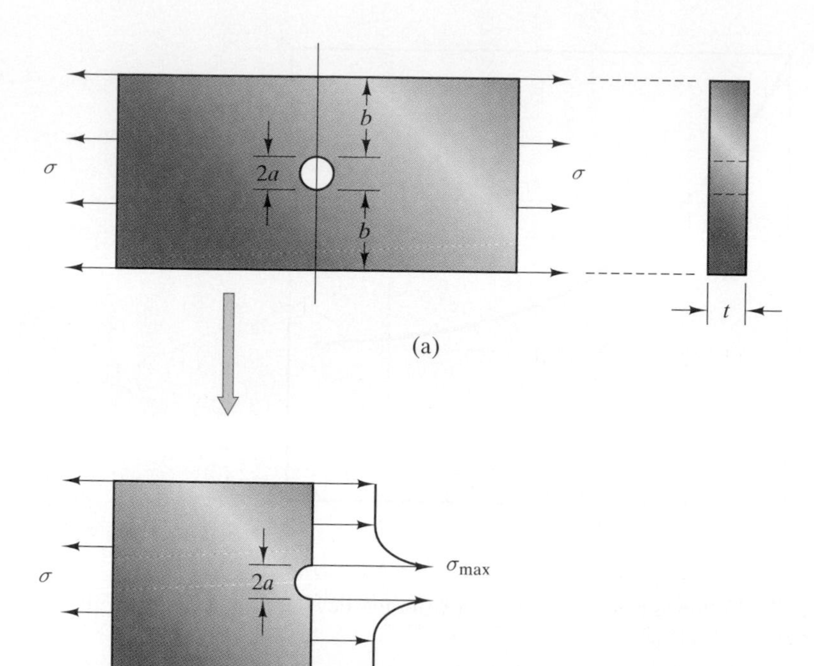

FIGURE 12-12 Subjecting a bar with a centrally located hole to tensile stress.

Stress concentration factors must be determined by advanced analytical or numerical methods. Under the assumption that the width $2(a + b)$ of the bar is much larger than its thickness t, values of C are shown as a function of a/b in Fig. 12-13. (Because the width of the bar is assumed to be much greater than its thickness, we could have anticipated that the stress concentration factor should depend only on the dimensionless quantity a/b.) Notice that even a microscopic hole resulting from a manufacturing flaw results in a maximum stress three times the nominal stress. This is one illustration of why designers must choose conservative factors of safety.

We have already seen (Fig. 12-11) that a stress concentration occurs in a bar with a stepped cross section. Figure 12-14 shows stress concentration factors for stepped circular bars subjected to tensile load as a function of the ratio of the radius a of the *fillet* to the smaller diameter d_2. Figure 12-15 shows equivalent stress concentration factors for thin rectangular bars. As the radius a approaches zero, the stress concentration factor approaches infinity, which emphasizes the importance of incorporating fillets (literally "rounding inside corners") in design.

Torsion

In Chapter 4 we used Eq. (4-9),

$$\tau = \frac{Tr}{J}, \tag{12-30}$$

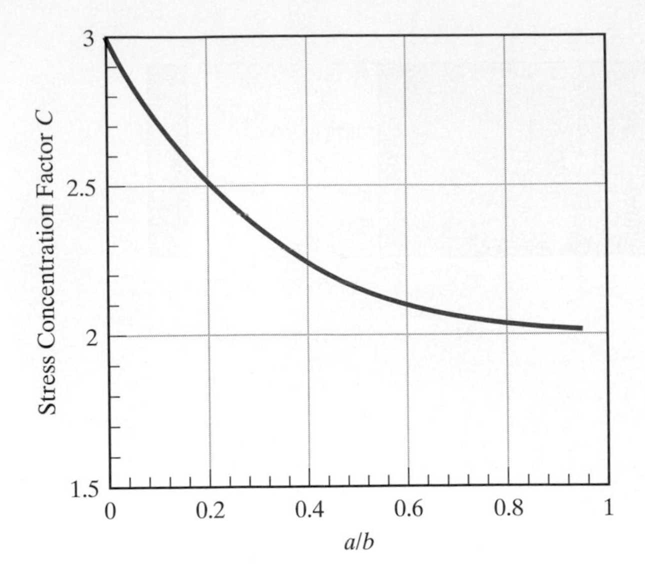

FIGURE 12-13 Stress concentration factor C for tensile loading of a thin bar with a centrally located hole.

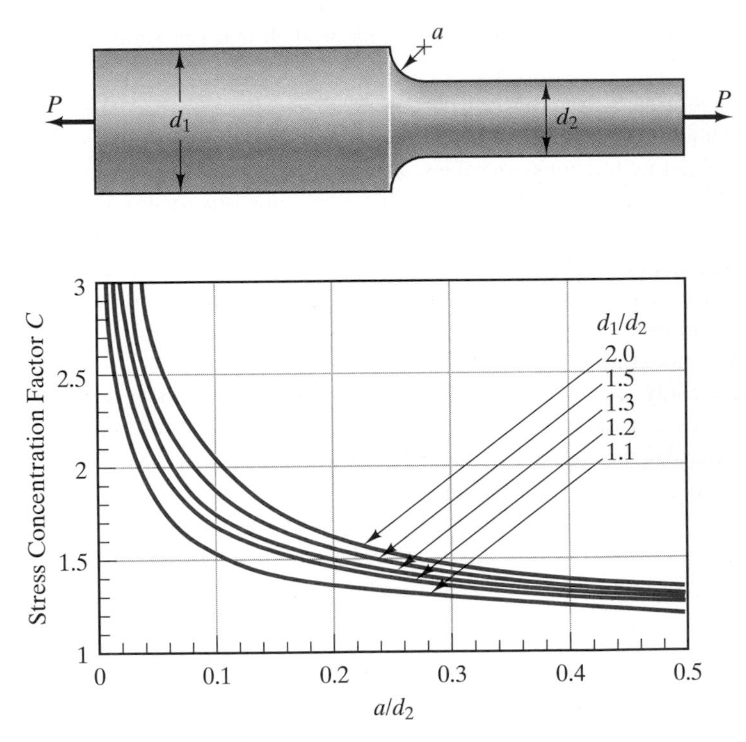

FIGURE 12-14 Stress concentration factor for a circular bar with a shoulder fillet under axial loading. The nominal stress is the normal stress in the part of the bar with the smaller diameter.

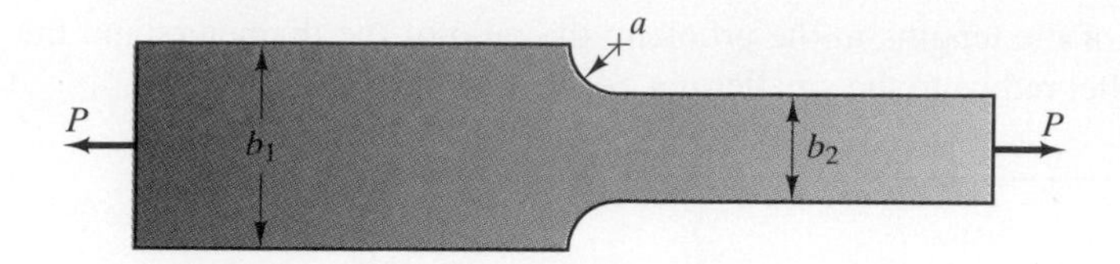

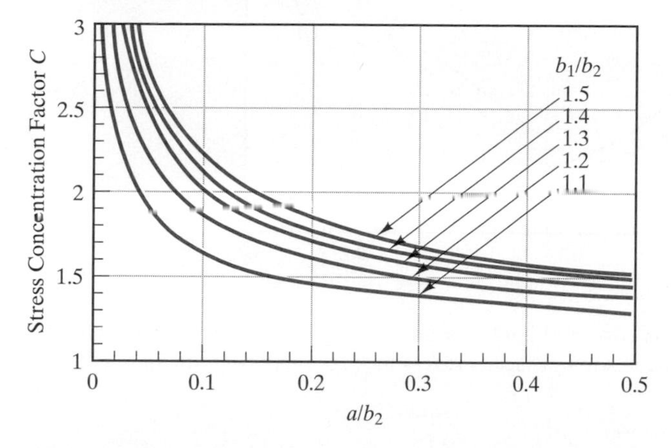

FIGURE 12-15 The stress concentration factor for a flat bar with shoulder fillets. The nominal stress is the normal stress in the part of the bar with the smaller height.

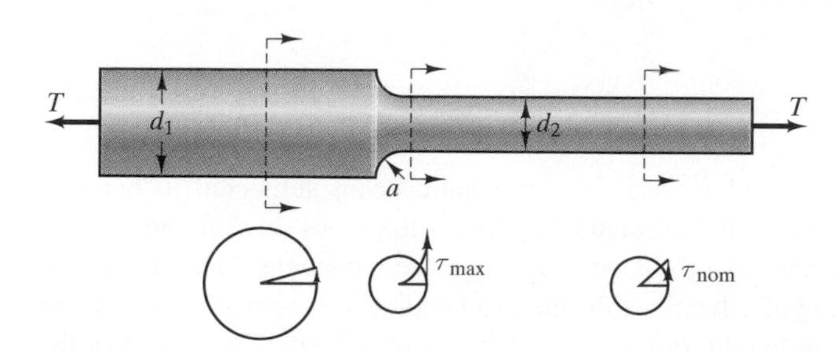

FIGURE 12-16 Stress distributions before, after, and at a step in diameter in a rod subjected to torsion.

to determine the shear stresses in a circular bar subjected to a torque T, where J is the polar moment of inertia of the cross section and r is the radial distance from the bar's axis. In a stepped bar (Fig. 12-16), the distribution of stress is approximated by Eq. (12-30) at cross sections that are not near the step, but a stress concentration exists at the step. The stress concentration factor is defined in terms of the shear stress,

$$C = \frac{\tau_{max}}{\tau_{nom}}, \tag{12-31}$$

where τ_{nom} is the maximum shear stress in the part of the bar with the smaller diameter. As can be seen in Fig. 12-17, the stress concentration factor depends on the ratios of the lengths in the problem: the ratio of the diameters and the ratio of the fillet radius to the smaller diameter.

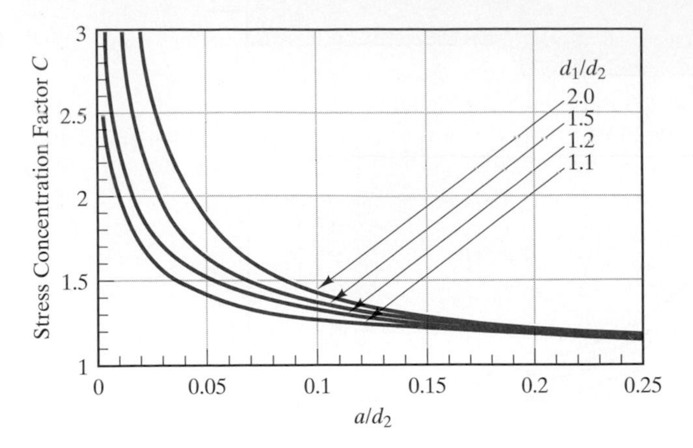

FIGURE 12-17 Stress concentration factor for a stepped bar subjected to torsion. The nominal stress is the maximum shear stress in the part of the bar with the smaller diameter.

Bending

In Chapter 8 we used Eq. (8-12),

$$\sigma_x = \frac{My}{I}, \tag{12-32}$$

to determine the normal stresses in a prismatic beam subjected to bending moment M, where I is the moment of inertia of the cross section and y is the distance from the neutral axis. As examples of stress concentrations in beams, we consider rectangular beams subjected to bending moment M that contain holes, have steps in height, or contain notches (Fig. 12-18). The height of the rectangular cross section is assumed to be large compared to its width.

For the case of a beam containing a hole (Fig. 12-18a), let the hole be centered at the neutral axis. When the hole diameter $2a$ is small compared to the beam's height, the stress is approximated by the nominal linear distribution

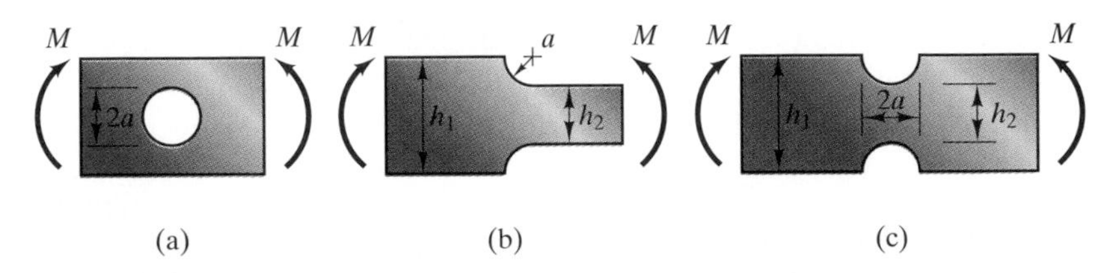

FIGURE 12-18 Examples of beam geometries that result in stress concentrations.

except near the hole (Fig. 12-19a). Although amplification of the stress occurs near the hole, the increased value does not exceed the maximum stress resulting from the nominal distribution. For larger holes ($2a/h_1 > \frac{1}{2}$), the stress concentration dominates the linear distribution and the resulting maximum stress is approximately twice the maximum nominal stress.

Figure 12-20 shows stress concentration factors $C = \sigma_{max}/\sigma_{nom}$ for a stepped beam subjected to bending moment as a function of the ratio of the fillet radius a to the smaller height h_2. The stress σ_{nom} is the maximum stress resulting from Eq. (12-32) in the part of the beam with the smaller height. Stress concentration factors for a notched beam are shown in Fig. 12-21.

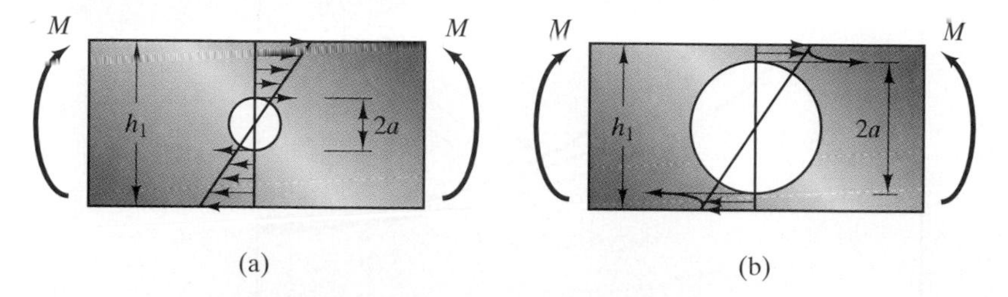

FIGURE 12-19 Stress distributions in beams with small and large holes.

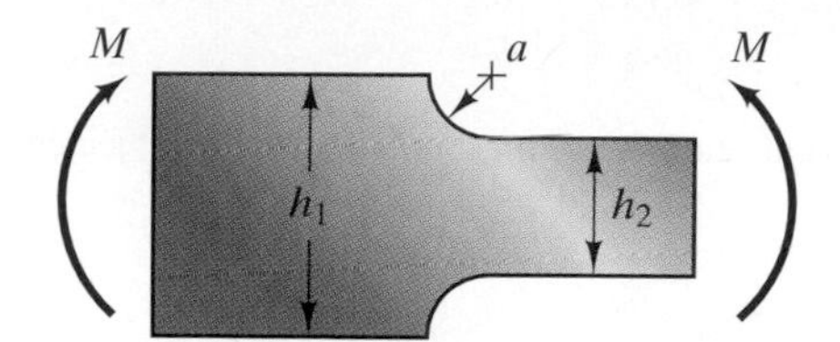

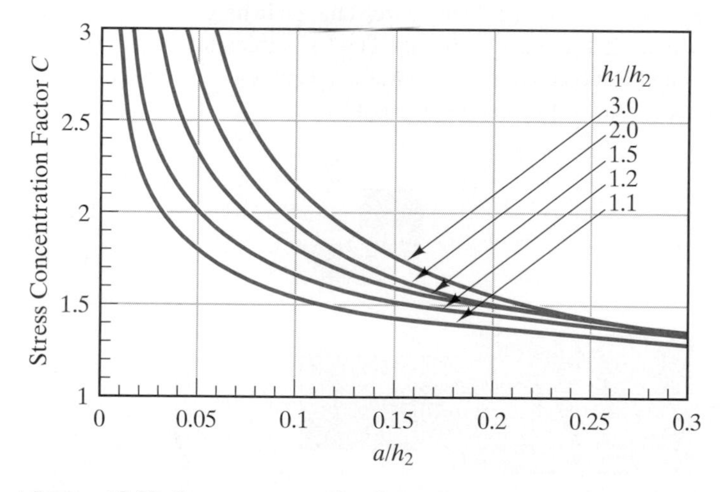

FIGURE 12-20 Stress concentration factor for a stepped beam subjected to bending. The nominal stress is the maximum stress in the part of the bar with the smaller height.

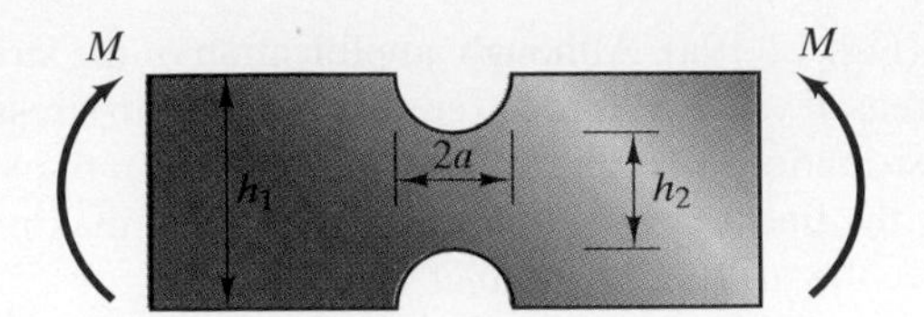

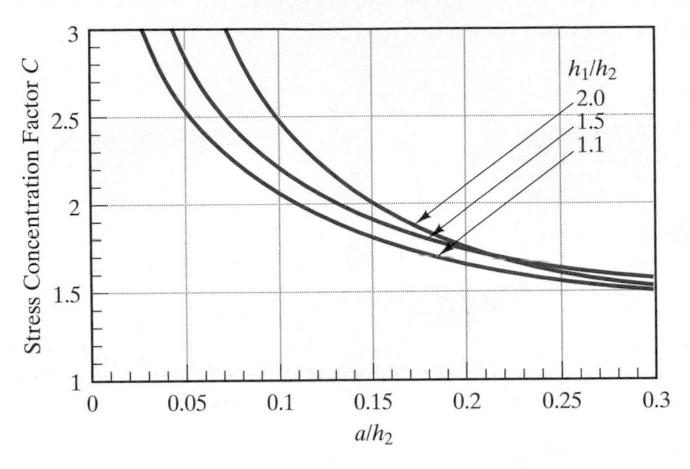

| **FIGURE 12-21** Stress concentration factor for a notched beam subjected to bending.

EXAMPLE 12-7

The thin rectangular bar in Fig. 12-22 has a step change in height from $b_1 = 3$ in. to $b_2 = 2$ in. with a fillet radius $a = 0.2$ in. The bar's thickness is $t = 0.01$ in. If the material can be subjected to an allowable stress $\sigma_{\text{allow}} = 40$ ksi, what is the maximum axial load P that can be applied?

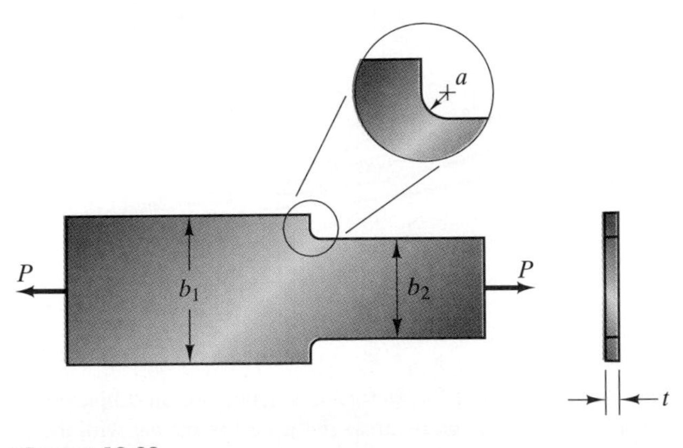

| **FIGURE 12-22**

Strategy

We must determine the stress concentration factor for an axially loaded bar with this geometry and fillet radius. Then by expressing the nominal stress (the normal stress in the part of the bar with the smaller height) in terms of P, we can determine the value of P for which the maximum stress due to the stress concentration is equal to the allowable stress.

Solution

From Fig. 12-15, the stress concentration factor for $b_1/b_2 = \frac{3}{2} = 1.5$ and $a/b_2 = 0.2/2 = 0.1$ is $C = 2.24$. The nominal stress in terms of P is

$$\sigma_{\text{nom}} = \frac{P}{b_2 t},$$

so the maximum stress resulting from the stress concentration in terms of P is

$$\sigma_{\text{max}} = C\sigma_{\text{nom}} = \frac{2.24P}{b_2 t}.$$

We equate this stress to the allowable stress.

$$\frac{2.24P}{b_2 t} = \sigma_{\text{allow}} :$$

$$\frac{2.24P}{(2)(0.01)} = 40{,}000.$$

Solving for P, the maximum permissible axial load is 357 lb.

12-3 | Fracture

Although the subdiscipline of mechanics of materials known as fracture mechanics essentially originated with a paper by A. A. Griffith in 1920, substantial growth of the field did not begin until the 1960s. This growth was motivated by a series of catastrophic structural failures, including the literal splitting in two of Navy T2 tankers while at anchor in calm but cold conditions and crashes of two Comet jet airliners, in which it became clear that the cause was growth of cracks from preexisting flaws. Such flaws had always been present in structural materials as a consequence of the manufacturing processes used, stress concentrations, and repeated loading. But their effects were accentuated in these more recent failures because the structures consisted of new high-strength materials that were much more sensitive to the presence of small cracks. As a result, the structures failed at much lower stress levels than had been anticipated by engineers following the procedures described in our discussion of failure in Section 12-1. In the materials science community, crack sensivity has led to a classic balancing act in the design of new materials in which high strength does not necessarily mean that a material is tough, or resistant to crack growth. The organization of this section mirrors to some extent the development in our discussion of failure, considering overloads and repeated loads separately.

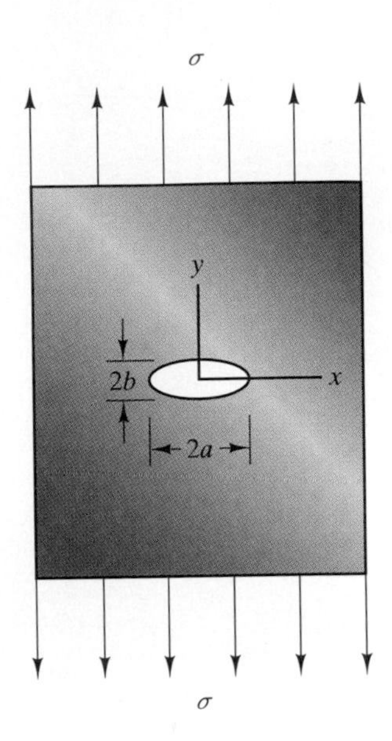

FIGURE 12-23 Applying tension to a sheet containing an elliptical hole.

Overloads and Fast Crack Growth

We consider an overload to be a one-time event in which load levels have exceeded a safe allowable value. This is in contrast to repeated loading, which can cause failure at much lower load levels. In first thinking about cracks in materials, you might think that as soon as there is a crack in a structure, it will fail. In fact, some structures can tolerate quite sizable cracks before failing, which makes it clear that severity of cracks can be ranked. We therefore need to develop a "crack meter," a quantitative measure of crack severity. To do so, we must return to the concept of stress concentrations.

STRESS CONCENTRATION DUE TO AN ELLIPTICAL HOLE

We have already seen that circular holes and fillets give rise to stress concentrations. The ellipse is a useful hole geometry for us to consider here, because at one extreme it can be made a circle while at the other extreme it becomes a crack. We can relate the former extreme to what we already know, and with the latter extreme we can explore new ground.

In Fig. 12-23, a uniform tensile stress is applied to the ends of a sheet containing an elliptical hole. The major axis of the ellipse ($2a$) is perpendicular to the direction of the applied stress. We assume that the width and height of the sheet are large compared to the hole. The maximum normal stress occurs at the root of the ellipse ($x = \pm a$) and is given by

$$\sigma_{\max} = \sigma\left(1 + \frac{2a}{b}\right). \tag{12-33}$$

Setting $a = b$ in Eq. (12-33), we recover the stress concentration factor for a circular hole: $C = \sigma_{\max}/\sigma = 3$. In terms of the *root radius* $\rho = b^2/a$ of the ellipse, the maximum stress is

$$\sigma_{\max} = \sigma\left(1 + 2\sqrt{\frac{a}{\rho}}\right). \tag{12-34}$$

When the crack length (the major axis of the ellipse) is large compared to the root radius, the stress concentration factor is approximately

$$C = 2\sqrt{\frac{a}{\rho}}. \tag{12-35}$$

For an infinitely sharp crack ($\rho \to 0$), the stress concentration factor is infinite, a result we will soon encounter again. Infinite stresses are predicted because our solution assumes elastic material behavior. In reality, the material would yield and the stresses remain finite.

Equation (12-35) provides justification for the common practice of drilling holes at the tips of sharp cracks (for example, in the metal skin of an airplane's wing) in order to arrest their growth. The radius of the hole drilled is much larger than the notch radius of the crack. As a result, the stress concentration is significantly reduced, arresting the crack.

Stress Distribution near a Crack Tip

Figure 12-24 is a schematic of the region very near a crack tip. With reference to the coordinate system shown, the components of plane stress are

$$\begin{aligned}
\sigma_x &= \frac{K}{\sqrt{2\pi r}} \cos\frac{\theta}{2}\left(1 - \sin\frac{\theta}{2}\cos\frac{3\theta}{2}\right), \\
\sigma_y &= \frac{K}{\sqrt{2\pi r}} \cos\frac{\theta}{2}\left(1 + \sin\frac{\theta}{2}\cos\frac{3\theta}{2}\right), \qquad \textbf{(12-36)} \\
\tau_{xy} &= \frac{K}{\sqrt{2\pi r}} \cos\frac{\theta}{2}\sin\frac{\theta}{2}\cos\frac{3\theta}{2},
\end{aligned}$$

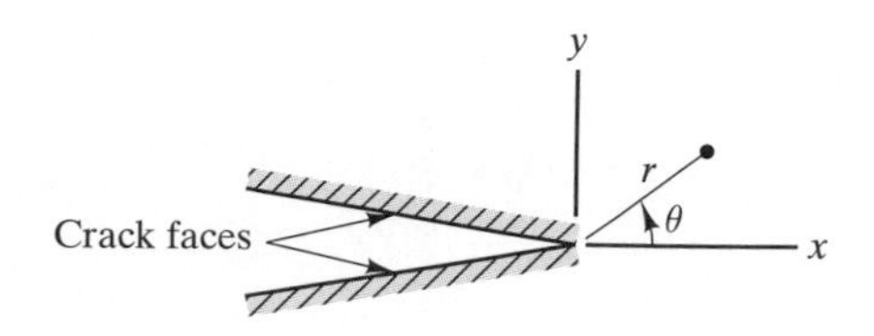

FIGURE 12-24 Region near the crack tip.

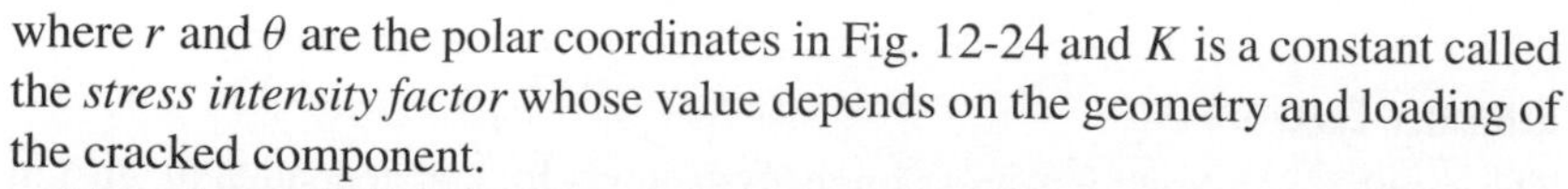

where r and θ are the polar coordinates in Fig. 12-24 and K is a constant called the *stress intensity factor* whose value depends on the geometry and loading of the cracked component.

Equations (12-36) imply that the stresses near all cracks have the same dependence on r and θ. These solutions predict that the stresses are infinite at the crack tip, just as we found for an elliptical hole when it reduces to an infinitely sharp crack ($\rho \to 0$), but yielding of the material will occur very near the crack tip. Let σ_Y be the yield stress of the material, and let us assume that the *crack opening stress,* the component σ_y, governs yielding. As the crack tip is approached along the x axis ($\theta = 0$), σ_y increases as the radial distance r decreases and is equal to the yield stress when the radial distance is

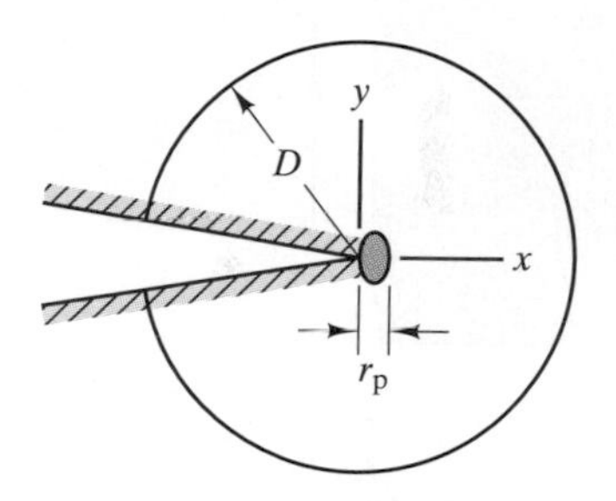

FIGURE 12-25 Yielded zone and region of dominance of Eqs. (12-36).

$$r_p = \frac{1}{2\pi}\left(\frac{K}{\sigma_Y}\right)^2. \qquad \textbf{(12-37)}$$

If r_p is much smaller than the smallest characteristic dimension of the cracked component (for example, the thickness of a cracked plate), the state of stress within a neighborhood of the crack tip is described by Eqs. (12-36) except very near the crack tip (Fig. 12-25). To avoid boundary effects, the radius D of this neighborhood must be significantly smaller than any in-plane dimensions of the cracked component. For example, in the cracked plate in Fig. 12-26, D must be small in comparison with the length L and width W of the plate.

Since the r and θ dependence is the same near all cracks, the only parameter distinguishing one crack from another is the stress intensity factor K. It is, in essence, the crack meter that we suggested must exist. We will show that it allows us to make quantitative comparisons between the severities of cracks in different structural components under different loads. Basically, the larger the value of K, the more severe is the crack.

Stress Intensity Factor Solutions

The solution leading to Eqs. (12-36) focuses on the region of the crack tip without concern for where the crack is located within a component, the overall shape or geometry of the component, or the applied loads. This explains why the stress intensity factor K appears as an undetermined constant. To determine K requires the development of solutions that are specific to each cracked component and its loading. There was a period in the development of fracture mechanics when a great deal of emphasis was placed on obtaining such solutions. Analytical (rather than numerical) solutions were preferred because varying parameters was simpler. However, it was found that analytical solutions could be obtained only for relatively simple geometries.

Numerical techniques and dimensional analysis have now extended the range of problems that can be solved, and handbooks exist in which many solutions are cataloged. We give examples of stress intensity factors in this section, and handbooks contain many others. When an existing solution cannot be found, finite element codes with special "crack elements" and fracture mechanics post-processing packages are available that can be used to solve very specific and complex crack problems.

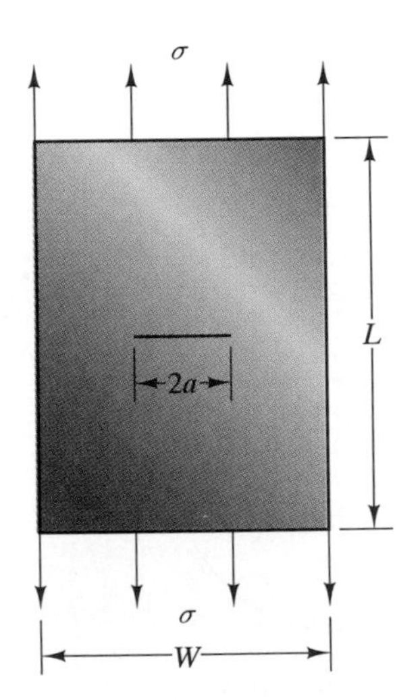

FIGURE 12-26 Plate with a central crack under tension.

CENTRALLY CRACKED PLATE

The simplest solution for a stress intensity factor is for a plate loaded in tension by a uniform normal stress σ and containing a centrally placed crack whose plane is perpendicular to the loading direction (Fig. 12-26). The crack length is assumed to be small in comparison to the plate's dimensions ($W, L \gg 2a$). This assumption means that the outer boundary is effectively at infinity, so the crack length $2a$ is the only geometrical parameter in the problem. The stress intensity factor is

$$K = \sigma\sqrt{\pi a}. \tag{12-38}$$

Thus the stress intensity factor is proportional to the applied stress and to the square root of the crack length. For a given level of stress, the longer the crack becomes, the more severe it becomes.

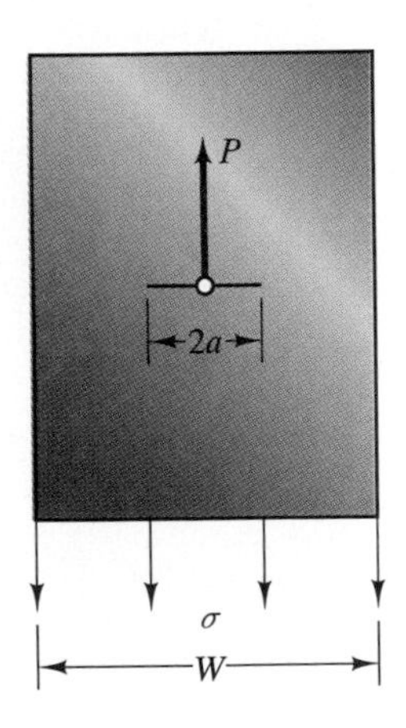

FIGURE 12-27 Central load P on a crack face.

CENTRAL LOAD ON A CRACK FACE

This problem can occur when cracks emanate from a bolthole. In Fig. 12-27, a concentrated force P is exerted by a bolt or rivet at the top of the bolthole, which is at the center of a crack of length $2a$. The diameter of the hole is assumed to be small in comparison to the crack length $2a$. Under these conditions the stress intensity factor is

$$K = \frac{P}{b\sqrt{\pi a}}, \tag{12-39}$$

where b is the thickness of the plate. We see once again that the stress intensity factor is proportional to the applied load. But in this case a appears in the denominator, and although K can be very large for short cracks, it decreases as the crack grows. This suggests the possibility of spontaneous crack arrest.

In many cases stress intensity factors can be expressed in the form

$$K = \sigma\sqrt{\pi a}\,Q\left(\frac{a}{W}\right), \tag{12-40}$$

where σ represents the stress applied to the component, a is the crack length, and $Q(a/W)$ is a nondimensional function of the crack length a normalized by an in-plane dimension W (such as the width of a plate) that is called the *configuration factor*. Comparing Eqs. (12-38) and (12-40), we see that $Q = 1$ for a centrally cracked plate, so this problem is the prototype for the expression (12-40). Many of the stress intensity factors given in handbooks are expressed in the form of Eq. (12-40).

Fracture Criterion

Up to this point we have dealt with the stress analysis of cracks. Doing so has allowed us to compare the severities of different crack configurations, but says nothing as to whether or not a crack will grow. Since this section deals with overloads, the type of crack growth we now consider is known as *fast crack growth*. We will show subsequently that slow crack growth can occur as a result of repeated loads and requires a different criterion.

The philosophy underlying the establishment of a fast crack growth criterion is very much akin to the use of the yield stress to establish the onset of yielding. Since we use the stress intensity factor to measure crack severity, an experiment can be conducted to measure the stress level and crack length at which a crack begins to grow (becomes a fast crack). These values can be substituted into the equation for the stress intensity factor, establishing the value of the stress intensity factor at which the crack begins to grow. This critical value of stress intensity factor is called the *fracture toughness* K_c. Once determined for a particular material, the fracture toughness can be used to predict the onset of fast crack growth in other structural components made of the same material. Since Eq. (12-40) is a general expression for the stress intensity factor that applies for any crack configuration, the criterion for fast crack growth is

$$\sigma\sqrt{\pi a}\,Q\left(\frac{a}{W}\right) = K_c. \tag{12-41}$$

From the left-hand side of this equation we see that there are two possible conditions for fast crack growth. The first one arises when the crack length a is fixed at a given value and the stress level for fast crack growth is to be established. Then Eq. (12-41) can be solved for the stress level σ at which a crack will become a fast crack before it can be seen by nondestructive inspection techniques or before it emerges from behind an overlapping component such as a washer covering the region around a hole. The second criterion for fast crack growth arises when the stress σ is fixed at some level. In this case it is often necessary to establish how long a crack can become before it becomes a fast crack by solving Eq. (12-41) for a. We will show in the following section that Eq. (12-41) can also be used to determine when a slow crack becomes a fast crack under cyclic loading.

The two criteria we have discussed bring out an important difference between yielding and fast crack growth. In the former we need only be concerned about stress levels. In the latter, both stress level and crack length must be considered. Another complication is that for many materials the fracture toughness is dependent on the thickness for thin sections, although it becomes constant as the thickness increases. We will assume that components are sufficiently thick that K_c does not depend on the thickness, in which case it is called the *plane strain fracture toughness*.

EXAMPLE 12-8

The steel plate in Fig. 12-28 has an edge crack of length $a = 2$ cm midway along its length. The plate is 5 mm thick, 20 cm wide, and has a length L sufficiently large that the loads P can be assumed to result in a uniform stress distribution across the width of the plate in the neighborhood of the crack. The stress intensity factor obtained from a handbook for this configuration is

$$K = 1.12\sigma\sqrt{\pi a}. \qquad (a/W \leq 0.13,\ L/W \geq 2) \qquad \textbf{(12-42)}$$

An alternative expression that applies for all values of $\alpha = a/W$ is

$$K = \sigma\sqrt{\pi a}\left[0.265(1-\alpha)^4 + \frac{0.857 + 0.265\alpha}{(1-\alpha)^{3/2}}\right]. \qquad (L/W \geq 2) \quad \textbf{(12-43)}$$

The fracture toughness of the plate is $K_c = 66$ MPa-m$^{1/2}$. Determine the load level that will cause the crack to grow.

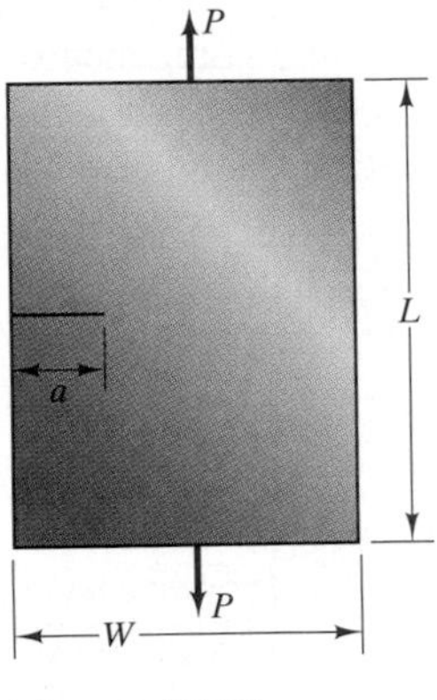

FIGURE 12-28

Strategy

We can use the simpler expression for the stress intensity factor since $a/W = 0.1$. The stress in the expression for the stress intensity factor is $\sigma = P/Wb$, where b is the thickness of the plate. Since the fracture toughness is known, we can solve for the value of P that will cause crack growth.

Solution

We substitute the fracture toughness and the expression $\sigma = P/Wb$ into Eq. (12-42),

$$K_c = 1.12\left(\frac{P}{Wb}\right)\sqrt{\pi a}:$$

$$66 \times 10^6 = 1.12\left[\frac{P}{(0.2)(0.005)}\right]\sqrt{\pi(0.02)}.$$

Solving, we obtain the critical load level $P = 0.235$ MN.

EXAMPLE 12-9

The container of weight W in Fig. 12-29 is supported by two parallel beams, each with semispan $L = 20$ ft and rectangular cross section of height $h = 10$ in. and width $b = 1$ in. The weight of the container and its contents is $W = 15{,}000$ lb. A crack of length a exists at the bottom of one of the beams at its midpoint as shown. The stress intensity factor obtained from a handbook for this configuration is

$$K = 1.12\left(\frac{6M}{h^2b}\right)\sqrt{\pi a}, \qquad (a/h \le 0.4) \qquad \textbf{(12-44)}$$

where M is the bending moment at the cross section containing the crack. An alternative expression that applies for all values of $\alpha = a/h$ and large values of h/L is

$$K = \frac{6M}{h^2b}\sqrt{\pi a}\sqrt{\frac{2}{\pi\alpha}\tan\frac{\pi\alpha}{2}}\left[\frac{0.923 + 0.199[1 - \sin(\pi\alpha/2)]^4}{\cos(\pi\alpha/2)}\right]. \qquad \textbf{(12-45)}$$

The fracture toughness of the beam material is $K_c = 120$ ksi-in$^{1/2}$. Determine the largest crack length a that could be tolerated before fast crack growth would occur.

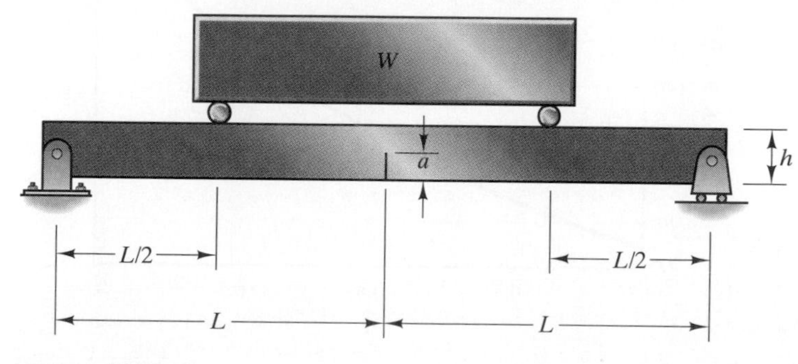

FIGURE 12-29

Strategy

We will first obtain a solution for the crack length using the simpler Eq. (12-44) for the stress intensity factor. If the resulting crack length is not within the range of validity of that expression, we must determine the crack length from Eq. (12-45).

Solution

Each beam supports half the weight of the container and its contents, so the bending moment at the cross section containing the crack is

$$\begin{aligned} M &= \frac{WL}{8} \\ &= \frac{(15{,}000)(240)}{8} \\ &= 450{,}000 \text{ in-lb.} \end{aligned}$$

Substituting this value and the fracture toughness into Eq. (12-44),

$$120{,}000 = 1.12\left[\frac{6(450{,}000)}{(10)^2(1)}\right]\sqrt{\pi a},$$

and solving, we obtain a critical crack length $a = 5.01$ in. This value is beyond the range of validity of Eq. (12-44), so we must obtain the solution from Eq. (12-45).

From a graph of the values of K given by Eq. (12-45) as a function of α (Fig. 12-30), we estimate that $K = 120$ ksi-in$^{1/2}$ at $\alpha = 0.40$. Using software designed to solve nonlinear algebraic equations, we obtain $\alpha = 0.406$. The critical crack length is $a = 4.06$ in.

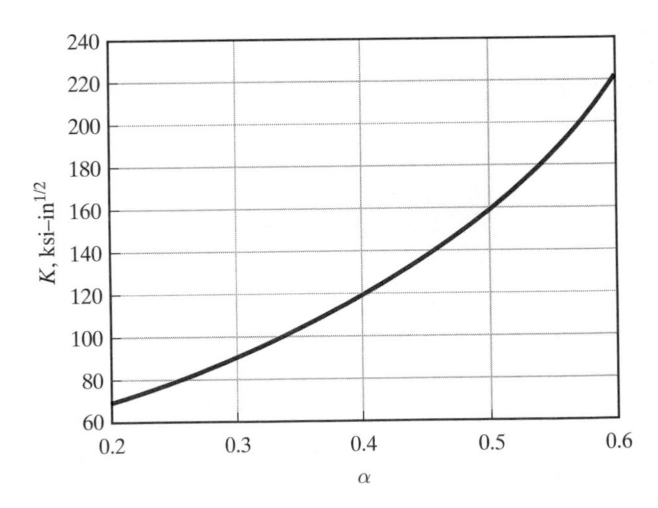

FIGURE 12-30 Graph of K as a function of α.

Repeated Loads and Slow Crack Growth

We have already seen in Section 12-1 that repeated loads can cause failure at stress levels that are lower than the yield strength of the material. When cracks are present, an analogous situation arises in the sense that crack growth may occur at stress intensity factor levels that are lower than the fracture toughness K_c. Repeated loading under these conditions can result in rates of crack growth that are much lower than those we discussed in the preceding section and are termed *slow crack growth*. Slow crack growth can be influenced by elevated temperatures and diffusive environments, but for this introduction we consider only the effects of repeated loading.

An Aloha Airlines Boeing 737 lands safely despite the loss of its fuselage.

The photograph of the Aloha Airlines Boeing 737 shows how multiple damage sites, combined with repeated loads, caused slow growth of fatigue cracks that eventually resulted in fast crack growth and the loss of part of the fuselage. In this case, the damage sites or initial flaws were induced by corrosion. The aircraft was used for short but frequent fights among the Hawaiian Islands so that the number of pressurization/depressurization cycles accumulated more quickly than usual.

Paris Law

The repeated application of loads is also called *cyclic loading,* and the ensuing crack growth is known as *fatigue crack growth*. In the simplest case the repeated loads have equal maximum and minimum values over time (Fig. 12-31), which is called *constant amplitude loading*. For example, constant amplitude loading can occur in the structure of an airliner whose cabin is pressurized to the same degree from flight to flight. In contrast, turbulent airflow and other loadings can subject the airliner to an irregular pattern of repeated loads. We first consider

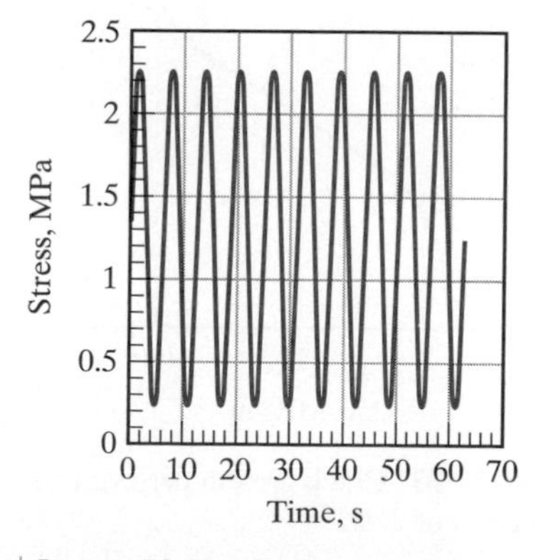

Figure 12-31 Example of the stress resulting from constant amplitude loading.

constant amplitude loading, which provides a basis for our subsequent analysis of more complex load histories.

Let us describe a fatigue crack growth experiment on the plate with a central crack shown in Fig. 12-26, whose stress intensity factor is given by Eq. (12-38). Instead of the monotonically increasing level of stress that would be used to measure K_c, the applied stress is varied sinusoidally as in Fig. 12-31. The change in applied stress over each cycle is

$$\Delta\sigma = \sigma_{\max} - \sigma_{\min}.$$

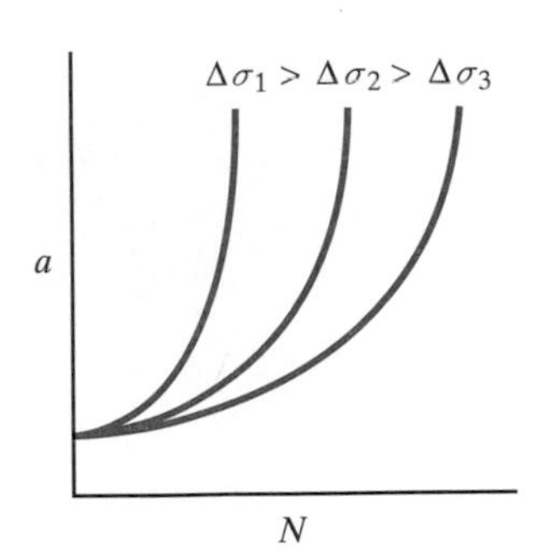

FIGURE 12-32 Crack growth history for various stress amplitudes.

In such experiments the crack length is measured as a function of the number of loading cycles N, yielding data that can be presented as shown in Fig. 12-32. Observe that increasing the stress amplitude causes the crack to grow faster.

However, Fig. 12-32 does not give a universal picture of a material's resistance to fatigue crack growth. A better measure is the rate at which a crack grows in a given material. We will see that rate information allows crack growth predictions to be made. From our discussion of fast cracks it can be anticipated that the crack growth rate da/dN depends not only on stress amplitude but also crack length, which brings the stress intensity factor to mind. The crack length does not increase very much over one loading cycle (usually $<10^{-4}$ in.), so from Eq. (12-38) the change in the stress intensity factor over one cycle is given approximately by

$$\Delta K = K_{\max} - K_{\min} = (\sigma_{\max} - \sigma_{\min})\sqrt{\pi a} = \Delta\sigma\sqrt{\pi a}.$$

When the results of a fatigue crack growth experiment are presented as crack growth per cycle da/dN as a function of the change in stress intensity factor per cycle ΔK, the curves corresponding to different levels of stress lie on a single universal curve that measures the material's resistance to fatigue crack growth (Fig. 12-33). This curve, which is sigmoidal in shape, has two limits. The limit at low ΔK values is the threshold toughness K_{th}. The upper limit is the fracture toughness K_c, at which fast crack growth begins.

A crack subjected to stress intensity factors lower than K_{th} will never grow, which makes it possible to design structural components so that the possibility of crack growth is very small. In less critical circumstances it may be permissible to tolerate slow crack growth in order to reduce structural weight.

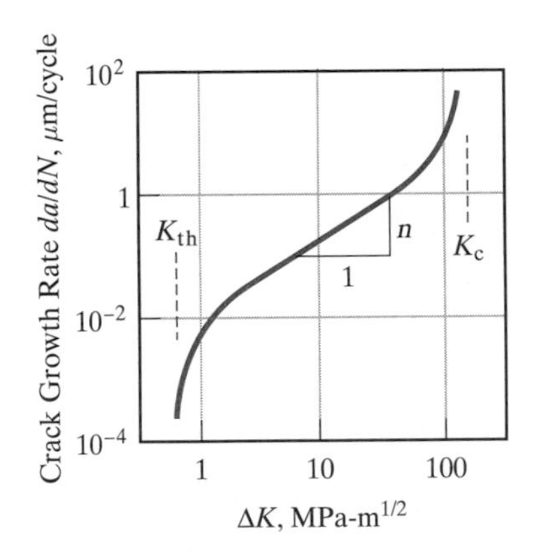

FIGURE 12-33 Crack growth per cycle as a function of ΔK.

Because Fig. 12-33 is a log-log graph, the linear portion of the curve indicates that da/dN and ΔK are related through a power law:

$$\frac{da}{dN} = A(\Delta K)^n. \tag{12-46}$$

This equation is named the *Paris law* after an early pioneer in fracture research. The constants A and n are material properies reflecting resistance to fatigue crack growth. For steels, $n = 3$, and for polymers, $n = 6$ to 10. We will show that once A and n are determined for a given material, Eq. (12-46) provides a basis for predicting crack growth.

Predicting Slow Crack Growth

Slow crack growth may take years to become fast, so if slowly growing cracks can be tolerated despite other reasons (appearance, leaks, etc.), a structure may be able to continue operating quite safely. Nevertheless, inspections must be carried out at regular intervals to monitor the progress of crack growth. The ability to predict paths and lengths of cracks helps establish the necessary frequency of inspections and also provides an estimate of the lifetimes of cracked components.

To predict the length, we write Eq. (12-46) as

$$dN = \frac{da}{A(\Delta K)^n}$$

and integrate, obtaining a relation between the number of cycles N and the crack length a:

$$\int_0^N dN = N = \frac{1}{A}\int_{a_0}^{a} \frac{da}{(\Delta K)^n}. \qquad \textbf{(12-47)}$$

The limit a_0 is the initial crack length from which the cycle count is made. We say it is where the "clock" starts running. Its value often results from the resolution of the equipment used to detect the presence of cracks. The time required for cracks to reach this length, which is often substantial, is not accounted for. In effect, this introduces a degree of conservatism into the design process.

Equation (12-47) determines the number of cycles N required to reach a given crack length a. The number of cycles at which failure occurs is obtained by setting a equal to the critical crack length for fast crack growth. The critical crack length is determined from Eq. (12-41) with $\sigma = \sigma_{max}$, the maximum magnitude of the constant amplitude loading. The following examples demonstrate these procedures.

EXAMPLE 12-10

A bolt with diameter $d = 0.5$ in. is subjected to a cyclic tensile service load that varies from zero to 10,000 lb. Due to a manufacturing flaw, the bolt has a crack initiating from a 0.01-in.-deep thread. The stress intensity factor obtained from a handbook for such a circumferential crack in an axially loaded bar is

$$K = 1.12\sigma\sqrt{\pi a}, \quad (2a/d < 0.21) \qquad \textbf{(12-48)}$$

where a is the crack depth and σ is the normal stress in the bar. The fracture toughness of the bolt material is $K_c = 23$ ksi-in$^{1/2}$. Its resistance to fatigue crack growth can be expressed by the Paris law with $A = 1 \times 10^{-18}$ in./cycle and $n = 3$. How many load cycles can the bolt withstand?

Strategy

By substituting the fracture toughness and the maximum stress due to the constant amplitude loading into Eq. (12-48), we can determine the critical crack length for fast crack growth. Then we can determine the number of load cycles to failure from Eq. (12-47).

Solution

The maximum normal stress due to the constant amplitude loading is

$$\sigma_{\max} = \frac{10{,}000}{\pi(0.5)^2/4} = 50{,}900 \text{ psi}.$$

Substituting this value and the fracture toughness into Eq. (12-48),

$$K_c = 1.12\sigma_{\max}\sqrt{\pi a}:$$
$$23{,}000 = (1.12)(50{,}900)\sqrt{\pi a},$$

we find that the critical crack length is $a = 0.052$ in. We also use Eq. (12-48) to obtain the change in the stress intensity factor over one cycle:

$$\Delta K = K_{\max} - K_{\min} = 1.12(\sigma_{\max} - \sigma_{\min})\sqrt{\pi a}$$
$$= (1.12)(50{,}900 - 0)\sqrt{\pi a}.$$

The initial depth of the crack is the depth of the thread: $a_0 = 0.01$ in. From Eq. (12-47), the number of cycles to failure is therefore

$$N = \frac{1}{A}\int_{a_0}^{a} \frac{da}{(\Delta K)^n}$$
$$= \frac{1}{1 \times 10^{-18}}\int_{0.010}^{0.052} \frac{da}{[(1.12)(50{,}900)\sqrt{\pi a}]^3}$$
$$= 10{,}800 \text{ cycles}.$$

EXAMPLE 12-11

The plate in Fig. 12-27 is 1 cm thick and 50 cm in width and the hole is 2 cm in diameter. The load P is cyclic and varies from zero to 30 kN. The stress intensity factor for this configuration is given by Eq. (12-39). The resistance to fatigue crack growth satisfies the Paris law with $A = 5 \times 10^{-9}$ m/cycle, $n = 4$, and $K_{th} = 5$ MPa-m$^{1/2}$. The crack emanating from the hole is initially obscured by a washer with a 4-cm outer diameter. How many cycles are required for a visible crack to stop growing?

Strategy

Equation (12-39) indicates that the stress intensity factor decreases with increasing crack length, which raises the possibility of crack arrest as the stress intensity factor drops below K_{th}. The length of the arrested crack is

$$a_{th} = \frac{1}{\pi}\left(\frac{P_{max}}{bK_{th}}\right)^2 = \frac{1}{\pi}\left[\frac{30 \times 10^3}{(0.01)(5 \times 10^6)}\right]^2 = 0.115 \text{ m},$$

so the crack stops growing before reaching the edges of the plate. We can use Eq. (12-47) to determine how many loading cycles occur from the time the crack becomes visible until it stops growing.

Solution

From Eq. (12-39), the change in the stress intensity factor over one cycle is

$$\begin{aligned}\Delta K = K_{max} - K_{min} &= \frac{P_{max} - P_{min}}{b\sqrt{\pi a}} \\ &= \frac{30 \times 10^3}{(0.01)\sqrt{\pi a}} \text{ Pa-m}^{1/2} \\ &= \frac{30 \times 10^{-3}}{(0.01)\sqrt{\pi a}} \text{ MPa-m}^{1/2}.\end{aligned}$$

The crack first becomes visible when it grows beyond the outer diameter of the washer, so $a_0 = 0.02$ m. From Eq. (12-47), the number of cycles to crack arrest is

$$\begin{aligned}N &= \frac{1}{A}\int_{a_0}^{a_{th}} \frac{da}{(\Delta K)^n} \\ &= \frac{1}{5 \times 10^{-9}}\int_{0.020}^{0.115}\left[\frac{(0.01)\sqrt{\pi a}}{30 \times 10^{-3}}\right]^4 da \\ &= 1.23 \times 10^4 \text{ cycles}.\end{aligned}$$

Chapter Summary

Failure

Failure in brittle materials occurs when a component breaks or fractures, whereas in ductile materials failure is usually defined to occur with the onset of yielding.

The *maximum normal stress criterion* states that fracture of a brittle material occurs when

$$\text{Max}(|\sigma_1|, |\sigma_2|, |\sigma_3|) = \sigma_U, \qquad \text{Eq. (12-1)}$$

where $\sigma_1, \sigma_2, \sigma_3$ are the principal stresses and σ_U is the ultimate stress. In plane stress, the values of σ_1 and σ_2 for which fracture will not occur are bounded by the square region shown in Fig. 12-1. It is bounded by the lines

$$\sigma_1 = \pm\sigma_U, \qquad \sigma_2 = \pm\sigma_U. \qquad \text{Eq. (12-2)}$$

When the ultimate stress of a brittle material subjected to plane stress differs in tension and compression, *Mohr's failure criterion* states that fracture occurs when σ_1 and σ_2 lie on the boundaries shown in Fig. 12-4a and described by Eqs. (12-4)–(12-7).

The *Tresca criterion* states that yielding of a ductile material occurs when the absolute maximum shear stress equals the yield stress in shear τ_Y. It can also be expressed in terms of the tensile yield stress σ_Y:

$$\text{Max}(|\sigma_1 - \sigma_2|, |\sigma_2 - \sigma_3|, |\sigma_1 - \sigma_3|) = \sigma_Y. \qquad \text{Eq. (12-12)}$$

The *Tresca factor of safety* is

$$S_T = \frac{\sigma_Y}{\sigma_T}, \qquad \text{Eq. (12-16)}$$

where

$$\sigma_T = \text{Max}(|\sigma_1 - \sigma_2|, |\sigma_2 - \sigma_3|, |\sigma_1 - \sigma_3|). \qquad \text{Eq. (12-15)}$$

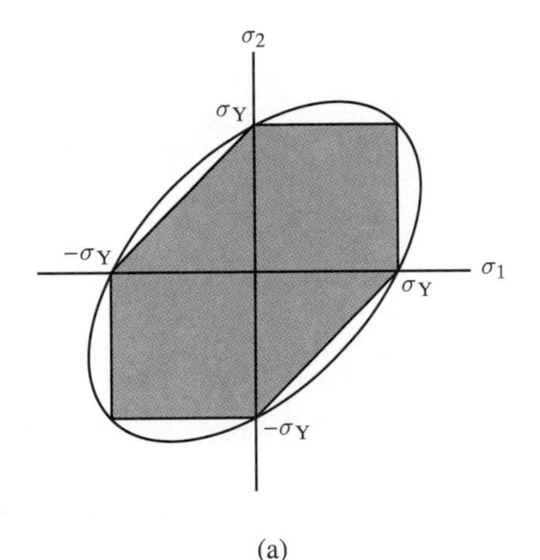

(a)

In plane stress, the safe region according to the Tresca criterion is bounded by the hexagon in Fig. (a).

The *von Mises criterion* states that yielding of a ductile material occurs when

$$\tfrac{1}{2}\left[(\sigma_1 - \sigma_2)^2 + (\sigma_2 - \sigma_3)^2 + (\sigma_1 - \sigma_3)^2\right] = \sigma_Y^2. \qquad \text{Eq. (12-17)}$$

The *von Mises factor of safety* is

$$S_M = \frac{\sigma_Y}{\sigma_M}, \qquad \text{Eq. (12-21)}$$

where σ_M is the *von Mises equivalent stress*

$$\sigma_M = \frac{1}{\sqrt{2}}\sqrt{(\sigma_1 - \sigma_2)^2 + (\sigma_2 - \sigma_3)^2 + (\sigma_1 - \sigma_3)^2}. \qquad \text{Eq. (12-19)}$$

In plane stress, the safe region according to the von Mises criterion is bounded by the ellipse in Fig. (a).

Fatigue is subdivided into low-cycle, high-cycle, and fatigue crack growth. In low-cycle fatigue, stress levels exceed the yield strength and the number of cycles to failure is relatively low ($<10^3$). High-cycle fatigue can occur when stress levels are lower than the yield strength, and failure may require 10^3 to 10^6 cycles.

Denoting the maximum and minimum values of the stress in constant frequency and amplitude loading by σ_{max} and σ_{min}, the *stress amplitude* σ_a

and *mean stress* σ_m are

$$\sigma_a = \frac{\sigma_{max} - \sigma_{min}}{2}, \qquad \sigma_m = \frac{\sigma_{max} + \sigma_{min}}{2}. \qquad \text{Eq. (12-22)}$$

An S–N curve, or *endurance curve,* is a graph of the stress amplitude or *fatigue strength* as a function of the number of cycles N to failure at zero mean stress. For some materials, there is a stress amplitude, the *fatigue limit* σ_{fat}, below which fatigue life is essentially infinite. Keeping stress levels below the fatigue limit is known as *safe life design*. An empirical expression used to fit S–N curves is

$$\sigma_a = \sigma_{fat} + \frac{b}{N^c}, \qquad \text{Eq. (12-23)}$$

Solving this equation for N yields

$$N = \left(\frac{b}{\sigma_a - \sigma_{fat}}\right)^{1/c}. \qquad \text{Eq. (12-24)}$$

Empirical equations that can be used to determine the effect of a constant level of mean stress on the S–N curve include the *Goodman relation* (12-25), the *Gerber parabola* (12-26), and the *Soderberg line* (12-27).

Suppose that a component is subjected to a sequence of blocks of loading. Let n_i be the number of cycles applied in the ith block, and let N_i be the number of cycles required to fail the undamaged component at that stress amplitude and mean stress. *Miner's law* states that the accumulated damage after k blocks of loading is

$$D = \sum_{i=1}^{k} D_i = \sum_{i=1}^{k} \frac{n_i}{N_i}, \qquad \text{Eq. (12-28)}$$

and that failure occurs when $D = 1$.

Stress Concentrations

Structural members may have geometric features (including holes, notches, and corners) that result in localized regions of high stress called *stress concentrations*. The ratio C of the maximum stress resulting from a stress concentration to a defined nominal stress is called the *stress concentration factor*. Stress concentration factors are given for an axially loaded bar in Figs. 12-13 to 12-15, for a stepped circular bar subjected to torsion in Fig. 12-17, and for a rectangular beam subjected to bending in Figs. 12-20 and 12-21.

Fracture

The *stress intensity factor* K is a measure of the magnitude of the stress in the neighborhood of a crack tip [see Eqs. (12-36)]. Stress intensity factors for a centrally cracked plate subjected to tension and a plate subjected to a central load on a crack face are given by Eqs. (12-38) and (12-39), respectively. In many cases stress intensity factors can be expressed in the form

$$K = \sigma\sqrt{\pi a}\,Q\left(\frac{a}{W}\right), \qquad \text{Eq. (12-40)}$$

where σ represents the stress, a is the crack length, and $Q(a/W)$ is a function called the *configuration factor*.

The value of the stress intensity factor at which *fast crack growth* begins is called the *fracture toughness* K_c. From Eq. (12-40), the criterion for fast crack growth is

$$\sigma\sqrt{\pi a}\,Q\left(\frac{a}{W}\right) = K_c. \qquad \text{Eq. (12-41)}$$

Let ΔK be the change in the stress intensity factor over one cycle in a cracked component subjected to cyclic loading. The derivative of the crack length a with respect to the number N of loading cycles is given by the *Paris law*

$$\frac{da}{dN} = A(\Delta K)^n. \qquad \text{Eq. (12-46)}$$

The constants A and n are material properties reflecting resistance to fatigue crack growth. Rearranging Eq. (12-46) and integrating give a relation between the number of cycles N and the crack length a:

$$\int_0^N dN = N = \frac{1}{A}\int_{a_0}^{a} \frac{da}{(\Delta K)^n}. \qquad \text{Eq. (12-47)}$$

The number of cycles at which failure occurs is obtained by setting a equal to the critical crack length for fast crack growth. The critical crack length is determined from Eq. (12-41) with $\sigma = \sigma_{max}$, the maximum magnitude of the constant amplitude loading.

PROBLEMS

12-1.1. A thin sheet of brittle material is subjected to the state of plane stress $\sigma_x = 125$ ksi, $\sigma_y = 75$ ksi, $\tau_{xy} = 100$ ksi. The ultimate stress of the material (in tension and compression) is $\sigma_U = 175$ ksi. Determine whether the sheet will fail according to the maximum normal stress criterion.

12-1.2. A load applied to a machine component results in the state of plane stress $\sigma_x = 80$ MPa, $\sigma_y = 100$ MPa, $\tau_{xy} = 60$ MPa. The component is made of a brittle high-strength steel that follows the maximum normal stress criterion with $\sigma_U = 200$ MPa. If increasing the load increases each stress component proportionally, determine the percentage increase that can be applied before the component fails.

12-1.3. A closed, thin-walled cylinder is to be pressurized internally at 600 psi. It is also subjected to a torque which results in a torsional stress equal to 0.0335 times the internal pressure. The mean radius and thickness of the cylinder are 2 and 0.1 in., respectively. The material being considered for the design has an ultimate stress (in tension and compression) $\sigma_U = 180$ ksi. Will it suffice? Explain.

12-1.4. A point on the surface of a material is subjected to the state of plane stress $\sigma_x = -50$ ksi, $\sigma_y = -35$ ksi, $\tau_{xy} = 40$ ksi. The material is brittle and has tensile and compressive ultimate stresses $\sigma_U^t = 80$ ksi, $\sigma_U^c = 140$ ksi. Is this a safe state of stress according to Mohr's failure criterion?

12-1.5. A material that follows Mohr's failure criterion has a compressive ultimate stress of 2 GPa, which is four times the tensile ultimate stress. A load is applied that causes the principal stresses to be in the ratio $\sigma_2/\sigma_1 = -4$, with $\sigma_1 \geq 0$ at all times. Determine the value of σ_2 at failure.

12-1.6. A thin sheet of material is loaded in such a way that $\sigma_x = -95$ ksi, $\sigma_y = 75$ ksi, $\tau_{xy} = 29.5$ ksi. It is made of a brittle material with tensile and compressive ultimate stresses of 150 and 300 ksi, respectively. If increasing the load multiplies each plane stress component proportionally, determine the percentage increase that can be applied before failure occurs according to Mohr's failure criterion.

12-1.7. Determine the factor of safety when a ductile material that follows the Tresca criterion is in the state of plane stress $\sigma_x = 50$ ksi, $\sigma_y = 30$ ksi, $\tau_{xy} = 25$ ksi. The yield stress of the material is $\sigma_Y = 80$ ksi.

12-1.8. A thin sheet of material is loaded in plane stress and yielding occurs when $\sigma_x = 80$ MPa, $\sigma_y = 35$ MPa, $\tau_{xy} = 45$ MPa. What is the yield stress σ_Y of the material?

12-1.9. A rod with a solid circular cross section is subjected to a bending moment M and a torque T. Obtain an expression for the radius of the rod according to the Tresca criterion if the yield stress of the material is σ_Y and a safety factor of 2.5 is required.

Worksheet 15 can be used to solve Problems 2-1.10 through 2-1.12.

12-1.10. The L-shaped bar has a solid circular cross section with 5-mm radius. The yield stress of the material is $\sigma_Y = 300$ MPa and it follows the von Mises failure criterion. If a safety factor of 2 is desired, determine the maximum value of the vertical load P that can be applied.

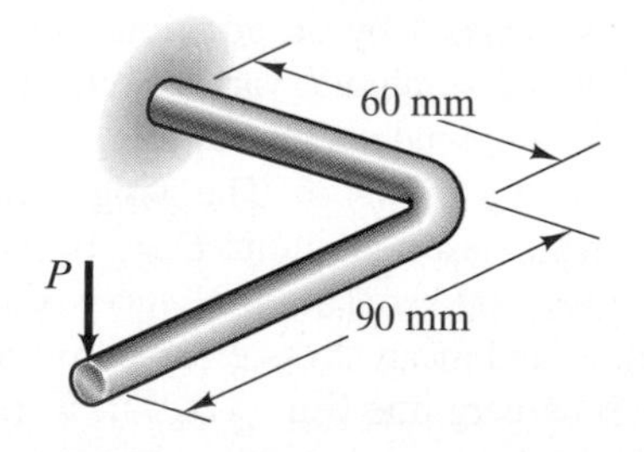

PROBLEM 12-1.10

12-1.11. The open cylindrical tank has height $h_S = 25$ ft, inner radius $R = 20$ ft, and wall thickness $t = 3$ in. It is made of steel that has weight density 490 lb/ft^3, yield stress $\sigma_Y = 110$ ksi, and follows the von Mises yield criterion. Based on the state of stress in the cylindrical wall due to pressure and the tank's weight, determine the factor of safety when the tank is filled with water (weight density 62.4 lb/ft^3) to a depth $h_W = 20$ ft.

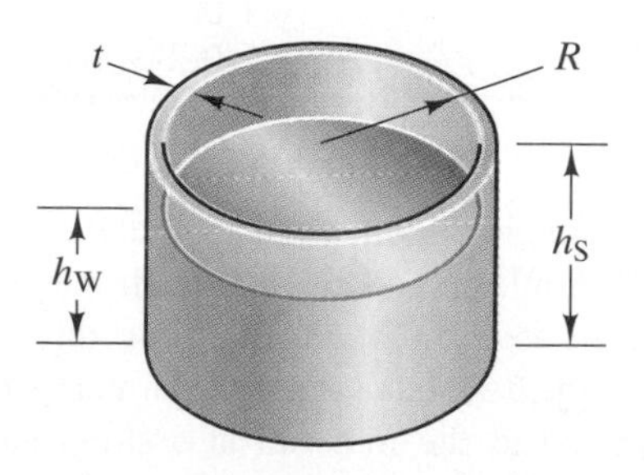

PROBLEM 12-1.11

12-1.12. The thin cylindrical shell is subjected to a compressive axial load P and an internal pressure p. It has internal radius R and thickness t and is made of material with yield stress σ_Y. For a fixed axial load P, determine the allowable pressure p so that the von Mises factor of safety is no less than 2.

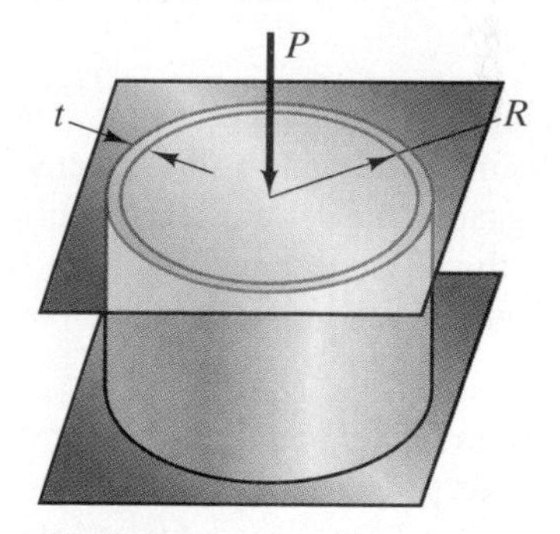

PROBLEM 12-1.12

12-1.13. A series of fatigue tests have been conducted on a material. The tests were conducted at zero mean stress level and resulted in the data tabulated below. If the endurance limit σ_{fat} was 15 ksi, find an empirical expression of the form of Eq. (12-23) that fits the data. How many cycles could be applied to the material at a stress amplitude of 35 ksi?

Stress Amplitude (ksi)	Cycles to Failure
52.7	1.000×10^3
38.8	1.000×10^4
30.0	1.000×10^5
24.5	1.000×10^6
21.0	1.000×10^7

12-1.14. A sample of undamaged material of the kind described in Problem 12-1.13 has been subjected to the blocks of cyclic loading tabulated below, all at zero mean stress. Determine how many more zero-mean-stress cycles could be applied to the material at a stress amplitude of 35 ksi.

Stress Amplitude (ksi)	Cycles Applied
42	7.938×10^2
28	5.318×10^4
25	8.353×10^4
22	1.536×10^6

12-1.15. A sample of undamaged material of the kind described in Problem 12-1.13 has been subjected to the blocks of cyclic loading tabulated below, all at a level of mean stress equal to 20% of the ultimate tensile strength. If the material follows the Goodman relation, determine how many more cycles could be applied to the material at a stress amplitude of 35 ksi and the same level of mean stress.

Stress Amplitude (ksi)	Cycles Applied
42	7.188×10^2
28	1.231×10^4
25	7.363×10^4
22	1.595×10^5

12-1.16. The titanium alloy rod has length $L = 15$ in. and a circular cross section with radius $R = 0.5$ in. The axial load $P_a = 98$ kip is constant. The load P_b is applied cyclically with zero mean value and the amplitudes tabulated below. The material has an ultimate stress of 250 ksi, an endurance limit $\sigma_{fat} = 100$ ksi, and follows the Gerber parabola for mean stress effects. Its S–N curve is described by Eq. (12-23) with constants $b = 200$ ksi, $c = 0.15$. Determine how much damage is caused by each block of loading. How many additional cycles of the load P_b could be applied at an amplitude of 548 lb before fatigue failure would occur?

Amplitude of P_b (lb)	Cycles Applied
610	1.000×10^3
560	3.850×10^4
518	2.000×10^6

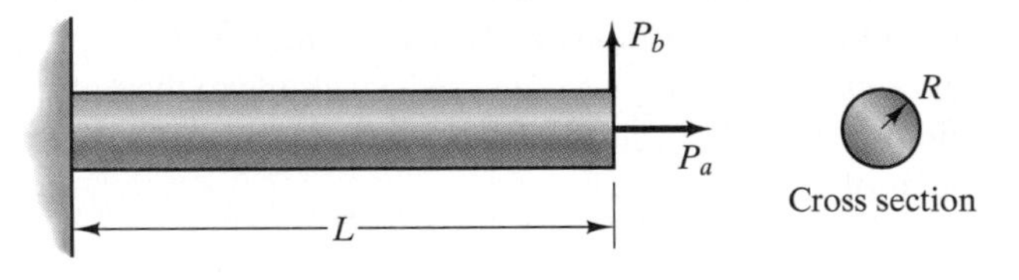

PROBLEM 12-1.16

12-1.17. In a preliminary design study of fatigue loading of an airplane's wing, the wing's structural response is modeled by a 20-ft cantilever beam with 20-in. height and 1-in. width. The wing's lift is represented by the constant 52 lb/in. uniformly distributed load. Fatigue due to gust loading is modeled by an additional uniformly distributed load whose amplitude varies cyclically with zero mean value. The amplitude of the resulting cyclic displacement of the wing tip is 8.64 in. The wing consists of aluminum alloy with elastic modulus $E = 10 \times 10^6$ psi and yield stress $\sigma_Y = 100$ ksi. The S–N curve of the material is shown below, and mean stress effects are accounted for through the Soderberg line [Eq. (12-27)]. If there are 355 cycles per flight, how much fatigue damage results from 5000 flights?

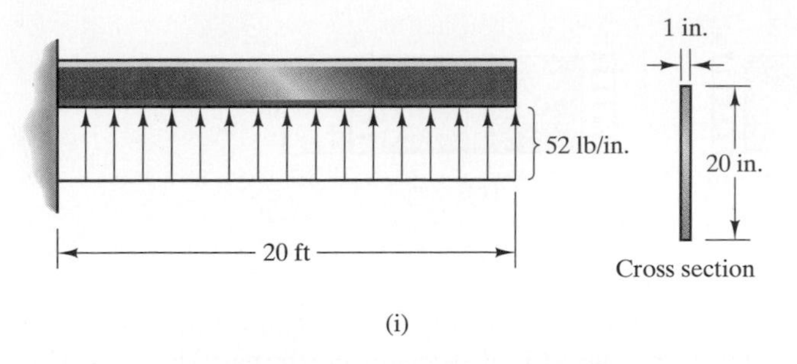

(i)

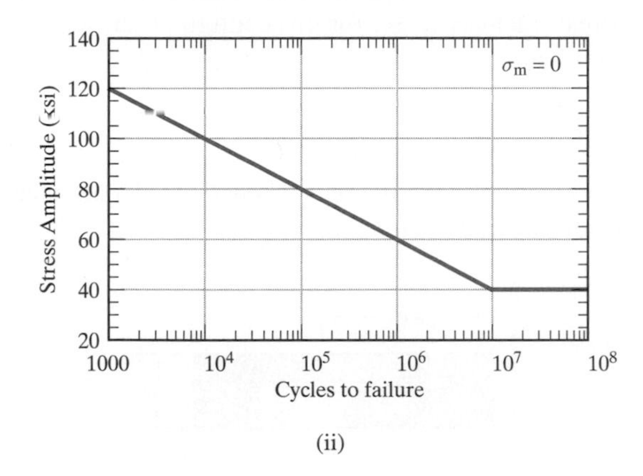

(ii)

PROBLEM 12-1.17

12-1.18. A bridge carries cars, trucks, and trains. It has been in service for three years and has carried the numbers of vehicles tabulated below on a daily basis. The main structural element of the bridge is a simply supported rectangular beam. The height, width, and span of the beam are 0.25, 0.1, and 250 m, respectively. The weight carried by the beam for each type of vehicle is tabulated. The S–N curve of the bridge material is described by Eq. (12-23) with $b = 11$ GPa, $c = 0.2$, and mean stress effects are accounted for by the Soderberg line [Eq. (12-27)]. Laboratory testing determines that $\sigma_{fat} = 500$ MPa and $\sigma_Y = 1200$ MPa. The level of traffic in the first three years was greater than originally anticipated, so the rail traffic was diverted. How much longer can the bridge be used by cars and trucks?

Vehicle	Weight W (kN)	Vehicles/Day
Auto	18.3	5000
Truck	26.0	100
Train	37.6	30

12-2.1. The plate is 0.005 m thick and has a central hole with 0.1-m diameter. If the axial load $P = 45$ kN, what is the maximum stress in the plate?

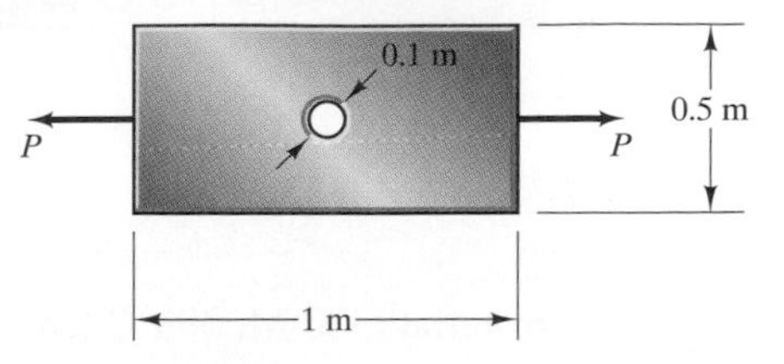

PROBLEM 12-2.1

12-2.2. Solve Example 12-7 if the two widths of the bar are 1.2 and 1.0 in., the thickness of the bar is 0.05 in., and the fillet radius is 0.3 in.

12-2.3. The diameters of the stepped rod are $d_1 = 2.625$ in. and $d_2 = 1.5$ in. A fillet radius of 0.225 in. is used to transition from one cross section to the other. If the rod is subjected to a 40,000-lb axial load, what is the resulting maximum stress? (It may be necessary to interpolate between d_1/d_2 values in Fig. 12-14.)

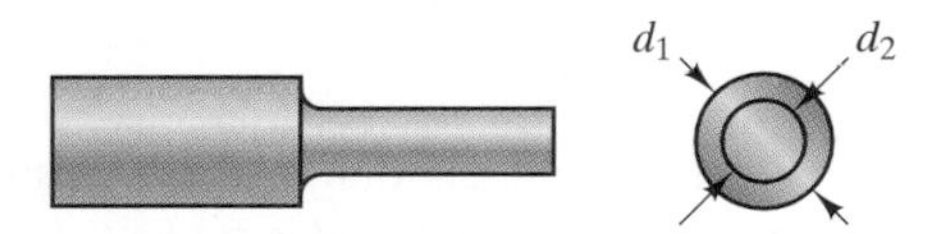

PROBLEM 12-2.3

12-2.4. Suppose that the diameters of the stepped rod shown in Problem 12-2.3 are $d_1 = 1.8$ in. and $d_2 = 1.5$ in. If the rod is to be subjected to a torque of 28.7 in-kip and the allowable shear stress is 65 ksi, what fillet radius should be used?

12-2.5. A double-notched beam with a height ratio $h_1/h_2 = 1.5$ is subjected to a pure bending moment (see Fig. 12-21). As a designer, you have the option of using either a semicircular notch or a deep notch with $a/h_2 = 0.05$. Show that the allowable bending moment is increased by 75% when the semicircular notch is used.

12-3.1. Three thin sheets with the same dimensions (5-in. width, 0.1-in. thickness) are subjected to tensile stress. Sheet (a) has a central hole with 0.5-in. diameter, sheet (b) has a 0.5-in. central crack, and sheet (c) has no defects.

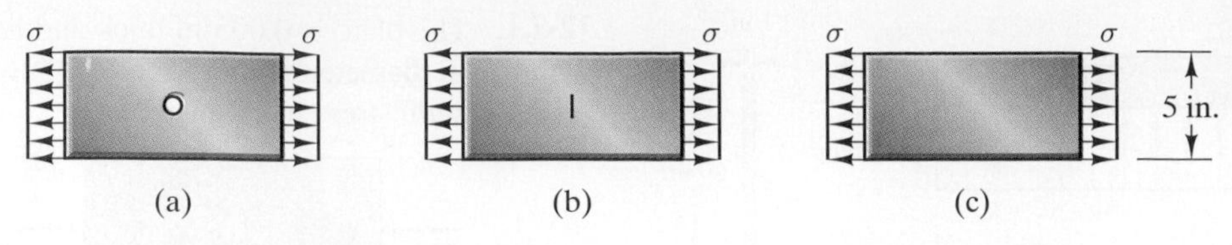

PROBLEM 12-3.1

They are made of a material (300-M, 300°C temper steel) whose yield stress and fracture toughness are 252 ksi and 59 ksi-in$^{1/2}$, respectively. Compare the failure loads in the three cases.

12-3.2. Compare the failure loads of the plates described in Problem 12-3.1 if they are made of 2024-T351 aluminum, which has yield stress and fracture toughness 47 ksi and 31 ksi-in$^{1/2}$, respectively. [You should find that the sheet with the hole has the lowest failure load in this case. This is because the aluminum is a relatively tough or crack-growth-resistant material. To understand why, compare the plastic zone sizes in the steel and the aluminum using Eq. (12-37). The plastic zone in the aluminum is almost eight times as large as in the steel, which helps resist crack growth.]

12-3.3. The titanium strut has radius $b = 10$ mm, a circumferential crack with depth a, and is subjected to tensile loads $P = 200$ kN. The fracture toughness of the titanium is 66 MPa-m$^{1/2}$ and the stress intensity factor for this configuration is given by

$$K = \frac{1.12P}{\pi b^2}\sqrt{\pi a}.$$

Nondestructive inspection techniques are used that can resolve a crack with depth $a = 1$ mm. Will inspection reveal any cracks before fast crack growth occurs? Explain.

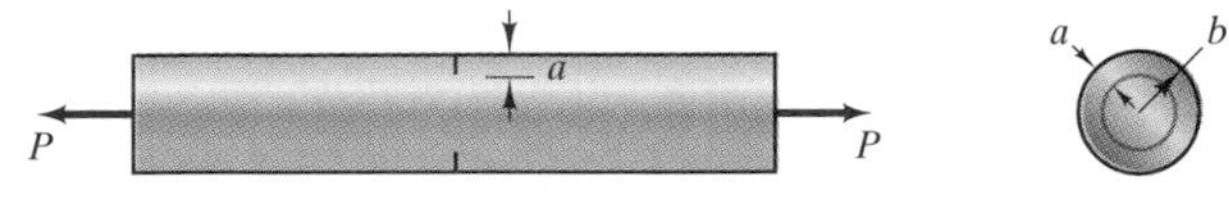

PROBLEM 12-3.3

12-3.4. In the fracture toughness test shown, a beam with a crack of length $a = 5$ in. (called a *double cantilever beam* specimen) is subjected to the load P. The height $2h = 1$ in. and the thickness (the dimension into the page) is $b = 0.5$ in. The stress intensity factor for this configuration is

$$K = \frac{2\sqrt{3}\,Pa}{bh^{3/2}}.$$

If fast crack growth initiates when $P = 460$ lb, what is the material's fracture toughness?

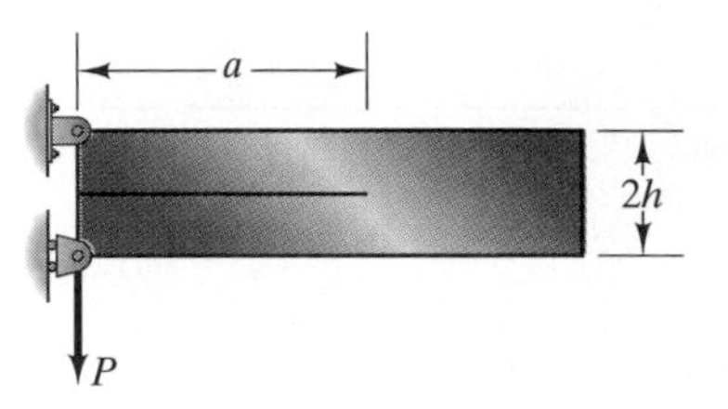

PROBLEM 12-3.4

12-3.5. A strip of the material tested in Problem 12-3.4 is bonded to rigid bars at its top and bottom edges and has an edge crack of length a. The length $L = 5$ in., height $2h = 1$ in., and thickness $b = 0.2$ in. A weight $W = 50{,}000$ lb is suspended from the bottom bar. The stress intensity factor of the configuration is

$$K = \frac{W}{bL}(1 - e^{-5a/2h})\sqrt{h(1 - \nu^2)},$$

where $\nu = 0.3$ is the Poisson's ratio of the aluminum. Determine the critical crack length for fast crack growth.

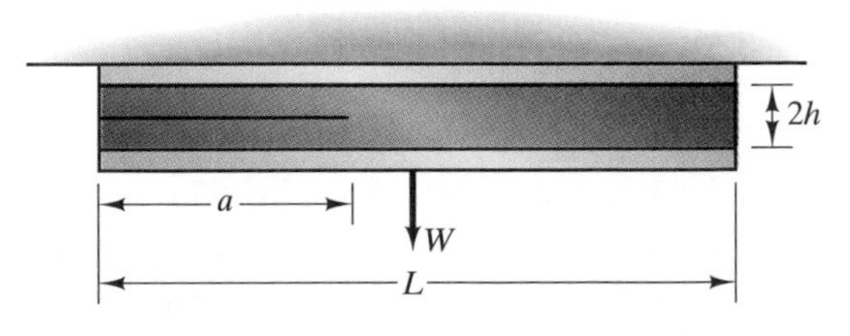

PROBLEM 12-3.5

12-3.6. The wedge-shaped blade of an axe is being driven into a piece of wood of thickness $2h = 1$ in. The blade imparts a displacement $\Delta = 0.25$ in. when the crack length $a = 6$ in. The stress intensity factor for this situation is

$$K = \frac{\sqrt{3}\, E\, \Delta h^{3/2}}{4a^2},$$

where $E = 20 \times 10^6$ psi is the elastic modulus of the wood. Determine how much more the crack will grow under this displacement if the fracture toughness of the wood is 8 ksi-in$^{1/2}$. (Notice that the stress intensity decreases with increasing crack length, which explains why further insertions of the blade would be required to split the wood completely.)

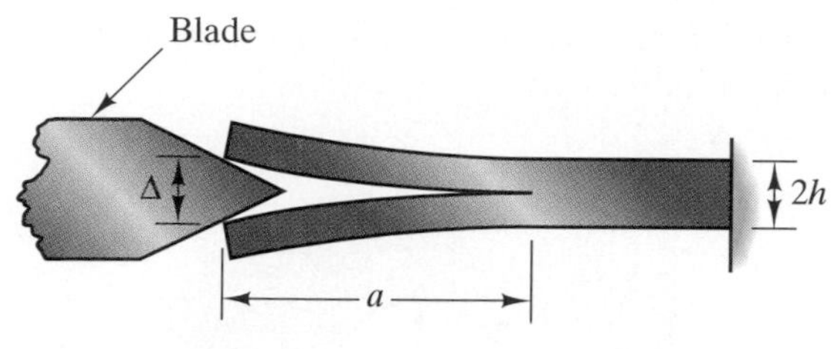

PROBLEM 12-3.6

12-3.7. The wall of a cylindrical pressure vessel contains an elliptical surface flaw with width $2c$ which may grow through the wall to produce a leak. Alternatively, it may become a fast crack before reaching the other surface of the wall, in which case it will grow axially in an explosive manner and destroy the vessel with other severe consequences. Thus it is preferable that the vessel leaks before it breaks. To design for this, we check whether the critical half-crack length c_0 under the action of the hoop stress is greater than the wall thickness, or

$$c_0 > \frac{1}{\pi}\left(\frac{K_c}{\sigma_h}\right)^2.$$

Derive this expression. The material being considered for a vessel with 2-m inner diameter and 15-mm wall thickness has a fracture toughness of 200 MPa-m$^{1/2}$. Determine whether the vessel satisfies this "leak before break" design philosophy if it is to be subjected to an internal pressure of 5 MPa.

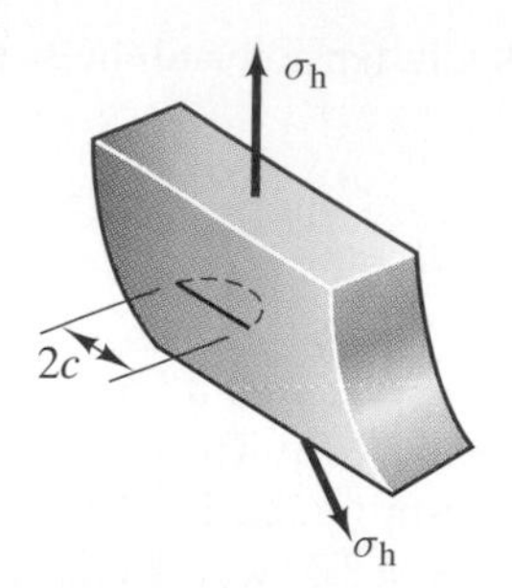

PROBLEM 12-3.7

12-3.8. A large sheet under constant-amplitude cyclic loading contains a central crack with length $2a = 2$ in. (Fig. 12-26). The change in applied stress over each cycle is 20 ksi. The material follows the Paris law for fatigue crack growth rate with constants $A = 1 \times 10^{-20}$ and $n = 3$, where crack growth rate is in inches per cycle and stress intensity factor is in psi-in$^{1/2}$. How many cycles are required for the crack to double in size?

12-3.9. In Problem 12-3.8, suppose that the sheet contains an edge crack with length $a = 1$ in. instead of a central crack (see Example 12-8). How many cycles are required for the crack to double in size? Compare your answer to the answer to Problem 12-3.8.

12-3.10. The titanium strut described in Problem 12-3.3 is subjected to tensile loads that cycle between 0 and 200 kN. The material follows the Paris law with constants $A = 1 \times 10^{-13}$ and $n = 3.5$, where crack growth rate is in meters per cycle and stress intensity factor is in MPa-m$^{1/2}$. If a 1-mm-deep circumferential crack is detected by nondestructive evaluation techniques, determine how many more cycles occur before the strut fails.

12-3.11. The strip described in Problem 12-3.5 is loaded cyclically by suspending and removing a weight W. The material follows the Paris law with constants $A = 1 \times 10^{-18}$ and $n = 3$, where crack growth rate is in inches per cycle and stress intensity factor is in psi-in$^{1/2}$. If microscopic measurements of the fracture surface indicate that the crack growth rate is 8×10^{-6} inches per cycle when the crack is 0.5 in. long, what is the weight W?

12-3.12. The sheet of width $W = 0.5$ m and thickness $b = 15$ mm contains a quarter-circular crack of radius a.

The sheet is loaded in tension and the stress intensity factor is

$$K = \frac{0.722P\sqrt{\pi a}}{bW}$$

as long as $a/b < 0.25$ and $a/W < 0.35$. The load P cycles between 0 and 3 MN. The fracture toughness of the material is 30 MPa-m$^{1/2}$ and its Paris law constants are $A = 3 \times 10^{-11}$ and $n = 3.7$, where crack growth rate is in meters per cycle and stress intensity factor is in MPa-m$^{1/2}$. Determine how many cycles it will take the sheet to fail if the initial crack radius is 0.5 mm and the crack maintains its quarter-circular shape until failure.

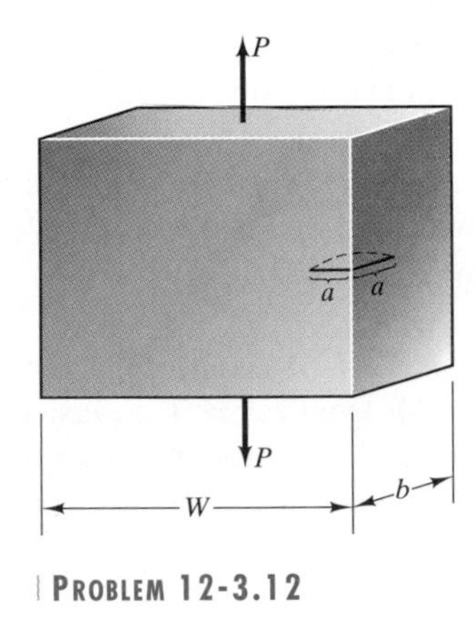

PROBLEM 12-3.12

12-3.13. The figure shows cracks at opposite sides of a 0.5-in.-diameter hole in a plate. A bolt inserted into the hole is larger than the hole and exerts the loads P. The stress intensity factor for this configuration is $K = P/b\sqrt{\pi a}$, where $b = 0.25$ in. is the thickness of the sheet.

(a) If the fracture toughness of the material is 35 ksi-in$^{1/2}$ and the bolt imparts loads $P = 6000$ lb, will fast crack growth occur?

(b) An external load applied to the plate causes the loads on the face of the hole to drop to $P = 3000$ lb. The external load is applied and released in a repeated manner, and when it is released the loads $P = 6000$ lb irrespective of crack length. If the bolt head obscures a 2-in.-diameter portion of the sheet, determine whether a fatigue crack will arrest before it becomes visible, given a threshold stress intensity factor of 12 ksi-in$^{1/2}$.

(c) Suppose that the diametrically opposed cracks are initially 0.005 in. in length. Determine the number of cycles at which they would stop growing if the material follows the Paris law with constants $A = 0.5 \times 10^{-21}$ and $n = 4$. Crack growth rates are in inches per cycle and stress intensity factor is in psi-in$^{1/2}$.

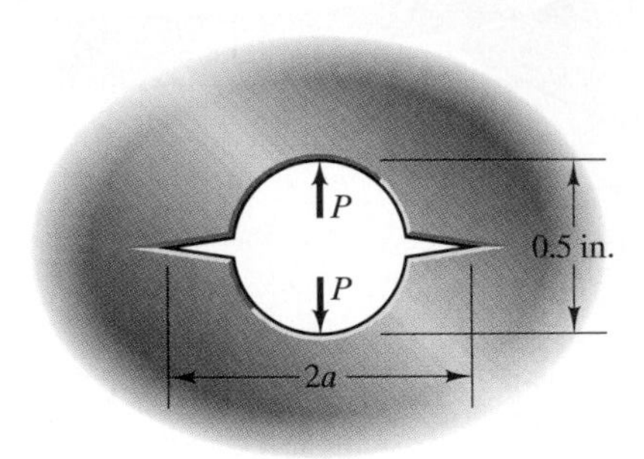

PROBLEM 12-3.13

APPENDIX A

Results from Mathematics

Algebra

Quadratic Equations

The solutions of the quadratic equation

$$ax^2 + bx + c = 0$$

are

$$x = \frac{-b \pm \sqrt{b^2 - 4ac}}{2a}.$$

Natural Logarithms

The natural logarithm of a positive real number x is denoted by $\ln x$. It is defined to be the number such that

$$e^{\ln x} = x,$$

where $e = 2.7182\ldots$ is the base of natural logarithms.

Logarithms have the following properties:

$$\ln(xy) = \ln x + \ln y,$$

$$\ln \frac{x}{y} = \ln x - \ln y,$$

$$\ln y^x = x \ln y.$$

Trigonometry

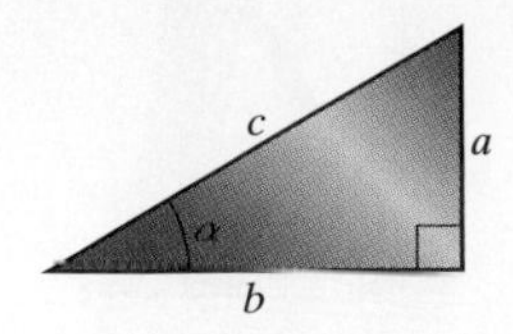

The trigonometric functions for a right triangle are

$$\sin\alpha = \frac{1}{\csc\alpha} = \frac{a}{c}, \qquad \cos\alpha = \frac{1}{\sec\alpha} = \frac{b}{c}, \qquad \tan\alpha = \frac{1}{\cot\alpha} = \frac{a}{b}.$$

The sine and cosine satisfy the relation

$$\sin^2\alpha + \cos^2\alpha = 1,$$

and the sine and cosine of the sum and difference of two angles satisfy

$$\sin(\alpha+\beta) = \sin\alpha\ \cos\beta + \cos\alpha\ \sin\beta,$$
$$\sin(\alpha-\beta) = \sin\alpha\ \cos\beta - \cos\alpha\ \sin\beta,$$
$$\cos(\alpha+\beta) = \cos\alpha\ \cos\beta - \sin\alpha\ \sin\beta,$$
$$\cos(\alpha-\beta) = \cos\alpha\ \cos\beta + \sin\alpha\ \sin\beta.$$

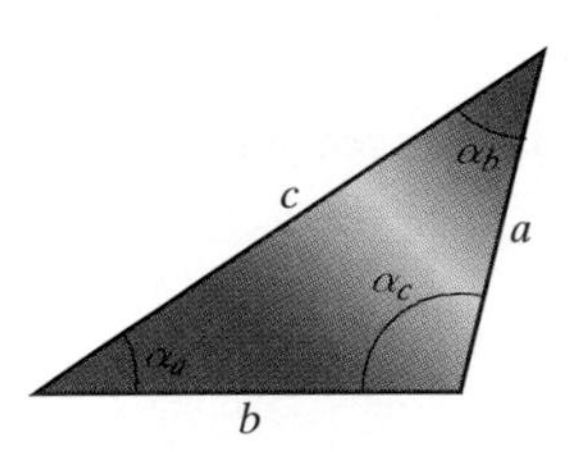

The *law of cosines* for an arbitrary triangle is

$$c^2 = a^2 + b^2 - 2ab\cos\alpha_c,$$

and the *law of sines* is

$$\frac{\sin\alpha_a}{a} = \frac{\sin\alpha_b}{b} = \frac{\sin\alpha_c}{c}.$$

Derivatives

$$\frac{d}{dx}x^n = nx^{n-1}.$$

$$\frac{d}{dx}e^x = e^x.$$

$$\frac{d}{dx}\ln x = \frac{1}{x}.$$

$$\frac{d}{dx}\sin x = \cos x.$$

$$\frac{d}{dx}\cos x = -\sin x.$$

$$\frac{d}{dx}\tan x = \frac{1}{\cos^2 x}.$$

$$\frac{d}{dx}\sinh x = \cosh x.$$

$$\frac{d}{dx}\cosh x = \sinh x.$$

$$\frac{d}{dx}\tanh x = \operatorname{sech}^2 x.$$

Integrals

$$\int x^n\,dx = \frac{x^{n+1}}{n+1} \qquad (n \neq -1).$$

$$\int x^{-1}\,dx = \ln x.$$

$$\int (a+bx)^{1/2}\,dx = \frac{2}{3b}(a+bx)^{3/2}.$$

$$\int x(a+bx)^{1/2}\,dx = -\frac{2(2a-3bx)(a+bx)^{3/2}}{15b^2}.$$

$$\int (1+a^2x^2)^{1/2}\,dx = \frac{1}{2}\left\{x(1+a^2x^2)^{1/2} + \frac{1}{a}\ln\left[x + \left(\frac{1}{a^2}+x^2\right)^{1/2}\right]\right\}.$$

$$\int x(1+a^2x^2)^{1/2}\,dx = \frac{a}{3}\left(\frac{1}{a^2}+x^2\right)^{3/2}.$$

$$\int x^2(1+a^2x^2)^{1/2}\,dx = \frac{1}{4}ax\left(\frac{1}{a^2}+x^2\right)^{3/2} - \frac{1}{8a^2}x(1+a^2x^2)^{1/2} - \frac{1}{8a^3}\ln\left[x+\left(\frac{1}{a^2}+x^2\right)^{1/2}\right].$$

$$\int (1-a^2x^2)^{1/2}\,dx = \frac{1}{2}\left[x(1-a^2x^2)^{1/2} + \frac{1}{a}\arcsin(ax)\right].$$

$$\int x(1-a^2x^2)^{1/2}\,dx = -\frac{a}{3}\left(\frac{1}{a^2}-x^2\right)^{3/2}.$$

$$\int x^2(a^2-x^2)^{1/2}\,dx = -\frac{1}{4}x(a^2-x^2)^{3/2} + \frac{1}{8}a^2\left[x(a^2-x^2)^{1/2} + a^2\arcsin\frac{x}{a}\right].$$

$$\int \frac{dx}{(1+a^2x^2)^{1/2}} = \frac{1}{a}\ln\left[x+\left(\frac{1}{a^2}+x^2\right)^{1/2}\right].$$

$$\int \sin x\,dx = -\cos x.$$

$$\int \cos x\,dx = \sin x.$$

$$\int \sin^2 x\,dx = -\frac{1}{2}\sin x\cos x + \frac{1}{2}x.$$

$$\int \cos^2 x\,dx = \frac{1}{2}\sin x\cos x + \frac{1}{2}x.$$

$$\int \sin^3 x\,dx = -\frac{1}{3}\cos x(\sin^2 x + 2).$$

$$\int \cos^3 x\,dx = \frac{1}{3}\sin x(\cos^2 x + 2).$$

$$\int \cos^4 x\,dx = \frac{3}{8}x + \frac{1}{4}\sin 2x + \frac{1}{32}\sin 4x.$$

$$\int \sin^n x\cos x\,dx = \frac{(\sin x)^{n+1}}{n+1} \qquad (n \neq -1).$$

$$\int \sinh x\,dx = \cosh x.$$

$$\int \cosh x \, dx = \sinh x.$$

$$\int \tanh x \, dx = \ln \cosh x.$$

Taylor Series

The Taylor series of a function $f(x)$ is

$$f(a+x) = f(a) + f'(a)x + \frac{1}{2!} f''(a)x^2 + \frac{1}{3!} f'''(a)x^3 + \quad ,$$

where the primes indicate derivatives.

Some useful Taylor series are:

$$e^x = 1 + x + \frac{x^2}{2!} + \frac{x^3}{3!} + \cdots$$

$$\sin(a+x) = \sin a + (\cos a)x - \frac{1}{2}(\sin a)x^2 - \frac{1}{6}(\cos a)x^3 + \cdots$$

$$\cos(a+x) = \cos a - (\sin a)x - \frac{1}{2}(\cos a)x^2 + \frac{1}{6}(\sin a)x^3 + \cdots$$

$$\tan(a+x) = \tan a + \left(\frac{1}{\cos^2 a}\right)x + \left(\frac{\sin a}{\cos^3 a}\right)x^2 + \left(\frac{\sin^2 a}{\cos^4 a} + \frac{1}{3\cos^2 a}\right)x^3 + \cdots$$

APPENDIX B

Material Properties

This appendix summarizes the properties of selected materials. These values may be used as approximations for preliminary design. However, in final design calculations you should try to use values obtained from the actual materials specified in your design. Material properties can sometimes be obtained from the sources supplying the materials, or it may be necessary to have measurements made from samples of the specified materials.

Table B-1 Elastic moduli of selected materials

Material	Modulus of elasticity E		Shear modulus G		Poisson's ratio ν
	10^6 psi	GPa	10^6 psi	GPa	
Aluminum	10	70	3.8	26	0.33
Aluminum alloys	10–12	70–80	3.8–4.4	26–30	0.33
2014-T6	10.6	73	4	28	0.33
6061-T6	10	70	3.8	26	0.33
7075-T6	10.4	72	3.9	27	0.33
Brick (compression)	1.5–3.5	10–24			
Cast iron	12–25	80–170	4.5–10	31–69	0.2–0.3
Gray cast iron	14	97	5.6	39	0.25
Concrete (compression)	2.6–4.4	18–30			0.1–0.2
Copper	17	115	6.2	43	0.35
Copper alloys	14–18	96–120	5.2–6.8	36–47	0.33–0.35
Brass	14–16	96–110	5.2–6	36–41	0.34
80% Cu, 20% Zn	15	100	5.5	38	0.33
Naval brass	15	100	5.5	38	0.33
Bronze	14–17	96–120	5.2–6.3	36–44	0.34
Manganese bronze	15	100	5.6	39	0.35
Glass	7–12	50–80	2.9–5	20–33	0.20–0.27
Magnesium	5.8	40	2.2	15	0.34

(Continued)

Table B-1 (*Continued*)

Material	Modulus of elasticity E		Shear modulus G		Poisson's ratio ν
	10^6 psi	GPa	10^6 psi	GPa	
Nickel	30	210	11.4	80	0.31
Nylon	0.3–0.4	2–3			0.4
Rubber	0.0001–0.0006	0.0001–0.004	0.00004–0.0002	0.0003–0.0014	0.44–0.50
Steel	28–32	190–220	10.8–12.3	75–85	0.28–0.30
Stone (compression)					
Granite	6–10	40–70			0.2–0.3
Marble	7–14	50–100			0.2–0.3
Titanium	16	110	5.8	40	0.33
Titanium alloys	15–18	100–124	5.6–6.8	39–47	0.33
Tungsten	52	360	22	150	0.2
Wood (bending)					
Ash	1.5–1.6	10–11			
Oak	1.6–1.8	11–12			
Southern pine	1.6–2	11–14			
Wrought iron	28	190	10.9	75	0.3

Table B-2 Yield and ultimate stresses of selected materials

Material	Yield stress σ_Y		Ultimate stress σ_U		Percent elongation (2-in. gauge length)
	10^3 psi	MPa	10^3 psi	MPa	
Aluminum	3	20	10	70	60
Aluminum alloys	6–75	40–520	15–80	100–560	2–45
2014-T6	60	410	70	480	13
6061-T6	40	270	45	310	17
7075-T6	70	480	80	550	11
Brick (compression)			1–10	7–70	
Cast iron (compression)			50–200	340–1400	
Cast iron (tension)	17–41	120–280	10–70	70–480	1
Gray cast iron	17	120	20–58	140–400	1
Concrete (compression)			1.5–10	10–70	
Copper					
Hard-drawn	49	340	55	380	10
Soft-annealed	8	55	33	230	50
Copper alloys					
Beryllium copper, hard	109	750	120	830	4

(*Continued*)

Table B-2 (*Continued*)

Material	Yield stress σ_Y		Ultimate stress σ_U		Percent elongation (2-in. gauge length)
	10^3 psi	MPa	10^3 psi	MPa	
Brass	12–80	80–540	36–90	240–600	4–60
80% Cu, 20% Zn, hard	67	460	84	580	4
80% Cu, 20% Zn, soft	13	90	43	300	50
Naval brass, hard	58	400	84	580	15
Naval brass, soft	25	170	59	410	50
Bronze	12–100	82–700	30–120	200–830	5–60
Manganese bronze, hard	65	450	90	620	10
Manganese bronze, soft	25	170	65	450	35
Glass (plate)			9	65	
Magnesium	3–10	20–68	15–25	100–170	5–15
Nickel	20–90	140–620	45–110	310–760	2–50
Nylon			6–10	40–72	50
Rubber	0.3–1	2–7	1–3	7–20	100–800
Steel, structural	30–104	200–720	50–118	340–820	10–40
ASTM-A36	36	250	60	400	30
ASTM-A572	50	340	70	500	20
ASTM-A514	100	700	120	830	15
Stone (compression)					
Granite			10–40	70–280	
Marble			8–25	50–180	
Titanium	60	400	70	500	25
Titanium alloys	110–125	760–860	130–140	900–960	10
Tungsten			200–600	1400–4000	0–4
Wood (bending)					
Ash	6–10	40–70	8–13	50–90	
Oak	6–8	40–55	8–13	50–90	
Southern Pine	6–8	40–55	8–13	50–90	
Wood (compression parallel to grain)					
Ash	4–6	30–40	5–8	30–50	
Oak	4–6	30–40	5–8	30–50	
Southern pine	4–8	30–50	6–10	40–70	
Wrought iron	30	210	48	330	30

Table B-3 Densities and coefficients of thermal expansion of selected materials

Material	Density ρ		Coefficient of thermal expansion α	
	slug/ft^3	kg/m^3	10^{-6} °F^{-1}	10^{-6} °C^{-1}
Aluminum	5.2	2700	13.3	23.9
Aluminum alloys	4.9–5.4	2500–2800	13–13.4	23–24
Copper	17.4	9000	9.2	16.6
Copper alloys				
Brass	16.3–16.9	8400–8700	10.6–11.8	19–21
Bronze	14.3–17.2	7400–8900	9.9–11.6	18–21
Cast iron	13.9	7200	5.6–6.7	10–12
Gray cast iron	13.6–13.8	7000–7100	5.6	10
Concrete	2.9–4.7	1500–2400	4–8	7–14
Glass	4.6–5.8	2400–3000	3–6	5–11
Magnesium	3.4	1740	14	25
Magnesium alloys	3.4–3.5	1760–1830	14–16	26–29
Nickel	16.7	8600	7.2	13
Rubber	1.7–3.9	900–2000	70–110	130–200
Steel	14.9–15.2	7700–7830	6–10	10–18
Stone	3.9–5.1	2000–2900	3–5	5–9
Titanium	8.8	4540	4.7	8.5
Tungsten	37.4	1930	2.4	4.3
Wrought iron	14–15	7400–7800	7	12

APPENDIX C

Centroids and Moments of Inertia

Centroids and moments of inertia of areas arise repeatedly in analyses of problems in solid mechanics. These quantities are defined and discussed in this appendix. Centroids and moments of inertia of specific areas are given in Appendix D.

C-1 | Centroids of Areas

Consider an area A in the x–y plane (Fig. C-1a). The coordinates of the centroid of A are defined by

$$\bar{x} = \frac{\int_A x\,dA}{\int_A dA}, \qquad \bar{y} = \frac{\int_A y\,dA}{\int_A dA}, \tag{C-1}$$

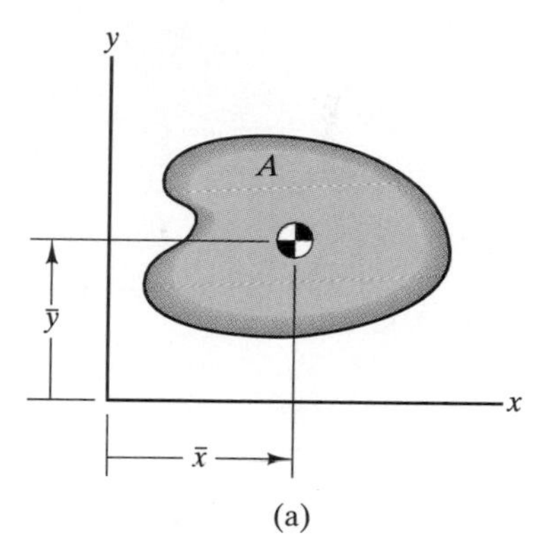

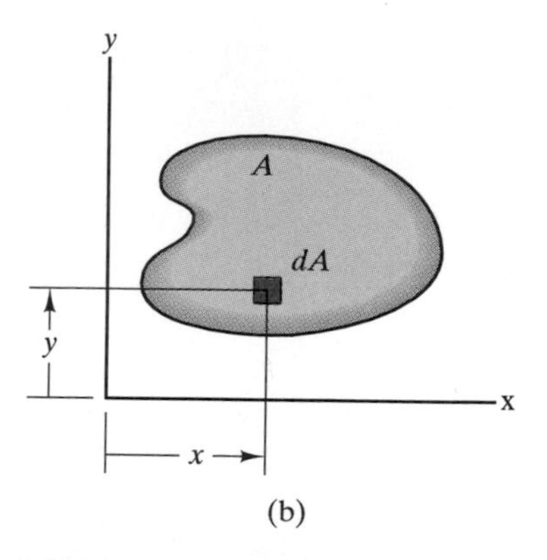

FIGURE C-1 (a) Centroid of an area A. (b) Coordinates of dA.

where x and y are the coordinates of the differential element of area dA (Fig. C-1b). The subscript A on the integral signs means that the integration is carried out over the entire area.

If an area is symmetric about an axis, its centroid lies on the axis (Fig. C-2a), and if an area is symmetric about two axes, its centroid lies at their intersection (Fig. C-2b).

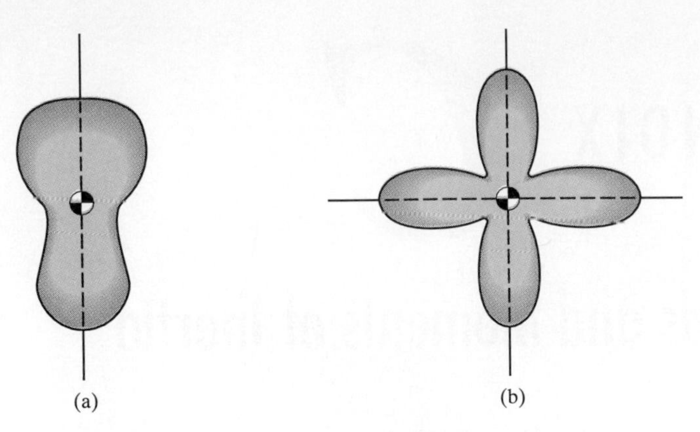

FIGURE C-2 (a) Area that is symmetric about an axis. (b) Area with two axes of symmetry.

EXAMPLE C-1

Determine the coordinates of the centroid of the triangular area in Fig. C-3.

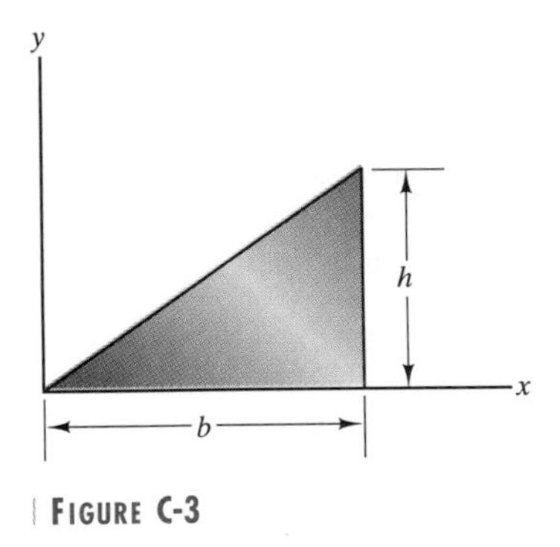

FIGURE C-3

Solution

Using the element of area $dA = dx\,dy$ in Fig. (a), the x coordinate of the centroid is

$$\bar{x} = \frac{\int_A x\,dA}{\int_A dA}$$

$$= \frac{\int_{x=0}^{b} \int_{y=0}^{(h/b)x} x\,dx\,dy}{\int_{x=0}^{b} \int_{y=0}^{(h/b)x} dx\,dy}$$

$$= \frac{\displaystyle\int_0^b x \left[\int_0^{(h/b)x} dy\right] dx}{\displaystyle\int_0^b \left[\int_0^{(h/b)x} dy\right] dx}$$

$$= \frac{\displaystyle\int_0^b (h/b)x^2\, dx}{\displaystyle\int_0^b (h/b)x\, dx}$$

$$= \frac{2}{3}b.$$

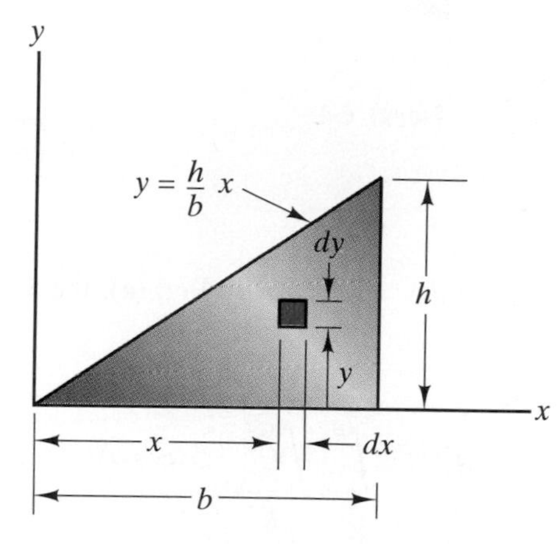

(a) Element dA for determining $\bar{x}$ and $\bar{y}$.

The y coordinate of the centroid is

$$\bar{y} = \frac{\displaystyle\int_A y\, dA}{\displaystyle\int_A dA} = \frac{\displaystyle\int_{x=0}^{b} \int_{y=0}^{(h/b)x} y\, dx\, dy}{\displaystyle\int_{x=0}^{b} \int_{y=0}^{(h/b)x} dx\, dy}$$

$$= \frac{1}{3}h,$$

where the integrals have been evaluated in the same way as in the determination of $\bar{x}$.

EXAMPLE C-2

Determine the coordinates of the centroid of the semicircular area in Fig. C-4.

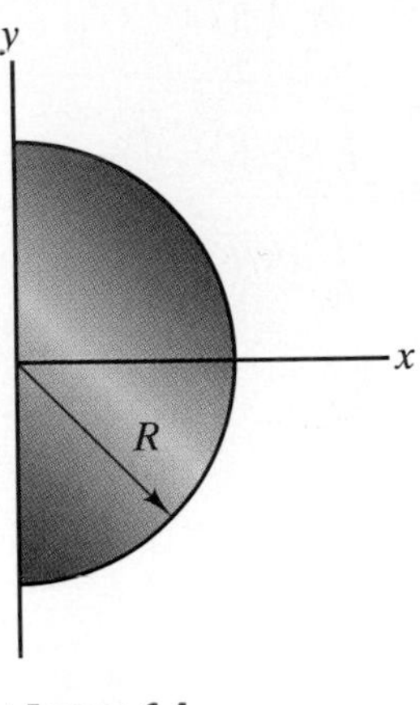

FIGURE C-4

Solution

Using the element of area $dA = r\,dr\,d\theta$ in Fig. (a), the x coordinate of the centroid is

$$\bar{x} = \frac{\displaystyle\int_A x\,dA}{\displaystyle\int_A dA} = \frac{\displaystyle\int_{r=0}^{R}\int_{\theta=-\pi/2}^{\pi/2} (r\cos\theta)(r\,dr\,d\theta)}{\displaystyle\int_{r=0}^{R}\int_{\theta=-\pi/2}^{\pi/2} r\,dr\,d\theta}$$

$$= \frac{\displaystyle\int_0^R r^2\left[\int_{-\pi/2}^{\pi/2}\cos\theta\,d\theta\right]dr}{\displaystyle\int_0^R r\left[\int_{-\pi/2}^{\pi/2}d\theta\right]dr}$$

$$= \frac{\displaystyle\int_0^R 2r^2\,dr}{\displaystyle\int_0^R \pi r\,dr}$$

$$= \frac{4R}{3\pi}.$$

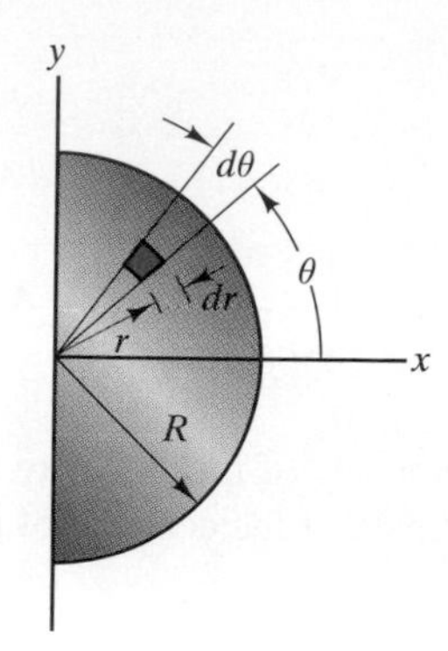

(a) Element dA for determining $\bar{x}$.

Because the area is symmetric about the x axis, $\bar{y} = 0$.

C-2 | Composite Areas

An area consisting of a combination of simple parts is called a *composite area*. Figure C-5a shows a composite area consisting of a triangle, a rectangle, and a semicircle, labeled 1, 2, and 3, respectively. The x coordinate of the centroid

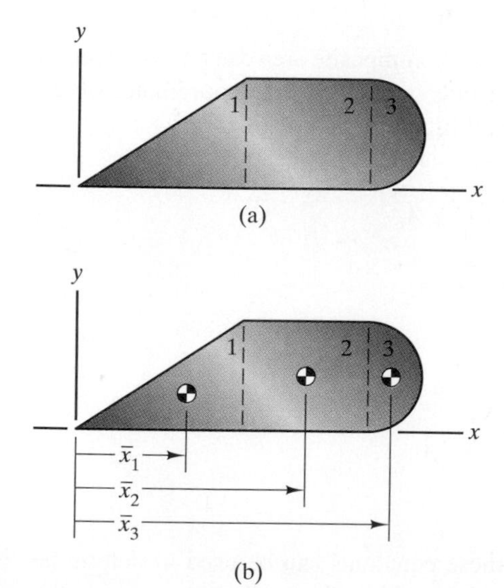

| **FIGURE C-5** (a) Composite area composed of three parts. (b) Centroids of the parts.

of the composite area is

$$\bar{x} = \frac{\int_A x\,dA}{\int_A dA} = \frac{\int_{A_1} x\,dA + \int_{A_2} x\,dA + \int_{A_3} x\,dA}{\int_{A_1} dA + \int_{A_2} dA + \int_{A_3} dA}. \tag{C-2}$$

The x coordinates of the centroids of the parts are shown in Fig. C-5b. From the definition of the x coordinate of the centroid of part 1,

$$\bar{x}_1 = \frac{\int_{A_1} x\,dA}{\int_{A_1} dA},$$

it is seen that

$$\int_{A_1} x\,dA = \bar{x}_1 A_1.$$

Using this equation and equivalent expressions for parts 2 and 3, Eq. (C-2) can be written as

$$\bar{x} = \frac{\bar{x}_1 A_1 + \bar{x}_2 A_2 + \bar{x}_3 A_3}{A_1 + A_2 + A_3}.$$

Thus the x coordinate of the centroid of the composite area can be expressed in terms of the x coordinates of the centroids of its parts. The coordinates of the centroid of a composite area with an arbitrary number of parts are

$$\bar{x} = \frac{\sum_i \bar{x}_i A_i}{\sum_i A_i}$$

$$\bar{y} = \frac{\sum_i \bar{y}_i A_i}{\sum_i A_i}. \tag{C-3}$$

As demonstrated in Example C-3, these equations can be used to determine the centroids of composite areas containing "holes," or cutouts, by treating the cutouts as negative areas.

EXAMPLE C-3

Determine the coordinates of the centroid of the area in Fig. C-6.

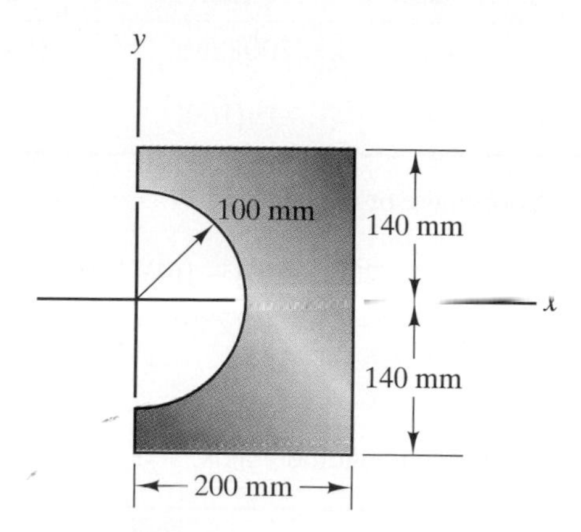

FIGURE C-6

Solution

The area can be treated as a composite consisting of the rectangle without the semicircular cutout and the area of the cutout, which will be called parts 1 and 2, respectively [Fig. (a)].

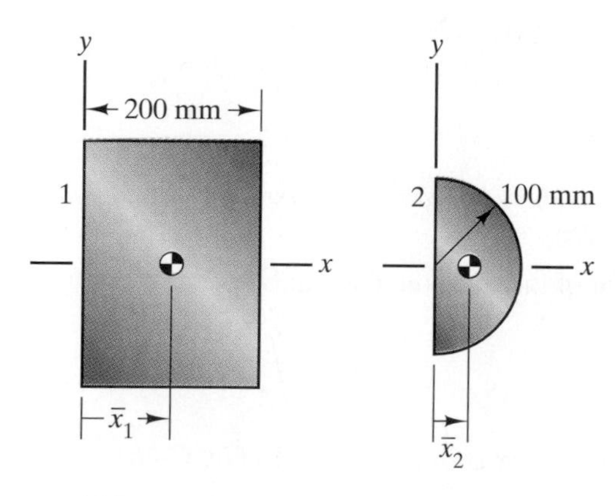

(a) Rectangle 1 and semicircular cutout 2.

From Appendix D the x coordinate of the centroid of the cutout is

$$\bar{x}_2 = \frac{4R}{3\pi} = \frac{4(100)}{3\pi} \text{ mm.}$$

The information for determining the x coordinate of the centroid of the composite area is summarized in Table C-1. Notice that the cutout is treated as a

Table C-1 Information for determining $\bar{x}$

	$\bar{x}_i$ (mm)	A_i (mm^2)	$\bar{x}_i A_i$ (mm^3)
Part 1 (rectangle)	100	(200)(280)	(100)[(200)(280)]
Part 2 (cutout)	$\frac{4(100)}{3\pi}$	$-\frac{1}{2}\pi(100)^2$	$-\frac{4(100)}{3\pi}\left[\frac{1}{2}\pi(100)^2\right]$

negative area. The x coordinate of the centroid is

$$\bar{x} = \frac{\bar{x}_1 A_1 + \bar{x}_2 A_2}{A_1 + A_2} = \frac{(100)[(200)(280)] - [(4)(100)/3\pi]\left[\frac{1}{2}\pi(100)^2\right]}{(200)(280) - \frac{1}{2}\pi(100)^2}$$

$$= 122.4 \text{ mm}.$$

Because the area is symmetric about the x axis, $\bar{y} = 0$.

C-3 | Moments of Inertia of Areas

The moments of inertia of an area are integrals similar in form to those used to determine the centroid of the area. Consider an area A in the x–y plane (Fig. C-7a). The moments and product of inertia of A are:

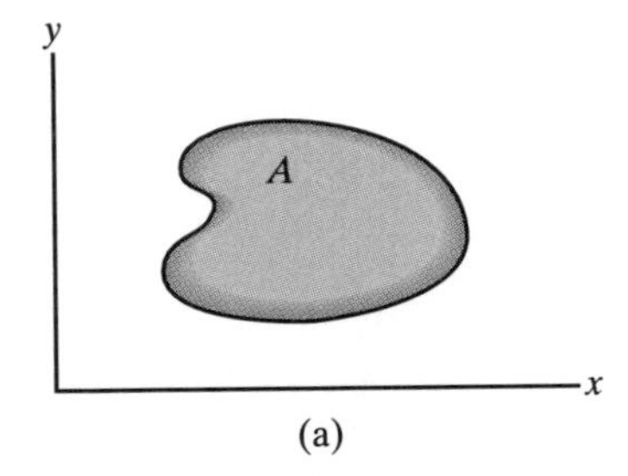

FIGURE C-7 (a) Area A in the x–y plane. (b) Coordinates of dA.

1. Moment of inertia about the x axis:

$$I_x = \int_A y^2\, dA, \qquad \textbf{(C-4)}$$

where y is the y coordinate of the differential element of area dA (Fig. C-7b).

2. Moment of inertia about the y axis:

$$I_y = \int_A x^2\, dA, \qquad \textbf{(C-5)}$$

where x is the x coordinate of dA (Fig. C-7b).

3. Product of inertia:

$$I_{xy} = \int_A xy\, dA. \qquad \textbf{(C-6)}$$

4. Polar moment of inertia:

$$J_O = \int_A r^2\, dA, \qquad \textbf{(C-7)}$$

where r is the radial distance from the origin O to dA (Fig. C-7b).

The dimensions of the moments of inertia of an area are (length)^4. The definitions of I_x, I_y, and J_O imply that they have positive values for any area. The polar moment of inertia about the origin is equal to the sum of the moments of inertia about the x and y axes:

$$J_O = \int_A r^2\, dA = \int_A (x^2 + y^2)\, dA = I_y + I_x.$$

The definition of I_{xy} implies that if A is symmetric about either the x axis or the y axis, the product of inertia is zero.

EXAMPLE C-4

Determine the moments of inertia of the triangular area in Fig. C-8.

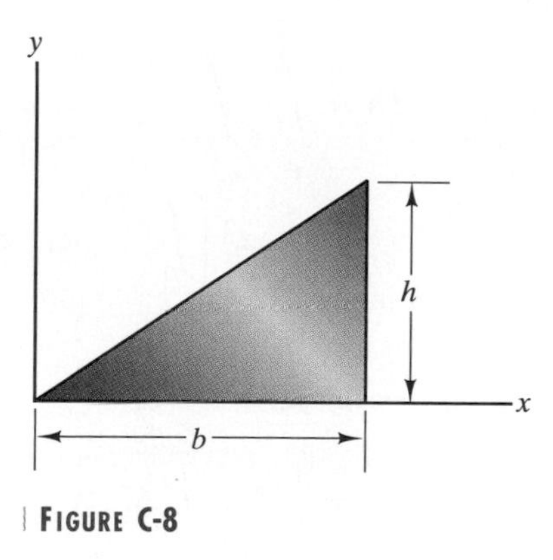

FIGURE C-8

Solution

The moments of inertia can be evaluated using the rectangular element $dA = dx\, dy$ shown in Fig. (a).

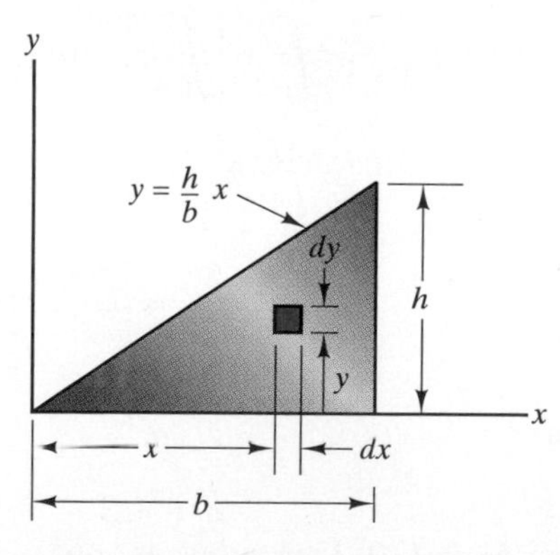

(a) Element dA for determining the moments of inertia.

Moment of inertia about the x axis

$$I_x = \int_A y^2\, dA = \int_{x=0}^{b} \int_{y=0}^{(h/b)x} y^2\, dx\, dy$$
$$= \int_0^b \left[\int_0^{(h/b)x} y^2 dy \right] dx$$
$$= \int_0^b \frac{h^3}{3b^3} x^3 dx$$
$$= \frac{1}{12} bh^3.$$

Moment of inertia about the y axis

$$I_y = \int_A x^2\, dA = \int_{x=0}^{b} \int_{y=0}^{(h/b)x} x^2\, dx\, dy$$
$$= \int_0^b x^2 \left[\int_0^{(h/b)x} dy \right] dx$$
$$= \int_0^b \frac{h}{b} x^3 dx$$
$$= \frac{1}{4} hb^3.$$

Product of inertia

$$I_{xy} = \int_A xy\, dA = \int_{x=0}^{b} \int_{y=0}^{(h/b)x} xy\, dx\, dy$$
$$= \int_0^b x \left[\int_0^{(h/b)x} y\, dy \right] dx$$
$$= \int_0^b \frac{h^2}{2b^2} x^3 dx$$
$$= \frac{1}{8} b^2 h^2.$$

Polar moment of inertia

$$J_O = I_x + I_y = \tfrac{1}{12} bh^3 + \tfrac{1}{4} hb^3.$$

EXAMPLE C-5

Determine the moments of inertia of the semicircular area in Fig. C-9.

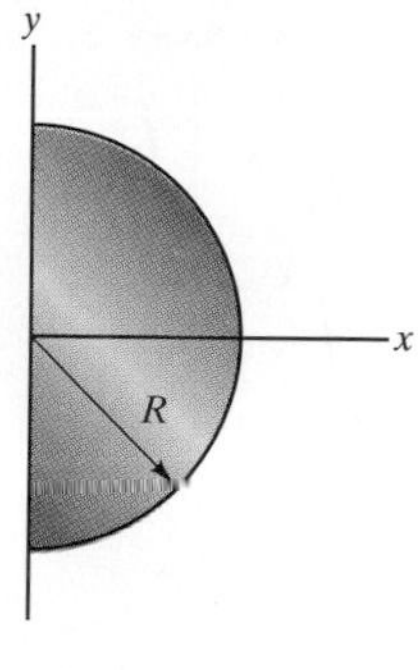

FIGURE C-9

Solution

The moments of inertia can be evaluated using the element $dA = r\,dr\,d\theta$ shown in Fig. (a).

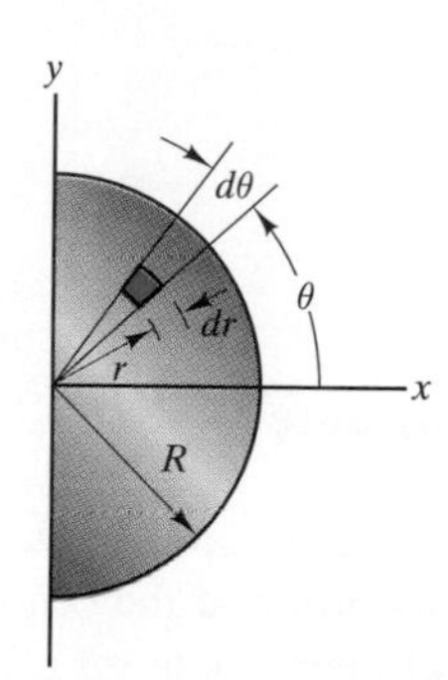

(a) Element dA for determining the moments of inertia.

Moment of inertia about the x axis

$$\begin{aligned} I_x = \int_A y^2\,dA &= \int_{r=0}^{R}\int_{\theta=-\pi/2}^{\pi/2} (r\sin\theta)^2(r\,dr\,d\theta) \\ &= \int_0^R r^3\left[\int_{-\pi/2}^{\pi/2}\sin^2\theta\,d\theta\right]dr \\ &= \int_0^R \frac{\pi}{2}r^3\,dr \\ &= \frac{1}{8}\pi R^4. \end{aligned}$$

Moment of inertia about the *y* axis

$$I_y = \int_A x^2\,dA = \int_{r=0}^{R}\int_{\theta=-\pi/2}^{\pi/2} (r\cos\theta)^2 (r\,dr\,d\theta)$$
$$= \int_0^R r^3 \left[\int_{-\pi/2}^{\pi/2} \cos^2\theta\,d\theta\right] dr$$
$$= \int_0^R \frac{\pi}{2} r^3\,dr$$
$$= \frac{1}{8}\pi R^4.$$

Product of inertia

Because the area is symmetric about the x axis, $I_{xy} = 0$.

Polar moment of inertia

$$J_O = I_x + I_y$$
$$= \tfrac{1}{8}\pi R^4 + \tfrac{1}{8}\pi R^4$$
$$= \tfrac{1}{4}\pi R^4.$$

C-4 | Parallel Axis Theorems

Suppose that the moments of inertia of an area A are known in terms of a coordinate system $x'y'$ with its origin at the centroid of A, and the objective is to determine the moments of inertia in terms of a parallel coordinate system xy (Fig. C-10a). This can be accomplished using results known as parallel axis theorems.

Let the coordinates of the centroid of A in the xy coordinate system be (d_x, d_y), and let $d = \sqrt{d_x^2 + d_y^2}$ be the distance from the origin of the xy coordinate system to the centroid (Fig. C-10b). Two preliminary results are needed to derive the parallel axis theorems: In terms of the $x'y'$ coordinate system, the coordinates of the centroid of A are

$$\bar{x}' = \frac{\int_A x'\,dA}{\int_A dA}, \quad \bar{y}' = \frac{\int_A y'\,dA}{\int_A dA}.$$

But the origin of the $x'y'$ coordinate system is located at the centroid of A, so

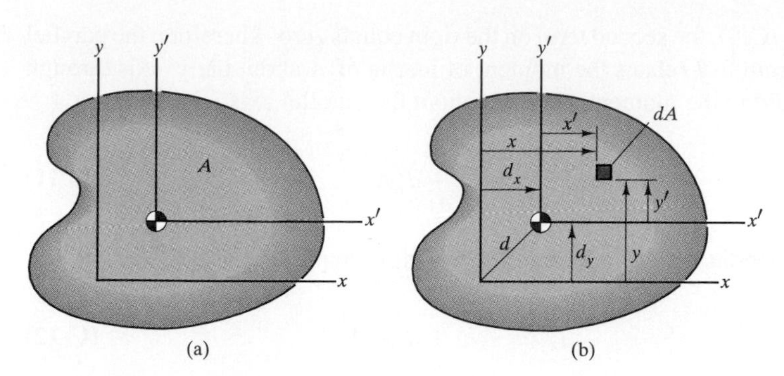

FIGURE C-10 (a) Area A and the coordinate systems $x'y'$ and xy. (b) Coordinates of the differential element dA.

$\bar{x}' = 0$ and $\bar{y}' = 0$. Therefore,

$$\int_A x'\,dA = 0, \qquad \int_A y'\,dA = 0. \tag{C-8}$$

Moment of Inertia about the *x* Axis In terms of the xy coordinate system, the moment of inertia of A about the x axis is

$$I_x = \int_A y^2\,dA, \tag{C-9}$$

where y is the coordinate of the element of area dA relative to the xy coordinate system. From Fig. C-10b it can be seen that $y = y' + d_y$, where y' is the coordinate of dA relative to the $x'y'$ coordinate system. Substituting this expression into Eq. (C-9) gives

$$I_x = \int_A (y' + d_y)^2\,dA = \int_A (y')^2\,dA + 2d_y \int_A y'\,dA + d_y^2 \int_A dA.$$

The first term on the right side of this equation is the moment of inertia of A about the x' axis. From Eq. (C-8), the second term on the right equals zero. Therefore,

$$I_x = I_{x'} + d_y^2 A. \tag{C-10}$$

This is a parallel axis theorem. It relates the moment of inertia of A about the x' axis through the centroid to the moment of inertia about the parallel axis x.

Moment of Inertia about the *y* Axis In terms of the xy coordinate system, the moment of inertia of A about the y axis is

$$I_y = \int_A x^2\,dA = \int_A (x' + d_x)^2\,dA$$

$$= \int_A (x')^2\,dA + 2d_x \int_A x'\,dA + d_x^2 \int_A dA.$$

From Eq. (C-8), the second term on the right equals zero. Therefore, the parallel axis theorem that relates the moment of inertia of A about the y' axis through the centroid to the moment of inertia about the parallel axis y is

$$I_y = I_{y'} + d_x^2 A. \tag{C-11}$$

Product of Inertia The parallel axis theorem for the product of inertia is

$$I_{xy} = I_{x'y'} + d_x d_y A. \tag{C-12}$$

Polar Moment of Inertia The parallel axis theorem for the polar moment of inertia is

$$J_O = J'_O + \left(d_x^2 + d_y^2\right) A = J'_O + d^2 A, \tag{C-13}$$

where d is the distance from the origin of the $x'y'$ coordinate system to the origin of the xy coordinate system.

EXAMPLE C-6

Determine I_x and I_{xy} for the area in Fig. C-11.

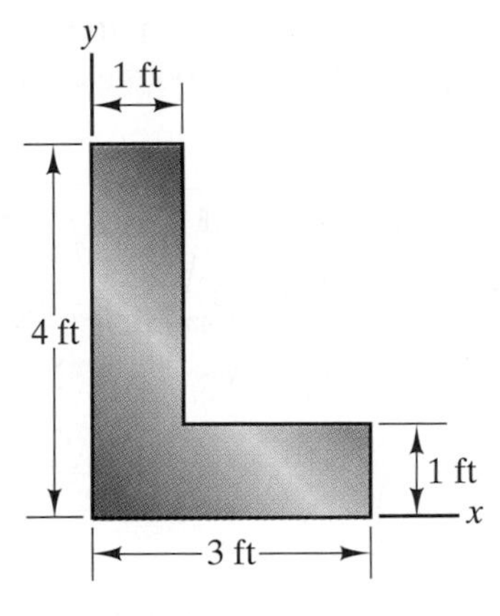

FIGURE C-11

Solution

The moments of inertia can be determined by treating the area as a composite consisting of the rectangular parts 1 and 2 shown in Fig. (a). For each part, introduce a coordinate system $x'y'$ with its origin at the centroid of the part [Fig. (b)]. The moments of inertia of the rectangular parts in terms of these coordinate systems are given in Appendix D.

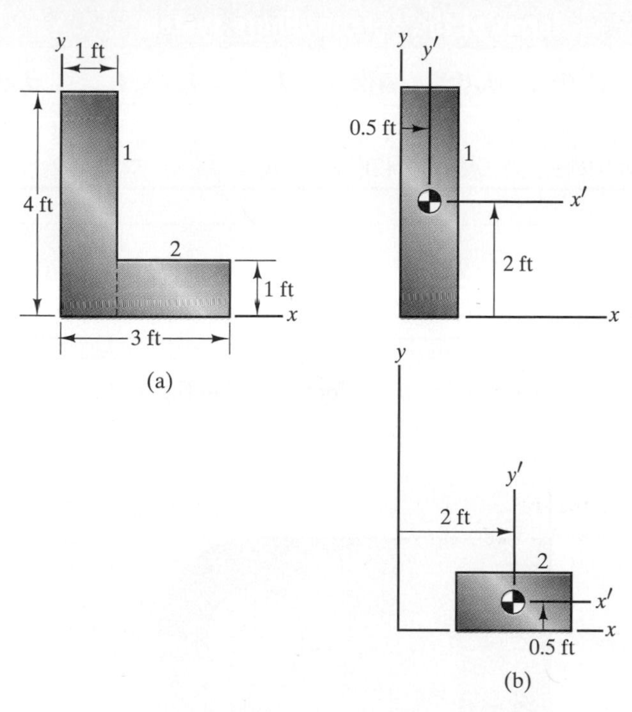

(a) Dividing the area into rectangles 1 and 2. (b) Parallel coordinate systems $x'y'$ with origins at the centroids of the parts.

Equation (C-10) can be used to determine the moment of inertia of each part about the x axis (Table C-2).

Table C-2 Determining the moments of inertia of the parts about the x axis

	d_x (ft)	d_y (ft)	A(ft^2)	$I_{x'}$ (ft^4)	$I_x = I_{x'} + d_y^2 A$(ft^4)
Part 1	0.5	2	(1)(4)	$\frac{1}{12}(1)(4)^3$	21.33
Part 2	2	0.5	(2)(1)	$\frac{1}{12}(2)(1)^3$	0.67

The moment of inertia of the composite area about the x axis is

$$I_x = (I_x)_1 + (I_x)_2 = 21.33 + 0.67 = 22.00 \text{ ft}^4.$$

Using Eq. (C-12), I_{xy} is determined for each part in Table C-3. The product of inertia of the composite area is

$$I_{xy} = (I_{xy})_1 + (I_{xy})_2 = 4 + 2 = 6 \text{ ft}^4.$$

Table C-3 Determining the products of inertia of the parts in terms of the xy coordinate system

	d_x (ft)	d_y (ft)	A(ft^2)	$I_{x'y'}$	$I_{xy} = I_{x'y'} + d_x d_y A$(ft^4)
Part 1	0.5	2	(1)(4)	0	4
Part 2	2	0.5	(2)(1)	0	2

EXAMPLE C-7

Determine the moment of inertia I_y for the area in Fig. C-12.

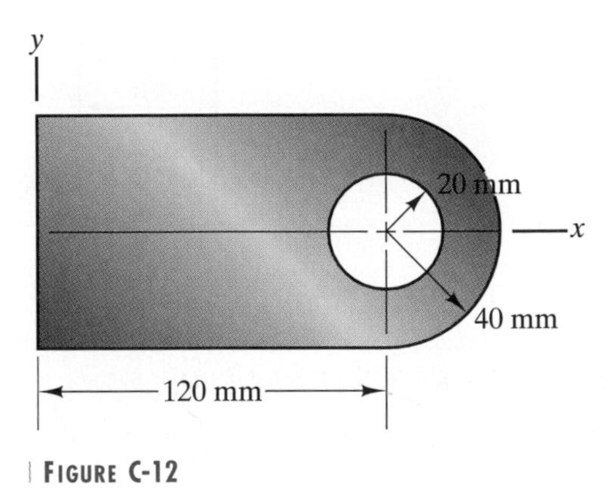

FIGURE C-12

Solution

The area can be divided into a rectangle, a semicircle, and a circular cutout, labeled parts 1, 2, and 3 in Fig. (a). The moments of inertia of the parts in terms of the $x'y'$ coordinate systems and the location of the centroid of the semicircular part are given in Appendix D. In Table C-4 Eq. (C-10) is used to determine the moment of inertia of each part about the x axis.

Table C-4 Determining the moments of inertia of the parts

	d_x (mm)	A(mm^2)	$I_{y'}$ (mm^4)	$I_y = I_{y'} + d_x^2 A$(mm^4)
Part 1	60	(120)(80)	$\frac{1}{12}(80)(120)^3$	4.608×10^7
Part 2	$120 + \frac{(4)(40)}{3\pi}$	$\frac{1}{2}\pi(40)^2$	$\left(\frac{\pi}{8} - \frac{8}{9\pi}\right)(40)^4$	4.744×10^7
Part 3	120	$\pi(20)^2$	$\frac{1}{4}\pi(20)^4$	1.822×10^7

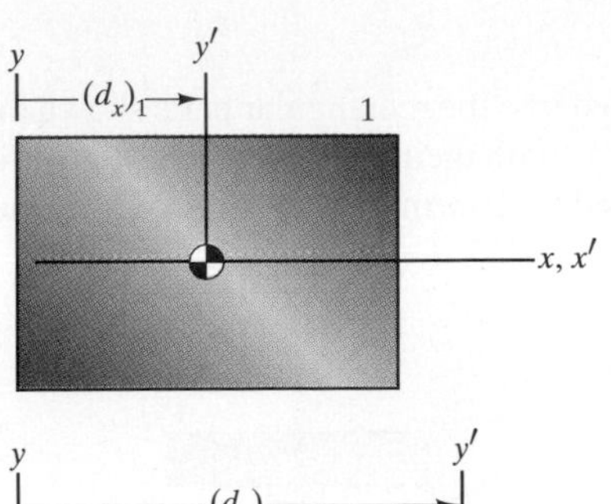

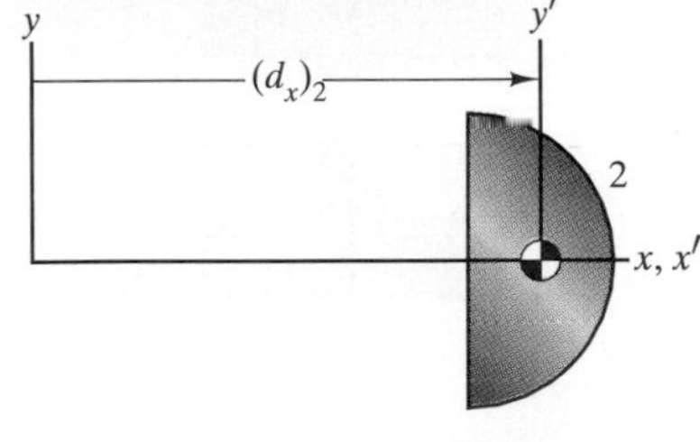

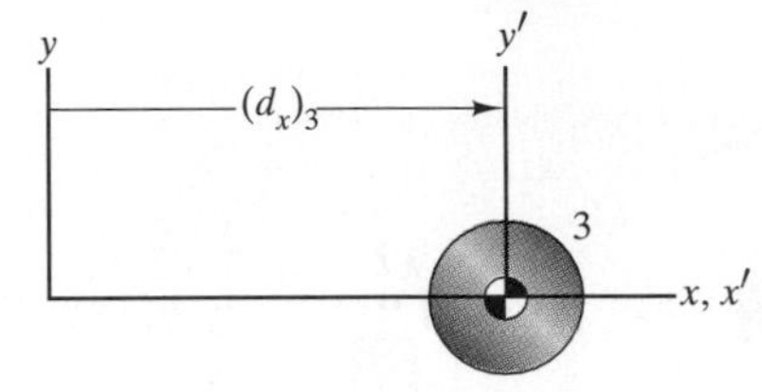

(a) Parts 1, 2, and 3.

The moment of inertia of the composite area about the y axis is

$$I_y = (I_y)_1 + (I_y)_2 - (I_y)_3 = (4.608 + 4.744 - 1.822) \times 10^7 = 7.530 \times 10^7 \text{ mm}^4.$$

EXAMPLE C-8

The cross section of an I-beam is shown in Fig. C-13. Determine its moment of inertia about the x axis.

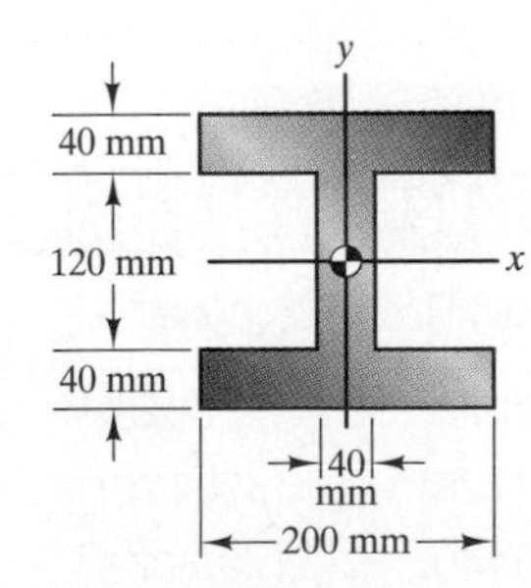

FIGURE C-13

Solution

The area can be divided into the rectangular parts shown in Fig. (a). Introducing coordinate systems $x'y'$ with their origins at the centroids of the parts [Fig. (b)], Eq. (C-10) can be used to determine their moments of inertia about the x axis (Table C-5).

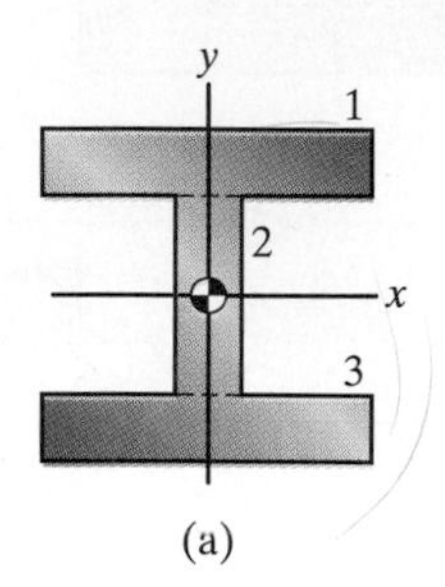

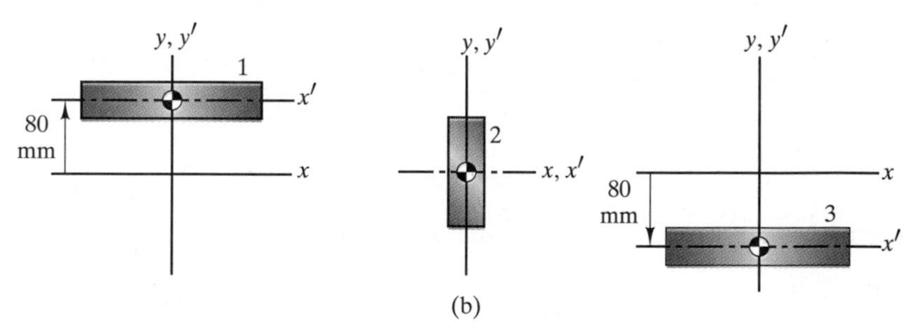

(a) Dividing the I-beam cross section into parts. (b) Parallel coordinate systems $x'y'$ with origins at the centroids of the parts.

The moment of inertia of the cross section is

$$I_x = (I_x)_1 + (I_x)_2 + (I_x)_3 = (52.27 + 5.76 + 52.27) \times 10^6 = 11.03 \times 10^7 \text{ mm}^4.$$

Table C-5 Determining the moments of inertia of the parts about the x axis

	d_y (mm)	A (mm²)	$I_{x'}$ (mm⁴)	$I_x = I_{x'} + d_y^2 A$ (mm⁴)
Part 1	80	(200)(40)	$\frac{1}{12}(200)(40)^3$	5.23×10^7
Part 2	0	(40)(120)	$\frac{1}{12}(40)(120)^3$	0.58×10^7
Part 3	−80	(200)(40)	$\frac{1}{12}(200)(40)^3$	5.23×10^7

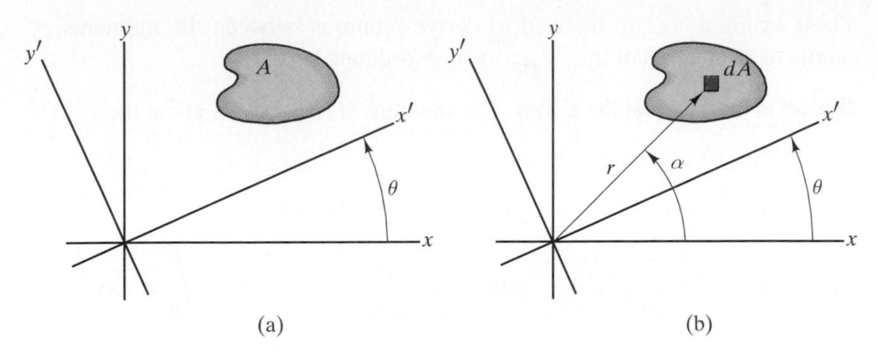

FIGURE C-14 (a) The $x'y'$ coordinate system is rotated through an angle θ relative to the xy coordinate system. (b) Differential element of area dA.

C-5 | Rotated and Principal Axes

In many engineering applications it is necessary to determine moments of inertia of areas for various choices of the angular orientation of the coordinate system relative to the area. Determining the angular orientation of the coordinate system for which the value of a given moment of inertia is a maximum or minimum is also frequently necessary. These procedures are discussed in this section.

Rotated Axes

Consider an area A, a coordinate system xy, and a second coordinate system $x'y'$ that is rotated through an angle θ relative to the xy coordinate system (Fig. C-14a). Suppose the moments of inertia of A are known in terms of the xy coordinate system. The objective is to determine the moments of inertia in terms of the $x'y'$ coordinate system.

In terms of the radial distance r to a differential element of area dA and the angle α in Fig. C-14b, the coordinates of dA in the xy coordinate system are

$$x = r\cos\alpha, \qquad \text{(C-14)}$$

$$y = r\sin\alpha. \qquad \text{(C-15)}$$

The coordinates of dA in the $x'y'$ coordinate system are

$$x' = r\cos(\alpha - \theta) = r(\cos\alpha\cos\theta + \sin\alpha\sin\theta), \qquad \text{(C-16)}$$

$$y' = r\sin(\alpha - \theta) = r(\sin\alpha\cos\theta - \cos\alpha\sin\theta). \qquad \text{(C-17)}$$

Substituting Eqs. (C-14) and (C-15) into Eqs. (C-16) and (C-17) yields equations relating the coordinates of dA in the two coordinate systems.

$$x' = x\cos\theta + y\sin\theta, \qquad \text{(C-18)}$$

$$y' = -x\sin\theta + y\cos\theta. \qquad \text{(C-19)}$$

These expressions can be used to derive relations between the moments of inertia of A in terms of the xy and $x'y'$ coordinate systems.

Moment of Inertia About the x' Axis The moment of inertia of A about the x' axis is

$$I_{x'} = \int_A (y')^2\,dA = \int_A (-x\sin\theta + y\cos\theta)^2\,dA$$
$$= \cos^2\theta \int_A y^2\,dA - 2\sin\theta\cos\theta \int_A xy\,dA + \sin^2\theta \int_A x^2\,dA,$$

which gives the result

$$I_{x'} = I_x\cos^2\theta - 2I_{xy}\sin\theta\cos\theta + I_y\sin^2\theta. \qquad \textbf{(C-20)}$$

Moment of Inertia About the y Axis The moment of inertia of A about the y' axis is

$$I_{y'} = \int_A (x')^2\,dA = \int_A (x\cos\theta + y\sin\theta)^2\,dA$$
$$= \sin^2\theta \int_A y^2\,dA + 2\sin\theta\cos\theta \int_A xy\,dA + \cos^2\theta \int_A x^2\,dA,$$

which gives the result

$$I_{y'} = I_x\sin^2\theta + 2I_{xy}\sin\theta\cos\theta + I_y\cos^2\theta. \qquad \textbf{(C-21)}$$

Product of Inertia In terms of the $x'y'$ coordinate system, the product of inertia of A is

$$I_{x'y'} = (I_x - I_y)\sin\theta\cos\theta + (\cos^2\theta - \sin^2\theta)I_{xy}. \qquad \textbf{(C-22)}$$

Polar Moment of Inertia From Eqs. (C-20) and (C-21), the polar moment of inertia in terms of the $x'y'$ coordinate system is

$$J'_O = I_{x'} + I_{y'} = I_x + I_y = J_O.$$

Thus the value of the polar moment of inertia is unchanged by a rotation of the coordinate system.

Principal Axes

The moments of inertia of A in terms of the $x'y'$ coordinate system in Fig. C-14a depend on the angle θ. Consider the following question: For what values of θ is the moment of inertia $I_{x'}$ a maximum or minimum? To consider this question, it is convenient to use the identities

$$\sin 2\theta = 2\sin\theta\cos\theta,$$
$$\cos 2\theta = \cos^2\theta - \sin^2\theta = 1 - 2\sin^2\theta = 2\cos^2\theta - 1,$$

to express Eqs. (C-20)–(C-22) in the forms

$$I_{x'} = \frac{I_x + I_y}{2} + \frac{I_x - I_y}{2}\cos 2\theta - I_{xy}\sin 2\theta, \quad \textbf{(C-23)}$$

$$I_{y'} = \frac{I_x + I_y}{2} - \frac{I_x - I_y}{2}\cos 2\theta + I_{xy}\sin 2\theta, \quad \textbf{(C-24)}$$

$$I_{x'y'} = \frac{I_x - I_y}{2}\sin 2\theta + I_{xy}\cos 2\theta. \quad \textbf{(C-25)}$$

Let a value of θ at which $I_{x'}$ is a maximum or minimum be denoted by θ_p. To determine θ_p, the derivative of Eq. (C-23) with respect to 2θ is equated to zero, obtaining the equation

$$\tan 2\theta_p = \frac{2I_{xy}}{I_y - I_x}. \quad \textbf{(C-26)}$$

Equating the derivative of Eq. (C-24) with respect to 2θ to zero to determine a value of θ for which $I_{y'}$ is a maximum or minimum also leads to Eq. (C-26). The second derivatives of $I_{x'}$ and $I_{y'}$ with respect to 2θ are opposite in sign:

$$\frac{d^2 I_{x'}}{d(2\theta)^2} = -\frac{d^2 I_{y'}}{d(2\theta)^2},$$

which means that at angles θ_p for which $I_{x'}$ is a maximum, $I_{y'}$ is a minimum, and at angles θ_p for which $I_{x'}$ is a minimum, $I_{y'}$ is a maximum.

A rotated coordinate system $x'y'$ that is oriented so that the derivative of Eq. (C-24) with respect to 2θ is equal to zero is called a set of *principal axes* of the area A. The corresponding moments of inertia $I_{x'}$ and $I_{y'}$ are called the *principal moments of inertia*. It can be shown that if x' and y' are principal axes, the product of inertia $I_{x'y'}$ equals zero. This is also a sufficient condition: If $I_{x'y'}$ equals zero, x' and y' are principal axes.

Once the orientation of the principal axes is determined by solving Eq. (C-26) for θ_p, the principal moments of inertia can be determined from Eqs. (C-23) and (C-24). Alternatively, by substituting Eq. (C-26) into Eqs. (C-23) and (C-24) it can be shown that the principal moments of inertia are given by

$$\text{principal moments of inertia} = \frac{I_x + I_y}{2} \pm \sqrt{\left(\frac{I_x - I_y}{2}\right)^2 + (I_{xy})^2}.$$

Because the tangent is a periodic function, Eq. (C-26) does not yield a unique solution for the angle θ_p. However, it does determine the orientation of the the principal axes within an arbitrary multiple of 90°. Observe in Fig. C-15 that if $2\theta_0$ is a solution of Eq. (C-26), then $2\theta_0 + n(180°)$ is also a solution for any integer n. The resulting orientations of the $x'y'$ coordinate system are shown in Fig. C-16.

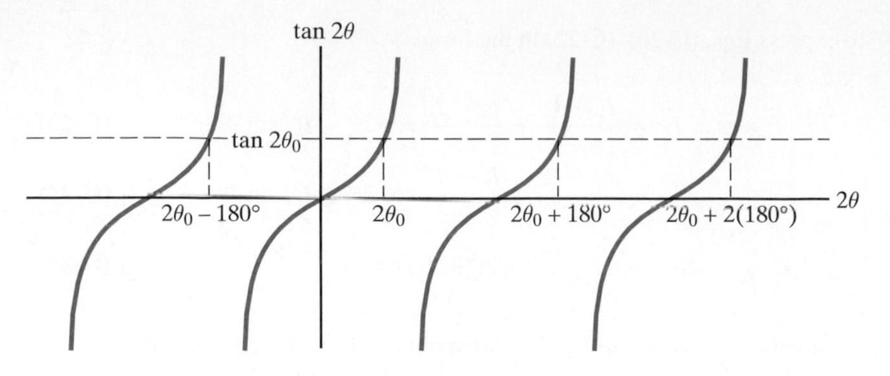

FIGURE C-15 For a given value of $\tan 2\theta_0$, there are multiple roots $2\theta_0 + n(180°)$.

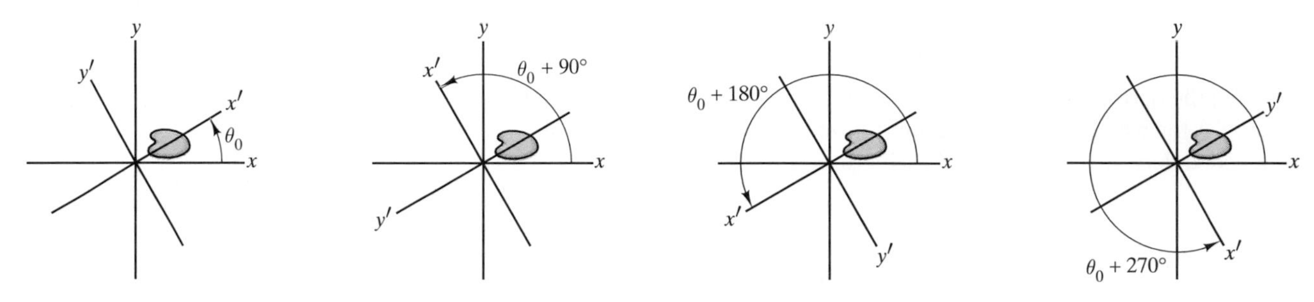

FIGURE C-16 The orientation of the $x'y'$ coordinate system is determined only within a multiple of 90°.

EXAMPLE C-9

Determine a set of principal axes and the corresponding principal moments of inertia for the triangular area in Fig. C-17.

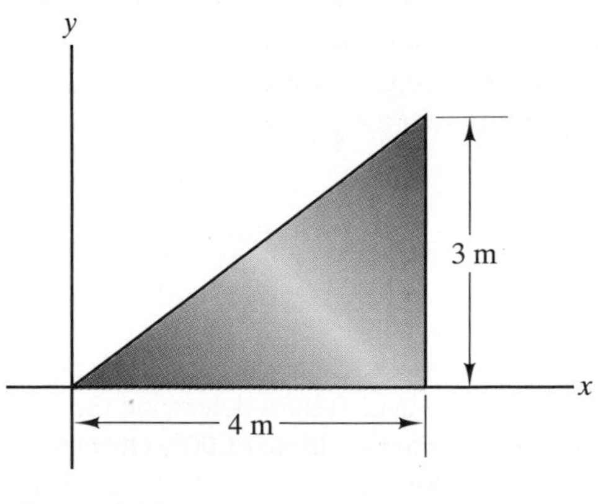

FIGURE C-17

Solution

From Appendix D, the moments of inertia of the triangular area are

$$I_x = \tfrac{1}{12}(4)(3)^3 = 9 \text{ m}^4,$$

$$I_y = \tfrac{1}{4}(4)^3(3) = 48 \text{ m}^4,$$

$$I_{xy} = \tfrac{1}{8}(4)^2(3)^2 = 18 \text{ m}^4.$$

From Eq. (C-26),

$$\tan 2\theta_p = \frac{2I_{xy}}{I_y - I_x} = \frac{2(18)}{48 - 9} = 0.923,$$

the angle $\theta_p = 21.4°$. Principal axes corresponding to this value of θ_p are shown in Fig. (a).

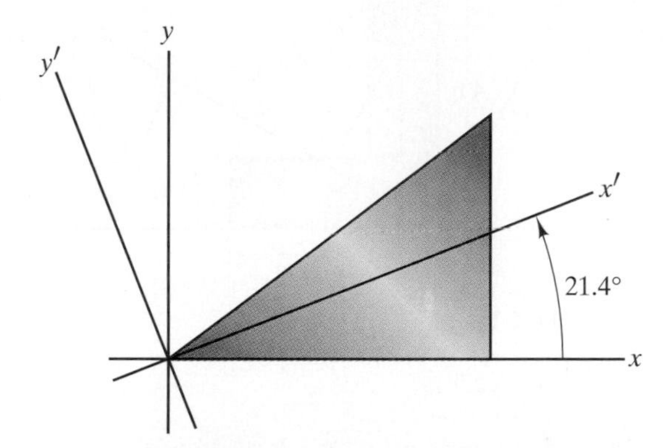

(a) Principal axes corresponding to $\theta_p = 21.4°$.

Substituting $\theta_p = 21.4°$ into Eqs. (C-23) and (C-24) gives the values of the principal moments of inertia:

$$I_{x'} = \frac{I_x + I_y}{2} + \frac{I_x - I_y}{2}\cos 2\theta - I_{xy}\sin 2\theta$$

$$= \frac{9 + 48}{2} + \frac{9 - 48}{2}\cos[(2)(21.4°)] - (18)\sin[(2)(21.4°)] = 1.96 \text{ m}^4,$$

$$I_{y'} = \frac{I_x + I_y}{2} - \frac{I_x - I_y}{2}\cos 2\theta + I_{xy}\sin 2\theta$$

$$= \frac{9 + 48}{2} - \frac{9 - 48}{2}\cos[(2)(21.4°)] + (18)\sin[(2)(21.4°)] = 55.0 \text{ m}^4.$$

These results can also be obtained from Eq. (C-5).

Discussion

The product of inertia corresponding to a set of principal axes is zero. In this example, substituting $\theta_p = 21.4°$ into Eq. (C-25) confirms that $I_{x'y'} = 0$.

EXAMPLE C-10

The moments of inertia of the area in Fig. C-18 in terms of the xy coordinate system are $I_x = 22\ \text{ft}^4$, $I_y = 10\ \text{ft}^4$, and $I_{xy} = 6\ \text{ft}^4$. **(a)** Determine $I_{x'}$, $I_{y'}$, and $I_{x'y'}$ for $\theta = 30°$. **(b)** Determine a set of principal axes and the corresponding principal moments of inertia.

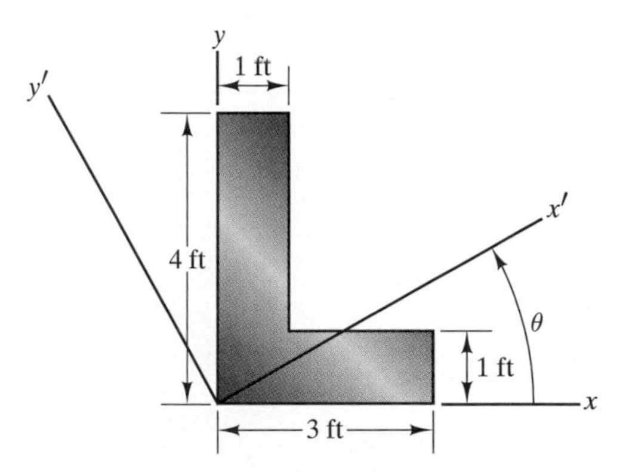

FIGURE C-18

Solution

(a) Setting $\theta = 30°$ in Eqs. (C-23)–(C-25) gives

$$I_{x'} = \frac{I_x + I_y}{2} + \frac{I_x - I_y}{2}\cos 2\theta - I_{xy}\sin 2\theta$$

$$= \frac{22 + 10}{2} + \frac{22 - 10}{2}\cos[(2)(30°)] - (6)\sin[(2)(30°)] = 13.8\ \text{ft}^4,$$

$$I_{y'} = \frac{I_x + I_y}{2} - \frac{I_x - I_y}{2}\cos 2\theta + I_{xy}\sin 2\theta$$

$$= \frac{22 + 10}{2} - \frac{22 - 10}{2}\cos[(2)(30°)] + (6)\sin[(2)(30°)] = 18.2\ \text{ft}^4,$$

$$I_{x'y'} = \frac{I_x - I_y}{2}\sin 2\theta + I_{xy}\cos 2\theta$$

$$= \frac{22 - 10}{2}\sin[(2)(30°)] + (6)\cos[(2)(30°)] = 8.2\ \text{ft}^4.$$

(b) Substituting the moments of inertia in terms of the xy coordinate system into Eq. (C-26),

$$\tan 2\theta_p = \frac{2I_{xy}}{I_y - I_x} = \frac{2(6)}{10 - 22} = -1,$$

gives the result $\theta_p = -22.5°$. The principal axes corresponding to this value of θ_p are shown in Fig. (a).

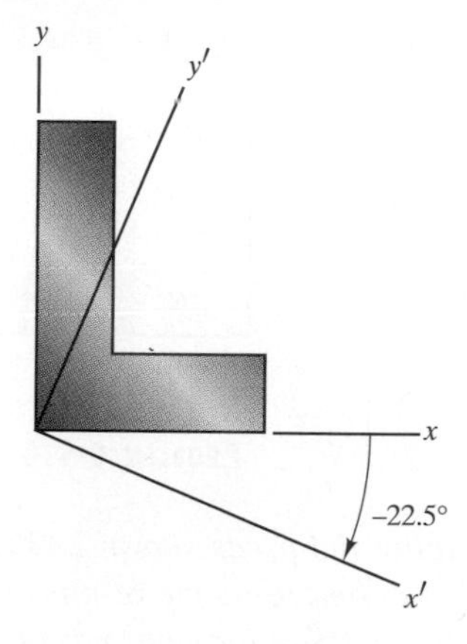

(a) Set of principal axes corresponding to $\theta_p = -22.5°$.

Substituting $\theta_p = -22.5°$ into Eqs. (C-23) and (C-24) gives the principal moments of inertia:

$$I_{x'} = 24.5 \text{ ft}^4, \qquad I_{y'} = 7.5 \text{ ft}^4.$$

PROBLEMS

For Problems C-1.1–C-1.8, use integration to determine the coordinates of the centroids of the areas shown.

C-1.1.

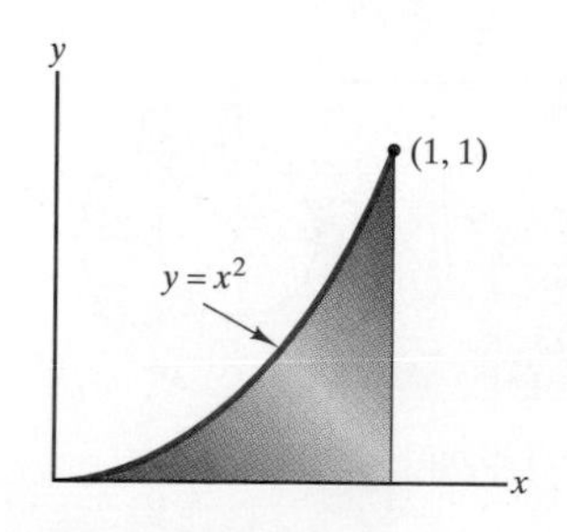

PROBLEM C-1.1

C-1.2.

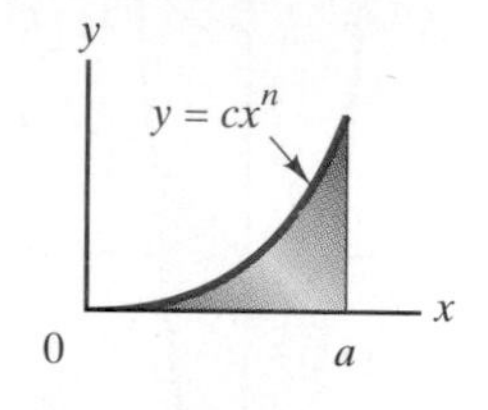

PROBLEM C-1.2

C-1.3.

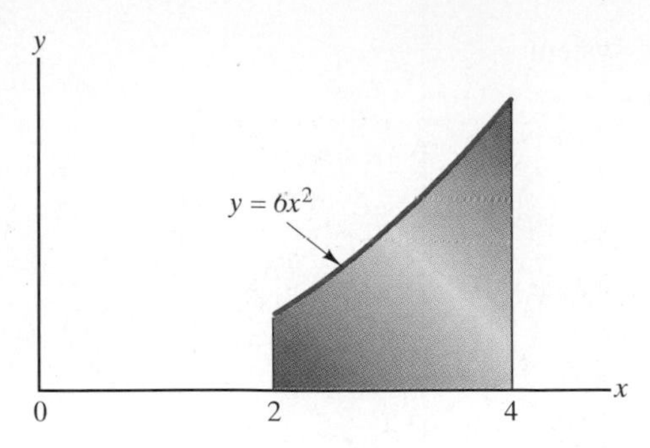

PROBLEM C-1.3

C-1.4.

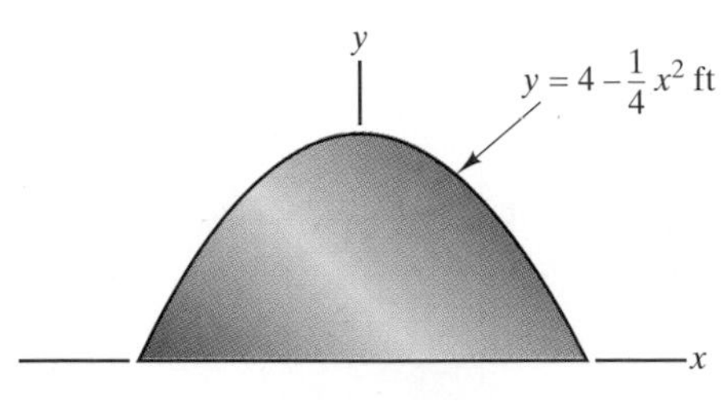

PROBLEM C-1.4

C-1.5.

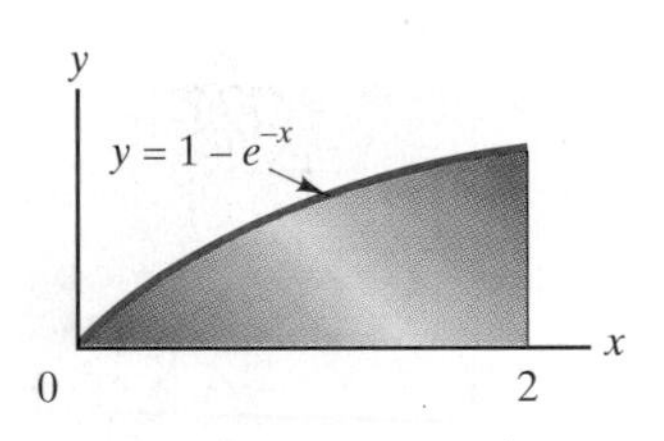

PROBLEM C-1.5

C-1.6.

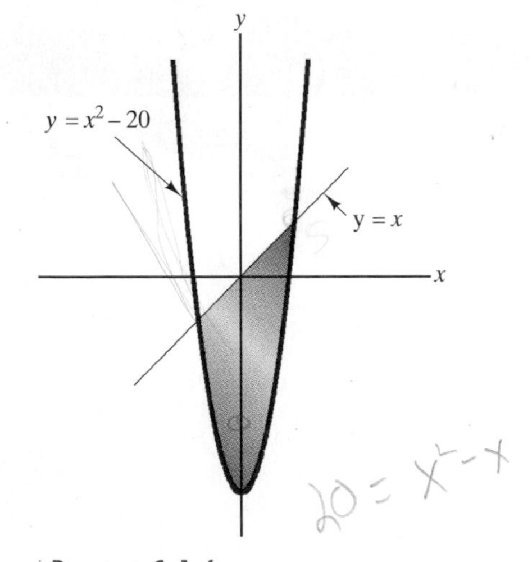

PROBLEM C-1.6

C-1.7.

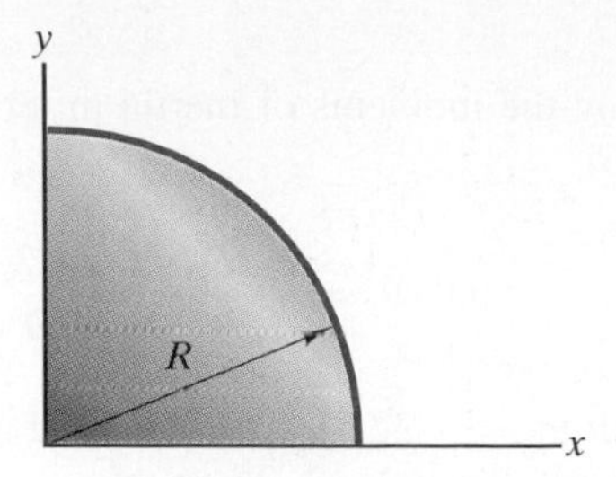

PROBLEM C-1.7

C-1.8.

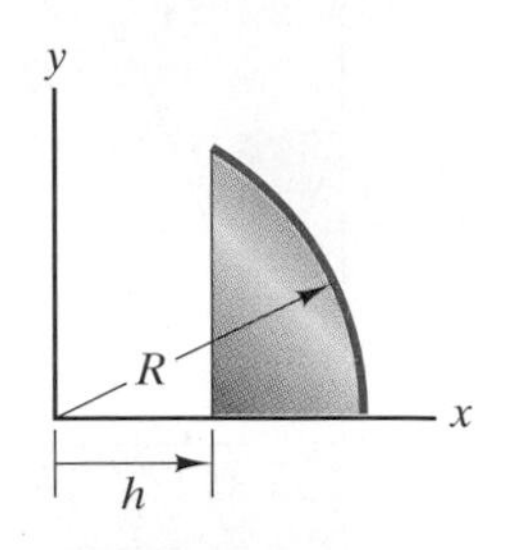

PROBLEM C-1.8

By treating the areas shown in Problems C-2.1–C-2.6 as composites and using the results in Appendix D, determine the coordinates of the centroids of the areas.

C-2.1.

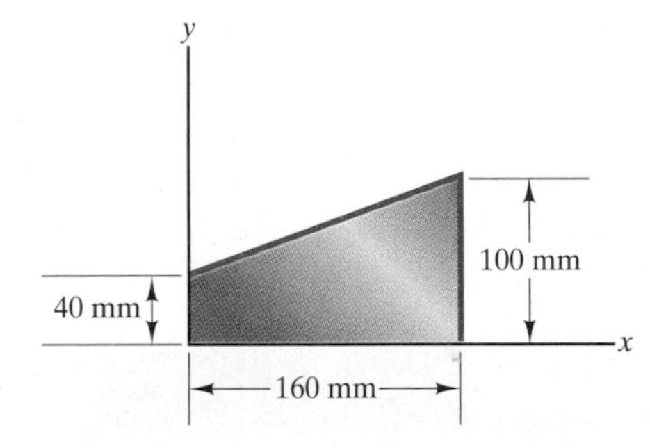

PROBLEM C-2.1

C-2.2.

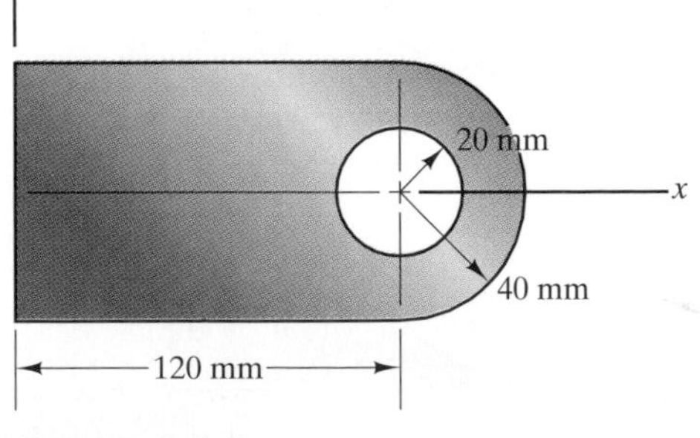

PROBLEM C-2.2

C-2.3.

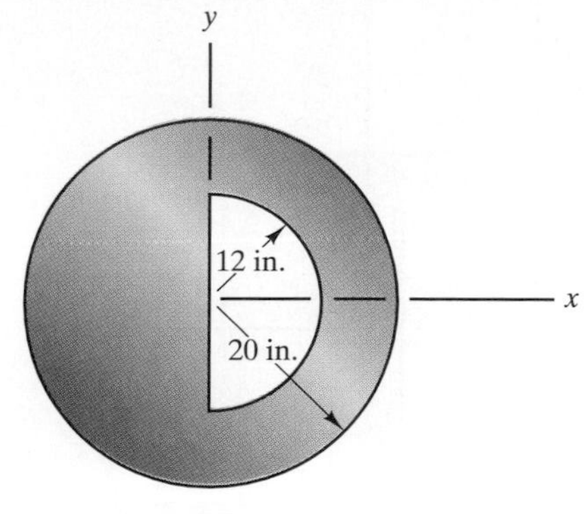

PROBLEM C-2.3

C-2.4.

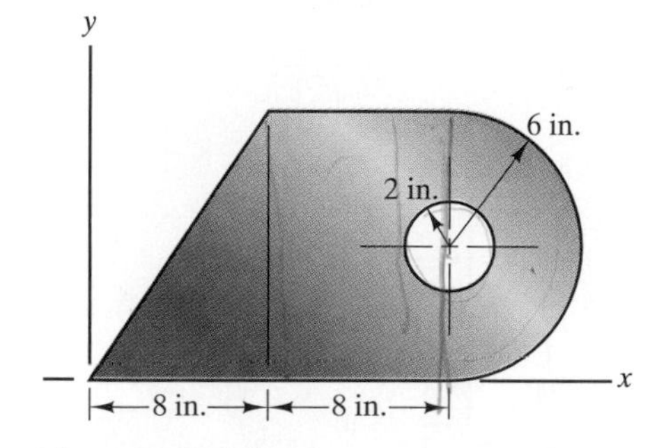

PROBLEM C-2.4

C-2.5.

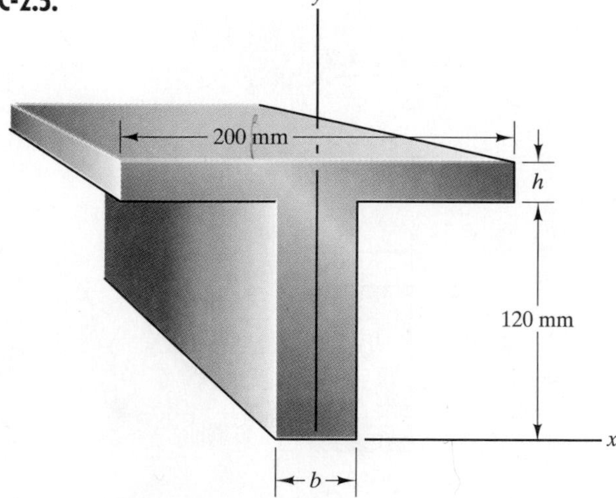

PROBLEM C-2.5

The dimensions $b = 40$ mm and $h = 20$ mm.

C-2.6.

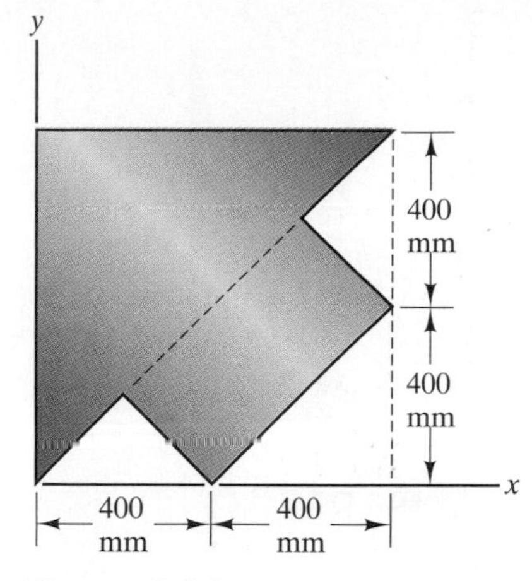

PROBLEM C-2.6

For Problems C-3.1–C-3.4, use integration to determine the moments of inertia I_x, I_y, and I_{xy} for the areas shown.

C-3.1.

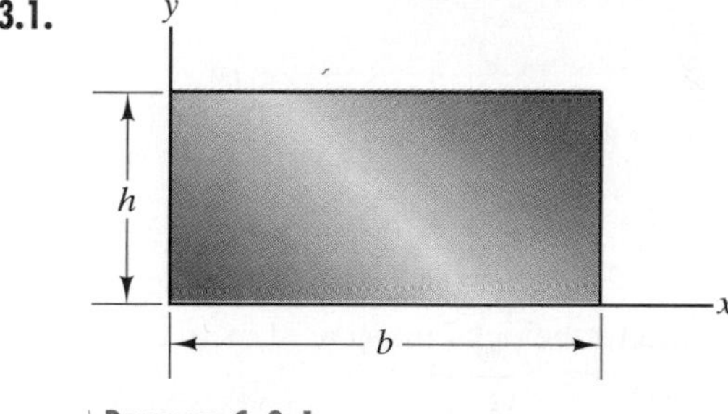

PROBLEM C-3.1

C-3.2.

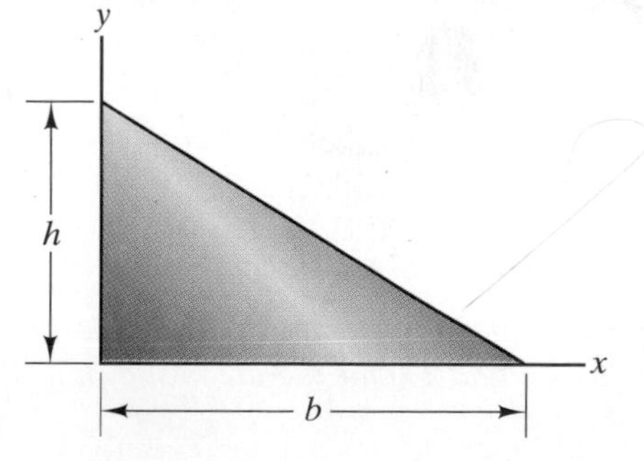

PROBLEM C-3.2

C-3.3.

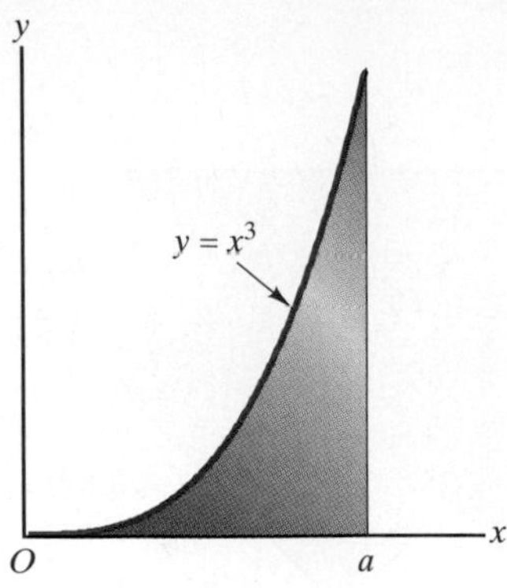

| PROBLEM C-3.3

C-3.4.

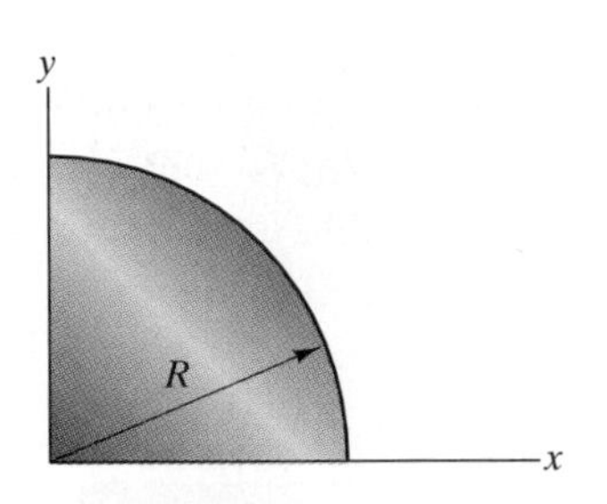

| PROBLEM C-3.4

C-3.5. Determine the polar moment of inertia J_O.

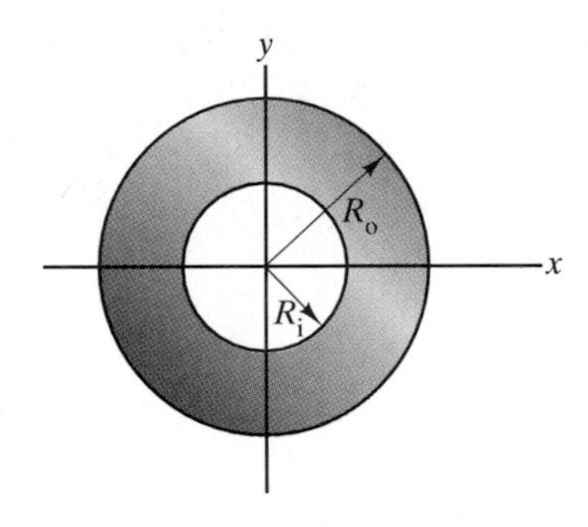

| PROBLEM C-3.5

By using the results in Appendix D, determine the moments of inertia I_x, I_y, and I_{xy} for the areas shown in Problems C-4.1–C-4.8.

C-4.1.

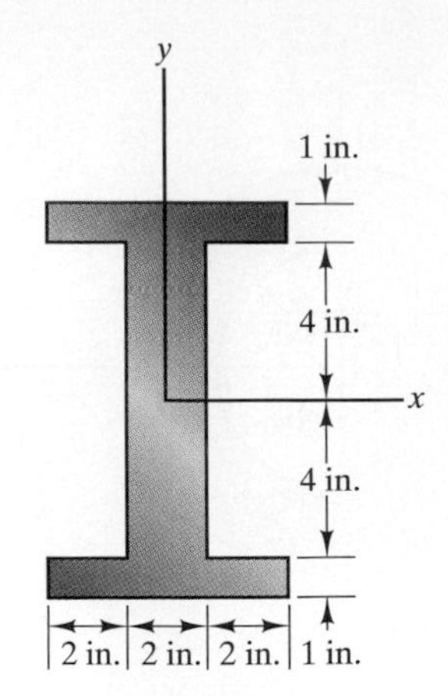

| PROBLEM C-4.1

C-4.2.

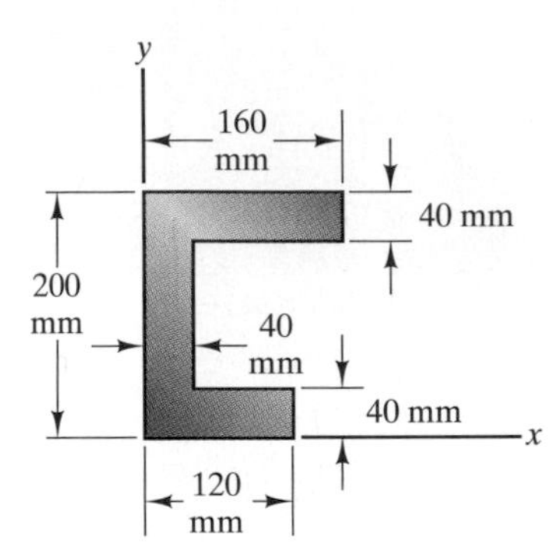

| PROBLEM C-4.2

C-4.3.

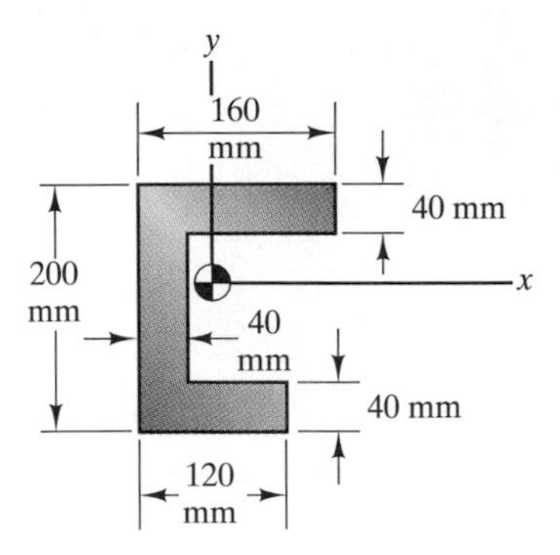

| PROBLEM C-4.3

C-4.4.

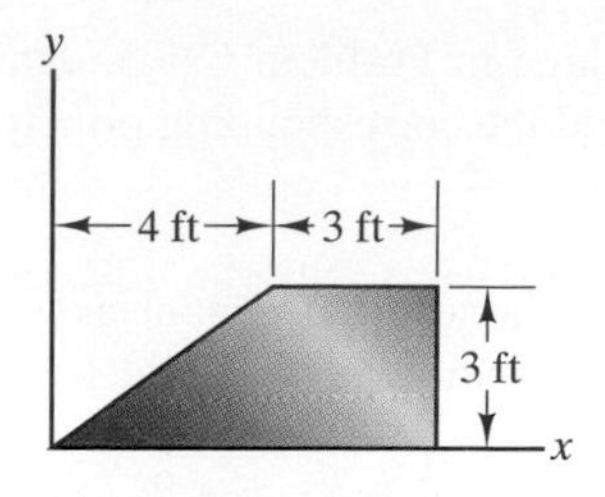

PROBLEM C-4.4

C-4.5.

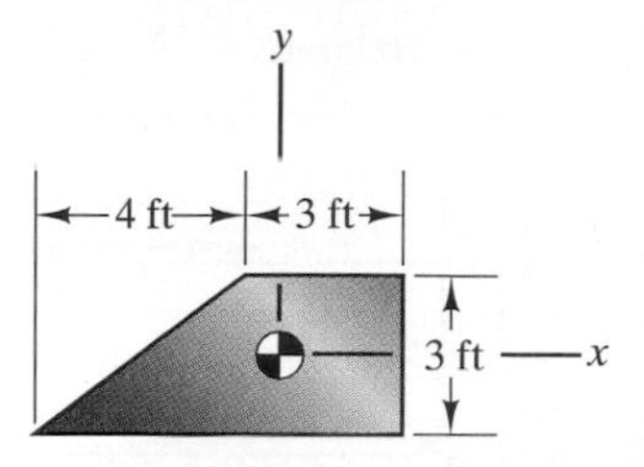

PROBLEM C-4.5

C-4.6.

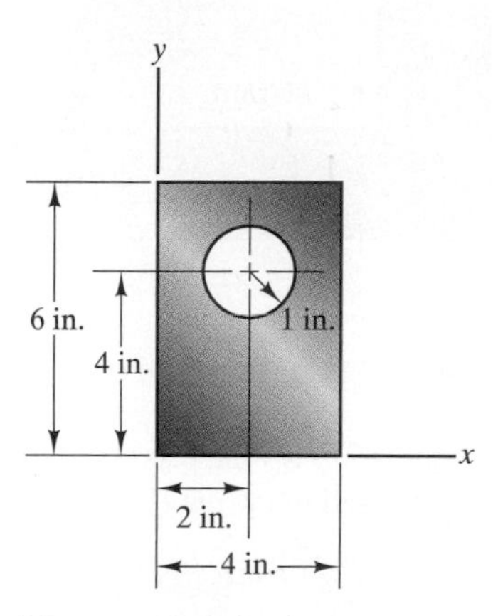

PROBLEM C-4.6

C-4.7.

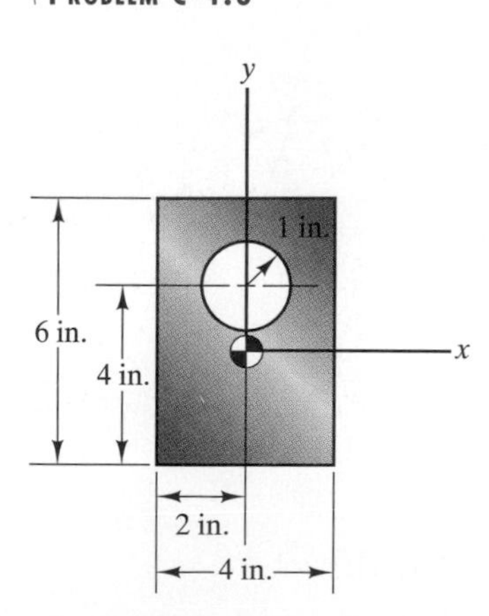

PROBLEM C-4.7

C-4.8.

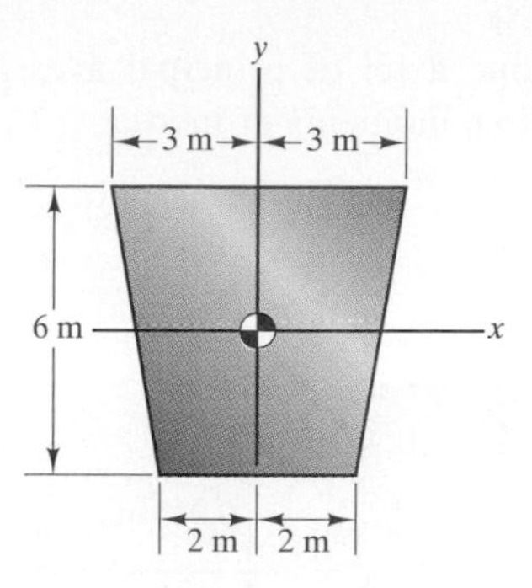

PROBLEM C-4.8

C-5.1. Determine $I_{x'}$, $I_{y'}$, and $I_{x'y'}$.

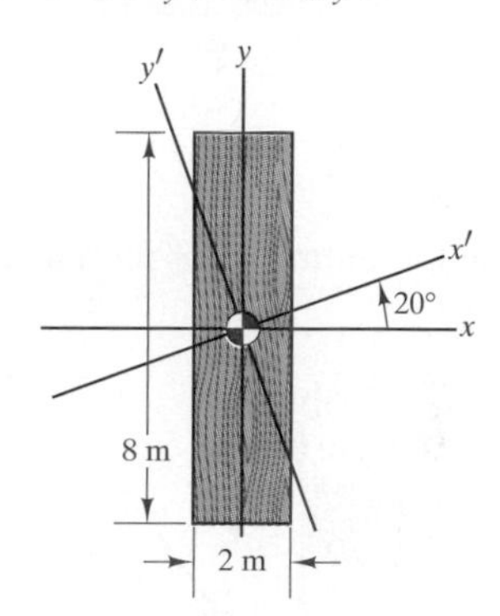

PROBLEM C-5.1

C-5.2. For the area in Problem C-5.1, determine a set of principal axes and the corresponding principal moments of inertia.

C-5.3. The moments of inertia of the rectangular area in terms of the xy coordinate system shown are $I_x = 76.0 \text{ m}^4$, $I_y = 14.7 \text{ m}^4$, and $I_{xy} = 25.7 \text{ m}^4$. Determine a set of principal axes and the corresponding principal moments of inertia.

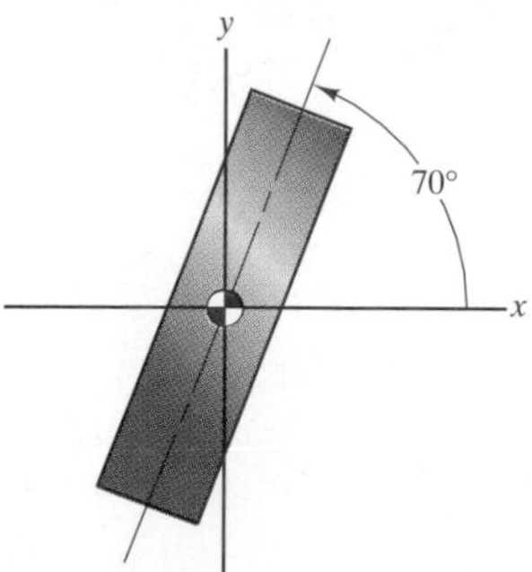

PROBLEM C-5.3

C-5.4. Determine a set of principal axes and the corresponding principal moments of inertia.

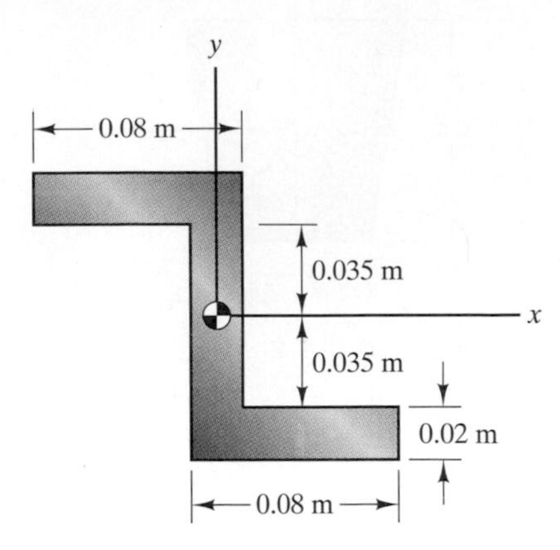

PROBLEM C-5.4

C-5.5. Determine the moments of inertia $I_{x'}$, $I_{y'}$, and $I_{x'y'}$ if $\theta = 15°$.

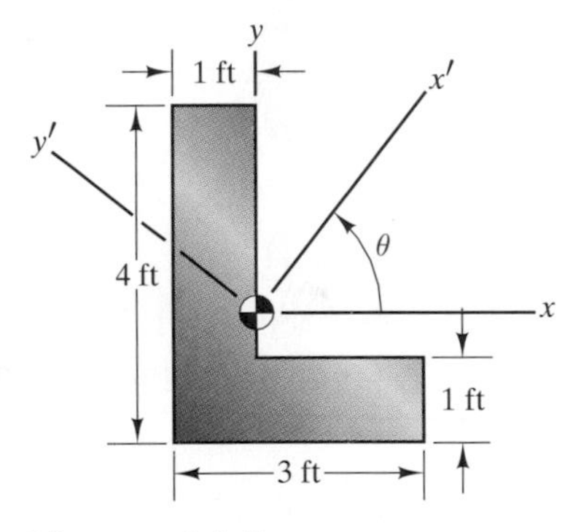

PROBLEM C-5.5

C-5.6. For the area in Problem C-5.5, determine a set of principal axes and the corresponding principal moments of inertia.

C-5.7. Determine a set of principal axes and the corresponding principal moments of inertia.

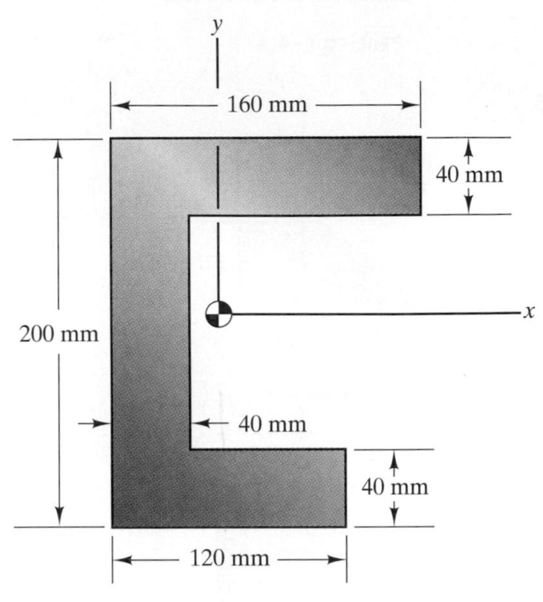

PROBLEM C-5.7

C-5.8. Derive Eq. (C-22) for the product of inertia by using the same procedure used to derive Eqs. (C-20) and (C-21).

APPENDIX D

Properties of Areas

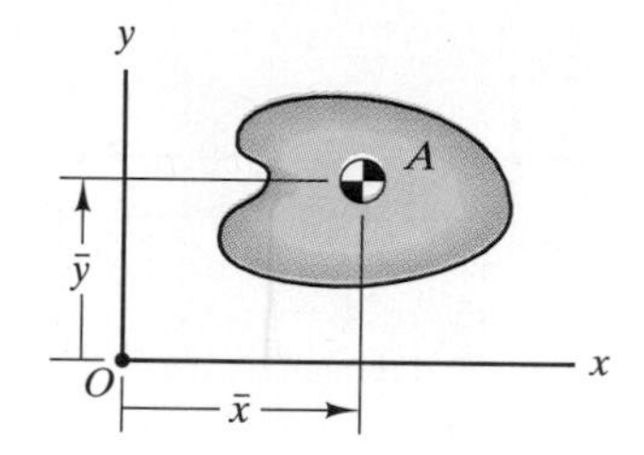

The coordinates of the centroid of the area A are

$$\bar{x} = \frac{\int_A x\,dA}{\int_A dA}, \qquad \bar{y} = \frac{\int_A y\,dA}{\int_A dA}.$$

The moment of inertia about the x axis I_x, the moment of inertia about the y axis I_y, and the product of inertia I_{xy} are

$$I_x = \int_A y^2\,dA, \qquad I_y = \int_A x^2\,dA, \qquad I_{xy} = \int_A xy\,dA.$$

The polar moment of inertia about O is

$$J_O = \int_A r^2\,dA = \int_A (x^2 + y^2)\,dA = I_x + I_y.$$

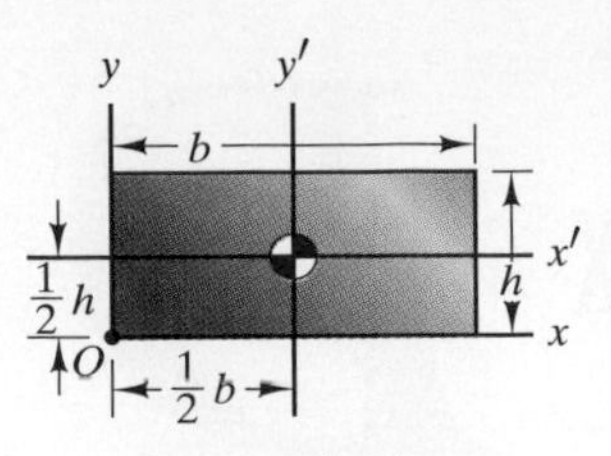

Rectangular Area

Area $= bh$.

$$I_x = \tfrac{1}{3}bh^3, \qquad I_y = \tfrac{1}{3}hb^3, \qquad I_{xy} = \tfrac{1}{4}b^2h^2.$$

$$I_{x'} = \tfrac{1}{12}bh^3, \qquad I_{y'} = \tfrac{1}{12}hb^3, \qquad I_{x'y'} = 0.$$

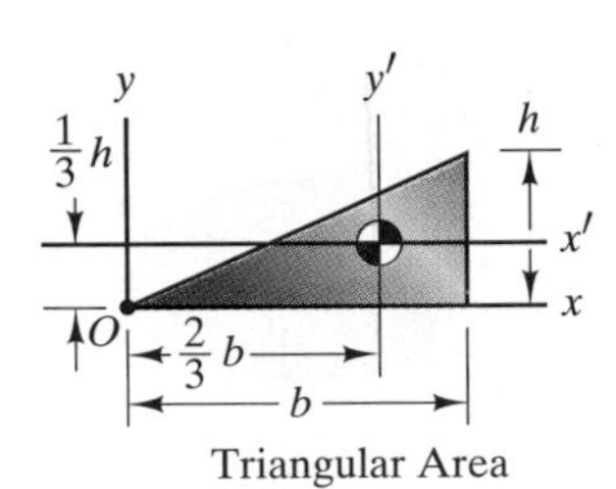

Triangular Area

Area $= \tfrac{1}{2}bh$.

$$I_x = \tfrac{1}{12}bh^3, \qquad I_y = \tfrac{1}{4}hb^3, \qquad I_{xy} = \tfrac{1}{8}b^2h^2.$$

$$I_{x'} = \tfrac{1}{36}bh^3, \qquad I_{y'} = \tfrac{1}{36}hb^3, \qquad I_{x'y'} = \tfrac{1}{72}b^2h^2.$$

Triangular Area

Area $= \tfrac{1}{2}bh$.

$$I_x = \tfrac{1}{12}bh^3, \qquad I_{x'} = \tfrac{1}{36}bh^3.$$

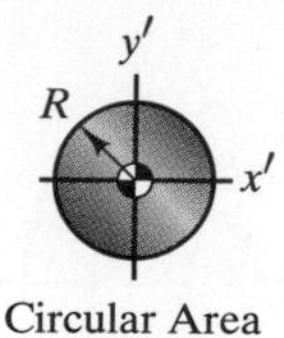

Circular Area

$$\text{Area} = \pi R^2.$$

$$I_{x'} = I_{y'} = \tfrac{1}{4}\pi R^4, \qquad I_{x'y'} = 0.$$

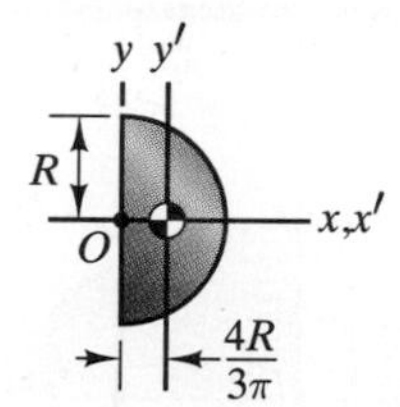

Semicircular Area

$$\text{Area} = \tfrac{1}{2}\pi R^2.$$

$$I_x = I_y = \tfrac{1}{8}\pi R^4, \qquad I_{xy} = 0.$$

$$I_{x'} = \tfrac{1}{8}\pi R^4, \qquad I_{y'} = \left(\frac{\pi}{8} - \frac{8}{9\pi}\right) R^4, \qquad I_{x'y'} = 0.$$

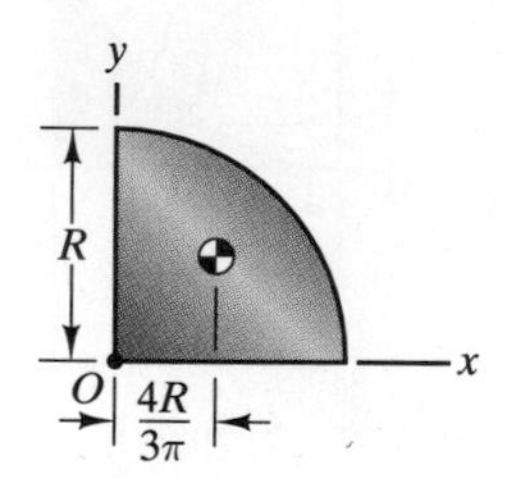

Quarter-Circular Area

$$\text{Area} = \tfrac{1}{4}\pi R^2.$$

$$I_x = I_y = \tfrac{1}{16}\pi R^4, \qquad I_{xy} = \tfrac{1}{8} R^4.$$

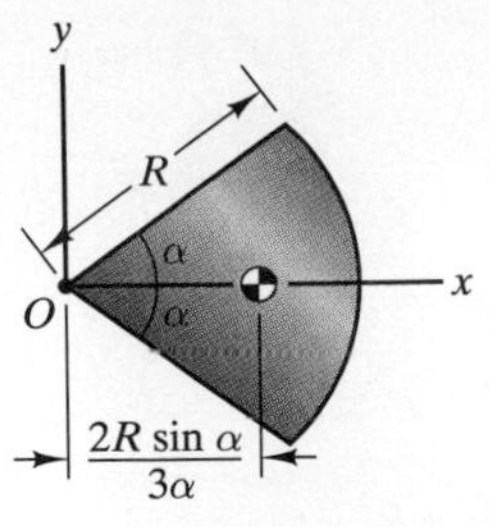

Circular Sector

Area $= \alpha R^2$.

$$I_x = \tfrac{1}{4}R^4\left(\alpha - \tfrac{1}{2}\sin 2\alpha\right), \qquad I_y = \tfrac{1}{4}R^4\left(\alpha + \tfrac{1}{2}\sin 2\alpha\right), \qquad I_{xy} = 0.$$

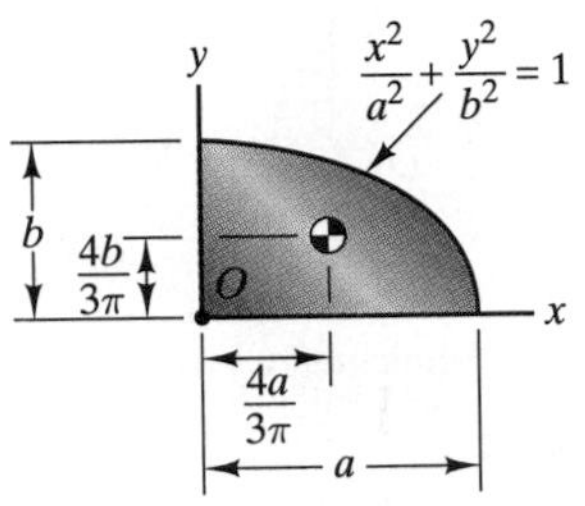

Quarter-Elliptical Area

Area $= \tfrac{1}{4}\pi ab$.

$$I_x = \tfrac{1}{16}\pi ab^3, \qquad I_y = \tfrac{1}{16}\pi a^3 b, \qquad I_{xy} = \tfrac{1}{8}a^2b^2.$$

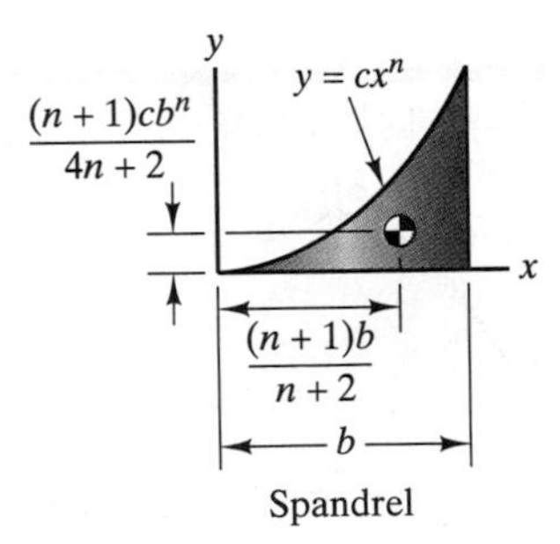

Spandrel

Area $= \dfrac{cb^{n+1}}{n+1}$.

$$I_x = \frac{c^3 b^{3n+1}}{9n+3}, \qquad I_y = \frac{cb^{n+3}}{n+3}, \qquad I_{xy} = \frac{c^2 b^{2n+2}}{4n+4}.$$

APPENDIX E

Deflections and Slopes of Prismatic Beams

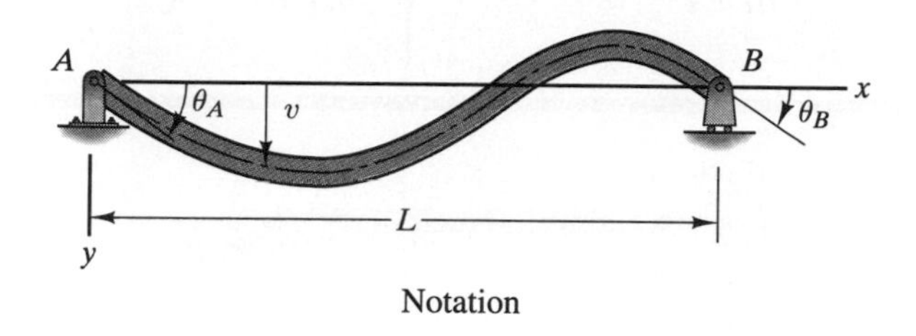

Notation

Simply Supported Beams

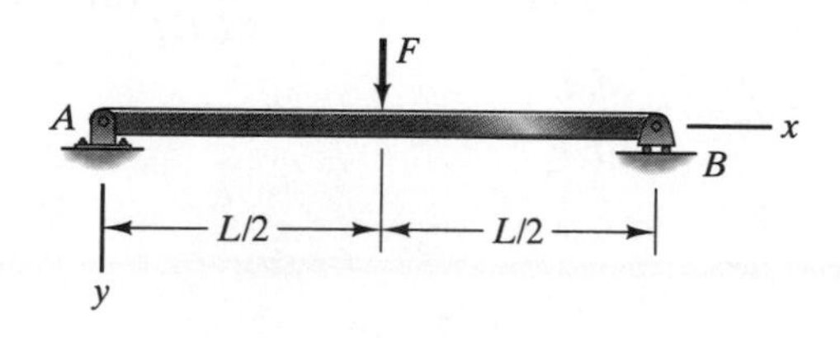

$$v = \frac{Fx}{48EI}(3L^2 - 4x^2), \qquad 0 \le x \le L/2$$

$$v' = \theta = \frac{F}{16EI}(L^2 - 4x^2), \qquad 0 \le x \le L/2$$

$$\theta_A = -\theta_B = \frac{FL^2}{16EI}.$$

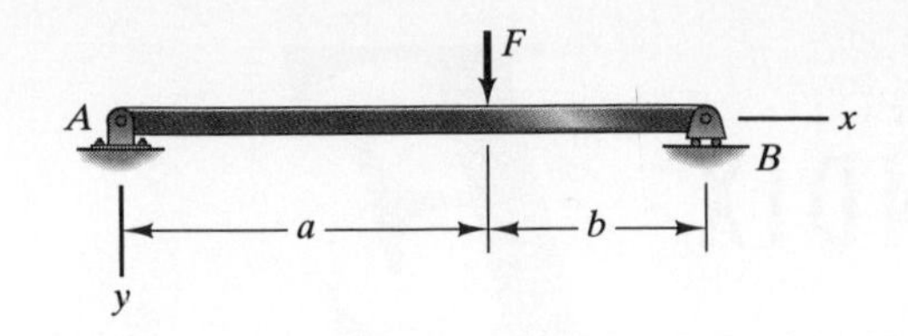

$$v = \frac{Fbx}{6LEI}(L^2 - b^2 - x^2), \qquad 0 \le x \le a$$

$$v' = \theta = \frac{Fb}{6LEI}(L^2 - b^2 - 3x^2), \qquad 0 \le x \le a$$

$$v = \frac{Fa}{6LEI}(L - x)[L^2 - a^2 - (L - x)^2], \qquad a \le x \le L$$

$$v' = \theta = -\frac{Fa}{6LEI}[L^2 - a^2 - 3(L - x)^2], \qquad a \le x \le L$$

$$\theta_A = \frac{Fab(L + b)}{6LEI}, \qquad \theta_B = -\frac{Fab(L + a)}{6LEI}.$$

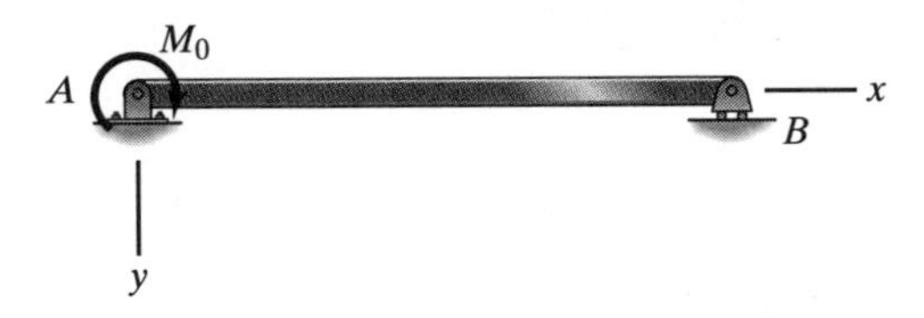

$$v = \frac{M_0 x}{6LEI}(2L^2 - 3Lx + x^2), \qquad v' = \theta = \frac{M_0}{6LEI}(2L^2 - 6Lx + 3x^2),$$

$$\theta_A = \frac{M_0 L}{3EI}, \qquad \theta_B = -\frac{M_0 L}{6EI}.$$

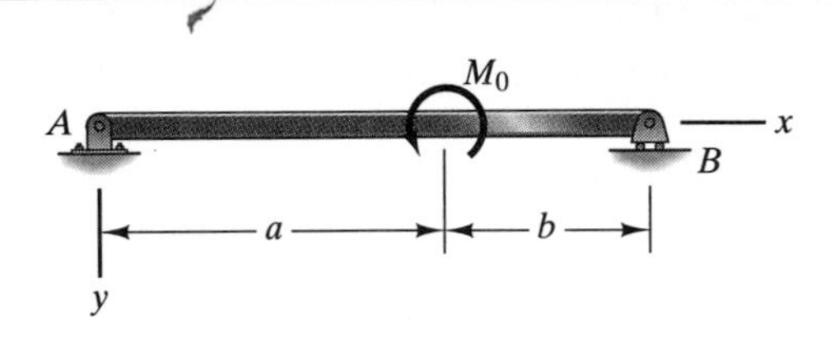

$$v = \frac{M_0 x}{6LEI}(6aL - 3a^2 - 2L^2 - x^2), \qquad 0 \le x \le a$$

$$v' = \theta = \frac{M_0}{6LEI}(6aL - 3a^2 - 2L^2 - 3x^2), \qquad 0 \le x \le a$$

$$v = \frac{M_0}{6LEI}(3a^2L - 3a^2x - 2L^2x + 3Lx^2 - x^3), \qquad a \le x \le L$$

$$v' = \theta = -\frac{M_0}{6LEI}(3a^2 + 2L^2 - 6Lx + 3x^2), \qquad a \le x \le L$$

$$\theta_A = \frac{M_0}{6LEI}(6aL - 3a^2 - 2L^2), \qquad \theta_B = -\frac{M_0}{6LEI}(3a^2 - L^2).$$

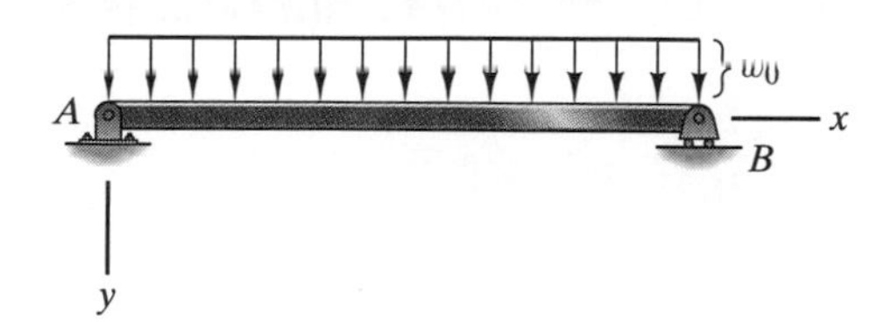

$$v = \frac{w_0 x}{24EI}(L^3 - 2Lx^2 + x^3), \qquad v' = \theta = \frac{w_0}{24EI}(L^3 - 6Lx^2 + 4x^3),$$

$$\theta_A = -\theta_B = \frac{w_0 L^3}{24EI}.$$

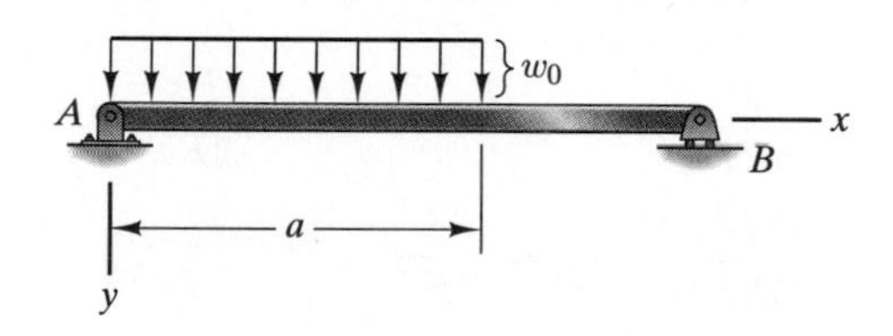

$$v = \frac{w_0 x}{24LEI}(a^4 - 4a^3L + 4a^2L^2 + 2a^2x^2 - 4aLx^2 + Lx^3), \qquad 0 \le x \le a$$

$$v' = \frac{w_0}{24LEI}(a^4 - 4a^3L + 4a^2L^2 + 6a^2x^2 - 12aLx^2 + 4Lx^3), \quad 0 \le x \le a$$

$$v = \frac{w_0 a^2}{24LEI}(-a^2L + 4L^2x + a^2x - 6Lx^2 + 2x^3), \qquad a \le x \le L$$

$$v' = \theta = \frac{w_0 a^2}{24LEI}(4L^2 + a^2 - 12Lx + 6x^2), \qquad a \le x \le L$$

$$\theta_A = \frac{w_0 a^2}{24LEI}(2L - a)^2, \qquad \theta_B = -\frac{w_0 a^2}{24LEI}(2L^2 - a^2).$$

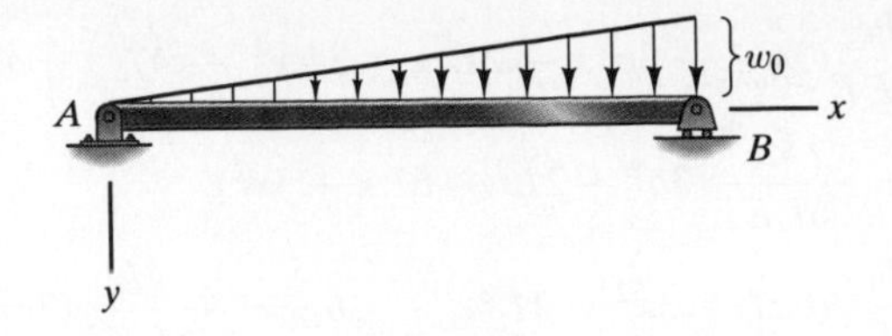

$$v = \frac{w_0 x}{360LEI}(7L^4 - 10L^2x^2 + 3x^4),$$

$$v' = \theta = \frac{w_0}{360LEI}(7L^4 - 30L^2x^2 + 15x^4),$$

$$\theta_A = \frac{7w_0L^3}{360EI}, \qquad \theta_B = -\frac{w_0L^3}{45EI}.$$

Cantilever Beams

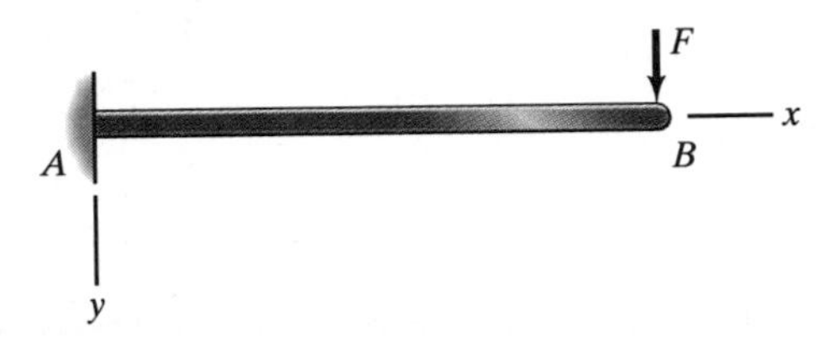

$$v = \frac{Fx^2}{6EI}(3L - x), \qquad v' = \theta = \frac{Fx}{2EI}(2L - x),$$

$$v_B = \frac{FL^3}{3EI}, \qquad \theta_B = \frac{FL^2}{2EI}.$$

$$v = \frac{Fx^2}{6EI}(3a - x), \quad v' = \theta = \frac{Fx}{2EI}(2a - x), \qquad 0 \le x \le a$$

$$v = \frac{Fa^2}{6EI}(3x - a), \quad v' = \theta = \frac{Fa^2}{2EI}, \qquad a \le x \le L$$

$$v_B = \frac{Fa^2}{6EI}(3L - a), \quad \theta_B = \frac{Fa^2}{2EI}.$$

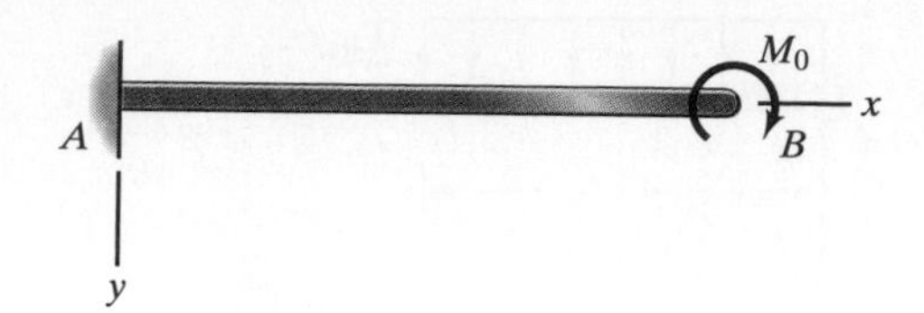

$$v = \frac{M_0 x^2}{2EI}, \qquad v' = \theta = \frac{M_0 x}{EI},$$
$$v_B = \frac{M_0 L^2}{2EI}, \qquad \theta_B = \frac{M_0 L}{EI}$$

$$v = \frac{M_0 x^2}{2EI}, \qquad v' = \theta = \frac{M_0 x}{EI}, \qquad 0 \le x \le a$$
$$v = \frac{M_0 a}{2EI}(2x - a), \qquad v' = \theta = \frac{M_0 a}{EI}, \qquad a \le x \le L$$
$$v_B = \frac{M_0 a}{2EI}(2L - a), \qquad \theta_B = \frac{M_0 a}{EI}.$$

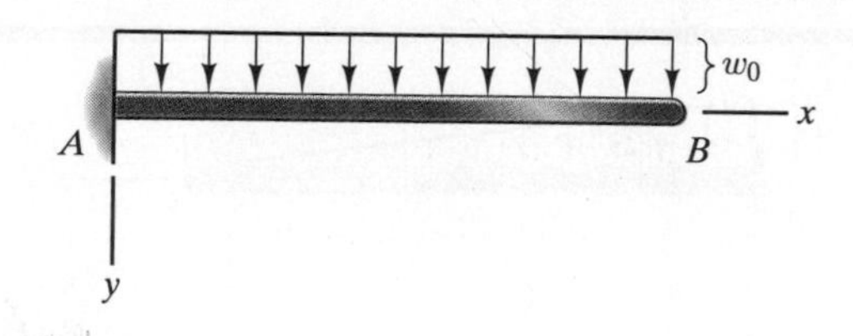

$$v = \frac{w_0 x^2}{24EI}(6L^2 - 4Lx + x^2), \qquad v' = \theta = \frac{w_0 x}{6EI}(3L^2 - 3Lx + x^2),$$
$$v_B = \frac{w_0 L^4}{8EI}, \qquad \theta_B = \frac{w_0 L^3}{6EI}.$$

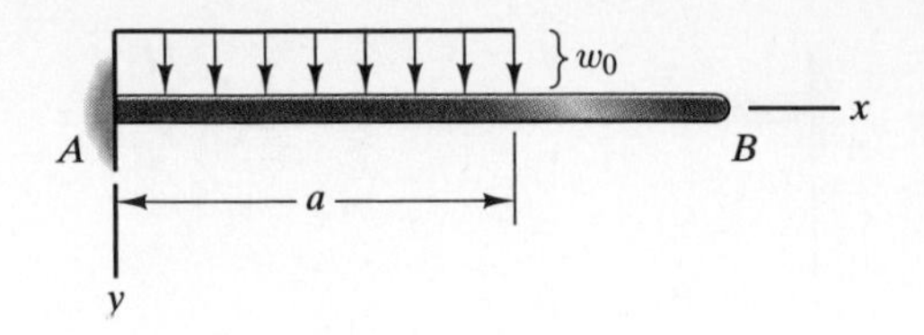

$$v = \frac{w_0 x^2}{24EI}(6a^2 - 4ax + x^2), \qquad 0 \le x \le a$$

$$v' = \theta = \frac{w_0 x}{6EI}(3a^2 - 3ax + x^2), \qquad 0 \le x \le a$$

$$v = \frac{w_0 a^3}{24EI}(4x - a), \qquad v' = \theta = \frac{w_0 a^3}{6EI}, \qquad a \le x \le L$$

$$v_B = \frac{w_0 a^3}{24EI}(4L - a), \qquad \theta_B = \frac{w_0 a^3}{6EI}.$$

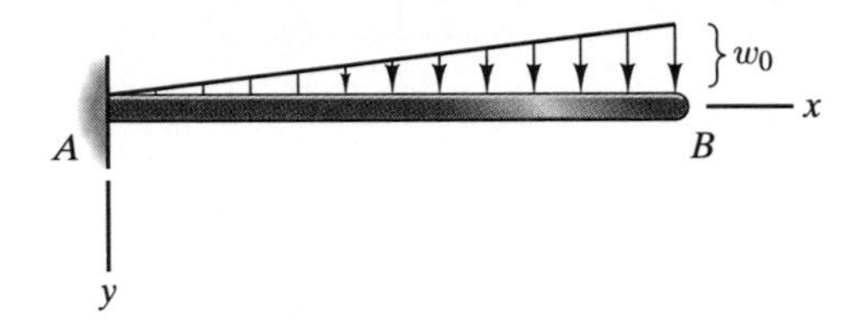

$$v = \frac{w_0 x^2}{120LEI}(20L^3 - 10L^2 x + x^3),$$

$$v' = \theta = \frac{w_0 x}{24LEI}(8L^3 - 6L^2 x + x^3),$$

$$v_B = \frac{11 w_0 L^4}{120EI}, \qquad \theta_B = \frac{w_0 L^3}{8EI}.$$

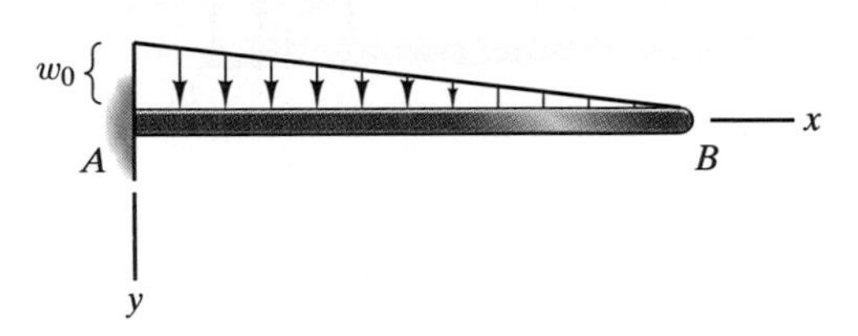

$$v = \frac{w_0 x^2}{120LEI}(10L^3 - 10L^2 x + 5Lx^2 - x^3),$$

$$v' = \theta = \frac{w_0 x}{24LEI}(4L^3 - 6L^2 x + 4Lx^2 - x^3),$$

$$v_B = \frac{w_0 L^4}{30EI}, \qquad \theta_B = \frac{w_0 L^3}{24EI}.$$

APPENDIX F

Isotropic Stress–Strain Relations

This appendix explains why the stress-strain relations for an *isotropic* linearly elastic material must have the forms given by Eqs. (6-28)–(6-33):

$$
\begin{aligned}
\epsilon_x &= b_{11}\sigma_x + b_{12}\sigma_y + b_{12}\sigma_z,\\
\epsilon_y &= b_{12}\sigma_x + b_{11}\sigma_y + b_{12}\sigma_z,\\
\epsilon_z &= b_{12}\sigma_x + b_{12}\sigma_y + b_{11}\sigma_z,\\
\gamma_{xy} &= b_{44}\tau_{xy},\\
\gamma_{yz} &= b_{44}\tau_{yz},\\
\gamma_{xz} &= b_{44}\tau_{xz}.
\end{aligned}
\qquad \textbf{(F-1)}
$$

The stress-strain relations for an arbitrary linearly elastic material are given by Eqs. (6-25):

$$
\begin{aligned}
\epsilon_x &= b_{11}\sigma_x + b_{12}\sigma_y + b_{13}\sigma_z + b_{14}\tau_{xy} + b_{15}\tau_{yz} + b_{16}\tau_{xz},\\
\epsilon_y &= b_{21}\sigma_x + b_{22}\sigma_y + b_{23}\sigma_z + b_{24}\tau_{xy} + b_{25}\tau_{yz} + b_{26}\tau_{xz},\\
\epsilon_z &= b_{31}\sigma_x + b_{32}\sigma_y + b_{33}\sigma_z + b_{34}\tau_{xy} + b_{35}\tau_{yz} + b_{36}\tau_{xz},\\
\gamma_{xy} &= b_{41}\sigma_x + b_{42}\sigma_y + b_{43}\sigma_z + b_{44}\tau_{xy} + b_{45}\tau_{yz} + b_{46}\tau_{xz},\\
\gamma_{yz} &= b_{51}\sigma_x + b_{52}\sigma_y + b_{53}\sigma_z + b_{54}\tau_{xy} + b_{55}\tau_{yz} + b_{56}\tau_{xz},\\
\gamma_{xz} &= b_{61}\sigma_x + b_{62}\sigma_y + b_{63}\sigma_z + b_{64}\tau_{xy} + b_{65}\tau_{yz} + b_{66}\tau_{xz}.
\end{aligned}
\qquad \textbf{(F-2)}
$$

The coefficients in Eqs. (F-2) are

$$
\begin{matrix}
b_{11} & b_{12} & b_{13} & b_{14} & b_{15} & b_{16}\\
b_{21} & b_{22} & b_{23} & b_{24} & b_{25} & b_{26}\\
b_{31} & b_{32} & b_{33} & b_{34} & b_{35} & b_{36}\\
b_{41} & b_{42} & b_{43} & b_{44} & b_{45} & b_{46}\\
b_{51} & b_{52} & b_{53} & b_{54} & b_{55} & b_{56}\\
b_{61} & b_{62} & b_{63} & b_{64} & b_{65} & b_{66}
\end{matrix}
\qquad \textbf{(F-3)}
$$

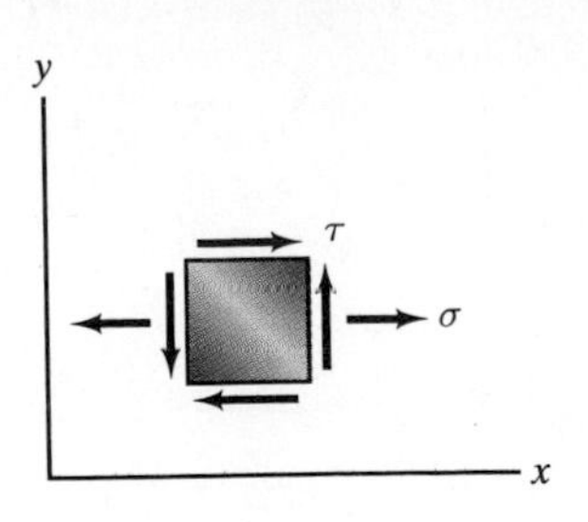

FIGURE F-1 Applying normal and shear stresses to an isotropic material.

Subjecting an isotropic material to a normal stress σ and shear stress τ as shown in Fig. F-1, Eqs. (F-2) are

$$\begin{aligned} \epsilon_x &= b_{11}\sigma + b_{14}\tau, & \gamma_{xy} &= b_{41}\sigma + b_{44}\tau, \\ \epsilon_y &= b_{21}\sigma + b_{24}\tau, & \gamma_{yz} &= b_{51}\sigma + b_{54}\tau, \\ \epsilon_z &= b_{31}\sigma + b_{34}\tau, & \gamma_{xz} &= b_{61}\sigma + b_{64}\tau. \end{aligned} \qquad \text{(F-4)}$$

Rotations about the z Axis

The coordinate system in Fig. F-2 is obtained by rotating the coordinate system in Fig. F-1 90° about the z axis. In terms of this coordinate system,

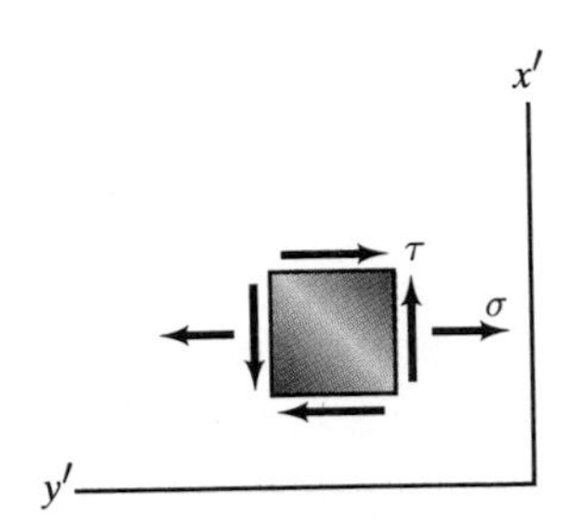

FIGURE F-2 Rotating the xyz coordinate system 90° about the z axis.

$$\begin{aligned} \epsilon'_x &= b_{12}\sigma - b_{14}\tau, & \gamma'_{xy} &= b_{42}\sigma - b_{44}\tau, \\ \epsilon'_y &= b_{22}\sigma - b_{24}\tau, & \gamma'_{yz} &= b_{52}\sigma - b_{54}\tau, \\ \epsilon'_z &= b_{32}\sigma - b_{34}\tau, & \gamma'_{xz} &= b_{62}\sigma - b_{64}\tau. \end{aligned} \qquad \text{(F-5)}$$

The complete strain correspondences between the coordinate systems in Figs. F-1 and F-2 and the resulting conclusions obtained by comparing Eqs. (F-4) and (F-5) are

$$\begin{aligned} \epsilon_x &= \epsilon'_y & &\Rightarrow & b_{11} &= b_{22}, & b_{14} &= -b_{24}, \\ \epsilon_y &= \epsilon'_x & &\Rightarrow & b_{21} &= b_{12}, & b_{24} &= -b_{14}, \\ \epsilon_z &= \epsilon'_z & &\Rightarrow & b_{31} &= b_{32}, & b_{34} &= -b_{34}, \\ \gamma_{xy} &= -\gamma'_{xy} & &\Rightarrow & b_{41} &= -b_{42}, & b_{44} &= b_{44}, \\ \gamma_{yz} &= \gamma'_{xz} & &\Rightarrow & b_{51} &= b_{62}, & b_{54} &= -b_{64}, \\ \gamma_{xz} &= -\gamma'_{yz} & &\Rightarrow & b_{61} &= -b_{52}, & b_{64} &= b_{54}. \end{aligned}$$

The coordinate system in Fig. F-3 is obtained by rotating the coordinate system in Fig. F-1 180° about the z axis. In terms of this coordinate system,

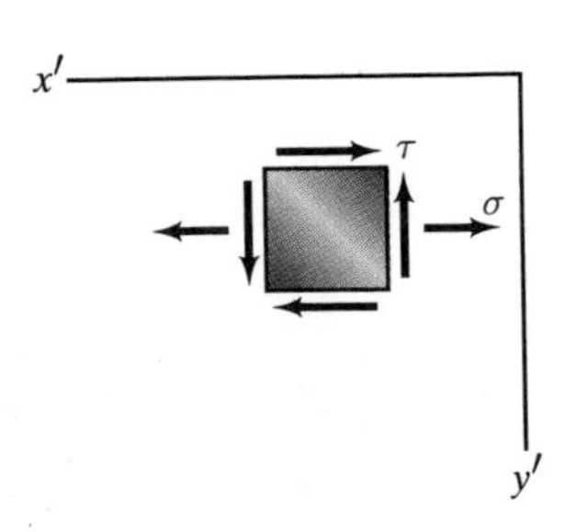

FIGURE F-3 Rotating the xyz coordinate system 180° about the z axis.

$$\begin{aligned} \epsilon'_x &= b_{11}\sigma + b_{14}\tau, & \gamma'_{xy} &= b_{41}\sigma + b_{44}\tau, \\ \epsilon'_y &= b_{21}\sigma + b_{24}\tau, & \gamma'_{yz} &= b_{51}\sigma + b_{54}\tau, \\ \epsilon'_z &= b_{31}\sigma + b_{34}\tau, & \gamma'_{xz} &= b_{61}\sigma + b_{64}\tau. \end{aligned} \qquad \text{(F-6)}$$

The complete strain correspondences between the coordinate systems in

Figs. F-1 and F-3 and the resulting conclusions are

$$\begin{aligned}
\epsilon_x &= \epsilon'_x && \Rightarrow && b_{11} = b_{11}, && b_{14} = b_{14},\\
\epsilon_y &= \epsilon'_y && \Rightarrow && b_{21} = b_{21}, && b_{24} = b_{24},\\
\epsilon_z &= \epsilon'_z && \Rightarrow && b_{31} = b_{31}, && b_{34} = b_{34},\\
\gamma_{xy} &= \gamma'_{xy} && \Rightarrow && b_{41} = b_{41}, && b_{44} = b_{44},\\
\gamma_{yz} &= -\gamma'_{yz} && \Rightarrow && b_{51} = -b_{51}, && b_{54} = -b_{54},\\
\gamma_{xz} &= -\gamma'_{xz} && \Rightarrow && b_{61} = -b_{61}, && b_{64} = -b_{64}.
\end{aligned}$$

Applying all the restrictions obtained so far, the coefficients (F-3) are

$$\begin{matrix}
b_{11} & b_{12} & b_{13} & b_{14} & b_{15} & b_{16}\\
b_{12} & b_{11} & b_{23} & -b_{14} & b_{25} & b_{26}\\
b_{31} & b_{31} & b_{33} & 0 & b_{35} & b_{36}\\
b_{41} & -b_{41} & b_{43} & b_{44} & b_{45} & b_{46}\\
0 & 0 & b_{53} & 0 & b_{55} & b_{56}\\
0 & 0 & b_{63} & 0 & b_{65} & b_{66}
\end{matrix}$$

Rotations about the y Axis

The coordinate system in Fig. F-4 is obtained by rotating the coordinate system in Fig. F-1 90° about the y axis. In terms of this coordinate system,

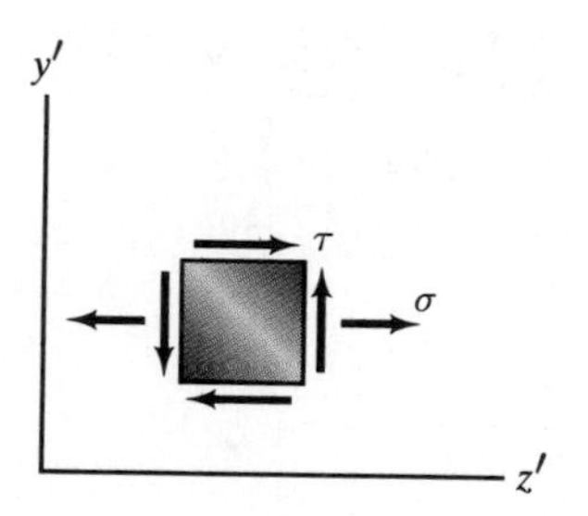

FIGURE F-4 Rotating the xyz coordinate system 90° about the y axis.

$$\begin{aligned}
\epsilon'_x &= b_{13}\sigma + b_{15}\tau, & \gamma'_{xy} &= b_{43}\sigma + b_{45}\tau,\\
\epsilon'_y &= b_{23}\sigma + b_{25}\tau, & \gamma'_{yz} &= b_{53}\sigma + b_{55}\tau,\\
\epsilon'_z &= b_{33}\sigma + b_{35}\tau, & \gamma'_{xz} &= b_{63}\sigma + b_{65}\tau.
\end{aligned} \tag{F-7}$$

The complete strain correspondences between the coordinate systems in Figs. F-1 and F-4 and the resulting conclusions are

$$\begin{aligned}
\epsilon_x &= \epsilon'_z && \Rightarrow && b_{11} = b_{33}, && b_{14} = b_{35},\\
\epsilon_y &= \epsilon'_y && \Rightarrow && b_{21} = b_{23}, && b_{24} = b_{25},\\
\epsilon_z &= \epsilon'_x && \Rightarrow && b_{31} = b_{13}, && b_{34} = b_{15},\\
\gamma_{xy} &= \gamma'_{yz} && \Rightarrow && b_{41} = b_{53}, && b_{44} = b_{55},\\
\gamma_{yz} &= -\gamma'_{xy} && \Rightarrow && b_{51} = -b_{43}, && b_{54} = -b_{45},\\
\gamma_{xz} &= -\gamma'_{xz} && \Rightarrow && b_{61} = -b_{63}, && b_{64} = -b_{65}.
\end{aligned}$$

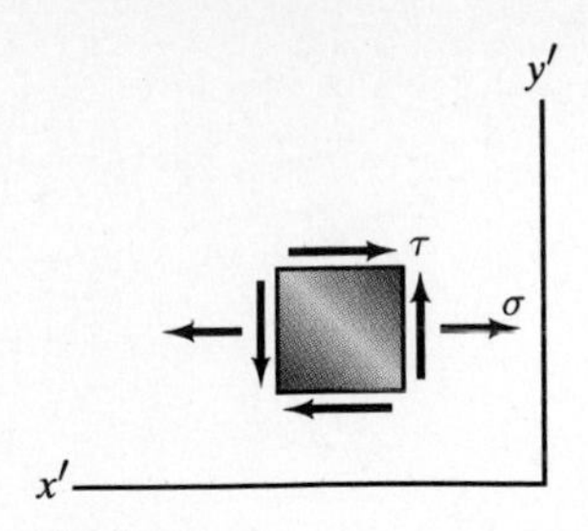

FIGURE F-5 Rotating the xyz coordinate system 180° about the y axis.

The coordinate system in Fig. F-5 is obtained by rotating the coordinate system in Fig. F-1 180° about the y axis. In terms of this coordinate system,

$$\begin{aligned} \epsilon'_x &= b_{11}\sigma - b_{14}\tau, & \gamma'_{xy} &= b_{41}\sigma - b_{44}\tau, \\ \epsilon'_y &= b_{21}\sigma - b_{24}\tau, & \gamma'_{yz} &= b_{51}\sigma - b_{54}\tau, \\ \epsilon'_z &= b_{31}\sigma - b_{34}\tau, & \gamma'_{xz} &= b_{61}\sigma - b_{64}\tau. \end{aligned} \tag{F-8}$$

The complete strain correspondences between the coordinate systems in Figs. F-1 and F-5 and the resulting conclusions are

$$\begin{aligned} \epsilon_x &= \epsilon'_x & &\Rightarrow & b_{11} &= b_{11}, & b_{14} &= -b_{14}, \\ \epsilon_y &= \epsilon'_y & &\Rightarrow & b_{21} &= b_{21}, & b_{24} &= -b_{24}, \\ \epsilon_z &= \epsilon'_z & &\Rightarrow & b_{31} &= b_{31}, & b_{34} &= -b_{34}, \\ \gamma_{xy} &= -\gamma'_{xy} & &\Rightarrow & b_{41} &= -b_{41}, & b_{44} &= b_{44}, \\ \gamma_{yz} &= -\gamma'_{yz} & &\Rightarrow & b_{51} &= -b_{51}, & b_{54} &= b_{54}, \\ \gamma_{xz} &= \gamma'_{xz} & &\Rightarrow & b_{61} &= b_{61}, & b_{64} &= -b_{64}. \end{aligned}$$

Applying all the restrictions obtained so far, the coefficients (F-3) are

$$\begin{matrix} b_{11} & b_{12} & b_{13} & 0 & 0 & b_{16} \\ b_{12} & b_{11} & b_{12} & 0 & 0 & b_{26} \\ b_{13} & b_{13} & b_{11} & 0 & 0 & b_{36} \\ 0 & 0 & 0 & b_{44} & 0 & b_{46} \\ 0 & 0 & 0 & 0 & b_{44} & b_{56} \\ 0 & 0 & 0 & 0 & 0 & b_{66} \end{matrix}$$

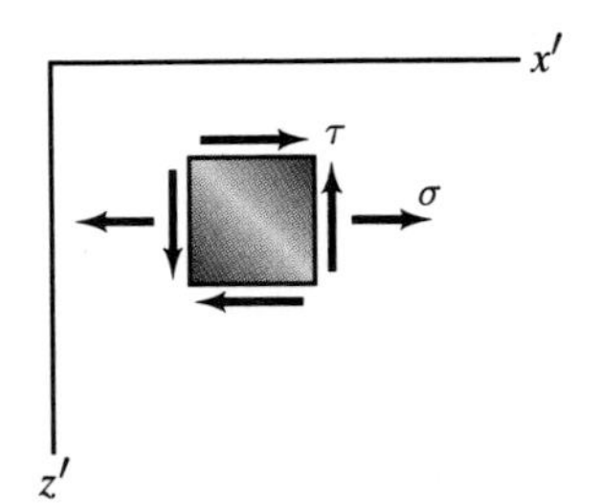

FIGURE F-6 Rotating the xyz coordinate system 90° about the x axis.

Rotation about the x Axis

The coordinate system in Fig. F-6 is obtained by rotating the coordinate system in Fig. F-1 90° about the x axis. In terms of this coordinate system,

$$\begin{aligned} \epsilon'_x &= b_{11}\sigma - b_{16}\tau, & \gamma'_{xy} &= b_{41}\sigma - b_{46}\tau, \\ \epsilon'_y &= b_{21}\sigma - b_{26}\tau, & \gamma'_{yz} &= b_{51}\sigma - b_{56}\tau, \\ \epsilon'_z &= b_{31}\sigma - b_{36}\tau, & \gamma'_{xz} &= b_{61}\sigma - b_{66}\tau. \end{aligned} \tag{F-9}$$

The complete strain correspondences between the coordinate systems in Figs. F-1 and F-6 and the resulting conclusions are

$$
\begin{array}{lll}
\epsilon_x = \epsilon'_x & \Rightarrow & b_{11} = b_{11}, \quad b_{14} = -b_{16}, \\
\epsilon_y = \epsilon'_z & \Rightarrow & b_{21} = b_{31}, \quad b_{24} = -b_{36}, \\
\epsilon_z = \epsilon'_y & \Rightarrow & b_{31} = b_{21}, \quad b_{34} = -b_{26}, \\
\gamma_{xy} = -\gamma'_{xz} & \Rightarrow & b_{41} = -b_{61}, \quad b_{44} = b_{66}, \\
\gamma_{yz} = -\gamma'_{yz} & \Rightarrow & b_{51} = -b_{51}, \quad b_{54} = b_{56}, \\
\gamma_{xz} = \gamma'_{xy} & \Rightarrow & b_{61} = b_{41}, \quad b_{64} = -b_{46}.
\end{array}
$$

Applying all the restrictions that have been obtained results in the coefficients in Eqs. (F-1):

$$
\begin{array}{cccccc}
b_{11} & b_{12} & b_{12} & 0 & 0 & 0 \\
b_{12} & b_{11} & b_{12} & 0 & 0 & 0 \\
b_{12} & b_{12} & b_{11} & 0 & 0 & 0 \\
0 & 0 & 0 & b_{44} & 0 & 0 \\
0 & 0 & 0 & 0 & b_{44} & 0 \\
0 & 0 & 0 & 0 & 0 & b_{44}
\end{array}
$$

Rotating the original coordinate system 180° about the x axis does not result in additional restrictions on the coefficients (see Problem 6-4.16).

APPENDIX G

Answers to Even-Numbered Problems

1-3.2 $A_x = 0$, $A_y = 10$ kN, $B = 10$ kN.

1-3.4 3.33 ft.

1-3.6 $P = 0$, $V = -144$ kN, $M = -288$ kN-m.

1-3.8 $\mathbf{R} = 173\mathbf{i} + 100\mathbf{j}$ (N), $\mathbf{M} = -566\mathbf{k}$ (N-m).

1-3.10 $m = 52.3$ kg.

1-3.12 35.4 kip (compression).

1-3.14 $F = 2.47$ kip.

1-3.16 Member CE, 4.72 kN (T). Member DE, 3.92 kN (C).

1-3.18 $C_x = 0$, $C_y = F$, $D_x = 0$, $D_y = -2F$, $E = F$.

1-3.20 4660 lb (compression).

1-3.22 6000 lb (tension).

1-3.24 1.80 kN (tension).

1-3.26 1650 N (compression).

1-3.28 $T_B = 49.2$ N-m.

1-3.30 $\mathbf{R} = 20\mathbf{i} - 20\mathbf{j} - 10\mathbf{k}$ (lb), $\mathbf{M} = -200\mathbf{i} - 20\mathbf{j} - 360\mathbf{k}$ (in-lb).

1-3.32 $\bar{x} = 90.3$ mm, $\bar{y} = 59.4$ mm.

1-3.34 Distance = 3.28 ft.

1-3.36 **(a)** $A_x = 0$, $A_y = 80.7$ N, $B = 209.8$ N. **(b)** $w = 223.4$ N/m.

1-3.38 $A_x = 0$, $A_y = 159$ lb, $B = 169$ lb.

1-3.40 $R_x = 0$, $R_y = -913$ N, $M_R = 1970$ N-m clockwise.

1-3.44 $A = c_0/2$, $B_x = 0$, $B_y = -c_0/2$.

2-1.2 $\tau_{av} = 0$.

2-1.4 **(a)** $\sigma_{av} = 318.3$ psi. **(b)** $\sigma_{av} = -636.6$ psi.

2-1.6 $\tau_{av} = -140$ kPa, $P = 7.15$ kN.

2-1.8 $\sigma_{av} = 27.5x$ kPa.

2-1.10 $P = 800$ lb, $\sigma_{av} = 240$ psi.

2-1.12 $\sigma_{av} = 33.3$ psi, $\tau_{av} = 57.7$ psi.

2-1.14 $\sigma_{av} = 64$ kPa, $\tau_{av} = 5.33$ kPa.

2-1.16 $\sigma_{av} = 200$ kPa, $\tau_{av} = 400$ kPa.

2-1.18 $\sigma_{av} = 5.96$ MPa (864 psi).

2-1.20 $\tau_{av} = 34.4$ psi.

2-1.22 $\tau_{av} = 216$ kPa.

2-1.24 $F = 9.6$ kN.

2-1.26 $\tau_{av} = 41.7$ MPa.

2-1.28 $\sigma_{av} = 102$ ksf.

2-1.30 $\tau_{av} = 3601$ psi.

2-1.32 $\sigma_{av} = 30.3$ MPa, $\tau_{av} = 10.0$ MPa.

2-1.34 $\sigma_{av} = -66.7$ psi, $\tau_{av} = 115.5$ psi.

2-1.36 $\tau_{av} = 2.32$ MPa.

2-1.38 $\tau_{av} = 18.75$ psi.

2-1.40 $\tau_{av} = 21.2$ MPa.

2-1.42 $\tau_{av} = F/bt$.

2-1.44 $\tau_{av} = 1.82$ psi.

2-1.46 $\sigma = 45$ kPa.

2-1.48 $\sigma = 3430$ psi.

2-1.50 **(a)** $70\mathbf{i} - 10\mathbf{j} - 5\mathbf{k}$ (kN). **(b)** $-70\mathbf{i} + 10\mathbf{j} + 5\mathbf{k}$ (kN).

2-1.52 $\sigma_{av} = 1.68$ MPa.

2-1.54 $\mathbf{t}_{av} = 1.955\mathbf{i} + 0.090\mathbf{j} + 1.060\mathbf{k}$ (MPa).

2-1.56 **(a)** Zero.

2-2.2 $dL' = 1.15dL$.

2-2.4 $L' = 0.206$ m.

2-2.6 $L' = 0.211$ m.

2-2.8 $\epsilon = 0.00625$.

2-2.10 $\delta = -0.028$ in.

2-2.12 $\epsilon = -0.240$.

2-2.14 $\delta = 0.04 \ln 2 = 0.028$ in.

2-2.16 $\epsilon_{AB} = 0.134$.

2-2.18 $\epsilon = 0.00454$.

2-2.20 $\epsilon = 0.002$.

2-2.22 $\epsilon_1 = 0.000333$, $\epsilon_2 = -0.008333$.

2-2.24 $\gamma = 0.698$.

2-2.26 The end rotates 139°.

2-2.28 $\gamma = 0.290$.

2-2.30 $\gamma = 0.0242$.

3-1.2 $P = 240$ kip.

3-1.4 $\sigma = 9$ ksi.

3-1.6 $\sigma = 2.31$ MPa.

3-1.8 $\sigma = -7.15$ MPa.

3-1.10 $\sigma = F/2A \sin\beta$.

3-1.12 $\beta = 45°$, $A = F/2\sigma_0 \sin\beta$.

3-1.14 $F = 49.2$ kN.

3-1.16 $\sigma_{BC} = -75.0$ MPa.

3-1.18 $\sigma = 1.50$ MPa.

3-1.20 $\sigma_\theta = 1170$ psi, $\tau_\theta = -3214$ psi.

3-1.22 $\theta = 50.2°$, $P = 61.0$ kN.

3-1.24 $F = 5196$ lb.

3-1.26 $\sigma = 120.0$ ksi, $|\tau| = 69.3$ ksi.

3-1.28 $\sigma = -1.90$ ksi.

3-1.30 $\sigma_3 = -63.1$ MPa.

3-2.2 10.02 in.

3-2.4 $E = 6.37$ GPa, $\nu = 0.30$.

3-2.6 Force $= 29.5$ kip, diameter $= 0.749$ in.

3-2.8 $D' = 0.7495$ in.

3-2.10 $\delta_{AC} = -0.0532$ in.

3-2.12 0.400 mm downward.

3-2.14 1.95° clockwise.

3-2.16 $\epsilon_{AB} = 0.000349$, $\epsilon_{CD} = 0.000698$, $\epsilon_{EF} = 0.001047$.

3-2.18 $\delta_{BE} = 0.0411$ in.

3-2.20 $\delta_{AB} = 5.28$ mm, $\delta_{AC} = -3.96$ mm.

3-2.22 $\delta_{AB} = 0.1949$ in, $\delta_{AC} = -0.0192$ in.

3-2.24 -0.0150 mm.

3-3.2 $\sigma = 2F/3A$.

3-3.4 900 kN.

3-3.6 $\sigma_B = -7.16$ MPa.

3-3.8 $F_1 = 97.1$ kN.

3-3.10 $b = 0.0681$ in.

3-3.12 $\sigma_{AB} = -F\cos^2\theta/[A(1+\cos^3\theta)]$, $\sigma_{AC} = -F/[A(1+\cos^3\theta)]$.

3-3.14 310 kN.

3-3.16 $h = 2.95$ mm.

3-3.18 9.92 kN.

3-3.20 $\sigma_{AB} = -1.74$ GPa, $\sigma_{AC} = -2.30$ GPa.

3-3.22 $\sigma_{AB} = 7.45$ ksi, $\sigma_{AC} = -10.17$ ksi, $\sigma_{AD} = 3.19$ ksi.

3-3.24 0.268 mm to the left, 0.247 mm upward.

3-4.2 $\delta = 0.0116$ in.

3-4.4 $\delta = 0.392$ mm.

3-4.6 $\sigma = 30.6$ ksi.

3-4.8 $\delta = 0.127$ mm.

3-4.10 $\delta = 6.13$ mm.

3-4.12 $\delta = 0.0441$ m.

3-4.14 $\delta = 0.0535$ m.

3-4.16 $x = L/2$, displacement $= qL^2/8EA$.

3-4.18 $x = 1.26$ m, displacement $= 0.101$ mm.

3-4.20 $\sigma_A = 57.6$ ksi, $\sigma_B = 28.8$ ksi.

3-4.22 $\delta = -F/\pi Ed \tan^2 \alpha$.

3-4.24 $\delta = 0.0131$ mm.

3-5.2 125°F.

3-5.4 30.024 mm.

3-5.6 $\sigma = 0$.

3-5.8 **(a),(b)** $\delta = 0.0111$ in.

3-5.10 134°F.

3-5.12 $\sigma = -7467$ psi.

3-5.14 $\sigma_A = -16.8$ MPa, $\sigma_B = -67.2$ MPa.

3-5.16 16,000 lb downward.

3-5.18 0.711 in. to the right, 0.377 in. upward.

3-5.20 1.20 mm to the left, 19.01 mm downward.

3-5.22 $\sigma_{AB} = \sigma_{AD} = 46.3$ MPa, $\sigma_{AC} = 79.8$ MPa.

3-5.24 $\sigma_{AB} = -1.73$ GPa, $\sigma_{AC} = -2.31$ GPa.

3-5.26 $\sigma_{AB} = 229$ MPa, $\sigma_{AC} = 162$ MPa, $\sigma_{AD} = 354$ MPa.

3-5.28 $\sigma_{AB} = 986$ psi, $\sigma_{AC} = -6665$ psi, $\sigma_{AD} = 4759$ psi.

3-7.2 Either 2014-T6 or 7075-T6.

3-7.4 ASTM-A514.

3-7.6 3.69 in^2.

3-7.8

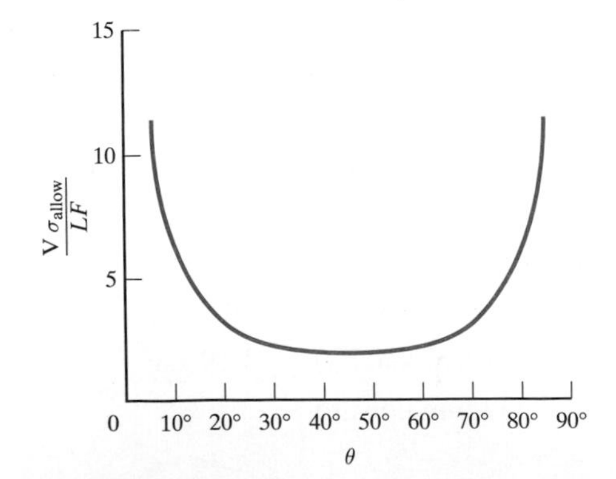

3-7.10 2014-T6 or 7075-T6.

3-7.16 $A = 0.00912$ in^2.

3-7.18 $A_3 = 1850$ mm^2.

4-1.2 $\beta = 89.9°$.

4-1.4 $\sigma_\theta = 10.4$ MPa, $|\tau_\theta| = 6$ MPa.

4-1.6 $\tau = 16.2$ MPa.

4-1.8 **(a)** $\sigma_\theta = -17.3$ ksi, $|\tau_\theta| = 10$ ksi. **(b)** 20 ksi.

4-1.10 $\gamma = 0.00346$.

4-2.2 $J = 23.6$ in^4.

4-2.4 $|\tau| = 17.0$ MPa.

4-2.6 $\tau = 19.9$ MPa.

4-2.8 $\tau = 11.7$ MPa.

4-2.10 $\phi = 18.9°$.

4-2.12 $\phi = 0.000382$ rad $(0.0219°)$.

4-2.14 **(a)** $|\tau| = 5093$ psi. **(b)** $|\phi| = 1.006°$.

4-2.16 $|\tau_A| = 19.89$ ksi, $|\tau_B| = 8.49$ ksi.

4-2.18 $|\tau_A| = 3.98$ ksi, $|\tau_B| = 8.49$ ksi, $|\phi| = 0.686°$.

4-2.20 $T = 1.99$ kN-m.

4-2.22 $\tau_{AB} = 37.7$ MPa, $\tau_{CD} = 28.3$ MPa.

4-2.24 $r_C = 108$ mm.

4-2.26 $\sigma_\theta = 20.5$ kPa, $|\tau_\theta| = 24.4$ kPa.

4-3.2 $|T_O| = 13.7$ in-kip.

4-3.4 $T_A = 1107.7$ N-m, $T_B = -92.3$ N-m.

4-3.6 $L_A = 186.7$ mm, $L_B = 93.3$ mm, $|\tau| = 84.9$ MPa.

4-3.8 $|\tau_A| = 8.13$ ksi, $|\tau_B| = 4.06$ ksi.

4-3.10 $|\phi_A| = 1.820°$, $|\phi_B| = 0.180°$.

4-3.12 $|\tau| = 0.656$ GPa.

4-3.14 $|\tau| = 21.9$ MPa.

4-4.2 $\phi = 0.0150$ rad $(0.861°)$.

4-4.4 $\phi = 4.60°$.

4-4.6 $T = 6.19$ N-m.

4-4.8 102 N-m.

4-4.10 $|\tau| = 40.7$ MPa.

4-4.12 $c_0 = 7200$ in-lb/in., $\tau = 18.7$ ksi.

4-4.14 $|\tau| = 325$ MPa.

4-4.16 $T_{\text{left}} = c_0 L/12$, $T_{\text{right}} = c_0 L/4$.

4-4.18 $T_{\text{left}} = 5c_0 L/192$, $T_{\text{right}} = c_0 L/64$.

4-5.2 $T = 6.49$ kN-m.

4-5.4 $\phi = 117°$.

4-5.6 $r_Y = 0.553$ in., $\phi = 74.6°$.

4-5.8 $T = 15.5$ kN-m.

4-5.10 $\phi = 142.3°$.

4-5.12 $T = 32.6$ kN-m.

4-6.2 $\phi = 0.00937$ rad (0.537°).

4-6.4 $\tau = 30.4$ MPa.

4-6.6 $\phi = 0.975°$.

4-6.8 $\tau = 5540$ psi.

4-6.10 $t = 0.0802$ in, $\phi = 1.95°$.

4-6.12 $\phi = 4.27°$.

4-6.14 $\tau = 69.9$ MPa.

4-6.16 **(a)** $\tau = 1591.5$ lb/in^2. **(b)** $\tau = 1590.6$ lb/in^2.

4-6.18 $|\tau_{max}| = 31.8$ MPa.

4-7.2 7075-T6.

5-2.2 $\sigma_x' = 25$ ksi, $\sigma_y' = -25$ ksi, $\tau_{xy}' = 0$.

5-2.4 $\sigma_x = 64.00$ MPa, $\sigma_y = 85.00$ MPa, $\tau_{xy} = 0.00$ MPa.

5-2.6 $\theta = -20.0°$ or $160°$.

5-2.8 $\sigma_x = 41.81$ MPa, $\sigma_y = -25.81$ MPa, $\tau_{xy} = -2.96$ MPa.

5-2.10 $\sigma = -2.23$ ksi, $|\tau| = 1.60$ ksi.

5-2.12 $\sigma = -352.6$ psi, $\tau = -44.5$ psi.

5-2.14 $\sigma = -7.86$ MPa, $\tau = 13.50$ MPa.

5-2.16 $\tau_{xy}' = 4.90$ MPa, $\theta = 19.5°$ or $\tau_{xy}' = -4.90$ MPa, $\theta = 40.2°$.

5-2.18 $\tau_{xy} = -78.4$ psi, $\tau_{xy}' = -114.1$ psi.

5-2.22 $\sigma_1 = \sigma_x$, $\sigma_2 = 0$, $\tau_{max} = |\sigma_x/2|$.

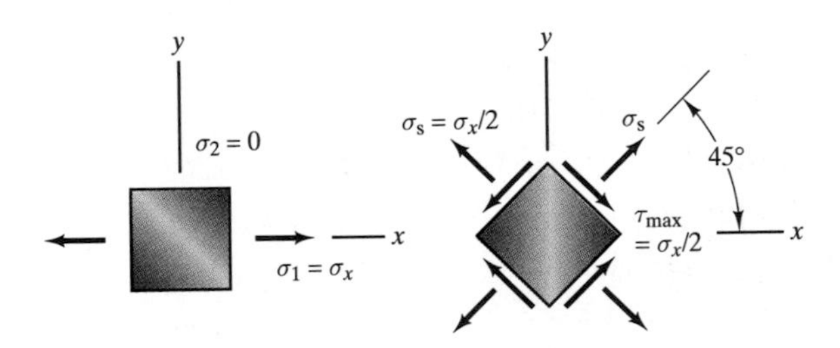

5-2.24 $\sigma_1 = 20$ MPa, $\sigma_2 = 10$ MPa, $\tau_{max} = 5$ MPa.

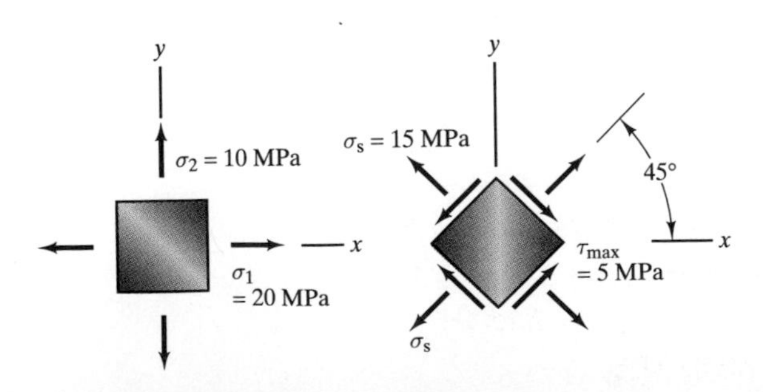

5-2.26 $\sigma_1 = 8.22$ ksi, $\sigma_2 = -10.22$ ksi, $\tau_{max} = 9.22$ ksi.

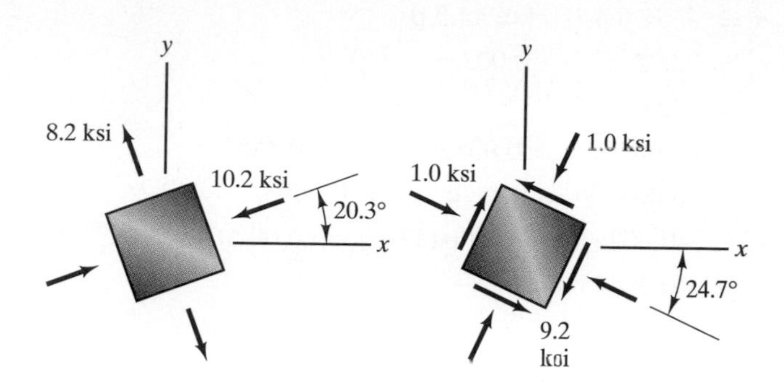

5-2.28 Absolute maximum shear stress = 10 MPa.

5-2.30 Absolute maximum shear stress = 6.54 ksi.

5-2.32 $\sigma_1 = 52.4$ MPa, $\sigma_2 = -32.4$ MPa, absolute maximum shear stress = 42.4 MPa.

5-3.2 $\sigma'_x = 25$ ksi, $\sigma'_y = -25$ ksi, $\tau'_{xy} = 0$.

5-3.4 $\sigma_x = 64.00$ MPa, $\sigma_y = 85.00$ MPa, $\tau_{xy} = 0.00$ MPa.

5-3.6 $\theta = -20°$.

5-3.8 $\sigma = 5.77$ MPa, $|\tau| = 4.89$ MPa.

5-3.10 $\sigma = 177.2$ psi, $\tau = -237.3$ psi.

5-3.12 See the answer to Problem 5-2.22.

5-3.14 See the answer to Problem 5-2.24.

5-3.16 See the answer to Problem 5-2.26.

5-3.18 $\sigma_1 = 52.4$ MPa, $\sigma_2 = -32.4$ MPa, $\tau_{max} = 42.4$ MPa.

5-3.20 Absolute maximum shear stress = 6.54 ksi.

5-4.2 $\sigma_1 = 40.45$ ksi, $\sigma_2 = 0$, $\sigma_3 = -15.45$ ksi, $\tau_{max} = 27.95$ ksi.

5-4.4 $\sigma_1 = 85.00$ MPa, $\sigma_2 = 65.00$ MPa, $\sigma_3 = 0$, $\tau_{max} = 42.50$ MPa.

5-4.6 $\sigma_1 = 240$ MPa, $\sigma_2 = 240$ MPa, $\sigma_3 = -120$ MPa, $\tau_{max} = 180$ MPa.

5-4.8 $-500 \leq \sigma_z \leq 600$ MPa.

5-4.10 $\sigma_1 = 409$ ksi, $\sigma_2 = 148$ ksi, $\sigma_3 = -257$ ksi, $\tau_{max} = 333$ ksi.

5-4.12 **(a),(b)** $\sigma_1 = 8.22$ ksi, $\sigma_2 = -10.22$ ksi, $\sigma_3 = 0$.

5-5.2 $\sigma = 20$ MPa, $\tau_{max} = 10.09$ MPa.

5-5.4 $\sigma = 142$ ksi, $\tau_{max} = 71.3$ ksi.

5-5.6 $\sigma_h = 250$ MPa.

5-5.8 $\sigma_h = 15$ MPa, $\tau_{max} = 7.65$ MPa.

5-5.10 $t = 6.02$ mm.

5-6.2 $\sigma = 7.10$ MPa, $|\tau| = 1.57$ MPa.

5-6.4 $\mathbf{t} = 2.667\mathbf{i} - 0.332\mathbf{j} + 3.002\mathbf{k}$ (ksi).

5-6.6 $\sigma = 6.86$ GPa, $|\tau| = 2.15$ GPa.

5-6.8 $|\tau| = 10.03$ MPa.

5-6.10 $\sigma = \sigma_1$, $|\tau| = 0$.

5-6.12 $\sigma = -352.6$ psi, $|\tau| = 44.5$ psi.

6-2.2 $\epsilon'_x = 0.002$, $\epsilon'_y = -0.002$, $\gamma'_{xy} = 0$.

6-2.4 $\theta = -20.0°$.

6-2.6 $\epsilon_x = 0.00640$, $\epsilon_y = 0.00850$, $\gamma_{xy} = 0.00000$.

6-2.8 $\gamma'_{xy} = 0.0098$, $\theta = 19.5°$ or $\gamma'_{xy} = 0.0098$, $\theta = 40.2°$.

6-2.10 $\epsilon'_x = -0.00257$, $\epsilon'_y = 0.00457$, $\gamma'_{xy} = 0.00207$.

6-2.12 $PQ = 0.99665$ mm.

6-2.14 $\gamma_{xy} = -0.00360$.

6-2.16 1.57626 rad (90.313°).

6-2.18 $\gamma_{xy} = 0$.

6-2.20 $\epsilon_x = -0.00160$, $\epsilon_y = 0.00100$, $\gamma_{xy} = 0.00201$.

6-2.22 $\epsilon_1 = 0.002$, $\epsilon_2 = 0.001$, $\gamma_{max} = 0.001$.

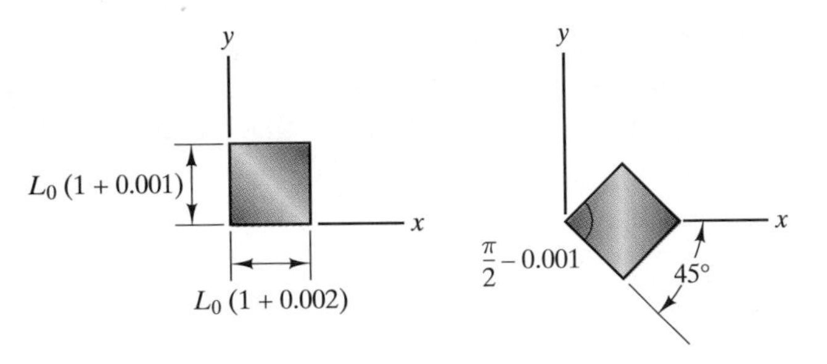

6-2.24 $\epsilon_1 = 0.00822$, $\epsilon_2 = -0.01022$, $\gamma_{max} = 0.01844$.

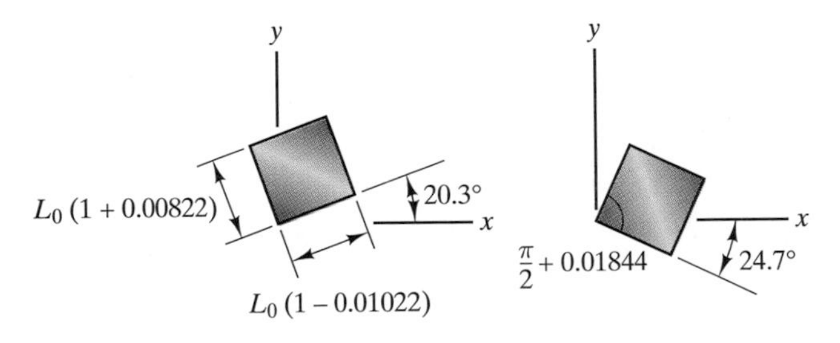

6-2.26 $\epsilon_1 = 0.00472$, $\epsilon_2 = -0.00273$, $\gamma_{max} = 0.00743$.

6-2.28 Absolute maximum shear strain = 0.00559.

6-2.30 Absolute maximum shear strain = 0.00314.

6-3.2 $\epsilon'_x = 0.002$, $\epsilon'_y = -0.002$, $\gamma'_{xy} = 0$.

6-3.6 $\gamma'_{xy} = 0.00175$, $\theta = 18.1°$ or $\gamma'_{xy} = -0.00175$, $\theta = 35.0°$.

6-3.8 $\epsilon_x = 0.0039$, $\epsilon_y = -0.0059$, $\gamma_{xy} = 0.0119$.

6-3.10 See the solution to Problem 6-2.22.

6-3.12 See the solution to Problem 6-2.24.

6-3.14 $\epsilon_1 = 0.00850$, $\epsilon_2 = 0.00640$, $\gamma_{max} = 0.00210$.

6-4.2 $\begin{bmatrix} 26.9 & 13.5 & 0 \\ 13.5 & 188.5 & -13.5 \\ 0 & -13.5 & 134.6 \end{bmatrix}$ MPa.

6-4.4 The required condition is that $\sigma_x + \sigma_y = 0$.

6-4.6 $\sigma_x = 64.0$ ksi, $\sigma_y = 83.5$ ksi, $\tau_{xy} = -93.8$ ksi.

6-4.8 **(a)** $\lambda = 16.2$ GPa, $\mu = 10.8$ GPa. **(b)** $K = 23.3$ GPa.

6-4.10 $\begin{bmatrix} 34.2 & 56.4 & -56.4 \\ 56.4 & -33.5 & 0 \\ -56.4 & 0 & -44.8 \end{bmatrix}$ ksi.

6-4.12 **(a)** $\epsilon_x = \sigma_x/E$, $\epsilon_y = \epsilon_z = -\nu\sigma_x/E$, other strain components equal zero. **(b)** Volume $= LA[1 + (1 - 2\nu)\sigma_x/E]$.

6-4.14 $\sigma_x = -253$ MPa, $\sigma_y = 809$ MPa, $\tau_{xy} = -234$ MPa.

6-4.16 $\sigma_x = -55.5$ MPa, $\sigma_y = 104.4$ MPa, $\tau_{xy} = -94.3$ MPa.

6-4.18 $\sigma_x = 412.6$ MPa, $\sigma_y = 444.2$ MPa, $\tau_{xy} = -74.5$ MPa.

7-1.2 $P_A = P_B = P_C = 0$, $V_A = V_B = V_C = 2$ kN, $M_A = 2$ kN-m, $M_B = 4$ kN-m, $M_C = 6$ kN-m.

7-1.4 $P_A = 866$ lb, $V_A = -500$ lb, $M_A = 3000$ ft-lb.

7-1.6 **(a)** $P_A = 0$, $V_A = -F$, $M_A = -LF/2$. **(b)** $P_A = 0$, $V_A = F$, $M_A = -LF/2$.

7-1.8 **(a)** $P_B = 0$, $V_B = -20$ N, $M_B = -5$ N-m. **(b)** $P_B = 0$, $V_B = -20$ N, $M_B = 5$ N-m.

7-1.10 **(a)** $P_A = 0$, $V_A = 4$ kN, $M_A = 4$ kN-m. **(b)** $P_A = 0$, $V_A = 2$ kN, $M_A = 3$ kN-m.

7-1.12 $P_A = 0$, $V_A = 16.7$ lb, $M_A = 575$ in-lb.

7-1.14 $P_A = 0$, $V_A = 225$ kN, $M_A = -375$ kN-m.

7-1.16 $P_A = 0$, $V_A = -475$ lb, $M_A = -1275$ ft-lb.

7-1.18 $P_A = 0$, $V_A = 4.8$ kN, $M_A = 13.6$ kN-m.

7-1.20 $P_C = 0$, $V_C = -3.7$ kN, $M_C = 14.1$ kN-m.

7-1.22 $P = 0$, $V = -7.81$ kN, $M = 4.56$ kN-m.

7-2.2 **(a)** $P = 0$, $V = -50x + 250$ kN, $M = -25x^2 + 250x$ kN-m.
(c)

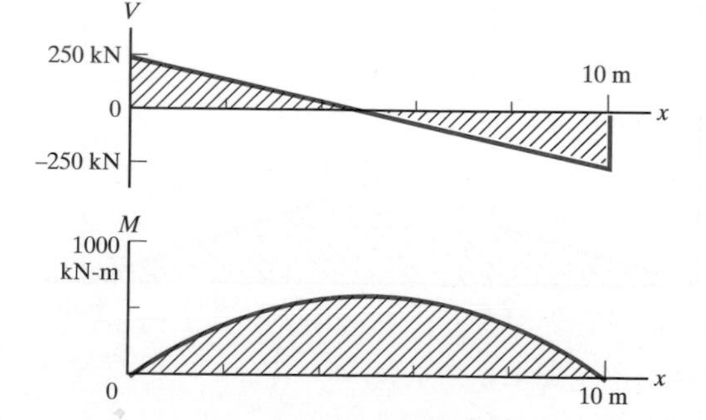

7-2.4 **(a)** $P = 0$, $V = 1080 - 10x^2$ lb, $M = 1080x - (10/3)x^3$ ft-lb.
(b) $M = 7482$ ft-lb at $x = 10.39$ ft.

7-2.6 $P = 0$, $V = -F$, $M = -Fx$.

7-2.8 $P = 0$, $V = -20$ N, $M = -20x$ N-m.

7-2.10 $P = 0$, $V = -100(4 + x^2/12)$ lb, $M = -100(4x + x^3/36)$ ft-lb.

7-2.12 **(a)** $0 < x < 6$ ft: $P = 0$, $V = 300$ lb, $M = 300x - 3000$ ft-lb.
$6 < x < 12$ ft: $P = 0$, $V = 100(x - x^2/12)$ lb,
$M = 100(-24 + x^2/2 - x^3/36)$ ft-lb.
(b)

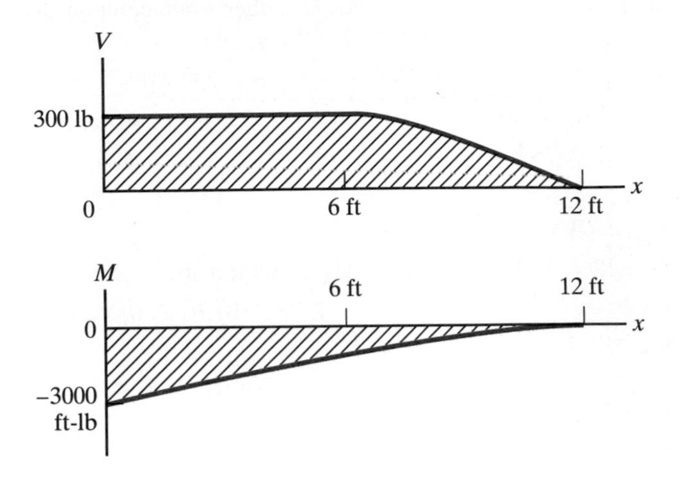

7-2.14 No. The resulting maximum bending moment magnitude is 8 kN-m.

7-2.16

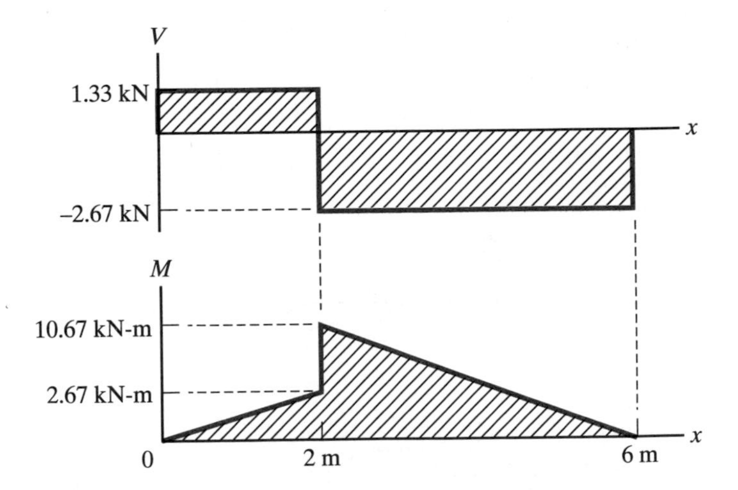

7-2.18 $M = 578$ in-lb at $x = 9.33$ in.

7-2.20

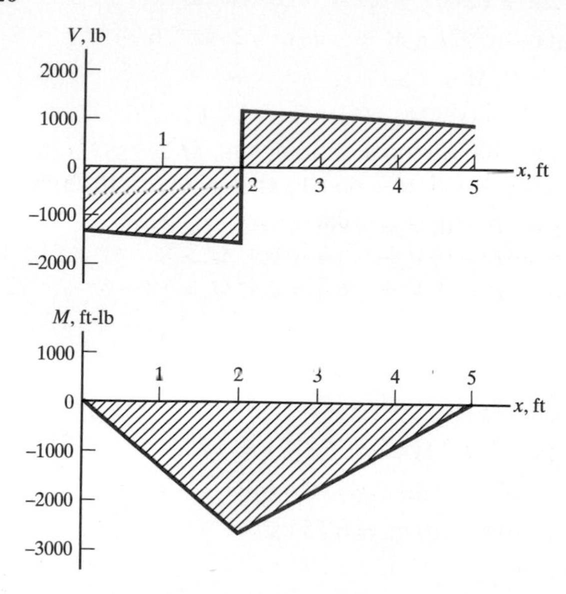

7-2.22

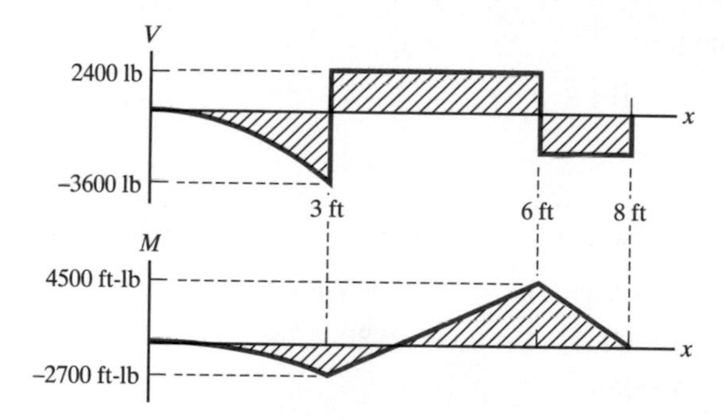

7-2.24

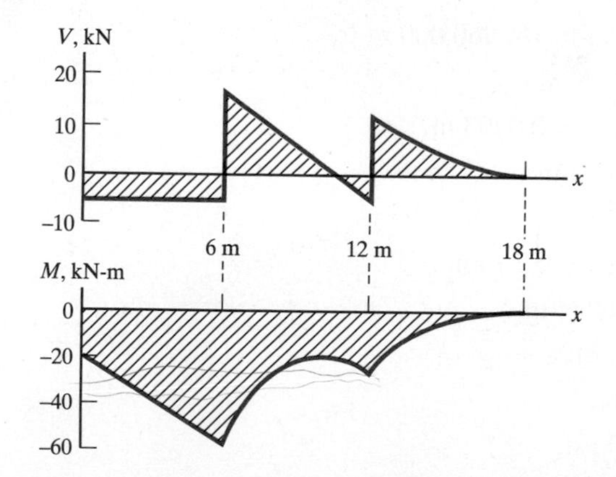

7-2.26 $V = -31.25 + 10.00x + 0.25x^2 - 0.20x^3$ kN.

7-3.2 $V = -w_0(x - x^2/2L)$, $M = -w_0(x^2/2 - x^3/6L)$.

7-3.4 $P = 0$, $V = F$, $M = Fx$.

7-3.6 $V = w_0L/6 - w_0x^2/2L$, $M = (Lx - x^3/L)w_0/6$.

7-3.8 $0 < x < 8$ ft: $P = -1500$ lb, $V = 225$ lb, $M = 225x$ ft-lb.
$8 < x < 10$ ft: $P = 0$, $V = 600$ lb, $M = 600(x - 10)$ ft-lb.

7-3.14 $0 < x < 2$ m: $P = 0$, $V = x$ kN, $M = x^2/2$ kN-m.
$2 < x < 5$ m: $P = 0$, $V = -4 + x$ kN, $M = 8 - 4x + x^2/2$ kN-m.
$5 < x < 6$ m: $P = 0$, $V = -6 + x$ kN, $M = 18 - 6x + x^2/2$ kN-m.

8-1.2 $M = 492$ kN-m.

8-1.4 **(a)** $\rho = -1170$ in. **(b)** $\sigma_x = 23{,}900$ psi.

8-1.6 $M = 176{,}700$ in-lb.

8-1.8 $\sigma_x = 2\sqrt{3}w_0L^2/9h^3$ at $x = L/\sqrt{3}$, $y = h/2$.

8-1.10 **(a)** $\sigma_x = 128.8$ MPa. **(b)** $\sigma_x = 78.6$ MPa.

8-1.12 **(a)** $\sigma_x = 13.10$ ksi. **(b)** $\sigma_x = 6.75$ ksi.

8-1.14 $\sigma_x = 1.24$ GPa.

8-1.16 $\sigma_x = 41.0$ kPa, $\sigma_x = -86.3$ kPa.

8-2.2 $S = 1.87$.

8-2.4 $h = 3.90$ in.

8-2.6 **(a)** $S = 2.01$. **(b)** $S = 4.69$.

8-3.4 $H = 0.02$ m.

8-3.6 $H = 2.67$ in.

8-3.8 $E_B = 216$ GPa.

8-3.10 **(a)** $\sigma_x = 26.3$ MPa. **(b)** $\sigma_x = 13.8$ MPa.

8-3.12 **(a)** $w_0 = 1320$ lb/in. **(b)** $w_0 = 2580$ lb/in.

8-3.14 $w_0 = 9.32$ kN/m.

8-3.16 $M_A = 180$ N-m, $M_B = 220$ N-m.

8-4.2 $c = 0.0274$ m.

8-4.4 **(a)** 640,000 in-lb. **(b)** 960,000 in-lb.

8-4.6 $w_0 = 50{,}300$ N/m.

8-4.8 $x = \sqrt{12}$ m, $c = 0.0293$ m.

8-4.10 $w_0 = 22{,}680$ N/m, $x = 0.6$ m.

8-4.12 $w_0 = 116$ lb/in., $x = 18$ in.

8-4.14 $x = 2.86$ m, $c = 50.7$ mm.

8-4.18 $M_U = 81{,}200$ N-m.

8-5.2 $\sigma_x = -31.3$ MPa.

8-5.4 $\sigma_x = 14$ Mpa.

8-5.6 $|\sigma_x| = 9200$ psi.

8-5.8 $|\sigma_x| = 21{,}400$ psi.

8-5.10 $|\sigma_x| = 3.14$ MPa.

8-5.12 $\sigma_x = 18{,}200$ psi.

8-5.14 $\sigma_x = 4.11$ MPa.

8-6.2 **(a)** $\tau_{av} = 4.44$ MPa. **(b)** $\tau_{av} = 2.78$ MPa.

8-6.4 $|\tau_{av}| = 576$ kPa at $x = 0$, $y' = 0$ and at $x = 8$ m, $y' = 0$.

8-6.6 **(a)** $\tau_{av} = -2810$ psi. **(b)** $\tau_{av} = -1230$ psi.

8-6.8 **(a)** $\tau_{av} = -19.9$ MPa. **(b)** $\tau_{av} = -11.1$ MPa.

8-6.12 $\tau_{av} = 6.86$ MPa.

8-6.14 $\tau_{av} = 188$ psi.

8-6.16 $\tau_{av} = 88.5$ psi on each joint.

8-6.18 $y' = 1$ in., $|\tau_{av}| = 4170$ psi.

8-6.20 **(a)** $\tau_{av} = 11.42$ MPa. **(b)** $\tau_{av} = 1.82$ MPa.

8-6.22 $\tau_{av} = 6.31$ MPa.

8-6.24 $\tau_{av} = 5.35$ MPa.

8-7.2 $\tau = 30.9$ MPa.

8-7.4 $\tau = 2250$ psi.

8-7.6 $\tau = 0.138\zeta$ GPa (ζ in meters).

8-7.8

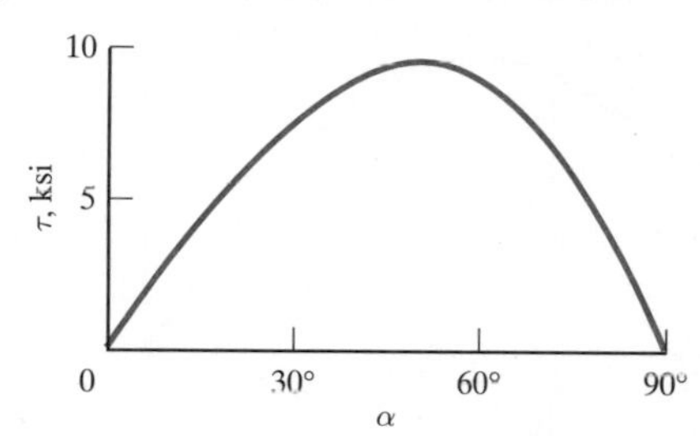

8-7.10

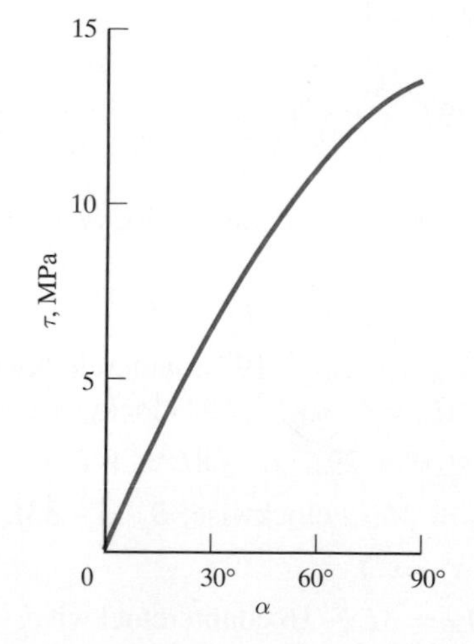

8-8.2 $e = 0.312$ m.

8-8.4 $e = 0.0332$ m.

8-8.6 $e = 7.54$ in.

8-8.8 $e = 0.0636$ m.

8-8.10 $e = 3.15$ in.

8-8.12 $e = 0.0394$ m.

9-1.2 $v = 6.22$ mm, $v' = 0$.

9-1.4 $v = 0.558$ in, $v' = 0.0106$ rad.

9-1.6 $|v| = 2.97$ mm at $x = 1.15$ m.

9-1.20 $\sigma_x = 5.04$ MPa.

9-1.22 $v = 3.02$ mm at $x = 1.63$ m.

9-1.24 $M_0 = 155$ kN-m.

9-1.26 $x = 36.4$ in.

9-2.2 $v = -(M_0/4LEI)(Lx^2 - x^3)$.

9-2.4 $A_x = 0$, $A_y = -9w_0L/40$, $M_A = 7w_0L^2/120$ counterclockwise, $B_y = -11w_0L/40$.

9-2.6 **(a)**

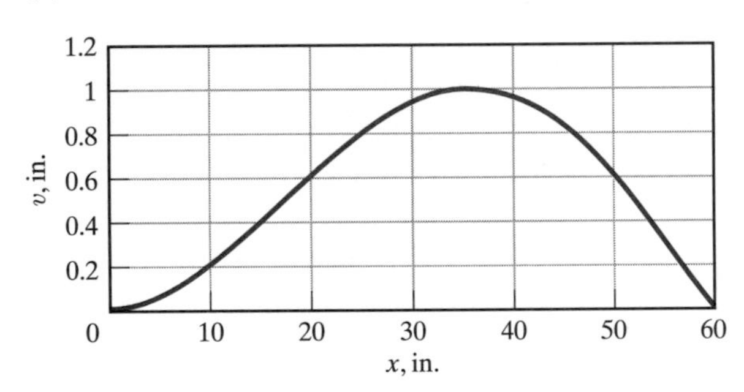

(b) $v = 0.988$ in. at $x = 35.9$ in.

9-2.8 $v = (w_0x^2/24EI)(L^2 - 2Lx + x^2)$.

9-2.10 $w_0 = 38.1$ kN/m.

9-2.12 $v = (w_0x^2/120LEI)(2L^3 - 3L^2x + x^3)$.

9-2.14 $v = -27.1$ mm.

9-2.16 $A_x = 0$, $A_y = -11F/16$, $M_A = 3LF/16$ counterclockwise, $B_y = -5F/16$.

9-2.18 $v = (F/96EI)(5x^3 - 15Lx^2 + 12L^2x - 2L^3)$.

9-2.20 $A_x = 0$, $A_y = -3w_0L/32$, $M_A = 5w_0L^2/192$ counterclockwise, $B_x = 0$, $B_y = -13w_0L/32$, $M_B = 11w_0L^2/192$ clockwise.

9-2.22 $v = (w_0/384EI)(16x^4 - 38Lx^3 + 29L^2x^2 - 8L^3x + L^4)$.

9-3.12 $A_x = 0$, $A_y = 3M_0/2L$, $M_A = M_0/2$ clockwise, $B_y = -3M_0/2L$.

9-3.14 $v = (w_0x^2/24EI)(L^2 - 2Lx + x^2)$.

9-3.16 $A_x = 0$, $A_y = -11F/16$, $M_A = 3LF/16$ counterclockwise, $B_y = -5F/16$.

9-4.2 $M_0 = 2FL/3$.

9-4.4 $M = M_0 - F(L - x)$.

9-4.6 $M_0 = w_0L^2/4$.

9-4.8 $v = (w_0x/24EI)(L^3 - 2Lx^2 + x^3) - (Fx/48EI)(3L^2 - 4x^2)$.

9-4.10 $A_x = 0$, $A_y = C_y = -3w_0L/16$, $B_y = -5w_0L/8$.

9-4.12 $v = (Fx/18EI)(2L^2 - 3x^2)$.

9-4.14 $v = (M_0x/2EI)(L - x)$.

9-4.16 $v = (w_0x^2/24EI)(L^2 - 2Lx + x^2)$.

9-4.18 $A_x = 0$, $A_y = 3M_0/2L$, $M_A = M_0/2$ clockwise, $B_y = -3M_0/2L$.

9-4.20 $A_x = 0$, $A_y = -11F/16$, $M_A = 3LF/16$ counterclockwise, $B_y = -5F/16$.

9-4.22 $B_y = -19w_0L/56$, $C_y = -12w_0L/56$.

9-4.24 $v = (w_0/1344EI)(13L^4 - 85L^3x + 192L^2x^2 - 176Lx^3 + 56x^4)$.

10-1.2 $P = 965$ kip.

10-1.4 $P = 202$ kN.

10-1.6 $P = 806$ kN.

10-1.10 $R = 25.5$ mm.

10-1.12 $F = 80.4$ kip.

10-1.14 $F = 3.02$ kN.

10-1.16 $m = 250$ kg.

10-1.18 $b = 3.10$ in., $F = 5.66$ kip.

10-1.20 $r_{AB} = 29.6$ mm, $r_{CD} = 0.892$ mm.

10-1.22 $F = 368$ N.

10-1.24 $F = 3.18$ kN.

10-2.2 $P = 31.7$ kip.

10-2.4 $P = 412$ kN.

10-2.6 $P = 7.20$ MN. It bends in the x–y plane.

10-2.8 $P = 1.66$ MN. It bends in the x–y plane.

10-2.10

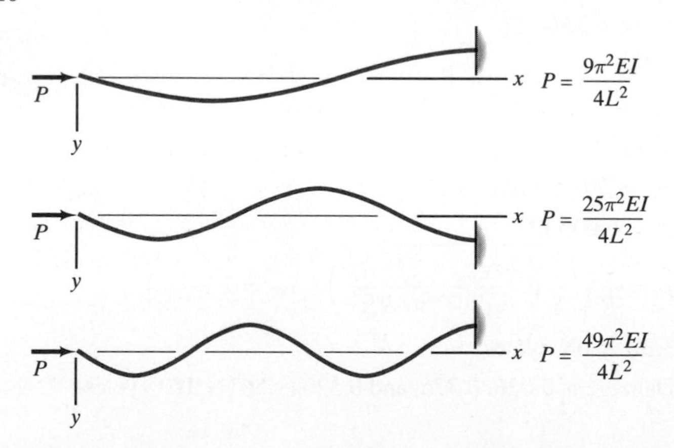

10-2.12 $P = \pi^2 EI/L^2$.

10-2.14 $L_e = L, L/2, L/3$, and $L/4$.

10-2.16 $L_e = 2L, P = \pi^2 EI/4L^2$.

10-3.2

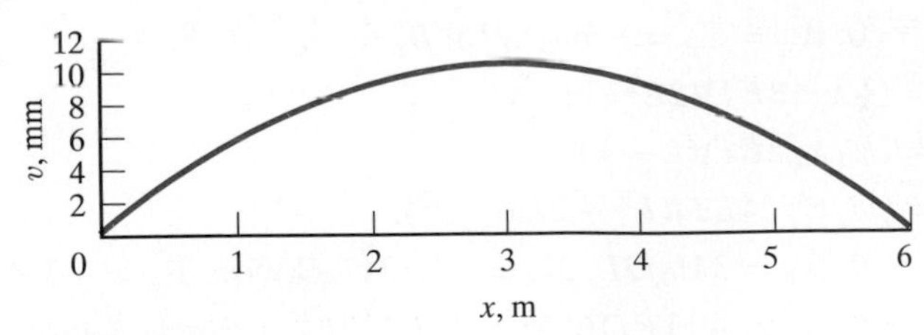

10-3.4 $v_{max} = 0.781$ in.

10-3.6 $v_{max} = 0.000738$ m.

10-3.8 $v_{max} = 0.00668$ m.

10-3.10 $P/A = 117$ MPa, $P = 258$ kN [obtained by numerical solution of Eq. (10.43)].

11-1.2 **(a)** $U = 0.00143$ J (joules, or N-m). **(b)** $W = 0.00143$ J.

11-1.4 **(a)** $U = 386$ in-lb. **(b)** $W = 386$ in-lb.

11-1.6 $\theta = 0.00321$ rad (0.184°) counterclockwise.

11-1.8 $F = 84.0$ kN.

11-1.10 $v = (1 + 2\sqrt{2})FL_0/EA$.

11-1.12 $v = 2.69$ mm.

11-1.14 $v = FL^3/48EI$.

11-1.18 $v = FL^3/12EI$.

11-1.20 $u = 1.24$ MJ/m^3.

11-2.2 **(a),(b)** $v = 0.0977$ in.

11-2.4 $v = 0.0111$ m.

11-2.6 $\theta_B = w_0 L^3/6EI$.

11-2.8 $\theta_B = M_0 L/EI$.

11-2.10 $v = FL^3/12EI$.

11-2.12 $B_y = -3M_0/2L$.

12-1.2 33%.

12-1.4 $\sigma_1 = -1.8$ ksi, $\sigma_2 = -83.2$ ksi; state of stress is safe.

12-1.6 15.4% increase.

12-1.8 $\sigma_Y = 107.8$ MPa.

12-1.10 $P = 6.033$ kN.

12-1.12 $p = \frac{t}{R}\sqrt{\frac{2}{3}\left(\frac{2\sigma_Y}{S_M} - \frac{P^2}{2\pi^2 R^2 t^2}\right)}$, $S_M = 2$.

12-1.14 3.32×10^3 cycles.

12-1.16 Damage is 0.036, 0.376, and 0.520. 9.561×10^3 cycles.

12-1.18 10.3 years.

12-2.2 $P = 1333$ lb.

12-2.4 $a = 0.075$ in.

12-3.2 **(a)** $P = 8.76$ kip. **(b)** $P = 17.48$ kip. **(c)** $P = 23.5$ kip.

12-3.4 $K_c = 45.1$ ksi-in$^{1/2}$.

12-3.6 Crack grows an additional 3.78 in.

12-3.8 1.315×10^6 cycles.

12-3.10 17,350 cycles.

12-3.12 1908 cycles.

C.1.2 $\bar{x} = a(n+1)/(n+2)$, $\bar{y} = ca^n(n+1)/(4n+2)$.

C.1.4 $x = 0$, $\bar{y} = 1.6$ ft.

C.1.6 $\bar{x} = 0.5$, $\bar{y} = -7.6$.

C.1.8 $\bar{x} = (R^2 - h^2)^{3/2}/3A$, $\bar{y} = [(2R^3/3) - R^2h + h^3/3]/2A$, where $A = (R/2)[(\pi R/2) - h(1 - h^2/R^2)^{1/2} - R \arcsin(h/R)]$.

C.2.2 $\bar{x} = 70.9$ mm, $\bar{y} = 0$.

C.2.4 $\bar{x} = 12.0$ in., $\bar{y} = 5.5$ in.

C.2.6 $\bar{x} = 344$ mm, $\bar{y} = 456$ mm.

C.3.2 $I_x = \frac{1}{12}bh^3$, $I_y = \frac{1}{12}hb^3$, $I_{xy} = \frac{1}{24}b^2h^2$.

C.3.4 $I_x = \frac{1}{16}\pi R^4$, $I_y = \frac{1}{16}\pi R^4$, $I_{xy} = \frac{1}{8}R^4$.

C.4.2 $I_x = 2.65 \times 10^8$ mm^4, $I_y = 0.802 \times 10^8$ mm^4, $I_{xy} = 1.08 \times 10^8$ mm^4.

C.4.4 $I_x = 36$ ft^4, $I_y = 327$ ft^4, $I_{xy} = 92$ ft^4.

C.4.6 $I_x = 237$ in^4, $I_y = 115$ in^4, $I_{xy} = 119$ in^4.

C.4.8 $I_x = 88.8$ m^4, $I_y = 65.0$ m^4, $I_{xy} = 0$.

C.5.2 $\theta_p = 0$, $I_{x'} = 85.33$ m^4, $I_{y'} = 5.33$ m^4.

C.5.4 $\theta_p = 36.9°$, $I_{x'} = 10.40 \times 10^{-6}$ m^4, $I_{y'} = 1.40 \times 10^{-6}$ m^4.

C.5.6 $\theta_p = 26.6°$, $I_{x'} = 10$ ft^4, $I_{y'} = 2.5$ ft^4.

INDEX

N

O

P

Q

R

S

T

U

Mathcad Engine User Agreement

NOTICE: MATHSOFT, INC. IS WILLING TO LICENSE THE SOFTWARE TO YOU ONLY UPON THE CONDITION THAT YOU ACCEPT ALL OF THE TERMS CONTAINED IN THIS LICENSE AGREEMENT. PLEASE READ THE TERMS CAREFULLY BEFORE DOWNLOADING THE SOFTWARE. IF YOU DO NOT AGREE TO THESE TERMS, YOU MAY NOT DOWNLOAD OR USE THE SOFTWARE.

MathSoft Incorporated ("MathSoft") grants you a free license to use the Mathcad Engine ("Software"). You may not use the Software for any development, commercial or production purpose. Any commercial reproduction or redistribution of the Software is expressly prohibited by law, and may result in severe civil and criminal penalties

The Software is protected under applicable copyright laws, international treaty provisions, and trade secret statutes of the various states. This Agreement grants you a limited non-exclusive, non-transferable license to use the Software. This is not an agreement for the sale of the Software or part thereof. Your right to use the Software is limited to the terms and conditions described herein.

You may use the Software solely for your own personal or internal purposes, for non-remunerated demonstrations (but not for delivery or sale) in connection with your personal or internal purposes on only one computer at a time and by only one user at a time. You may not make copies of the Software.

MathSoft, Inc. reserves all rights not expressly granted to you by this License Agreement. The license granted herein is limited solely to the uses specified above and, without limiting the generality of the foregoing, you are NOT licensed to use or to copy all or any part of the Software in connection with the sale, resale, license, or other for-profit personal or commercial reproduction or commercial distribution of computer programs or other materials without the prior written consent of MathSoft, Inc. In particular, the DLL interface specifications, the HBK file format and other confidential information and copyrighted materials may not be used for creating computer programs or other materials for sale, resale, license, or for remunerated personal or commercial reproduction or commercial distribution without the prior written consent of MathSoft, Inc.

Your license to use the Software will automatically terminate if you fail to comply with the terms of the Agreement. If this license is terminated, you agree to destroy/delete and/or remove all copies of the Software in your possession.

DISCLAIMER OF WARRANTY. THE SOFTWARE IS PROVIDED "AS IS" WITHOUT WARRANTY OF ANY KIND. TO THE MAXIMUM EXTENT PERMITTED BY APPLICABLE LAW, MATHSOFT FURTHER DISCLAIMS ALL WARRANTIES, INCLUDING WITHOUT LIMITATION ANY IMPLIED WARRANTIES OF MERCHANTABILITY, FITNESS FOR A PARTICULAR PURPOSE AND NONINFRINGEMENT. THE ENTIRE RISK ARISING OUT OF THE USE OR PERFORMANCE OF THE SOFTWARE REMAINS WITH YOU. TO THE MAXIMUM EXTENT PERMITTED BY APPLICABLE LAW, IN NO EVENT SHALL MATHSOFT, OR ITS SUPPLIERS BE LIABLE FOR ANY CONSEQUENTIAL, INCIDENTAL, DIRECT, INDIRECT, SPECIAL, PUNITIVE, OR OTHER DAMAGES WHATSOEVER (INCLUDING, WITHOUT LIMITATION, DAMAGES FOR LOSS OF BUSINESS PROFITS, BUSINESS INTERRUPTION, LOSS OF BUSINESS INFORMATION, OR OTHER PECUNIARY LOSS) ARISING OUT OF THE USE OF OR INABILITY TO USE THE SOFTWARE, EVEN IF MATHSOFT, INC. HAS BEEN ADVISED OF THE POSSIBILITY OF SUCH DAMAGES. BECAUSE SOME STATES/JURISDICTIONS DO NOT ALLOW THE EXCLUSION OR LIMITATION OF LIABILITY FOR CONSEQUENTIAL OR INCIDENTAL DAMAGES, THE ABOVE LIMITATION MAY NOT APPLY TO YOU.

This Agreement shall be governed by the laws of the State of Massachusetts, and you further consent to jurisdiction by the state and federal courts sitting in the State of Massachusetts. If either MathSoft or you employ attorneys to enforce any rights arising out of or relating to this Agreement, the prevailing party shall be entitled to recover reasonable attorneys' fees.

The Software is provided with RESTRICTED RIGHTS. Use, duplication, or disclosure by the Government is subject to restrictions as set forth in subparagraph (c)(1)(ii) of The Rights in Technical Data and Computer Software clause of DFARS 252.227-7013 or subparagraphs (c)(i) and (2) of the Commercial Computer Software – Restricted Rights at 48 CFR 52.227-19, as applicable. Manufacturer is MathSoft, Inc., 101 Main Street, Cambridge, MA 02142.

You acknowledge that the Software acquired hereunder are subject to the export control laws and regulations of the U.S.A., and any amendments thereof. You confirm that with respect to the Software, it will not export or re-export them, directly or indirectly, either to (i) any countries that are subject to U.S.A export restrictions (currently including, but not necessarily limited to, Cuba, the Federal Republic of Yugoslavia (Serbia and Montenegro), Iran, Iraq, Libya, North Korea, Sudan, South Africa (military and police entities), Syria, and Vietnam); (ii) any end user who you know or have reason to know will utilize them in the design, development or production of nuclear, chemical or biological weapons; or (iii) any end user who has been prohibited from participating in the U.S.A. export transactions by any federal agency of the U.S.A. government. You further acknowledge that the Software may include technical data subject to export and re-export restrictions imposed by U.S.A. law.

Should you have any questions concerning this Agreement, or if you desire to contact MathSoft, Inc. for any reason, please write: MathSoft, Inc., 101 Main Street, Cambridge, MA 02142.

LICENSE AGREEMENT AND LIMITED WARRANTY

READ THE FOLLOWING TERMS AND CONDITIONS CAREFULLY BEFORE OPENING THIS DISK PACKAGE. THIS LEGAL DOCUMENT IS AN AGREEMENT BETWEEN YOU AND PRENTICE-HALL, INC. (THE "COMPANY"). BY OPENING THIS SEALED DISK PACKAGE, YOU ARE AGREEING TO BE BOUND BY THESE TERMS AND CONDITIONS. IF YOU DO NOT AGREE WITH THESE TERMS AND CONDITIONS, DO NOT OPEN THE DISK PACKAGE. PROMPTLY RETURN THE UNOPENED DISK PACKAGE AND ALL ACCOMPANYING ITEMS TO THE PLACE YOU OBTAINED THEM FOR A FULL REFUND OF ANY SUMS YOU HAVE PAID.

1. **GRANT OF LICENSE:** In consideration of your payment of the license fee, which is part of the price you paid for this product, and your agreement to abide by the terms and conditions of this Agreement, the Company grants to you a nonexclusive right to use and display the copy of the enclosed software program (hereinafter the "SOFTWARE") on a single computer (i.e., with a single CPU) at a single location so long as you comply with the terms of this Agreement. The Company reserves all rights not expressly granted to you under this Agreement.
2. **OWNERSHIP OF SOFTWARE:** You own only the magnetic or physical media (the enclosed disks) on which the SOFTWARE is recorded or fixed, but the Company retains all the rights, title, and ownership to the SOFTWARE recorded on the original disk copy(ies) and all subsequent copies of the SOFTWARE, regardless of the form or media on which the original or other copies may exist. This license is not a sale of the original SOFTWARE or any copy to you.
3. **COPY RESTRICTIONS:** This SOFTWARE and the accompanying printed materials and user manual (the "Documentation") are the subject of copyright. You may not copy the Documentation or the SOFTWARE, except that you may make a single copy of the SOFTWARE for backup or archival purposes only. You may be held legally responsible for any copying or copyright infringement which is caused or encouraged by your failure to abide by the terms of this restriction.
4. **USE RESTRICTIONS:** You may not network the SOFTWARE or otherwise use it on more than one computer or computer terminal at the same time. You may physically transfer the SOFTWARE from one computer to another provided that the SOFTWARE is used on only one computer at a time. You may not distribute copies of the SOFTWARE or Documentation to others. You may not reverse engineer, disassemble, decompile, modify, adapt, translate, or create derivative works based on the SOFTWARE or the Documentation without the prior written consent of the Company.
5. **TRANSFER RESTRICTIONS:** The enclosed SOFTWARE is licensed only to you and may not be transferred to any one else without the prior written consent of the Company. Any unauthorized transfer of the SOFTWARE shall result in the immediate termination of this Agreement.
6. **TERMINATION:** This license is effective until terminated. This license will terminate automatically without notice from the Company and become null and void if you fail to comply with any provisions or limitations of this license. Upon termination, you shall destroy the Documentation and all copies of the SOFTWARE. All provisions of this Agreement as to warranties, limitation of liability, remedies or damages, and our ownership rights shall survive termination.
7. **MISCELLANEOUS:** This Agreement shall be construed in accordance with the laws of the United States of America and the State of New York and shall benefit the Company, its affiliates, and assignees.
8. **LIMITED WARRANTY AND DISCLAIMER OF WARRANTY:** The Company warrants that the SOFTWARE, when properly used in accordance with the Documentation, will operate in substantial conformity with the description of the SOFTWARE set forth in the Documentation. The Company does not warrant that the SOFTWARE will meet your requirements or that the operation of the SOFTWARE will be uninterrupted or error-free. The Company warrants that the media on which the SOFTWARE is delivered shall be free from defects in materials and workmanship under normal use for a period of thirty (30) days from the date of your

purchase. Your only remedy and the Company's only obligation under these limited warranties is, at the Company's option, return of the warranted item for a refund of any amounts paid by you or replacement of the item. Any replacement of SOFTWARE or media under the warranties shall not extend the original warranty period. The limited warranty set forth above shall not apply to any SOFTWARE which the Company determines in good faith has been subject to misuse, neglect, improper installation, repair, alteration, or damage by you. EXCEPT FOR THE EXPRESSED WARRANTIES SET FORTH ABOVE, THE COMPANY DISCLAIMS ALL WARRANTIES, EXPRESS OR IMPLIED, INCLUDING WITHOUT LIMITATION, THE IMPLIED WARRANTIES OF MERCHANTABILITY AND FITNESS FOR A PARTICULAR PURPOSE. EXCEPT FOR THE EXPRESS WARRANTY SET FORTH ABOVE, THE COMPANY DOES NOT WARRANT, GUARANTEE, OR MAKE ANY REPRESENTATION REGARDING THE USE OR THE RESULTS OF THE USE OF THE SOFTWARE IN TERMS OF ITS CORRECTNESS, ACCURACY, RELIABILITY, CURRENTNESS, OR OTHERWISE.

IN NO EVENT, SHALL THE COMPANY OR ITS EMPLOYEES, AGENTS, SUPPLIERS, OR CONTRACTORS BE LIABLE FOR ANY INCIDENTAL, INDIRECT, SPECIAL, OR CONSEQUENTIAL DAMAGES ARISING OUT OF OR IN CONNECTION WITH THE LICENSE GRANTED UNDER THIS AGREEMENT, OR FOR LOSS OF USE, LOSS OF DATA, LOSS OF INCOME OR PROFIT, OR OTHER LOSSES, SUSTAINED AS A RESULT OF INJURY TO ANY PERSON, OR LOSS OF OR DAMAGE TO PROPERTY, OR CLAIMS OF THIRD PARTIES, EVEN IF THE COMPANY OR AN AUTHORIZED REPRESENTATIVE OF THE COMPANY HAS BEEN ADVISED OF THE POSSIBILITY OF SUCH DAMAGES. IN NO EVENT SHALL LIABILITY OF THE COMPANY FOR DAMAGES WITH RESPECT TO THE SOFTWARE EXCEED THE AMOUNTS ACTUALLY PAID BY YOU, IF ANY, FOR THE SOFTWARE.

SOME JURISDICTIONS DO NOT ALLOW THE LIMITATION OF IMPLIED WARRANTIES OR LIABILITY FOR INCIDENTAL, INDIRECT, SPECIAL, OR CONSEQUENTIAL DAMAGES, SO THE ABOVE LIMITATIONS MAY NOT ALWAYS APPLY. THE WARRANTIES IN THIS AGREEMENT GIVE YOU SPECIFIC LEGAL RIGHTS AND YOU MAY ALSO HAVE OTHER RIGHTS WHICH VARY IN ACCORDANCE WITH LOCAL LAW.

ACKNOWLEDGMENT

YOU ACKNOWLEDGE THAT YOU HAVE READ THIS AGREEMENT, UNDERSTAND IT, AND AGREE TO BE BOUND BY ITS TERMS AND CONDITIONS. YOU ALSO AGREE THAT THIS AGREEMENT IS THE COMPLETE AND EXCLUSIVE STATEMENT OF THE AGREEMENT BETWEEN YOU AND THE COMPANY AND SUPERSEDES ALL PROPOSALS OR PRIOR AGREEMENTS, ORAL, OR WRITTEN, AND ANY OTHER COMMUNICATIONS BETWEEN YOU AND THE COMPANY OR ANY REPRESENTATIVE OF THE COMPANY RELATING TO THE SUBJECT MATTER OF THIS AGREEMENT.

Should you have any questions concerning this Agreement or if you wish to contact the Company for any reason, please contact in writing at the address below.

Robin Short
Prentice Hall PTR
One Lake Street
Upper Saddle River, New Jersey 07458